TERMS RELATED TO FARM ANIMALS

	Cattle	Dog	Goat	Horse	Donkey	Chicken	Sheep	Swine
Genus	Bos	Canis	Capra	Equus	Equus	Galus[a]	Ovis	Sus
Group	Herd	Pack	Band	Band	Band	Flock	Flock	Herd
Newborn	Calf	Pup	Kid	Foal	Foal[b]	Chick[c]	Lamb	Pig
Young male	Bull calf	Pup	Buck kid	Colt	Colt	Cockerel	Ram lamb	Boar pig
Young female	Heifer calf	Pup	Doe kid	Filly	Filly	Pullet	Ewe lamb	Gilt
Mature male	Bull	Dog	Buck	Stallion	Jack	Cock	Ram	Boar
Mature female	Cow	Bitch	Doe	Mare	Jennet	Hen	Ewe	Sow
Male castrate	Steer		Wether	Gelding		Capon	Wether	Barrow
Birth of newborn	Calving	Whelping	Kidding	Foaling	Foaling	Hatching	Lambing	Farrowing

Gallus is the genus for chicken and Aves is the class for chickens.

The mule is a hybrid cross resulting when a mare is bred to a jack; the "hinny" is a hybrid cross resulting when a jennet is bred to a stallion; both are sterile.

A poult in turkeys, a gosling in geese, and a duckling in ducks.

SELECTED ABBREVIATIONS

US

Bushel	bu
Cup	c
Degrees Fahrenheit	°F
Fluid ounce	fl oz
Foot	ft
Gallon	gal
Hundredweight	cwt
Inch	in.
Parts per million	ppm
Peck	pk
Pound	lb
Quart	qt
Square	sq
Tablespoon	T or Tbsp
Teaspoon	t or tsp
Yard	yd

METRIC

Centimeter	cm
Degrees Celsius	°C
Gram	g
Hectare	ha
Kilogram	kg
Kiloliter	kL
Kilometer	km
Liter	L
Meter	m
Milligram	mg
Milliliter	mL
Millimeter	mm

SELECTED WEIGHT, AREA, AND VOLUME EQUIVALENCIES

1 mg	0.001 g
1 g	1,000 mg
1 kg	1,000 g
1 lb	453.5 g or .4535 kg
1 kg	2.205 lb
1 ppm	1 microgram/g
	1 mg/L
	1 mg/kg
1 in.	2.54 cm
1 cm	0.3937 in.
1 m	39.37 in.
1 mi	5,280 ft
1 km	0.6214 mi
1 acre	43,560 sq ft
1 sq mi	640 acres (1 section)
1 acre	0.4047 ha
1 ha	2.471 acres

DIRECTIONAL TERMS RELATED TO THE BODY

TERM	PLANE OR DIRECTION
Anterior, cranial	Before or in front of
Bilateral	On both sides
Distal	Farthest from the vertebral column
Dorsal	Toward the vertebral column
Lateral	To the side; away from the medial plane
Medial	Middle or center; closest to the middle
Posterior, caudal	Behind, in back of; toward the tail
Proximal	Closest to a given point; opposite of distal
Ventral	Toward the abdominal wall

Animal Science and Industry

Seventh Edition

Animal Science and Industry

Merle Cunningham

Professor Emeritus of Animal Science, Purdue University

Mickey A. Latour

Associate Professor of Animal Science, Purdue University

Duane Acker

President Emeritus, Kansas State University and
Collaborative Professor of Animal Science,
Iowa State University

PEARSON
Prentice
Hall

Upper Saddle River, New Jersey 07458

Library of Congress Cataloging-in-Publication Data

Cunningham, Merle, 1932-
    Animal science and industry / Merle D. Cunningham,
    Mickey A. Latour, Duane Acker.— 7th ed.
        p. cm.
    Includes bibliographical references and index.
    ISBN 0–13–046256–X /hardcover : alk. paper)
    1. Livestock. 2. Animal industry. I. Acker, Duane. II. Title.

SF61.A3 2005
636—dc22

                                            2004001902

Executive Editor: *Debbie Yarnell*
Development Editor: *Kate Linsner*
Production Editor: *Cindy Miller and Holly Henjum, Carlisle Publishers Services*
Production Liaison: *Janice Stangel*
Director of Manufacturing and Production: *Bruce Johnson*
Managing Editor: *Mary Carnis*
Manufacturing Manager: *Ilene Sanford*
Manufacturing Buyer: *Cathleen Peterson*
Creative Director: *Cheryl Asherman*
Marketing Manager: *Jimmy Stephens*
Cover Design Coordinator: *Christopher Weigand*
Composition: *Carlisle Communications, Ltd.*
Printing and Binding: *Courier Westford*

Cover photos: Photograph of dairy cows and swine courtesy of Purdue University; photograph of Polled Hereford cow-calf pair and Beef Feedlot courtesy of *Hereford World;* photograph of sheep courtesy of The Continental Dorset Club, Inc.; photograph of two young children holding a cat courtesy of Mickey A. Latour; photograph of horses courtesy of Dorling Kindersley Media Library, photographed by Bob Langrish; photograph of aquaculture courtesy of Purdue University; photograph of domestic chicken courtesy of Dorling Kindersley Media Library.

Pearson Education Ltd.
Pearson Education Singapore, Pte. Ltd.
Pearson Education Canada, Ltd.
Pearson Education—Japan

Pearson Education Australia PTY, Limited
Pearson Education North Asia Ltd.
Pearson Educacíon de Mexico, S.A. de C.V.
Pearson Education Malaysia, Pte. Ltd.

10 9 8 7 6 5 4 3 2 1
ISBN: 0-13-046256-X

Contents

Preface

FOCUS OF THIS TEXT: THE STUDENT AND THE INSTRUCTOR

The principal objective of *Animal Science and Industry* is to make the introduction to animal science an enjoyable and rewarding experience for the student—one that provides a sound foundation for further learning and stimulates the student's interest in pursuing additional courses and experiences in animal sciences. This seventh edition of *Animal Science and Industry* also has two principal goals:

- To provide students—who are *not majoring* in animal science and those with *nonfarm backgrounds*—with appropriate information to understand animal agriculture and its role in our society and economy. This is a challenging goal that cannot be accomplished by use of this text alone but requires special efforts from both instructor and student. The instructor can help ensure student success with well-prepared lectures, assignment of appropriate reading from the text to support lecture material, providing well-structured and organized laboratory sessions where possible, and offering timely help sessions. The student should make every effort to attend each class session, to develop an organized set of notes and supplement them with specific subject matter from the text, and to become acquainted with upcoming lecture topics by reading ahead in the text. With effort from both student and instructor, every student should be able to successfully progress through the introductory course. *Then, both instructor and student will be rewarded.*
- To provide students *majoring* in animal science—as well as those in preprofessional science programs—material, principles, and concepts that will better prepare them for more specialized and advanced courses.

The authors greatly appreciate the role of the instructor in teaching an introductory course in animal science to students of diverse backgrounds and interests. We hope instructors will find the material in the text to be of sufficient depth and breadth and to be presented in a manner that is easy to read and to understand. We have provided both basic principles of animal science and realistic examples of their application to efficient animal care and production in an attempt to make instruction easier and more exciting.

This revised edition places emphasis upon the *principles of animal science* and their relationship to livestock and poultry production. Comprehensive information and discussion focuses on (1) biological principles of animal function and management; (2) characteristics of the different animal species production systems; (3) production of high-quality animal products—meat, milk, eggs, wool—and the use of the horse for work and pleasure; and (4) systems and concepts in marketing animals and the processing of their products.

Most animal science instructors have found that learning in introductory courses is more effective and the principles presented are more usable when there is a comparative basis for learning. Therefore, this book was developed to provide an integrated and comparative presentation of the scientific principles that apply to most species involved in animal agriculture, and to describe major differences among species.

Animal Science and Industry, Seventh Edition, continues to present material in considerable scientific depth in the major disciplines of animal science, and provides coverage and updating of a broad number of topics. Research and technology have led to significant advances in animal production practices, marketing, and merchandising of products. Discussion of the more pertinent of these advances has been incorporated into selected chapters.

Throughout the revision of this text, the needs of the student have been our highest priority. The authors recognize the diverse backgrounds of students enrolled in most introductory animal science courses. *Animal Science and Industry* does not assume extensive acquaintance with all domesticated species or their enterprises. Therefore, it provides appropriate descriptions and vocabulary in current usage in the animal industry. This edition also contains new and challenging material for the student with a strong animal production background.

MAJOR CHANGES IN THIS EDITION

Every effort has been made in this seventh edition to improve upon the strengths of earlier editions. Some of those acquainted with previous editions have expressed appreciation for the larger page size and this feature is continued in this seventh edition. This change provides for more pleasing organization and appearance of illustrations.

This textbook resolves a problem that often confronts the instructor during the initial portion of the course as it provides an early overview of the breadth of animal science and related industries. With this approach, later discussions relating to the principles of animal science will be better appreciated and understood. Each of the major animal enterprises is briefly described in a new section of Chapter 1 so that initial coverage of specific species in separate chapters is not required. Also added to the introductory chapter is the topic of companion animals.

The role of companion animals in animal science is rapidly developing, providing students an alternative animal to study and careers beyond traditional agriculture. These animals are a significant part of human society and the economy, thus Chapter 37 has been devoted to this new topic.

LEARNING AIDS

To help students focus on important concepts and terminology, several learning aids have been incorporated into this seventh edition, including:

- *Chapter Introduction.* Each chapter begins with a brief introduction and overview of the chapter topic. Also provided is a list of objectives the reader should be able to accomplish—define, describe, explain—upon completion of that chapter.

- *Review Questions.* In order to emphasize important material and review of concepts and facts within each chapter, questions for study and discussion are provided at the end of each chapter. These questions are designed to encourage students to think independently, and to reinforce and retain information they have studied and encountered in lecture and laboratory sessions.

- *Margin Glossary.* This feature has been included to place further emphasis and understanding on important terms and their definitions contained in the nearby text. The student only needs to refer to the *Glossary* for further description and terms not found in the margins.

- *Vocabulary.* Important new terms within each chapter are boldfaced and are presented in the Glossary at the end of the book.

- *Color Insert and Endpaper Tables.* Review of the descriptive color plate and inside cover information enables the student to become acquainted with the various animal enterprises and their characteristics, as well as with common terminology. More effective instruction can then proceed into a discussion of the major disciplines of animal science.

- *Information Boxes.* Also continued in this seventh edition are information boxes in numerous chapters. These boxes enhance the student's appreciation and understanding of topics related to the material covered in the chapters. They are also provided to increase the student's interest in particular areas of animal science.

- *Other Aids.* Several chapters refer to additional information found in Appendix A that provides sketches of market grades of livestock and Appendix B that provides illustrations of animals, their external parts, and related terms. Throughout the text, brief footnotes provide clarification and additional information.

SEQUENCE OF TOPICS

Many options are available to the instructor because of the organization and content of this text. Each chapter can be presented independently. The *Instructor's Manual* provides supporting materials such as topic overviews, additional materials, and questions for quizzes and examinations.

The first two chapters introduce the reader to animal agriculture and the animal industry. Several chapters on nutrition follow: nutrients, ruminant and nonruminant digestion, and appropriate feeds and feeding of common species. These are areas in which most students who have managed animals for food and pleasure want to become more proficient. Following the chapters on nutrition are chapters related to animal comfort, responsiveness, and performance; animal growth; the feeding enterprise; and animal environment, health, and behavior.

Equally important to success in breeding and management of herds and flocks are the topics of reproduction and lactation (Chapters 14–16). Emphasis is placed upon the structures and normal function of the male and female reproductive systems and the need for synchronization of events for successful fertilization, pregnancy, and parturition. The topic of lactation follows logically. An understanding of the mammary system of all farm mammals, milk synthesis, and milk let-down is a prerequisite to achievement of desired quantity and quality of milk for thrifty, growing offspring, or for sale for human consumption.

Chapters 17–20 on genetics, genetic improvement, animal evaluation, and mating systems enable the student to learn and easily understand the basic principles. Chapter 21 introduces the student to the more popular breeds and their historical, current, and future roles in animal improvement. We hope it will instill appreciation for breeds as sources of DNA needed for competitive production of consumer-desired animal products in the twenty-first century.

Few animal science students are acquainted with the marketing of animals and their products. Therefore, Chapters 22–26 provide a wealth of information on marketing terminology and procedures, humane handling and slaughtering, and important aspects of meat processing and merchandising.

Application of the basic principles of animal science is emphasized in the chapters discussing animal production and management (Chapters 27–36). Chapter 37 emphasizes the broad world of companion animals. These topics include an overview of the industry of specific species and current practices, including recent technological developments and recommendations to improve efficiency of production.

The authors greatly appreciate the ability and desire of college and university instructors to implement the most efficient structure in their introductory course. This text permits flexibility. For example, an instructor may prefer to teach a section on animal products or species production systems earlier in the course, prior to the more in-depth study of animal genetics, reproductive and environmental physiology, nutrition, and behavior. This edition is arranged so that chapters or specific sections can be easily referenced by both instructor and student.

INTERNET RESOURCES

AGRICULTURE SUPERSITE

This site is a free online resource center for students and instructors in the field of Agriculture. Located at http://www.prenhall.com/agsite, this site contains numerous resources for students including additional study questions, job search links, photo galleries, PowerPoint slides™, *The New York Times eThemes* archive, and other agriculture-related links.

On this supersite, instructors will find a complete listing of Prentice Hall's agriculture texts as well as instructor supplements that are available for immediate download. Please contact your Prentice Hall sales representative for password information.

The New York Times eThemes of the Times for AGRICULTURE and The New York Times eThemes of the Times for AGRIBUSINESS

Taken directly from the pages of *The New York Times*, these carefully edited collections of articles offer students insight into the hottest issues facing the industry today. These free supplements can be accessed by logging onto the Agriculture Supersite at: http://www.prenhall.com/agsite.

AGRIBOOKS: A CUSTOM PUBLISHING PROGRAM FOR AGRICULTURE

Just can't find the textbook that fits your class? Here is your chance to create your own ideal book by mixing and matching chapters from Prentice Hall's agriculture textbooks. Up to 20% of your custom book can be your own writing or come from outside sources. Visit us at: http://www.prenhall.com/agribooks.

ACKNOWLEDGMENTS

This text is the result of a combined effort of the principal authors; many knowledge-able scientists, teachers, and industry leaders who provided assistance and suggestions; and the editors and support staff of Prentice Hall.

Special appreciation is extended to those who outlined, drafted, and reviewed chapters and sections of the book, and who were individually acknowledged in earlier editions. These individuals include colleagues and associates from the following uni-versities: Iowa State University, Kansas State University, Louisiana State University, Louisiana University at Monroe, Ohio State University, Oregon State University, New Mexico State University, North Carolina State University, Oregon State University, Purdue University, South Dakota State University, Texas A & M University, Texas Tech University, University of Arkansas, University of California—Davis, University of Florida, University of Illinois, University of Nebraska, University of Wisconsin, and Virginia Polytechnic Institute.

Several colleagues using the text in the classroom provided important feedback that was enormously helpful to us in preparing this seventh edition. We are grateful to Gordon F. Jones, Western Kentucky University; R. Kraig Peel, Colorado State University; and Michael S. Roeber, Northeast Community College, for their valuable assistance in this endeavor.

The authors also appreciate the help of colleagues in the animal industries, fields of animal production, and other universities and colleges who provided photos and illustrations. The authors particularly appreciate the guidance and assistance of Dr. LaDon Swan, Aquaculture Specialist, Auburn University, and the Illinois-Indiana Sea Grant Program in the development of Chapter 36—*Aquaculture*. Also, personnel in nu-merous USDA agencies were especially helpful in providing the most current statistical information.

An effort has been made to properly cite sources of illustrations, data, and other materials, and to provide proper credit to individuals, institutions, and organizations. If proper credit has not been given, the omission is unintentional and sincerely regretted.

The dedicated editorial services of Cindy Miller and Holly Henjum and the staff of Carlisle Communications to ensure a well-edited and attractive seventh edition are greatly appreciated.

Finally, the authors would like to offer special thanks to the many students they have taught and counseled, and who inspired them to develop a popular text well suited for course work and later reference.

1

Animal Agriculture

Food is the primary product of agriculture and animals are a major provider of food. Natural fibers, such as wool, mohair, and cotton; leather; alcohol; and certain pharmaceuticals, oils, and plastics are also important agricultural products, valuable to society and the economy, both nationally and globally.

Plant and animal products—cereals, vegetables, fruits, meat, milk, and eggs—provide nutrients essential for human life in a form that also gives enjoyment and satisfaction. It would be difficult to duplicate the pleasures that come from waffles with maple syrup, fresh strawberries, a grilled T-bone steak, chicken breast, or pork or lamb chops, smooth chocolate frozen yogurt, or a fresh garden salad.

This chapter is largely devoted to the importance of domesticated animals in providing food, fiber, and recreation for people not only in the United States but also in other countries of the world.

An additional objective of this chapter is to provide the reader with some of the more important characteristics of the various animal enterprises prior to subsequent chapters on the principles of genetics, reproduction, and nutrition so that the application of the principles are better understood.

A third objective of this introductory chapter is to acquaint the reader with some of the environmental challenges that face animal agriculturists, particularly in the areas of animal welfare and waste management. Upon completion of this chapter, the reader should be able to

1. Describe the general trends in per capita consumption of animal products in the United States, other developed countries, and less-developed countries of the world.

2. Relate the relationship between family income and consumption of animal products in developing countries.

3. Note the region(s) of the world where the greatest increase in human population is expected.

4. Explain the shifts in U.S. farm population, farm numbers, and farm size.

5. Describe the availability of land in the United States well suited for livestock.

6. Better understand why U.S. farmers are highly productive.

7. Provide examples of how domesticated farm animals benefit people.

8. Trace the zoological order of our most prevalent domesticated species.

9. Compare differences between ruminants and monogastric animals in their ability to utilize feeds.

10. Give examples of how animals differ in ability to adapt to climatic conditions and diseases.

11. Know several factors essential for all animal agriculture enterprises.

12. Better understand specific species enterprises and be able to characterize their labor and capital investment and degree of business risk.

13. Discuss vertical integration in animal agriculture, its degree, and where it prevails.

14. Relate changes in animal production systems that have made both producers and consumers more conscious of their welfare.

15. Better understand how animal agricultural units have expanded and necessitated improved management of animal waste for control of air, water, and soil pollution.

16. Appreciate the measures taken by breeders and producers of farm animals to improve the welfare of their animals and the quality of our environment.

1.1   GLOBAL TRENDS IN ANIMAL PRODUCT CONSUMPTION

Per capita consumption of animal products in the United States steadily increased during the twentieth century. This increase accompanied steadily increasing purchasing power, and, since 1970, the average consumer has been eating about as much meat and other animal products as he or she desires. In America, meat consumption is more influenced by desire as opposed to economics, i.e., some developing countries.

In Table 1–1, note the significant changes in meat consumption since 1970. After peaking in 1978, beef consumption decreased until about 1990, then leveled off at about 66 to 68 pounds per person. In contrast, broiler and turkey consumption doubled during this period, largely because of promotion and retail prices and choices. Currently, chicken consumption is the highest per capita meat consumed (77 lbs/yr). For many years, American consumers have continued to include only small amounts of veal and lamb in their diets.

Less fluid milk and cottage cheese are being consumed, but the demand for other cheeses (such as ripened and pizza varieties) has increased dramatically. Also, the popularity of frozen dairy products has remained high.

The number of eggs consumed per person today is considerably lower than in 1970, but consumption has been on the increase in recent years after reaching a low in the mid-1990s. Total egg consumption for 2002 was estimated at 250 eggs per person, about 7 more than was consumed in 1998. Note that increasing amounts of liquid, frozen, or dried eggs are being consumed each year.

In recent years people in the United States have been changing the way they spend money on food. During the early 1980s consumers spent nearly twice as much money on food prepared at home versus away from home (Figure 1–1). It is estimated that by 2010 money spent on food will be split equally between food prepared at home versus away from home. This trend is not surprising because one of the strongest investments during the economic downturn of 2001–2002 was restaurants. In addition, people are spending money at a faster rate on food prepared away from home versus at home (note the rate of increase for both food prepared at home and away from home over the 30-year period).

TABLE 1–1 — Consumption of Animal Products, 1970–2002

Product	1970	1978	1986	1998	2002
Meats, boneless trimmed equiv. lb					
Beef	79.6	82.2	74.4	64.9	65.7
Pork	48.0	42.3	45.2	49.1	51.6
Veal	2.0	2.0	1.6	0.7	0.6
Lamb	2.1	1.0	1.0	0.9	1.1
Broilers	24.9	28.8	35.8	50.6	77.6
Turkeys	6.4	6.9	10.2	14.2	18.2
Chicken	2.5	1.5	1.4	0.2	1.5
Total, meats	165.5	164.8	169.5	180.5	216.3
Dairy Products, lb					
Fluid milk and cream	275.0	254.0	240.0	219.0	209.4
Butter	5.4	4.4	4.7	4.2	4.6
Cheeses, exc. cottage cheese	11.4	17.0	23.1	28.5	29.8
Cottage cheese	5.2	4.7	4.1	2.7	2.6
Evap. and cond. milk	7.2	4.1	3.6	2.2	2.1
Skim milk, bulk and canned	5.0	3.5	4.3	4.2	5.8
Frozen products	28.5	27.3	27.8	29.5	27.8
Dry products	7.2	6.0	6.8	7.3	3.1
Eggs					
Shell eggs, number	275.9	237.2	214.4	175.8	176.7
Products, egg equivalent[a]	33.0	34.3	39.1	67.8	73.0
Total, egg equivalent	308.9	271.5	253.5	243.6	249.6

[a]Products include "broken-out" pasteurized egg mixtures; liquid, dried, frozen; whole, whites, yolks.
Sources: *Adapted from USDA, ERS, Statistical Bulletin 965, July 1999 & January 2003.*

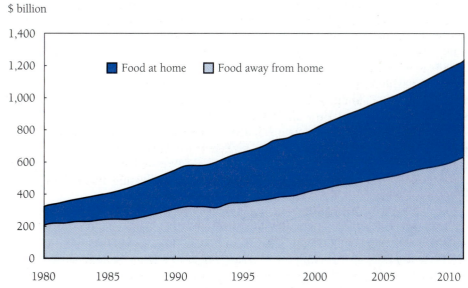

$ billion

Figure 1–1
Food Expenditures in the United States
Source: USDA Agricultural Baseline Projections to 2011, February 2002.
USDA Economic Research Service.

Million metric tons

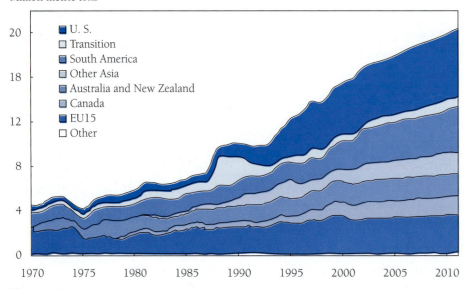

Figure 1–2
Meat exports
Source: USDA Agricultural Baseline Projections to 2011, February 2002.
USDA Economic Research Service.

At the same time that a major portion of American dollars are being spent on food away from home, there is an increasing trend to export meat (Figure 1–2). Compare the amount of meat exported in 1970 (less than 1 million metric tons) to projected levels in 2010 (approximately 7 million metric tons). The trend to export is expected to be considerably higher by 2010 for all countries, but the percentage increase is largest for America—suggesting that a major portion of meat produced in the United States is shipped to other countries. This trend to export may be explained in two ways: (1) the amount of meat produced in the United States exceeds consumption and alternative sources to sell the product must be investigated so American farmers can sell their products, and (2) other countries are becoming more industrialized and thus have the economic means to begin purchasing more meat.

Import of meat is highest for east, south, and Southeast Asia, followed by China and transition. The transition is a staging of meat that waits to be purchased and delivered to a country; that is, during the transition meat can be held in cold storage until a country wants to purchase it for consumption. Also, note the amount of imports has drastically decreased for western Europe (Figure 1–3).

In other developed countries, such as Japan and those of western Europe, animal product consumption rose rapidly after the reconstruction that followed World War II. Rapidly growing family income allowed increased purchase of more nutritious and desired animal products. Though the increase has slowed somewhat in recent years, the general trend continues slightly upward (Figure 1–4).

The countries of eastern Europe and the former Soviet Union, then with "controlled economies" and fewer incentives for human productivity, did not make significant economic progress during the 1950-to-1990 period. Standard diets in these countries, therefore, continued to carry high proportions of cereals and root crops.

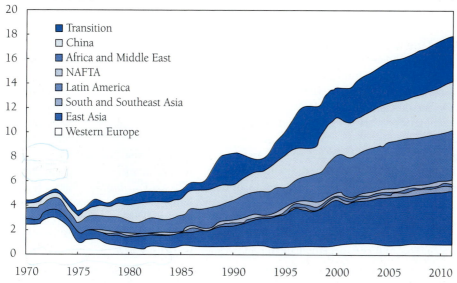

Million metric tons

Legend:
- Transition
- China
- Africa and Middle East
- NAFTA
- Latin America
- South and Southeast Asia
- East Asia
- Western Europe

Figure 1–3
Meat imports
Source: USDA Agricultural Baseline Projections to 2011, February 2002.
USDA Economic Research Service.

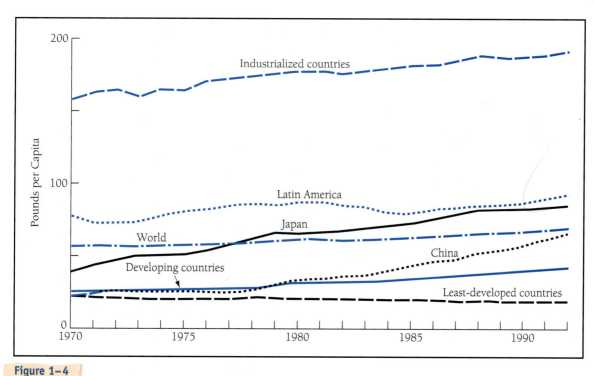

Figure 1–4

Pounds-per-capita consumption of meat is much higher for industrialized countries and is rising for developing countries, in contrast to least-developed countries.
(*Source:* USDA.)

5

Animal products were less available and less affordable. With the switch to a market economy in the early 1990s, increased family incomes and demand for animal products would be expected, but economic recovery has, in fact, been very slow.

Much of the globe, however, consists of countries with very low family incomes, where the diet is composed largely of cereals (rice, sorghum, maize), root crops (yams, sweet potatoes, peanuts, cassava), and pulses (beans, chick peas). In these countries, animal product consumption is very low. These countries are largely in tropical and semi-tropical regions of Central and South America, Africa, south Asia, and the Near East. Population, in relation to productive natural resources, is large. There is little Class I soil; land may be mountainous, subject to periodic flooding, or too dry for significant production.

These regions of the world will be greatly challenged to provide sufficient food as their human population increases. A greater priority will continue to be placed on simply meeting nutrient needs, with little opportunity for selection of plant or animal sources. This will be especially true in regions such as Africa where the growth rate continues to be unusually high, as shown in Table 1–2.

The United States and other developed countries have provided money, technology, and food aid to many of these less-developed countries (LDCs) to help them develop, and there have been some extraordinary successes. Taiwan, South Korea, Thailand, and Indonesia have made rapid progress since 1950. Their successful economic development generally began with government policies that gave incentive to human productivity, plus large investments in agriculture, the basic enterprise of most societies.

In most LDCs, 50 to 85 percent of people who work are farmers, perhaps on two- to three-acre farms. If their productivity is increased, a significant increase in country gross national product (GNP) results. Perhaps as important, many people in these countries have such inadequate diets that their productivity, both physical and mental, is severely limited. When the farmers produce a bit more, the farm families eat better; any surplus food moves on to villages and towns and the people there eat better. All become more productive.

The United States and other donors helped these countries develop short-strawed rice varieties that would use fertilizer better and produce more grain without falling over. They developed experiment stations, extension services, and colleges of agriculture. It took a long time. Roads were needed to move produce to market and bring production inputs—seed, fertilizer, and equipment—to the farmers. Agribusinesses,

TABLE 1–2	Predicted Human Population Increase for World Regions and World[a]			
	1994	**2010**	**Total Period Change**	**Yearly Change**
Regions	*Millions*	*Millions*	*%*	*%*
Europe	728	734	0.8	0.05
North America	290	334	15.2	0.95
Latin America	470	584	24.3	1.52
Asia	3,392	4,253	25.4	1.59
Africa	700	1,078	54.0	3.38
World	5,607	7,022	25.2	1.58

[a]*The world population was estimated at 6.3 billion in 2003.*

Source: *Adapted from* Proceedings of 1995 Maryland Nutrition Conference, E. Brown.

credit systems, and markets had to develop. In the meantime, surplus food from the United States and other countries often was provided to improve diets while the production system was being strengthened.

As people throughout the LDCs became more productive, incomes went up and demand for consumer goods—bicycles, food of higher quality, clothing, and other items—increased. The total economy expanded and more jobs resulted, which, in turn, brought more family income. Food consumption patterns changed from largely rice toward wheat, then more meats and vegetables.

Of special interest to U.S. agriculture, which depends heavily on exports to market much of its output, is that, even as agricultural production expands in LDCs, family income and food demand grow faster once the economy gets moving. GNP often grows 5 to 9 percent per year, as in Asia (Table 1–3), whereas agricultural production can rarely grow more than 3 percent per year.

South Korea illustrates the latter point. Between the early 1970s and early 1980s, whereas per capita agricultural production increased 27 percent, annual imports of purchased agricultural commodities from the United States increased from about $200 million to about $1.7 billion. As income went up, South Koreans ate not only more food but also a higher proportion of meat and other animal products, which require feed grains and plant protein sources for production.

During the last several decades, developing countries were the largest growth markets for U.S. agricultural exports, and that is where the most growth potential lies in the years ahead. In western Europe and Japan, like the United States, population growth has plateaued and people are consuming about as much good food as they want or as is recommended. Not much growth potential remains.

In the rest of the world, however, over 4.5 billion people—primarily in Asia, Latin America, and Africa, as shown in Table 1–2—would like to eat more food and food of higher nutritional quality and palatability. In most such countries, about 60 cents out of every dollar of *increased* family income is spent on more food or higher quality and more nutritious food.

TABLE 1–3	Economic Growth for World Regions and the World Based upon Gross Domestic Product, Annual Percentage Change				
	1990	**1994**	**1998**	**2000**[a]	**2002**
Developed					
United States	1.2	3.5	3.9	1.0	2.2
Canada	(−0.3)	4.7	2.8	3.3	3.4
Japan	4.9	0.7	(−3.1)	2.4	(−.60)
European Union	3.0	3.0	2.8	2.5	1.30
Former Soviet Union	(−3.2)	(−13.9)	(−5.3)	0.0	3.70
Developing					
Asia	5.7	9.4	2.3	5.8	4.83
Latin America	0.0	5.1	2.0	2.3	(−.84)
Middle East	2.9	0.3	1.2	3.5	3.47
Africa	1.4	2.7	3.5	3.7	3.91
World	**2.1**	**3.2**	**2.0**	**2.6**	**1.72**

[a]*Forecast*

Source: *Adapted from USDA, ERS, Agricultural Outlook, April 1996, March 1999, and February 2003.*

With continued economic growth in LDCs, as well as in eastern Europe and other formerly "controlled economies," the outlook for animal agriculture is very strong. The demand for inputs to animal agriculture systems—feed grains and protein sources; seedstock, semen, and embryos; equipment; and expertise—should also be strong.

1.2   CHANGES IN AMERICAN FARM POPULATION, NUMBERS, AND SIZE

Changes in both U.S. plant and animal agriculture have been dramatic when viewed from 100 years ago, 50 years ago, or just 6 years ago. These changes have resulted in fewer farmworkers working on fewer and larger farms (Table 1–4 and Figure 1–5).

Today, there are less than 1.7 million farms as compared to nearly 5.5 million farms 50 years ago. During that time interval, the average size of farms increased from about 220 acres to almost 490 acres (Figure 1–6)[1].

Major shifts occurred in the mid-1940s and 1950s, from largely diversified crop-livestock farms to larger, more specialized farms. Overall farm efficiency steadily improved through specialization in fewer enterprises, increased volume of production, and use of newer and more automated equipment.

The efficiency of the American farmer is the envy of the world. Fewer and fewer farmers have met the challenge to produce sufficient food for the increasing U.S. population. For many years, the percentage of American farmers steadily decreased. However, not only have they kept pace with domestic food needs but they have also made food available to other people in the world.

TABLE 1–4	Changes in U.S. Farm Numbers, Farm Population, and Farm Size			
		Farm Population		
Year	**Farms (thousands)**	*Thousands*	*% of U.S Population*	**Farm Size (acres)**
1951	5,428	21,890	14.2	225
1956	4,515	18,712	11.1	
1961	3,825	14,803	8.1	
1966	3,257	11,595	5.9	
1971	2,902	9,425	4.6	384
1976	2,497	8,253	3.9	
1981	2,434	5,850	2.6	425
1986	2,212	5,226	2.2	447
1991	2,105	4,361	1.8	467
1996	2,075	2,842		470
1997	1,900	2,670		487

Note: Blank cells in tables indicate "not applicable."

Source: *Adapted from USDA data.*

---

[1] An *acre* of land contains 43,560 square feet. A *section* of land represents 1 square mile, or 640 acres. The average U.S. farm consists of about three-quarters of a section of land.

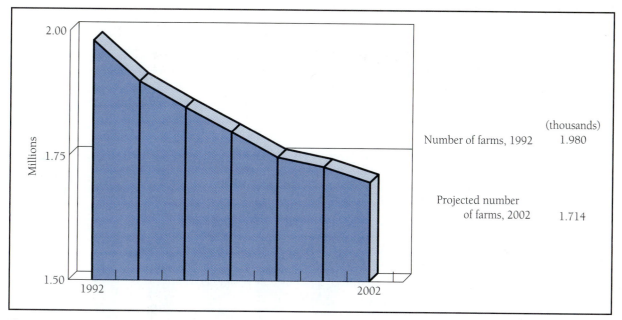

	(thousands)
Number of farms, 1992	1.980
Projected number of farms, 2002	1.714

Figure 1–5
Projected changes in U.S. farm numbers, 1992–2002.
(*Source:* U.S. Census.)

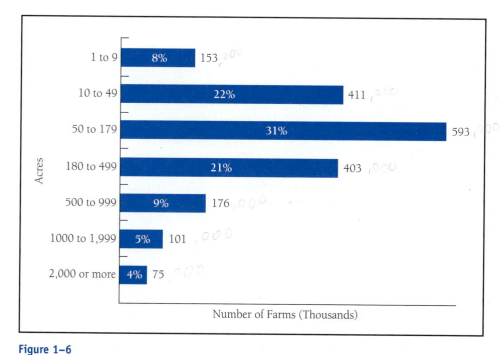

Figure 1–6
U.S. farm size in 1997 averaged 487 acres. Note the percentages and total number of farms and average farm size within acreage categories. For example, 593,000, or 31 percent of all U.S. farms are between 50 and 179 acres in size. Total acreage equals about 1.9 million acres.
(*Source:* USDA, NASS data.)

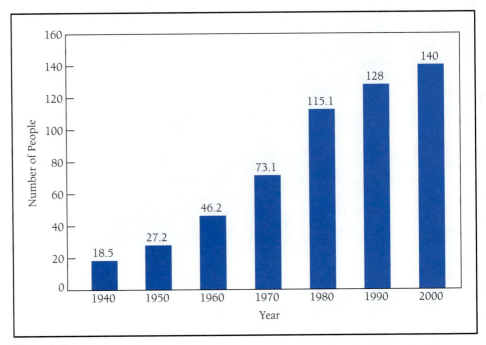

Figure 1–7
Approximate number of people supported by the average U.S. farmer from 1940 to 2000 in food and fiber.
(*Source:* Adapted from USDA data.)

By 2000, the U.S. farm population represented less than 2 percent of the people in the country. The amount of food and fiber produced by one U.S. farmer in 1940 was enough to support about 18.5 people. In 2000 that number was estimated to be 140 people, about 98 in the United States and 42 abroad (Figure 1–7).

1.3   U.S. LAND RESOURCES

Soil is the basis of all agricultural production. The amount, topography, fertility, and location of land available for production are of concern as the world human population continues to expand.

U.S. farmers have become more conservation-minded and have steadily increased the use of conservation tillage practices (crop residue management, minimum tillage, strip tillage, grassed waterways, etc.) to reduce the rate of soil erosion and to preserve land productivity. Of some concern is that cities, towns, rural residences, industrial plants, highways, airports, reservoirs, and recreational facilities have removed much U.S. land from agricultural production. Land diverted to nonagricultural uses, from 1959 to 1992, increased 147 million to 340 million acres, or over 2.3 times. Loss of agricultural land has occurred in other countries, especially where populations are growing more rapidly.

Of the estimated 2.26 billion acres of U.S. land, about 460 million acres, or slightly more than 20 percent, are classified as *cropland*. Of this amount, about 15 percent is actually utilized for crops, the balance being idle acres or used for pasture.

Grassland, rangeland, and *forest land* represent about 1.24 billion acres, or almost 55 percent of total U.S. land, where little human food can be produced unless the grass and other vegetation are consumed by meat-, milk-, or fiber-producing animals.

Privately owned farms accounted for only about 40 percent of all U.S. land in the mid-1990s. The remaining land was largely range and forest land managed by the federal government. Use of some of this land for grazing is permitted, for a fee per animal unit, by ranchers.

1.4   WHY ARE AMERICAN FARMERS SO PRODUCTIVE?

American farmers deserve much credit for their efficient production of agricultural products. Other factors that have enhanced their capability include an economic system of free enterprise, favorable agricultural land and climate, and strong educational and agribusiness support. These are briefly described in this section.

Our Economic System

The free enterprise system in the United States encourages the establishment and attainment of family goals, and encourages its citizens to be innovative and industrious. These are major reasons why American farmers are so efficient in the production of high-quality products. What and how much is produced is largely determined by product marketability and profit potential.

Farmers place a high value upon freedom and quality of life obtained from farming, the beauty of nature, openness of space, and greater independence of life. Most farms are family operated and provide many opportunities for children to become involved.

Land and Climatic Conditions

In relation to most countries, the United States has an abundance of land well suited for production of fruits, vegetables, grain, forages, and other agricultural commodities. Farming practices throughout the United States differ and are influenced by the existing soil type, topography, and environment of the region.

The Midwest is characterized by rich soils and high crop yields. Some of the drier semiarid areas of the West and Southwest have land that becomes very productive with sufficient irrigation. The northern states are better suited for forage crops and shorter season plant varieties. Marginal cropland (rolling, hilly land) found throughout the United States is best suited for livestock systems that utilize the production of forage crops.

Our Educational Systems

The efficiency of American agriculture has been enhanced through information obtained via the various educational services. There are 68 "land-grant" universities and colleges in the United States. Their establishment was made possible by passage of the Morrill Act of 1862 that granted the land for and established at least one university or college in each state. About 44 of every 100 farmworkers have attended college.[2]

The Hatch Act, passed in 1887, provided for the establishment of experiment stations in each of the states. Here, researchers solve problems, develop new concepts, and bring about the latest advances in technology and biotechnology.

---

[2] USDA agricultural statistics, 1998.

Research information is readily available to the public and is especially sought by farmers and agribusiness people. Its dissemination is greatly facilitated by the Cooperative Extension Service found in every county and was made possible by the Smith-Lever Act in 1914.

The teaching of principles, gaining of knowledge through research, and dissemination of new information (extension education) have contributed greatly to past agricultural accomplishments. An even stronger educational system will be needed to help the American farmer of the future become still more efficient.

Our Specialized Agribusiness

Today's successful farmer depends upon high-quality products and services provided through many agribusiness firms (Chapter 2). Each segment of farming depends upon specialization—whether it involves production, harvest, and preservation of crops—or the breeding, feeding, housing, and management of animals. Agribusiness also provides well-trained and experienced people specifically trained in providing service to the farmer.

Our Specialized Research

Agricultural productivity input has remained steady over the course of nearly 50 years, with a drop occurring around 1980 through 1996 (Figure 1–8). At the same time, the amount of output and productivity has been on a steady rise; more specifically, the amount of output and productivity has doubled during this period. The "doubling effect" can be attributed to several factors: (a) an increase in genetic selection of animals, (b) refinement of management practices on farms, (c) increased quality of grains, (d) advancements in machinery, and (e) overall advancements through research and development both within the private and public sector.

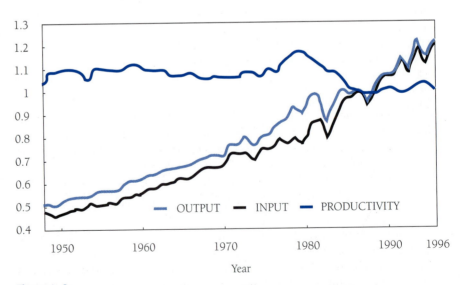

Figure 1–8
Agriculture productivity: 1948–1996.
(*Source:* USDA-ERS.)

1.5   ANIMAL AGRICULTURE: THE ROLE OF ANIMALS

Since prehistoric times, the domestication of animals has contributed greatly to humanity. Their major role has been in providing a dependable source of food and fiber. Modern-type animals of today provide other benefits and pleasures to people throughout the world:

Food protein of high biological value.   Animal products contain protein of high biological value. This term refers to the ability of these protein sources to supply amino acids in the proper proportion needed for body functions. These products contain essential amino acids in almost the same ratio as body tissues. Therefore, animal protein sources are used more efficiently than most plant proteins deficient in one or more essential amino acids. Plant proteins can provide an assortment of amino acids when several plant sources are combined.

Most people in the United States are fortunate to have an abundance of animal products such as milk, cheese, meats, fish, and poultry as primary protein sources in their diets. It may be difficult for some to realize that much of the world's population still depends upon vegetable diets and that protein deficiency continues to be widespread in the developing countries.

Enhancement of diet palatability.   Well-prepared animal products, in addition to providing nourishment, contribute greatly to a person's sense of meal satisfaction. The family meal, special dinner occasions, and refreshments at social gatherings often center around well-prepared animal products (see color insert). Palatability is enhanced by the aroma, flavor, and juiciness of foods that stimulate salivary and gastric juice flow.

Soil conservation and enhancement.   Grazing animals can be efficient converters of vegetation into nutrients for useful purposes (see color insert). Much of this vegetation, not otherwise usable, comes from marginal cropland, pastures, and rangeland not suitable for cultivation. Well-managed grazing also promotes soil conservation by reducing soil and moisture loss. The components of animal waste enhance the soil's tilth and organic matter content.

Stabilizers of our food economy.   Countries utilizing domesticated animals for meat purposes generally have lower risk of food shortage during difficult climatic and economic times. During periods of drought and scant vegetation, an increased number of animals can be diverted into the human food chain, thus reducing the risk of famine. In times when vegetation is abundant, more animals can be kept for further growth and development to provide food when they attain the most desired market condition and weight.

Companion animals and pleasure.   Animals provide enjoyment to many people. Farm animals, especially the horse and dog, are used extensively in sports and recreation. Much pride exists in the breeding, training, and showing of well-disciplined and superior animals. Many persons, less interested or unable to work directly with animals, attend events where animals provide entertainment through their display or performance.

1.6   THE GENE POOL AND ANIMAL CLASSIFICATION

Over 15,000 species of mammals are known to exist. Cattle, sheep, and goats are major sources of animal food and fiber worldwide. In the far North, reindeer serve the same function. These **ruminants** (cud-chewing animals with a multicompartmented stomach), adapted to the grazing lands, efficiently utilize forage. There are other ruminants that are similarly adapted, though not generally used for domestic food consumption—the prong-horn antelope of the northern U.S. plains, the gazelle of India, the waterbuck, the eland, the pigmy antelope of Africa (13 inches tall and the smallest known ruminant), the giraffe (tallest of living mammals at up to 18 feet), the bison, the deer, the Rocky Mountain goat, the elk, the moose, and others. These species, and others, represent a gene pool for potential use in domestic food production, at least in grassland regions.

> **Ruminants**
> Cud-chewing animals with a multicompartmented stomach.

Hundreds of species of ruminants exist. By natural selection, and because of human action in recent centuries, certain species have become common or predominant in specific regions on each continent. Humans and domestic animal species have developed a synergistic relationship as humans have advanced to become the dominant force on the globe. The domesticated animal species have flourished and in turn have provided humans their companionship and their products—meat, milk, eggs, fiber, and, in some cases, draft power and transportation. As we have confined, tended, nourished, and selectively mated individuals or groups of individuals within each of certain species, relatively true-breeding groups, called breeds, have developed.

The presence of 16 breeds of milk goats in Switzerland and 25 or more breeds of beef cattle in the United States suggests the intense degree to which we humans have selected and refined groups within species to achieve a maximum synergistic relationship.

Humans have also studied the many forms of zoological life and have classified them according to body structure in an attempt to better understand the evolutionary process and the relationships among species existing today or known to have existed. Both goats and cattle are species within the family Bovidae, one of the five families in the suborder Ruminantia.

Systems used by zoologists to classify animal life vary. In one of the systems in common use (see Figure 1–9), the suborder Ruminantia is one of many within the order Ungulata (hoofed mammals). The ungulates are among 11 orders within the subclass Eutheria (animals with a placenta). Eutheria and two other classes comprise the class Mammalia (mammals). This also illustrates that the number of different kinds of forage-consuming mammals on the face of the globe is awesome.

The domestic pig is another species in the mammalian subclass Eutheria. Birds form the class Aves, which is considered second to mammals in importance among land-living vertebrates. More than 8,600 species of birds are known to exist.

Domestic chickens, turkeys, pheasants, quail, grouse, and guinea hens, all used routinely for meat in the United States, are part of the worldwide group (order) Galliformes, the henlike, ground-nesting birds. Several of the 17 other orders of birds contain species that often contribute to the nutritional well-being of humans. These include Columbiformes (pigeons), Anseriformes (ducks and geese), and Struthioniformes (ostriches).

Today's commercial broiler, a young chicken fed for meat production, has been adapted by humans from the early chicken by selection and planned mating systems, to (1) live well in confinement, (2) grow rapidly, (3) utilize high-energy rations, and (4) yield a high proportion of edible meat. Similar adaptation has resulted in efficient egg-laying birds. Perhaps others of the 8,600 bird species carry the genes, among individual birds within species, that would allow their use as a major food source in decades or centuries ahead.

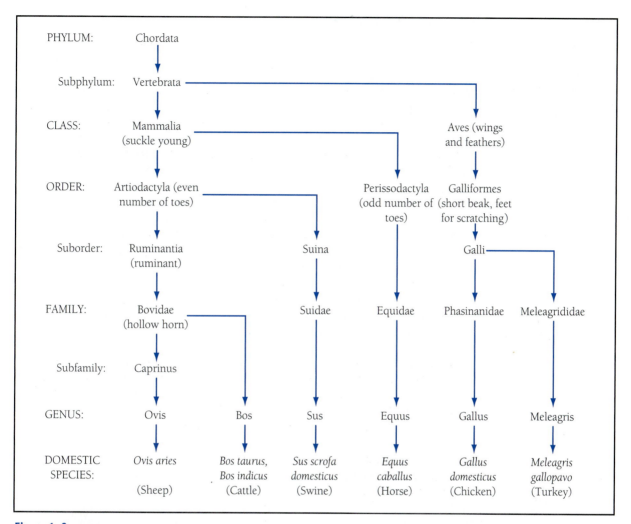

Figure 1–9
Zoological classification and comparison of certain domestic species.
(*Source:* Plimpton, R. F., and J. F. Stephens, *Animals and Science for Man: A Study Guide,* 2nd ed. Minneapolis, MN: Burgess Publishing, 1979.)

Seventy-one percent of the globe is covered with saltwater; freshwater streams and lakes cover another major segment. These vast aquatic environments contain an infinite pool of genetic material in the form of fish and other water-adapted species. Humans have been primarily gatherers of marine products until recent decades, when intensive aquaculture (oyster, shrimp, and fish) systems have been developed in a number of countries.

As the world population increases, fish and seafood will provide an increasing proportion of the protein in human diets. (See Chapter 36.) Aquaculture involves the propagation, rearing, and marketing of aquatic animals (and plants) in selected or controlled environments. This nontraditional form of agriculture, which requires specialized facilities and expertise, shows promise as an efficient provider of human protein.

1.7   ANIMAL ADAPTATION

Species characteristics influence their adaptation to geographic areas or production systems. The kinds of feed they can utilize, their size in relation to product volume, and their susceptibility to stresses—extreme temperatures, parasites, and diseases—largely determine where each species will flourish or the type of land or facility where they can be efficiently produced.

Poultry and swine digestive systems are similar to that of the human. Their relatively small, single-compartment stomach cannot handle large volumes of bulky forages, so these animals (Figure 1–10) depend primarily on concentrated grains and similar feeds. Mature sows can use considerable forage, but growing swine and poultry and laying hens cannot consume enough or effectively digest forage to provide the energy they need.

Cattle, sheep, and goats are designed for forage utilization. As ruminants, their large four-compartment stomach accommodates bulky grass and hay. The first three compartments prepare fibrous forage, by means of microbial fermentation and thorough grinding, for efficient utilization. Although not a ruminant, the horse with its well-developed cecum and colon is well suited for forage utilization.

Whereas cattle and sheep can grow, mature, and reproduce on forage alone, those being fed for slaughter are usually given a higher proportion of grain for several months to speed up growth and bring about "marbling," the deposition of fat within muscle tissue. Dairy cattle fed for high milk production also need higher proportions of grain in their rations.

Goats have the ability to graze over rough terrain, thereby utilizing forage almost inaccessible to other species. They are agile and sure-footed and are "browsers," often preferring brushy species and weeds rather than the more typical pasture or range

Figure 1–10
Hogs and poultry get most of their energy from concentrated feeds.
(*Sources:* HY-Line Poultry Farms and Purina Mills, Inc.)

grasses. Sheep rank next in this ability, followed by cattle. Sheep and cattle are often handled by the same rancher because the combination makes efficient use of grazing land. Cattle prefer tall grass in open areas (Figure 1–11); sheep will consume short grass and will forage on steep slopes.

Because sheep can adapt to higher altitudes than cattle, flocks are seen in high mountain ranges of the western United States. Goats are more common in certain mountain areas and in the dry, rocky areas of west-central Texas.

Sheep breeds, especially, vary in their foraging ability. The Merino and Rambouillet breeds, which were developed in rough, hilly country, are popular in range areas. Mutton-type (or meat type) breeds, however, which are shorter legged, more compact, and less agile, are most practical on farms where forage is lush and more accessible.

Extremely hot or cold weather is a stress to any animal, but the effects are more severe in certain cases. Dairy cattle are extremely sensitive to hot, humid weather; production often goes down markedly during late summer and is lower, on the average, in the southern states. Cold, dry weather is not as severe a stress for dairy cattle because they consume large volumes of feed and generate much body heat as the feed is utilized.

Temperature effects on most breeds of beef cattle are similar. Beef cows consuming large quantities of forage, or fattening cattle on full feed, are not severely affected by cold, dry weather. But because animals of most breeds lose little heat by perspiration, extremely hot weather is a *severe* stress. Cattle of Zebu-type breeds, such as American Brahman and their crosses, are exceptions. They do perspire considerably and because of this and other traits are more adapted to hot climates.

Where poultry or swine are maintained continually in confinement, temperature and humidity are usually controlled by extensive ventilation systems, sprinkler devices, and heating and cooling systems.

Figure 1–11
Cow herds gain most of their nutrients from grass.
(*Source:* National Livestock Producers Assn.)

All young and small animals—poultry, calves, lambs, and pigs—are extremely sensitive to cold, damp weather. Because they have a large surface area per unit weight, relatively more body heat is lost by such means as conduction, radiation, and evaporation. Also, in extremely young animals, built-in temperature control mechanisms do not yet operate effectively.

Among the other temperature influences is the reduction in feed consumption of animals on full feed during periods of hot and humid weather.

Animals also vary in their susceptibility to disease organisms and parasites. Young animals, for example, are more susceptible to and adversely affected by disease and parasite infestation than are older, more mature animals.

Poultry and swine can be more troubled with disease outbreaks. Because they are more commonly in confinement, a disease outbreak can intensify and spread rapidly.

Sheep and horses are very susceptible to parasite infestation and must be treated often for parasites during the summer months. This is another reason why sheep are more adapted to higher altitudes, where parasites are less numerous.

Beef and dairy cattle are less troubled with disease and parasite infestations than other species. This is partially because they are larger, more rugged animals and, in the case of beef herds, are usually kept on grazing land, where disease organisms are not so concentrated. More problems are encountered with dairy cows or beef cattle in drylot, and include such things as mastitis, foot rot, and respiratory diseases.

1.8   ESSENTIAL FACTORS OF ALL ANIMAL ENTERPRISES

Food production businesses compete with other businesses, such as retailing, mining, and manufacturing, for management talent, labor, and capital investment. In other words, the animal producer is in business to make a profit from effective use of management talent, labor, and capital.

Successful production of high-quality meat, milk, or eggs demanded by the consumer requires an extremely high level of business and animal management skills. Each species enterprise has its own characteristics and specific needs (Section 1.9). The following paragraphs review components somewhat common to all successful animal enterprises.

Quantity and Quality of Labor

The adoption of modern technology in almost all confinement housing systems requires skilled labor and management. The potential loss in animal productivity or life demands that managers employ the best available personnel. Inexperienced persons are sometimes hired, but they must be diligent and willing to quickly learn important aspects of the enterprise.

All employees must be interested in the welfare of the animals and the success of the enterprise. They must be alert to building and equipment conditions that appear to be abnormal, and if not able to correct them, quickly report them. Similarly, those working with animals must understand normal and abnormal behavior, know signs of good health, and take proper action when unusual signs are detected.

Today's managers, some of whom are recent college graduates, must be able to attract and manage employees. High-quality, dedicated employees are more likely to stay with an operation that provides a competitive salary, fringe benefits, and incentives for excellence in performance.

Capital Investment

Most beginning farmers or ranchers are short on capital, need quick return, and cannot afford large risks. All enterprises need a minimum of capital to provide sufficient volume to be profitable. In some cases, beginners are able to work into a partnership with the owner, substituting labor for a portion of the required capital. After successful establishment, partnerships or family corporations often consider expansion to generate additional income. However, it is wise to first evaluate the entire enterprise and concentrate on those areas that can be improved with little additional capital investment. "Getting good *before* bigger" is usually a good rule to follow.

Acreage Needs—Pasture or Confinement

Many livestock and poultry production units have moved toward partial or total confinement where limited acreage is needed for animals, animal buildings, sufficient feed storage, and adequate animal waste management structures.

Where land is suitable for crop production, it is often more advantageous to confine animals. In this way, more homegrown feeds (forage and grain) can be grown, harvested, stored in permanent year-round structures, and fed via automated systems. Both low- and high-investment confinement systems permit greater labor efficiency, closer observance of animals, and better control of the animal's environment.

However, there are several examples where dairy and beef producers utilize forage through the efficient use of intensive grazing systems. This system involves the rotation of cattle on small lots of young, lush pasture. Each lot is sized to provide about 1 to 3 days' allowance, intensively grazed, then allowed to recover for about 40 days prior to regrazing.

Confinement systems are utilized extensively in swine, dairy, and poultry production units, the finishing of feedlot cattle and lambs, specialized veal production, and in some sheep operations where accelerated lambing is practiced.

With grass and hay as the prevalent feeds, ranching in the range areas is restricted to cattle herds, yearling beef and steer (stocker) enterprises, sheep or goat flocks, range horse herds, and yearling beef steer and heifer enterprises.

Shelter needs of livestock and poultry are closely associated with sensitivity to extreme temperatures, discussed in Section 1.7. Baby pigs, lambs, and calves born during cold weather need a warm, dry, draft-free environment. Baby chicks and poults also need to be confined to areas of the facility that are warm, dry, and free of drafts.

Semiconfined shelters for older animals do not need to be elaborate. However, they must be properly designed so that warm, moist air is moved upward and away from the animals. High environmental temperatures are common in the southern and southwestern states. In these areas, the large operations usually provide supplemental shade. By reducing the ambient heat load, cattle are more comfortable and more productive.

Totally confined shelter ventilation systems are more elaborate and must function properly at all times. These systems must be designed to minimize the build-up of high humidity, excessive moisture, stagnant air, cold drafts, and airborne organisms and particulates.

Stability of Product

Perishability or stability of product is less important in considering adaptation of species to a region, but does influence enterprise costs and returns. Refrigeration, other preservation techniques, and rapid, timely transportation are available for milk, meat, and eggs.

Red meat and poultry meat are perishable but not until the animal is slaughtered. Market animals can be rapidly shipped long distances with today's modern trucking and highway systems. With freezing, canning, and other effective methods of preservation, the distance that meat products can be moved is limited only by cost.

Wool is a rather stable product, not perishable if stored under good conditions. It can be stored many months and shipped via relatively inexpensive carriers.

1.9   CHARACTERISTICS OF SPECIFIC ENTERPRISES

The magnitude of the entire livestock and poultry industry in the United States, including all species, can be assessed by the total dollar value of livestock and poultry sold each year. In recent years, it has approached $100 billion (Figure 1–12). Each type of animal enterprise, several within each species, has its distinct characteristics, advantages, and disadvantages. However, each provides opportunity for success for those with the desire and management skills, and adequate start-up capital.

The purpose of this section is to provide a brief summary of several enterprises—to give the reader an introductory acquaintance and/or review. Some initial familiarity with the different enterprises should help in better understanding how the principles of animal science apply to their production and management. A more comprehensive coverage of the numerous animal enterprises is provided in Chapters 28 through 36.

Dairy Production

The business of dairying requires dedicated managers and workers who are well trained and experienced (see color plate A). The volume of labor required is high; milking requires more attention and time than any other activity. Cows are milked twice daily, and three times per day in some cases. Dependable labor must be available every day of the year.

The production of high-quality milk requires a large investment in facilities and equipment (see Figure 1–13). Modern dairy farmers are quick to adopt the latest technological improvements to maintain an efficient operation. In addition, very high skills in areas of breeding, reproduction, nutrition, housing, and waste management are essential for profitable dairying. Proper growing and harvesting of high-quality forages receives a high priority on most dairy farms.

Figure 1–12
The total value of all livestock and poultry and their products produced in the United States in 1997 amounted to about $99 billion. (*Source:* USDA, NASS.)

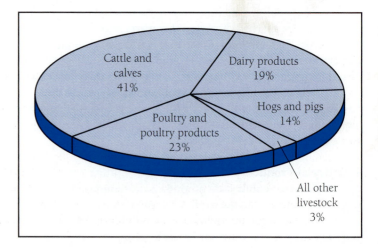

Financially, the daily production of milk and at least every-other-day delivery of milk to the processing plant results in a good cash flow, and dairying can be accurately characterized as a relatively low-risk long-term business. (See Chapter 30.)

Every dairy operation depends upon an adequate number of 2-year-old first lactation cows to replace about one-third of the milking herd each year. The major reasons for cows leaving the herd are mammary system and reproductive problems. Most dairies replace cull cows with heifers that were born and raised on their own farm. However, some larger dairies contract out their heifers to people who specialize in rearing heifers from about 6 months of age until a few weeks before their first calving.

Very few, if any, bull calves are raised on most dairy farms; instead they are sold at 5 to 10 days of age to those who specialize in either veal calf or dairy-beef production. Also, very few dairy producers raise bull calves for breeding purposes. In fact, over 75 percent of our U.S. dairy farmers take advantage of genetically superior sires available through the services of the artificial insemination (AI) industry. On dairies that breed AI, an inventory of carefully selected semen is maintained in a nitrogen tank, and cows are bred artificially by a designated person who has mastered the proper AI techniques.

Sheep Production

Most U.S. sheep flocks are bred for their meat-producing ability rather than for wool production; the sheep owners depend upon the sale of feeder lambs ready to enter the feedlot for finishing or the sale of market slaughter lambs finished in their own feedlot. The value of wool sales amount to about 10 percent of the annual sales from the typical sheep operation.

The labor needs of a ewe farm flock are not usually considered high but are somewhat seasonal, especially during the lambing period. In many flocks of less than 100 head, considerable labor is required in relation to the investment; however, facilities and equipment do not need to be elaborate (see color plate B).

To make the farm flock the major enterprise, some owners have increased the ewe flock to several hundred head. Others have adopted the practice of accelerated lambing—a practice of breeding ewes more often than once a year. Winter lambing requires more attention to proper housing and management to prevent lambs from chilling. Some sheep operations have adopted the practice of confinement lamb production. In these cases, high management skills and greater investment in facilities are required. (See Section 29.8.)

Market lamb production in the southwestern and northwestern states, where flock size usually exceeds 1,000 head, accounts for most of the lambs sold in the United States. These range flocks require special skills in range forage management, coping with predators, and care of the flocks during extreme weather conditions.

The business of lamb finishing consists of the growing, developing, and finishing of lambs coming off pasture or range grass in the fall. They remain in the feedlot for about 60 days until the desired weight (about 110 pounds) and finish are reached, then are sold as market lambs. Considerable skill is needed to properly feed and manage feedlot lambs, and due to the uncertainties of market, disease outbreaks, cost of weight gain, and so on, finishing lambs has been a relatively high-risk business. (See Chapter 10.)

Beef Cattle

The cow-calf operations across the United States vary greatly in herd size, choice of breeds, and business objectives. However, all generally share the common goals of economical production, a high percentage of calf crop weaned, and excellent average weaning weights for each calf crop (see color plate C).

Many cow-calf owners depend upon other enterprises as a major source of income. For example, some may own land well suited for cattle, live in town, and hire workers to manage the daily routines of their operation. Others may own or operate farms with some land suited for cow-calf operations and other acreage that provides a major source of income from crop production.

Cow-calf operations generally prevail in states where land is best suited for grazing. These include the leading states of Texas, Missouri, Oklahoma, Nebraska, South Dakota, Montana, Kansas, Kentucky, and Tennessee. Also, those cow-calf operations located west of Missouri generally consist of ranches with herd sizes that far exceed the typical cow herd of less than 100 head located in the midwestern and eastern states.

Labor demands of the cow herd are somewhat seasonal. Most owners prefer a breeding season of about 60 to 80 days; thus, considerable devoted attention is needed about 280 days later as the calving season begins and extends to another 60 to 80 days. Then, some 200 to 240 days later, the calf crop will be weaned and properly managed to ensure that they are healthy, consume proper amounts of feed and water, and receive the appropriate vaccinations. (See Chapter 29.)

Most newly weaned steers and heifers are grown and developed prior to feedlot finishing. They may stay on the same ranch or farm, or be purchased by operators who continue their growth on available forages. Typically these feeder calves will be placed on a wintering program on a high-roughage, limited-grain ration. They then are placed on spring and summer pastures for continued economical gains prior to feedlot placement (Figure 1–14).

Several factors determine the profitability of growing and developing feeder cattle, the major one being the price margin (the difference in purchase price and weight and the sale price and weight). In addition, the operator must be especially skilled in providing the proper health and nutrition practices needed for continued profitable growth.

Figure 1–14
Yearling beef steers and heifers can make economical weight gains on a spring and summer grazing program before being finished in the feedlot.
(*Source:* Purdue University.)

The feedlot enterprise involves the placement of feeder cattle into feedlots to finish them to the body condition and weight desired by the beef-processing plants. Operations consist of the farm feedlot or the commercial feedlot.

The farmer-feeder may choose to retain ownership of feeder cattle, utilizing homegrown feed and labor. However, most cattle are finished in commercial feedlots with the capacity to feed thousands of head at one time. These operations depend upon skillful financing, maintaining feedlot capacity, economical purchase of feeds, proper nutrition, and a sound health program.

Most of the larger feedlots are located in the northern and southern plains states. These areas are more remote from populated areas, have excellent sources of feeder cattle, a relatively dry climate, nearby sources of energy feeds, and the availability of processing plants. (See Chapter 10.)

Typically, feeder cattle (e.g., steers) enter the feedlot at about 700 to 730 pounds and are fed for about 140 to 160 days. At that time they are ready for slaughter at about 1,150 to 1,250 pounds. Heifers usually enter the feedlot at about 60 to 90 pounds lighter than steers and are ready for market at about 1,075 to 1,100 pounds.

Swine Production

The United States is one of the leading pork-producing countries. Over 19.4 billion pounds was processed in 1999. In recent years the United States exported considerably more pork than it imported (see Figure 1–15 in poultry production discussion). Japan has provided the largest market for U.S. pork.

The swine industry has experienced significant changes in recent years. These include the concentration of fewer but larger production units, shifts in genetic selection toward leaner hogs, new sources of genetic stock through breeding companies, and increased efficiency of production (see color plate D).

Recent changes in swine management include greater use of the all-in, all-out practice to better control disease, segregated early weaning, split-sex feeding, multiple-site systems, and producer networking. (See Chapter 28.)

Most of the U.S. pork is produced in the midwestern states where abundant grain supplies provide a low-cost feed source. However, North Carolina and other southeastern states have experienced a dramatic increase in relation to the midwestern states in recent years. Their increases have been due mainly to higher concentrations of large-scale swine operations and development of nearby packing facilities.

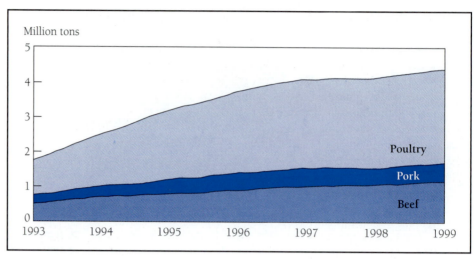

Figure 1–15
U.S. meat exports have been led by poultry meat (largely broilers), followed by beef and pork. (*Source:* USDA, ERS.)

Most hogs are produced in farrow-to-finish operations where pigs are farrowed and then finished to a slaughter weight of 230 to 270 pounds. The different phases of production require facilities to accommodate sow and gilt breeding, gestation, and farrowing; pig nursery; and pig growing and finishing. Although some operations utilize less elaborate facilities on pasture systems, most are all-in, all-out confinement operations.

Some producers specialize in farrow-to-feeder operations. These operations function very much like farrow-to-finish except pigs are sold at or after weaning at weights of 40 to 60 pounds. In most cases, the producer of feeder pigs does not have ready access to economical feed sources.

Because profitability of farrow-to-feeder operations depends heavily upon number of pigs weaned and high weaning weights per litter, great emphasis is placed upon obtaining high conception rates, sufficient pigs per litter, good mothering ability, and a low mortality rate. Operations that farrow pigs have a high investment per sow because of the specialized building and equipment needs. Also, considerable high-quality labor is needed during the farrowing and caring for baby pigs.

Some people specialize in the growing and finishing of pigs. This enterprise contrasts greatly with farrow-to-finish because the operator generally has available sufficient energy feeds to grow and finish feeder pigs to the 230- to 270-pound market weight, but requires only limited facilities and equipment. However, considerable capital is needed for the purchase of pigs, and purchasing from varied sources increases the risk of disease and economic loss.

Poultry Production

The U.S. poultry industry is highly efficient in the production and processing of eggs and meat for its consumers. Almost all eggs consumed are white and are produced by very productive strains of the White Leghorn chicken breed. The predominant poultry meats are derived from meat-type chickens such as broilers, turkeys, and ducks.

The poultry industry has become vertically integrated in both egg and meat production. Today, the egg industry is owned by a few companies with complete control of all phases of production. Although highly integrated, poultry meat production usually involves contract growing. This arrangement allows the grower, who usually provides the land, to concentrate on providing labor and management. The poultry company provides the baby birds, feed, medicines, and veterinary or managerial assistance. Also, marketing and processing of meat birds (or eggs) are performed by the company.

Vertical integration reduces the capital risk of the grower and allows the company to maintain better production and quality control throughout the stages of production. Since the start of integration in the mid-1940s, the poultry industry has made dramatic advancements in genetics, nutrition, health management, and engineering of buildings and equipment. As a result, high-quality meat and egg products are available to the consumer at an economical cost.

The commercial egg layer operation consists of several phases of management. The most critical phase is the brooding of chicks, which requires advance preparation of the house (such as cleanliness, air temperature, provision of feed, and water). Then, the young pullets are carefully managed through the starter, growing and development, and prelaying phases (see color plate E).

Layer performance is greatly influenced by management, especially during the first 6 to 20 weeks of age. Proper nutrition (feed and water), vaccination and sanitation, air (ventilation and temperature), and lighting are major factors in pullet development.

At about 18 weeks of age, the pullets are moved to the lay house where both intensity and day length of light are increased. The objective of light control is to bring each group of pullets into egg production together. It also helps ensure production of eggs of uniform size and quality.

Most operations keep each group of laying hens for two lay cycles, a total of about 90 weeks. After approximately 50 weeks of lay, molting (interruption of lay) is induced and a second lay cycle follows. Many layer operations attain a high egg production average per hen of 290 to 300 eggs per year. Currently, when all U.S. laying flocks are included, the average per hen is about 264 eggs per year. (See Chapters 33 and 34.)

As a result of strong and growing demand for poultry meat (primarily broilers), U.S. consumption increased more than 5 percent each year during the 1990s. Also, global demand for U.S. poultry meats resulted in double-digit increases in exports—especially to China, Russia, and Mexico. Currently, the United States supplies over 50 percent of all poultry meats imported by other countries in the world and leads all other U.S. meat exports (Figure 1–15).

Broiler production consists of large production units that have the capability of rearing up to 120,000 to 130,000 birds per group with up to six groups per year.[3] The broiler firm and grower predetermine the market date for each group of birds, which is usually 45 to 50 days after arrival at the production unit. Birds are managed in an all-in, all-out system that begins with properly cleaned and equipped floor-type housing. When each group of birds is marketed, the facilities are prepared for another cycle of broiler chicks that are scheduled to arrive within about 10 to 12 days (see color plate F).

Modern commercial strains of meat-type breeds grow efficiently and rapidly. A broiler should attain the desired fleshing and weight (about 4.8 to 5.0 pounds) within the 45- to 50-day period. Also, well-grown broilers attain excellent feed conversion ratios of 1.9 to 2.0 pounds of feed per pound of gain. Today's large and highly automated broiler farms are very labor efficient and require high management skills. (See Chapter 35.)

---

[3] Lance, G. Chris, *Poultry Sci,* 74 (1995): 398.

Turkey production is somewhat more seasonal but similar to that of broilers. Vertical integration approaches the level observed in broiler production. However, turkeys can be reared on open range or in pen shelters, so facility costs are considerably lower (see color plate D). In 1994, the typical tom (male) was marketed at about 127 days and weighed almost 30 pounds. The average hen was marketed at 100 days and weighed slightly over 15 pounds.

Some breeding companies are moving toward strains of turkeys that would be marketed in two categories: medium-heavy and super-heavy weight. Their goals are to market medium-heavy hens at 14 pounds in 84 days and toms at 24 pounds in 99 days. Super-heavy strains would be marketed at about 22 pounds for hens in 122 days and 32 pounds for toms in 125 days.

Some processors specializing in deboned products anticipate even heavier (36- to 38-pound) toms in the future.[4] (See Section 35.6; also Section 35.7 for other meat-producing birds.)

Horse Management

For the most part, horses and mules are no longer necessary for draft purposes in the United States. However, well-bred and well-trained horses are considered invaluable on numerous farms and ranches throughout the country.

Although horse numbers are about one-third of their peak 80 years ago, they continue to be an important segment of animal agriculture. Today, the pleasure provided by horses and ponies as well as mules is reflected in a fairly steady national population of about 6.5 million head.

Many adult owners continue a tradition of horse or pony ownership established as youths, and have one or more for their family. They value the pleasures and responsibilities of family ownership, especially as their children become involved. Children develop a sense of responsibility as they care, train, and ride their pony or horse. In contrast to most other farm livestock, the well-trained pony or horse often remains with the same family for a dozen years or more.

Some less experienced owners, with ample leisure time and money, have chosen the horse to be a part of their recreational activities. Others have chosen the breeding, training, showing, and merchandising of horses to provide both enjoyment and a profession.

Although horses, ponies, and mules comprise a small portion of the U.S. animal population, purchases for their care and management are economically significant. When all economic factors related to the U.S. horse industry are considered, it is estimated to be a $25 billion plus business. For example, annual feed purchases for horses and mules amounts to about 1.4 million tons. Also, their economic contribution to other states is considerable. In Kentucky, horses and mules were the second leading agricultural commodity in 1996, with cash receipts estimated at $628 million.

The horse industry contributes greatly to the economy of many areas of the country. Events such as racing, rodeos, and horse shows are popular and attract large numbers of spectators. The dollar flow into the general economy becomes significant when aspects of ownership are considered. These include the costs of land, fencing, stables, trucks and trailers, tack, as well as the animal(s). Related needs include feed, medicine and veterinary care, breeding, feed, breed association fees, and sometimes training and boarding.

---

[4] *Turkey World,* July–August 1995.

The student of animal science should appreciate the equine species as an important part of our society. Although not a provider of food or fiber in the United States, most of the principles of animal science that apply to the providers of meat, milk, and eggs also apply to the health and well-being of the horse.

Knowledge of the principles of genetics, nutrition, reproductive and environmental physiology, lactation, behavior, and equine care and management is essential for ownership or when providing products or services to others involved in the horse industry. (See Chapter 32.)

Aquaculture

The growing, rearing, and marketing of aquatic organisms such as fish or shrimp is a rapidly growing industry in the United States. The term *aquaculture* is commonly used and is derived from the Latin word *aqua* or water and *culture* which means to grow or cultivate. It is a form of animal agriculture where cold-blooded organisms are grown for food and cultured in a water environment.

Although fish and other seafood are not consumed by the average American family to the extent of beef, pork, or poultry, per capita consumption still averages about 15 pounds each year. The demand for high-quality fish, and the probability that wild fish will not meet consumer demand, will likely result in well-managed aquaculture operations being profitable in the foreseeable future.

Aquaculture, such as fish farming, is somewhat labor intensive and requires close attention to water quality, regular feeding, and health of the fish. In pond culture, initial capital output can be quite high for construction of the ponds and levees. Other methods of production include raceways, cages, and water recirculation systems. Aquaculture operations can be established in the northern region of the United States through the use of recirculation systems located within confined structures, although these require a high dollar investment to establish and maintain.

The more common fish species grown in the United States are channel catfish, tilapia, hybrid striped bass, and rainbow trout. Channel catfish are the most prevalent species. Other aquaculture operations include oysters, crayfish, or freshwater shrimp.

Much of the success of aquaculture is dependent upon having a clean fresh source of water and maintaining high quality of water. Water temperature must be within the range of requirement for the species. For examples, hybrid striped bass, tipalia, and channel catfish require warmer water than rainbow trout or walleye. High water quality requires the maintenance of adequate dissolved oxygen, low levels of carbon dioxide, and a pH range between 6.5 and 9.0. Also, high nitrogenous wastes (e.g., ammonia and nitrites), if allowed to build up excessively, are detrimental to fish health.

Catfish operations usually have stocking rates up to about 6,000 pounds of fish per acre. Feed must be formulated to best suit the species being cultured—the protein requirement ranging from 30 percent for catfish to 45 percent for trout. All rations should be formulated according to current National Research Council (NRC) recommendations.

When fish reach market size they must be harvested and quickly processed. Prior to harvest they should be sampled to ensure that no off-flavors are present in their meat. Much of the success of fish farming depends upon the availability of a market and its continuous need for a high-quality finished product. To be profitable, every aquaculture producer should develop a sound marketing plan. (See Chapter 36 for more details.)

Over the past 10 to 15 years, there has been considerable interest in the area of companion animals, specifically within the discipline of agriculture. The level of interest may be attributed to the large number of animal science students entering veterinary school and many of these students are expressing a desire to work with small animals, such as cats and dogs. In addition, there is a growing population of students and faculty members who are engaging in companion animal research, which in turn funds graduate student projects and then offers employment following graduation. The trend is becoming clear that students have an interest and desire to study alterative animals in agriculture.

1.10   ANIMAL AGRICULTURE, ANIMAL WELFARE, AND COMPANION ANIMALS

When one asks or speaks of the welfare of others, very likely he or she is referring to their well-being, that is, their health, prosperity, or happiness. Today, animal welfare, or animal well-being, is frequently the topic of discussion. However, how animal welfare is defined or viewed depends largely upon each individual's or interest group's background, their degree of involvement with animals, their social beliefs, and how relevant animal welfare is to their interests.

The federal government is also involved in animal well-being. The U.S. Department of Agriculture (USDA) provides the following mission statement: "The Mission of the Animal Well-Being and Stress Control Systems Program is to provide information needed to assess farm animal well-being scientifically, reduce animal stress, produce fit animals, and improve animal production systems for enhanced well-being." Within the umbrella of the USDA, research programs are targeted toward (a) scientific measures of well-being, (b) adaptation and adaptedness, (c) social behavior and spacing, (d) cognition and motivation, (e) evaluation of practices and systems to improve well-being, and (f) bioenergetic criteria for environmental management. Currently, the USDA has animal well-being centers at the following locations: Livestock Behavior Research (West Lafayette, IN), Animal Physiology Research (Columbia, MO), U.S. Meat Animal Research Center: Biological Engineering Research (Clay Center, NE), and Poultry Research (Mississippi State, MS). To find out more about the specific projects being conducted at these four USDA-supported centers, visit them online at http://www.usda.gov/

Why should animal welfare be discussed in a text devoted to animal science and industry? Actually this topic cannot be excluded if we look at the past trends, current practices, and anticipate the future direction of animal agriculture. In today's world, animal welfare is of concern to all—the breeder and producer of animals, the consumer and general public, as well as others such as animal rightists. In the past, concern about the treatment of animals and their welfare was essentially limited to the occasional incidence of animal cruelty. Today our society insists that ethical concerns extend beyond the cruelty of animals to include their comfort as well as productivity.

For some readers, it might be helpful to reflect upon some examples of changes that have occurred in animal production and management to better relate to the welfare of farm animals. Some of these are presented below.

- For every 100 farms in operation some 60 years ago, there are now only 25 much larger and more efficient farms. Today, less than 2 percent of the U.S. population is actively engaged in farming. Those families who moved to the urban and suburban areas, their children, and perhaps grandchildren, now have little if any contact with farm animals or their management.

- Several decades ago one would have seen several different species of live-stock and poultry on these smaller but more diversified farms. Today, most farms specialize in only one species for their major enterprise because of the need for sufficient size, increased technology and knowledge, and greater capital investment.

- In those days, herds of milk cows could be seen grazing pasture or cooling themselves in streams or ponds on a summer day. Today, very few dairy herds are allowed to graze. Instead, they are housed in well-designed facilities throughout the year that provide a cleaner, more sanitary environment for the production of high-quality milk. Improved genetics and management has resulted in increased milk production from about 4,500 pounds to over 18,000 pounds per cow per year.

- In the past, sows and pigs could be observed in a pasture setting with their individual farrowing hutches, whereas today most are confined in labor-efficient facilities and give birth to pigs in carefully designed farrowing crates. Today's hogs reach a heavier market weight more quickly and pro-duce a much leaner carcass than the lard-type hogs produced several decades ago.

- The egg-laying flock has evolved from a family-size operation where hens laid their eggs in individual nests and received some grain to supplement their diet (as they were free to roam about in search of food) to very large groups of birds housed in highly specialized environmentally controlled fa-cilities. Most are confined to cages and receive a carefully programmed diet. Today's highly efficient laying hen produces about 300 eggs per year as com-pared to about 200 eggs produced by the average hen some 60 years ago.

The above examples illustrate how animal agriculture has shifted to much larger, more concentrated, and more efficient operations. Greater demands for animal productiv-ity have been and will continue to be made. Higher quality products as demanded by con-sumers are now produced because of improved genetics and excellence in animal care and management. Today's livestock and poultry producers recognize that the humane care of animals is essential for excellent performance, and that stress must be kept to a minimum.

As viewed by many in animal agriculture, *especially producers*, in addition to the animal's well-being, welfare refers to the presence of suitable conditions and manage-ment most relevant to the animals' productivity and profitability.[5] This encompasses all sound practices of animal science such as excellence in nutrition, housing, protection from diseases and parasites, as well as protection from predators, and most importantly, excellence in care and management by the owners or managers.

A similar definition of animal welfare has been adopted by the American Vet-erinary Medical Association (AVMA) as follows: "Animal welfare is a humane responsi-bility that encompasses all aspects of animal well-being, including proper housing, management, nutrition, disease prevention and treatment, responsible care, humane handling, and when necessary humane euthanasia."[6]

---

[5] Rollin, B. E. *Farm Animal Welfare. Social, Bioethical, and Research Issues*. Ames: Iowa State University Press. 1995.
[6] DeHart, H. "Animal rights position allows no room for Compromise." *Feedstuffs* 65, no. 1 (1993): 8.

Few people would disagree with the above definitions as expressed by producer or veterinarian groups. However, it should be noted that a very high percentage of the U.S. population is no longer associated with farming or well acquainted with modern animal agricultural practices. This *consumer* group, or general public, tends to relate the welfare of animals to how they would want their companion animals (dog, cat, or horse) to be treated. They also are more apt to be concerned about possible boredom of animals, perceived unhappiness, social deprivation, and animal suffering that could be associated with animal productivity and efficiency.

A third and relatively small percentage of people—the animal rights activists—express stronger views regarding animal welfare. Extremists in this group advocate that animals should enjoy the same rights as humans and should not be exploited for any reason. However, most animal rightists are more conservative in their views and recognize differences between the rights of humans and animals but stress the rights of animals to freedom, sufficient space, and social contact for expression of their natural instincts.

A major concern among many in the animal industry is that the differences between animal welfare and animal rights need to be more clearly understood. Some also fear that lobbied actions of animal rights extremists could possibly result in unrealistic regulations that might greatly alter the efficiency and cost of animal production. Therefore, it is apparent that the animal rights and welfare groups be recognized and the various issues of animal rights and welfare be addressed in a cooperative effort with members of the educational institutions and animal industry groups. Through communication of thoughts, education, better understanding, and collaboration between interest groups, additional and beneficial guidelines for humane care and management of most farm animal species will likely evolve.

Most Americans, animal agriculturists as well as animal welfare advocates, are probably neither extreme (abolitionists) nor conservative (dominionists) in their views regarding animal rights. Instead they can be categorized as *utilitarian believers*.[7] As such these people contend that animals have rights, but that human needs such as food, fiber, work, and research outweigh possible violations of animal rights. Utilitarian believers advocate the compassionate use of animals for the benefit of humans.

Those most involved with animal production and their welfare must continually minimize conditions that might contribute to pain, suffering, or excessive stress. Their thorough understanding and observation of animal behavior (as discussed in Chapter 13) can be beneficial through the implementation of practices that enhance animal welfare as well as overall productivity of the enterprise.

1.11   THE ENVIRONMENT AND ANIMAL WASTE MANAGEMENT

Animal waste management problems have steadily intensified from the days when animals were first domesticated. As discussed in the previous section, animal production was much less concentrated several decades ago and much of the manure was distributed by the animal onto pasture and rangeland. Table 1–5 illustrates the enormous amounts of manure currently produced annually by meat animals (cattle, swine, lambs, and poultry), dairy cows, and laying hens—fed and housed in *confined operations* in the United States. (See also Section 11.11.) The yearly production of over 422

---

[7] *Instructional Materials about Animal Welfare*, National FFA Foundation, Madison, WI.

TABLE 1–5	Estimated Yearly Manure Production of Confined Animals in the United States			
Species	**Number of Animals** (Million)	**Daily Yield per Animal[a]** (Lb)	**Days on Feed** (Days)	**Total Yearly Production** (Million tons)
Meat Animals[b]				
Feedlot Cattle	22.0	58.00	150	95.70
Swine	101.8	6.50	170	56.24
Feedlot Lambs	5.6	3.30	130	1.20
Broilers	7,934.3	0.28	50	55.54
Turkeys	301.3	0.95	114	16.32
Dairy Cows[c]	9.2	105.00	365	176.30
Laying Hens	318.1	0.36	365	20.90
Total				422.20

[a]Wet weight basis of fecal waste per day; excludes urine, bedding, or added water.

[b]Animals marketed for slaughter per year; excludes breeding stock, other poultry, and exotic animals.

[c]Excludes herd replacements.

Source: Adapted from Council for Agricultural Science and Technology (CAST) Task Force Report No. 128 Integrated Animal Waste Management, November 1996 and USDA sources.

million tons of manure would be considerably higher if the wastes from all farm animals (e.g., horses and other animals on pasture) were considered.[8] Concentrated feeding operations are probably the greatest risk for air, water, and soil pollution because larger volumes of manure are produced within a small geographic area.

The USDA suggests that nutrients from livestock and poultry manure are key sources of water pollution. The Clean Water Act was developed in 1972 as a result of livestock and poultry operations becoming more concentrated and specialized. Since 1972, the overall number of farms has decreased, but, conversely, they have increased in size and thus the numbers of livestock or poultry per farm and per acre have increased the risk of water pollution, with manure being disposed of in ways not adequately addressed in the original regulations. In order to keep up with the ever-changing size of livestock and poultry operations, the U.S. Environmental Protection Agency (EPA) proposed new regulations that would compel operations with the largest number of animals to manage their manure according to a nutrient management plan. These new regulations were implemented in 2003 and called for in the *Unified National Strategy for Animal Feeding Operations,* which was developed by the USDA and EPA. The overall goal was to minimize water quality and public health impacts from improperly managed animal manure. Much of the focus has been on the largest animal-feeding operations. For smaller operations a nutrient management plan would be recommended but not required. Table 1–6 depicts the number of animals that constitute a small, medium, and large animal operation unit.

---

[8] In addition to the fecal waste of animals, animal waste includes a variety of materials such as urine, used bedding, spilled feed, etc.

TABLE 1–6	Number of Animals Defined in Small, Medium, and Large Operations		
Operation	**Large**	**Medium**	**Small**
Cattle or cow-calf pairs	1,000 or more	300–999	Less than 300
Swine (weighing less than 55 pounds)	10,000 or more	3,000–9,999	Less than 3,000
Swine (weighing over 55 pounds)	2,500 or more	750–2,499	Less than 750
Mature dairy cattle	700 or more	200–699	Less than 200
Turkeys	55,000 or more	16,500–54,999	Less than 16,500
Laying hens or broilers (liquid manure-handling system)	30,000 or more	9,000–29,999	Less than 9,000
Chickens other than laying hens (other than a liquid manure-handling system)	125,000 or more	37,500–124,999	Less than 37,500

Source: *Adapted from 7191 Federal Register 68, no. 29, Wednesday, February 12, 2003, Rules and Regulations.*

The following examples illustrate how animal production units have expanded, and continue to expand, in the United States.

- New dairies, ranging from 700 to 10,000 or more cows, are becoming established in areas where feed, climate, and other resources best suit their needs.
- Today there are about 400 commercial cattle feedlots in the United States, largely in the Plains states, each containing over 8,000 head; about one-fourth of these have the capacity to feed over 32,000 head per day.
- A commercial swine operation may consist of 4,000 to 6,000 sows or more. Some companies anticipate having as many as 100,000 sows in different units within a small geographic area.[9]
- Many of the laying hen facilities have the capacity for several hundred thousand birds, with some units that house over 2 million.

As the advantages of confinement rearing have become more apparent, many farm operations have shifted their production to larger, more concentrated animal production units. However, as animal numbers have become more concentrated so has the volume of waste—and the potential for increased pollution problems.

As animal production units have enlarged, so has the human population with expansion of homes into the suburban and rural areas. Potential problems of coexistence of animal production operations and their new nonfarm neighbors have increased. Today's animal producers are challenged not only to produce animals more efficiently and profitably but also to properly manage their operations so that the risks of pollution from

---

[9] Herrick, J. B. *Large Animal Veterinarian*, January–February 1995.

animal waste are minimized. If not properly managed and utilized, the large amounts of animal waste can cause significant water, air, and soil pollution.

Potential air pollution or odor problems can occur because animal waste is primarily organic and largely biodegradable through microbial action. As it decomposes, odors are liberated, especially when agitated, with some that may be offensive enough to cause neighbors alarm, particularly during field application. Some odors are characteristic of livestock production, and even though they may not be unpleasant to the owner, often are offensive to neighbors. Careful facility site location and design, and consideration of neighbors when hauling and spreading animal waste on the land, reduces the possible risk of alienation and litigation.

Water quality must be maintained to ensure its safety and use for humans, as well as for animal consumption, agriculture, industry, fisheries, wildlife, and recreational activities. The EPA is responsible for the establishment of standards to prevent harm to the human, plant, animal, and water populations and to protect human health from harmful contaminants (e.g., pathogens and toxic chemicals) in drinking water. These standards have been legislated through the 1972 federal Clean Water Act. All animal feeding operations must prevent contamination of *any* water source. If not, they are violating water pollution regulations. A permit is required for all operations that exceed a certain number of animals (e.g., 1,000 slaughter and feeder cattle, 2,500 swine, 700 mature dairy cattle) after approval of the operation's manure management plan. More recently, the USDA has initiated a plan that focuses on proper management and disposal of livestock waste and requires all animal feeding operations to implement nutrient management planning to minimize their impact on water quality.[10]

Careless waste management practices, or accidental waste spillage, can result in groundwater (underground wells and aquifers)—and surface water such as lakes and streams—being contaminated from animal sources. Water pollution can result from runoff water from open feedlots, after field application of manure, or leaching into soil water from leaking manure storage.

Soil pollution and changes in soil properties can be adversely affected through excessive applications of animal waste. For example, due to its soluble salt content, the application of animal waste at high rates can increase soil salinity enough to reduce crop growth. Also, high amounts of organic nitrogen, as found in manure, upon decomposition releases ammonium that can accumulate in the soil—and high levels of ammonia in concentrated areas—resulting in reduced seed germination and plant seedling vigor.

Some differences exist in the composition of manure of different animal species. For example, poultry manure is high in uric acid, which is rapidly converted to ammonia, and is considered a greater hazard to crops than livestock manure when applied at similar rates.

Livestock producers should develop a system and strategies for managing animal waste that considers the following resource areas:[11]

- *Soil.*   Waste must be applied to the soil at levels not to exceed its capacity to absorb and store them.

---

[10] The USDA plan is referred to as the *Draft Strategy for Addressing Environmental and Public Health Impacts from Animal Feeding Operations.*
[11] Boyd, W. H. *Supplement to the Journal of Soil and Water Conservation* 49, no. 47 (1994): 53.

- *Water.*   The quality of surface water and groundwater must be maintained. Polluted water, such as runoff water from open lots, must be retained for treatment or storage for later field application when conditions are appropriate.

- *Air.*   Producers must consider potential objectionable odors from confined animals and waste storage areas, including lagoons and field application, and must employ methods that minimize degradation of air quality. Air movement, temperature, humidity, and intensity of odors must be considered in deciding the best time for application of manure to fields.

- *Plants.*   Plants can effectively recycle waste nutrients. Application rates must not be toxic or exceed the plant's needs. Tree plantings, such as evergreens, also can be helpful in screening areas of the operation, providing wind direction or wind breaks, reducing noise, and reducing soil erosion.

- *Animals.*   A healthy safe environment is essential and a good waste management system protects animals from the potential hazards of toxic gases, diseases, parasites, and insects. Contaminated water must not be accessible to animals.

Certain nutritional practices can be beneficial in reducing the risk of soil, water, and air pollution. Both odor and nitrogen pollution are primarily produced by the crude protein contained in the ration. Many odor compounds are associated with protein and amino acid degradation. Some research has focused on the source and conditions promoting odors by (1) altering the type of intestinal microbial populations; (2) changing the structure of odoriferous materials through changes in the composition of the ration, and through additives to feed and waste storage (e.g., underground pits or lagoons); and (3) trapping offensive odors through specially designed exhaust systems.[12]

Excessive amounts of nitrogen and phosphorus can be a major cause of soil and water pollution. For example, runoff of either nutrient from cropland or feedlots into ponds can result in rapid growth of algae (or pond scums). Eventually algae die and the decomposition process utilizes large quantities of dissolved oxygen in the water—and results in poor growth of fish and other aquatic animal life. (See also Chapter 36.)

Changes in rations can reduce the risk of water and soil pollution due to excessive nitrogen and phosphorus. In the traditional formulation of nonruminant rations, more than one high-protein feed is sometimes included to ensure proper balance of amino acids. However, their amounts often can be reduced by the inclusion of commercially available amino acids such as lysine, methionine, tryptophan, and threonine, with significant reduction in nitrogen excretion into the waste. Feed preservatives containing propionic acids also have been shown to reduce up to 25 percent of the ammonium nitrogen level in the gastrointestinal tract of the pig.

Much of the phosphorus stored in plant seeds is present as phytate-bound phosphorus that cannot be utilized by monogastric animals and is excreted in the waste—and ultimately returned to the soil. Supplementing diets with enzymes, such as microbial phytase, has resulted in better use of phytate-bound phosphorus with up to 25–30 percent reduction of phosphorus in fecal waste. Recently more farmers are formulating rations that conform closely to nutritional requirements, and utilizing phase feeding (by age groups) to further reduce fecal nitrogen and phosphorus. Today's farm-

---

[12] Coelho, M. B. *Feedstuffs,* June 20, 1994.

ers also pay considerable attention to soil tests to better utilize nitrogen, phosphorus, and potassium, and to minimize the risk of soil and water pollution.

An increasing number of farmers are utilizing a *Global Positioning System (GPS)* to assist them in nutrient applications to fields. This system is mounted on their applicator equipment and receives the precise geographical position from a satellite. The signal is received by an onboard computer and indicates the position in the field where the equipment is operating. With this information, the operator can more accurately apply fertilizer according to the soil fertility (or yield data) at specific locations within a field. This system helps to minimize overapplication of fertilizers, whether from commercial or animal sources, and to reduce excessive applications that contribute to soil and water pollution.

Several methods are used to return animal waste to the land. The conventional manure spreader was once the method of choice because of the solid or thick slurry consistency of manure. Today most confinement systems use tank wagons for liquid manure, or an increasing number are utilizing a tank wagon equipped with injection knives (Figure 1–16). By incorporating waste into the soil, fewer nutrients are lost, less runoff occurs, and fewer odors are emitted into the environment during application. Today's farmers must keep accurate records of the quantity of manure produced on their farm, and how, where, when, and the quantities applied to the land. Land application should be timely with consideration to the weather, especially prevailing winds, and to possible

Figure 1–16
Tank wagons are commonly used to efficiently and uniformly apply animal waste. The use of injection knives (lower photo) can reduce nutrient losses and odors. (*Source:* Purdue University.)

offensive odor for adjoining neighbors. The rate of application must not exceed the capacity of the soil and planned crops to properly assimilate the nutrients and to prevent soil or water pollution.

Other practices that farmers employ to properly manage animal waste and reduce the risk of air, water, and soil pollution include

- Providing grassed filter strips to receive liquids and solids draining from settling basins, and buffer strips between fields to help control runoff
- Storing animal waste away from environmentally sensitive areas
- Using air "scrubbers" that remove odors and dust particles from animal confinement buildings
- Diverting water away from lots and waste storage
- Placing synthetic, permeable cover materials that float on top of the waste lagoons that reduce the release of gases
- Using solids-liquid separators and composting manure, and, in some cases, selling the compost to commercial businesses such as yard and garden shops
- Using manure to produce gases for power generation.

Each animal production unit system will differ to some extent in a waste management plan, but the handling, storage, and application of animal waste should be environmentally and economically sound. Also, every system must fulfill the requirements of state and federal regulations.

QUESTIONS FOR STUDY AND DISCUSSION

1. Which animal products increased significantly in per capita consumption in the United States during the 1980s and 1990s?

2. Briefly discuss progress being made by less-developed countries (LDCs) and the aid provided to them by the United States.

3. What region of the world is well above average in population growth?

4. What percent of the U.S. population is represented by farmers, and how many acres does the average farm contain?

5. Explain how our forefathers aided agricultural education by certain law enactments.

6. Define *palatability* and *biological value of protein*.

7. What species belongs to the genus *Ovis? Bos? Sus? Equus? Galus?*

8. Compare how species characteristics affect adaptation to the environment and kind of feed consumed, such as swine as compared to cattle.

9. Which of the major enterprises have very high labor and investment needs per animal unit? Which provide good cash flow? Which may have greater financial risks?

10. At what age do most dairy heifers calve and enter the milking herd?

11. What is the fate of most male dairy calves?

12. What is the major source of income derived from the U.S. farm sheep flock?

13. At what weight range are most market hogs sold?

14. Briefly describe how the system of vertical integration of the poultry industry works.

15. At what age are most egg-laying birds brought into production?

16. If 1-day-old broiler chicks arrive at the rearing facility on March 1, at what date and weight should they be scheduled for delivery to the processor?

17. Explain how humane care of animals benefits the producer and consumer, as well as animal rights and welfare groups.

18. Calculate the estimated amount of manure produced by a 400-cow dairy for 1 year. (Utilize Table 1–5 and exclude herd replacements.)

19. What changes can be made in nutrition and feeding of animals to reduce the potential for soil, water, and air pollution?

20. List several of the additional management practices farmers have taken to reduce environmental pollution.

2

The Animal Industry

All producers of farm animals rely heavily upon products and services provided by related industries, colleges and universities, and governmental agencies for their products, equipment, or services. The many segments of agribusiness also depend upon the patronage of animal producers in the purchases of their products or services. This chapter will illustrate the breadth and depth of the total animal industry and the related sciences.

The gross income from animals and animal products sold from or utilized on U.S. farms and ranches has ranged from $90 to $100 billion in recent years, the variation due to both volume and price. For individual animal products or species, plus the major crops, gross incomes are shown in Table 2–1. Note that many crops of lesser dollar volume are not listed. Also, because a considerable amount of the corn is fed on the farm in essence, it is counted twice in these data.

Meat processors have been a significant part of American agriculture since soon after the Pilgrims landed. Woolen mills and dairies also developed early. Veterinarians

TABLE 2–1	U.S. Gross Income for Selected Animals, Animal Products, and Major Crops		
	Millions of Dollars		
	2002	2001	2000
Cattle	38,302	40,804	
Hogs	9,651	12,462	53,395
Sheep	439	405	
Broilers	13,400	16,693	13,988
Turkeys	2,700	2,790	2,822
Eggs	4,200	4,446	4,345
Milk	20,688	24,848	20,771
All meat and animal products[a]	89,445	102,514	95,407
Cotton	3,594	3,122	4,260
Corn for grain	21,213	18,888	18,499
Wheat	5,863	5,440	5,782
All hay	12,433	12,603	11,407
Soybeans	14,755	12,606	12,467
Potatoes	3,151	3,058	2,591
Tobacco	1,726	1,952	2,002

[a]Includes estimated figures for sheep, goats, wool, and mohair.

Source: USDA, NASS.

have been an identified segment of the animal industry since the early 1900s because of their special academic training and skills.

Over time, scientific and technological developments, specialization, and desire to improve production and lower on-farm labor requirements resulted in differentiation from farming and ranching of several input industries, such as feed manufacturing, the animal health–pharmaceutical industry, and livestock and poultry equipment. Food retailers and restaurants are part of the broad animal industry; they are the ultimate product outlet to consumers.

An industry still in the process of differentiating from traditional farming and ranching is the animal genetics industry. A number of purebred herds and flocks, suppliers of breeding stock to commercial producers, are on general farms and ranches. However, production of swine breeding stock has moved considerably into specialized companies and that movement is virtually complete in poultry genetics.

University colleges of agriculture and veterinary medicine, agricultural experiment stations, and extension education programs, as well as the U.S. Department of Agriculture and some other federal research units, provide valuable information to growers and feeders as well as to the allied professions and businesses.

If the total livestock and poultry industry is to successfully provide consistently high-quality and wholesome food and other products to the consumer at competitive prices, the industry must be staffed by men and women well trained, educated, and experienced in animal science. Nutritionists, sales and field service workers, buyers, production managers, geneticists, veterinarians, physiologists, food scientists, and other specialists are essential. Journalists, advertising specialists, accountants, office managers, chemists, economists, engineers, and others are also employed in the industry, and they can benefit from experience and education related to livestock and poultry.

The sciences related to animal production and the processing of animal products utilize some of the world's most intelligent and ingenious people. They may be educated as nutritionists, mathematicians, population geneticists, biotechnologists, physiologists, economists, microbiologists, or in some other discipline. Their lives are devoted to searching for clues to the mysteries of animal life, growth, reproduction, and product quality. Beneficiaries of their problem-solving careers are all of society—the consumers.

Upon completion of this chapter, the reader should be able to

1. Identify the major areas of the United States where livestock and poultry density is high, and explain why.

2. Provide examples of shifts in animal population density during this century, and list some reasons for these shifts.

3. Describe the feed manufacturing, meat-processing, and animal product processing industries in terms of general size, location, concentration, and change.

4. List three different roles veterinarians play in the total animal industry.

5. Describe the panorama of sciences that relate to animal production and processing.

6. List several functions related to the animal industry that are performed by universities and governmental agencies.

Memorization of the numerous statistics provided in this chapter is not intended or necessary. They are provided to better illustrate the magnitude and changes that have occurred in animal science and related industries.

2.1   LIVESTOCK AND POULTRY DENSITY

The geographic density of animal agriculture is heavily dependent on feed supply. It is generally more costly to transport the feed required to produce the animal product than to transport the product. In recent years, however, because of concerns about odor and other features of large and intensive animal production units, some have been located in less populous areas, more distant from feed production.

Rainfall and soil fertility are important for feed production. Water is essential for plant metabolism and growth, and the soil must supply most plant nutrients. Soil structure, which is partially dependent upon temperature and decay of organic matter, must be loose enough to provide aeration and good drainage yet store enough moisture for plant growth between rains.

More grain and forage can be grown on flat or gently rolling land than on hilly or mountainous land. Because steep or rough land is subject to erosion and usually has more shallow soil, it is usually devoted to grazing or timber.

A combination of these factors—level or gently rolling topography, adequate rainfall, well-drained and friable, fertile soil—is essential for maximum livestock and poultry feed production (Figure 2–1). Such a situation exists in the central region of the United States, especially in the Corn Belt. Iowa continues to have the densest population of cattle and hogs, producing more red meat per square mile than any other state in the country. Adjoining states rank high in production of feed and animals because of suitable soil fertility, topography, rainfall, and other climatic conditions. Other regions

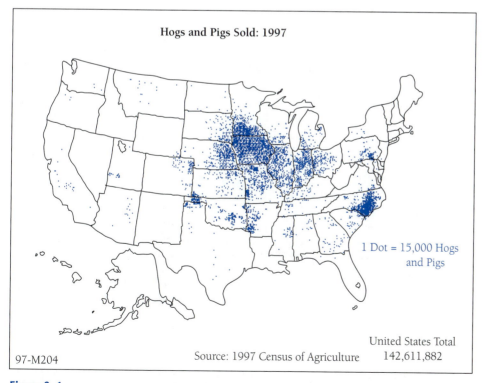

Figure 2–1
Relative density of hog marketings illustrates the predominance of the Corn Belt in providing grain for meat production. USDA, NASS at http://www.nass.usda.gov.

of the United States where land may be less suited for crop production can be competitive in animal agriculture, depending on labor, utilities, grain transportation, and other costs.

Intensive farming exists in the irrigated valleys of New Mexico, Arizona, California, and other western and Plains states. Where water is supplied by irrigation, but the soil is rich and level intense crop production is conducive to high-level meat, egg, and milk production. The same is true for good soil areas of the Northeast, Appalachia, and Southeast regions and Washington, Idaho, and the Canadian provinces, where there is adequate rainfall or water for irrigation.

Direct effects of climate and location on the animals are also important. Certain species or breeds of livestock and poultry are more adapted to the temperate conditions in the central United States than to the semitropical or tropical climate near the Gulf of Mexico or in Central or South America.

Geographic density is not static. During the mid-twentieth century the center of the beef cow industry shifted toward the Southeast, prompted in part by the conversion of cotton land to grass, improved insect control, and availability of cattle more tolerant to heat. Egg production shifted from the Corn Belt and Northeast toward the Southeast and Arkansas. Feed company and processor financing was offered in the Southeast to underemployed farmers short on capital, and that coincided with specialization and other attractive enterprises in the Corn Belt.

But egg production moved back toward the Corn Belt states in the 1980s. Iowa dropped from number 1 in egg production in the 1950s to 15, then came back to number 1 in 2002. Some facilities in the Southeast had become obsolete and, before rebuilding, major egg-producing companies studied comparative grain costs and saw cheaper corn in the Corn Belt.

Cattle shifted both geographically and in type of operations, from the traditional farmer or feeder of the central Corn Belt of Indiana, Illinois, and Iowa to large commercial and custom feedlots in the western Corn Belt and southern Plains.

Dairy production dispersed somewhat from the traditional "dairy belt," Wisconsin to New York, during the latter part of the twentieth century, and, at this writing, South Dakota, Iowa, and some other states are recruiting dairy producers from both California and northern Europe. Environmental regulations and land cost may make the Midwest attractive to these dairy producers, and the Midwest processing industry needs more milk.

Although these and other shifts in geographic density—and type of operation—may be responses to feed supply, transportation, labor availability, or other external factors, the changes are invariably initiated by individual *people* who see *opportunity* in a different location or a different form of business. The changes don't "just happen." Investors, people in other industries or professions who see opportunity for themselves or their communities in animal production—do some calculations, study feasibilities, and make or bring about investment.

The same process no doubt occurred decades ago, in the beginning stages of industry differentiation, to the industries described in the next sections.

2.2   THE ANIMAL GENETICS INDUSTRY

Although Chapter 21 illustrates many breeds of all major livestock and poultry species, these are primarily of historical significance for commercial poultry production and that trend is apparent in swine.

In the general evolution of domestic animals for food production, by selection, herds and flocks met specific needs of an area. Continued selection over generations resulted in "true breeding," offspring much like their parents. Color markings were important and usually became the "trademark," for easy identification. There was pride in "my breed," and breeds have contributed tremendously to productivity improvements.

During the latter half of the twentieth century, focus in poultry improvement moved to selection and development of lines within the breeds that were most efficient and popular for egg or meat production. Because such lines are generally the same color, lines are identified by name or number. More rapid genetic change was possible in poultry than in other animal species because of higher prolificness (more than 200 eggs per hen per year) and shorter generation time.

Accompanying this process was the industrialization of broiler and egg production—units with thousands of birds under precision management and responding to consumer demand for uniform quality.

The result in the early twenty-first century is a few poultry genetics companies that develop and provide the male and female grandparents or parents for chicks that enter broiler or pullet growing houses. Four companies, all global in their marketing, provide most of the layer lines and a similar number provide most of the broiler lines used in the United States.

These poultry genetics companies usually sell line-cross females and males to selected franchise growers who, in turn, produce the eggs and hatch and deliver the chicks (a cross from those male and female parents) to the broiler or layer producer.

Even though a similar trend clearly exists in swine, with several global companies developing and selling male and female parents as described for poultry, separate breeds and individual breeders yet play a prominent role in the swine industry. However, a prominent swine breeder must maintain a sufficiently large herd of parent stock, perhaps several hundred females, to produce enough uniform boars or gilts to satisfy the needs of large farrow-to-finish or feeder pig producers.

There remain some individual breeders with limited volume, who may sell one or several boars to individual producers, but continued industry concentration suggests a limited opportunity for such breeders.

It is apparent that relative prolificness is a primary influence on the speed of genetic change described above. Some line development within breeds has occurred in beef cattle, but most focus continues to be on selecting bulls within breeds for crossbreeding. Consequently, beef breeding stock, the genetic source for the beef industry, is provided by many farm-based or ranch-based breeders either at auction or by private treaty.

In dairy cattle, breeds have retained their full identity and significance, but the high-volume-producing-Holstein is far more numerous because of the industry's focus on milk volume. Detailed production record systems, beginning early in the last century, allowed rather accurate prediction of the breeding value of young bulls. That, plus the relative ease of artificially inseminating constrained dairy cows, led to "bull studs," companies or cooperatives that purchase very highly rated young bulls and collect, process and sell semen for the nation's dairy herds. Here, as in the poultry genetics field, considerable consolidation of these businesses has resulted in a small number of companies now providing most of the input genetics for dairy production.

More recent components of the animal genetics industry also speeding genetic improvement, are embryo transfer and cloning. **Embryo transfer (ET)** enables the production of several genetically similar calves, pigs, or lambs from genetically superior

> *Embryo transfer (ET)*
> Five- to 6-day-old embryo from an elite female is transferred to one of the uterine horns of a recipient female—both horns if two embryos are transferring; most commonly done with cattle.

parents. ET involves superovulation of the donor female by hormone administration and artificial insemination. Embryos are harvested at an early stage and transferred into several female recipients whose estrus cycles have been synchronized with that of the donor animal. Prior to the transfer, the embryos can be sorted by sex, frozen for later use, or split. Microsurgical division of an embryo into two, three, or four groups of cells can result in the production of identical siblings (e.g., twins, triplets, quadruplets).

In contrast to embryo transfer, cloning of animals results in animals with identical genetic makeup, including the sex chromosome, to the animal cloned. Cloning makes possible the development of genetically uniform herds or flocks. In a herd or flock, successive "generations," as well as the animals within a generation, could be identical in their genetic makeup. (See Box 2–1.)

Both embryo transfer and cloning are highly delicate processes that require skilled technicians and well-equipped facilities. Veterinarians, physiologists, and endocrinologists, many with advanced degrees, are involved in these businesses. Most clientele are owners of superior breeding stock who see a good market for more animals than their flock or herd could provide under traditional reproduction schedules.

2.3 THE FEED INDUSTRY

The American Feed Industry Association indicated in 2003 that 181 companies, members of that association, manufacture 75 percent of the commercial feed and pet food produced in the United States. These tend to be the larger companies, many with several manufacturing plants in this country and abroad. In addition, several thousand other companies, including cooperatives, operate feed mixing facilities that serve limited geographic areas or a specific clientele (see Figure 2–2).

In addition to the feed manufacturers, 25 companies that produce and sell microbial or enzyme products, largely for improving value of forage, 73 companies that provide major ingredients to feed mixers, and 66 suppliers of microingredients (vitamins, trace minerals, etc.) were also members of this association. This illustrates the breadth of the feed industry, and suggests diverse career opportunities in research, ration formulation, quality control, mill operation, advertising, consultation with customers, and sales.

With a high proportion of poultry and swine in large confinement facilities, most of their manufactured feed is a complete ration, usually delivered daily in bulk (not bagged) by truck and augered or blown into bins at the feeding site. For small and hobby operations, and for baby chicks and calves as well as pets, complete rations are usually sold in 20- or 40-pound bags. For horses, beef cow herds, dairies without their own mixing facilities, and for smaller commercial livestock operations where farm-grown grain is fed, the industry provides mixed "supplements," in bag or bulk, higher in protein and with vitamins and minerals to complement the nutritional features of the grains and roughage.

The U.S. feed industry in the early twenty-first century is described as a "mature industry," not rapidly growing in terms of dollar volume or personnel. This does not mean an unchanging industry; a 2003 industry survey indicated a number of changes.[1] For example, 27 percent now produce some antibiotic-free feeds and 6 percent produce one or more organic feeds.

Although some companies are expanding their business, there is some surplus manufacturing capacity in the total industry. The reason is the rapid movement from

---

[1] *Feedstuffs* 75, no. 28 (2003): 12.

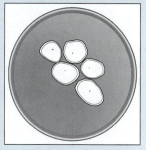

1. Cells from a 30-day-old calf fetus serve as the genetic source in the cloning process. These 30-day-old cells are also known as primordial germ cells.

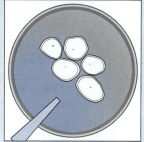

2. In a step that involves a patented process, growth-promoting proteins are introduced to get 30-day-old primordial cells to become stem cells.

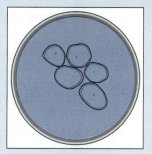

3. Permanent embryonic stem cells develop. These cells are different from 30-day-old primordial cells because they are now capable of developing into a calf.

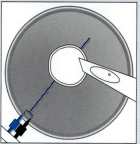

4. The nucleus of an unfertilized egg is removed in a process called enucleation. The unfertilized egg "shell" will become the home of an embryonic stem cell.

5. An embryonic stem cell is introduced into an unfertilized egg "shell," scientifically known as the zona pellucida. The stem cell is fused into the zona pellucida by applying a low-voltage current.

6. Now the cell has the genetic material from the 30-day-old calf fetus and is wrapped by the zona pellucida of the unfertilized egg. Growth in the cell is activated with a protein media.

7. Cells grow rapidly, just as if sperm and egg united to create a new calf. This step creates more duplicates or clones of the original, extending the genetic source. Cell growth occurs for seven days.

8. This step is the same as Step 4. After the nucleus of another unfertilized egg is removed, it will be used to grow a cloned cell from Step 7, inside the empty egg's zona pellucida.

9. This step is similar to Step 5, except a copy of the cell in Step 5 is placed inside the zona pellucida of the unfertilized egg. Low voltage is used to join the two components.

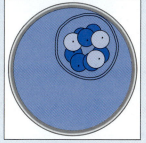

10. The cloned cell multiplies for seven days, just like a normal embryo. After Day 7, lab technicians look at the embryo to see if it's growing properly and can be transplanted.

11. Until this point, all the entire cloning process is done in the lab. Now the 7-day-old embryo is transplanted into a recipient cow where it develops into a calf.

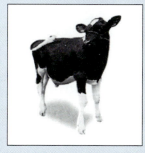

12. After a gestation period of 280 days, the cloned calf is born. To verify the process, DNA from the calf, dam, and sire are used to verify his genetics are the same.

Source: *Hoards Dairyman,* August 25, 1997.

Figure 2–2
Aerial view of portions of a north-central company's research and demonstration farm, above, and a midwestern feed manufacturing plant, below.
(*Sources:* Cargill Animal Nutrition Division, Inc. and Kent Feeds, Inc.)

smaller animal production operations, which depended on others to manufacture their rations or supplements, to a preponderance of large poultry, swine, and cattle feeding units, many with their own mixing facilities on site and purchasing ingredients directly in large volume.

2.4   VETERINARIANS AND ANIMAL HEALTH

The animal industry and owners of companion animals are very dependent upon veterinarians and suppliers of animal health products to help maintain the health, productivity, and vitality of individual animals, herds, and flocks, and in the treatment and recovery of sick or injured animals.

The American Veterinary Medical Association (AVMA) reported 69,000 members in 2003. Seventy-four percent of them work in private or corporate clinical practice and more than half of those treat exclusively horses, cats, dogs, birds, hamsters, reptiles, or other companion animals. A significant number serve in federal or state governmental agencies, including the Agricultural Research Service (ARS), Food Safety and Inspection Service (FSIS), Animal and Plant Health Inspection Service (APHIS) of the USDA; the Centers for Disease Control and Prevention (CDC); the Environmental Protection Agency (EPA); the Fish and Wildlife Service (FWS) of the Department of Interior; and parallel state agencies. In these agencies, a veterinarian's work may include research, monitoring of animal diseases, inspection of animal processor facilities and products, negotiations with other countries on certification of animal or animal product imports and exports, or interactions between animal health and human health.

Most members of AVMA are graduates of 1 of the 17 U.S. or 4 Canadian colleges of veterinary medicine accredited by AVMA, and 1,000 or more serve on college faculties in teaching, research, or clinical education.

For most commercial poultry, swine, and dairy operations, the work of the veterinarian, whether in private practice or as an employee of an input supplier, is to design a complete herd or flock health program, then consult or troubleshoot when problems develop. Such tasks as castration, vaccination, pregnancy checking, or help with birthing are usually handled by animal unit personnel. Individual animal treatment by a veterinarian is more common with companion animals or in smaller herds and flocks.

Twenty-eight animal health and pharmaceutical companies hold membership in the American Feed Industry Association. These companies employ veterinarians and persons trained in other disciplines to develop, test, and sell vaccines, feed additives, and other animal health products to clinics, livestock and poultry producers, and owners of companion animals.

2.5   FACILITIES AND EQUIPMENT

To serve the needs of the livestock industry, hundreds of companies manufacture and sell equipment designed especially for that industry, such as gates, portable corrals, working chutes, trailers, farrowing crates, scales, feed bunks or pans, feeder wagons, silage loaders, waterers, and manure loaders and spreaders. Some equipment is highly specialized and innovative, such as "nose pumps," designed for cattle to pump water from a reservoir with their noses. A panel at the back of the water bowl is a "pump handle." The cow puts her nose in the bowl, pushes at the panel to get at the water in the bottom of the bowl, and that action brings up more water.

Some of the companies are national, but many serve only a relatively small area. They may range from a small welding shop producing a single item to a company that employs several hundred to design, fabricate, and sell a full line of equipment.

Relatively few companies supply the poultry industry, and a higher proportion are national or at least cover much of the country. The poultry industry is less diverse than the livestock industry. A higher proportion are total confinement enterprises, and many of those need similar chain or flexible auger feed distribution systems and nipple waterers. For young birds, most use similar floor-level feeders.

Three or four companies provide most of the cages and the egg gathering and packing systems for layer houses. They not only manufacture such equipment but also usually provide trained and experienced installation crews.

A critical piece of equipment for many confinement systems is a computer, plus specially written software for automatic operation of ventilation fans or curtains, heating (and sometimes air conditioning), feed distribution and, in the case of layers, delivery of eggs from the cage area to the packer. In response to temperature or other sensors and to parameters entered by the operator, computers can keep temperature within a specific range in building sections, deliver fresh feed to animals at set times in 24 hours, and even blend eggs from two or more banks of cages to achieve a predetermined egg weight per dozen at the packer.

Several companies in given areas specialize in constructing poultry, swine, or other buildings. Repetitive construction experience results in lower costs and design details that serve both the animal and operator needs.

There are hundreds of small items used by the animal industry and for companion animals—show halters, brushes, combs, bridles, saddles, etc.—that no doubt comprise a multimillion-dollar business.

2.6   MARKETS

In the early twenty-first century a high proportion of animals and animal product marketing is being done in the absence of the animals or the product. It is done by negotiated contract, usually covering a period of time or by electronic (phone, fax, e-mail) or face-to-face buy/sell negotiation for specific lots. Even in the case of livestock auctions (Figure 2–3), where tradition has had the bidders looking at animals before them and responding to an auctioneer's chant, recent years have brought electronic auctions. Animals are seen on a computer or TV screen at any location and bidding is electronic.

Futures markets, described in detail in later chapters, provide a mechanism to buy and sell "contracts for future delivery," transactions that can provide price protection for either a buyer or seller. Their use by producers can be part of a marketing plan.

Figure 2–3
A market hog auction in the Northwest.
(*Source:* USDA.)

Before electronic capability and prevalence of long-term contracts, many of the cattle and sheep, and a significant proportion of fed hogs, were marketed at "terminal markets," consisting of large acreages of livestock pens located in major cities of livestock regions. There, commission agents, representing producers, would negotiate the sale of lots to processors or, in the case of feeder cattle or lambs, to feedlot operators. Although some finished hogs went through terminal markets, most were delivered to "packer buying stations" in rural towns, the producer having checked the day's prices at several stations by phone before delivering. Such terminal markets and buying stations virtually disappeared in the latter part of the twentieth century.

The most common livestock auctions today, where buyers see the animals and an auctioneer's chant is heard, involve feeder cattle, breeding animals, horses, and the animals shown by 4-H and (FFA) members at county and state fairs throughout the nation. At the latter, local citizens and businesses may purchase the animals at premium prices to support the youth, often simultaneously reselling the animals at a base price to a processor.

A nationwide association, the Livestock Marketing Association, serves the interest of about 800 member auctions, most of which hold weekly or biweekly livestock auctions. The largest volume auctions, through which many feeder cattle move, are located from western Iowa through Nebraska and the Dakotas to eastern Montana. A few auctions in that area also handle sufficient numbers of fed cattle to attract competitive buyers.

Since well before 1990, more than 90 percent of turkeys, broilers, and eggs were marketed through some form of long-term agreement with a processor, or might have been owned by the processor during production. In contrast, through virtually all of the twentieth century, most marketed cattle, hogs, and sheep were priced on what is called the "spot market," in which the price was negotiated or set that day. Milk from a dairy herd went daily to a single processor, usually through a producer-owned cooperative, and was priced each day according to some "blend" or "formula" system. In the 1990s, however, the proportion of hogs produced under some form of long-term contract—either a sale contract or a contract where the producer was paid to feed hogs owned by the processor—increased sharply to approach 80 percent. Long-term contracting of milk, whether or not the processor was a cooperative, became rather common. But in marketing of finished cattle, many buyers continue to negotiate the purchase at the pen in commercial or farmer feedlots. More detailed discussion of markets and marketing are provided in several later chapters.

2.7 PROCESSORS

Cattle, hogs, sheep, and poultry go in one side of a processing plant, and boxed beef and pork, hams, sliced bacon, legs of lamb, wieners, sausages, packaged broilers, lard, and hundreds of other meat products come out the other. A meat processor is a "disassembler" of meat animals, but also prepares some of the disassembled cuts for cooking or, in some cases, for serving (Figure 2–4). Chapter 25 is devoted entirely to meat processing.

Eggs go in one side of a plant and a variety of items, such as liquid egg whites, a seasoned omelet mix, cooked and sliced eggs for salads, or cartoned eggs for retail come out the other. From the nation's milk-processing plants come about a billion pounds of butter, seven billion pounds of cheese, 1.5 billion pounds of yogurt, a billion gallons of ice cream and, of course, containers of fluid milk to drink.

Figure 2–4
A modern meat-processing plant in southwestern Kansas. Note cattle in pie-shaped pens at the right. Slaughter area is at the right, cutting area in the middle, and boxing area to the left. (*Source:* Iowa Beef Processors. Vol 75, No. 27)

More than 7,000 meat-processing plants exist in the United States and most of the larger plants are devoted to one species: broilers, turkeys, beef, lamb, or swine. There is considerable ownership concentration. About 75 percent of the beef processing is handled by the top five companies. For pork and broilers, the top five handle about 65 and 60 percent, respectively.[2]

Some plants process the animals to retail-ready packages; some move carcasses to other facilities for further steps in preparation for retail or the food service industry. A majority of the meat plants, ranging from more than 1,000 to less than 10 employees, operate under a federal inspection program. The same is true for about 70 egg-processing plants. Twenty-five states provide a state meat inspection program, and the plants that process under these inspection programs—more than 2000, generally smaller businesses—are limited to selling their product within the state.

For both humane treatment of animals (avoidance of extended time in holding pens) and efficiency, delivery of animals to the larger processing plants is carefully scheduled. Minimum elapsed time for eggs or milk from the production unit is also important. In large production dairies, milk may be rapid chilled enroute from the cow directly to a refrigerated tanker trailer, which, when loaded, will be taken 10, 100, or 1,000 miles for processing.

A meat-, milk-, or egg-processing plant may be owned by a private corporation or a producer cooperative, or may be a joint venture, with a producer cooperative providing the animals or other input, milk or eggs.

A hundred or more woolen manufacturers, most in the eastern United States, produce more than a billion dollars' worth of woolen cloth and fabric annually. Considerable wool is imported for these mills; U.S. sheep and wool production has decreased in recent years. The fascinating story of the wool fiber and how it is made into cloth, as well as a brief discussion of wool marketing, is presented in Chapter 27.

[2]*Feedstuffs*

2.8   RESTAURANTS AND FOOD RETAILERS

Restaurants and food retailers are not only the ultimate outlet for nearly all animal products but also the animal industry's major information link *from* the consumer. The restaurants' patrons and the food retailers' customers demonstrate, by the meals they order and the products they choose, consumer preferences and volume demand. It can therefore be said that the animal industry is "consumer controlled." The industry is successful only to the extent it meets consumer desires, in quality, quantity, and the form in which it is provided to the consumer.

Historically and through most of the twentieth century, most of U.S. agriculture, especially animal agriculture, was presumed to be controlled by seasonality and convenience of production. Most calves and lambs were born in the spring, so their mothers' milk production would benefit from lush grass, and the young animals' early months would be in favorable weather. Pigs were farrowed in the spring for the same reason, and some in the fall. Those fall-farrowed pigs and the calves and lambs could be finished over the winter on newly harvested grain. Transportation cost and time precluded volume imports of animal products from other climates.

Such constraints today are minimal. More animal production is in temperature-controlled facilities. Product refrigeration and rapid transportation are commonplace, and imported products compete. Consumers now expect year-round availability of what they like.

There are changing ethnic preferences: a higher proportion of the current U.S. population was born in or their ancestry is of Southeast Asia, the Caribbean, and Central America. About 25 percent of the U.S. first graders, at this writing, are Hispanic.

More than half the U.S. meals are eaten outside the home, and franchise fast-food restaurants provide a major share of these meals. That volume allows them to prescribe the type, quality, portion size, and form in which the product is delivered and, indirectly through processors, they can exert influence on how the animals are produced. Long-term contracts for hamburger, sliced meats, scrambled egg mix, or other products are common and may include detailed specifications on processing and production.

Large-volume retailers, through electronic means, record and aggregate checkout-counter product sales data and transmit the next day's or next week's needs to suppliers. Their volume gives large retailers extraordinary leverage in the food chain and similarly allows them to prescribe product features to their suppliers.

2.9   THE ANIMAL SCIENCES

Animal agriculture is a competitive business based in significant part on scientific principles, concepts, and data. The more facts that are known about nutrition, animal behavior, product quality, and consumer behavior, the more successful the managers and workers in the industry can be. Scientists seek the facts; they are, in essence, seeking to "read successive pages in Mother Nature's Book of Knowledge" and apply what they read.

It takes good training, facilities, equipment, and time to ascertain the facts, such as the nutrient levels for the most productive rations, the most productive mating systems, the most efficient housing or the preservation techniques that ensure food safety. Animal scientists, perhaps trained in genetics, animal management, or other subdisciplines, work closely with equally well-trained chemists, physicists, mathematicians, economists, and others in this search for knowledge.

Figure 2–5
Research journals, experiment station and USDA reports, and extension service pamphlets are available to keep researchers, instructors, extension educators, and livestock breeders and producers abreast of new developments in the animal sciences. Many are available on the Internet. (*Source:* Purdue University.)

Scientific journals, such as the *Journal of Dairy Science, Journal of Nutrition, Poultry Science, Cytology,* and *Journal of Animal Science* (Figure 2–5), serve as official records of research completed. Research done at state agricultural experiment stations, USDA laboratories, and other public and private research installations is reported in these volumes and in bulletins, and is increasingly available on the Internet.

The following paragraphs, grouped by subject, illustrate the breadth of scientific endeavors directed to serve the broad animal industry. They also provide some historical perspective.

Nutrition

Animal nutrition knowledge has developed rapidly since the early 1900s. Even though the need for carbohydrates, proteins, and fats in animal diets was recognized, it was not until about 1910 that Babcock and others demonstrated that other "factors" were also needed.[3] Their studies led to the use of purified diets in nutrition studies and an understanding of the need for vitamin A and its precursor, carotene.

Other vitamins, amino acids, and minerals have been identified and characterized by similar chemical and physical procedures, such as determining molecular weight, reactivity with known compounds, color reactions, and transmission of light. Once the chemical composition and structure of a nutrient is established, a manufacturing chemist may try to reproduce it. An architect with atoms, the chemist may duplicate the functions of plants, chemically manufacturing the pure nutrient. Or, the item may be produced biologically, in bacteria or animals.

Because nutrient utilization in all species is influenced by microorganisms in the digestive tract, nutritionists have directed much attention toward the characteristics

---

[3] Babcock conceived of the idea of feeding only one feed at a time to better determine its nutritional contribution. He and his Wisconsin Experiment Station associates found that rations composed entirely of wheat (plus salt) failed to support normal growth and reproduction in dairy heifers. Heifers receiving only corn (and salt), however, were much sleeker and more vigorous, produced normal calves, and produced three times more milk the first month after calving. These dramatic differences were later determined to be due to the carotene content found in yellow corn but devoid in wheat grain.

and physiology of such microorganisms. In some studies, a portion of the digestive tract, such as the rumen or small intestine, is cannulated and samples removed for **in vitro** studies and laboratory analyses.

Environmental Physiology and Animal Management

This field of study deals with the effects of the surrounding environment upon animals and their performance. How much shelter does an animal need for top performance in the winter? How much air movement is needed for removal of moisture and animal comfort in confinement? Can air conditioning in the summer be practical? If not, what practical environmental modifications can be made? For layers housed in cages, how much space is needed per bird?

Though livestock and poultry shelters were formerly built to protect animals from the extremes of heat and cold so apparent to the producer, research has been directed to determining the environment needed by the animal for top performance. Basic research in fully insulated chambers at Pennsylvania, Missouri, California, and other experiment stations has measured the effects of controlled temperatures and humidity on milk production, growth, and reproduction in animals. With such data, the practicability of providing shelter and artificial heat or cooling to counteract adverse climatic conditions has been determined.

In the late 1990s, with many animals raised in confinement, more research focused on the influence of space, types of floors or bedding, and other features of confinement facilities on both the comfort and productivity of animals. Animal confinement continues to refined in terms of productivity vs. animal density. Currently there are many investigators attempting to understand the overall welfare vs. productivity of these highly concentrated animal facilities.

Animal Health

Effective control of animal disease involves (1) recognition of the causative agent, (2) learning the growth habits and physiology of the virus or organism involved, (3) developing chemicals, vaccines, and/or management practices that will prevent or cure the disease and (4) achieving resistance by genetic change.

Bacteriologists and virologists study microorganisms and viruses and classify them according to their growth habits, the growth media on which they thrive, and their appearance. Continued dilution of solutions known to contain disease-causing agents, and characterizing the agents separated in this fashion, often allow the identification of causative agents.

Chemicals manufactured or produced by other living organisms and known to have certain inhibitory properties are then tested for effectiveness against the disease-producing organism or agent. Often, new chemicals must be developed for specific diseases. A special serum, usually produced by animals (or another species) infected with low levels of the disease, may be developed. Dosages must be determined for various ages and weights of animals, and the effects of these disease-combatting chemicals and serums on other aspects of animal physiology and performance must be checked.

Genetics and Animal Reproduction

During the twentieth century scientists established the roles of genes (now known to be segments of the **deoxyribonucleic acid** or DNA, molecule) in transmitting characteristics, such as color, leanness of meat animals, or the speed of a horse, to the next generation and they determined the heritability ranges (relative influence of heredity vs.

In vitro
In an artificial environment such as in a laboratory rather than in the living body (in vivo).

Deoxyribonucleic acid (DNA)
Double-stranded nucleic acid that is a component of chromosomes and that contains coded genetic information within its nucleotides. The genome would be all the DNA for that animal.

environment) for most important animal traits. They documented the benefits of heterosis, or hybrid vigor, the tendency of offspring of genetically diverse parents to perform better than the average of their parents. (Offspring of genetically similar parents will tend to perform at the average of their parents.)

Research disclosed how the endocrine system controls the reproduction cycle, including the laying of eggs, and techniques for diluting, storing, and handling collected semen, so that inseminations will be more successful.

As the twenty-first century begins, scientists are trying to achieve a complete genome map for each species, the segment or segments (genes) that influence individual traits and their location in the DNA molecules (chromosomes) of that species. This is a mammoth task; the number of significant segments (genes) may be in the millions. Such precise information gathered to date has allowed dramatic benefit, allowing the transfer of specific genes from one species to another to augment product quality or disease resistance.

Late in the twentieth century, animal scientists learned how to synchronize estrus in a group of females and to transfer embryos from one female to another, and how to clone animals. These developments opened the door to allow production of several genetically similar offspring from a genetically superior female and male, as discussed in Section 2.2.

In the late 1990s, the term *biotechnology* emerged as a popular term to embrace gene transfer, embryo transplanting, cloning, and a multiplicity of other recently developed biological techniques and processes. In truth, all animal processes are part of biology and animal scientists have always pursued new technologies, so the term could be more embracing. However, as currently used, *biotechnology* can be defined as the use of biological organisms or processes in the development of *new* products (consumer products or intermediary "products," such as enzymes, organisms with new and different properties, etc.).

Other Biotechnology Examples

Through molecular biotechnology, genes can be transferred from one organism (such as the cow) to another unrelated organism (a bacterium) to produce a transgenic organism. An excellent example is the development and use of **bovine somatotropin** (BST) for increased milk production in dairy cows. This process involves the transfer of the gene responsible for BST production from the dairy cow to a common bacterium, E. coli (Figure 2–6).

The newly engineered *Escherichia coli* (E. coli) organisms are multiplied in large quantities through **fermentation** techniques. Then they are harvested, killed, and their protein separated, purified, and formulated for use in dairy cows. With most cows and with excellent management practices, milk yield increases of 10 to 20 percent and feed efficiency improvement of 5 to 10 percent can be expected.

BST is a peptide hormone consisting of 191 amino acids, so if consumed, it is digested like other proteins. Therefore, it must be introduced systematically (by injection or implant). Because the level of BST in the milk of BST-treated cows does not increase when consumed but is broken down in the digestive tract, it has been deemed safe for human consumption.

Similar genetic engineering techniques can result in the production of recombinantly produced **porcine somatotropin** (PST). A natural swine protein growth hormone, PST administered to hogs can improve feed efficiency, increase lean muscling, and reduce fat deposition.

Another example of advances in biotechnology is the development of beneficial products using transgenic animals. These are animals that have DNA from humans or

---

Somatotropin
Growth hormone; a protein hormone produced by the pituitary which promotes body growth.

Escherichia coli (E. coli)
A bacterium that lives in the intestinal tract of most vertebrates. Much of the work using recombinant DNA techniques has been conducted with this organism because it has been genetically well characterized.

Fermentation
The conversion of one of more substances into more desirable substances through the actions of microorganisms (i.e., rumen fermentation).

Porcine somatotropin (PST)
The type of somatotropin produced by swine.

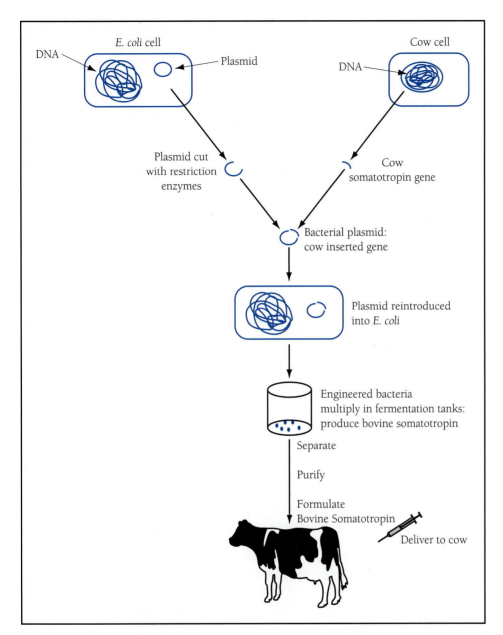

Figure 2–6
Diagram illustrating the processes of bovine somatotropin production.
(*Source:* Monsanto.)

another animal inserted into their gene makeup. By introducing a new segment of DNA into both the somatic and reproductive genes of an animal's cell, the new DNA may be transmitted to the offspring as a continuing trait. Transgenic animals can be used to produce beneficial products such as pharmaceuticals containing human vaccines, medically beneficial foods and dietary supplements, and even organs for transplant that would not require antirejection drugs.

Biotechnology also has resulted in the production of field crops with improved agronomic qualities—such as resistance of plants to selected herbicides and protection against insect pests, and in plants that better meets the needs of livestock and poultry producers.

For example, soybean meal is the most important protein source for livestock and poultry in the United States. Through genetic engineering, soybeans with increased levels of the amino acids lysine and methionine have been developed that reduce the quantities of higher-cost protein feeds needed in the ration.

The development of low-phytate corn has increased the availability of phosphorus. Phytates are compounds that may bind phosphorus and reduce its absorption from the digestive tract. As corn is a major component in livestock and poultry rations, its greater digestibility should result in lowered feed costs because of increased absorption of the phosphorus in corn grain and reducing the need for phosphorus supplements. Low-phytate corn also means that animals will pass less phosphorus in their waste and reduce the risk of soil and water pollution.

Products

Whether it is milk, eggs, meat, or a resultant processed product from one of these, the technologies of preservation and the maintenance of purity and quality during handling and storage involve the chemistry and microbiology of the product. The potential chemical and microbiological risks were well elucidated and the technologies of canning, freezing, dehydration, curing, and irradiation were rather well developed and refined during the twentieth century.

In addition, methods of quality assessment received much research attention. The Babcock test to determine butterfat content of milk was developed early in the twentieth century and electrophoresis and other technologies later in the century allowed rapid and low-cost quantification of most milk components.

Scientists learned early that egg candling (holding a shell egg before a limited light source) would detect interior abnormalities and allow estimating egg freshness. Later research developed more precise and automated assessment techniques, in both shell eggs and egg products. Yellow corn imparts a pigment to egg yolk, and other pigment sources are sometimes used in layer rations.

With livestock, scientists learned in midcentury to estimate the fatness of a pork carcass by measuring the fat thickness over the back with a thin metal probe, and soon thereafter adapted for that purpose the sonoscope, which emits sound waves and measures the time it takes for them to bounce back from the sheath over the loin muscle. Also, there was much research attention to the influence of feeding regime on product characteristics. For example, the carotene and other pigments in a high-forage diet result in a yellow tinge to deposited fat, not desired in hamburger. Leanness of a hog can be increased by a high-protein diet and by certain feed additives.

Eating qualities of meat, eggs, or other products are not so easily assessed. It is difficult to electronically or mechanically replicate the "eating sensation," the flavor, aroma, or other palatability features (Figure 2–7). A shearing device to measure meat tenderness and a device to measure juiciness by forcing out fluid under pressure were developed early in the twentieth century and there were some refinements, but for full palatability assessment, "taste panels" of discriminating people continue to be used.

There was also much research effort invested in product development. Examples are (1) replacing some butterfat in low-fat dairy spreads with other milk ingredients while retaining the color and "spreadability" of butter; (2) mechanically restructuring

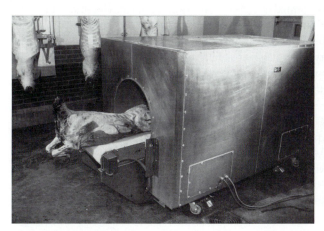 

Figure 2–7
Electromagnetic scanning utilizing the TOBEC™ provides a high degree of precision in measurement of carcass and primal cut lean meat.
(*Source:* Purdue University.)

less tender pork muscles into formed, more tender product but retaining much of the loin chop texture; and (3) curing turkey as a competitor for ham.

A major current focus of animal product research is on food safety, prompted in part by consumer concern over several well-publicized illnesses from inadequately cooked meat products, some product recalls, and the risks of terrorist attacks on the food system. This focus puts the research emphasis on how animals and products are routinely handled, and the sites and processing steps where there may be risk of bacterial contamination. Several changes and refinements in animal and product handling have resulted and will result from ongoing research.

Economists and sociologists also play roles in animal product research. Economists trace and project product demand and price, and the influences of the general or local economy, employment, or other factors on demand. Sociologists may structure questions for in-depth interviews of consumers to determine geographic, income, or ethnic preferences, how consumers may react to a new or modified product, or how it might be best packaged or advertised.

2.10 GOVERNMENT AGENCIES AND PRIVATE ORGANIZATIONS

Federal regulations concerning animals, animal production systems, and animal products are written and enforced by the U.S. Department of Agriculture, the Environmental Protection Agency, the Food and Drug Administration, and perhaps other agencies in response to congressional action. Specific issues related to these regulations are discussed in several chapters.

As a guide to the consumer or retailer, the USDA has developed relative quality and product yield grades for many animals and animal products, and employs trained personnel to do the grading. The same is true for inspection of processing plants and monitoring the procedures in those plants.

Most land-grant universities and the USDA have intensive research and extension education programs in the animal and related sciences. In the universities, the work is organized in an agricultural experiment station and a cooperative extension service, financed by both state and federal funds. In the case of extension, field specialists, working in a county or multicounty area, may be partially financed by county tax funds. Staff organize field days, write reports, provide informational material via the Web and public media, and often work with individual producers, processors, or others in the broad animal science arena.

Many privately financed organizations serve the animal industries. Commodity organizations promote their products to encourage consumption and use, support research in universities, and lobby federal and state legislatures and agencies on behalf of their members' interests. Most funds for these organizations are provided by membership fees and by "check-off" deductions at the point of animal or product sale. At this writing, however, the constitutionality of several of the check-off programs have been challenged in court, and their continuation is yet to be determined.

Livestock breed and other industry organizations, such as equipment or feed manufacturers, also carry out educational, research, and lobbying efforts to serve their respective industry sectors.

QUESTIONS FOR STUDY AND DISCUSSION

1. In what region of the country is animal production more heavily concentrated and why?

2. Describe an example of a significant geographic shift in the livestock or poultry industry, and suggest some reasons for that shift.

3. List four different career opportunities for veterinarians.

4. In what species has the development and marketing of breeding stock moved from the farm or ranch to specialized genetics companies?

5. Who "controls" the animal industry—has the most influence on its long-term viability and success?

6. List several scientific disciplines within animal science where information is constantly being gathered by dedicated researchers.

7. Define *biotechnology* and describe how BST was developed and made available to dairy producers.

8. How can cloning of animals benefit the animal producer and society?

9. Define *genome* and give some reasons why animal genome maps are being developed.

10. List three services provided by governmental and educational agencies to producers, consumers, or others.

3
Nutrients and Their Sources

Efficient production of meat, milk, and eggs from farm animals, as well as the vitality and performance of companion or working animals, requires sound nutrition and nutritional management. This involves proper feeding and management of breeding stock, their production of strong, healthy offspring, and the steady growth and development of young animals from birth to market or into the breeding herd or flock.

Productivity, herd or flock health, and costs of production are greatly influenced by nutrition. Nutrition deals with the kind and amounts of feeds, their composition of nutrients, the animal performance desired, and digestion and metabolism of nutrients. Animal productivity and herd or flock health are greatly influenced by nutrition. The use of nutritionally sound **rations** and recommended practices also can reduce livestock and poultry production costs—which typically amount to 50 to 70 percent of the total cost of production.

A **nutrient** is a food constituent that aids in the support of life. It may be a single element such as iron or copper, or it may be a large, complex chemical compound composed of many different units, such as starch or protein.

About 100 different nutrients are known to have value in livestock and poultry rations. Many are individually required for normal body metabolism, growth, and reproduction; others either are not essential or can be replaced by other nutrients.

This chapter provides in-depth material on nutrients, their general and specific roles, their formation in plants, and how feeds are classified according to their nutrient content. Upon completion of this chapter, the reader should be able to

1. List the six classes of nutrients and describe the general role each plays in body metabolism.
2. Identify the major component units of carbohydrates, proteins, and fats.
3. Understand differences in nutrient needs and use according to the purpose of the animal.
4. Know differences in cellulose, starch, and lignin, and their usefulness to animals.
5. Compare the roles of carbohydrates and fats in providing energy, and how they are similar and dissimilar.
6. Understand how saturated and unsaturated fatty acids affect properties of fat.
7. List the essential amino acids.
8. Explain why nonprotein nitrogen can be a partial protein substitute for some species.
9. Know how water requirements are related to feed intake and other factors.

> **Ration**
> A combination of feeds, perhaps mixed together, that are fed to animals to meet nutrient requirements. May vary in quantity.

> **Nutrient**
> A chemical element or compound that is essential for normal body metabolism.

10. Name the macro and micro minerals.

11. Know the classification of vitamins and important differences related to their solubility.

12. Describe how nutrients are formed in plants.

13. Differentiate between roughages and concentrates.

14. List, define, and differentiate among four methods of measuring the energy value of feeds.

15. List at least three feeds that are relatively rich sources of protein, carotene, calcium, and the B vitamins.

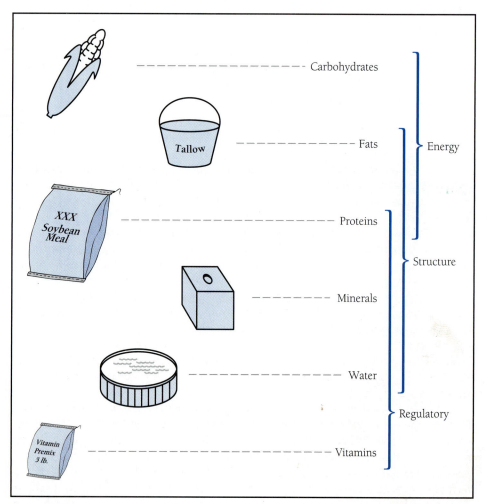

Figure 3–1

The six classes of nutrients grouped according to their functions. There are exceptions to this generalization. Protein supplies energy only when fed in excess quantities or when rations contain insufficient carbohydrates or fats. Classes of nutrients that are considered regulatory may (1) carry nutrients, enzymes, or hormones; (2) form a part of enzymes or hormones; or (3) stimulate or catalyze the activity of an enzyme or hormone in metabolism.
(*Sources:* Iowa State University and Purdue University.)

3.1   CLASSES OF NUTRIENTS

The many individual nutrients are grouped into six classes—*carbohydrates, fats, proteins, water, minerals,* and *vitamins*—according to chemical or functional similarities. Space does not permit a complete listing and discussion of all the nutrients within these classes. The main functions of the classes are illustrated in Fig 3–1.

Carbohydrates are related both chemically and functionally; they all contain the same elements and all yield energy. Proteins also have chemical similarities; all are composed primarily of amino acids. Most proteins eventually become a part of new tissue, a hormone, or an enzyme.

Minerals are not closely related chemically except that they are individual, inorganic elements. Functions of most minerals, however, are similar. Most become a part of body structure and also influence the rate of certain chemical reactions in body metabolism.

Further evidence of chemical or functional similarities among the nutrients within a class is disclosed in later sections. No attempt is made to enumerate all functions of all nutrients. Rather, examples are given that illustrate the types of functions that the nutrients or classes of nutrients perform.

3.2   HOW NUTRIENTS MAY BE USED

To perform well, animals must receive *balanced rations.* This is defined as the quantity of feed that provides an animal all of the nutrients needed in desired proportion. On most farms there are animals of different sizes and in different phases of life, or animals being kept for different purposes. Therefore, they require different rations and are usually grouped and fed according to their needs. These nutrient needs can be grouped into the areas of maintenance, reproduction, lactation, growth and fattening, and work.

Animals fed a *maintenance ration* are usually in a nonproductive period of their life. For example, the feed requirement of mature beef cows would be close to maintenance just after weaning of their calves but before significant needs for pregnancy occur. Adult horses not being worked or exercised, or nonactive breeding stock, should also be fed a ration approximating their maintenance need.

Maintenance requirements refer to the total nutrients necessary for normal body function (for example, tissue repair, respiration, circulatory system activity, metabolic activity). Of course, these needs will vary with type of species and age and weight of the animal, as well as season of the year.

Reproductive needs must be met during pregnancy as the nutrient needs increase for development of the reproductive system, placenta, and fetus. These requirements, over and above the daily maintenance needs, steadily increase during pregnancy and are most critical during the last **trimester** of pregnancy (Figure 3–2). Although body reserves of the mother help provide nutrients for the fetus during periods of inadequate nutrition, extended underfeeding can result in weak or undersized offspring and adversely affect the mother's performance in the next period of lactation and reproduction. The additional reproductive needs of males during periods of breeding can be considered insignificant; they should perform well if nutrients for maintenance and increased activity related to breeding are properly provided.

Lactational needs of the mother increase dramatically after **parturition** in order to provide nutrients for synthesis of milk to be used for initial growth and development of the offspring. In the case of high-producing dairy cows, the nutrient requirements to

> **Trimester**
> A period of 3 months.

> **Parturition**
> The act of giving birth—calving, lambing, farrowing, or foaling.

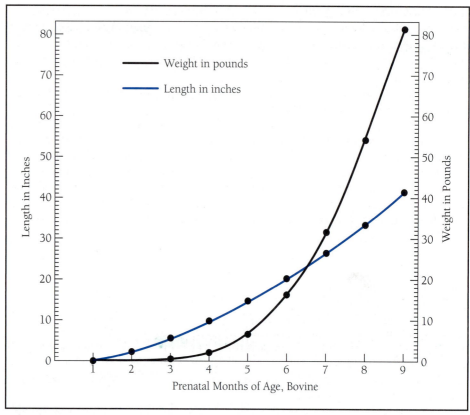

Figure 3–2
Growth of the fetal calf, showing greatly accelerated development in the last trimester of the 280-day gestation period.
(*Source:* Texas A&M University.)

Lactation
The period of milk secretion. Usually begins at parturition and ends when offspring are weaned or, in the case of dairy cattle, when milking is stopped.

provide for maintenance and high yields of milk are especially critical during the first few months of **lactation**. Again, nutrient needs for lactation will vary with species. The sow nursing her litter of 9 or 10 pigs provides milk of high protein and fat content. Thus, she must be fed a lactation ration that allows for her body size, number of pigs, and composition of milk.

All healthy young animals are building protein into body tissues and organs. In addition, their skeleton is growing rapidly. This growth, as measured in increased body weight and skeletal size, occurs most rapidly in early life. Therefore, young animals must receive a ration highly concentrated in nutrients, such as milk, or a highly digestible ration to help ensure early and proper growth and development. The nutrient requirements, expressed as percent of nutrients per pound of feed, is greatest during rapid growth and becomes less as the animal approaches maturity. For example, growing pigs usually receive a ration containing 16 to 18 percent protein, whereas finishing hogs on full feed need a ration of about 14 percent protein.

Other productive purposes of the animal and its species must be considered when determining the nutritional needs of the animal. The production of eggs for hatching and commercial use requires consideration of both nutrient requirements and the effects of the ration on egg and eggshell quality. For example, a very high calcium requirement, about 2.75 percent of the ration, is needed for proper calcium deposition in the structure of the eggshell. Certain feeds, even though they are good sources of nutrients, are unacceptable because they cause undesirable flavors or darkened color to the yolk of the egg.

Wool production also has special nutrient requirements because of the high content of protein in the wool fiber.

The same principles of nutrition apply when managing animals involved in *work* as in other productive functions. That is, allowances in the ration must consider maintenance needs and the demands for light, moderate, or intense work. Increased energy demands are proportional to the amount of work performed. Although protein requirements increase little, if any, during periods of work, animals usually need more carbohydrates or fat for energy. Loss of minerals and dehydration of body fluids are special concerns during periods of sustained work because of the increased losses related to sweating and respiration.

3.3   CARBOHYDRATES

Carbohydrates, the main source of energy in most livestock and poultry rations, make up 65 to 80 percent of the dry weight of nearly all grains and roughages. Energy supplied by carbohydrates is used for (1) maintenance, (2) growth, which includes formation of meat tissue and wool, (3) reproduction, (4) production of animal products such as eggs or milk, and (5) work. Energy for maintenance includes all that is used for voluntary and involuntary muscle action, such as in walking, eating, blood circulation, and movement of food through the digestive system.

> *Carbohydrate*
> A class of nutrients, each composed of carbon, hydrogen, and oxygen, the latter elements present in a 2:1 ratio.

Most carbohydrate energy that is not used soon after absorption is stored permanently or semipermanently in the animal after conversion to glycogen or fat. All carbohydrates contain the elements carbon, hydrogen, and oxygen; the latter elements are present in the same ratio as in water (H_2O). This chemical similarity and the fact that all carbohydrates yield energy justify their classification as a group (Table 3–1). As shown in this table, carbohydrates are classified as (1) monosaccharides, (2) disaccharides, and (3) polysaccharides.

Carbohydrates consist of from two to a large number of 5-carbon or 6-carbon sugars (**monosaccharides**) with an empirical formula of $C_5H_{10}O_5$ or $C_6H_{12}O_6$. The number of monosaccharides and the manner in which they are linked together influence the solubility and digestibility of the carbohydrate.

> *Monosaccharide*
> A single-sugar molecule; a carbohydrate.

Sucrose (table sugar), lactose (milk sugar), and maltose (the more soluble part of the carbohydrate in grains, and shown in Figure 3–3) consist of two units each, and so are called *disaccharides*. Because their linkages are easily ruptured, these carbohydrates are very soluble, and their sweet taste is apparent as soon as they touch the moist tongue. The internal structures of the component monosaccharides vary, however, and this explains the variation in the nature of the sweetness.

Large polymers (*polysaccharides*) make up most of the carbohydrate of livestock feeds. When 6-carbon sugars combine, a molecule of water is released at every linkage, so the empirical formula of a polysaccharide thus formed is $(C_6H_{10}O_5)_n$. The symbol n indicates the number of monosaccharide units in the polysaccharide. The starch of corn, wheat, and other grains is a polysaccharide consisting largely of maltose units (each made up of two 6-carbon hexoses), linked together in a manner that permits rather quick and complete rupture by digestive enzymes.

TABLE 3–1	Common Carbohydrates, Sources, and Functions	

Name	Sources	Functions
Monosaccharides		
Pentoses (5-carbon sugars)		
Hexoses (6-carbon sugars)		
Glucose	Corn syrup, fruit	Principal blood sugar
Fructose	Fruit, honey	Converted to glucose in liver
Galactose	One of the two molecules in milk	(same)
Disaccharides		
Lactose	Milk	Yields glucose and galactose in digestion
Maltose	Germinating seeds	Yields two units of glucose in digestion
Sucrose	Cane and sugar beets	Yields glucose and fructose in digestion
Polysaccharides		
Pentosans		
Hexosans		
Starch	Seeds, tubers	Principal energy source for monogastrics
Cellulose	Plant leaves and stems (cell walls)	Major source of energy for ruminants; requires microbial enzyme cellulase
Glycogen	Limited supply in animal tissues	Reserve form of glucose in muscles and liver

Note: *This abbreviated table provides only a partial list of the carbohydrates, but includes the more common ones that the student should know.*
Source: (Iowa State University and Purdue University.)

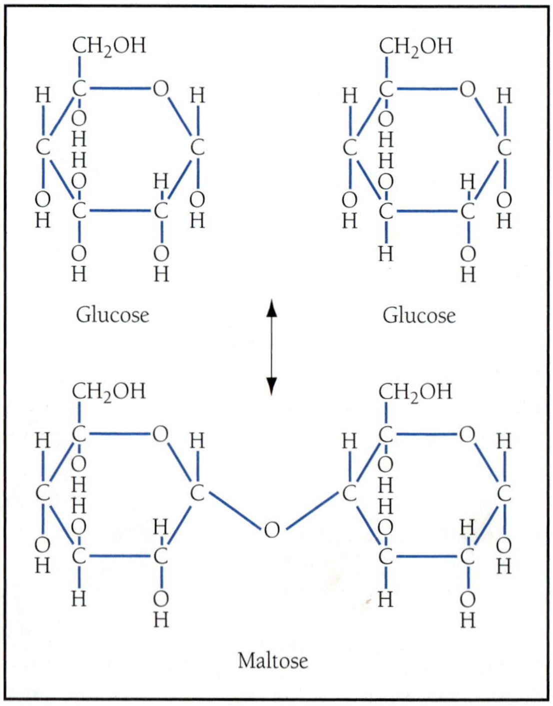

Figure 3–3
Two units of glucose, when combined, yield the disaccharide maltose and a molecule of water. Further consolidation in plant growth results in starch formation. During digestion, the opposite process occurs, yielding monosaccharides and water.
(*Sources:* Iowa State University and Purdue University.)

Cellulose is built by the plant to provide structure to the stems and leaves. Therefore, it is found largely in the cell walls of the plant. It has hexose (or 6-carbon sugar) component units similar to starch, but the units are linked in such a way to cause stiffness and strength in the plant structure. Because of this difference in the molecular arrangement of its glucose units, cellulose is *indigestible when consumed by* **nonruminants**.[1] However, cellulose provides an excellent source of energy when the enzyme *cellulase* is present, as found in the digestive system of all ruminants.

Lignin, although not a carbohydrate, is almost always associated with the fibrous portion of plant material. It is a noncarbohydrate fibrous component of more mature plants and is indigestible to both ruminants and monogastrics. The lignin content rapidly increases as plants mature. Because it is indigestible and prevents some of the more digestible carbohydrates from being digested, the practice of harvesting forages when they are immature is recommended. Both cellulose and lignin provide bulk and dietary fiber when included in the ration of animals and thus can be beneficial when present in small amounts.

For carbohydrates to be absorbed from the intestinal tract into the circulatory system of the animal body, they must be digested essentially to the monosaccharide stage. Very few disaccharides or larger units are absorbed.

When the chemist or nutritionist appraises the carbohydrate content of feeds, he or she divides this class of nutrients into two groups, *fiber* and **nitrogen-free extract (NFE)**. Fiber remains undissolved after a sample of feed has been washed alternately with dilute acid and dilute alkali. Because this washing simulates, to some degree, the action of digestive enzymes in the intestinal tract, the fiber content roughly indicates the amount of carbohydrates that are poorly digested and therefore inefficient as an energy source, largely cellulose and lignin.

Hay of average quality contains about 28 percent fiber and 38 percent NFE. Corn contains only about 2 percent fiber and almost 70 percent NFE. Most grains have less than 12 percent fiber, and the NFE content is usually about 60 percent. These figures, plus the fact that grains are not as bulky as hay, explain why grains are valuable for fast gains in birds or livestock and for maximum milk or egg production. They supply a high proportion of efficiently digested carbohydrate.

Although feed analysis for crude fiber and NFE continues to be used, methods are now available for providing more accuracy in evaluation of the useful energy in forages. For example, methods are currently being used to determine the **neutral detergent fiber (NDF)** and **acid detergent fiber (ADF)** components of the plant. The NDF-ADF method distinguishes between the more digestible cell contents and the fibrous portion of the plant cell wall.

Also, scientists have made possible a more rapid technique for obtaining estimates of forage composition and quality, known as **near-infrared reflectance (NIR)**. After calibration with known samples, a near-infrared spectrometer equipped for diffuse reflectance and connected to a minicomputer provides forage quality prediction values for each forage sample tested.

Because cellulose and other fibrous carbohydrates are inefficient sources of energy for livestock and poultry, perhaps the animal nutritionist should recommend that

---

[1] Nonruminants are animals with a single stomach compartment, such as swine and poultry (and humans). Cattle, sheep, and certain other animals have several stomach compartments and are called ruminants because one of the compartments is a rumen that becomes relatively large with maturity (see Chapter 1 for a definition).

> *Cellulose*
> A prevalent polysaccharide found in the fibrous portion of plants; digestible because of cellulase produced by rumen, cecum, and colon microbes.

> *Nonruminant*
> An animal without a functional rumen. Sometimes called a monogastric.

> *Lignin*
> Indigestible material; not a carbohydrate, but a component of the cell wall of plants and woody materials; steadily increases with plant maturity, thus reducing nutritive value.

> *Nitrogen-free extract (NFE)*
> Part of a feed consisting largely of sugars and starches.

> *Neutral detergent fiber (NDF)*
> A measure of the cellulose, hemicellulose, lignin, and insoluble ash content of forages; an indicator of the intake potential of forages.

> *Acid detergent fiber (ADF)*
> An indicator of relative digestibility of fibrous feeds.

> *Near-infrared reflectance (NIR)*
> A rapid method of obtaining quality estimates of forages; utilizes an NIR spectrometer.

forages with a low cellulose and lignin content be developed. The plant breeder would like to comply, except that a certain amount of these components is needed to keep plants standing erect for easy harvesting, exposure to needed sunlight, and maturation of seed heads.

NFE and fiber are rarely used in ration formulation. Megacalories per 100 pounds of dry matter, a more precise measure of the digestible energy value of feeds for maintenance, gain, and other functions, is used. This is further discussed in Section 3.10 and Chapters 5 and 6.

3.4   LIPIDS: FATS AND OILS

If a producer or nutritionist wants to increase the energy intake of animals without requiring the animals to eat more, he or she can replace some of the carbohydrates in the ration with fats or oils. This is possible because fats are the most potent energy source in rations. A pound of digested fat has about 2.25 more times the energy value of a pound of digested carbohydrate or protein.

> **Lipids**
> A term that denotes fats, oils, and fatlike materials.

Fats and oils, also called **lipids**, are almost synonymous when used in reference to energy value. They are insoluble in water but soluble in organic solvents. Because an ether-extraction process can be used to measure fat or oil in a feed sample, a feed analysis may list either **ether extract** or "crude fat" on the feed analysis tag.

> **Ether extract**
> Common expression for lipids (fats, oils) that are extractable with ether from feedstuffs. Also referred to as crude fat.

Most grains and roughages contain less than 5 percent fat, so the total energy supplied by fat in most rations is not nearly as great as that supplied by carbohydrates. Although the calories per pound can be increased by substitution of fats, the amounts must be restricted to about 15 percent of the ration in swine and poultry rations. Even less must be added to ruminant rations because of the risk of altered rumen function and reduced digestion. Animal nutrition studies indicate that the level of fat should be restricted to 6 to 8 percent in ruminant rations; increasing the calcium in the ration enhances its utilization.

During times when such fats are plentiful and the price is low, they often compete favorably on a "cost-of-energy" basis with grain. Because fats and oils contain no other nutrients, their addition tends to dilute the other nutrients in the ration that must be adjusted upward to keep the ration balanced.

> **Triglyceride**
> A compound composed of glycerol and three fatty acids.

A **triglyceride**, or true fat molecule, is composed of three fatty acid units chemically united with glycerol (Figure 3–4). The kind and size of the fatty acids attached to glycerol can vary in size and chemical properties. For example, the oils are generally smaller in molecular weight and are liquid at room temperature. Also, the oils are said to be "unsaturated" because of the presence of double bonds between some of the carbons. Unsaturated fats tend to be more reactive at the double bond locations. Therefore, use of fats containing largely unsaturated fatty acids requires the addition of antioxidants to reduce the risk of the feed becoming rancid before being used.

When the fat is made up largely of saturated fatty acids, it is not only solid at room temperature but also more stable and less likely to become oxidized and rancid upon storage.

True fats are composed of carbon, hydrogen, and oxygen, as are carbohydrates, but the proportion of oxygen is less. The chemical formula for the disaccharide sucrose is $C_{12}H_{22}O_{11}$. In contrast, the composition of tristearin, a fat in beef tallow, is $C_{57}H_{110}O_{6}$. This is the primary reason that fats yield about 9 calories per gram as compared to about 4 calories per gram for carbohydrates. The basis for this higher energy yield can be explained by the metabolism in the body cells, where energy-yielding nutrients are used.

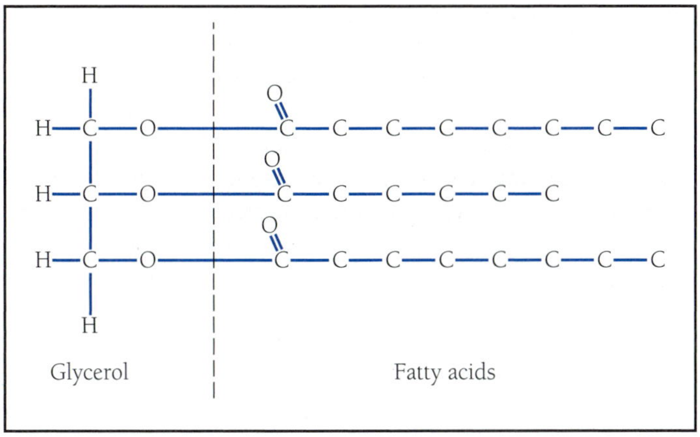

Figure 3–4
Fats are normally composed of glycerol and three fatty acid units. The fatty acid units vary in length, size, and other characteristics. All three fatty acid units that are part of a fat need not be identical.
(*Sources:* Iowa State University and Purdue University.)

In this process of oxidation, oxygen is inhaled through the lungs and transported via the blood to body cells that react with the nutrient. This oxygen combines with the carbon and hydrogen to form the end products—carbon dioxide and water—with the simultaneous release of energy. Because fats contain low amounts of oxygen, more oxidation occurs and more energy is released.

One interesting aspect about feeding of fats to animals is that monogastrics tend to deposit the type of fat consumed in their ration. If a polyunsaturated oil, such as peanut oil, is fed to pigs, their body fat tends to be softer, a feature not desired by companies that slaughter and process meat, nor by the consumer. This does not occur when fats are included in low amounts in the ration of ruminants, because the microorganisms in the rumen bring about saturation of the fatty acids. This also explains the greater firmness of beef fat, or tallow, as compared to pork fat, or lard.

In addition to providing energy, fats and oils serve other useful functions.

- Fat in the ration aids the absorption of vitamins A, D, E, and K from the digestive system. These are therefore termed the fat-soluble vitamins.
- Fat cushions and protects vital organs in the body. This fat, which physiologically is an energy storehouse, helps suspend vital organs but also allows **peristalsis** and absorbs shock and sudden movement.
- The deposits of fat in lean muscle, which are also energy stores, add to the juiciness and flavor of steaks, chops, and roasts. The term **marbling** is used to describe this dispersion of fat in the lean tissue.
- Small amounts of added fat can have beneficial effects on the physical characteristics of the feed, as this tends to enrich the color and reduce dustiness due to fine particle size. Feed manufacturers also like it because of its lubrication properties and reduced friction in mixing and handling equipment.

> **Peristalsis**
> Wavelike contractions of the muscles in the walls of the hollow digestive organs. These contractions force contents onward.

> **Marbling**
> The interspersion of fat particles in lean meat; intramuscular fat.

> **Protein**
> A class of nutrients containing amino acids that may be present in feeds. Contains the elements carbon, hydrogen, oxygen, nitrogen, and usually sulfur.

3.5   PROTEINS (AMINO ACIDS)

All livestock and poultry producers are aware of the importance of **protein**. First, they know that protein-deficient rations adversely affect animal performance. Second, they purchase much of the protein feeds and thus keep abreast of feed price relationships and seek the best feed buy. Because most rations, with the exception of those for cultured fish, will contain from 12 to 20 percent protein, the total quantity purchased on many production units represents a significant portion of the total feed cost. Skillful managers keep feed costs as low as possible without jeopardizing the efficiency or total production of their animals.

Chemically, a protein is an organic compound composed of carbon, hydrogen, and oxygen, but unlike carbohydrates and fats, it also contains about 16 percent nitrogen. In addition, most proteins contain the element sulfur. The actual makeup of a protein is determined by the kind and number of amino acids and the sequence in which the amino acids are bonded together.

An estimate of the total protein in a feed can be made by chemically measuring the nitrogen content and (1) dividing by 16 percent or (2) multiplying by 6.25 (the reciprocal of 16). Another procedure utilizes the "protein analyzer." Although more expensive equipment is required, this method is more rapid and labor saving. Both methods result in a calculated value called **crude protein**.

Some nitrogen in feeds, especially forages, is not incorporated into a protein molecule. When immature plants are eaten or harvested, there is considerable nitrogen in the plant in the form of nitrates, amines, amides, or individual amino acids not converted into protein. This form of nitrogen is referred to as **nonprotein nitrogen (NPN)**, usable by ruminants but of little, if any, value to nonruminants. Therefore, crude protein as measured by nitrogen content is not a precise appraisal of true protein. The quantity of NPN present in grain, or in animal byproduct feeds, is relatively insignificant.

Some 22 different amino acids may be present in feed or animal proteins. All amino acids are needed for body functions, but there are some amino acids that the body tissues cannot make, or cannot make rapidly enough to meet their needs. These are classified as the *essential amino acids* and can be remembered by learning the acronym "PVT MAT HILL." The 10 essential amino acids for most animals are

Phenylalanine	Methionine	Histidine
Valine	Arginine	Isoleucine
Threonine	Tryptophan	Leucine
		Lysine

A *nonessential amino acid* also is one needed by the animal, but because it can be formed from other amino acids, it does not have to be contained in the proteins in the ration of an animal. The nonessential amino acids include

Alanine	Cystine	Hydroxyproline
Aspartic acid	Glutamic acid	Proline
Citrulline	Glycine	Serine
		Tyrosine

For optimal growth, three of these (glutamic acid, glycine, and proline) are required by the chick, in addition to the 10 essential amino acids listed earlier.

Nutritionists use the terms *first-limiting amino acid* in discussing or evaluating the protein efficiency of rations for nonruminants. The first-limiting amino acid is the one present in the diet in the least amount, *in relation to* the animal's need for that specific amino acid. The other amino acids present will be used by the nonruminant animal only to the extent of the presence of the limiting amino acid. In most poultry diets, these are the sulfur-containing amino acids (methionine and cystine), whereas in the typical corn-soybean meal diet for growing pigs, lysine is most likely to be first-limiting.

Crude protein
Nitrogen, present in feed, multiplied by 6.25. This factor is used because amino acids contain about 16% nitrogen. Crude protein includes nonprotein nitrogen compounds.

Nonprotein nitrogen (NPN)
Refers to components in feed that contains nitrogen not incorporated into protein, such as urea.

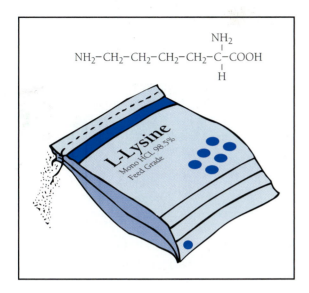

Figure 3–5
Lysine, an essential dietary amino acid for nonruminants. NH_2 is an amino unit and COOH is an acid unit. The presence of these units on such compounds is the basis of the name *amino acid*.
(*Source:* Purdue University.)

Several amino acids are manufactured chemically or by cultured strains of organisms and may be added to rations low in a specific amino acid. In some diets, the addition of a commercial source of amino acid enables the lowering of the total protein level. Synthetic lysine (Figure 3–5), methionine, or methionine hydroxy analogue (MHA), threonine, and tryptophan are commercially available and commonly used in poultry and swine feeds.

It is fortunate that ruminants have microorganisms in their rumen that manufacture all of the essential amino acids. Most natural feeds do not contain amino acids in the exact proportions needed by the animal. In some cases, a feed may be almost devoid of a certain essential amino acid. The ability of the microorganism to synthesize the essential amino acids prevents their deficiencies from occurring in ruminants and also allows the protein in the ration to be used efficiently. Without these microorganisms, the dietary protein would probably be used only in proportion to the level of the first-limiting amino acid.

The ability of ruminants to synthesize amino acids allows the use of certain protein substitutes in some livestock rations, especially when plant protein is quite expensive. About one-third of the protein needs of mature cattle and sheep can be supplied by NPN if careful procedures are followed in mixing and time is allowed for the animal and its microbes to adapt. Commercial sources include the ammoniated products, biuret, protein supplements containing urea, and feed-grade urea (Figure 3–6).

Microbial amino acids are formed when rumen bacteria and protozoa incorporate the nitrogen from the NPN source with carbon, hydrogen, and oxygen obtained from the carbohydrate in grain or other energy components of the ration. The amino acids are then built into bacterial and protozoal protein. The ruminal microbes, along with other protein that escapes breakdown in the rumen, move into the abomasum and small intestine to provide protein for digestion and absorption.

The mature horse can utilize some NPN in its ration, although it is not used as efficiently as in ruminants. Poultry, swine, and other nonruminants cannot efficiently utilize NPN for amino acid synthesis.

Figure 3–6
Urea. Atomic weights of the elements indicate that urea contains approximately 46 percent nitrogen. Commercial urea contains about 45 percent nitrogen. Because crude protein contains about 16 percent nitrogen, one might say that urea has the equivalent of 281 percent crude protein (45 × 6.25 = 281) if incorporated into protein.
(*Sources:* Iowa State University and Purdue University.)

Cell walls of animals are composed primarily of protein; hence, the primary function of protein is often listed as the building and repair of tissue. Individual amino acids have many other specific and important functions. Only a few examples are given here. Enzymes that aid digestion in the stomach and small intestine, as well as those that catalyze metabolic reactions in body cells, are composed of amino acids. One or more amino acids are essential constituents of some hormones that help regulate body processes. Certain amino acids are needed for antibody formation; others are concerned with hair and skin pigmentation.

3.6   WATER

Water, the largest single component of the body that is present in every living cell, must be considered as a nutrient critical for optimum performance. Although the cheapest of the nutrients, the importance of water quality and availability to the animal is sometimes overlooked until signs of water deprivation are noted and animal productivity has suffered.

Water represents about 90 percent of the developing embryo and 70 percent of the newborn animal. Body composition steadily decreases to about 60 percent water as the animal grows and matures or is fed for slaughter, although the requirement for water increases with body size, activity, and productive performance.

Animals suffer quickly from a deficiency of water. It is so vital to biological functions that a loss of 10 percent results in disturbance of body functions, and if loss continues to 20 percent or more, death of the animal is likely. Limiting water intake reduces feed consumption and, therefore, productivity and efficiency of farm animals.

An important role of water is that it serves as a medium for chemical reactions—enzymatic digestion in the digestive tract and metabolism in the cells. In addition to providing the medium where these reactions occur, water actively participates in many chemical reactions. The digestion of fats, carbohydrates, and proteins is primarily through the process of hydrolysis, where these substances are broken down into smaller units and the ions OH^- and H^+ from water are united with these units. The

Figure 3–7
Water is the cheapest nutrient and should be readily available for maximum animal performance. It is the major constituent of body tissues and products such as milk and eggs.
(*Source:* Bohlmann, Inc.)

splitting of the water molecule enables the bond of the more complex compound to be broken.

For example, sucrose is hydrolyzed to two simple sugars, glucose and fructose, when water is added to its molecule.

$$C_{12}H_{22}O_{11} + H_2O \rightarrow C_6H_{12}O_6 + C_6H_{12}O_6$$
$$\text{sucrose} \qquad \text{water} \quad \text{glucose} \quad \text{fructose}$$

Water is an excellent transport medium, moving nutrients to the tissue cells and waste products from the cells to be eliminated from the body. Water is extremely important in the *regulation of body temperature* because of its capacity to gain and lose heat slowly, thus preventing sudden body temperature changes. Also, as evaporative cooling (panting, sweating) occurs, a great deal of heat dissipates as the moisture evaporates from the surfaces of the lungs and skin.

The animal derives its daily water needs mainly from three sources: metabolic water, water in its feeds, and drinking water (Figure 3–7). About 10 percent of the water used in body processes is produced in the cells during oxidation of energy-yielding nutrients. Metabolic water is the water resulting from hydrogen and oxygen atoms released during metabolism.

Water contained in the feed merely replaces water the animal otherwise would drink. Immature grass may contain 80 percent or more water and silages usually contain from 50 to 75 percent, whereas hay and grain stored in an air-dry condition usually contain less than 14 percent water.

The term *dry matter* will be used in later discussion of ration formulation and feed consumption. It is important to recognize that a feed containing 80 percent water, such as immature grass, contains but 20 percent dry matter. Grain containing 14 percent water contains 86 percent dry matter. Therefore, water content of feeds has a direct effect upon the amount of dry matter consumed per pound of feed.

The most common sources of drinking water for farm animals are wells and farm ponds or reservoirs. Generally, the quality of these sources is satisfactory for animal production unless they become contaminated from external sources. Some organisms, such as salmonella and coliforms, can be dangerous if they have polluted the drinking water. High sulfate levels may affect availability of certain nutrients. Also, high nitrate or nitrite levels may adversely affect the health of animals.

Inadequate water intake can be due to several causes. Water supplies may be greatly reduced during severe drought conditions. Faulty water valves can reduce water flow. Stray voltage from the watering equipment that is detectable to the animal but often not to the caretaker can be responsible for reduced intake and unusual animal behavior. Therefore, daily checking of equipment and animal appearance and behavior should be done. Dehydration of horses should be prevented by allowing them to drink frequently during continued and strenuous work. Diarrhea in very young animals can be responsible for severe dehydration, and in such cases restoration of fluids and electrolytes is essential.

Water consumption by animals is especially influenced by air temperature. For example, a 1,000-pound beef cow nursing a calf that drinks 100–110 pounds of water daily at 50°F air temperature needs more than 150 pounds when temperatures approach 90°F. The dairy cow requires 4 to 5 pounds of drinking water per pound of milk. Therefore, a cow producing 80 pounds of milk daily can be expected to need at least 40 gallons of water.

The approximate daily water intake needs of adult livestock in temperate climate are as follows:

	Gallons[2]	Pounds
Beef cattle—finishing	7.5–10.8	50–150
Dairy cattle	10–42	84–350
Horses	8–12	67–100
Swine	3–5	25–42
Sheep and goats	1.1–4	9–33

Laying hens—91 to 325 mL per bird per day, or 1.1 to 3.6 mL water intake per gram of feed

As shown above, water intake varies considerably and will depend upon factors such as the age, weight, and body surface area of the animal, as well as the kind of feed, environmental temperature, and performance of the animal.

The amount of water needed by an animal also is related to the amount of dry matter consumed. Most monogastric animals require from 2 to 3 pounds of water per pound of air-dry feed consumed. The requirement for ruminants is about 3 to 4 pounds of water per pound of feed consumed. It is important to note that the amount of dry matter consumption and consequent daily performance is greatly influenced by the amount of water consumed.

The magnitude of daily water requirements on large production systems can be quickly appreciated by estimating the amounts needed by all animals.

3.7   MINERALS

Minerals are solid, inorganic crystalline elements that cannot be decomposed or synthesized by the body. Organic compounds, which contain carbon (such as carbohy-

---

[2] A gallon of pure water weighs 8.34 pounds.

drates, fats, proteins, and vitamins), will burn. When a sample of feed is burned at 600°C until it ceases to lose weight, the minerals remain as ash. Hence, the word *ash* on a feed analysis refers to the total mineral content.

Minerals known to perform essential body functions can be classified into either *macro* (major) minerals or *micro* (trace) minerals. The macro minerals are normally present in the body at levels greater than 100 parts per million (ppm) and are usually expressed as a percent of the ration. The micro minerals are usually present in the body at levels less than 100 ppm and are typically expressed as parts per million of the ration.

Macro Minerals

Calcium (Ca)	Magnesium (Mg)
Phosphorus (P)	Potassium (K)
Sodium (Na)	Sulfur (S)
Chlorine (Cl)	

Micro Minerals

Copper (Cu)	Chromium (Cr)
Iron (Fe)	Fluorine (F)
Manganese (Mn)	Selenium (Se)
Molybdenum (Mo)	Silicon (Si)
Zinc (Zn)	

The structural role of minerals is best illustrated by the fact that about 99 percent of the body calcium and 80 percent of the body phosphorus is contained in the skeleton. Magnesium, fluorine, and certain other mineral elements are also integral parts of the bone structure and teeth. When the need arises, calcium and other minerals can be mobilized from the skeleton and used for other body functions. Examples are the removal of calcium for eggshells and for milk production following parturition of the cow. Excess calcium, consumed before these production needs exist, is temporarily stored in the skeleton.

Also, minerals are important in regulation of certain body processes and energy metabolism. A specific mineral may be involved in several different roles. For example, *calcium* also is essential for transmission of nerve impulses, performs a necessary role in blood coagulation, and serves as a catalyst in oxidation of nutrients in cells to yield energy.

Phosphorus, next to calcium in greatest quantity in the body, has other specific roles in addition to its need for proper bone strength and rigidity. It is actively involved in cellular transfer of energy, and necessary for proper function of the B vitamins and numerous enzymes.

Fifty to 80 percent of the phosphorus in feeds is in a form, phytin, that is not well utilized by animals, especially nonruminants. Much of the phosphorus consumed is therefore excreted, and this becomes a liability in handling manure from concentrated animal feeding operations. Manure management plans in most states must be based on the phosphorus need of the land to which the manure will be applied—both the phosphorus level in the soil and the amount of phosphorus that will be used by the crop to be grown.

Poultry and swine producers increase phytin phosphorus utilization by adding the enzyme phytase to the ration. Today's research focuses on genetic modification of grain crops to increase the proportion of phosphorus that is in a form other than phytin.

If all the phosphorus in grain crops were nonphytin, supplemental phosphorus sources, such as bonemeal or dicalcium phosphate, would likely not be needed in rations.

Simple sugars that result from carbohydrate digestion require the presence of *sodium* in order to be absorbed through the intestinal and capillary walls into the bloodstream. Sodium from salt and potassium helps stabilize osmotic pressure inside and outside blood cells and muscle cells, to keep them from shrinking or bursting.

The *chlorine* ion, in addition to being necessary in regulation of the acidity (**pH**) of body fluids, is contained in hydrochloric acid (HCl) which maintains the low pH of the stomach.

> **pH**
> An index of the acidity of a substance, the value of 7.0 being neutral, values above 7.0 being alkaline, and values below 7.0 being acid.

Certain mineral elements are integral parts of key body compounds. *Iron, copper,* and *cobalt,* for example, are components in synthesis of hemoglobin, the red blood pigment responsible for transport of oxygen from the lungs to body tissues. *Iodine* is an integral part of the hormone thyroxine. *Zinc* is a component of insulin, which controls the rate of carbohydrate utilization. Other examples could be given where minerals are components of body compounds.

Microorganisms functioning in the digestive system of ruminants have need for certain minerals. Phosphorus and iron, for example, have been shown to be critical in bacterial digestion of roughages[3]. Certain minerals are also needed for synthesis of amino acids and B vitamins. *Sulfur* must be available for microbial synthesis of methionine; *cobalt* is an essential constituent of vitamin B_{12}.

The above does not include all of the functions of minerals in livestock and poultry rations. Instead, it merely illustrates the types of roles and specific functions that minerals perform. Minerals that serve as catalysts in digestion and metabolism are needed in very small quantities. Skeletal requirements for minerals are large, however, as indicated by the proportions of body calcium and phosphorus contained in the skeleton.

Many minerals are present in natural feedstuffs. The content depends on plant species, type of soil, stage of harvest, and other factors. Most rations composed of plant materials and various byproduct feeds are short in sodium, calcium, and phosphorus. Salt should therefore be supplied to livestock, either by free choice or by adding about 0.5 percent of the ration as a source of sodium.

Limestone, containing 38 percent calcium, is commonly used to provide supplemental calcium. Bonemeal, dicalcium phosphate, defluorinated rock phosphate, or other products that supply both calcium and phosphorus are commonly added to livestock and poultry rations. To balance rations already high in calcium, sources of phosphorus such as monosodium phosphate, containing about 22 percent phosphorus, are sometimes used.

Other minerals, such as *potassium, manganese,* and *magnesium,* are sometimes considered critical in rations. See Box 3–1 for a description of the deficiency symptoms of magnesium in cattle and sheep. Extra iodine is needed for all livestock and poultry, especially in areas where soils are deficient. Because all animals need supplemental salt, it is the most common carrier for commercially available trace minerals. For example, a swine mineral supplement might contain specified and guaranteed amounts of minerals such as manganese, zinc, iron, copper, and cobalt in addition to salt (Figure 3–8).

Certain minerals, although considered essential, can be toxic to animals when consumed in relatively small amounts. Selenium, for example, needed at about 0.1 ppm

---

[3] *Critical* is used to mean not only that the nutrient is essential but also that normal rations may often supply insufficient quantities.

BOX 3-1   Grass Tetany in Lactating Cattle (Hypomagnesemia)

Magnesium is a cofactor of enzymes involved in nutrient metabolism. It also is essential for proper neuromuscular function. Grass tetany is caused by insufficient levels of magnesium in extracellular fluids.

Grass tetany generally occurs during early lactation in older cattle on lush early spring pasture. In some cases, it occurs when the dietary intake of magnesium is not low and is related to factors that reduce its availability (e.g., high levels of potassium and ammonia).

In acute cases of grass tetany, normally quiet behaving cows become restless, walk with a stiff gait, or may even run excitedly. This is followed by falling, tetanic spasms, and convulsions.

Affected animals must be treated quickly. An IV injection of both calcium and magnesium is slowly administered. All animals on pasture should receive supplemental magnesium with energy feeds or molasses licks. Magnesium "bullets" placed in the rumen of cattle and sheep have been developed to provide slow magnesium release.

Source: Purdue University.

Figure 3-8
The addition and careful blending of relatively small amounts of prescribed trace mineral and vitamin premixes into a mixture of several tons of feed can prevent nutritional deficiencies and can be the difference between profit and loss. (*Source*: Purdue University.)

of the ration, is toxic when the ration contains amounts above about 5 ppm. Fluorine and molybdenum can also be toxic in relatively low amounts, even though a small amount of each is essential for certain body functions.

3.8   VITAMINS

Vitamins are essential organic compounds that are not a source of energy or usable for protein. However, certain vitamins are essential in the proper utilization of the carbohydrates, fats, and proteins; thus their role is mainly **catalytic**, or they are ingredients such as coenzymes that help enzymes function properly. Vitamins are required in very small amounts. About 0.01 gram (22 millionths of a pound) of vitamin B_{12}, for example, is enough to meet the requirements in a ton of swine ration.

> **Catalyst**
> A substance that speeds up a reaction without permanently entering into the reaction itself.

Vitamins are not necessarily related chemically or functionally, nor connected to each other, as glucose and amino acid molecules may be. Although classed together because their functions are mainly of a regulatory nature, each has specific functions.

About 15 different vitamins are recognized to exist and to have specific functions in animal metabolism. Not all must be present in livestock or poultry rations, however, because synthesis of certain vitamins can occur in various parts of the animal body.

Vitamin Solubility

Vitamins are classified according to their solubility either in fats (vitamins A, D, E, K) or in water (the B complex vitamins and vitamin C). There are major differences in the absorption, transport, and storage of the fat- and water-soluble vitamins. These differences are discussed here.

The *absorption* of fat-soluble vitamins involves their movement with fatty acids across the intestinal wall into the lymph system. In contrast, the water-soluble vitamins are absorbed directly into the bloodstream, where they move freely in the blood. To be *transported* to the cells via the bloodstream, fat-soluble vitamins must be made soluble. This is accomplished by their attachment to **lipoproteins** circulating in the blood. These lipoproteins, because of their structure, allow both lipids and fat-soluble vitamins to move freely in the bloodstream.

> **Lipoprotein**
> A compound of lipid and protein; lipids in blood are transported as lipoproteins.

Another important difference in fat- and water-soluble vitamins is how they are *stored* in the body. Body reserves of the fat-soluble vitamins tend to be greater than the water-soluble vitamins because, when supplied in excess, they are stored in the areas of fat deposition. However, water-soluble vitamins, when consumed in excess, are largely excreted from the body, and more attention, at least for the B vitamins, must be directed toward providing the animal's requirement on a daily basis.

Sources of Vitamins

Green forage and animal products, especially liver, are good sources of *vitamin A*. Dry, weathered grass and hay are very low in vitamin A potency. Deficiencies of this vitamin often occur, therefore, during periods of prolonged drought in range areas. *Carotene* and other plant pigments are the source of vitamin A potency in green forages. After the forage is ingested, these precursors are converted to vitamin A in the wall of the small intestine and in other tissues. Carotene in hay is partially destroyed by exposure to the sun during curing. Other factors, such as extreme heat and high concentration of minerals in feed, will destroy some of the vitamin A potency.

The rearing of animals such as broilers, layers, dairy calves, and pigs in confinement increases the need for *vitamin D* in rations. As long as animals are exposed to the sun, ultraviolet rays catalyze the formation of vitamin D from certain compounds present just under the skin. Animals raised or kept inside, however, must depend on their rations to supply adequate vitamin D.

There are several forms of vitamin D, the most significant being D_2 and D_3. Vitamin D_2 is formed in plants after the plant has matured or has been cut and lies exposed to the ultraviolet rays of the sun, which catalyze conversion of a plant sterol, ergosterol, to D_2. Growing pastures, therefore, are low in vitamin D_2, whereas sun-cured hay is relatively high. (It is interesting to note that cut hay, exposed to the sun for curing, loses some vitamin A potency, but becomes a more potent source of vitamin D.) In animals, ultraviolet rays catalyze conversion of the sterol 7-dehydrocholesterol, located in animal tissues near the skin surface, to Vitamin D_3.

Vitamins D_2 and D_3 are equally potent for humans and livestock (and other four-footed animals), but only D_3 has significant potency for poultry. Concentrated sources of vitamin D_2 for supplementing animal feeds, as well as human diets, are manufactured by exposure of yeast (single-celled plants) to ultraviolet rays. Vitamin D_3 is manufactured by exposing animal sterols to ultraviolet light.

Vitamin E (primarily tocopherol) is abundant in whole cereal grains, especially in the germ portion. Green forages and leafy materials also are good sources. Vitamin E is easily oxidized and deteriorates when feeds are stored for long periods.

Vitamin K is produced by bacteria in the digestive system and a dietary source is not usually needed, except in newborn animals or animals with extreme digestive disturbances. Because newborn animals developed in and were born from a sterile reproductive system, several days elapse before their digestive systems contain a bacterial population for sufficient vitamin K production. Some is stored in the liver before birth, and milk, or the yolk sac, supplies additional amounts until intestinal synthesis becomes adequate. Green vegetation provides rich sources of vitamin K. Also, the nutritionist may supplement the ration by the addition of menadione compounds, as vitamin K_2.

The *B vitamins* are supplied by green, good-quality forage, animal byproducts, and milk products. Fermentation byproducts, from the cheese, brewing, and distilling industries, are also good sources (Figure 3–9). The healthy, cud-chewing ruminant animal should have sufficient B-vitamin production from its rumen microflora.

Figure 3–9
B vitamins and antibiotics are produced by biological fermentation in stainless steel tanks. Note pipes for heating and paddles for agitation in right photo. The fermentation product is dried, then standardized for potency, and sold for mixing in livestock feeds. (*Source:* Merck and Company, Inc.)

TABLE 3–2	Some Important Vitamins, Their Functions, and Deficiency Symptoms	

Fat-Soluble	Specific Role	Likely Deficiency Symptoms
A	Needed for vision Health of epithelial cell lining of body	Poor vision in dim light Increased infections Weak offspring
D	Aid calcium absorption	Bone disorders, rickets
E	Antioxidant Aids selenium	Encephalomalacia in chicks White muscle disease in calves or "stiff lamb disease"
K	Needed for blood clotting	Increased clotting time and possible hemorrhaging

Water-Soluble B-Complex	General Roles as a Group	Some Multiple Vitamin Deficiency Symptoms[a]
Thiamin Riboflavin Niacin Pyridoxine Folacin Panothenic acid Biotin	Act as coenzymes in release of energy Health of skin	Weakness Paralysis Anorexia Dermatitis Lesions of lips, tongue Neutral degeneration
Vitamin B_{12}	Red blood cell production	Poor oxygen transport, anemia Poor growth

[a]If a deficiency of one occurs, it is likely that several will be deficient. All B-complex vitamin requirements should be met in the healthy ruminant by rumen microorganisms without need for dietary supplementation.

Source: Adapted from National Academy of Science publications, Nutritional Requirements of Domestic Animals, 1985, 1988, 1996.

Farm animals rarely need a dietary source of *ascorbic acid (vitamin C)*. Normally, it is synthesized in ample quantities in animal tissues, except chickens.

Most vitamins function primarily as regulators of body metabolism. Examples of some known vitamin functions are shown in Table 3–2.

Specific Roles of Vitamins

In addition to its function in transmission of light received in the eye, vitamin A is concerned with the nervous system, growth and maintenance of the epithelial cells, and health of the reproductive tracts of the male and female, as well as with fat metabolism.

Vitamin D is essential for many functions, including (1) calcium absorption, (2) calcium deposition in bones, (3) storage of carbohydrates in the liver and cells, (4) excretion of protein end products through the kidney, and (5) oxidation of fats and carbohydrates to supply energy.

Coagulation of blood requires vitamin K in addition to the mineral calcium. Ample vitamin K must be in the liver to cause formation of certain materials that later participate in the coagulating process.

Ascorbic acid (vitamin C) is involved with mineral deposition in the bones and also with healing of wounds. It is also an antioxidant, thus sparing other nutrients from

TABLE 3–3	Recommended Vitamin Additions for Pigs, Weaning to Market (amount per pound of diet)	
Vitamin	**Starter Diet**	**Grower-Finisher Diet**
Vitamin A	2,000 IU[a]	1,200 IU
Vitamin D	200 IU	120 IU
Vitamin E	5 IU	5 IU
Vitamin K	1 mg	1 mg
Riboflavin	1.5 mg	1.2 mg
Pantothenic acid	8 mg	6 mg
Niacin	10 mg	8 mg
Vitamin B$_{12}$	10 mcg	6 mcg
Choline	85 mg	50 mg

[a]IU=International Unit

Source: *Pork Industry Handbook, PIH-2.*

oxidation, and in some manner it enhances the absorption of iron. In most cases vitamin C is not added to the ration because it is synthesized by the animal in adequate amounts, although it is commonly added to fish diets. Vitamin C is often added to chicken diets during periods of intense stress (e.g., while being transported or in excessively hot weather).

Vitamin E is also an antioxidant and is apparently needed to maintain the integrity of the blood capillaries. Vitamin E and selenium work together in prevention of white muscle disease in calves, or "stiff lamb disease," a form of muscular dystrophy.

Members of the B-vitamin group are generally employed in animal tissues for releasing energy from the carbohydrates, fats, and proteins, breaking down ingested proteins and amino acids and recombining these units into body protein. They also help maintain the nervous system and the epithelium tissues. Each member of the group has separate and distinct functions, some of which are not included above. Certain B vitamins stimulate appetite and are needed to maintain good feed consumption in nonruminant animals. Supplemental levels of B vitamins are sometimes used for this purpose during extremely hot weather, especially with poultry, when feed consumption normally declines.

Vitamins probably have many specific functions that are not known. Clues to some of these are provided by symptoms that occur during a deficiency of the specific vitamin.

Several synthetic vitamins are produced by commercial companies. These are as effectively used as vitamins from natural feedstuffs, and when natural sources are not available, a commercial vitamin supplement should be added. As an example, the recommended amounts of vitamin additions needed for pigs are shown in Table 3–3.

3.9   NUTRIENT ABSORPTION AND FORMATION BY PLANTS

Nutrients that are a single element are absorbed from the soil by the many fine, penetrating plant roots. These elements may be in free form in the soil or they may be contained in simple compounds. Absorption by plants is described in Figure 3–10.

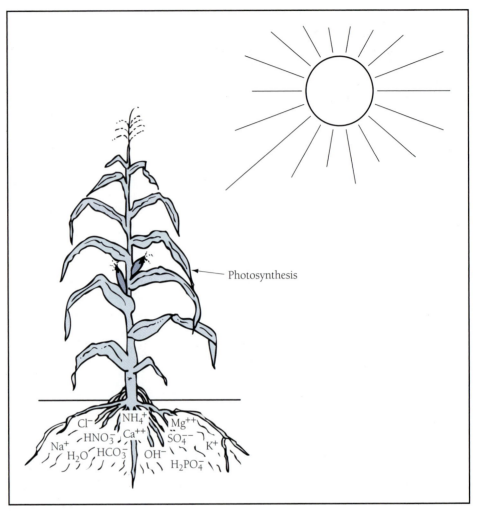

Figure 3–10
Plants such as corn obtain most of their nitrogen and other elements from the soil by absorption through the roots. Photosynthesis in leaves, with energy from the sun's rays, uses CO_2 from the air and H_2O from the air and soil to build basic organic compounds such as carbohydrates. (*Sources:* Iowa State University and Purdue University.)

Carbohydrates, fats, and proteins, which are rather complex nutrients, are manufactured by photosynthesis and other chemical processes in plant tissues. Catalyzed by chlorophyll present in green plant cells, photosynthesis combines carbon, hydrogen, and oxygen from the soil, air, and water into carbohydrate molecules. Energy, provided by the rays of the sun, is captured in these molecules. The net process of photosynthesis is usually described as follows:

$$6CO_2 + 6H_2O + \text{Radiant Energy} - \boxed{\text{Photosynthesis}} \rightarrow C_6H_{12}O_6 + 6O_2$$
$$\text{(673 calories)} \qquad \text{Chlorophyll}$$

Carbohydrates, the main form of energy storage in plants, form most of the plant structure. A large part of the plant fat (formed from carbohydrates after photosynthesis) accumulates in the seeds, apparently to supply a concentrated source of energy for germination and early growth in the new generation.

Amino acids (which later combine to form proteins) are synthesized by incorporation of nitrogen, previously absorbed from the soil, with an organic unit containing carbon, hydrogen, and oxygen. In some plants this process has been demonstrated to be an integral part of photosynthesis. Various amino acids are then linked together to form proteins that participate in plant metabolism and growth as well as in seed production.

As introduced in Section 3.8, vitamins are essential organic compounds in animals. Photosynthesis must occur for their formation in plants. Further anabolic steps fix nitrogen and other elements into the compounds.

Vitamins also serve important catalytic functions in plants as well as in animals, promoting nutrient formation, tissue growth, and reproduction. Their presence in the plant means that harvested plants and plant products are sources of these vital nutrients.

3.10   FEEDS AND THEIR NUTRIENT CONTENT

Most animal feeds are plant materials, such as grass, hay, grain, soybean meal (residue of the soybean after the oil is removed), and distiller's grains (residue of grain after fermentation to produce alcohol). A few, such as meat and bone scraps, fish meal, and whey result from the processing of animal products.

Feeds are generally classified as concentrates or roughages. **Concentrates** are low in fiber and high in usable energy and include grains and animal and plant byproducts. **Roughages** are high in fiber and are relatively low in usable energy. The National Research Council (NRC) has developed a system of classification of feeds (Table 3–4) that is helpful in understanding how they are best used and how they can be substituted for each other.

> *Concentrate*
> A feed high in digestible energy and low in fiber.

> *Roughage*
> A feed low in digestible energy and high in fiber.

Tables 3–5 and 3–6 show the nutritive value of some common animal feeds. Note especially that the values for silages and pastures are presented on a dry matter basis so that they can be fairly compared and so that the data can be used in ration formulation. For proper preservation, most silages should contain about 50 to 65 percent moisture. Succulent pastures usually will contain from 75 to 90 percent moisture. Ensiled high-moisture grains will contain about 25 to 32 percent moisture, whereas the moisture content of dried grains and forages is normally about 10 to 14 percent. The moisture content of stored feeds, such as corn silage or legume-grass silages, should be monitored regularly. Vegetative moisture content tends to vary considerably with advancing maturity and from one crop cutting to the next. As illustrated in Box 3–2, failure to correct for moisture changes can drastically affect the actual dry matter intake by the animals.

The energy value of feeds is measured in calories.[4] A **calorie** is the amount of heat required to raise 1 gram (g) of water 1° Centigrade. A *kilocalorie* (kcal) equals 1,000 calories and is sometimes expressed as a *calorie*. A *megacalorie* (Mcal) equals 1,000 kcal, and is also sometimes called a *therm*.

> *Calorie*
> The amount of heat required to raise 1 kg of water 1°C or 1 lb of water approximately 4°F.

---

[4] Some publications include TDN as another measure of the energy value of feeds. The term was developed early in the 20th century and stands for "total digestible nutrients," so the term is a misnomer. TDN is calculated by adding digestible fat × 2.25 (because fat is relatively more potent as an energy source) to digestible carbohydrate (NFE and fiber) and digestible protein. It is an index of the energy value of feeds, considerably less precise than measures used in this chapter, and its use has decreased significantly.

TABLE 3–4	Classes of Feeds According to the NRC System, Their Traits and Examples	

Class	Trait(s)	Examples
1—Dry forages or roughage	> 18 percent fiber[a]	Hay, straw, seed hulls, fodder, stover
2—Succulent forages or roughage	> 18 percent fiber	Pasture, green chop, cannery residues
3—Silages	> 18 percent fiber	Wholeplant grain crops, wilted or low-moisture grasses or legumes
4—Energy feeds	< 20 percent protein[b] and < 18 percent fiber	Cereal grains, milling byproducts, roots and tubers, brewery byproducts
5—Protein supplements	> 20 percent protein and < 18 percent fiber	Animal byproducts (meat scraps) Marine byproducts (fish meal) Avian byproducts (hydrolized feathers) Plant byproducts (soybean meal, cottonseed meal, linseed meal, corn gluten meal)
6—Mineral supplements	Guaranteed analysis	Steamed bone meal Dicalcium phosphate Iodized salt Trace mineralized salt
7—Vitamin supplements	Guaranteed potency	Vitamin A acetate Vitamins A, D, E B-complex vitamins
8—Additives	Specific	Antibiotics (chlortetracycline, oxytetracycline, tylosin) Coloring materials Flavors Hormones Medicants

[a]The symbol > = greater than.

[b] The symbol < = less than.

Source: *Adapted from National Academy of Science publications.*

Measures of energy in feeds can be determined and expressed as follows:

- *Gross energy:* heat yielded by burning.
- *Digestible energy:* that which is absorbed through the wall of the gastrointestinal tract (gross minus the energy in the feces).
- *Metabolizable energy:* gross minus energy in the feces, urine, and combustible gases. Energy available for use in the body cells.
- *Net energy:* energy available for maintenance and growth or lactation (metabolizable minus the energy required for the consumption, mastication, movement, and digestion of the feed).
- *Megacalories:* is 1000 kcal of energy.

For swine and poultry, energy requirements and the energy values of feeds (Table 3–6) are usually expressed as metabolizable energy. These species consume primarily concentrates, so there is not a large variance among feeds in the energy contents (for example, corn grain, 3,325 vs. sorghum grain, 3,229 kcal/kg).

TABLE 3–5	Nutritive Value of Feeds Commonly Used in Ruminant Rations, as a Proportion of Dry Matter

Feed	NE_m[a] (Mcal/kg)	NE_g[b] (Mcal/kg)	NE_l[c] (Mcal/kg)	Protein (%)	Calcium (%)	Phosphorus (%)	Vitamin A (IU/kg)
Roughages:							
Alfalfa hay	1.24	0.59	1.30	17.2	1.25	0.30	72,000
Alfalfa meal, dehydrated	1.31	0.73	1.40	19.7	1.32	0.24	27,000
Alfalfa silage	1.24	0.59	1.30	17.2	1.25	0.30	72,000
Beet pulp	1.79	1.19	1.79	8.0	0.75	0.11	—
Bermuda grass hay	1.03	0.19	1.05	6.0	0.46	0.18	25,000
Bluegrass pasture, early	1.60	1.03	1.64	17.3	0.56	0.47	179,000
Bluegrass pasture, bloom	1.52	0.95	1.57	14.8	0.46	0.39	112,000
Bluestem pasture, mature	1.00	0.11	—	4.5	0.40	0.11	—
Bluestem pasture, early	1.23	0.57	—	11.0	0.63	0.17	219,000
Brome hay	1.33	0.73	1.40	10.5	0.30	0.35	26,000
Corncobs	1.01	0.15	1.03	2.8	0.12	0.04	—
Corn silage	1.54	0.97	1.59	8.0	0.27	0.20	18,000
Corn stover	1.26	0.62	1.32	5.9	0.60	0.09	2
Cottonseed hulls	0.86	—	0.81	4.3	0.16	0.73	—
Cowpea hay	1.36	0.76	1.42	18.4	1.34	0.32	14,000
Fescue pasture	1.33	0.73	1.40	12.4	0.61	0.42	135,000
Grama grass pasture, early	1.39	0.79	—	13.1	0.53	0.19	—
Oat silage	1.33	0.73	1.40	12.8	0.50	0.10	65,000
Soybean straw	0.96	0.01	0.96	5.2	1.59	0.06	—
Sudan grass pasture	1.54	0.97	1.59	16.8	0.43	0.41	79,000
Vetch hay	1.33	0.73	1.40	19.0	1.18	0.34	154,000
Wheat pasture	1.64	1.07	—	28.6	0.42	0.40	520,000
Concentrates:							
Barley	1.96	1.31	1.91	13.9	0.05	0.37	—
Brewers dried grains	1.44	0.86	1.52	26.0	0.29	0.54	—
Corn, grain	2.15	1.42	2.03	10.0	0.03	0.31	1,000
Corn, ground ear	1.86	1.24	1.84	9.3	0.05	0.26	3,000
Cottonseed meal	1.69	1.11	1.72	44.8	0.17	1.31	—
Molasses, beet	1.69	1.11	1.72	8.7	0.21	0.04	—
Oats	1.73	1.14	1.74	13.6	0.07	0.39	—
Rye	1.86	1.24	1.84	13.8	0.07	0.36	—
Sorghum grain	1.86	1.24	1.84	11.7	0.03	0.33	—
Soybean meal	1.89	1.26	1.86	49.6	0.36	0.75	—
Wheat	2.15	1.42	2.03	14.4	0.05	0.45	—

[a]NE_m=net energy for maintenance.

[b]NE_g=net energy for growth or gain.

[c]NE_l=net energy for lactation.

Source: *Adapted from National Research Council,* Nutrient Requirements of Dairy Cattle, *1988; bluestem grama grass and wheat pasture values adapted from National Research Council,* Nutrient Requirements of Beef Cattle, *1996.*

TABLE 3–6

Nutritive Value of Feeds Commonly Used in Nonruminant Rations, on an Air-Dry Basis

Feed	Metabolizable Energy (kcal/kg)	Protein (%)	Lysine (%)	Methionine (%)	Tryptophan (%)	Calcium (%)	Phosphorus (%)	Sodium (%)
Alfalfa meal, dehydrated	2,270	17.5	0.73	0.2	0.28	1.44	0.22	0.08
Barley	2,870	11.6	0.42	0.2	0.14	0.05	0.36	0.04
Blood meal	1,927	85.0	8.10	1.5	1.10	0.30	0.25	0.33
Corn, grain	3,325	8.8	0.24	0.2	0.05	0.02	0.28	0.02
Cottonseed meal	2,555	41.4	1.71	0.5	0.47	0.15	0.97	0.04
Fish meal, Menhaden	2,230	60.5	4.83	1.8	0.68	5.11	2.88	0.41
Meat and bonemeal, 50%	2,434	50.4	2.60	0.7	0.28	10.10	4.96	0.72
Oats	2,668	11.4	0.40	0.2	0.16	0.06	0.27	0.06
Peanut meal, solvent	2,920	47.0	1.76	0.4	0.48	0.20	0.65	0.10
Rye, grain	2,712	12.6	0.49	0.2	0.12	0.08	0.30	0.02
Skim milk, dried	3,360	33.5	2.40	0.9	0.44	1.28	1.02	0.44
Sorghum, grain (milo)	3,229	8.9	0.22	0.1	0.10	0.03	0.28	0.01
Soybean meal, solvent	3,090	44.0	2.93	0.7	0.62	0.29	0.65	0.34
Wheat, hard, winter	3,220	14.1	0.40	0.2	0.18	0.05	0.37	0.04
Whey, dried	3,190	13.6	0.97	0.2	0.19	0.97	0.76	2.00
Mineral supplements:								
Bonemeal	—	—	—	—	—	28.0	13.0	—
Dicalcium phosphate	—	—	—	—	—	26.0	20.0	—
Limestone	—	—	—	—	—	39.0	—	—

TABLE
3–6

Feed	Manganese (mg/kg)	Iron (mg/kg)	Zinc (mg/kg)	Vitamin A (IU/kg)	Riboflavin (mg/kg)	Niacin (mg/kg)	Pantothenic Acid (mg/kg)	Choline (mg/kg)	Vitamin B$_{12}$ (mcg/kg)
Alfalfa meal, dehydrated	28.0	310	17	27,000	15.7	38	28.4	1,097	0.004
Barley	8.0	50	17	—	1.2	63	9.2	990	—
Blood meal	6.4	3,000	306	—	1.3	22	1.1	749	0.440
Corn, grain	5.0	35	10	1,000	1.0	34	7.5	530	—
Cottonseed meal	20.2	110	—	—	4.0	40	9.9	2,933	—
Fish meal, Menhaden	33.0	440	147	—	4.9	55	9.0	3,056	0.150
Meat and bonemeal, 50%	14.2	490	93	—	4.4	46	4.1	1,996	0.070
Oats	43.2	70	1	—	1.1	15	29.2	1,100	—
Peanut meal, solvent	29.9	—	—	—	11.0	165	50.6	1,960	—
Rye, grain	66.9	100	31	—	1.5	16	9.2	—	—
Skim milk, dried	2.0	50	40	—	22.0	12	33.0	1,250	0.010
Sorghum, grain (milo)	12.9	40	14	—	1.1	41	12.0	678	—
Soybean meal, solvent	29.3	120	27	—	2.9	60	13.3	2,794	—
Wheat, hard, winter	62.2	50	14	—	1.4	56	1.35	1,090	—
Whey, dried	6.1	130	—	—	27.1	10	44.0	1,980	0.015
Mineral supplements:									
Bonemeal	—	—	—	—	—	—	—	—	—
Dicalcium phosphate	—	—	—	—	—	—	—	—	—
Limestone	—	—	—	—	—	—	—	—	—

Source: *Adapted from National Research Council, Nutrient Requirements of Swine, 1979.*

BOX 3–2

Effects of Moisture in Feed upon Dry Matter and Nutrient Intake by the Animal

Feed always contains some moisture—as little as 3–5%, or as much as 90–92%. Nutrients are contained only in the dry matter (DM) portion of a ration. The moisture contributes water but no other nutrients.

Note a typical protein content of 17.2% for medium-quality alfalfa silage shown in Table 3–5. This amount is contained in the dry matter portion.

$$100 \text{ lb} \times 0.172 = 17.2 \text{ lb per 100 lb DM}$$

However, an animal consumes it on an "as fed" basis (containing both moisture and DM).

Assume that an animal eats 50 pounds of this alfalfa silage, and a moisture test showed it contained 40% dry matter.

Answer the following:

(1) How much DM is consumed?

$$50 \text{ lb} \times 0.40 = 20 \text{ lb}$$

(2) How much protein is consumed?

$$20 \text{ lb} \times 0.172 = 3.44 \text{ lb}$$

because protein is contained only in the dry matter portion.

Because moisture contents of feeds vary considerably, they must be monitored regularly and the ration contents adjusted accordingly.

Source: Purdue University.

With ruminants, however, there is much variation between concentrates and roughages and also among the roughages in energy required for eating, chewing, and digesting. Roughages vary considerably in coarseness, bulkiness, and the amount of work required of the animal to utilize the feed. Therefore, the energy requirements and the energy values of feeds (Table 3–5) are usually expressed as net energy.

Research has disclosed that the net energy value of a feed for maintenance (breathing, walking, normal body heat loss, etc.) is higher than the energy value of that same feed when the maintenance needs have been met and the energy is used for growth. That is the reason for two net energy values for each feed listed in Table 3–5. NE_m is the value when the feed is being used for maintenance, and NE_g is the value when the feed is being used for growth (or gain).

Research with beef cattle has resulted in *slightly* different energy values for maintenance and for gain, in the case of many feeds, from values established in research with dairy cattle. Space does not permit inclusion of separate NE_m and NE_g data for both. It is sufficient to indicate that most of the data in Table 3–5 is based on work with dairy cattle and is used here to illustrate ration formulation for several species. In the case of dairy cattle, note in Table 3–5 that the efficiencies of energy use by a dairy cow for maintenance and for milk production are similar.

Roughages are used primarily for ruminant animals (such as mature cattle and sheep) and by horses, and in some instances are a more economical nutrient source than grains. A certain proportion of roughage is usually necessary in ruminant rations for proper functioning of the digestive system, including rumen and reticulum motility and health of epithelial tissues lining these compartments. Other benefits and values are discussed in later chapters.

Roughages (or forages) may be classed as high, medium, or low quality. An increasingly common and more precise measure of roughage quality is relative feed value

(RFV), a calculated figure currently based on NDF digestibility, crude protein, NFE, and ether extract. A forage of high quality is one that is high in nutrients, nutrient digestibility, and palatability. The nutrient content and digestibility of most grasses and legumes are inversely related to the stage of maturity of the plant when harvested.

Alfalfa or comparable forage with an RFV of 150 to 200 is needed for high-producing dairy cows. To achieve such high values, the first cutting of alfalfa should be harvested at or before the early bud stage, baled or chopped without rain and with no leaf loss, and stored under cover. Subsequent cuttings, depending on growing conditions, generally should be harvested every 22 to 24 days during the growing season. Such high-quality legumes will exceed 20 percent protein and be highly palatable.

Forage with an RFV of 100 to 125 is recommended for 1,000-pound pregnant beef cows. Such cows need less energy and protein than dairy cows and the protein content of such forage would generally range from 10 to 15 percent. Higher RFV forage would provide more digestible energy and other nutrients than needed.

Some suggest that to be fully efficient in alfalfa production and use, a producer needs to have both a dairy herd and a beef cow herd outlet, or a hay market for both. The alfalfa harvested on time and stored without rain can go to dairy cows; the crop that gets rained on after cutting and before baling, or where cutting was unduly delayed, can go to the beef cows.

Mature cattle and sheep can also make use of *low-quality roughages,* such as corn-cobs or stalks, straw, hulls, or similar coarse feeds can be supplemented with roughages from the high-quality group and/or concentrate mixtures. Low-quality roughages usually provide 0 to 4 percent protein, 1 Mcal/kg or less net energy for maintenance, negligible net energy for gain, negligible quantities of minerals, and little or no vitamin A.

Researchers have been successful in increasing the digestibility, palatability, and protein-equivalent content of low-quality forages for ruminants by use of such chemicals as anhydrous ammonia. When forages are scarce, such as during extended drought, the straw of cereal grains can be ammoniated by carefully covering the bales with plastic to provide a gas-tight seal and introducing about 60 pounds of anhydrous ammonia per ton of dry matter. About 3 weeks is necessary for proper ammoniation, at which time the plastic is removed to allow remaining ammonia gas to escape prior to feeding. The protein-equivalent content and energy value of the ammoniated straw is about equal to medium-quality grass hay.

Few roughages are used in rations for nonruminants. Silage and ground hay are sometimes fed to sows, and rations for some poultry and growing swine may contain small proportions of dehydrated alfalfa meal as a source of critical vitamins.

Though good quality grass hay is usually adequate for horses, most horse owners will buy the highest quality alfalfa they can find. Forages for horses are discussed in detail in Chapter 7.

Concentrates (e.g., grains) are used primarily by poultry, swine, and other non-ruminants, as well as by cattle and sheep being fed for processing and dairy cows fed for high milk production. They are high in usable energy and may be high or low in protein.

Grains are fed primarily for the energy they contain. Corn and grain sorghum, the most popular grains, are highly digestible because they contain high levels of starch and very little fiber. They are relatively low in protein, and the protein they contain is low in certain amino acids. Corn is the only grain with considerable carotene content.

See the latter paragraphs of section 2.9 for a discussion of genetic modification of grains and soybeans for higher feeding value.

> **Byproduct**
> A product of less value than the major product. In oil-bearing seeds, the high-protein feeds sold as animal feeds after oil extraction.

Wheat is a potent energy feed but is often too expensive for livestock. Barley and oats, being higher in fiber and lower in usable energy, are less valuable.

Soybean, cottonseed, linseed, and peanut meal—all **byproducts** of oil extraction—are the main plant sources of supplemental protein. Their economy and use in rations are greatest in the areas where they are produced. Soybean meal, because of its excellent palatability and high content of the critical amino acid lysine, is the most popular supplemental protein source in nonruminant rations. It contains 44 to 50 percent protein, depending on whether or not hulls are added back following oil extraction and toasting.

Byproducts of the meat industry—meat and bone scrap, tankage, liver meal, and blood meal—and fish byproducts are potent sources of supplemental protein. Such byproducts of bovine processing are not allowed in bovine rations because of the risk of spreading diseases unique to the bovine, such as bovine spongiform encephalopathy (BSE), commonly known as mad cow disease. For nonruminants, these products supply some of the amino acids that are relatively low in the protein of grain and plant oil meals, plus higher levels of certain minerals and vitamins.

Dried skim milk, brewer's dry yeast, fish solubles, whey, and some other feed ingredients supply supplemental protein and are valuable in nonruminant rations for the same reasons.

Quantity of protein does not disclose the total value of a protein source. Levels of specific amino acids, as well as processing techniques that might influence the digestibility and availability of these amino acids, are important. Purchasers of protein ingredients use certain laboratory tests to check processing methods and protein quality before accepting delivery.

QUESTIONS FOR STUDY AND DISCUSSION

1. What are the six classes of nutrients? Which provide energy? Which perform structural roles? Which are involved in regulatory roles?

2. Define *maintenance ration*. Give an example of an animal that might be on a maintenance ration.

3. When is growth of an animal most rapid?

4. Give several examples of the need for nutrients for productive purposes.

5. What is the main source of energy in livestock and poultry rations?

6. Give two examples of polysaccharides that are commonly present in livestock feeds.

7. Which major polysaccharide cannot be efficiently used by nonruminant animals? Why?

8. Name three 6-carbon sugars. Which is the most prevalent sugar of the body?

9. What effect does the amount of lignin in a feed have upon its nutritive content?

10. Define *crude fiber* and NFE. Name a feed high in fiber and low in NFE. Name a feed low in fiber and high in NFE.

11. How much more concentrated are fats as an energy source than most carbohydrates? What is a typical amount of fat in most forages and cereal grains?

12. What is the composition of a triglyceride? What is the major reason some fats are liquid and others are solid at room temperature?

13. Is it true that a nonruminant (pig, chicken) will tend to have the same type of body fat as that consumed in its feed? Is this also true for ruminants? Explain.

14. In addition to providing energy, name two other reasons fats might be included in a ration.

15. Which elements do amino acids always contain?

16. Explain the difference in the definition of *essential* and *nonessential amino acids*.

17. Which of the following would not need all essential amino acids in their ration: cattle, chickens, sheep, horse, swine, goats, young calves, lamb, foals?

18. Define *NPN* and indicate a possible use for it.

19. In addition to water per se, indicate two other sources of water for use by the animal's body.

20. Give examples where animals might not receive adequate amounts of water.

21. Indicate how water and feed dry matter intake are related.

22. Distinguish between macro and micro minerals.

23. List three important functions of minerals.

24. List three minerals considered toxic at relatively low levels.

25. Distinguish between fat-soluble and water-soluble vitamins and how they are absorbed and stored in the body.

26. What are two examples where certain vitamins can be provided from sources other than the feed?

27. How do protein supplements, energy feeds, and forages and silages differ in their amounts of fiber and protein?

28. Distinguish between gross and net energy of a feed.

The Digestive and Metabolic Systems 4

How does an animal utilize the nutrients in the feed it consumes? What are the mechanisms for releasing specific nutrients from complex feeds and moving these nutrients into organs and cells of the body where they can become part of the animal structure, contribute to animal life, or help form animal products? Meat, milk, and eggs, as well as such nonedible products as wool and leather, are the eventual result of nutrient utilization—digestion, absorption, circulation, and metabolism.

In the previous chapter, we discussed the nutrients necessary for proper animal health and performance and the typical feeds used in production of farm animals. Now we direct our attention to the important aspects of the anatomy and function of the digestive tract of common farm animals and their utilization of nutrients.

Digestion includes the physical and chemical changes that feeds undergo in the gastrointestinal tract (mouth, esophagus, stomach, and intestines) and the release of individual nutrients for absorption. Chewing, swallowing, crushing, and rhythmic movements of the stomach and small intestine cause physical breakdown of feed particles into smaller pieces, increasing the surface area exposed for chemical digestion. Sometimes chemical action can enhance digestion through physical breakdown of nutrient-containing substances. For example, the emulsification of fats by bile results in more but smaller fat globules; thus the total surface area exposed to enzymatic action is greatly increased.

Chemical changes in feeds are accomplished by secreted enzymes[1] and also, to varying degrees, by bacteria and other microorganisms present in these digestive organs. Enzymes catalyze hydrolysis of carbohydrates, fats, and proteins into smaller units that can be absorbed.

Most **absorption** of nutrients in nonruminants (and much in ruminants) occurs from the small intestine, through the intestinal wall and blood or lymph capillary walls, to the circulatory system. Absorbed nutrients are *circulated* to individual cells where they are finally utilized.

A *cell* may refer to a typical muscle cell, liver cell, kidney cell, or cell of some other tissue. Cells that comprise organs have specific functions and use absorbed nutrients for these functions.

> **Digestion**
> The chemical and physical breakdown of nutrients in the gastrointestinal tract preparatory to absorption.

> **Absorption**
> Movement of nutrients through the wall of the gastrointestinal tract and capillary walls into the circulatory system.

---

[1] As mentioned in Chapter 2, *enzymes* are organic catalysts. They promote changes in other organic compounds without themselves being changed. All known enzymes that have been studied are composed of amino acids.

> **Metabolism**
> All physical and chemical processes occurring within a biological system.

In these cells **metabolism** occurs. Carbohydrates and fats are further degraded to yield energy. Amino acids are recombined to form protein and eventually tissue, hormones, or enzymes. Vitamins and minerals also function here in the utilization of carbohydrates, fats, and proteins.

It is important to realize that use of a ration by livestock and poultry involves four steps: (1) digestion, (2) absorption, (3) circulation of absorbed nutrients, and (4) cellular metabolism. The following sections provide a brief discussion of these steps, considering the organs and tissues involved as well as the differences among species. Digestion and absorption in nonruminants are emphasized in the first sections; then the unique characteristics of the ruminant digestive system are discussed.

Upon completion of this chapter, the reader should be able to

1. Differentiate between the ruminant and nonruminant digestive systems.

2. Identify the key parts of each system.

3. Describe the unique characteristics of the digestive systems of the horse and of poultry.

4. Define *prehension, mastication, digestion, absorption, metabolism, emulsification,* and other key terms.

5. Identify just where the metabolism of nutrients occurs.

6. Describe the role of the circulatory system, as related to nutrient use.

7. Describe where and in what form nutrients are stored in the animal body.

4.1   NONRUMINANT DIGESTIVE SYSTEM

Nonruminants, such as swine and poultry (and humans), have a single stomach compartment. Young ruminants such as calves and lambs are functionally nonruminants until the accessory stomach compartments develop and begin to actively aid digestion. This is a gradual process; the additional stomach compartments are not completely functional until the animal is several months of age. The digestive system of the pig, including accessory organs (the liver and pancreas), are diagrammatically illustrated in Figure 4–1.

> **Monogastrics**
> The term denoting animals with a single stomach, in contrast to ruminants with four stomach compartments.

In **monogastrics** such as the pig, upon prehension (eating) each bolus of feed formed in the *mouth* is pushed back into the pharynx region and passes rapidly down the *esophagus* by **peristalsis** into the stomach. After digestion in the stomach, the material (chyme) is moved into the small intestine and progresses through its three sections, the *duodenum, jejunum,* and *ileum.* These three areas receive secretions of the liver, pancreas, and cells within the walls of the small intestine and so are major sites of nutrient digestion and absorption.

> **Peristalsis**
> A process by which muscular contractions propel food material through the digestive tract.

The undigested material will move from the ileum directly into the *large intestine.* It is shorter in length than the small intestine but of greater diameter and volume. It is divided into the large and small colon. The last portion of the intestine is the *rectum,* and the entire digestive tract is terminated by the *anus,* or opening to the outside of the body.

To summarize, the digestive pathway in monogastrics, such as the pig, is

mouth $\longrightarrow$ esophagus $\longrightarrow$ simple stomach $\longrightarrow$

$\longrightarrow$ small intestine (duodenum, jejunum, ileum) $\longrightarrow$

$\longrightarrow$ large intestine $\longrightarrow$ rectum $\longrightarrow$ anus

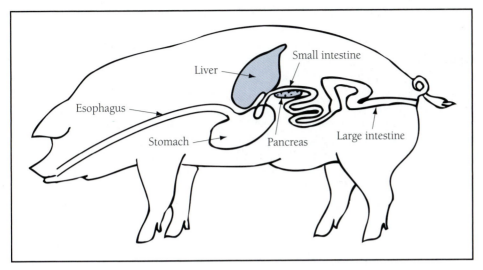

Figure 4–1
A diagrammatic outline of the digestive system of the pig. The intestines are much longer than illustrated.
Source: Iowa State University.

In Figure 4–1, note the small stomach volume in relation to the animal's size. This small size, a characteristic of all monogastrics, dictates frequent feeding and the use of a concentrated ration low in bulkiness and fiber. A 2-week-old pig has a stomach capacity of only about 0.7 pound; the stomach of a 200-pound pig will hold about 8.5 pounds. Total capacity of the four stomach compartments of a 100-pound lamb, by contrast, is about 3 gallons (24 pounds).

In poultry the feed particles move quickly from the *mouth* into the *esophagus* (or gullet), shown in Figure 4–2. This tubular structure directs the feed into a temporary storage pouch, termed the *crop*. The feed-saliva mixture moves from the crop into an enlarged section of the esophagus known as the *proventriculus*. This enlargement is also referred to as the glandular stomach because it produces both hydrochloric acid (HCl) and digestive enzymes. Material moves rapidly from the proventriculus into the *gizzard*, a structure with powerful muscles that reduces the particle size of coarse feed. Fine particles of feed aren't retained by the gizzard but move rapidly into the small intestine.

The *small intestine* in the bird is a major site of digestion and absorption. The first portion, the duodenum, is arranged in a looplike structure and is referred to as the duodenal loop. Within this loop is the pancreas, which pours its secretions into the duodenum. The typical small intestine of the chicken is about 5 feet in length. The avian species possesses two blind pouches between the small and large intestines. Known as the *ceca,* they play a very small role in digestion but can harbor unfavorable organisms (such as coccidia). This inspection occurs prior to slaughter, so producers can observe the overall health status. In contrast to the small intestine, the bird's large intestine is very short—only 4 to 5 inches in the chicken—so digestive residue passes quickly through this area. The large intestine empties waste material into a bulbous area called the **cloaca**, which also serves as a receiving area for the egg in the hen. The external opening of the cloaca, which allows both the egg and waste material to be expelled from the body, is termed the *vent.*

> *Cloaca*
> The bulblike structure at the end of the digestive tract of birds; empties the digestive, urinary, and reproductive tracts.

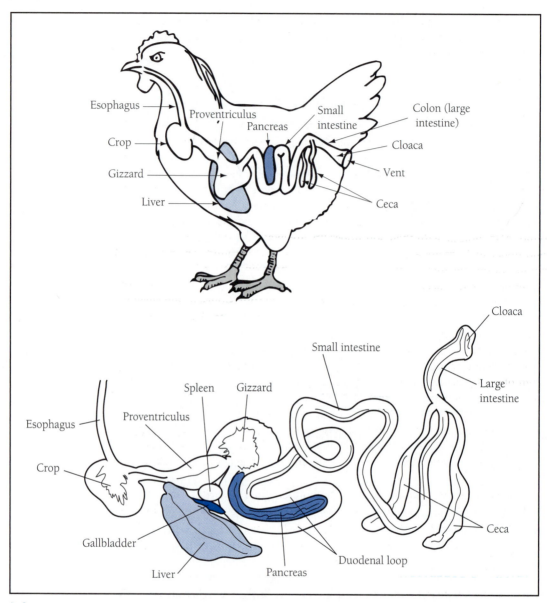

Figure 4–2

Diagrammatic outlines of the digestive system of the chicken. Note the pancreas and its location within the duodenal loop. Also note the blind ceca and the very short large intestine.

Source: Iowa State University and Purdue University.

To summarize, the digestive pathway in poultry is

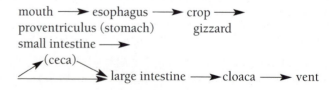

In livestock, excretion of solid wastes and of urine are separate; in poultry, urine is released into the cloaca and excreted with feces. The urine (uric acid) appears as a whitish coating on the droppings.

4.2 PREHENSION, MASTICATION, AND SALIVATION

Prehension, or the act of obtaining food, differs between animals and is affected by the shape of the lips, type and arrangement of teeth, and differences in tongue structure.

Differences in prehension and eating are particularly noticeable in grazing animals. For example, cattle, without soft, flexible lips, do not nibble like sheep and goats. Instead, they use their strong, rough tongue to grasp vegetation and shear it against the incisors of their lower jaw. The shape of the upper lip of sheep and goats enable them to nibble, and because they prefer short, tender vegetation, they tend to graze closer to the ground. Also, horses, with their soft upper lip, are excellent nibblers and with both lower and upper incisors are able to graze plants close to the ground. The rather hard and rigid lips of the pig are not much help in their prehension of food.

Mastication, the mechanical breakdown of feeds into smaller particles, increases the area exposed for more complete enzyme action in the mouth, stomach, and small intestine. A pig has a full set of incisors and molars for complete and effective chewing. Cattle and sheep have only a hard dental pad instead of top incisors. However, whereas molars do most of the grinding, mastication of rations by these animals can be very effective.

Because poultry do not have teeth, they depend on the gizzard for mechanical breakdown of feed particles. A rough inner lining, presence of grit as grinding stones, and strong muscular action ensure thorough grinding.

Saliva secretion is greatest during feeding and mastication (or remastication when ruminants chew their cud). Saliva contains salivary amylase[2] in pigs and chickens. It also acts as an important lubricant in forming the bolus and swallowing. The large quantities produced by ruminants also help maintain the rumen contents in the desired pH range of 5.8 to 7.0.

Following mastication and ensalivation, the tongue moves the bolus into the pharynx. Once here, the act of swallowing is involuntary, and the bolus moves down the esophagus by peristaltic action.

4.3 ENZYMATIC DIGESTION

An enzyme, as previously defined, is a complex substance, composed of amino acids, that increases by catalytic action the rate of chemical change in organic substances. As a catalyst, an enzyme is not used up in the process and is not itself a reactant. Enzymes occur and function at many locations in plant and animal metabolism, including nutrient formation in plant tissue cells and nutrient utilization in animal tissue cells. This

---

[2] *Salivary amylase:* An enzyme that catalyzes or promotes the hydrolytic digestion of carbohydrates. The term denotes a class of enzymes, all of which attack carbohydrates. The carbohydrase in saliva, amylase, is a relatively nonspecific carbohydrase and causes hydrolysis of most carbohydrates. Certain other amylases will cause hydrolysis of specific types of carbohydrates. Amylases are often called carbohydrates.

Digestive enzymes generally are named according to the type of compound they work on, plus the "ase" suffix. Proteases catalyze digestion of proteins and lipases catalyze digestion of lipids.

section is devoted to the function of digestive enzymes that are secreted by and into the digestive tract and that aid in the chemical breakdown of animal feed to permit nutrient absorption.

Enzymatic digestion of feeds, releasing nutrients, begins in the mouth. Saliva elaborates from several sets of glands there. As earlier stated, saliva may contain salivary amylase, which catalyzes digestion of certain carbohydrates, especially sugar and starch. This digestion is not completed in the mouth, but enough does occur in swine to release a few simple sugars, giving a sweet sensation with certain feeds. The saliva of cattle and sheep apparently contains no salivary amylase.

Because saliva and ingested feed are mixed thoroughly during chewing, salivary amylase, if present, continues to hydrolyze carbohydrates after the feed enters the stomach.

The presence of food in the stomach stimulates secretion of *gastric juice* from cells in the stomach lining. Gastric juice contains hydrochloric acid, enzymes, and water. The hydrochloric acid plays two important roles in initiation of protein digestion: (1) it creates an acid environment (pH of 2 or lower) necessary for the hydrolysis of protein, and (2) it activates the inactive proteolytic enzyme pepsinogen to pepsin, which breaks down protein into polypeptides.

Lipase
An enzyme present in gastric juice and pancreatic juice that acts upon fats to produce glycerol and fatty acids.

Some lipase capable of digesting fats is contained in the gastric juice, but its effect in the stomach is minimal. The reasons for its limited action are (1) very few fats are yet in the emulsified form in the early stages of digestion, and (2) the low pH created by hydrochloric acid limits the action of lipase. However, those fats that are emulsified may be broken down to glycerol and individual fatty acids. A third enzyme, *rennin,* may be found in gastric juice and is discussed in Box 4–1.

Chyme
A mixture of consumed feed, saliva, other enzyme-containing digestive juices, and/or microorganisms present in the digestive tract.

The cells of the stomach wall secrete a protective substance, mucus, that protects these cells from the enzymes and acids.

Once partially digested material (**chyme**) moves into the small intestine, it is subjected to an alkaline environment. Secretions into the small intestine from the pancreas and liver elevate the pH, allowing the lipase plus pancreatic and intestinal wall enzymes to work effectively.

BOX 4–1	The Milk Coagulating Enzyme—Rennin

The enzyme rennin, which causes milk to coagulate, is found in the stomach of calves, and perhaps other young ruminants. However, it is not present in monogastric animals.

The discovery of cheese, according to ancient tales, occurred when milk—contained in a traveler's canteen made from the lining of a sheep stomach—changed to curds and whey.

For many years rennin was used in the process of cheese making, its source being the stomachs of suckling calves and the principle based upon its coagulating ability. Today, a synthetic form of rennin is available for cheese making.

Rennin is important in the digestion of the principal milk protein, casein, contained in milk. Rennin converts casein to paracasein which then combines with free calcium ions. The result is a coagulum, calcium paracaseinate, that slows the rate of passage of milk through the stomach, allowing more time for protein digestion and reducing the risk of digestive upsets.

Source: Purdue University.

Bile also aids fat digestion. Produced continually by the liver, bile is stored in the gallbladder and enters the duodenum when fatty materials are in the intestine. Upon contact with bile, fats become emulsified and the fatty acids become more soluble in water. Now, with the surface area of the fatty acids greatly increased and a more favorable higher pH, pancreatic lipase is highly effective in digestion of fats in the small intestine.

Efficient absorption of nutrients depends on the continuation of digestion until the smallest units, which comprise carbohydrates, fats, and proteins, are eventually broken apart. Amino acids apparently are absorbed *most* efficiently through the wall of the small intestine, but *some* dipeptides, especially those composed of amino acids small in molecular weight and therefore small in size, are also absorbed.

Proteases in the small intestine, produced there and in the pancreas, continue the protein hydrolysis initiated in the stomach. Carbohydrates produced by the pancreas continue the carbohydrate digestion, releasing monosaccharides (simple sugars) for absorption.

The small intestine plays a relatively larger role in digestion in poultry than in livestock. In fact, stomach enzymes do much of their work after the chyme passes into the small intestine. Secretions from the avian pancreas are less alkaline, so the digestive material is in a more acidic environment throughout the digestive tract of the bird.

> **Bile**
> Secretion of the liver stored in the gallbladder, if present, that emulsifies fats in the small intestine.

4.4 ABSORPTION

Nutrient absorption into the circulatory system is not a simple filtration process for most nutrients. It is an active process occurring only in living tissues and requiring certain specific conditions. Efficient absorption of many nutrients is dependent on the type of nutrients present, the rate that chyme is moving through the tract, and other factors.

Most absorption in nonruminants (and much in ruminants) occurs from the small intestine, where enzymatic digestion reaches a climax. Because the small intestine is a long organ, nutrients released there can be absorbed before chyme is moved on to the large intestine, especially in the ileal portion. The small intestine of a 100-pound pig, for example, is about 60 feet long. Little absorption, except for water and some small fatty acids, takes place from the stomach or large intestine.

Villi, minute proliferations of the small intestine lining, provide an extremely large surface area for active absorption and engulf the nutrient molecules with their fingerlike structure. Each villus contains an elaborate network of capillaries, so that only two cellular membranes separate the digested nutrients in the intestine from the circulating fluids in the blood or lymph capillaries.

Absorption of monosaccharides, the simple sugars resulting from carbohydrate digestion, requires energy from **adenosine triphosphate (ATP)**, a phosphorus-containing molecule, and also the presence of a certain level of sodium. Evidence indicates that absorption of amino acids also involves sodium. Absorption of fatty acids and glycerol involves ATP and, in the case of glycerol, interim combination with a phosphorus unit.

Minerals, being single elements, apparently are absorbed by simple filtration or through a process controlled by osmotic pressure. In certain cases minerals in feeds are held in complex compounds that are not completely digestible, and thus are relatively

> **Adenosine triphosphate (ATP)**
> A universal energy-transfer molecule. Upon oxidation of organic structures, the energy is captured and stored as ATP. A reverse reaction results in a release of energy for useful purposes.

unavailable for absorption. Over half of the phosphorus in grains is contained in such a compound, called *phytin*. It is not absorbed and utilized by poultry, and only partially utilized by swine. Bacteria in the rumen apparently produce enzymes capable of breaking down the phytin complex, releasing the phosphorus for absorption. Certain feeds contain high levels of oxalates, organic compounds that tie up calcium and thereby render it unavailable for absorption.

Mechanisms for vitamin absorption are not completely understood. It is likely that the water-soluble vitamins are absorbed directly into the blood, whereas fat-soluble vitamins are absorbed into the lymph system and enter the bloodstream later.

The large intestine performs an important role in absorption of water from the contents of the cecum and colon, thereby conserving water for body functions. Also, as noted in the next section, absorption of end products of microbial digestion occurs in this part of the digestive tract of the horse.

4.5   DIGESTIVE SYSTEM OF THE HORSE

Horses and mules are not classified as ruminants (cattle, sheep, goats) or nonruminants (pigs, chickens). However, mastication and other physical processes, enzymatic digestion, and absorption of end products of digestion from the small intestine are similar to that of the pig.

The horse has a full set of upper and lower incisors and molars. These permit easy grazing and also provide rather thorough grinding or mastication before the forage or grain, mixed with saliva, is swallowed. Enzymatic digestion then proceeds in the stomach and small intestine.

It should be noted that although the horse is a herbivorous species and can obtain much of its nutrient needs from forages, digestion of fibrous material cannot occur in the stomach because fermentation breakdown has not occurred, as in the rumen compartment of ruminant animals. Also, the capacity of its stomach is relatively small in relation to the size of the horse and its feed needs (Figure 4–3).

The large intestine of the equine animal consists of the cecum, large colon, small colon, rectum, and anus. These provide a combined structure of about 30 feet in length and 35-gallon capacity in the adult horse. The **cecum**, which consists of a large blind sac, is located between the small intestine and large colon. Considerable microbial digestion occurs in both the cecum and large colon.

Because of the enzymatic digestion in the stomach and small intestine and the absorption of nutrients from the small intestine, the fermentation that occurs in the large intestine is largely of fibrous residues and other materials that escaped enzymatic action. **Volatile fatty acids (VFAs)** are the principal end product of bacterial fermentation and provide up to 25 percent of the animal's energy needs.

The B vitamins and certain amino acids are formed by the bacterial action in the large intestine. Because of this, the absorption that takes place from the large intestine, and the normal levels of these nutrients in horse rations, B-vitamin deficiencies are not common in mature horses and amino acid balance is not a significant concern. For young horses, supplemental B vitamins are usually provided and amino acid levels are of concern in ration formulation.

Cecum
A large pouch that is the forward part of the large intestine of a horse.

Volatile fatty acids (VFAs)
The group of volatile organic acids resulting from rumen fermentation; most prevalent are acetic, propionic, and butyric acids.

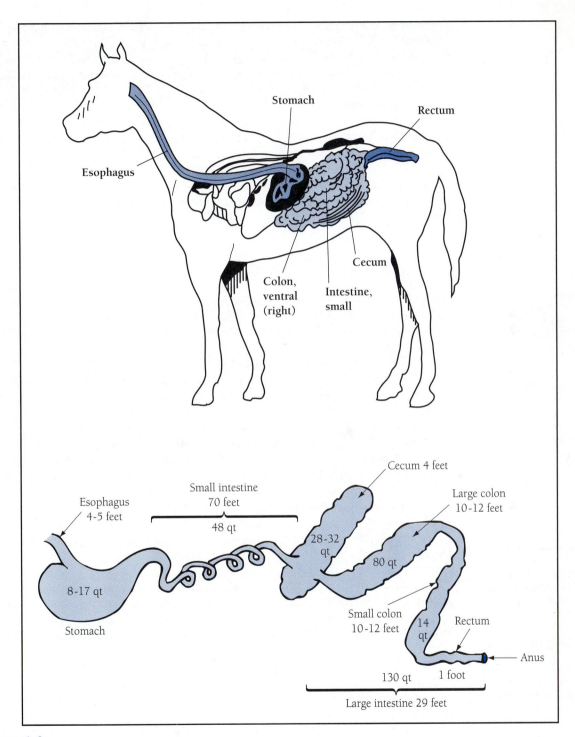

Figure 4–3

Diagrammatic outlines of the digestive system of the horse. The upper diagram shows portions (labeled) of the digestive tract as viewed from the left side, partially surrounded by other organs, and their approximate location within the body. The lower diagram shows the entire digestive tract and approximate capacities of each of the structures.
Source: Purdue University.

TABLE 4–1	Approximate Relative Capacity of Ruminant Stomach Compartments (percent)		
Compartment	**At Birth**	**At 4 Months**	**At Maturity**
Rumen	25	75	80
Reticulum	5	5	5
Omasum	10	9	7
Abomasum	60	11	8
Total	100	100	100

Source: Church, D. C. *The Ruminant Animal, Digestive Physiology and Nutrition.* Upper Saddle River, NJ: Prentice Hall, 1988.

Rumen
The largest of the four stomach compartments in the adult ruminant. The site of active microbial digestion.

Reticulum
The second compartment of the ruminant stomach, where bacterial digestion continues. Has a honeycomb-textured lining, so is often called the "honeycomb," or "hardware" stomach.

Omasum
The third compartment of the ruminant's stomach which receives material from the reticular-omasal orifice.

Abomasum
The fourth compartment of the ruminant stomach where enzymatic digestion occurs. Often called the "true stomach."

4.6 RUMINANT DIGESTIVE SYSTEM

Ruminants are especially important to humanity because they can convert fibrous feeds into meat, milk, and fiber (wool, mohair) without directly competing with the human food supply. This ability of both domesticated cattle, sheep, and goats, as well as non-domesticated species (such as antelope, buffalo, deer, and elk) is possible because of the complex digestive system of ruminants and the symbiotic relationship with the microorganisms that flourish within the digestive system.

Mature ruminants have four functional stomach compartments (Figure 4–4). The largest is the **rumen,** which is located predominantly on the left side. Immediately in front of the rumen is a smaller structure called the **reticulum.** The third compartment is the **omasum,** located to the right of the reticulum and slightly right of the median plane. The **abomasum** is the fourth structure, also located on the right side, somewhat ventral to (beneath) the omasum and extending caudally (to the rear) on the right side of the rumen. The abomasum corresponds to the stomach of nonruminants; thus, it is sometimes referred to as the "true stomach."

Development of the ruminant digestive system is gradual. When calves and lambs are born they are functionally nonruminants because the first three compartments are relatively small. These compartments grow faster than the true stomach, as indicated in Table 4–1, and gradually begin to function.

Before the various compartments and their functions are discussed in detail, other differences between ruminants and nonruminants should be mentioned. Cattle and sheep do not have incisors on the upper jaw, only a dental pad. There is no apparent advantage to this; these animals simply do not seem to need upper incisors. Forage, natural feed for ruminants, is easily severed from the growing plant by the shearing action of the lower incisors when grasped by the tongue.

Cattle and sheep swallow ingested feed almost immediately, with very little chewing.[3] They graze or consume harvested forage for long periods without apparent interruption. Later they lie down in a pleasant spot to initiate the process of rumination, or "chewing their cud."

---

[3] Because ruminants, especially cattle, eat rapidly without thorough chewing, they are less capable of sorting out hard materials and sometimes swallow foreign material contained in the feed. This can result in "hardware disease" when sharp objects (wire, nails) penetrate the walls of the rumen or reticulum. Animal owners sometimes place magnets in the reticulum to accumulate metal and reduce the risk of puncture.

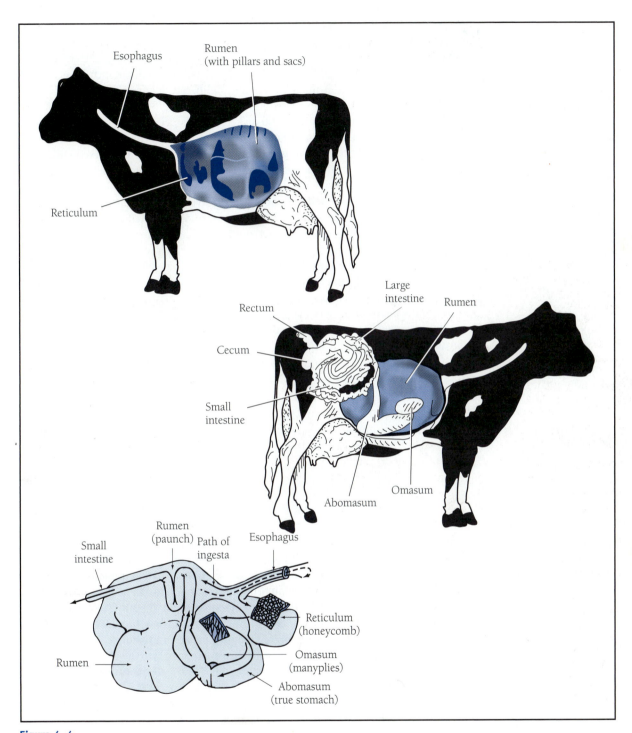

Figure 4–4

The digestive system of the ruminant. The reticulum and rumen are located primarily on the left side, and the omasum and abomasum (true stomach) are located primarily on the right side. Note the approximate location of the small intestine, cecum, large intestine, and rectum. Also shown is a schematic of the four compartments of the stomach with a cutaway showing the lining of the reticulum and omasum.

Source: Purdue University.

Rumination involves the processes of regurgitation, reensalivation, remastication, and reswallowing of earlier ingested rumen materials. Rumen contractions move the feed mass forward and into contact with the *cardia* (lower opening into the esophagus), and it moves up the esophagus by reverse peristalsis into the mouth. This bolus is rechewed and salivated in a much slower and deliberate manner than when first ingested. During the entire process of rumination, involving 8 hours or more each day, large quantities of saliva (as much as 100 to 120 pounds per day in a mature cow) with its sodium bicarbonate enters the rumen to maintain the pH near 7.0.

Because each bolus of feed is swallowed into the rumen and mixed with other ingesta, it is probable that individual feed particles are regurgitated, chewed, and swallowed several times.

Regurgitation described here does not involve the bitter sensation experienced by humans and other nonruminants. The bolus of feed regurgitated by cattle comes from the rumen, which is almost neutral in pH, not from the acidic true stomach.

The rumen, the largest of the four compartments and the one to which most bulky feed first goes, functions as a large fermentation vat. No digestive enzymes are produced in the walls of the rumen and the saliva of cattle and sheep apparently contains no carbohydrase, therefore enzymatic digestion occurring in the rumen is almost completely the result of bacterial and protozoal enzymes.

The rumen provides a near-ideal environment for enormous numbers of microorganisms to live and multiply: a warm temperature, adequate moisture, nutrients from the feed, a suitable pH, largely anaerobic conditions, adequate mixing, and almost continual removal of the products of fermentation.[4] In turn, the organisms

1. Convert large amounts of starch and cellulose to volatile fatty acids as sources of energy.
2. Convert both preformed protein and nonprotein nitrogen to microbial or protozoal protein, which serves as a source of protein containing all the essential amino acids for the host.
3. Synthesize vitamin K and all of the B vitamins.

Most of the digested carbohydrate in ruminant rations is degraded by organisms in the rumen to VFAs, primarily acetic, propionic, and butyric acids. These are absorbed directly through the rumen wall into capillaries of the circulatory system. A small amount of the carbohydrate is incorporated into organisms as their energy store. When these organisms reach the small intestine, enzymatic digestion releases this carbohydrate and the microbial protein comprising the organism structure for absorption. Other items—minerals, water, ammonia, and glucose, an intermediate in rumen digestion of carbohydrate—are absorbed through the rumen wall into the circulatory system.

Because the smaller reticulum is completely open to the rumen, fermentation occurs here, too, and ingesta moves back and forth with the rhythmic movements of rumination. The reticulum lining, however, appears like the surface of a honeycomb. Rumination and contractions cause movement of the smaller food particles into the omasum, the third compartment.

---

[4] The number of bacteria is estimated to be near 10^{10} per gram of rumen contents, whereas protozoa numbers approximate 10^6 per gram.

As indicated in Figure 4–4, the omasum is composed of a mass of suspended, adjacent, and parallel leaves with very coarse surfaces. These leaves, with the aid of regular movements of the organ, may cause a limited amount of grinding or crushing of ingesta, but also permit absorption of considerable water.

To summarize, the digestive pathway in ruminants is

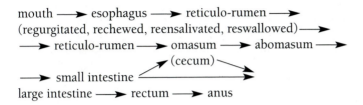

The pathway of liquid diets in young ruminants is different and unique. On the right side of the reticulum is a groovelike structure. It extends from the cardia of the esophagus to the opening into the omasum, the *reticulo-omasal orifice*. This groove also consists of lips extending its length so that when closed, it forms a tube that directs milk (or other liquids) into the omasum. Thus, this reticular groove, closed during nursing or drinking, is beneficial in directing milk into the abomasum for digestion rather than allowing it to accumulate in the undeveloped reticulo-rumen.

4.7   NUTRIENT TRANSPORT

Absorbed nutrients are carried primarily by the blood to organs and tissues throughout the body. Because lymph capillaries carry plasma from tissue spaces back to blood system veins, the lymph system plays a modest role in nutrient transport. In fact, the lymph system is poorly developed in birds, and transportation occurs mainly through the portal system.

Most of the blood leaving the absorptive area of the intestine eventually enters the portal vein leading to the liver. The liver is not only an active site of metabolism—degrading and rebuilding amino acids, detoxifying excretion products, building blood cells, and performing dozens of other vital functions—but is also a storehouse for all classes of nutrients.

The complete circulatory system, which reaches and bathes every cell in circulatory fluid, is a most efficient nutrient distribution system. In a healthy animal the need for a nutrient in any cell is almost immediately met, if the nutrient is available in adequate quantities. The communication medium for this distribution system is primarily the profuse nervous system. Hormones and other factors contribute to the regulation of nutrient use and distribution by providing stimuli for the nervous system.

4.8   NUTRIENT STORAGE

Intake of nutrients in the form of livestock and poultry rations is not continuous, nor is it always uniform through succeeding days, weeks, or months. Nor is the *demand* for nutrients within the animal body continuous or uniform. Just as cities build reservoirs to guarantee a water supply between rains, the animal body is equipped to store, temporarily or semipermanently, essential nutrients.

Calcium, phosphorus, and other minerals are stored in bones and teeth. These deposits accumulate while a dairy cow is nonlactating or before a pullet begins to lay, and are mobilized when the need arises. Each eggshell contains about 2 grams of calcium, so the extra bone storage is needed.

The liver is also an important site of nutrient storage. This storage function is reflected in the high nutritional value of liver meal, a byproduct of the meat-processing industry. Human nutritionists also recognize the value of liver in the diet.

Fat-soluble vitamins, essential amino acids, and carbohydrates are among the nutrients temporarily stored in the liver. Liver storage of vitamin A, accumulated by a cow during the summer on green pasture, may help supply the nutrient during the winter when she may receive roughage low in carotene. Glycogen stored in the liver is a "quick-energy" reserve that can be mobilized rapidly. Newborn animals depend on liver nutrients stored during embryonic development for survival.

Every body cell maintains some nutrient storage, which may be small or large, depending on the type and function of the cell and on the particular nutrient being stored. Active muscle cells in the arms, legs, and jaws must maintain carbohydrate stores plus certain B vitamins, vitamin D, and minerals employed in the energy-releasing oxidative process needed for muscle contraction. Bone marrow cells, which manufacture hemoglobin for red blood cells, store and use amino acids, iron, copper, cobalt, folacin, and other nutrients, in addition to carbohydrates.

> **Adipose fat**
> Fat stored in loose connective tissue throughout the body.

Adipose and subcutaneous tissue cells are primarily energy storage cells where little active metabolism occurs. Fat, from absorption or converted from excess carbohydrate, occupies these cells as the main form of energy storage in animals.

4.9   NUTRIENT UTILIZATION IN THE CELL

Nutrients do their job inside a cell. Most energy-yielding nutrients, such as monosaccharides and fatty acids, are oxidized or burned inside body cells. This oxidation obviously does not involve burning at high temperatures. Rather, enzymes within each cell (different enzymes from those present in the digestive system, but with somewhat related functions) promote this oxidation, permitting it to proceed at body temperature. This oxidation, the reverse of plant photosynthesis, releases energy, primarily from ATP, and gives off carbon dioxide and water as end products.

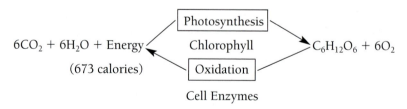

$$6CO_2 + 6H_2O + Energy \quad \xrightarrow{\text{Photosynthesis}} \quad C_6H_{12}O_6 + 6O_2$$

(673 calories)   Chlorophyll

Oxidation

Cell Enzymes

The released energy is then utilized by the tissue for maintenance, growth, reproduction, and production of animal products. In an animal being fattened for slaughter on a full feed of grain, much of the energy is not released by oxidation but is deposited in cells as fat molecules.

Amino acids, released from whole protein by digestion in the small intestine, then absorbed and transported to cells, become a part of the cell structure, allowing it to grow or divide. Muscle development, therefore, is primarily amino acid or protein deposition. Certain amino acids become part of the pigmentation molecules in hair or skin cells. Others are used in glands and organs to build hormones or enzymes.

Excess protein in the diet, above that needed for specific protein functions, is usually converted in the liver to an energy-yielding compound. Such protein is not necessarily harmful, but is wasteful. Protein in animal rations, from oil meals of animal byproducts, invariably is more expensive than the normal energy sources, carbohydrate and fat in grains.

Most other nutrients, besides calcium, phosphorus, and others that become part of the skeletal structure, play their role in the various cells by helping release energy from carbohydrates or fats, or by promoting the build-up of amino acid–containing tissues and products. Some examples of these functions have been previously cited.

QUESTIONS FOR STUDY AND DISCUSSION

1. Which domesticated farm animals are able to graze plants close to the ground? Why?
2. What process aids the passage of food as it moves down the esophagus?
3. Name the three areas of the small intestine.
4. Name the structure located at the junction of the small and large intestine.
5. Compare the length of the large intestine of poultry to that of swine when body size is considered.
6. What portion of the small intestine receives secretions from the pancreas?
7. Name two actions that are examples of the physical processes of digestion.
8. Name two chemical effects of HCl secreted in the stomach.
9. What are the benefits of bile upon entering the small intestine?
10. Although not a true ruminant, why can the horse utilize considerable forage in its ration?
11. Name the four compartments of the stomach of the ruminant. Which two are located primarily on the left side? Which is the largest? Which is the true "chemical" stomach?
12. Name and describe the processes of rumination.
13. What are important functions of saliva, especially in ruminant digestion?
14. Describe conditions in the rumen that make it a favorable structure for microorganisms to flourish.
15. Name three favorable contributions made by the rumen microbes that benefit the host animal.

5

Nutrition of Nonruminants

Animal owners or managers have the responsibility of ensuring that all animals receive properly formulated rations. For desired animal performance and economical production, several steps must be fulfilled. These are discussed in this chapter and include

1. Determining the nutrient requirements based upon desired performance, weight and age, environmental conditions (such as climate and stress), and, in some cases, gender.
2. Locating dependable sources of feed and determine or estimate its nutrient content and digestibility.
3. Determining the need for feed additives such as medicants and antibiotics.
4. Considering ration alternatives, their costs, and palatability.
5. Formulating, mixing, and delivering the rations to the animals.

Animals with one stomach compartment—swine and poultry, as well as calves, lambs, and kid goats up to a few months of age—usually need a concentrated ration. The ration must be low in fiber and highly digestible because these animals have a small stomach capacity and are without the large numbers of microorganisms needed for bacterial digestion of fibrous feeds. Nonruminants, especially swine and poultry, grow and mature rapidly, so the requirements of energy and other nutrients are high in relation to feed consumption and capacity.

The most common feeds for nonruminants[1] are *concentrates,* defined as feeds low in fiber (usually under 12 percent) and high in digestible energy. As previously introduced (Section 3.10), concentrates may be high or low in protein. Grains usually contain only 8 to 14 percent protein and are fed primarily to supply energy. Molasses, sugar, and animal or vegetable fats also are good energy sources and almost devoid of protein. Most other concentrates, such as oil meals and most meat, dairy, and fish byproducts, contain 20 to 60 percent protein. A few, such as blood meal, have 80 percent or more protein.

Concentrated rations, amply fortified with needed nutrients and feed additives, permit rapid gains. Swine, poultry, and veal produced for meat are fed to the most profitable market weight as rapidly as possible. To attain this goal, each group of animals must have a high **average daily gain (ADG)** and excellent feed conversion. Extra days

---

[1] Nonruminants are also called monogastrics, as mentioned, but some avoid that term because both ruminants and nonruminants have but one stomach compartment that produces gastric enzymes for digestion.

> **Average daily gain (ADG)**
> A measure of weight response of an animal or group of animals based on per-day performance.

needed to attain market weight and condition usually means increased costs of production, especially in feed costs, because the daily nutrient needs for maintenance of each animal must be met before nutrients can be used for growth and conditioning. Risk of sickness and death continues until animals are marketed, as does the payment of interest on money borrowed to finance the livestock or poultry operation. Faster gains and quicker marketing also mean more efficient use of facilities because more hogs, broilers, poults or fish can be fed to market weight during the year.

What about sows and gilts that are not necessarily fed for most rapid gains? When offered feed free choice, they may consume more energy than required for maintenance and pregnancy and become too fat. A logical management strategy for extended reproductive efficiency in the mature sow is to minimize weight and fat loss of the sow during lactation, and control excessive weight gain during gestation. *Roughages,* feeds high in fiber and low in digestible energy, are often used *with* concentrates in sow rations. Such roughages include ground alfalfa hay or meal, silage, pasture, and other forages. The same result can be accomplished by hand-feeding a limited quantity of a concentrated ration, but this requires extra labor. Roughage is sometimes a cheaper source of energy than is grain. With laying hens, broilers, and turkeys, top production at lowest cost is desired, and this requires that high-energy, highly digestible rations be used (Figure 5–1).

Upon completion of this chapter, the reader should be able to

1. Differentiate between nutrient requirements and recommended nutrient levels, and explain why they are different.

Figure 5–1
Modern turkey facility. Most poultry such as these poults are fed complete rations, with the ingredients ground, mixed, and often pelleted. *Source:* Purdue University.

2. Describe how the protein requirement as a percent of the ration changes as an animal grows and matures.

3. Differentiate among protein, individual amino acids, calcium, a trace element, and one of the B vitamins in the relative amount required in a ration.

4. Calculate by algebraic method a ration mix based on protein, given the animal's protein requirement and the protein content of ration ingredients.

5. List the approximate daily feed consumption of a baby pig, mature sow, broiler, laying hen, and baby calf, as a percent of body weight.

6. Define and list three examples of feed additives.

5.1   NUTRIENT REQUIREMENTS

Nutrient levels for nonruminant rations are usually stated as a proportion of the ration. Protein and macromineral requirements are stated as a percentage of the ration, and the requirements for microminerals and vitamins are expressed as units, milligrams, or micrograms per pound of ration (Table 5–1).

The need for most nutrients is influenced by volume of feed consumed and most nonruminant rations are fed in blended or mixed form, so expressing requirements as a proportion of the ration is convenient and functional.

It is noted that vitamins A and D and, in some cases, other nutrients or additives have significant functions not related to feed intake. Therefore, recommended allowances for these are usually stated as a proportion of the ration *for certain weights, ages, or stages* of production. The volume of feed provided per day for layers during molt or sows after pigs are weaned may be reduced, but the need for certain nutrients or additives may continue at the same level (perhaps for disease or parasite control), so the proportions in the mixed ration would need to be modified accordingly.

How are nutrient requirements of animals established? There are several techniques used by university experiment stations and by federal and private research agencies, usually involving large numbers of animals. Animals may be fed individually, or in small groups, rations containing graded levels of the nutrient in question—for instance, protein. Precautions include that the rations be adequate in all other nutrients; that all animals, or groups of animals, be as uniform in breeding, weight, and nutritional history as possible; and that all be handled similarly. These precautions will ensure that differences noted in performance of animals receiving different levels of protein will be due specifically to the level of that nutrient, rather than to some other influence.

The lowest level of each specific nutrient that provides normal growth and performance is considered that nutrient's requirement under the conditions of that particular experiment. This does not necessarily mean that different animals, under different environmental conditions, will do as well on the same level. Animals, seasons, and facilities vary, as does quality of ration ingredients.

Fortunately, different research teams have studied the nutrient requirements of the same species, conducting their research at several locations, with animals having different inheritance and nutritional history, and probably with rations composed of a variety of feeds. Similarity in a requirement established under these different conditions indicates that nutritive requirement is relatively stable and not greatly influenced by the different circumstances. Wide discrepancies in requirements established indicate that these different conditions do have a marked influence and should be considered in ration formulation.

TABLE 5–1	Examples of Minimum Requirements for Certain Nonruminants as a Proportion of the Diet

Nutrient	Chickens		Swine		Young Calves[a]
	Broilers, 0–21 days	Laying Hens	5–10 kg (11–22 lb)	50–110 kg (110–242 lb)	45 kg (99 lb)
Metabolizable energy (kcal/kg)	3,200	2,900	3,240	3,275	3,780
Crude protein (%)	23[b]	14.5	20	13	22
Lysine (%)	1.2	0.64	1.15	0.60	—[c]
Methionine cystine (%)	0.93	0.55	0.58	0.34	—
Tryptophan (%)	0.23	0.14	0.17	0.10	—
Calcium (%)	1.00	3.40	0.80	0.50	0.70
Phosphorus available (%)	0.45	0.32	0.65	0.40	0.60[d]
Sodium (%)	0.15	0.15	0.10	0.10	0.10
Manganese (mg/kg)	60	30	4	2	40
Iron (mg/kg)	80	50	100	40	100
Zinc (mg/kg)	40	50	100	50	40
Selenium (mg/kg)	0.15	0.1	0.30	2	0.30
Vitamin A (IU/kg)	1,500	4,000	2,200	1,300	3,800
Vitamin D (IU/kg)[e]	200	500	220	150	600
Riboflavin (mg/kg)	3.6	2.2	3.5	2	(5)[f]
Niacin (mg/kg)	27	10	15	7	—
Pantothenic acid (mg/kg)	10	2.2	10	7	—
Choline (mg/kg)	1,300	4,000	500	300	—
Vitamin B_{12} (mcg/kg)	9	3	17.5	5	(15)

Note: *Lambs are omitted because little research has been done on nutrient requirements of young lambs. Also, the 1994 NRC amino acid and macromineral calculations for broilers were modified based on recommendations regarding the energy content of the diet. Today's rations should be calculated using the most current National Research Council's nutrient requirements.*

[a]*Calves on milk replacer diet.*

[b]*Drops to 20 percent at three weeks and 18 percent at six weeks. Requirements for component amino acids drop similarly.*

[c]*A dash indicates lack of sufficient information.*

[d]*Phosphorus need for calves expressed as total phosphorus.*

[e]*Vitamin D_3, expressed as ICU (International Chick Units) for poultry; Vitamin D_2 or D_3 for swine and calves.*

[f]*Figures in parentheses are estimates based on available research.*

Source: *Adapted from National Academy of Sciences,* Nutrient Requirements of Poultry, 1984, Nutrient Requirements of Swine, 1988, *and* Nutrient Requirements of Dairy Cattle, 1988.

5.2   RECOMMENDED NUTRIENT LEVELS

Minimum nutrient requirements listed in Table 5–1 for various species presumably were established with healthy, sound animals under good environmental conditions and with rations adequate in all other nutrients. Requirements given are interpreted as sufficient to prevent deficiency symptoms from occurring and usually permitting normal health and productivity. This does not mean that these stated levels will permit maximum performance in all animals under all circumstances. Nor does it mean that nutrients present when a ration is mixed will maintain their potency until the ration is fed. Safety factors are added to minimum requirements to allow for such conditions. Table 5–2 provides some recommended nutrient levels for nonruminants, including reasonable safety factors suitable for most production situations.

TABLE 5–2	Recommended Nutrient Levels in Nonruminant Rations					
	Egg-type Chickens		**Swine**		**Young Calves**	**Lambs[a]**
Nutrient	*0–8 Weeks*	*Laying Hens*	*11–22 lb*	*110–240 lb*	*75–100 lb*	*25 lb*
Crude protein (%)	22.0	16.0	24.2	14.3	24.20	22.50
Calcium (%)	1.10	3.74	0.88	0.55	0.77	0.85
Phosphorus (%)	0.50	0.35	0.72	0.44	0.66	0.70
Salt (%)	0.42[b]	0.42	0.28	0.28	0.25	0.10
Iron (mg/lb)	40.0	25.0	50.0	20.0	50.0	—[c]
Manganese (mg/lb)	30.0	15.0	2.00	1.00	20.0	—
Iodine (mg/lb)	0.18	0.15	0.07	0.07	0.12	—
Zinc (mg/lb)	20.00	25.0	49.9	25.0	20.0	—
Vitamin A (IU/lb)	750	2,000	1,100	649	3,800	—
Vitamin D (IU/lb)[d]	100	250	110	75	600	—
Thiamin (mg/lb)	0.90	0.40	0.50	0.50	—	—
Riboflavin (mg/lb)	1.80	1.10	1.75	1.00	—	—
Niacin (mg/lb)	13.5	4.99	7.50	3.50	—	—
Pantothenic acid (mg/lb)	4.99	1.10	4.99	3.50	—	—
Choline (mg/lb)	649	?	250	150	—	—
Pyridoxine (mg/lb)	1.50	1.50	—	—	—	—
Vitamin B_{12} (mcg/lb)	4.50	2.00	8.75	2.50	—	—
Folic acid (mg/lb)	0.28	0.12	0.15	0.15	—	—
Biotin (mg/lb)	0.075	0.05	0.02	0.02	—	—
Additives (when used):						
Antibiotics (mg/lb)	0–10	0	25.00	5.00	20.00	—
Ethoxyquin (mg/lb)	50.5	50	—	—	—	—

[a]*The figures given, except salt, are calculated by converting approximate nutrient content of ewe's milk to an air-dry basis.*
[b]*Inorganic.*
[c]*A dash indicates lack of sufficient information.*
[d]*Vitamin D_3 for chickens; Vitamin D_2 or D_3 for swine and calves.*
Source: *Compiled by the authors and colleagues.*

Lack of complete knowledge of nutrient requirements is a limiting factor in using the minimum requirement figures in Table 5–1 for ration formulation. Knowledge is sparse and sometimes completely lacking on the needs of animals for certain nutrients. Deficiencies of these nutrients may not have been suspected with normal rations, and research on requirements, therefore, may not have been done.

Extra levels of relatively unstable nutrients are often added to rations. A good example is vitamin A in rations with forages that have lost some provitamin A (**carotene**) during maturation, curing, or storage (Table 5–3).

Both vitamins and feed additives can lose potency during processing or storage. Such loss can be promoted by sunlight, high concentration of minerals, moisture, temperature, and pelleting. Destruction is proportionally greater when large quantities have been added to the ration. **Antioxidants** and stabilizers are sometimes added to rations to deter vitamin destruction, and sources of the vitamins or additives may vary in product stability.

> **Carotene**
> A precursor of vitamin A found largely in the yellow and green pigments of plants.

> **Antioxidant**
> A compound that prevents oxidation. Used in mixed feeds to prevent rancidity or loss of vitamin potency.

TABLE 5–3	Approximate Vitamin A Activity of Yellow Corn and Alfalfa (International Units per Kilogram)		
Corn		**Alfalfa**	
New	1,920	Dehydrated:	
Stored:		22% protein	101,000
4 months	1,440	17% protein	52,000
		Sun-cured:	
8 months	1,000	Hay, vegetative stage	81,000
12 months	720	Hay, mid-bloom stage	46,000
24 months	680	Hay, full-bloom stage	26,000

Source: *Alfalfa data adapted from National Academy of Sciences,* Nutrient Requirements of Dairy Cattle, *1988; corn data adapted from Watson.* Proceedings of 17th Corn Research Conference, *American Seed Trade Association, Washington, D.C., 1962.*

Micronutrients
Nutrients added in extremely small quantities to rations.

A safety factor may be added because of the danger of incomplete mixing of **micronutrients.**[2] A baby pig eats less than a pound of feed per day. A 3-day-old chick consumes less than a tenth of a pound daily. Unless mixing is almost perfect, this small portion of feed may not contain sufficient quantities of all nutrients.

Hot weather often reduces feed consumption, so less of each nutrient is consumed each day. A high safety factor should be considered, therefore, for those nutrients whose requirements are not necessarily proportional to feed intake and that may be needed on a daily basis.

Extra levels of such critical nutrients as vitamins, trace minerals, and protein are often included in rations fed to animals that have a poor nutritional history or that have been subjected to disease and other stresses. Runt pigs or feeder pigs shipped a great distance are examples (Figure 5–2). These animals are more susceptible to disease and infections, so a high safety factor in critical nutrients, antibiotics, and other such additives for these animals is usually justifiable.

In some instances certain added ingredients may improve performance of pigs or chicks even when all known nutrient requirements have been otherwise met. Nutritionists refer to this phenomenon as the presence of some *unidentified growth factor* in the added ingredients. This suggests there is yet more to learn through continued research.

5.3   FORMULATING RATIONS

Feed represents the largest expense in the production of animals, especially in poultry and swine enterprises. Therefore, selection of ingredients and formulation of efficient rations are necessary for these enterprises to be profitable.

---

[2] One example is manganese in the ration of laying hens (Table 5–2). The 30 mg/lb of manganese, or 30 parts per million, represents almost the same proportion as that of 2 inches to 1 mile. These small amounts of micronutrients must be thoroughly dispersed in premixes before being incorporated into large quantities of ration mixes.

Figure 5–2
It is worthwhile to provide high quality nutrients to feeder pigs.
Source: Purdue University.

Formulation requires knowledge of feed ingredients and the nutrient requirements for desired performance at a particular age or weight. One must also consider the
palatability of individual feeds, their commercial availability, their freedom from contaminants or toxins, and the quality of the nutrients each contains.

Protein is usually the criterion for determining the proportion of major ingredients in rations for nonruminants. Protein is the most expensive of the macronutrients
(carbohydrate, fat, and protein) and cannot be replaced, as carbohydrate can be replaced
by fat to provide energy. Enough high-protein ingredients must be used to supply ample
protein. Although excess protein can be deaminated in the liver and utilized as a source
of energy, the process is inefficient, and high-protein feeds usually are too expensive to
serve as a major energy source.

When corn or grain sorghum is combined with a "complete" supplement (one
which supplies protein, minerals, vitamins, and additives), a simple algebraic procedure
discloses the correct and most economical mixture that will meet the animal's requirements (Figure 5–3, Example A). This same technique can be used for more complex
mixtures where a certain ratio of grains is to be used in the ration or where predetermined quantities of certain feeds are to be incorporated (Figure 5–3, Example B). These
illustrations indicate the precision that is possible with the use of simple algebra and a
hand calculator.

All feed manufacturers and most poultry and swine producers who formulate
their own rations use computers. Nutrient requirements of the animal, ingredient characteristics and costs per pound, and "limitations" to be imposed, such as maximum
percent of molasses, are entered. The computer formulates the ration that meets specifications at the lowest cost. This is especially valuable where large quantities of feed are
mixed, where a variety of ingredients are available, and where prices of major ingredients fluctuate considerably (Figure 5–4). It also permits easy formulation changes as
animals mature and nutrient requirements change.

Example A (Feeder wants a 16 percent ration):

$$x = \text{lb corn, 8.8\% protein}$$
$$100 - x = \text{lb "complete" supplement, 35\% protein}$$
$$x + 100 - x = 100 \text{ lb rations}$$
$$.088x + .35y = 16 \text{ lb protein}$$

Now solve for x:

$$.088x + 35(100 - x) = 16.0$$
$$.088x + 35 - 35x = 16 - 35$$
$$-.262x = -19$$
$$x = 72.52 \text{ lb corn}$$
$$100 - x = 27.48 \text{ lb supplement}$$

Now check the work:

72.52 lb corn	× 8.8% protein	=	6.38
27.48 lb supplement	× 35.0% protein	=	9.62
			16.00 lb protein

Example B (Feeder wants to use half barley and half corn for the grain portion of the ration):

$$x = \text{lb grain, 10.20\% protein (av. of corn and barley)}$$
$$y = \text{lb "complete" supplement, 35\% protein}$$
$$x + 100 - x = 100 \text{ lb ration}$$
$$.102x + .35(100 - x) = 16 \text{ lb protein}$$

You carry it from here—

Figure 5–3

Using simple algebraic equations to determine the correct proportions of a commercially available "complete" supplement and available energy feeds on the farm.

In addition to meeting the total protein needs of the animal, precise monogastric nutrition requires consideration of *protein quality*—the amounts and balance of essential amino acids needed to meet the needs of the animal. For example, specific amino acid levels are recommended in broiler rations, expressed as a percentage of the metabolizable megacalories per pound of feed. In swine, the concept of *ideal protein*—the essential amino acid requirements expressed in a proportional relationship to the requirements of the amino acid lysine may be used.

Palatability, cost, mechanical conditions, and bulkiness of ingredients are often as important in ration formulation as nutrient content. There is a limit to the amount of molasses, fish solubles, or animal fat that other ingredients will absorb. An excess will cause the feed to cake or "set up" in storage or in a feeder. Pellets made from the mixture will crumble. Also, feeds that are too dusty or otherwise unpalatable will not be consumed well.

Because protein is the most expensive macronutrient, the prices of protein sources have a great influence on the cost of a ration per pound and therefore on the cost of weight gain. During seasons when protein sources such as oil meals[3] are relatively cheap, it is usually advantageous to feed as high a protein level as will be beneficial to

---

[3] Oil meals usually refer to high-protein supplements such as cottonseed meal (41% protein), peanut meal (48% protein), and dehulled soybean meal (48.5% protein).

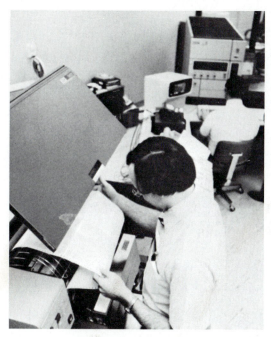 

Figure 5-4 *(continued)*

Up-to-date computer software programs are used extensively by animal production managers and feed manufacturers to formulate livestock and poultry rations. The O.B.S. (Outcome-Based System) model™ employed by Cargill, Inc., estimates the performance of specific pigs raised in a particular environment, provided with a defined nutritional program, and consuming feed at a given rate. An example of factors considered in ration formulation and the calculated ration is shown on the following page.
Sources: Pfizer, Purina Mills Inc., and Cargill Inc.

animal performance. In such cases, the most rapid gains will be the cheapest gains because the animals are maintained fewer days and a larger proportion of consumed nutrients is used for growth.

If, however, oil meals increase in price or grain decreases in price, it may be practical to formulate a lower protein ration containing less oil meal. The slower gains that would result from this lower protein ration mean that relatively more of the feed consumed is being used for maintenance and less for growth, so more pounds of feed are consumed per pound of gain and more days are required to reach market weight. This may be practical, though, if the low-protein ration is cheap enough so that the cost of gain is less (assuming no change in price of market hogs).

5.4 FEEDS AND THEIR EFFECT ON CONSUMPTION

Maximum feed consumption is usually desired for young animals—chicks, pigs, calves, and lambs—so they will be sure to consume enough nutrients to supply their needs (Figure 5–5). Availability of fresh feed helps feed intake. Animals prefer freshly ground or mixed rations. Most large poultry and swine units have mechanical feeding systems that bring fresh feed to the feeders or troughs multiple times in 24 hours. These systems are usually computer controlled, so the times can be easily changed as needed. Where

Figure 5–4

(continued)

```
                        OUTCOME SUMMARY


                    Old MacDonald's Farm

                 OVERALL PERFORMANCE SUMMARY

     Outcome Start Date:  11/27/96

     Avg Daily Gain, Lb:     1.54    ¹ Avg Daily Lean Gain:    0.64
            Feed/Gain, Lb:   3.29          Feed/Lean Gain:     7.85
            Cost/CWT Gain:  $22.06     Cost/CWT Lean Gain:   $52.77
            Days on Feed:    131       ¹ Est Percent Lean:     54.7


        FEEDING PROGRAM ASSUMPTIONS      ¹Calculated on a fat-free lean basis


           Location: Farm 1          Avg Feed Intake:    5.03 Lbs/Head/Day
           Genotype: MIXED              Begin Weight:   50.00 Lbs
             Gender: MixedSex            End Weight:   251.16 Lbs
    Avg Temperature: 65.00 Degrees F  Avg Pen Density:    8.00 Sq Ft/Pig


       RESULTS BY WEIGHT RANGE


       RATION:              1        2        3        4       5       6

    ASSUMPTIONS:
      Weight In, Lb:     50.00   101.32   150.29   200.58
      Weight Out, Lb:   101.32   150.29   200.58   251.16
      Facility:            FIN      FIN      FIN      FIN
      Avg Temp, Degrees F: 65.00    65.00    65.00    65.00
      Feed Type:       GrindMix GrindMix GrindMix GrindMix
      Feed Wastage:     6-10%    6-10%    6 10%    6-10%
      Pen Density Sq Ft Pg: 8.00     8.00     8.00     8.00
      End Date:    12/29/96 01/29/97 03/03/97 04/06/97
      Opt Criteria:      Days     Days     Days     Days
    PERFORMANCE:
      Feed Intake, Lb:    3.63     4.72     5.50     6.22
      Days on Feed:         33       31       33       34
      Avg Daily Gain, Lb:  1.56     1.58     1.52     1.49
      Avg Daily Lean Gain,Lb: 0.60  0.65     0.65     0.65
      Feed/Gain:          2.33     2.99     3.61     4.18
      Feed/Lean Gain:     6.02     7.27     8.47     9.57
      Cost/CWT Gain:    $16.97   $20.68   $23.68   $26.68
      Cost/CWT Lean Gain: $43.74  $50.31   $55.57   $61.05
      Est Percent Lean:  54.60    54.41    54.52    54.70


           Location: Farm 1          Feed Intake:    6.22 lbs/head/day
        Ration Name: Mixed 4        Begin Weight:  200.00 lbs
              Class: Finisher         End Weight:  250.00 lbs
           Genotype: MIXED          Feed Wastage:  6-10%
             Gender: MixedSex          Feed Form:  GrindMix


                    BATCH SIZE: 6000 lbs


                                         LBS/       ACCUMULATED
       INGREDIENT               PERCENT   BATCH          LBS

       Corn Ground               78.02   4681.20      4681.20
       Soybean ML 48%            19.98   1198.80      5880.00
       OPTIMIX PREMIX*            2.00    120.00      6000.00

       TOTALS                   100.00   6000.00      6000.00

       RATION PRICE      $6.38 /cwt    $127.60 /Ton

       NUMBER OF PIGS:    200    BATCHES PER FEEDING PERIOD:   6.22
          DAYS/PERIOD:     30    DAYS TO CONSUME ONE BATCH:    4.82
                                       BATCHES PER WEEK:       1.45


       *Optimix Premix is a product of Walnut Grove Techmixes™

    NOTE: Walnut Grove does not warrant the accuracy of these estimates. This is
          not a guarantee of performance to be achieved (see Cover Sheet Note).

    Version 1.03                    (C) Copyright 1994-1996 Cargill, Incorporated
```

Figure 5–5
To permit maximum feed consumption, ample feed
and feeder space must be provided for young animals.
Sources: Purdue University, Hy-Line Poultry Farms,
and Geo. A. Hormel & Company.

batch feeders are used, poultry producers may stir the feed several times daily, and some
let the feeders remain empty for an hour or two each day. Then, when fresh feed is supplied, the birds really eat.

In some instances, for pigs or poultry, sugar, molasses, saccharin, or flavors may
be added to encourage feed intake.

The normal daily feed consumption for nonruminants of various weights
given in Table 5–4 indicates that consumption on full feed ranges from approximately
2 to 13 percent of body weight. Feed consumption in swine and poultry, expressed as a
percentage of body weight, declines as the animal grows because volume consumed does
not increase as fast as body weight. In calves and lambs, however, feed consumption increases faster than body weight because of the development of the rumen and other
stomach compartments, and because the animals consume increasing proportions of
less efficiently digested roughages.

Feeds vary greatly in palatability. New corn is preferred to old corn that has
been stored several years. For poultry, corn is considered more palatable than grain
sorghum or most small grains, and among the small grains, wheat is most palatable to
swine and poultry. Cracked, rolled, or crushed grain is usually more desired than whole
kernels but must not be ground too fine.

TABLE 5–4	Approximate Daily Feed Consumption		

| Species and Age or Weight | Air-dry Feed Consumption | | Gain (lb) |
	Pounds	Body Weight	
Swine:			
25 lb	2.00	8.00	0.80
50 lb	3.20	6.40	1.20
100 lb	5.30	5.30	1.60
150 lb	6.80	4.53	1.70
200 lb	7.50	3.75	1.90
250 lb	8.30	3.32	1.90
Pregnant gilts, 300 lb	4.50	1.50	0.70
Pregnant sows, 500 lb	4.50	1.00	0.50
Lactating gilts, 350 lb	11.00	3.14	—
Lactating sows, 450 lb	12.50	2.78	—
Mature boars, 500 lb	5.0	1.50	—
Poultry:			
0.5-lb broiler	0.065	13.00	
2.5-lb broiler	0.190	7.60	0.04
4.0-lb layer	0.241	6.00	0.045
Calves:			
50 lb	0.90	1.80	0.50
100 lb	2.00	2.00	0.90
150 lb	4.00	2.67	1.40
200 lb	6.00	3.00	1.60
Lambs:			
15 lb	0.25 (plus milk)	1.60	0.25

Note: *A dash in the gain column assumes that weight is maintained, although there could be some weight loss in lactating gilts and sows.*

Source: *Adapted from National Research Council publications and other sources.*

Soybean meal is the most palatable of common protein feeds. Complete supplements sold in the Midwest are usually composed largely of soybean meal because of its economy, availability, and good, uniform quality.

Many valuable ingredients are relatively unpalatable so must be used in limited quantities or in combination with especially well-liked feeds. Meat and bonemeal is relatively unpalatable, especially for young animals. Milk byproducts are unpalatable when fed alone but seem to improve the palatability of supplements or complete rations.

Excess amounts of minerals are unpalatable to all farm animals. Alfalfa meal, ground oats, and distiller's solubles are not especially well-liked because of their small particle size and tendency to be dusty. Therefore, they may be used to limit feed intake of sows or boars.

Pelleting improves palatability and feed consumption especially when relatively unpalatable ingredients are used. Crumbles, which are usually broken pellets, are well-liked by nonruminants unless they include too much fine material.

Availability of feed that is clean and fresh helps feed intake. Animals prefer freshly ground or mixed rations. Self-feeders should be closely adjusted to prevent feed accumulation in the trough where it may become stale or moldy. Most modern poultry units include mechanical feeding troughs that provide fresh feed to the birds on a continuing basis.

More than 300 feet between feed and water will reduce feed consumption and pig gains, according to South Dakota research. This is obviously important for other species as well, especially in adverse weather.

Animals eat less in extremely hot weather, an unfortunate consequence because the need for certain vitamins is actually increased by high temperature. For example, if animals consume less total ration, they will consume less of the B vitamins. These same B vitamins are recognized as being important for maintaining normal feed intake. Therefore, some poultry producers in hotter climates sprinkle a high-potency "top feed" in the feeders several times each week during prolonged hot weather. This top feed usually contains 10 or more times the normal level of vitamins and extra-high levels of trace minerals. It stimulates consumption and helps prevent slumps in egg production.

Artificial cooling, by fan or water mist, or at least good ventilation and air movement may be worthwhile to maintain feed consumption and utilization during the summer months.

5.5 FEED ADDITIVES

Feed additives, including certain antibiotics, are often used in poultry and swine rations to improve growth rate and feed conversion, and also to reduce mortality and morbidity caused by clinical and subclinical infections. An antibiotic is synthesized by microorganisms and inhibits the growth of one or more other microorganisms. Chemotherapeutics, also sometimes used, have the same function as antibiotics, but are produced chemically.

At this writing, however, there is some consumer concern that human infectious agents may have become immune to certain antibiotics because of long-time heavy antibiotic use in both human medicine and in animal feeding regimes. Therefore, some major fast-food companies have considered or implemented limitations on antibiotic use in the production of animal products they purchase. In addition, there has been growth in the demand for and production of "natural pork," production systems where, among other features, antibiotics are not included in the rations.

Other additives, such as **anthelmintics** to control parasites, may be used, depending on need.[4]

> **Anthelmintics**
> A substance that destroys or expels intestinal parasites.

The U.S. Food and Drug Administration (FDA) conducts a comprehensive review of each new feed additive, and new products must be shown to be both "safe" and "effective" before they are approved by FDA. Each year, a feed additive compendium[5] is published that provides information on what additives, what levels, and what combinations of approved additives can be used in animal feeds. For most medicated feeds, there is a required period of withdrawal from the feed prior to slaughter.

---

[4] One antibiotic found to be a very effective anthelmintic is *Ivermectin*. This broad-spectrum antibiotic is effective against both internal and external parasites.

[5] *Feed Additive Compendium*, The Miller Publishing Co., 12400 Whitewater Drive, Minnetonka, MN 55343.

5.6   NUTRITIONAL IMBALANCES AND DEFICIENCIES

In recent years, it has been recognized that numerous interrelationships exist among nutrients. For example, excess calcium in the ration increases the zinc requirement and affects phosphorus utilization. If additional zinc is not supplied, a classical zinc deficiency—rough haircoat and skin disorders—may occur. The vitamin E requirement will depend on the selenium level in the ration. Adequate calcium and phosphorus nutrition is dependent upon a sufficient supply of each element, a suitable ratio between them, and the presence of vitamin D.

Beyond this, specific deficiencies are relatively rare. Insufficient levels of vitamin D will cause poor eggshell quality, and other classical deficiency symptoms could be cited. But most swine and poultry operations use high-quality rations formulated according to current knowledge regarding nutrient levels needed for high-level production.

5.7   OTHER SPECIES

Nutrition of ducks, pheasant, quail, turkeys, dogs, mink, and immature game animals such as deer and antelope is very similar to that described in previous sections.

The National Research Council (NRC) publication on poultry (see Table 5–1) includes information on the nutrient needs of turkeys, geese, ducks, pheasants, and quail. Companion publications are also available for other monogastrics such as dogs, rabbits, cats, nonhuman primates, laboratory animals, fish, and mink. Generally, the smaller the animal, the higher the metabolic rate (heartbeat) and the higher the requirement for protein and vitamins, expressed as percentages of the ration. Whereas young egg-type chickens require an 18 percent protein ration, young bobwhite quail require a 26 percent protein ration. An exception is turkeys, which, although larger than chickens, have higher amino acid requirements.

It is important to recognize the similarities among nonruminant species, however, in digestive systems, kinds of feeds on which they thrive, and methods of computing and mixing rations.

5.8   NUTRITION AND THE NEWBORN

Colostrum, produced and secreted from the mammary gland the first 3 or 4 days postpartum, is valuable to the newborn calf, lamb, pig, or foal. First-day colostrum is very high in protein, vitamin A, and minerals, and also contains large quantities of antibodies that help newborn animals resist disease and infection (see nutrient composition in Table 6–5). Newborn animals have little antibody material in their blood at birth, so without colostrum very soon after birth they are more susceptible to disease (Figure 5–6).

Many dairy farmers routinely save all first-day colostrum not used by calves and freeze it for later use. It may be valuable for a later calf, should its mother die or develop udder infection, and can also be used for orphan lambs, pigs, or foals.

Energy requirements for dams in all species of farm animals increase markedly at the beginning of lactation (see Tables 6–1, 6–2, and 7–2). More energy and also more protein, minerals, and vitamins are needed for milk production than for development of the embryo in the uterus.

As discussed in Sections 15.1 and 15.2, health and survival of newborn animals can be greatly influenced by the feeding of the dam during gestation. Animals that are

Figure 5–6
These newborn piglets are receiving essential antibodies and nutrients through the consumption of their mother's colostrum very soon after birth.
Source: Iowa State University.

small and weak at birth because of inadequate feeding during gestation have less chance of survival and often grow more slowly. On the other hand, excessively fat females are more likely to experience dystocia (difficult birth) because of greater fetal size and excessive internal fat.

QUESTIONS FOR STUDY AND DISCUSSION

1. Briefly define the terms *concentrate* and *roughage feeds*.
2. Name four concentrates and four roughages normally fed in your area of the country.
3. Give an example where nutrient requirements are expressed on a "per-day" basis and as a "proportion of the ration."
4. Explain the difference between minimum nutrient requirements and recommended nutrient levels.
5. List several factors that may destroy certain nutrients and thus require higher amounts of nutrient addition to the ration.
6. What factors in addition to nutrient content help determine if specific feeds may be added to the ration?
7. Why are excessive amounts of protein in the ration unwise?
8. What is meant by protein quality?
9. List some ration specifications easily and rapidly performed by use of the computer.
10. Give a realistic example of feed intake (air-dry basis) expressed as a percent of animal body weight for a nonruminant animal. (For help, see Table 5–4.)
11. Give three classifications of nonnutritive feed additives.
12. Provide three examples of NRC publications that provide nutritional requirements for species other than poultry, swine, cattle, sheep, and horses.
13. Provide an example where the relationship between two nutrients is important.
14. Have you tried, or do you know someone who tried, to raise a young orphan mammal without the benefit of colostrum? Why is it extremely difficult?

6

Nutrition of Ruminants

Many of the principles of nonruminant nutrition discussed in the previous chapter also apply to the nutrition of ruminant animals. However, the anatomy and physiology of the adult ruminant's digestive system require special discussion regarding their nutritional needs. The first portion of this chapter hopefully brings about a greater appreciation of ruminants and their role in providing food and fiber. The chapter then concentrates on nutritional requirements, factors affecting feed intake, and specific nutritional needs of beef cattle, sheep, and dairy cattle.

Almost all of the milk and about 50 percent of the meat consumed by people in the world is provided by ruminants.[1] In the United States, cattle and sheep are our most important species in converting grasses and other forage into marketable products—meat, milk, wool, hides, and other byproducts. In some countries, important domesticated species of ruminants are the buffalo, camel, goat, llama, reindeer, and yak. When fed largely concentrate diets, ruminants are somewhat less efficient in conversion of feed into meat than are swine and poultry. However, because of their ability to take in large amounts of fibrous feeds, which are digested by the large population of microorganisms they host within their reticulo-rumen, they are much more efficient than nonruminants in utilization of roughages.[2] Ruminants also can be confronted with serious problems related to their unique digestive system (see Box 6–1).

Over 50 percent of the agricultural land in the United States is in permanent pasture or timber that can also be used for grazing. Such land varies in productivity, but the total annual yield of forage is tremendous. A portion of the tilled or "crop" land grows forage—temporary pasture or forage for hay or silage—each year. In addition, much roughage is produced along with grains. Corn stalks and cobs are roughage byproducts of corn production; straw remains when small grains or beans are harvested; and beet tops and cottonseed hulls are also byproduct roughages available for ruminant consumption. It is easy to visualize, therefore, that a very large proportion of the energy stored by photosynthesis in plants is stored in what nutritionists call roughage, efficiently utilized *only* by ruminants (Figure 6–1).

---

[1] The world population of domesticated ruminants that provide food and other uses to humans totals nearly 2.8 billion. This number includes about 1 billion sheep, 1 billion cattle; 400 million goats, 125 million buffalo, and 30 million camel, yak, llama, and reindeer. An estimate of wild ruminants is several hundred million. (*The Role of Ruminants in Support of Man,* 1978. Winrock International Livestock Research and Training Center, Petit Jean Mountain, Morrilton, AR.)

[2] In our discussions, roughages include all fibrous plant material edible by livestock, generally characterized as being bulky and high in cellulose and lignin (major components of plant cell walls) and poorly digested. The term "forage" refers to that portion of roughages that are high quality, such as that produced on improved pastures and managed meadows. Forage also includes silages from meadows or whole-plant grain crops.

BOX 6–1

Ruminants—Sometimes Victims of Their Digestive System

Ruminant animals, because of their digestive physiology and anatomy, are fascinating animals in their ability to rapidly consume large amounts of feed and later regurgitate it for further chewing. Their symbiotic relationship with rumen microbes is equally interesting. As host animals, they reap the benefits of cellulose digestion of fibrous materials and production of microbial protein and B vitamins. However, at times they suffer from being a ruminant. The threat of hardware disease was described in Section 4.6. Two additional potential maladies are described here—bloat and ketosis.

Bloat sometimes results in tragic and sudden losses in ruminants. It occurs most commonly in cattle grazing some legume pastures.

Bloat is characterized by distention of the left side of the animal and is caused by excessive amounts of gases that accumulate in the reticulo-rumen that cannot be released. These gases are the natural products of rumen fermentation and are normally released through the belching process. Anything that interferes with this process can result in bloat. One known cause is the entrapment of the gases into foamy bubbles or froth brought about by the saponins in certain legumes.

As bloat advances, intraruminal pressure increases, respiration is impaired, and death may occur. In acute cases, treatment requires immediate release of the pressure, preferably by stomach tube. However, in some cases actual puncture into the rumen with a trocar may be necessary to avoid death.

Prevention of bloat involves careful pasture management, such as not turning hungry animals into legume pastures and strip-grazing to force whole-plant grazing. Also, a surfactant, such as poloxalene, can be effective if fed regularly in the feed or mineral supplements.

Ketosis is a metabolic problem most common in lactating dairy cows. It is characterized by elevated levels of ketone bodies and generally occurs within a few weeks after calving. Many high-producing cows have borderline ketosis. Advanced cases result in lack of appetite, depression, and loss in body weight and milk production.

Because much of the ruminal digestion of carbohydrates results in volatile fatty acids rather than glucose, the glucose level of ruminants is normally lower than that of nonruminants. Glucose levels are further lowered by production demands, with consequent excessive fat mobilization. In ketosis, insufficient glucose may prevent the complete oxidation of the resulting fatty acids, causing a build-up of the toxic ketone bodies.

Treatment of ketosis involves IV administering of glucose, and/or glucocorticoid injection. Problem cows may also receive oral doses of propylene glycol, a glucose processor not destroyed in the rumen.

Reduced incidence of ketosis involves careful management of the cow, especially during the week before calving. Cows should not be allowed to become overconditioned, as most become sluggish in their appetites that predisposes them to ketosis.

Ketosis also can occur in ewes in late pregnancy. Overconditioned ewes carrying twins or triplets are most susceptible, especially when stressed by short periods of inadequate feed or hot conditions.

Source: Purdue University.

Concentrates are used when needed to supplement the roughages, for more rapid growth, quicker marketing, maximum milk production during lactation, or higher quality carcasses. Dairy cows in high production consume large proportions of concentrates, but the concentrate mixture is formulated according to the quality and amount of forage fed. Lambs fed for meat often receive 50 or 60 percent concentrate so that their daily energy intake will be high enough to promote finishing at desired weights. The same is true in cattle fed for meat when the goal is the choice or prime grade. A few cattle and lambs are finished on all-concentrate or "high-concentrate" (80 to 90 percent) rations.

Figure 6–1
Cattle make good use of corn stalks or hay.
Sources: Purina Mills, Inc. and Union Pacific Railroad.

Upon completion of this chapter, the reader should be able to

1. Explain the significance of ruminant animals in converting roughage and other materials humans cannot utilize into meat or animal products they can use.

2. Explain why some rations are calculated on a per-day basis and others on the basis of dry matter proportion.

3. Explain why the energy value of feeds and the energy requirements for ruminants need to be measured and expressed so precisely, and also which of several energy measurements (gross, TDN, digestible, net) is the most useful in formulating ruminant rations.

4. Calculate the nutrient level of the ration, given the composition of each ingredient and the pounds used of each ingredient.

5. Explain the significance of concentrate-to-roughage ratios in feeding cattle and lambs.

6. Describe the influence on mature milk production of varying energy levels for growing dairy heifers.

7. List three separate factors that determine the daily nutrient requirement of a dairy cow.

6.1 NUTRIENT REQUIREMENTS

The unique aspects of the ruminant's digestive system have been previously discussed and are relevant in understanding their nutritional needs and how they differ from non-ruminants (for a brief review, see Section 4.6).

Once their digestive system becomes fully functional, ruminants require fewer specific nutrients in their rations than nonruminants. The B vitamins required for healthy and thrifty ruminants are synthesized by rumen microorganisms. The same is true for individual amino acids. Rumen microorganisms can build the essential amino acids from protein, other amino acids, and other nitrogenous material. Cellulolytic rumen bacteria break down the cellulose in plant material and convert it into useful energy, as volatile fatty acids.

Requirements for some cattle and sheep are stated on a per-day basis (Table 6–1). This is because the nutrient intake of grazing animals is largely controlled by the volume of roughage dry matter they consume per day and also because some ruminants fed in drylot are fed roughage and concentrate separately or twice per day.

However, because of the common use of complete rations for dairy cows and for beef cattle and lambs in commercial feedlots, nutritive requirements are often stated as a "percentage" or "proportion" of the dry matter in the ration (Table 6–2).

Note that the requirements listed in Tables 6–1 and 6–2 are dependent not only on animal weight but also on desired performance; that is, maintenance, finishing for slaughter, growth of dairy heifers, pregnancy, or lactation.

As stated in Chapter 5, National Research Council requirements, summarized in publications, are available not only for domesticated farm animals but also for other species including cats, dogs, fish, laboratory animals, mink, foxes, and nonhuman primates. These requirements are considered the levels of each nutrient satisfactory for good performance under average conditions.

Maximum performance is not guaranteed when requirement levels are used; no margins of safety are included. One must consider, therefore, providing additional levels of nutrients when formulating rations, depending on such factors as

- environmental stresses, such as intense heat and high humidity, cold and high wind velocity (wind chill factor), and lack of shelter
- genetic potential
- herd or flock health
- previous nutritional history
- ration palatability
- nutrient stability

Requirements stated as a proportion of the ration provide opportunity to be compared at different stages in the life cycle and among species. Additional requirements are given in the references cited in the source note of Table 6–2.

Section 3.9 described the nutritive values of selected feeds and provided a brief discussion of appraising the energy value of feeds. Except for wintering young cattle or handling mature beef cows or ewes, most ruminant feeding programs are directed toward rapid growth, where the animal will consume and use as much feed energy as possible. The more rapid the growth, the lower the cost of production in most instances. Feedlot cattle or lambs will reach market weight and condition in fewer days. Also, yard fees, interest charges, and labor will apply to fewer days, and there are fewer days that illness or death will be risked.

This means much attention must be paid to achieving a high energy intake. On the other hand, feed is expensive, and if the energy level is so high as to cause digestive problems, inefficiency results.

TABLE 6-1

Examples of Daily Nutrient Requirements of Certain Ruminants

	Daily Gain Expected kg	lb	Daily Dry Matter Consumption kg	lb	Rough-age (%)	NE$_m$ (Mcal)[a]	NE$_g$ (Mcal)	Crude Protein (g)	Calcium (g)	Phos-phorus (g)	Vitamin A (IU)	Vitamin D (IU)
Steer, medium frame, 300 kg (660 lb)												
Low gain	0.2	0.44	6.9	15.1	100	5.55	0.69	499	14	12	15,200	1,900
Moderate gain	1.0	2.2	7.3	16.0	20–25	5.55	3.68	944	32	16	16,100	2,000
Dry beef cow, middle third of pregnancy, 545 kg (1,200 lb)	0	0	8.8	19.4	—	8.87	N/A	960	16	13	24,640	2,420
Growing Holstein dairy heifer, 100 kg (220 lb)	0.8	1.76	3.0	6.6	—	2.72	1.66	483	18	10	4,240	660
Dairy cow:												
Maintenance: 600 kg (1,320 lb)	0.72	0.33	11.4	25.1		9.70		404	24	17	46,000	18,000
Production: 30 kg (66 lb) 3.5% milk			8.6	19.0		20.70	—	2,520	89	55	—	—
Total dairy cow			20.0	44.1		30.40		2,924	113	72	46,000	18,000
Ewe, 70 kg (154 lb), last 4 weeks of gestation, 180–225% lambing rate	225	0.51	1.9	4.2				214	7.6	4.5	5,950	388

[a] A portion of the net energy will be used for maintenance and the balance for growth, milk production, or development of the fetus. In the case of the 300-kg steer gaining 1.1 kg per day, about 5.5 Mcal will be used for maintenance and 4.90 Mcal for growth.

Source: Adapted from National Academy of Sciences, Nutrient Requirements of Beef Cattle, 1996; Nutrient Requirements of Dairy Cattle, 1988; and Nutrient Requirements of Sheep, 1985.

TABLE
6–2

Examples of Nutrient Requirements of Certain Ruminants as a Proportion of the Dry Matter in Rations

	Daily Gain Expected		Daily Dry Matter Consumption		Rough-age (%)	NE_m (Mcal/ kg)	NE_g (Mcal/ kg)	Crude Protein (g)	Calcium (g)	Phos-phorus (g)	Vitamin A (IU/kg)	Vitamin D (IU/kg)
	kg	lb	kg	lb								
Finishing yearling steer, 400 kg (880 lb)	1.4	3.0	8.2	18	15	2.09	1.41	10.14	0.41	0.23	2,200	275
Finishing lamb, 40 kg (88 lb)	0.275	0.60	1.6	3.5	25	(3.3 Mcal/ kg DE)		11.6	0.42	0.21	1,175	147
Growing Holstein dairy heifer, 100 kg (220 lb)	0.8	1.76	3.0	6.60	60	1.69	1.08	16.0	0.52	0.31	2,200	308
Dairy cow, 600 kg (1,320 lb), producing 30 kg (66 lb) 3.5% milk	0.15	0.33	20.0	44.1	60	1.62[a]	—	16.0	0.58	0.37	3,200	1,000

[a]Shown here as a single figure because the efficiency of energy used for milk production is about the same as that used for maintenance.

Source: Adapted from National Academy of Sciences, Nutrient Requirements of Beef Cattle, 1996; Nutrient Requirements of Dairy Cattle; 1988; and Nutrient Requirements of Sheep, 1985.

Much research has been directed toward precise measurements or estimates of net energy required for each function—maintenance, growth, lactation, development of the fetus—and at successive stages of life. The maintenance requirement is higher, of course, for a 1,000-pound steer (7.65 Mcal) than for a 500-pound steer (4.55 Mcal), but note that the difference is not in direct relationship to weight. This is true, in part, because the larger animal has less surface per unit weight and so loses relatively less energy by convection or radiation than does the smaller animal.

Net energy values on feeds and data on animal requirements permit precise formulation of ruminant rations for maximum energy efficiency or for specific levels of animal production.

Requirements for some nutrients are not included in Tables 6–1 and 6–2. Such nutrients are generally considered to be adequate in most typical rations. Exceptions do exist, however, such as the lack of iodine and selenium in feeds grown in certain central regions.

Salt may be provided free choice, preferably in granular form. However, adequate intake is ensured by adding salt to the concentrate mix or complete ration at levels suggested by the NRC.

Though corn is the only grain with considerable carotene (provitamin A), nearly all good-quality forages contain high levels. Requirements given in Tables 6–1 and 6–2 are usually met if animals are grazing lush grasses or legumes, or are being fed considerable silage or green, leafy hay. Deficiencies of vitamin A occur during periods of prolonged drought, when grass turns dormant, or when animals in confinement receive feeds low in carotene, such as straw, weathered hay, grain other than corn, or corn or hay that has been stored for a long time. Most formulated rations include supplemental vitamin A because of its low cost and as insurance against possible deficiency.

Ruminants exposed to sunlight synthesize adequate vitamin D_3 through conversion of cholesterol in their skin to active vitamin D. Less synthesis occurs in winter months when sunlight exposure is less. Either sun-cured forages or supplemental vitamin D usually is recommended during the winter or for animals confined under roof. This is especially true in dairy herds, where cows have a high vitamin D need for both its content in milk and its role in mobilization and utilization of the calcium necessary for high milk production.

6.2   FACTORS AFFECTING FEED INTAKE

Bulkiness, moisture content, and palatability of feeds must be considered in ration formulation. A ration formulated for one day may contain the quantities of nutrients the animal needs, but if they are supplied only by bulky ingredients totaling more than the animal can or will eat, the ration will not provide adequate nutrition. The daily dry matter consumption figures suggested in Tables 6–1 and 6–2 serve as examples in formulating rations.

The tremendous volume of the ruminant stomach compartments does allow a greater volume of feed intake and greater use of bulky feeds than is true for nonruminants. Consumption of feed dry matter usually ranges from 1.5 to 2 percent of body weight in older animals on a maintenance ration to about 2.5 to 3.5 percent for finishing lambs and beef cattle, or as much as 3.8 to 4.0 percent for dairy cows during peak production.

Daily consumption of pasture forage, and the nutrients it contains, can be estimated if the composition of the forage (including moisture) is known. Unless the pasture is extremely high or low in palatability, ruminants usually consume about as much "dry matter" in the form of pasture as they would in equally palatable air-dry feed.

Relative dry matter is also used as a guide in predicting silage consumption, except where the silage is extremely wet and soggy and therefore less palatable. Because most silages contain about 33 to 35 percent dry matter, and hay contains about 80 to 90 percent, 2.5 to 3 pounds of silage is considered about equivalent to 1 pound of hay. Some refer to this as *hay-equivalent basis*. Today, most rations are calculated on a *total-dry-matter basis* and then adjusted to an **as-fed** basis for weighing ration ingredients for mixing.

> **As-fed**
> An expression to denote nutrient content of a ration or feed when fed.

Corn, sorghum, or small cereal grain silages, however, should be considered mixtures of both roughage and concentrate, because they contain grain as well as forage and supply considerably more energy on a dry-matter basis than does grass or legume silages (Table 3–4).

6.3   FORMULATING DAILY RATIONS FOR COWS AND EWES

Designing a ration for a ruminant, using daily requirements such as those in Table 6–1 or current NRC recommendations as a guide, can be done by trial and error or by established methods. Individuals should use the most current NRC recommended levels as Tables 6–1 and 6–2 are examples. A computer and appropriate software, however, enable calculation of rations more precisely and rapidly. This also allows changing rations more frequently to hold down ration cost as relative prices of ingredients change.

It is obvious that a ration must be palatable, economical, and easy to handle. Consideration of these factors sometimes may be more important than meeting the exact nutrient level listed. Also, as was previously pointed out, nutrient requirements must be interpreted according to the condition of the animals, feed prices, and other factors. A 1-day ration for a pregnant ewe is given in Table 6–3.

6.4   RATIONS FOR FINISHING CATTLE AND LAMBS

Most feeders use complete rations for finishing lambs and cattle. Research results suggest quicker adaptation to finishing rations and more efficient utilization of feed when the ration is completely mixed and sometimes for lambs, pelleted. Complete rations can be formulated with extreme precision, using NRC requirements such as those shown in Table 6–2.

Ratio of concentrate to roughage must be considered when complete rations are formulated, and provision must be made for supplemental minerals, vitamins, and feed additives. The relative amounts of grains and oil meal used in the concentrate portion of the ration are usually dependent on protein content. Algebraic procedures, similar to those used for formulating nonruminant rations (Fig. 5–3), can be adapted to formulate ruminant rations and mixtures. A typical complete steer-finishing ration is listed in Table 6–4.

Liquid feeds, such as molasses, a byproduct of sugar refining, or stillage, a byproduct of ethanol production, may also be used and are excellent energy sources. Both research and feedlot experience suggest stillage may have feeding value beyond energy and known nutrients it contains. The economics of these liquid feeds may depend on transportation cost, or proximity of the feedlot to the processing plant.

<table>
<tr><th colspan="2">TABLE
6–3</th><th colspan="10">Example of One-Day Wintering Ration for a 140-Pound (63-Kg)
Pregnant Ewe in the Last Six Weeks of Gestation[a]</th></tr>
<tr><th></th><th colspan="2">Daily Feed</th><th rowspan="2">Crude
Protein
(lb)</th><th colspan="2">Digestible
Energy</th><th rowspan="2">Calcium
(g)</th><th rowspan="2">Phosphorus
(g)</th><th rowspan="2">Vitamin A
(IU)</th><th rowspan="2">Vitamin D
(IU)</th></tr>
<tr><th>Feed</th><th>As-Fed
Basis
(lb)</th><th>Dry-Matter
Basis
(lb)</th><th>TDN
lb</th><th>Mcal</th></tr>
<tr><td>Corn silage</td><td>6.0</td><td>1.8</td><td>0.13</td><td>1.10</td><td>2.22</td><td>2.72</td><td>1.91</td><td>13,922</td><td>324</td></tr>
<tr><td>Alfalfa hay</td><td>1.0</td><td>0.9</td><td>0.15</td><td>0.50</td><td>1.02</td><td>5.10</td><td>1.10</td><td>3,946</td><td>905</td></tr>
<tr><td>Corn, yellow</td><td>1.0</td><td>0.9</td><td>0.08</td><td>0.80</td><td>1.62</td><td>0.09</td><td>1.22</td><td>520</td><td>—</td></tr>
<tr><td>Total</td><td>8.0</td><td>3.6</td><td>0.36</td><td>2.40</td><td>4.86</td><td>7.91</td><td>4.23</td><td>18,388</td><td>1,229</td></tr>
<tr><td>Requirement</td><td></td><td></td><td>0.36</td><td>2.40</td><td>4.80</td><td>4.60</td><td>3.50</td><td>3,160</td><td>350</td></tr>
</table>

[a]This ration is nutritionally adequate, palatable, and usually economical. Surpluses of calcium, Vitamin A, and Vitamin D in such a ration have not been demonstrated to be harmful.
Source: Adapted from National Academy of Sciences, Nutritional Requirements of Sheep, 1996.

TABLE 6–4

Example of Complete Ration for 750-lb (340-kg) Yearling Cattle[a]

Feeds	Dry-Matter Basis (%)	As-Fed Basis		Daily Feed Consumption			Feed Nutrients	Calculated Analysis	Requirement
		%	lb/ton	Dry-Matter Basis (lb)	As-Fed Basis (lb)				
Sorghum, grain	70.0	50.3	1,006	12.6	14.2		Crude protein (%)	11.64	11.25
Corn silage	20.0	42.6	852	3.6	12.0		TDN (%)	80.3	85.0
Supplement[b]	10.0	7.1	142	1.8	2.0		NE_m (Mcal/kg)	1.84	2.09
							NE_g (Mcal/kg)	1.22	1.41
Total	100.0	100.0	2,000	18.0	28.2		Calcium (%)	0.32	0.26
							Phosphorus (%)	0.32	0.25
							Vitamin A (IU/kg)	3,670	2,250

[a]Proportions of concentrate to roughage would vary depending on quality of cattle and how long they have been on feed. Quantities of nutrients supplied by each ingredient illustrate differences in feed values. Premixes in the supplement may supply other vitamins, trace minerals, and/or additives. Since the premix "carrier" may be grain or oil meal, it may also supply small quantities of other nutrients. See Section 4.5 on protein equivalent supplied by urea.

[b]Ingredients in the supplement (lb/ton): grain sorghum, 1,320; soybean meal, 270; urea, 93; limestone, 112; cane molasses, 80; salt, 100; and premixes, 25.

Source: Adapted from National Academy of Sciences, Nutritional Requirements of Beef Cattle, 1996.

Enterotoxemia also occurs in cattle. This disorder is usually preceded by **acidosis,** a condition where the rumen pH decreases from above 7.0 to 6.5 or lower. Enterotoxemia is caused by a specific strain of organism in the digestive system. This organism, *Clostridium perfringens,* which produces a toxin that makes cattle or lambs "go off feed" and that can cause eventual death, often develops and flourishes when animals receive a high-concentrate ration. Vaccination is often used to reduce risk of enterotoxemia; a bacterin is sometimes used before lambs are placed on "full feed," and an antitoxin may be effective in preventing death loss.

When lambs are started on grain too rapidly, some will overeat. The first symptom is probably related to acidosis, with failure of lambs to come up to the bunk at eating time. They move about slowly, if at all. Usually they stand in a corner or near the fence, with their back arched and head down. Many feedlot operators maintain a "hospital" lot to handle these lambs. They are taken off all grain immediately, given only hay and water for several days, then started back on grain very slowly.

Because of the universal occurrence of this organism and the sensitivity it has to concentrates, lambs traditionally are started on grain slowly. When 60- to 70-pound feeder lambs are purchased from range areas, they usually have never eaten grain, having grown on grass and milk. Careful feeders usually offer the lambs only hay after arrival, then feed a small quantity of oats (more bulky than corn) each day. Corn or sorghum then gradually replaces the oats and is increased to about half the ration in 3 or 4 weeks. Experienced feeders in some areas are able to increase the proportion of concentrate faster and sometimes go as high as 60 or to 65 percent of the ration. Mixing a complete ration, using ground ingredients, or mixing and pelleting the ration often helps prevent enterotoxemia. This may be because mixing and pelleting prevents certain lambs from "sorting out" a higher proportion of concentrates, possible when ingredients are fed separately.

Although, as mentioned above, lambs can be vaccinated as an aid in preventing enterotoxemia, caution in feeding still must be practiced. New feeder lambs need fresh water, rest, and a ration consisting of predominantly high-quality forage to reduce health problems and help bring them on to full feed gradually. This is economically important because consumers prefer retail cuts from lambs that are well finished by the time they weigh 100 pounds.

Though most lamb feeders use a high proportion of concentrates, some have used pelleted or cubed rations containing as much as 70 or 80 percent roughage. Research suggests that the optimum "physical balance" between roughage and concentrate may be different in pelleted rations than when rations are not pelleted. There is evidence that pelleting improves the utilization of some feeds. The improvement is apparently larger in rations with a high proportion of roughage.

When cattle are being fed, the desired proportion of grain depends on the weight and quality of the cattle and the length of time they are expected to be fed. Relatively more grain may be fed high-quality cattle because they can potentially produce high-grade carcasses that will sell for a relatively high price (Figure 6–2).

Cattle can be finished on all-concentrate rations. This may reduce feed cost in areas where roughage is scarce, reduce the volume of feed handled, and make mechanical or automatic feeding easier. The practice is not common and requires especially close management because of bloat, acidosis, founder, and other problems that sometimes occur. A more practical approach is to incorporate 10 percent roughage into the ration.

Enterotoxemia
An acute disease within the intestine caused by the organism *Clostridium perfringens;* highly fatal and most common in sheep and cattle on concentrate rations.

Acidosis
A decrease of alkali in body fluids in proportion to the acid content; common in ruminants with sudden ration change and lowered rumen pH.

Figure 6–2
Commercial feedlots use total mixed rations (TMR) for feeder cattle to be grown and finished to market weights.
Source: American Angus Association.

6.5   DAIRY CALVES AND GROWING HEIFERS (HERD REPLACEMENTS)

Whereas beef calves and lambs are raised primarily for production of quality meat, dairy heifers are grown to replace culled cows, so the goal of the feeding system is well-grown heifers to enter the milking herd at about 24 months of age. The same nutrients are needed and, in general, the same feeds are used. There are several items, however, that might be considered unique in feeding dairy cattle.

Calves in most dairy herds are not allowed to nurse their mothers after the colostrum period because the cow's milk then becomes ready for sale (Figure 6–3). Fluid milk is valuable enough that most calves are shifted quickly to a dry, soluble milk replacer. Colostrum is highly nutritious (Table 6–5). When more colostrum is available than can be utilized by a calf, it is sometimes frozen and held for later use. In feeding such colostrum, it should be diluted with one to two parts water because of its high nutrient content.

A high-quality milk replacer, to be mixed with water for feeding morning and evening, should contain 20 to 24 percent protein, preferably from milk byproducts, although a nonmilk source such as special processed soybean concentrate can provide some of the protein. Much of the milk replacer energy is derived from high-quality animal fats, 10 to 15 percent in most formulations. Lecithin is added to aid fat digestibility, and an additional emulsifier is included to aid in mixing the high-fat replacer with water. The young calf has little ability to utilize nonmilk sources of carbohydrate, such as sucrose or starch, so the major carbohydrate must be lactose from milk products.

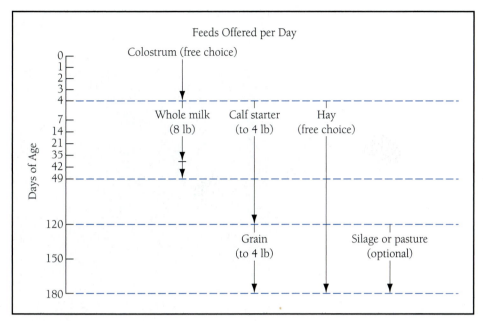

Figure 6–3
The limited whole milk–dry calf starter–hay system of raising dairy calves. A milk replacer may also be used, and weaning will range from 42 to 49 days. Amounts of each feed depend on size of calf.
Source: Adapted from *Iowa State University Extension Pamphlet 258.*

TABLE 6–5	Typical Composition of Colostrum and Transitional Milk Postcalving			

| | Colostrum Milk | | | Normal |
Component	1st	2nd	3rd[a]	Milk
Total solids (%)	23.9	14.1	13.6	12.9
Fat (%)	6.7	3.9	4.4	4.0
Protein (%)	14.0	5.1	4.1	3.1
Lactose (%)	2.7	4.4	4.7	5.0
Vitamin A (µg/dl)	295	113	74	34
Immunoglobulins (%)	6.0	2.4	1.0	0.1

[a]*Composite of fifth and sixth milkings.*
Source: Feeding the Newborn Dairy Calf, *Special Circular 311, Pennsylvania State University.*

A high-quality milk replacer, fed at the rate of 1 percent (dry material) of body weight, should provide all needed nutrients, including the essential amino acids, be highly digestible, be high in energy, and be less than 0.5 percent fiber.

The large-breed calves—Holstein and Brown Swiss—should receive about 1 pound of dry replacer daily. Small-breed calves—Ayrshire, Guernsey, and Jersey—should receive proportionally less, or approximately 0.7 to 0.8 pounds daily. Normally

Figure 6–4
Young calves will begin to eat a palatable calf starter within 7 to 10 days of age and readily consume 1.5 to 2.0 pounds by weaning at about 42 days. *Source:* Purdue University.

7 to 8 pounds of warm water is mixed with each pound of replacer to provide a dry-matter content of 12.5 to 14.5 percent in the final mixture.

The daily allowance is usually divided into equal proportions and fed twice daily, morning and evening. Because calves are functionally nonruminants, the principles expressed in Chapter 5 are applicable to dairy calf nutrition. (Nutrient requirements and recommendations are provided in Tables 5–1 and 5–2.)

In addition to milk replacer, calves should receive a high-quality concentrate (dry starter feed) and hay from 4 to 7 days of age (Figure 6–4). Because of the limited capacity of the calf's digestive system, fresh hay is the preferred forage.

The calf starter should be palatable to ensure early intake and 1.5 to 2.0 pounds daily consumption by the time the calf is weaned from the milk replacer at about 42 to 49 days of age. It should be relatively dust-free and contain coarsely cracked grains, about 18 percent high-quality protein, and supplemental vitamins and minerals. If hay is not provided free choice, as much as 10 percent of the starter can be ground, chopped, or pelleted hay.

The change from a nonruminant to a functional ruminant begins relatively early in the calf's life and is continuous (Section 4.6). A high-quality starter and hay will encourage early development of the rumen. As the rumen develops and makes up a larger proportion of the total digestive system, the calf is better able to utilize the more succulent forms of silage and pastures.

As a calf matures, it can also use cheaper concentrates as energy sources. As a ruminant, it begins to manufacture the essential amino acids and B vitamins. Protein requirements as a percentage of the ration decline slightly. The concentrate mixture that the more mature calf receives, therefore, can be formulated with less-expensive ingredients.

In raising dairy calves, care and management must receive high priority. Where raised in confinement, any disease can be easily spread. In some cases, ventilation may be inadequate. Also, the calf may not benefit from sunlight exposure to ensure against vitamin D deficiency. Therefore, especially good management and clean, relatively dry, and draft-free housing must be provided. Where disease or other stresses are high, nutrition is even more important. Milk replacers, starters, and forages must be of especially high quality.

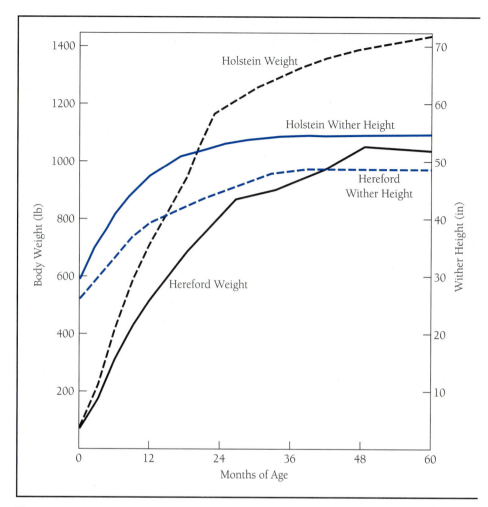

Figure 6–5

Growth of Holstein and Hereford heifers compared, both raised under typical systems of management. Dairy heifers must be fed for rapid and maximum growth without overconditioning. *Sources:* Adapted from Davis, H. P., and I. L. Hathaway, *Nebraska Agr. Exp. Sta. Bull. 179;* and Brown, C. J., et al., *Arkansas Agr. Exp. Sta. Bull. 570.*

A second unique consideration in dairy cattle nutrition is that growing heifers are fed to *grow, not fatten.* Note in Figure 6–5 that Holsteins, a dairy breed, grow in weight more rapidly than do Herefords, a popular beef breed. Naturally, dairy breeding stock are automatically selected for their inability to fatten when selection is directed toward high milk production. To hold down costs, to avoid fattening, and to cause maximum rumen development, heifers are raised primarily on roughage after a few months of age.

Overfeeding energy, especially in the 6- to 12-month-old dairy heifer, will lower lifetime production. In Cornell University studies, trios of Holstein heifers were assigned at random to three levels of energy at birth—65, 100, and 140 percent of the levels considered "normal"—and fed at these relative levels until they calved. Then all

TABLE 6–6	Influence of Energy Intake (From Birth to First Calving) on Subsequent Production		

| | Feeding Level | | |
	Low	Medium	High
Number started	33	34	34
Number culled	16	24	22
Pounds milk per cow:[a]			
First lactation	8,840	9,083	9,226
Second lactation	10,299	10,509	9,752
Fourth lactation	11,657	11,083	10,696
Sixth lactation	12,402	11,419	11,275

[a]*Averages are per cow and per lactation. Some, but not all, of the cows had completed six lactations.*

Source: *Reid, J. T., 1960. J. Dairy Sci. 43:103; and private communication, 1969. See also Table 15–2.*

were fed liberally, according to size and production. A summary of available results is shown in Table 6–6.

Fat deposition in the udder during growth, and therefore less development of secretory tissue, of heifers fed high levels of energy cause low production when they enter the milking herd. Though normal milk yields are considerably higher in the early twenty-first century, the principles established by this classic research hold true.

Between 6 and 12 months of age, roughage alone usually is not adequate for maximum growth in heifers. Even though the rumen has developed considerably, the energy needs are great enough that 2 to 5 pounds of concentrates must also be fed per day for rapid growth. As with other ruminants, the protein level of the concentrate is determined by the protein content and consumption of the roughage.

By the time heifers are 1 year of age, they are usually able to consume and utilize adequate levels of nutrients in the form of hay, silage, and/or pasture. Concentrates—grains or protein supplement—are needed only if the roughage is low quality or in limited supply.

6.6   DAIRY COWS

Examples of nutrient requirements for dairy cows were given in Tables 6–1 (per day) and 6–2 (proportion of the ration). These examples were given only to illustrate (1) how requirements are stated and (2) the magnitude of the difference in requirements for maintenance and for maintenance plus production (Table 6–1 only). The main factors that influence requirements are body weight, production, milk composition, and stage of gestation.

Principles of formulation mentioned previously for other ruminants apply to dairy cattle. Because forage is usually a more economical energy source, and because protein and energy requirements for good milk producers are so high, high-quality forage that is palatable and well utilized (RFV of 150 or higher; see Section 3.10) is important. Cows in milking herds are often provided all the forage they will eat as hay, silage, pasture, and/or freshly chopped forage (Figure 6–6). Most cows will eat 2 to

Figure 6–6
Good-quality forage—grass, hay, or silage—is the foundation of all lactating dairy cow rations, left. Vertical silos are still used on some farms but are being replaced by bunker silos, or large plastic bag containers, below, because of lower labor and maintenance costs. *Sources:* Holstein Assn. and Purdue University.

3 pounds of hay per 100 pounds of body weight. Where silages or greenchop material are fed, comparable dry-matter intake will occur if the moisture content of the total ration doesn't exceed 55 to 60 percent.

High-producing dairy cows can't get enough forage to provide all the energy and nutrients needed for milk production. Therefore, some concentrate must be provided, with the ratio of forage to concentrate determined by the milk production level

Total mixed ration (TMR)
All-in-one ration. Ration components combined; theoretically, each bite of food would contain all nutrients in the correct proportions.

desired. Also, high-producing cows aren't in the milking parlor long enough to eat the volume of concentrate they need. In larger herds, therefore, the cows often are grouped by production levels, and a forage-concentrate complete ration is provided. A high-producing group may require a 50:50 ratio of forage to concentrate, whereas a lower producing group late in lactation may receive a 70:30 ratio. A **total mixed ration (TMR)** ensures that each cow receives the proper proportions of desired nutrients. An example of a 1-day ration for a large, high-producing cow is given in Table 6–7.

As an alternative to TMR rations, some dairy producers provide a partial ration, largely concentrates, in computer-controlled concentrate feeders located in the housing area of the cows. Sensing devices triggered by ear or neck tags dispense concentrate mix when the cow enters, with the amount for each cow programmed according to production records.

During mid to late lactation (usually 150 days plus), milk yield and energy needs decline, so the amount of concentrate mix fed or the proportion of concentrate in a complete ration may be decreased.

Adequate forage in the ration is necessary to maintain desired percent of fat in the milk and to maintain good rumen and digestive tract function.[3] Guidelines include (1) daily roughage dry matter equivalent to 1.5 percent of the animal's body weight; (2) a minimum of 17.0 percent crude fiber and 21.0 percent acid detergent fiber in the complete ration; and (3) at least one-third of the total ration dry matter in the form of long hay, silage, or other coarse forage.

Milk fever, characterized by low blood calcium which causes temporary paralysis, is more prevalent among high-producing cows shortly after parturition because heavy milk production requires large quantities of calcium. High-ration calcium level during the nonlactating period contributes to milk fever after parturition. Adequate levels of vitamin D, the right ratio of calcium to phosphorus, and *modest* levels of calcium and other cations during the preceding dry period appear to help high producers maintain adequate blood levels of calcium.

A liberal supply of clean, fresh water is especially important for dairy cows. Milk contains about 87 percent water. Also, considerable water is necessary for digestion and metabolism of the tremendous volume of feed most cows consume. Cows in full production will usually drink 4 to 5 pounds of water for every pound of milk produced, depending, of course, on temperature, humidity, water content of roughage consumed, and other factors. That would be about 30 to 40 gallons per day for the cow featured in Table 6–7.

In late pregnancy, when the cow is nonlactating, energy is needed primarily for maintenance and for the developing fetus. During the first two-thirds of this 40- to 60-day period, good-quality roughage usually will provide adequate energy and nutrients. Hay or silage comprised of grasses, mixtures of grasses and legumes, or small grains such as oats are common forages. Corn silage is usually rather high in energy and may cause too much fattening; alfalfa is high in calcium and potassium, and when fed as the exclusive roughage source may contribute to the incidence of milk fever at the onset of lactation.

During the last 2 to 3 weeks prior to calving, the lactation ration should be introduced and the amount increased gradually. This will provide for (1) adaptation of rumen organisms to a high-energy ration and (2) energy needed for lactation.

---

[3] National Academy of Sciences, *Nutrient Requirements of Dairy Cattle,* 1988.

TABLE 6–7

One-Day Total Mixed Ration (TMR) for a 1320-lb (600-kg) Dairy Cow Producing 66 lb (30 kg) of 3.5 Percent Butterfat Milk[a]

Feed	Daily Feed lb	kg	Daily Dry Matter (kg)	NE[b] (Mcal)	Crude Protein (g)	Calcium (g)	Phosphorus (g)	Vitamin A (IU)	Vitamin D (IU)
Corn silage	58.7	26.7	8	12.4	640	21.6	16.0	144,000	3,172
Alfalfa hay, chopped	12.2	5.55	5	6.2	860	62.4	15.0	360,000	11,050
Sorghum grain	12.2	5.55	5	9.3	585	1.5	16.5	—	—
Soybean meal	4.4	2.0	1.8	3.4	893	6.5	14.0	—	—
Dicalcium phosphate	0.12	0.05	0.05	—	—	13.0	10.0	—	—
Trace-mineralized salt	0.24	0.1	—	—	—	—	—	—	—
Total	87.86	39.95	19.85	31.3	2,978	105.0	71.5	504,000	14,222
Requirement			20[c]	30.4	2,949	99	69.5	46,000	6,000

[a]The surpluses of calcium, Vitamin A, and Vitamin D are not harmful.

[b]Includes both energy used for maintenance and energy used for milk production.

[c]A guide, not a requirement; TMR fed according to cow size and weight and desired production; group-fed according to the number of cows in the group. Feeds must be regularly monitored for moisture and amounts adjusted accordingly.

Source: Adapted from National Academy of Sciences, *Nutritional Requirements of Dairy Cattle*, 1988.

QUESTIONS FOR STUDY AND DISCUSSION

1. What are examples of feeds that are more efficiently utilized by ruminants than nonruminants?

2. How important is protein quality in ruminant nutrition? Explain.

3. Give examples where animals should receive more nutrients than stated in the NRC requirements.

4. On what basis, dry matter or as-fed, are most rations calculated? On what basis are the final pounds in the ration determined?

5. Why are (a) calculation of the ratio of concentrate-to-roughage and (b) adaptation to the ration especially important when feeding ruminants?

6. What is meant by the terms *enterotoxemia* and *acidosis*?

7. How do the goals of feeding heifers for herd replacement differ from those for growing finishing cattle?

8. What are typical percentages of protein and fat in a milk replacer for dairy herd replacements?

9. At what ages are most dairy heifers weaned from liquid milk or milk replacer? How is calf starter intake related to age at weaning?

10. What are undesirable effects of overfeeding herd replacement heifers?

11. Why is forage quality so important when feeding lactating dairy cows?

12. What is meant by (a) "TMR" and (b) a "50:50" or "70:30" forage-to-concentrate ratio?

13. Explain how computer feeders can be effectively used in feeding of dairy cows.

14. List three guidelines that help ensure that a dairy cow ration is not milkfat depressing.

Nutrition of Horses and Ponies

7

Proper nutrition is essential for horses (or mules and ponies) to be in the best condition for optimum appearance and performance, whether kept for pleasure, show, work, breeding, or racing. Their care can come from the most novice to most professional owner. Where horses are kept also is an important factor in their nutrition. Most are kept at home, others on rented pastures or confined in exercise lots or stables. Therefore, the quantity and quality of the forage portion of their ration can vary greatly. This chapter is devoted to the nutritional needs of the horse, providing suitable feeds and other nutritional discussion unique to the horse.

As previously described in Section 4.5, the horse is a nonruminant herbivore and, like the pig, has a relatively small stomach. Only about 10 percent of the horse's gastrointestinal tract capacity is in the stomach. Therefore, the ability of the horse to utilize large amounts of forage is due to its cecum and colon. Digestion and absorption that occurs in this portion of the horse's digestive tract is similar to digestive processes occurring in the reticulo-rumen of ruminants. This enables the horse to gain nutritional value from larger amounts of forage than is the case with swine or poultry.

Because horses can use moderate quantities of roughage, most feeding programs are based on good-quality pasture or hay (Figure 7–1). Horses seem to crave such forage, and pasture or hay in the diet tends to minimize the incidence of management problems such as wood chewing, tail chewing, and ingestion of undesirable materials.

Upon completion of this chapter, the reader should be able to

1. List typical components of a horse ration.
2. Contrast the nutrient requirements of horses at leisure and at work.
3. Describe differences between a mare's energy needs for maintenance, reproduction, and nursing of her foal.
4. Compare the differences in cereal grains as feed for horses.
5. List three common and desirable forages for horses.
6. Describe the typical feeding pattern for a colt from birth to 1 year of age.
7. List four common problems in horse nutrition and feeding.

7.1 NUTRIENT REQUIREMENTS

Horses, like other farm animals, require adequate protein, energy, vitamins, minerals, and water. The quantity of nutrients required depends largely on

- Age and stage of growth (younger horses have higher nutrient needs)
- Size

Figure 7–1

Where possible, good-quality pasture should form the basis of a feeding and management system. Horses will be healthier and have fewer digestive problems when good-quality forage is included in the diet.

Source: Quarter Horse Journal.

- Activity (leisure or work)
- Special needs (reproduction and lactation)
- Environmental conditions (extremely hot or cold)
- Individual characteristics ("easy or hard keepers," as affected by metabolic rate)

Energy and protein are required in larger amounts, so rations are usually balanced for these, with vitamins and minerals added as needed. Most energy is provided in the form of carbohydrates, though small amounts of fats may be used to provide some energy. For rations high in grains, the principal carbohydrate digestion product is glucose, which is absorbed through the small intestine. If large amounts of roughage are fed, volatile fatty acids produced by microbial digestion in the cecum and colon also provide energy.

Energy requirements vary greatly in mature horses, depending on their level of work (Table 7–1). Also, the requirements increase as pregnant mares approach parturition and lactation. The careful feeder will increase the proportion of concentrate to forage to meet the nutritional needs of individual horses. Unfortunately, underfeeding and overfeeding of energy occur too frequently because of lack of adequate individual attention by many horse owners.

Growing horses and brood mares in gestation or lactation require higher levels of protein than do other mature horses. Crude protein content of horse rations ranges from about 8.0 percent for mature horses to 13.2 percent for lactating mares and as high as 18 percent for the young foal.

Protein quality, as well as quantity, is important in horse nutrition. All protein supplements, especially those fed to growing horses and brood mares, should contain protein with adequate levels of the essential amino acids. *Lysine* is most often the

TABLE 7–1

Nutrient Concentrations in Diets for Horses and Ponies on 100 Percent Dry-Matter Basis[a]

	Digestible Energy		Diet Proportions		Lysine (%)	Crude Protein (%)	Cal-cium (%)	Phos-phorus (%)	Vitamin A Activity	
	Mcal/kg	Mcal/lb	Concen-trate (%)	Rough-age (%)					IU/kg	IU/lb
Mature horses and ponies at maintenance	2.0	0.9	0	100	0.28	8.0	0.24	0.17	1,830	830
Mare, last 10 months of gestation	2.25	1.0	20	80	0.35	10.0	0.43	0.32	3,650	1,660
Lactating mare, first 3 months	2.6	1.2	50	50	0.46	13.2	0.52	0.34	2,750	1,250
Lactating mare, 3 months to weaning	2.45	1.15	35	65	0.37	11.0	0.36	0.22	3,020	1,370
Creep feed for foal	3.5	1.6	100	0	0.60	18.0	0.85	0.60		
Foal (4 months of age)	2.9	1.4	70	30	0.60	14.5	0.68	0.38	1,580	720
Weanling (6 months of age, moderate growth)	2.9	1.4	70	30	0.61	14.5	0.56	0.31	1,870	850
Yearling (12 months of age, moderate growth)	2.8	1.3	60	40	0.53	12.6	0.43	0.24	2,160	980
Long yearling (18 months of age, light training)	2.65	1.2	50	50	0.50	12.0	0.36	0.20	1,800	820
Two-year-old, light training	2.65	1.2	50	50	0.45	11.3	0.34	0.20	2,040	930
Mature working horses:										
Light work[b]	2.45	1.15	35	65	0.35	9.8	0.30	0.22	2,690	1,220
Moderate work[c]	2.65	1.2	50	50	0.37	10.4	0.31	0.23	2,420	1,100
Intense work[d]	2.85	1.3	65	35	0.40	11.4	0.35	0.25	1,950	890

[a]Values assume a concentrate feed containing 3.3 Mcal/kg and hay containing 2.0 Mcal/kg.

[b]Examples are horses used in western pleasure, bridle path hack, equitation, etc.

[c]Examples are ranch work, roping, cutting, barrel racing, jumping, etc.

[d]Examples are race training, polo, etc.

Source: Adapted from National Academy of Sciences, Nutrient Requirement of Horses, 1989.

NRC Standards

limiting essential amino acid in horse rations. Only limited protein synthesis and absorption of amino acids occur in the cecum and colon of the horse. More rapid and efficient gains can be expected when weanlings receive the proper balance of amino acids in the ration.

Trace-mineralized salt should be fed free choice or included in the concentrate to provide about 1 percent of the total ration. Prolonged exercise, especially in warm weather, increases the need for sodium chloride lost through sweating. The calcium and phosphorus levels should meet NRC standards (Table 7–1). Also, the ratio of calcium to phosphorus should be within a range of 1:1 to 3:1. Note that the suggested levels for foals and weanlings are 0.85 percent Ca and .60 percent P, when formulated into the creep feed, at a ratio of 1.42 to 1. If horses are on pasture or fed good-quality hay, vitamin supplements are usually not necessary.

Plenty of fresh, clean water should be available. Water intake varies with activity. Nonworking horses generally consume 4 to 8 gallons per day; hard-working horses, as much as 15 to 16 gallons. Water intake is highly correlated to dry-matter intake; about 2 to 4 pounds of water is needed per pound of feed dry matter. Also, the lactating mare may require twice as much water as required for maintenance alone.

Note the nutrient requirements of horses and ponies at different levels of work, shown in Table 7–1. Recommended concentrate-to-roughage ratios in the diet are also shown. A minimum of 0.75 to 1 percent of a horse's body weight should be provided as long-stemmed roughage. This amount should provide for normal activity of the digestive system and helps prevent the horse from developing undesirable vices such as chewing wood or eating bedding.

7.2   FEEDS

Energy Feeds

Concentrates used for horses can be considered in two categories: *energy feeds* and *protein supplements*.

Many grains can be used as energy sources, provided they are fed correctly (Figure 7–2). Oats have been preferred by most who work with horses. Oats are very palatable and are lower in digestibility and higher in fiber than other grains, so "over-

Figure 7–2
Horses should be fed individually. If kept in groups, each horse should have an individual feeder or adequate feeder space to ensure that it gets its share of the feed.
Source: Texas A&M University.

feeding" is less likely to cause digestive problems. However, oats are usually more expensive than corn or other grains in cost per unit of energy and their sources vary greatly in quality. Oats or other small grains (barley or wheat) are usually fed whole to mature horses with good teeth, and rolled or crimped to younger animals. Finely ground oats can contribute to respiratory problems.

Corn is lower in crude protein, higher in energy, and usually cheaper per unit of energy than oats. Corn can be fed to mature horses as ear corn or as whole or cracked grain. A pound of corn will contain about 20 percent more energy than a pound of oats because of its lower fiber and higher starch content.

Grain sorghum (milo) can be fed to horses but because of the smell and hard kernels, it should be ground or rolled. It is usually cheaper per unit of energy than oats, higher in energy content, and only slightly lower in protein.

Molasses is an energy feed commonly included in horse rations to increase palatability. When included at 3 to 5 percent of the grain mixture, it aids in controlling dust and binding the feed ingredients together. Liquid molasses is especially low in protein (4 to 5 percent). Therefore, if considerable molasses is used, proportions of other ingredients would need to be changed to provide adequate protein in the ration. Also, molasses at higher levels has some laxative effect.

Other grains such as barley and limited amounts of wheat are acceptable energy sources for horses. Rye grain, because of its poor palatability and risk of ergot infestation, is not a preferred feed source. Grain by-products, such as brewer's grain, beet pulp, or citrus pulp, are sometimes fed to horses when available and economical. Also, the addition of up to 10 percent high-quality fat to the concentrate portion of working horses is becoming more popular. It results in increased caloric density and palatability of the grain mixture as dustiness is reduced.

Protein Supplements

Soybean meal is the most common protein supplement for horses. It is an excellent source of protein because of its amino acid content, especially lysine, which is often limited in other vegetable protein sources. Linseed meal or cottonseed can also be used; if needed lysine is provided by other means. However, the risk of gossypol toxicity from a phenolic pigment in cottonseed makes it less popular. Also, the expense and availability of linseed meal limit its use.

Forages

The feeding program for horses should begin with good-quality forage. This can consist of grass and legume pastures and hay. Typical legumes include alfalfa, red clover, lespedeza, and birdsfoot trefoil. Popular grasses include orchardgrass, bromegrass, and timothy, and Kentucky bluegrass if available. Tall fescue is a very hardy grass that takes traffic well but is less palatable and nutritious in summer months. Older strains of tall fescue may also contain fungal endophyte that can cause reproductive problems in mares.

Hay should be palatable and free from dust, mold, and foreign material. Horses are more susceptible to colic from dusty or poor-quality hay than are other species (see Section 7.6). Most legume hays are higher in protein and calcium than grass hay but are sometimes dusty, especially if exceedingly dry when baled or stored. Because of its higher energy and protein content compared to grass hays, alfalfa is usually the hay of choice.

Several grain mixture options and horse rations are shown in Tables 7–2 and 7–3, respectively.

TABLE 7–2	Grain Mixtures for Horses			
Ingredients[a]	Weanlings and Yearlings (%)	Broodmares, Late Gestation (%)	Lactation (%)	Maintenance (%)
Oats	58.5	90	72.5	97.5
Corn	25		15.0	
Soybean meal	10	5	7.5	
Molasses	2	2	2	
Dicalcium phosphate	2		0.5	1.0
Limestone	1	1.5	1	
Salt (trace mineral)	1	1	1	1.0
Vitamin premix	0.5	0.5	0.5	0.5
Total	100	100	100	100

[a]Other recommended cereal grains or protein supplements as discussed in Section 7.2 could be used if adjusted for protein and energy contents.

Source: *Adapted from* Horse Industry Handbook, *HIH 790-7*

TABLE 7–3	Typical Daily Rations for Horses

(All ages and classes should have free access to iodized salt and a calcium-phosphorus supplement)

A. Light horse, 1,100 pounds, light use: 4 lb oats, 2 lb corn, 14 lb mixed hay

B. Pregnant mare, 1,100 pounds: 4.5 lb barley, 4.5 oats, 1 lb wheat bran, 12 lb mixed hay

C. Yearling, 800-pound, during summer: luxuriant pasture

D. Yearling, 800-pound, during winter: 4 lb oats, 1 lb wheat bran, 12 lb mixed hay

Source: *Adapted from USDA, ARS Bull. 353, 1972.*

Many horses are kept in urban areas where a variety of feeds, minerals, and vitamin supplements are not readily available, so the owner may need to purchase a concentrate mix to supplement the hay or pasture.

7.3   CONSUMPTION AND INFLUENCES

The major influence on feed consumption is size of horse. Larger horses eat more because they have higher maintenance requirements for protein and energy. For example, the suggested energy allowance for a 1,100-pound mature horse on a maintenance ration is 16.38 megacalories digestible energy; a 1,320-pound horse's allowance is 18.78 megacalories. Obviously, high-performance horses, such as active racehorses and those that work livestock, or mares during late pregnancy or lactation, require more feed to provide more energy and other nutrients. Mature horses in light work need about 0.5 percent of their body weight as concentrate mix and about 1.5 percent as forage dry matter.

7.4   NUTRITION AND PERFORMANCE

Nutrition has a significant influence on horse performance. Adequate levels and ratios of nutrients, including vitamins and minerals, are essential for hard-working horses. Considerable energy is used for muscle contraction during movement. Current NRC nutrient requirements suggest that for ponies and light horses, light, moderate, and intense work increases the energy requirements 25, 50, and 100 percent above maintenance needs, respectively. Thus, strenuous exercise—racing or working livestock—requires much more energy than does walking or trotting.

It is not necessary to increase the percentage of protein in the concentrate mix for working horses. Other than for the small amount of nitrogenous compounds lost in perspiration, little protein is used for physical work. Also, because feed intake increases to meet the energy requirement, additional protein is automatically provided. Excessive levels of protein may be detrimental to a horse's performance.

Though mineral balance is important in rations of working horses, it does not appear that the requirement for minerals as a proportion of the ration is increased during work. The same appears true for vitamins, though some trainers provide supplemental B vitamins to increase feed consumption in racehorses.

In summary, as work increases, feed intake (especially grain concentrate) by horses should increase, largely to meet the energy requirement. With a good ration, as intake is increased, quantities of needed protein, vitamins, and minerals are automatically provided. Other "rules of thumb" in feeding horses according to their body weight, activity, forage quality, and body condition are provided in Box 7–1.

7.5   PREGNANCY AND THE FOAL

The nutritional state of the mare during the last few months of gestation and the early weeks of lactation can have a significant effect on the health of the **foal**. Most of the fetal growth occurs during the last third of pregnancy, so protein, energy, calcium, phosphorus, and vitamin A needs are high (see Table 7–1).

> **Foal**
> A newborn horse or pony.

During the first 90 days of lactation, nutrient requirements of the mare are even higher as a proportion of the ration and on a per-day basis. A typical mare requires approximately 28 megacalories of digestible energy, 1.4 kilograms of protein, 56 grams of calcium, and 36 grams of phosphorus per day. A 1,000-pound mare in early lactation will produce about 30 pounds of milk daily, containing about 2.2 percent protein and 475 kilocalories digestible energy per kilogram. Such milk production is necessary to support normal foal growth of 2 to 3 pounds per day.

Milk production by mares peaks at about 2 months of lactation, then steadily decreases. It is, therefore, beneficial to start foals on a concentrate feed or mix as soon as possible. Foals usually will eat rolled oats at about 2 weeks of age, and consumption should increase steadily to accommodate the foal's needs when milk production begins to decline (about 3 months into lactation). A typical concentrate mix for young foals contains about 18 percent protein, 0.8 percent calcium, and 0.6 percent phosphorus, and would include ingredients such as corn, rolled oats, soybean meal, molasses, trace-mineralized salt, supplemental vitamins, and sources of calcium and phosphorus. It should be available in creep feeders, which enable foals to eat freechoice but which exclude the mares.

By the time foals are weaned, at about 6 months of age, they should be consuming enough of a balanced ration to fully sustain growth. This will include good-quality

BOX 7–1 Rules of Thumb in Feeding Horses

- The *amount of total ration* is related to *body weight* and *activity* of the horse.

For example, a mature 1,000-pound horse fed a maintenance ration needs about 1.5 to 2.0 percent of its weight as air-dry feed (average = 1.75 percent).

$$1,000 \times .0175 = 17.5 \text{ pounds air-dry feed}$$

The same horse if working needs from 1.5 to 3.0 percent of its weight, depending upon work intensity. The above horse at moderate work would need about:

$$1,000 \times .0225 = 22.5 \text{ pounds air-dry feed}$$

- Also, the *amount of forage* in the ration depends upon activity of the horse (e.g., maintenance, work). A greater percentage of the ration should come from concentrate feeds when the horse is working.

For example, this same horse, at maintenance and fed fair-quality forage, could receive all of its ration as forage (1.5 to 2.0 percent of body weight).

$$1,000 \times .0175 = 17.5 \text{ pounds air-dry feed as forage}$$

However, at moderate work, the same horse should receive only half of its ration as forage (half of 2.25 percent of body weight), the balance coming from concentrate feed.

$$1,000 \times .0225 = 22.5 \text{ pounds air-dry feed needed}$$
$$1,000 \times .01125 = 11.25 \text{ pounds as forage}$$
$$11.25 \text{ pounds as concentrate}$$

The basis for less forage and more concentrate is to provide less bulk in the ration and better meet the extra nutrient needs, particularly energy and protein required for work.

- The *quality of forage* (excellent, fair, poor) also must be considered. As quality decreases, the proportion of concentrate in the ration should increase. However, not more than 0.75 percent of body weight in concentrate should be fed at one feeding. Thus, feeding the 11.5 pounds of concentrate in the above ration divided into two equal feedings should pose no problem.

- Many experienced horse owners or managers also evaluate the horse's "condition score" and make adjustments in the feeding program. The procedure of condition scoring (e.g., poor = 1, extremely fat = 9) involves visual observation and feeling for fat at several body locations.

Source: Compiled by the authors.

pasture or hay in addition to a 15 to 16 percent protein concentrate mix. As the foal matures, more feed is required per day, but the proportion of protein and energy in the mix can be reduced. Mature size is reached at about 3 years.

Nutritional needs of breeding stallions are not significantly different from those of other mature horses (Figure 7-3). They should receive good-quality forage, concentrate mix as needed, salt and mineral supplement, and plenty of exercise. If in heavy use, the stallion will have an energy need somewhat above his maintenance requirements.

7.6 OTHER FEEDING MANAGEMENT PRECAUTIONS

Both under- and overfeeding are the most common problems of horse owners. Obese horses are lower in breeding efficiency, their performance is reduced, and their general health and longevity will likely be reduced.

Figure 7–3
Stallions should be kept in individual paddocks where they can get plenty of exercise. Stalls or shelters plus adequate forage are also desirable for good stallion management.
Source: Quarter Horse Journal.

Acute overeating—often resulting from a horse gaining access to a feed room or bin—usually causes **founder**, a severe digestive malfunction with many side effects. Usually preceded by acidosis, founder results in rectal temperature rising to 103° to 106°F. There is inflammation of the fleshy laminae (soles) of the feet, and the animal is usually reluctant to move because of extreme pain (Figure 7–4). Permanent foot deformity and laminitis can result. Founder can also be caused by overeating lush, green plants, by intense work on hard surfaces, by cooling horses too rapidly after an intense workout, or by retained placenta and uterus inflammation in mares that have recently foaled. If founder is recognized in the early stages and properly treated, horses may return to a serviceable life, but recurrence is common.

Another common problem is **colic**—any pain closely associated with the digestive system. Parasites are the usual cause of colic; however, it may also be caused by sudden changes in the ration, overeating, eating immediately before or after exercise, eating moldy or spoiled feed, or consumption of foreign materials. Intestinal motility usually decreases with colic; this can lead to accumulation of fluid, gases, and digesta in the intestines, and related discomfort. A horse's pain tolerance is low; even slight pain may cause a horse to show anxiety, crouch, lie down, or roll.

To avoid colic, managers should not feed old or moldy feed or hay and should feed similar amounts at standard times each day. They should make changes in the diet gradually and not allow the horse access to foreign material that it might ingest.

Respiratory problems, including what some call *heaves*, sometimes occur when horses are fed dusty grain or hay. Horses on pasture or horses fed rations pelleted, containing molasses, low in dust, or not finely ground have less risk of respiratory problems.

"Tying up" or *azoturia* sometimes occurs in horses that are worked hard for several days, rested a day or more without reducing feed intake, then put back to work. In

> **Founder**
> Digestive malfunction in a horse, usually caused by excessive overeating. Symptoms include high temperature, foot deformity, and pain.

> **Colic**
> Acute abdominal pain; pertaining to the colon.

Figure 7–4

Schematic showing a horse with acute laminitis and painful stance—with front legs extended and weight shifted back onto the hind feet which are drawn underneath the body for support. *Source:* Purdue University.

draft horses, it has been termed "Monday morning disease." Soon after exercise or muscular activity, the symptoms appear. Muscles may become rigid, excessive sweating occurs, and the animal is reluctant to move. It usually can be prevented by reducing feed intake and providing exercise on days the horse is not worked. *Selenium deficiency* symptoms resemble those of azoturia, and feeds containing selenium or selenium injections may be beneficial.

All horse owners need to be aware that monensin, an antibiotic that encourages weight gain in cattle, is deadly to horses. This is especially dangerous in situations where mineral supplements containing monensin (approved by the FDA) are made available to cattle on pastures or ranges shared with horses.

Horses should be wormed regularly, teeth should be examined on a periodic basis, and plenty of water should be provided. Because horses have relatively small stomachs and rapid passage rate of feed through the digestive tract, they should be fed at least twice and preferably three times a day. Because requirements are usually given in weights of certain nutrients, feed should be provided by *weight*, not by *volume*. A gallon of corn has almost twice the energy content of a gallon of oats, whereas a pound of corn has only about 15 percent more energy than a pound of oats.

Most nutritional or other horse disorders are related to mismanagement. Keen observation by the manager can permit early detection of illness or injury, so the horse may be treated or the management regime may be changed promptly.

7.7 NUTRITIONAL DEFICIENCIES AND IMBALANCES

Most metabolic disorders can be avoided by providing water, exercise, roughage, and adequate levels and ratios of required nutrients, especially for young, growing animals, lactating mares, and severely worked horses.

Energy- or protein-deficient diets cause poor performance, weight loss, and poor appearance in mature horses and will severely reduce growth rate in foals. Deviation in the leg structure of foals can be caused by excess energy intake or imbalances of other nutrients. This problem has become more prevalent in recent years as some owners have tried to get young horses to mature size at 2 years of age.

Although ammonia toxicity resulting from urea has been reported in horses, they are less susceptible than are cattle. Because conversion of urea to ammonia occurs in the horse's cecum, near the end of the digestive tract, less ammonia is absorbed than with cattle, in which urea conversion occurs in the rumen.

Calcium or phosphorus deficiency can hinder bone growth, cause weak bones, and may even cause **rickets** in young animals. In older animals, low levels of phosphorus can cause softening of the bones, low appetite, and poor overall performance.

> *Rickets*
> Abnormal bone development; bones easily bent or distorted.

The calcium-to-phosphorus ratio seems to be more critical than actual amounts, as long as the levels of each are within reason. Excess phosphorus accompanied by a low level of calcium is known to cause "big head"—connective tissue is formed in sections of the skull bone from which calcium has been mobilized to maintain blood calcium levels; the result is an apparent swelling of the head along the facial bones.

Other deficiencies may result in geographic areas where the soil and feeds are low in certain minerals, but with the addition of trace-mineralized salt to the ration most potential problems are avoided. Some deficiencies that have been observed include *goiter,* from iodine excess or deficiency in foals; *anemia,* from iron deficiency; *parakeratosis,* from low levels of zinc; and *white muscle disease* in foals with selenium deficiencies.

Horses fed good-quality grass or hay seldom have vitamin deficiencies, but supplemental vitamin A is often provided horses not on lush, growing pasture. Insufficient carotene or vitamin A can cause impaired vision or even blindness. The B vitamins are ample in good-quality hay or are produced by bacteria in the cecum, so supplementation is usually not needed, except for horses under severe stress.

It is important to learn about common deficiencies in areas where the horse is grazed or the feed is grown. Abnormalities, symptoms, or signs of poor health should be brought to the attention of an expert who might identify the problem and suggest a solution.

Owners or trainers seeking maximum performance from their horses sometimes provide supplemental nutrients at excessive levels, some of which can be toxic. Nutrients most likely to cause toxicity at high levels include selenium, iodine, iron, and vitamin D.

QUESTIONS FOR STUDY AND DISCUSSION

1. What structures of the horse's digestive system enable it to utilize large amounts of forage?
2. Give examples where the crude protein requirement of the horse will be greater than its maintenance needs.

3. Is a balance of amino acids needed in the diet of the horse? Explain.

4. How is water intake related to dry-matter intake of the horse?

5. Although not necessary, why would horse owners like oats as feed grain in the ration?

6. Compare the concentration of energy (Mcal/lb) needed in the ration of mature horses at maintenance and for the lactating mare (Table 7–1).

7. What effects does addition of liquid molasses have upon the ration?

8. What is usually an advantage of using soybean meal in the ration of the horse?

9. What is a good estimate of feed required for a working horse per 100 pounds of body weight (or percent of body weight)?

10. What nutrient requirements increase significantly as the work of the horse increases?

11. When are increased nutrient requirements of the pregnant mare most important?

12. When would creep feeding of the foal be recommended?

13. What are consequences of overeating resulting in founder?

14. Describe two additional disorders of horses related to feeding.

Evaluation 8
of Feeder Animals

Feeders refer to animals that are raised or sold for the purpose of going into an intensive feeding program. The term applies to pigs, lambs, and cattle. Most feeder pigs are raised by the operator who intends to feed them until slaughter, but an increasing number are raised for sale to another party at or soon after weaning. Feeder cattle[1] are usually calves or yearlings that have grown mainly on their mothers' milk and on grass in the range and pasture areas. Most feeder cattle are primarily of beef breeding, though dairy-beef has become increasingly popular. Feeder lambs are sold soon after weaning, usually in the Autumn months and to someone who will feed and finish them for slaughter. The main flow of calves, pigs, and lambs are shown in Fig. 8–1.

The term *feeder* also is sometimes used to describe the feedlot operator who feeds purchased or raised animals until slaughter. Most lamb and cattle feedlots are near irrigated areas of the West and Southwest and in other areas where a plentiful supply of grain and other high-energy feeds is available.

This chapter focuses on the characteristics to look for in selecting feeders. The variation that exists among feeders in their ability to gain, utilize feed, and produce high-quality carcasses will be emphasized and illustrated. The relationship of chronological and physiological growth to the quality characteristics of feeder animals is discussed in Chapter 9. Where and when most feeders are bought and sold will be treated in Chapter 22. Chapter 24 provides a discussion of market classes and grades. Sources of broiler chicks and turkey poults for poultry feeding units are discussed in Chapter 35.

Upon completion of this chapter, the reader should be able to

1. Describe the degree to which ability to gain and to perform efficiently in feeder animals is inherited.
2. Identify and describe the areas of the United States where most feeder cattle, lambs, and pigs are raised.
3. Contrast rate of gain and efficiency of gain among animals of different weights.
4. Describe the influence of sex and sex condition on the value of animals for feeding.
5. List U.S. feeder cattle grades and differentiate between frame and thickness grades.
6. List six factors that may influence rate and efficiency of gain of lambs, cattle, and pigs.

---

[1] The term *stocker cattle* is sometimes used to describe feeders that move into high-roughage programs *before* going to a feedlot. Principles discussed in this chapter also apply to stockers.

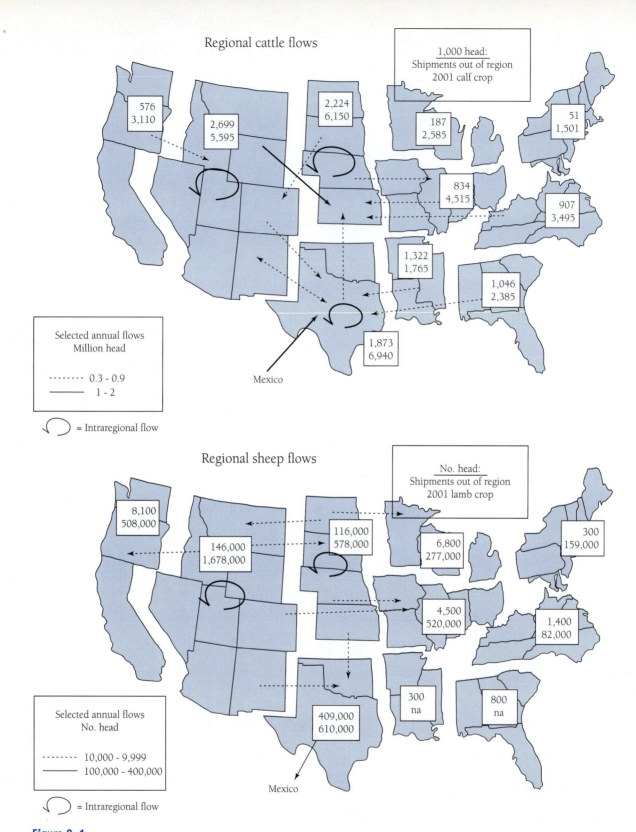

Figure 8–1

Major sources of feeder calves, lambs, and pigs are indicated by regional flow of animals.
Source: Economic Research Service, USDA 2003.

156

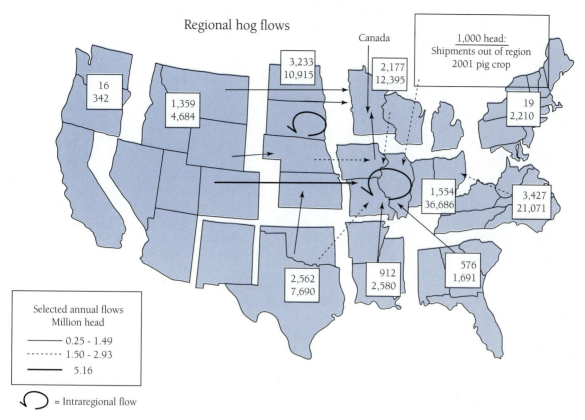

Figure 8–1
(continued)

8.1   INHERITANCE AND RANGE IN PERFORMING ABILITY

There are large ranges in performance characteristics of meat animals—rate and efficiency of gain, carcass grade, and percentage of lean. The following example illustrates the wide differences in average daily gain and feed efficiency that can occur in beef bulls, and the effect they can have upon dollar income. Calves sired by one polled Hereford bull, on a 140-day feeding test at the Arkansas Experiment Station, Fayetteville, Arkansas, 1987–88, gained an average of 3.65 pounds per day on 603 pounds of feed per 100 pounds of gain.[2] Calves on the same performance test sired by a different bull gained 2.60 pounds per day and required 677 pounds of feed per 100 pounds of gain. Assume you purchased the latter calves at 700 pounds and fed them until slaughter at 1,200 pounds. If they sold for the same price and the ration cost 7 cents per pound, you would need to pay at least $5.00 less per hundredweight for these calves as compared to the better performing calves just to break even on feed costs. Also, the calves of lower daily gain and efficiency would require about 2 months longer in the feedlot to reach equal market weight and factors such as labor, interest on investment, and risk would be higher. In addition, cattle that gain more slowly usually grade lower at market time.

---

[2] Arkansas Cooperative Beef Bull Performance Test, Research Series 375, Arkansas Agricultural Experiment Station, Fayetteville, September 1988.

> **Boar**
> A sexually mature,
> (uncastrated) intact male
> hog porcine).

Offspring of different **boars**, on feed from 70 to 230 pounds at the Iowa Swine Testing Station, have differed by more than 100 pounds of feed required per 100 pounds of gain.[3] Average feed efficiency of boars in one test was 257 pounds of feed per 100 pounds of gain. Typically, the range from highest to lowest boars on test will be 220 to 300 pounds of feed per 100 pounds of gain. If ration costs had been 7 cents per pound, each one of the most efficient feeder pigs (those with the feed efficiency of 2.2:1) would have required almost $9.00 less in feed cost than the pigs averaging a 3:1 feed efficiency. As 70-pound feeders, the most efficient pigs would be worth about 13 cents more per pound.

It pays a cattle breeder to practice careful selection of bulls and replacement heifers, considering these important production traits in addition to prolificacy of parents. Some progressive cattle producers use only performance-tested bulls, whose rate and efficiency of gain in a standard feeding trial have been outstanding and which may have sired calves that produced top-quality carcasses. Certain of these producers have followed their calves to commercial feedlots to check gains and efficiency and finally to processing plants where they measured the rib eye muscle and other carcass traits. They use this information to evaluate herd bulls or for selecting replacements closely related to certain good-performing animals.

> **Ram**
> A sexually mature, intact
> (uncastrated) male sheep.

Selection of swine breeding stock for rate and efficiency of gain, as well as for carcass merit, is commonplace. Though selection of **rams** and **ewes** on these bases is as important to the sheep industry as the previously mentioned programs are to the swine and cattle industries, sheep testing programs generally followed development of swine and beef cattle testing programs. Because each lamb is on feed a shorter period and consumes much less total feed than a steer, the benefit to be gained by rigid selection of a sire for feed efficiency is less. The same principles apply, however.

> **Ewe**
> A female sheep of any age.

An increasing number of producers are improving their herds and flocks by techniques mentioned above and by other methods, in order to establish a top reputation as a source of good feeders. Therefore, they can demand and receive top prices.

It is not always possible, however, to know the performance background in selecting feeders. Later sections, therefore, discuss other factors important in selection and the value of certain traits as indicators of performance and profitability.

8.2 FEEDER CATTLE

Feeder cattle are produced in all sections of the country, but most are raised in grazing areas, where mature cows can do the best job of utilizing available forage, i e., southeast. As shown in Fig 8–1, cattle are moved across the corn belt and reside in many feedlots throughout the plains. These states produce much grass, and producers make good use of breeds and lines of cattle that are relatively resistant to heat and insects. The Rocky Mountain states and the grass regions of the Plains states, Texas through the Dakotas, have long been important feeder-producing areas. Other pasture areas, in Missouri, Iowa, Kentucky, southern Illinois, Tennessee, and other states, also account for a considerable number of feeder cattle, usually produced in smaller herds.

Some feeder calves are raised in the Corn Belt where cow herds can be maintained on roughage that is a byproduct of grain production and on land not used for crop production. The former may involve feeding the roughage to the cows in confinement most or part of the year, which presents numerous management and labor problems.

---

[3] Iowa Swine Testing, Box A, Iowa State University, Ames, Fall 1984 Summary.

Figure 8–2
Limousin-Hereford feeder calves on a Colorado ranch in midwinter, left. Many crosses are represented in the right photo.
(Courtesy of North American Limousin Foundation and Iowa Beef Processors.)

Breed Differences

Quality in feeder cattle is not necessarily related to breed. Good and poor animals are found in all breeds. The range within breeds is much larger than the range in characteristics among averages of breeds.

The Hereford was the predominant breed in western U.S. range beef herds for many years. Until recent years, many animals in the Angus breed have been considered too small to cope with the rugged conditions prevailing in range areas; however, selection effort has been directed to larger, more rugged animals. Though Hereford and Angus are still popular and common in western range herds and there are some Shorthorn range herds, other European or "Continental" breeds, such as Limousin and Simmental introduced into the United States during the 1970s, are now widely used in crossbreeding for production of feeder cattle and replacement animals (Fig. 8–2).

The benefits of crossbreeding (see Chapter 20) in commercial cow herds have been demonstrated repeatedly, and a high proportion of feeder cattle entering feedlots are crossbred.

It is apparent that a cow-calf owner who consistently produces good-quality crossbred calves will develop a valuable reputation and will profit both from better performance before weaning and from the good demand for a vigorous calf that will perform well in the feedlot. Both result from hybrid vigor. The demonstrated merit of some of the breeds introduced into the United States during the 1970s contributed considerably to the growth of crossbreeding in cow herds.

Brahman, Santa Gertrudis, and Charbray, as well as animals that are crosses between these and the English breeds (Hereford, Angus, and Shorthorn), are used heavily in the southwestern and Gulf Coast states because of their tolerance to extreme heat and resistance to certain insects prevalent in those areas. Other breeds adapted to the tropics have been imported and used in recent years.

Cattle of the larger dairy breeds should not be ignored in this discussion. States such as California, Wisconsin, Minnesota, New York, and Pennsylvania produce

TABLE 8–1	Growth and Carcass Data on Beef and Dairy Breed Steers		
	Hereford	**Angus**	**Brown Swiss**
Average daily gain (lb)	2.38	2.30	2.52
Slaughter weight (lb)	1,000	993	1,179
Dressing percent	59.5	59.7	58.0
USDA quality grade[a]	11.2	12.7	11.3
Marbling score[b]	10.1	13.7	10.8
USDA yield grade	3.5	3.6	2.2
Rib eye area (in.^{2})	10.2	10.4	12.4
Fat thickness (in.)	0.60	0.58	0.19

[a]11 = High Good, 12 = Low Choice, 13 = Average Choice.
[b]9 = slight +, 10 = small −, 21 = slight abundant +.

Source: *USDA, SEA, Germ Plasm Evaluation Program, Progress Report No. 2, U.S. Meat Animal Research Center, Clay Center, Nebraska, 1975.*

significant numbers of male dairy calves that are finished as veal calves, or eventually are finished in commercial beef feedlots. Many feeders market considerable roughage and grain through Holsteins or Brown Swiss steers, or crosses involving these breeds. Such cattle normally gain faster, produce more carcass per pound of feed, and have higher percentage of lean in the carcass than the traditional English breed (Table 8–1). Carcasses may grade lower because of less marbling in the muscle.

Cattle of this larger type do not accelerate in fattening until heavier weights are reached, so gains to any given weight may be more efficient. A larger percentage of lean and a smaller percentage of fat is deposited with each pound of gain.

In countries where meat-producing capacity is taxed because of dense population, a larger percentage of beef consumed is from animals of dairy breeding. In these countries, milk is a cherished food that is produced more efficiently than meat, and dairy herds serve a dual function, with cull cows, bulls, and steers providing meat.

In the United States, we have been willing to pay some premium for cattle of beef breeding because of the desired marbling and other quality factors in the lean.

Age and Weight

In addition to species, the age and weight of feeders purchased usually depend on the kind and amount of feed, labor, and capital available. Feeder calves purchased soon after weaning typically average about 550 to 600 pounds (Fig. 8–3). Yearling steers or heifers usually weigh between 700 and 800 pounds. Older and heavier cattle, if high quality, are put on a concentrate ration rapidly in order to produce desirable-quality carcasses before the animals pass the ideal market weights, generally between 1,100 and 1,300 pounds. Because lower quality cattle are usually fed to lighter weights, fewer concentrates and more roughage are used.

Very few weaned calves are placed directly into the feedlot but are grown and developed, or **backgrounded,** on a feeding program that provides modest but economical gains. Their winter rations consist largely of roughages such as hay and limited grain, whereas in summer they generally are placed on range or farm pastures. The reasons are to provide a market for feeds that might otherwise be wasted and to let the stocker calves grow on a feed cheaper than concentrate. A later feeding period on a high-grain ration,

Backgrounding
Growing of animals, typically young beef animals, for an extended period prior to placing them into a feedlot.

Figure 8–3
Feeder calves sorted and penned for shipment.
(Courtesy of American Hereford Association.)

TABLE 8–2					
Influence of Initial Weight on Rate and Efficiency of Gain in Cattle					

Initial Weight, lb	Average Daily Gain, lb		Dry Matter/Gain	
	Angus	*Holstein*	*Angus*	*Holstein*
649–748	2.16		6.60	
748–849	2.42		7.01	
849–948	1.98	2.49	8.47	7.46
948–1,047	1.76	2.33	9.63	8.22
1,047–1,148	1.67	2.22	10.34	8.95
1,148–1,247		1.69		10.66
1,247–1,346		1.72		11.64

Source: *Thonney, M. L., et al., J.Animal Sci. 53(1981): 354.*

usually ranging from 120 to 180 days and depending on quality of cattle and feed, will put good-quality calves in the choice grade at about 1,100 pounds.

It may be less profitable for the cow-calf operator to finish livestock for market at an extremely light weight. When a calf is sold for slaughter, the sale price must pay not only for the feed and labor that went into the calf but also all expenses of growing and maintaining the cow and the calf's share of the bull cost (both included in the cost of the feeder calf). When a 700-pound finished calf is sold for slaughter, the price per pound obviously must be higher to pay these costs than if the animal is sold at a heavier weight.

Countering this effect, however, is the fact that a growing calf (or other animal) becomes less efficient as it grows and fattens. So each additional pound of weight is added at a greater feed cost than the previous pound. For example, the data in Table 8–2 indicate that the Angus cattle averaging 1,000 pounds gained about two-thirds of a pound less per day than 800-pound Angus cattle, but required 2.62 pounds more feed per pound of gain.

Lighter and younger animals need more shelter and are more susceptible to disease and other stresses so demand more attention. Less financial risk is involved with calves, however, because a smaller animal is purchased. Calves will also be fed longer, therefore, fewer total animals (meaning less investment in animals) are needed to utilize a given quantity of feed.

The ability to gain fast is important because feed efficiency (pounds of gain per pound of feed) is usually better in faster gaining animals. The obvious reason is that a faster gaining animal will be on feed fewer days to reach the desired market weight. In addition, fast-growing young animals provide more tender meat at the same target weight. Because a large part of the ration consumed is used for maintaining the animal and only the remainder contributes to increased growth, maintaining an animal fewer days means that a larger proportion is used for gain.

Sex and Sex Condition

Steer
A male bovine castrated before puberty.

Most cattle in feedlots are **steers** or **heifers**, though a few **bulls** and **cows** are fed. A majority are steers, because 30 to 50 percent of heifers are retained as herd replacements.

Steers gain faster than heifers on similar rations, but heifers fatten more quickly so are ready for slaughter at a lighter weight. (Table 23–2 compares heifers with steers and bulls in percent of lean and fat in the carcass.) Some prefer heifers because of the shorter feeding period and lower purchase price. If feeding raises the grade and value of an animal, the feedlot operator receives a positive "margin," selling for a price per pound higher than the purchase price. Because of the shorter feeding period (and less feed consumed per animal), more heifers than steers can be finished with a given amount of feed. Therefore, the feeder might realize a margin on more animals.

Heifer
A female bovine before calving. Some use the term until the second calving.

The shorter feeding period for heifers may sometimes be considered a disadvantage, depending on feed supplies and anticipated prices of slaughter cattle. Heifers not fed estrus-suppressing hormones often "come in heat" during the feeding period, becoming restless, riding each other "walking the fence," and gaining little for several days. This results in an increased cost per hundred pounds of gain of about $2.50 to $3.00 more than similar quality steers.

Bull
Most commonly refers to a sexually mature, intact (uncastrated) male bovine.

Also, because of the possibility of slaughter heifers being pregnant, causing a lower dressing percentage, processors often pay less for heifers. Though this differential is usually reflected in the purchase price of feeders, the gain put on in the feedlot also sells for less.

Certain feed additives such as melengestrol acetate (MGA) can be used to prevent feedlot heifers from showing signs of "heat," and may therefore result in faster and more efficient gains.

Cow
A female bovine, usually after first pregnancy and parturition.

Some bulls are fed for slaughter (the terms "bullock" or "intact male" are sometimes used for these animals). Bulls generally gain 5 to 9 percent faster and 3 to 7 percent more efficiently than steers and produce a carcass that has a higher percentage of lean. Table 8–3 shows the results of bull feeding research at the Fort Hays, Kansas, Agricultural Experiment Station. When bulls are weaned at a relatively early age and full-fed on a high-concentrate ration, they can be slaughtered at or before 15 months of age. This reduces ownership time, and that, combined with the higher proportion of lean, results in lower cost production of lean meat.

Older cows, such as cull beef or dairy cows, are sometimes confined to a feedlot for a very short feeding period before slaughter, to increase the fat in the carcass and the juiciness of the meat. This is not routine but is practical if cows are of desirable con-

TABLE 8–3	Performance and Carcass Traits of Steers and Bulls Full-Fed from Weaning to Slaughter at 15 Months	
	Steers	**Bulls**
Number	20	23
Weaning weight (lb)	476	496
Feedlot gain, 247 days (lb)	674	747
Feed per cwt gain (lb)	677	630
Slaughter weight (lb)	1,146	1,256
Dressing percent	64.7	65.5
Backfat (in.)	0.64	0.50
Rib eye area (in.^{2})	11.51	13.05
Yield grade, avg.[a]	3.66	3.04
Marbling score[b]	5.24	4.26

[a]*USDA yield grades. A grade of 3 indicates a higher yield of trimmed, valuable cuts than does a grade of 4.*

[b]*A higher score indicates more marbling in the lean.*

Source: *Brethour, J. R., Progress Report 384, Fort Hays Branch, Kansas Agr. Exp. Sta., 1980.*

formation and are young when culled so they may yield relatively high-quality carcasses. There have been times when drought has forced heavy culling of beef cows in range areas. Market price of these cows (which normally go directly to slaughter) dropped so low that feeders with available feed could afford to purchase them to utilize roughage and some grain.

Conformation and Performance

Feeder cattle that eventually will produce carcasses with high "cutability"—a high proportion of lean and valuable cuts—show their natural muscling clearly. Muscle is lean meat, and the most valuable muscle is in the rib, loin, and round.

As a high-cutability feeder walks, you can see the muscles of the *round* move and ripple, those of the *forearm* bulge. From the rear, the animal has a rounded "quonset" shape over its back and is widest through the middle of the round. Along the underline, a high-cutability feeder is trim; the brisket, twist, or flank is not heavy or deep.

To provide a common basis for trading, even when both buyer and seller cannot see the cattle, and to ensure meaningful market reports of supplies and prices, grades have been established for grouping feeder cattle. Grades and specifications that became effective August 1, 1979, take into account variations in frame size, thickness, and thriftiness (Fig. 8–4). When grouping animals by size categories (large, medium, or small), they are selected according to their skeletal size (height and body length) in relation to their age.

The weight gain of a large-frame feeder animal, of a given degree of thickness, normally will consist of more muscle and bone but less fat than the weight gain of a smaller frame animal. On the average, larger frame animals after feeding will not produce the desired quality of carcass until live weight exceeds about 1,200 pounds for steers (and bullocks), and about 1,000 to 1,050 pounds for heifers. (See the discussion of carcass grades in Chapter 24.)

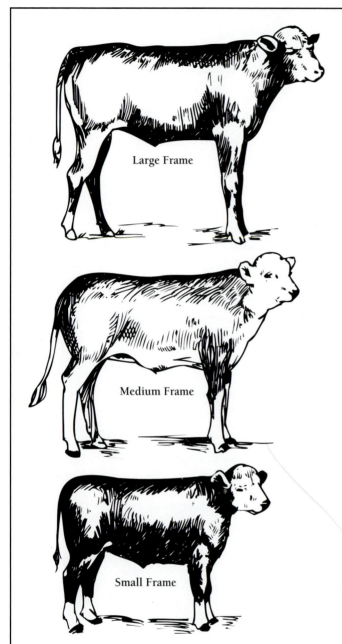

Large Frame Size. Cattle qualifying for the minimum of this grade have large frames, are thrifty, and are tall and long-bodied for their age. Steers and heifers would not be expected to produce carcasses with the amount of external (subcutaneous) fat opposite the 12th rib (usually about 0.5 in.) normally associated with the U.S. Choice grade until live weight exceeds about 1,200 pounds for steers and about 1,000 pounds for heifers.

Medium Frame Size. Cattle meeting minimum qualifications for this grade have slightly large frames, are thrifty, and are slightly tall and slightly long-bodied for their age. Cattle would be expected to have the amount of external fat (0.5 in.) normally associated with U.S. Choice beef carcasses at live weights of about 850 pounds for heifers and 1,000 pounds or more for steers.

Small Frame Size. Feeder cattle included in this grade have small frames, are thrifty, and are shorter-bodied and not as tall as specified as the minimum for the medium frame grade. Cattle would be expected to produce carcasses with the amount of external fat (0.5 in.) normally associated with the U.S. Choice grade at live weights of less than 1,000 pounds for steers and less than 850 pounds for heifers.

Figure 8–4

USDA feeder cattle grades. The 10 grades are: large frame, Nos. 1, 2, and 3; medium frame, Nos. 1, 2, and 3; small frame, Nos. 1, 2, and 3; and inferior.

Source: Adapted from *Cooperative Extension Service Bulletin,* Iowa State University.

No. 1

No. 1 Thickness. Feeder cattle which possess typical minimum qualifications for the No. 1 thickness grade are thrifty and are slightly thick throughout. They are slightly wide through the chest and are slightly thick and full in the crops, back, and loin. The rounds and forearms are slightly thick. Their legs are set moderately wide apart, and they usually show a high proportion of beef breeding. Feeder cattle produced by crossing English breeds with heavily muscled breeds, such as Charolais, Simmental, Limousin, and others, would produce No. 1 thickness.

No. 2

No. 2 Thickness. Feeder cattle which possess typical minimum qualifications for the No. 2 thickness grade are thrifty and narrow throughout. They are narrow through the chest and over the crops, back, and loin. The rounds and forearms are narrow, and their legs are set close together. Crossing English breeds with dairy-type Holstein would result in feeder cattle with No. 2 thickness.

No. 3

No. 3 Thickness. Feeder cattle included in the No. 3 thickness grade include thrifty animals that have less thickness than specified for the No. 2 grade. The thin-type dairy animals fall into this category.

Figure 8–4
(continued)

A thicker feeder animal normally will produce a carcass with a more desired yield grade (1 or 2 rather than 3, 4, or 5) after feeding than a feeder of the same frame size that is thinner.

This emphasizes that feeder grades are related largely to conformation and the potential yield grade or cutability of the finished animal or carcass. It also means that when cattle entering a feedlot have been sorted by frame size and thickness grades, cattle coming from the feedlot will be more uniform in cutability. Finish is not, of course, a factor in feeder grades.

Handling for Good Performance

Movement of feeder cattle, especially calves, from ranch to feedlot can cause much stress and result in illness, weight loss, and sometimes death. Some feeder calves are weaned the day they are loaded for the trip—perhaps a thousand or more miles and extending over several days—to the feedlot. Some go through several markets, experience climatic changes, and are exposed to miscellaneous diseases. Illness, loss of weight, and decreased appetite are costly to the purchaser.

Calves can be handled to minimize these effects. Nonstop movement on interstate highways or feeding and watering en route may help. Some feeder producers follow a system of "preconditioning" feeder calves for the trip and the feedlot environment, and such calves usually will bring several dollars more per hundredweight. Preconditioning is described in Section 29.5.

8.3   FEEDER LAMBS

Because sheep are well adapted to cool climates, rough terrain, and rather sparse vegetation, a majority of the sheep are raised in the mountain states and arid plains states (see Fig. 8–1), where lambs are cycled throughout the United States. Few concentrates are available in the grazing areas, so most lambs are shipped out as feeders after weaning and placed in feedlots.

In recent years, more and larger lamb feedlots have been developed near the lamb producing areas (Fig. 8–5). Although much of the feed must be shipped in, the transportation and handling costs of the lambs, as well as their stress, are less.

Figure 8–5
Lambs on feed in a western state.
(Courtesy of *Feedlot.*)

Lambs raised in midwestern "farm flocks" usually are sold for slaughter relatively soon after weaning. Such lambs normally have access to a creep ration in addition to their mothers' milk and pasture, so are not sold as "feeders" but as slaughter lambs that reach market weight of 100 to 110 pounds within 120 to 140 days of age.

Breeding and Color Markings

Lamb feeders prefer lambs that show evidence of some meat-type breeding—sired by rams of the mutton-type breeds (Chapter 21). Sheep of fine-wool breeding are narrower, rangier, and produce a less desirable carcass. Most ewe flocks in range areas are of fine-wool breeding because of the heavy yield of high-quality wool these breeds produce, and also because their type and conformation make them more adapted to range conditions. Many sheep ranchers have found it practical to maintain a fine-wool ewe flock, breeding enough ewes to fine-wool rams to provide flock replacements. They then use mutton-type rams on the rest of the ewes to produce the meatier crossbred feeder lambs. The crossbred lambs usually are more vigorous and gain faster, both on the range and in the feedlot.

Hampshire and Suffolk rams—both with black markings—are popular for such crossbreeding programs because of their large mature size, ruggedness, and muscling. Therefore, feeder-lamb purchasers have developed the habit of looking for black-faced or dark-faced feeder lambs, knowing they probably would gain faster and produce meatier carcasses than white-faced, straightbred, fine-wool lambs. In recent years, however, certain white-faced mutton-type rams, such as Columbia and Corriedale, have been used, so the black markings have become less reliable as an indicator of crossbreeding, faster gains, and potential carcass merit.

Age and Weight

Essentially all lambs in the range areas are sold to feedlot operators after weaning at 5 to 8 months of age and 65 to 95 pounds. Consumer desire for lightweight cuts and the slow fattening that results from about 50 percent roughage in the ration mean that light feeder lambs, weighing 60 to 65 pounds, are desired. Light lambs will more likely grade Choice by the time they reach market weight.

There are other reasons why light lambs may be more profitable for the feeder. Whenever lambs are bought and sold there is marketing expense and considerable transit shrink. If lambs are purchased at 80 pounds and sold at 100 pounds, shrink and marketing costs are high in relation to the profit that might be made on the gain. A feeder would need to buy, feed, and sell about twice as many 80-pound feeder lambs to market a certain quantity of feed as would be necessary if 60-pound feeder lambs were purchased. Also, the heavier lambs probably would gain less efficiently, on the average, than lighter lambs because a large proportion of the gain is fat and because their average daily maintenance requirement is higher.

As is true with cattle, heavier feeder lambs of the same age will gain faster, but because of the other factors mentioned this is less important in lambs. Older sheep—beyond yearlings—yield carcasses with stronger flavor, which is not desired by the consumer.

Sex and Sex Condition

Lambs seldom are sorted according to sex, although sex does influence feedlot performance. Male lambs will gain faster in the feedlot than ewe lambs, partly because they are heavier at weaning time and upon entering the feedlot. Though essentially all male

TABLE 8–4	Growth and Carcass Traits of Ram and Wether Lambs	
	Rams	**Wethers**
Gain (lb/day)	0.475	0.398
Slaughter weight (lb)	160.45	143.79
Hot carcass weight (lb)	80.70	75.34
Leg conformation[a]	12.42	12.42
Fat thickness (mm)	6.46	8.58
Yield grade	3.50	4.43
Quality grade[a]	11.73	12.40
Flavor score[b]	5.04	4.58

[a]Low Choice = 10, Choice = 11, High Choice = 12.

[b]Based on a 9-point scale, with 1 = imperceptible and 9 = very pronounced.

Source: Crouse, J. D., et al., J. Animal Sci. 53 (1981): 376.

Wether
A young male sheep or goat castrated before puberty.

lambs are castrated soon after birth and sold as **wethers**, considerable interest has developed in feeding young intact rams. Such would avoid the shock, weight loss, and risk associated with castration. Rams gain slightly faster and produce leaner carcasses than wethers, so gains usually are more efficient. Because lambs (especially those from farm flocks) are slaughtered at an early age, secondary sex characteristics have not developed and there is little evidence that carcass quality is impaired (Table 8–4). Most processors dock (discount) ram lambs $1.00 or more per hundredweight, regardless of weight. Evidence to support this practice, especially in young lambs, is not available, although heavy ram lambs (over 140 pounds) produce carcasses that are sometimes soft and oily and have undesirable aroma and flavor characteristics.

Leaving male range lambs intact probably would be less practical because range lambs are older when slaughtered. The possibility exists, however, that certain hormones could be used to inhibit development of the secondary sex characteristics, and the effect of castration could also be avoided.

Grades and Conformation

Market Grades
Grading refers to animals according to relative merit. Grading of live animals is unofficial and aids in establishing the value of an animal destined for market and slaughter.

Although official grades of feeder lambs have been designated and described, they have less significance in appraising feeder lambs than have the grades for feeder cattle. The main reason is that there is less total variation in feeder lamb quality, and so less need for grading. Because of (1) the heavy seasonal influence on lamb production, (2) the economic restriction of feeder lamb production to grazing areas, and (3) the limitation on age and weight for lambs entering feeding programs, nearly all feeder lambs are produced under very similar conditions and are sold at very similar ages and weights. About the only factors that contribute greatly to variation in feeder lamb quality are breeding and health.

In most years, healthy feeder lambs that carry some mutton breeding grade either Choice or Good-to-Choice (see Chapter 24 regarding grading). Lower grading feeders are usually of straight fine-wool breeding, are late lambs that did not get the benefit of early, lush grass (and their dams probably produced less milk for the same reason), or are from a drought area short on grass.

Because feeder lamb production is so seasonal and so dependent on grass production, especially in the mountain states, there is much yearly variation in lamb weights and quality. In some years lambs are thin, light, and look poor, though they may be highly profitable to a feeder for these very reasons. In other years rainfall and temperature are such that all lambs are heavy at the end of the grazing season; many carry enough finish to grade High Good or Choice and so are sold directly to packers. Whereas the lambs are in excellent condition, such years are usually bad for lamb feeders. The lambs are too heavy for an economical feeding program and feeder prices are high, with packers competing for those lambs that carry some finish.

Other Influences on Performance

Lambs born as singles are heavier at weaning than lambs born as twins or triplets, especially in range flocks where feed supply and milk production are less; so singles probably will continue to perform better in the feedlot. The same is true for lambs produced by mature ewes versus lambs born to 2-year-old ewes.

8.4   FEEDER PIGS

There is less movement of feeders in the swine industry than in the cattle and sheep industries, and those that do move travel less distance. Because sows depend on a rather highly concentrated ration rather than on range grasses, most are maintained in grain-producing regions where pigs are fed to slaughter. Major pig-producing states are presented in Fig. 8–1 and listed in Table 28–2.

With more highly specialized swine production and "off-site" production, movement of feeder pigs has increased, with significant numbers being farrowed and weaned at one location and finished at another. Farrowing stalls, heat lamps, insulated housing, controlled temperature, and other items used for farrowing require considerable capital and must be well utilized in order to be profitable.

Producers with facilities, experience, and personal desire to work with sows and young pigs have found it profitable to specialize in the production and sale of feeder pigs (Fig. 8–6). The feeder pig enterprise has advantages for some producers:

- It is well suited for grain-deficit areas.
- It fits the producer with excess labor for care and management of sow herd.
- Lower initial investment is needed than for farrow-to-finish operations.
- There is a rapid turnover of volume.

However, the feeder pig enterprise requires a high level of management and a dependable market to be successful.

Feeding weaned pigs to market weight requires relatively less labor, capital, and equipment, so many farmers prefer to buy feeder pigs. These farmers have ample grain for feeding but have specialized in other phases of the farming operation and lack labor, equipment, capital, or desire to farrow sows.

Several factors inhibit further or more rapid development of this specialization. Pigs are affected by and are potential carriers of many contagious and serious diseases. In fact, this has led to highly developed biosecurity plans to protect the animals from foreign diseases. Most farmers resist bringing pigs from other premises onto their farms, especially if they might have been hauled in several different trucks or routed

Figure 8–6
Feeder pigs, immunized for disease prevention, leave the yards of a feeder pig producer. (Courtesy of *National Hog Farmer.*)

through various markets. Also, marketing costs for feeder pigs are high in relation to animal value. Essentially as much labor, skill, and record keeping are involved in selling a 40-pound pig worth $40 as in selling a 500-pound calf worth $400 or a 700-pound yearling worth $525.

Because of relatively high marketing costs, most feeder pigs are sold directly by producers to feeders, or are handled by dealers. Certain cooperative units have developed in feeder pig-producing areas. Relatively few feeder pigs are sold through central public markets, but a significant number are sold through local auctions. Nearly all feeder pigs sold by regular producers or dealers are marketed soon after weaning, at about 40 pounds. Essentially all feeder pigs are crossbreeds.

Health and Preparation
State laws usually require that pigs brought into the state be vaccinated for certain diseases and be certified in good health. Such laws usually do not control shipments within a state, however, so it is important that producers and dealers develop a reputation for selling only healthy, disease-free pigs. Sanitation is extremely important, not only in buildings but also in scales, pens, and trucks occupied by pigs from a number of farms.

Male pigs will have been castrated, and most pigs will have their tails docked to prevent tail biting. Tails are cut off about one-fourth to one-half inch from the body with side cutting pliers or a chicken beak trimmer.

Appraisal Systems and Weight
USDA grades for feeder pigs exist, but purchasers are most concerned with health, reputation of grower or dealer, and uniformity in size of the pigs.

Prices asked and paid for feeder pigs depend primarily on the price of slaughter hogs, because of the relatively short feeding period. The price of corn or grain sorghum and other ration ingredients, in comparison to the current and anticipated slaughter hog prices, is also considered.

If pigs are healthy, there are few visual criteria that can be used to predict rate or efficiency of gain among feeder pigs. Reputation of the grower or dealer is much more important. Conformation, however, does give an indication of the quality of carcass that can be produced.

Sex and Sex Condition

Barrow pigs will gain faster than gilts and will be slightly fatter at the same market weight, though there is usually little difference in feed efficiency. Feeders are generally not sold according to sex.

Boars will gain faster, produce leaner carcasses, and gain more efficiently than either barrows or gilts. Consumer prejudice and the problem of an offensive odor that is sometimes, though not usually, present in boar carcasses prevent boar feeding from becoming common.

> *Barrow*
> A male pig castrated before puberty.

> *Gilt*
> A female pig before farrowing. Some use the term until the second farrowing.

QUESTIONS FOR STUDY AND DISCUSSION

1. What geographical areas in the United States produce the most beef calves?

2. List five states that lead in lamb production. Why?

3. Why are the above states well suited for cattle and sheep production?

4. Give examples of beef breeds introduced into the United States to increase skeletal size. Have the traditional breeds changed in size due to selection emphasis?

5. What are typical weights for steers and heifers entering the feedlot to be fattened for slaughter? At what weights should they typically be ready for market?

6. Explain why feeding extremely heavy cattle is inefficient.

7. Explain why purchase price, feeding performance, and slaughter price of heifers usually differ from those of steers or similar-quality cattle.

8. Because young bulls gain rapidly and are efficient in body weight gain, why aren't more fed for meat production in the United States?

9. Name the three frame sizes for feeder cattle. How would you describe large-frame feeder cattle?

10. At what age and weight are most midwestern feedlot lambs sold for slaughter? How does this differ from range-raised lambs?

11. List five states that lead in pig production. Compare the location of these states to the leading states in calf and lamb production.

12. What are the principal characteristics of the feeder pig enterprise?

Animal Growth and Carcass Composition 9

All breeders and producers of livestock poultry, as well as fish and companion animals, have an appreciation and knowledge of the growth and development of their animals. Each breeder has a special interest in the appearance and performance of animals resulting from each planned mating. Producers also regularly monitor the rate and efficiency of growth of each group of market animals as they increase in size and weight. The many processes occurring at the cellular level that bring about growth and development of the complete animal are less obvious. However, they also must be understood and appreciated.

Growth is defined, measured, and described in many ways. Growth could be simply defined as an increase in body size. In biological systems, the term *hyperplasia* is used to describe an increase in the number of cells and the term *hypertrophy* to describe an increase in the size of cells. Some scientists suggest that growth is the excess of **anabolism** (protoplasmic construction) over **catabolism** (protoplasmic destruction). In other words, body tissues undergo continuous building and destruction; growth occurs when tissues are built faster than they are broken down. In animal production, growth is usually defined as the increase in animal tissue—muscle, fat, bone, and related tissues.

> *Anabolism*
> The building up of body tissues.

> *Catabolism*
> The breaking down or degradation of body tissues.

This chapter is devoted to the biological process of growth, how it occurs, and the components of growth in the lifetime of an animal. Upon completion of this chapter, the reader should be able to

1. Define growth, chronological growth, physiological growth, and other key terms.
2. List six or more ways in which growth may be measured in livestock and poultry.
3. Calculate average daily gain and weight per day of age, given data needed for the calculations.
4. Chart a typical growth curve and mark key times—birth, puberty, maturity—on that curve.
5. Describe how hormones influence growth.
6. Chart or describe the deposition rates for lean, fat, and bone as an animal grows and matures.
7. Describe how growth may be modified during the feeding process to influence meat quality.

9.1   CHRONOLOGICAL VERSUS PHYSIOLOGICAL GROWTH

To help describe the increase in an animal's ability to perform, growth is separated into two types: chronological and physiological. *Chronological growth* refers to increases in size or body function due to an animal growing older. *Physiological growth* refers to increases in size or body function due to increase in tissue and organ growth and development.

It is difficult to separate these two types of growth under normal conditions. Naturally, as an animal grows older until adulthood, it will increase in size, and, at the same time, its tissues and organs will mature and develop. However, body functions in one species may be related more to physiological growth, whereas in another species they are related more to chronological growth. In swine, the appearance of first estrus depends more on the age of the female (chronological growth); in cattle, the appearance of first estrus is more dependent on the size of the female, indicating a relationship to tissue and organ development (physiological growth).

9.2   MEASURES OF GROWTH

Growth can be measured in a variety of ways. The type of measurement used should depend on the intended use of the animal. For meat animals intended for slaughter, the growth measurement should provide an indication of the edible portion of the carcass that meets minimum quality standards. Objective measurements such as animal weight, loin eye area, and backfat thickness are indicators of the edible portion of the carcass. In general, the heavier the animal (within limits), the larger the loin eye. Thinner backfat indicates a higher proportion of lean tissue.

Growth of breeding animals is related to their ability to produce offspring. Linear body measurements, weight, age at puberty or first estrus, and size of the testicles could all provide indications of an animal's ability to produce offspring. Measurements of milk, wool, and egg production of young females are used to evaluate growth at the cellular level necessary for producing products for human consumption. These generally will indicate how the female will perform later in life.

Measurements of growth in horses would include height at the withers, weight, and even speed or strength. Choice of an objective measurement that is related to the intended use of the horse is most important. The intended use may include strength for pulling, speed for racing, or a combination of speed, strength, and coordination for working livestock or recreation.

In addition to objective measurements of growth, as noted above, subjective or visual appraisal of growth is also used by producers. Visual appraisal helps estimate the nature (quality) of growth that has occurred and complements objective measurements. For example, yearling weight and testicular size and development would serve as excellent objective measurements of a bull's ability to grow and produce offspring. But visual appraisal of the straightness of hind legs or the strength of pasterns helps assess the quality of growth, in this instance the likelihood that a bull will be free of lameness or impaired movement through life—its ability to perform an intended use.

Quality grades and conformation scores of growing livestock may be based on both objective and subjective measurements. Several grading and scoring systems are used to evaluate growth rate potential. Cattle may be ranked by height at the hooks (hips) and a muscling score. Taller, longer bodied animals with moderate muscling and a minimum of external fat covering tend to grow faster. Shorter bodied cattle with an excess of external fat covering will grow more slowly.

The *rate* at which growth occurs is a better measurement of an animal's ability to perform than is weight itself. *Average daily gain (ADG)* measures growth rate. Average daily gain measures the average weight increase each day between the first weight and the last weight. It is determined by the formula

$$ADG = \frac{W_2 - W_1}{t_2 - t_1}$$

where W_1 is the beginning weight, W_2 is the ending weight, t_1 is the beginning time, and t_2 is the ending time. This is the way most producers measure growth rate of animals destined for slaughter.

Another useful measurement is *weight per day of age (WDA)*. For example, poultry for slaughter are marketed at standard weights following a relatively short growing period, so that growth rate is recorded as number of days to reach market weight. Broilers generally should reach the standard market weight of 4.7 to 4.9 pounds in 52 days following hatching. Growth rate in pigs usually is recorded as days to 230 pounds, or as average daily gain. WDA also measures growth rate and is commonly used with livestock raised for breeding purposes. It is determined by the formula

$$WDA = \frac{Weight}{Age\ in\ days}$$

Cattle producers frequently use weight at 205 days of age to evaluate growth rate in weaning animals. Weights of calves weaned at various ages may be adjusted mathematically to a common age of 205 days so their growth rates can be compared on an equal basis. The 205-day adjusted weaning weight is determined by the formula

$$205 - day\ adjusted\ weight = \frac{Actual\ weight\ -\ Birth\ weight}{Age\ in\ days} \times 205 + Birth\ weight$$

9.3   THE GROWTH CURVE AND ITS PHASES

Regardless of how growth is defined, it is not a straight-line function. The age-growth curve, or plot of weight versus time, is the common way of representing growth (Fig. 9–1). The S-shape is similar for all species of livestock and represents the continual process of growth and development of an animal from birth to maturity. Growth rate is represented by the steepness of the curve at any particular time. It is much more rapid near the middle of the inflection point of the curve than just after birth or just before maturity. This period of rapid growth is near puberty and corresponds to the rapid growth that humans experience during adolescence.

The growth curve for all farm animals except poultry can be separated into three phases: (1) prenatal, (2) preweaning, and (3) postweaning. **Prenatal** growth is the increase in weight from conception to birth. It is divided into the ovum, embryonic, and fetal phases. Students of growth usually describe the ovum phase as beginning with fertilization and continuing until attachment to the uterine wall, about 11 days in most meat animals.

Tissues, organs, and various biological systems form or differentiate during the embryonic phase, but the embryo increases very little in weight. This period covers 25 to 45 days, depending on the species.

Prenatal
Prior to birth.

Figure 9–1
A generalized age-growth curve for mammals from conception to maturity. Growth of specific parts of the body will vary from this generalized curve.
(*Source:* Iowa State University.)

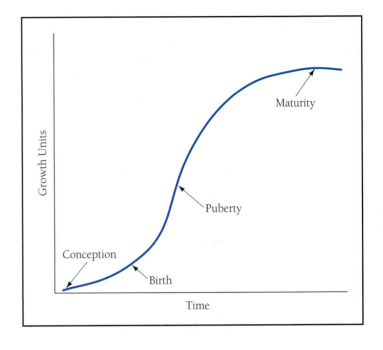

During the fetal phase, which lasts until birth, there are rapid increases in weight and dramatic organ and tissue development. Sixty to 70 percent of an animal's birth weight is amassed during the fetal phase. Organs and systems—including the circulatory and nervous systems—develop and increase in size and function. The digestive tract develops primarily during this phase, though some additional changes occur following birth, when the animal begins to ingest solid food.

> **Preweaning**
> Time period from birth to weaning of offspring.

Preweaning growth occurs from birth to weaning (when the mothers are nursing their young). Increases in size and organ development are highly dependent on quality and quantity of milk produced by the mother. The amount and quality of feed the mother receives affects her milk production. Young, aged, or unusually small mothers within a breed usually produce less milk than larger, mature mothers, so the preweaning growth rate of their offspring is lower. When sows produce more pigs than they have functional nipples, growth of the pigs is hampered. Single lambs grow more rapidly than twin lambs raised by the same ewe, and this is related to milk supply.

Sex of the young also affects growth rate. Intact males grow more rapidly than castrated males and castrated males more rapidly than females when the mother receives sufficient feed. These differences are relatively small (perhaps 10 to 20 percent) during the preweaning phase and usually do not occur when the feed supply to the mother is limited.

> **Postweaning**
> Time period after offspring is weaned from dam until slaughter (or selection for entering the breeding herd).

The **postweaning** phase covers the period from weaning to slaughter (or selection for a breeding herd). During this period the animal continues to increase in size, but growth rate declines as maturity is approached. The growth rate early in this phase is influenced significantly by the treatment the animal received during preweaning. Animals with superior genetic potential that have received limited milk or other feed prior to weaning often will gain at a rate more rapid than true growth (as shown later in Fig. 9–4) when fed a well-balanced ration after weaning. Animals fed excessively, to the extent of causing fattening, in the preweaning phase or early postweaning phase usually will not attain their full postweaning growth potential. Fattening early in life may prevent maximum development of the skeleton and muscle tissue.

The growth rate late in the postweaning phase is more dependent on genetic potential, sex, environment, and the quality of nutrition the animal receives. It is the phase with which most animal production processes are concerned. These factors will be discussed in detail in Section 9.5.

9.4 HORMONAL CONTROL OF GROWTH

Hormones, as chemical messengers, serve as the regulating mechanism for growth. Collectively they coordinate growth and development of body tissues, structures, and organs, so that an animal's size will complement its organ development. In addition, for growth to occur with ease and effectiveness, different organs and tissues must become functional at different times. This process of establishing priorities for tissue function occurs many times as the animal increases in age. In addition, the process of **homeostasis** regulates physiological body function on a somewhat constant and established basis.

> *Homeostasis*
> Maintenance of physiological stability even though environmental conditions may change.

Hormones are secreted from endocrine glands. Each of these ductless glands empties its secretion into the bloodstream where the hormone is carried to its "target organ." The primary hormones responsible for the control of growth are somatotropin, thyroxine, glucocorticoids, androgens, and estrogens.

Somatotropin is secreted from the anterior pituitary, which is located just below the brain. As mentioned, it is commonly called "growth hormone" because it is the primary hormone regulating increases in body size. It regulates development of bone and muscle. Growth hormone has considerable potential in increasing both yield and efficiency of meat and milk production now that technology has been developed to produce adequate quantities through recombinant DNA procedures. For example, **bovine somatotropin (BST)** has resulted in substantial increases in milk yield of lactating dairy cows, possibly due to the repartitioning of nutrients from the fat tissues to the mammary gland.

> *Bovine somatotropin (BST)*
> A peptide hormone produced through molecular biotechnology; when injected into lactating dairy cows generally brings about a response in increased milk yield.

Porcine somatotropin (PST) is the major hormone controlling protein and lipid metabolism in the pig. When administered as an *exogenous* (outside) source to pigs in their growing and finishing stages, PST increases protein deposition about 30 percent. During this same period, it decreases fat deposition as much as 45 percent without adversely affecting meat quality.

> *Porcine somatotropin (PST)*
> The type of somatotropin produced by swine.

The thyroid gland, located in front of and on either side of the larynx, secretes *thyroxine*. Thyroxine regulates the basal metabolic rate, which controls protein synthesis and, therefore, lean tissue growth and body weight of the animal. When the thyroid does not produce sufficient thyroxine, usually because of insufficient iodine, animals lose weight; they eat less and muscle and bone growth are impaired. This condition is called *hypothyroidism*. Hyperthyroidism occurs when the thyroid secretes too much thyroxine. The animal's metabolic rate increases so that nutrients are used less efficiently. This condition also generally results in weight loss or lack of normal growth rate.

Thyroxine compounds may be fed to animals in the form of thyroproteins. Such compounds will not increase weight gain; rather, they have been used in a limited number of instances to increase feathering in poultry, wool and milk production in sheep, and milk production and fat content of the milk in dairy cows. In the case of feeding thyroprotein to cows, the initial response is increased milk yield; however, upon removal from the ration, rather severe production decreases may occur.

Androgens are a group of hormones that are associated with the development of secondary sex characteristics in males, but that also have dramatic effects on bone and muscle growth in both sexes. *Testosterone,* the primary androgen, is secreted primarily from the testes in males and the adrenal glands in females. Androgen secretion increases

> *Androgen*
> A hormone associated with development of secondary sex characteristics in the male.

> **Testes**
> The paired male sex organs that produce sperm after sexual maturity. Sing., testicle.

> **Estrogen**
> A female hormone that promotes development of the female reproductive tract and mammary tissue.

markedly just prior to puberty and is partly responsible for the rapid growth that occurs at that time. Males have a more rapid growth rate than do females because the **testes** produce more androgens than do the adrenal glands.

Estrogens are secreted primarily from the ovaries and serve to develop the female reproductive tract in all species. Estrogens also increase muscle growth in ruminants and fat deposition in poultry. In other species, estrogens tend to slow growth of these tissues.

The adrenal glands, small kidney-shaped glands just above the kidney, secrete *glucocorticoids*. These hormones may cause removal of nutrients from body stores to produce energy, resulting in weight loss or decreased rate of gain, though this is not a problem in normal, healthy animals.

The functioning of each of the hormone-producing glands and the degree of harmony with which they function appear to be influenced largely by the genes animals carry. In other words, heredity appears to affect animal growth and development largely through hormonal activity.

9.5   FACTORS AFFECTING GROWTH

As mentioned earlier in the chapter, the S-shape of the age-growth curve is similar for all animal species. The actual rate at which growth occurs or the final size of an animal is determined by several factors.

Inheritance determines the potential size and growth rate of an animal. Vast differences in genetic potential exist among and within breeds. When two animals are compared, one may have a higher mature weight, a more rapid growth rate, or both, as illustrated in Fig. 9–2.

Favorable conditions are necessary for full expression of an individual's genetic potential. Inheritance, to be discussed in Chapter 18, has more influence on some growth traits or stages of growth than on others. Postweaning weight gain is highly dependent on the animal's genetic potential, whereas preweaning growth is more dependent on the mother's milking ability.

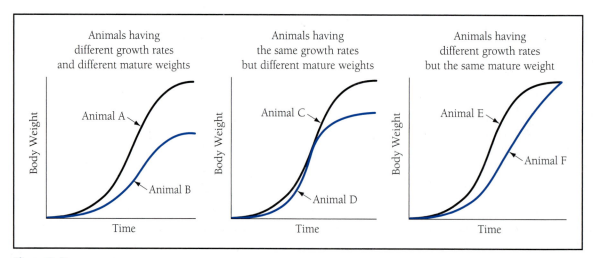

Figure 9–2
Theoretical growth curves of three pairs of animals. Note the captions for each pair of growth curves. (*Source:* Iowa State University.)

Genetic imbalances of growth hormone or thyroxine markedly affect growth. For example, a specific recessive gene expression results in dwarfism, sharply reduced growth in cattle.

Gender, or sex of an animal, plays an important role in growth of most meat animal species. High androgen levels initiate muscle development. Males therefore have more rapid growth rates, especially in protein mass, and higher mature weights than females (Fig. 9–3). Sex differences usually are smaller in swine than in cattle. Most heifers gain from 0.2 to 0.3 pound less per day than steers and are usually slaughtered at about 1,000 to 1,100 pounds, whereas steers are usually slaughtered at 1,100 to 1,300 pounds.

The plane of nutrition of the growing animal affects both immediate and subsequent growth traits and requirements. Generally, the more an animal eats, the faster its growth rate. The amount and quality of feed selected should match the growth desired. Maximum growth rate is not always desirable. Excessive feeding early in life will cause fattening and may inhibit bone growth and organ development. A plane of nutrition providing enough energy to promote maximum bone growth and organ development is best for females that are to be kept for breeding purposes. Very high energy rations for rapid weight gain and fattening are recommended only for animals destined for slaughter.

Many antibiotics improve growth rate and feed efficiency by reducing the incidence of disease. Also, in sheep and cattle, they alter the microbial population of the rumen and may improve feed utilization. Small amounts may be added to the feed, mineral supplement, or drinking water. Low levels of tranquilizers reduce stress susceptibility in cattle and sheep. Although tranquilizers normally are not given to meat animals, they have been shown to improve growth rate and feed efficiency.

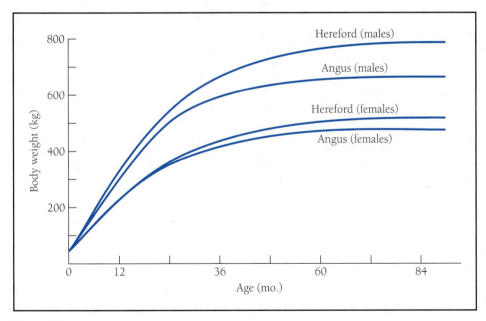

Figure 9–3
Effect of sex on growth curves of Hereford and Angus cattle. Note the higher growth rate of the males.
(*Sourc:* Adapted from Brown, et al., *Arkansas Agr. Exp. Sta. Bull. 570* and *571,* 1956.)

Figure 9–4
Normal growth curve of
Animal A versus the growth
curve of Animal B experiencing
compensatory gain.
(*Source:* Iowa State University.)

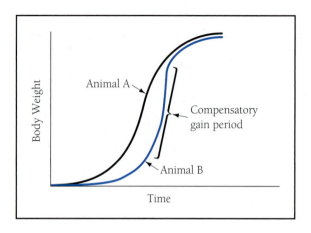

Other feed additives function in a variety of ways to improve growth rate or feed efficiency. In ruminants, ionophores such as Rumensin (monensin sodium) and Bovatec (lasalocid) alter rumen microflora causing an increase in production of the volatile fatty acid propionic acid, resulting in improved feed utilization and feedlot performance. Also, the additive MGA (melengestrol acetate) inhibits estrus and ovulation in heifers, thereby preventing the associated weight loss or lack of gain.

Implants such as those commonly referred to as Ralgro, Synovex, MGA, Revalor H, and Finaplix, in the form of small pellets that are implanted at the base of the ear in cattle, promote muscle and bone development. Active ingredients in the pellets are released gradually to the circulatory system. These compounds are considered to be *repartitioning agents* in that they redirect nutrients from fat to the development of protein.

Physical environment—shelter, type of flooring, animals per pen or cage, sound, or climatic conditions—can alter growth rate in livestock and poultry. (See Chapter 11.)

Sheep and cattle sometimes are raised in range conditions where energy intake is relatively low during the preweaning and the early part of the postweaning growth phases. Because feed eaten by the animal is used first for maintenance, then for bone growth, muscle development, and fat deposition, in that order, skeletal growth of an animal may be near normal and muscle development and fat deposition considerably less than would occur with an average-energy-level diet. This allows for compensatory gain to occur when the animal is fed a well-balanced, high-energy diet in the postweaning phase. **Compensatory gain** is the very rapid growth that occurs when an animal that has been on an energy-deficient diet is placed on an energy-rich diet (Fig. 9–4), such as lush spring grass, fall wheat pasture, or a finishing ration of corn and corn silage.

> ***Compensatory gain***
> Gain at an above-normal
> rate following a period of
> little or no weight gain.

9.6   CARCASS COMPOSITION

An animal's carcass has three main constituents: muscle, fat, and bone. All three increase in total amount as growth occurs. However, as one tissue comprises a higher percentage of the carcass, the others comprise proportionately less. As an animal grows older and larger, the proportion of bone and muscle tissue decreases whereas the percentage of fat of the carcass increases. The proportion of muscle and bone is high and the proportion

TABLE 9–1	Efficiency of Feed Utilization Related to Age in Broilers					
Weeks of Age	**Live Weight (lb)**		**Feed Consumption (lb)**		**Feed Conversion (lb feed/lb gain)**	
	End of Week	*Weekly Gain*	*Weekly*	*Cumulative*	*Weekly*	*Cumulative*
1	0.29	0.19	0.31	0.31		
2	0.61	0.32	0.48	0.79	1.50	1.30
3	1.07	0.46	0.78	1.57	1.69	1.47
4	1.64	0.57	1.05	2.62	1.84	1.60
5	2.31	0.67	1.28	3.90	1.91	1.69
6	3.13	0.82	1.67	5.57	2.04	1.78
7	3.96	0.83	1.84	7.41	2.22	1.87
8	4.81	0.85	2.03	9.44	2.38	1.96
9	5.65	0.84	2.21	11.65	2.63	3.06
10	6.47	0.82	2.28	13.93	2.78	2.15
11	7.22	0.75	2.14	16.07	2.85	2.22

Source: *North, M. O., Commercial Chicken Production Manual, 2nd ed., AVI Publishing Company, Westport, Conn., 1978.*

of fat is low in young animals. Older animals generally have a higher proportion of fat and a lower proportion of muscle and bone.

Fat thickness, determined by a probe or ultrasonic equipment, and loin eye area, measured by ultrasonics, are indicators of the proportions of muscle, fat, and bone in the live animal. The larger the loin eye and the less external fat, the more lean.

As an animal grows, fat deposition increases and muscle accretion (growth) slows. Also, daily nutrient requirements for maintenance increase. Therefore, the feed or nutrients needed to produce a pound of growth increases as the animal approaches mature size. Table 9–1 illustrates how the pounds of feed needed for a pound of gain increase as broilers increase in size and accumulate fat.

The rate at which proportions of bone, muscle, and fat tissues change vary with species, breed, animal type, ration, and sex. Larger cattle types are usually leaner than smaller cattle types at the same age or weight. Males are usually leaner than females of the same age.

Higher energy rations result in a faster rate of fattening than do lower energy rations. They increase the amount of external fat and the amount of marbling in a carcass. The longer an animal is fed a high-energy ration, the more external fat and marbling accumulates. Thus, an animal must be fed adequate energy long enough to reach the minimum acceptable level of marbling but then be slaughtered before the amount of external fat increases to the point that it lowers carcass cutability.

Protein is the most important nutrient in meat, and most of the protein in the carcass is present in muscle tissue. Therefore, the meat animal industry expends effort to maximize the protein content of carcasses. It's also the most expensive part of a carcass. Figure 9–5 illustrates that total protein in an animal's carcass increases to a maximum and remains about stationary, even though the animal continues to grow. At that point, additional growth is largely fat.

Figure 9–5
Graph of the amount of protein versus empty body weight of swine. (*Source:* Data from a series of studies summarized by Bailey and Zobrisky, National Academy of Sciences, 1968.)

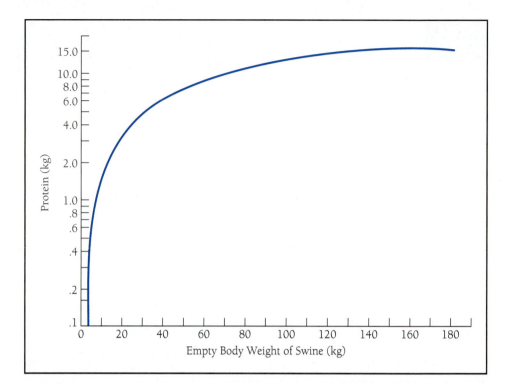

QUESTIONS FOR STUDY AND DISCUSSION

1. Define the terms *hyperplasia* and *hypertrophy*.
2. Distinguish between chronological and physiological growth.
3. Provide four examples of objective measurements of growth.
4. Calculate the average daily gain of 5-month feeder lambs that entered the feedlot on September 10 at 64 pounds live weight each and were shipped to market on November 29 at 112 pounds each. *Note:* September 10 is the 253rd day of the year; November 29 is the 333rd day of the year.
5. Explain how beef calf weights are adjusted to a standard day of age, such as 205-day weaning weight.
6. Draw the typical growth curve of an animal from birth to maturity.
7. Explain why the growth rate usually declines as the animal approaches maturity.
8. Define the term *homeostasis*.
9. What are the typical effects of the following hormones or hormonelike compounds on growth of meat animals: thyroxine, somatotropin, estrogens, androgens?
10. Name four materials that can be implanted or added to the feed to enhance growth rate.
11. Explain how gender may affect growth of meat animals.
12. What is "compensatory" growth?
13. Does feed efficiency usually increase or decrease as meat animals increase in maturity? Explain.

10

The Feeding Enterprise

Most cattle and lamb feeding is done in commercial and custom feedlots. In some areas the two terms may be synonymous, but generally the *custom feedlot* is defined as one where the feedlot operator feeds animals that belong to someone else, receiving payment on a per-day or per-unit or gain basis. In contrast, the *commercial feedlot* feeds its own animals.

This chapter describes the business of commercial and custom cattle feeding. Upon completion of this chapter, the reader should be able to

1. Differentiate between commercial and custom feeding.
2. Indicate where most cattle feeding businesses are located and explain why.
3. List five items that contribute to cost in a feeding enterprise.
4. List five factors that influence animal performance in feedlots.
5. Define and calculate margin, given data needed for the calculation.
6. Describe how feedlot operators reduce risk by price protections.
7. Explain how income tax considerations may influence the management of a feeding enterprise.

10.1 GEOGRAPHIC CONCENTRATION

The largest concentration of cattle and lamb feeding is on the Great Plains and western Corn Belt. The reasons are

- Desirable climate
    —drier feedlots resulting in fewer mud problems
    —lower humidity resulting in better animal comfort and appetites
- Proximity to grain supply—such as corn, corn byproducts, sorghum, and wheat
- Sparse population and more open space—fewer concerns about feedlot odor or dust by neighbors or cities and towns

As the density of animal feeding increased in these areas in the latter half of the twentieth century, allied and supporting industries and professionals such as meat processors, equipment suppliers, feedlot consultants, and veterinarians increased in numbers.

TABLE 10–1	Leading States in Cattle on Feed and Feedlot Capacities, 1998			

State	Annual Number on Feed (1,000)	Number of Commercial Feedlots by Capacity		
		Under 1,000	*1,000–15,999*	*16,000-Plus*
1. Texas	5,774	8	56	73
2. Kansas	5,210	45	92	58
3. Nebraska	4,430	270	363	32
4. Colorado	2,550	54	101	19
5. Oklahoma	907	3	23	9
6. Iowa	584	200	110	—
7. California	575	4	8	12
8. Idaho	554	19	41	—
9. Arizona	398	—	11	6
10. South Dakota	307	50	57	—
Total, 10 states	21,289	653	862	209
Other states	1,500	189	141	10
Total United States	22,789	842	1,003	219

Source: *Adapted from Cattle Final Estimates, 1994–98, NASS, USDA, January 1999.*

Although this chapter focuses on both cattle and lamb feeding, most discussion and illustrations pertain to cattle. The relative volume of lamb feeding is small and the concepts presented for cattle feeding apply as well to the lamb feeding business.

In 1998, almost 23 million head of cattle, about 93 percent of the cattle fed nationally, were fed in the 10 leading states (Table 10–1). Note the leading states of Texas, Kansas, Nebraska, and Colorado. They accounted for about 79 percent of the cattle fed and marketed from feedlots in 1998 (Figure 10–1).

Also note in Table 10–1 and in Fig. 10–2 that the large feedlots of 16,000-head-plus capacity have largely replaced the farmer-feeder once common in the Corn Belt. Well over *80* percent of the feeder cattle are fed and marketed from feedlots of 1,000-head capacity or greater. Lots under 1,000-head capacity are considered farm feedlots, largely located in the three States of Iowa, Minnesota, and Illinois.

Most meat-processing plants are in or near the major feeding areas. Meat cuts, contained in 60- to 80-pound boxes, can be moved to population centers in refrigerated trucks more cheaply than live animals can be moved to population centers for slaughter (see Section 25.6).

10.2 FEEDLOT MANAGEMENT

In feeding enterprises, profit is the goal. Major factors that affect profit are price margin (the difference between the purchase price per hundred pounds of feeders and the selling price of animals when finished), feed-to-gain ratio, death loss, and the costs of health-related treatments. Many individuals not personally employed in the industry invest in cattle feeding enterprises and expect a return on their invested capital. Most feedlot ownership is in the form of a corporation (there are a few cooperatives) with

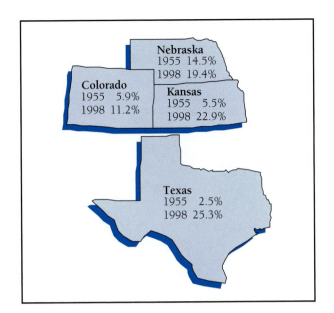

Figure 10–1
Percentages of all fed cattle marketed in the United States. (*Source:* Adapted from 1998 USDA data.)

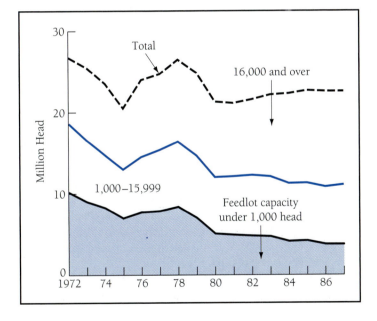

Figure 10–2
Since the 1970s, the number of cattle fed in large-capacity feedlots has increased whereas the number fed in smaller lots has decreased. (*Source:* USDA.)

experienced managers and many workers employed. Less than half of the cattle in most feedlots are owned by the feedlot owner; most are fed for other owners—farmers and ranchers with cattle but without grain and other investors or speculators. Cattle may be bought, managed, and sold on behalf of the investor by the feedlot managers. Grains and other feeds are usually purchased (Fig. 10–3).

Figure 10–3
Midwest and central feedlots usually consist of a few hundred to 1,000 head, left, whereas commercial feedlots have capacities for 1,000 to 30,000-plus head, right.
(Courtesy of American Hereford Association and Gray County Feedyards.)

Managers of commercial feedlots usually feed yearling cattle that require a feeding period of 4 to 5 months, whereas farm feedlot managers tend to feed younger and lighter cattle requiring longer feeding periods.

Feeding cattle is big business. Risk in purchased cattle is high and management must be effective. Decisions may determine a profit or loss difference of thousands of dollars.

Consumers demand an appealing product at competitive prices. Research developments prompt changes in ration formulation, disease prevention, and labor management. Cattle and feed prices fluctuate daily. Health problems arise overnight; weather may diminish feed intake. Feed-processing equipment may malfunction, necessitating a change in ration; this may affect the animal's appetite or performance.

Some commercial feedlot operations now seek to obtain most of their cattle from select suppliers (or alliances of producers). More uniform size and quality of cattle are obtained through this system. Also, with a guarantee of preconditioned cattle (prescribed vaccinations and other preparations for the feedlot) there are fewer medical problems and improved performance in the feedlot. (See Chapter 29.)

The feedlot manager may draw on many services and people, such as nutrition consultants to formulate rations, veterinarians to organize and manage health programs, market consultants and reporters, and accountants to help with financial records, cash flow, limited partnerships, taxes, and other issues. Extension services may provide some of these services. Most managers use satellite or e-mail-fed monitors for immediate market news and computer software for ration formulation and financial management.

Under federal Confined Animal Feeding Operation legislation and regulations, the Environmental Protection Agency (EPA) and a state agency with related responsibility may monitor manure handling facilities and operations, nutrient discharge or runoff, odor, or other environmental factors.

The Food and Drug Administration (FDA) regulates and may monitor use of feed additives or animal treatments. Other federal agencies are charged by legislation with regulating and monitoring other elements of a feeding business. The Occupational Safety and Health Administration (OSHA), for example, specifies certain working conditions, and the Internal Revenue Service (IRS) requires periodic deposits and summary reports for workers' social security.

10.3   CAPITAL AND COSTS

Large amounts of capital are necessary in feeding enterprises, perhaps $150 per head for open dry lots and $300 or more for partially or totally enclosed structures, plus feed inventory and several months' operating expense for labor, utilities, and fuel. Institutions with loan capital—banks, the Farm Credit System, and others—require detailed analysis of projected costs, returns, and cash flow before loans are approved.

Generally, costs are divided into two constituents: (1) the cost of the feeder animals and (2) the cost of gain. The *cost of the feeder animal* includes all expenses incurred in the purchase and transport of the animal to the feedlot. In addition to the initial purchase cost, there may be commissions paid to order buyers, trucking costs, shrinkage,[1] and death losses. Differences in grade, sex, size, time of year, health, and distance from the feedlot all influence cost. Seasonality of feeder cattle numbers and prices will be discussed later in the chapter.

Feeder heifers are cheaper than steers of the same size and quality. A higher death loss is expected with heifers; a small percentage may be pregnant and will abort or be induced to abort. Processors usually pay less per pound for heifers at the end of the feeding because of the lower cutout value of the carcasses.

Lighter feeder cattle usually cost more per pound, but heavier cattle require more investment per animal. Lightweight does not necessarily mean small-framed. Large-framed, lightweight cattle have a potential for compensatory gain, if healthy, and may well be a good buy. (See Section 8.5.)

Death loss during transport or in the feedlot can be significant. Lighter weight animals generally have a higher death loss than heavier animals.

Cattle purchased and moved directly from a ranch or through one market will usually have a lower death loss than cattle gathered from several markets over a period of weeks in groups of 20 to 30 head. The latter usually have been exposed to a variety of diseases and have undergone more stress.

Transporting feeder animals long distances is expensive. Shrink and death losses increase with length of time animals are in transit. Table 10–2 illustrates that weight losses increase with time in transit or distance traveled. The greatest losses occur in the first 3 to 4 hours or first 100 miles.

Loading and hauling are critical and are the responsibility of the trucker. Trucks should not be overcrowded and should move quickly to their destination. Trucks should be equipped with tall exhaust stacks to prevent the inhalation of fumes by the animals.

Cost of gain includes any cost incurred in producing a market animal from the lighter weight feeder animal. Expenses during this period generally include (1) feed costs, including markup or margin; (2) direct interest charges; (3) death losses; (4) yardage

---

[1] Shrinkage in this discussion refers to animal weight loss during transportation from point of origin to the feedlot.

TABLE 10–2	Typical Percentage Weight Loss in Transported Feeder Cattle	

Travel Time or Distance	Shrinkage (%)
Hours in transit:	
3–4	1.00/hr
4–14	0.150/hr
14+	0.085/hr
Miles traveled:	
0–100	3.0
100+	1.0/100 miles

Source: *Riley, J. G., Kansas State University, 1976.*

fees, which cover supervision and routine veterinary expense; and (5) medicines and other specific charges.

Feed is the major cost. Each animal will consume from 5 to 6 pounds of feed per pound of gain, or as much as 1.5 tons, during the finishing period, a good reason to locate feeding enterprises in grain-producing areas.

Feedlots usually charge $4 to $6 per ton to process feed (grind, mix, and perhaps steam or flake). Animals generally gain more rapidly and more efficiently midway to late in the feeding period on highly processed grain rations. Feed additives also contribute to ration cost.

In custom feedlots, markup on feed is usually the largest source of profit to the lot. The cattle owner has the interest cost of capital for (1) purchasing feed and (2) purchasing the feeder animal. In 2004 excluding interest costs on feed, the total interest charges for one yearling steer were as follows:

Purchase of 680-lb steer at $78/cwt × Interest at 8.3% for 160 days
$530.40 × .085 × 160/365 = $19.76

The owner may have adequate capital and not obtain a loan on cattle or feed but must take into account the interest the capital would have earned if invested elsewhere.

An example of animal performance is shown in Table 10–3, based on steers and heifers fed in large commercial feedlots. Note the average weights of animals as they enter and leave the feedlot, the lighter weights of heifers, and average daily gains. The average feed conversions shown for June 1998 and 1999 approximate 6.65:1 (90 percent air-dry feed per pound of weight gain). Death losses were 1.31 percent and cost per hundred pounds of gain approximated $54.58.

Death losses can play a big role in determining cost of gain. Any death loss incurred during the feeding period reduces the final selling weight of the pen. Many animals will have consumed considerable feed before they are lost. Fewer cattle marketed means fewer dollars to pay the total cost of feeding and the initial purchase price of the feeder animals. Cattle feeders assume that 0.5 to 1.5 percent death loss will occur during the feeding period.

Yardage fees charged the investor or cattle owner by the feedlot include veterinary expense, labor, interest on the feedlot investment, depreciation, operating and

TABLE 10–3	**Feedlot Performance**[a]					
	Steers			**Heifers**		
	June '99	May '99	June '98	June '99	May '99	June '98
Weight in, lb	662	685	701	608	621	638
Weight out, lb	1,191	1,199	1,198	1,090	1,082	1,103
Days fed	172	169	172	174	165	178
Daily gain, lb	3.08	3.04	2.89	2.77	2.79	2.61
Dry conversion, lb	6.30	6.57	6.73	6.53	6.74	7.05
Death loss, %	1.17	1.49	1.22	1.19	1.25	1.67
Cost of gain, $/cwt.	47.26	48.92	58.65	50.35	51.74	62.05

[a]*Data are based on 25 feedlots located in a four-state area, serviced by several nutritional and veterinarian consulting groups.*

Source: Feedstuffs, *June Feedlot Analysis, 1999, provided by Dr. Marcus Hoelscher, special consultant, Hereford, Texas.*

maintenance costs, and some margin for profit. Yardage fees are commonly 12 to 15 cents per head per day. Because yardage charges are on a per-head per-day basis, feedlots usually try to operate near capacity. Full-time employees are required year-round. Handling facilities and pens depreciate, and interest must be paid whether or not facilities are in full use.

10.4   INFLUENCES ON ANIMAL PERFORMANCE AND FEEDLOT PROFIT

Many factors affect feeding profitability, directly or indirectly. The genetic growth-rate potential of the feeder animal is a factor. The faster animals gain, the cheaper the cost of gain. Animals reach market weight in fewer days, consume less total feed, require less yardage expense, and incur less interest per animal. Genetic potential of feeder calves is difficult to estimate, but experienced feedlot operators contract for feeder animals year after year from producers whose cattle previously have performed well.

As discussed in Section 9.5, the previous treatment of the feeder animal often affects feedlot performance. Animals grown on low-energy, adequate-protein diets have a high probability of compensatory growth, usually rapid and at lowest cost. On the other hand, creep-fed calves usually will not perform as well in the feedlot. Feeder calves that have been "roughed" or moved frequently may appear to have potential for compensatory gain but may not perform well due to respiratory or other health problems that impair or retard growth.

Heifers generally gain more slowly and less efficiently than steers, for two reasons. Heifers mature at lower weights so have lower growth-rate potential than steers. Also, they are fatter than steers at the same weight, and more feed is required to produce a pound of fat than is required to produce a pound of lean. In addition, weight loss during **estrus** is frequent, unless an estrus inhibitor is added to the ration.

Environmental stress during the feeding period results in less-efficient gains. More of the feed consumed is used for maintenance and less for growth and tissue deposition. During cold weather, animals eat more but maintenance requirements increase,

> **Estrus**
> The time when a female is "in heat" and will breed readily. Also called *estrus period.*

so less feed is used for gain. Hot summer weather also increases the maintenance requirement but decreases feed consumption and gain. Managers increase the roughage content of feedlot rations during the winter and decrease it in the summer. More body heat per pound of ration is produced by roughage digestion and at a lower cost. Environmental effects on animal performance and stress will be discussed more thoroughly in Chapter 11.

Low feed consumption is usually a problem in young or lightweight cattle that have not previously been confined or fed grain rations. Starter rations usually include forages, such as grass hay, with which the cattle may be familiar. Two to 3 percent molasses or feed intake stimulant may be added to starter rations. Once animals are adjusted to the feedlot environment and to starter rations, energy content of the ration is increased quickly. Animals fed high-energy rations consume more energy, increasing tissue deposition and resulting in more rapid and more efficient gains.

Changes in weather affect feed consumption. Cattle significantly increase feed consumption before a storm but will not eat or will decrease consumption during and following a storm. Top "bunk management"—avoiding feed accumulation in bunks—is necessary to reduce off-feed conditions in cattle and to reduce feed spoilage.

Feed processing—including grinding, flaking, and steaming—usually increases ration digestibility and feed efficiency 2 to 3 percent especially in Great Plains Feedlots where grain sorghum is the major grain. However, energy and labor costs must be considered.

With high commercial fertilizer costs, manure removed from feedlots can be a valuable commodity. Manure can be applied to suitable cropland by spreading or through irrigation systems and provides nitrogen, phosphorus, and potassium. Although as much as 50 percent of the nitrogen in manure can volatilize before field application, little if any of the potassium and phosphorus is lost. A credit of 7 to 8 cents per 100 pounds of beef produced is sometimes allowed for the value of the feedlot manure.

10.5   PROFIT MARGIN

In livestock feeding as in other businesses, margin is the difference between cost and sale price. It is not synonymous with profit, as it can be either positive or negative. Two types of margin relate to the feeding enterprise. *Price margin* is the difference between the purchase price of the feeder and the sale price of the fed animal per hundredweight (cwt). *Feeding margin* is the difference between the cost of gain per hundredweight and the sale price of the fed animal per hundredweight. The following examples are illustrated in Fig. 10–4.

If a 700-pound steer is purchased for $70 per cwt and the price of fed market steers weighing 1,180 pounds at the end of the feeding period is $80 per cwt, the cattle feeder has a positive price margin of $10 per cwt (Example 1). This $10 difference, applied to the 700 pounds purchased, would contribute $70 to profit. If the selling price of fed market steers is $65 per cwt, there would be a negative price margin of $5 per cwt (Example 2). This would contribute a $35 loss.

The cattle feeder would have a $10 per cwt positive feeding margin if the cost of gain on a pen of cattle averages $50 per cwt and the price of fed market steers at the end of the feeding period is $60 per cwt. The cattle feeder would sell each 100 pounds of gain for $10 more than it cost. The feeding margin of $10 per cwt applied to 480 pounds of gain would contribute $48 to potential profit (Example 3). As shown in Example 4,

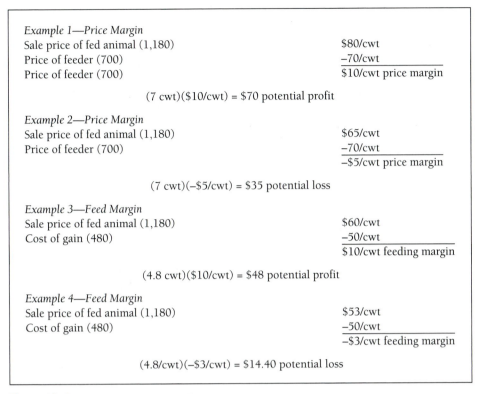

Example 1—Price Margin
Sale price of fed animal (1,180) $80/cwt
Price of feeder (700)   –70/cwt
Price of feeder (700) $10/cwt price margin

(7 cwt)($10/cwt) = $70 potential profit

Example 2—Price Margin
Sale price of fed animal (1,180) $65/cwt
Price of feeder (700)   –70/cwt
                                                                                      –$5/cwt price margin

(7 cwt)(–$5/cwt) = $35 potential loss

Example 3—Feed Margin
Sale price of fed animal (1,180) $60/cwt
Cost of gain (480)   –50/cwt
                                                                                      $10/cwt feeding margin

(4.8 cwt)($10/cwt) = $48 potential profit

Example 4—Feed Margin
Sale price of fed animal (1,180) $53/cwt
Cost of gain (480)   –50/cwt
                                                                                      –$3/cwt feeding margin

(4.8/cwt)(–$3/cwt) = $14.40 potential loss

Figure 10–4
Examples of price margin and feeding margin.

if the cost of gain is $50 per cwt and the price of fed market steers is $53 per cwt, the $3 negative feeding margin would contribute a $14.40 loss.

It is possible for a positive price margin to more than offset a negative feeding margin, resulting in a net profit to the enterprise. Or, a negative price margin could more than offset a positive feeding margin, resulting in a net loss. Both price and feeding margins are watched carefully by feedlot managers.

10.6   PRICE PROTECTION

Some feeders, especially farm owners, purchase feeder calves for backgrounding or grazing in the fall and winter them on pasture or available crop residues to avoid paying high spring prices. Cattle on summer pastures may be forward-contracted during the summer for fall delivery. Some cattle producers use an "integrated approach," placing their own cattle in a custom feedlot in the fall and fattening for slaughter at spring prices. Low fall prices are thus avoided and the owner may also profit from positive feeding margin.

Most feedlot operators are so specialized and have such large investments that they are very vulnerable to rapid cattle and feed price fluctuations. A sharp decline in cattle prices or a sharp increase in feed prices could cause financial stress or even bankruptcy. The futures market and contracts between the feeder and the processor, feeder cattle supplier, or feed supplier can help ensure financial stability and reduce risk.

In the case of custom feeding, the arrangements between the cattle owner and feedlot are usually in a written contract. Contracts may prescribe the weight gain to be achieved or the days cattle are to be fed, the cost of that gain, and an acceptable death loss. Some contracts prescribe only the cost of feed and daily yardage cost. The cattle owner guarantees payment for feeding the cattle.

Forward contracting of cattle being fed is a price protection option. A contract between the cattle owner and the processor specifies the number, quality, price, and delivery date of the fed cattle. The owner is afforded price protection and the processor is guaranteed both price and cattle numbers on the delivery day. This is a reasonable form of protection for the smaller cattle feeder who chooses not to enter the futures market. Forward contracts deal with real cattle and real prices.

A feedlot operator may also forward-contract for grain, forage, and protein supplement purchases, establishing the price for these feeds to be delivered to the feedlot on a certain date.

The **futures market,** which also provides a form of price protection, deals in "contracts," rather than real cattle or real grain, and reduces risk by letting speculators and others share the risk. The price of contracts reflects what traders—speculators and others—feel the cattle or grain price will be at some future date. A "commodity futures market" in Kansas City or Chicago (Fig. 10–5) acts as liaison between buyers and sellers. Each has separate agreements with the commodity market, often through a broker.

Future contracts state the quantity (per contract), quality, delivery time and place, and all other factors that might influence commodity value. A buyer buys a contract of "live cattle" or a seller sells a contract of "live cattle." No more contracts can exist than there are sellers, and every contract must have one buyer and one seller. The buyer has an agreement (contract) with the commodity market to accept delivery of the commodity at the price and date specified by the contract. The seller has an agreement with the commodity market to deliver the commodity for the price and on the date specified by the contract.

> *Futures market*
> A market at which contracts for future delivery of a commodity are bought and sold.

Figure 10–5
Trading floor of the Chicago Mercantile Exchange, where futures contracts in live cattle, carcasses, and other commodities are traded. Note banks of phones, foreground, and display of trades and prices in background.
(Courtesy of Chicago Mercantile Exchange.)

Because the futures market deals in the sale of contracts and not real cattle, a contract buyer may *sell* the contract before the delivery date. Once that is done, he or she is out of the market. The net price change during the time the contract was held determines profit or loss.

With the option of getting into the market or out of the market at any time, speculators may also enter the cattle and grain futures markets and own neither commodity in its physical state. Rather, they buy and sell contracts and receive profit or loss on price changes. Such *speculators* are valuable because they provide some of the market volume; they are always ready to buy or sell—at a price.

Because futures markets are separate from the current live cattle and grain markets, but closely follow them, cattle owners and feedlot operators use them for price protection. Speculators assume much of the price fluctuation risk.

The practice of buying or selling futures contracts in commodity markets for price protection is known as **hedging**. The person hedging cattle or grain is called the hedger. Cattle owners planning on selling fed, market-weight cattle in the fall may hedge in June by selling "October live cattle" futures contracts to counter any decrease in slaughter cattle price over the summer and fall months.

For example, the economics of the cattle market in June may suggest a projected profit in October of $75 per head. The feeder may prefer to accept that amount, to "lock it in" and achieve protection against possible price drops by giving up extra profit that might result from possible further price increases. Assuming that 400 head are in the feedlot and that they will be sold at about 1,200 pounds for slaughter in October, the feeder sells 12 October live cattle contracts (40,000 pounds per contract) on the futures market in June. Any change in the price paid by the processors is usually accompanied by simultaneous changes in the market value of the 12 contracts. When the cattle are sold to the processor in October at market price, the feeder buys back 12 futures contracts. If the price paid by the slaughter plant decreased between June and October, the price of the contracts also likely decreased, so the feeder buys them back at a lower price than when they were sold. Money lost on live cattle, relative to the projected profit in June, would be offset by profit on the contract transaction. Also, any increase in the market price of live cattle would likely be offset by a loss on the contract transaction. Therefore, little money was likely made or lost after the contracts were hedged in June. The selling price of the market-weight cattle was "protected" and the projected profit of $75 per head was almost guaranteed for the cattle feeder in June although the cattle were not sold until October. The costs for this price protection include brokerage fees, interest on margin money (a deposit for each contract to serve as collateral), and the risk that the price of live cattle and futures contracts may not parallel perfectly.

Feedlot operators may also reduce their risks in feed price fluctuations by trading grain futures. Grain prices and the price of grain futures contracts usually are lowest at harvest time because of large grain supplies. Managers purchase and store as much feed as possible at this time and buy futures contracts equal to the amount of additional grain needed later in the year. Any increase in the price of grain will be offset by the increase in the value of the futures contract.

Another form of price protection is the options market, where one may buy or sell an *option* to buy or sell a specific futures contract at a specific price during a future time period. Detailed discussion is not warranted here.

Frequently, feeding contracts and forward marketing contracts are used in combination with hedging to give maximum risk protection and maximize profit.

> **Hedge**
> A form of price protection, usually involving a commodities futures contract (e.g., selling beef carcass futures at the time feeder cattle enter a feedlot).

10.7   TAX CONSIDERATIONS

Cattle feeding provides some opportunities for income tax planning. Tax laws and interpretations change over time. Usually feed purchased for cattle is deductible as an expense in the year in which it is paid for, not the year in which it is fed. This assumes there is a bona fide business reason for the date of purchase, such as price or guarantee of feed supply. Feeders can purchase large quantities of grain near the end of high-income years to reduce taxable income for that year.

Tax liability is never avoided; cattle are eventually sold and taxable income received, so tax liability is only delayed. For those taxpayers who report their income on a cash rather than accrual basis, cost of feeders is charged in the year the finished animals are sold, not necessarily in the year they are purchased.

In those cases where animals are ready for slaughter near the end of December, taxable income for the year may influence the date chosen for sale. Should taxable income that year be unduly high, the sale date might be delayed to January and some tax liability delayed for a year.

Tax issues influence other management decisions. Investment credit, a net reduction in taxes payable, may be granted for certain investments. Facilities and equipment have a specified life span, and extra first-year depreciation or accelerated depreciation schedules may be used. These factors may influence when facilities investments are made or equipment purchased.

QUESTIONS FOR STUDY AND DISCUSSION

1. Distinguish between commercial and custom feedlots.
2. Name three of the top four states in commercial feedlot cattle production.
3. About what percentage of cattle are fed in commercial and farm feedlots?
4. Give several reasons for the location of the larger cattle feedlots in the United States.
5. Which usually cost more per pound to purchase as feeder cattle: heifers or steers? light or heavy feeders?
6. List four or five factors in addition to feed costs that affect the cost of production of feedlot cattle.
7. Explain how environmental stress or changes can affect feedlot animal performance.
8. What is meant by (a) price margin and (b) feeding margin? Give an example of each.
9. What strategies might be used to reduce the risk of cattle or feed price changes?

Animal Environment and Adaptation 11

Performance and production of domesticated animals are influenced by both genetics and environmental factors. The environment of an animal affects the degree to which its genetic potential is expressed (Fig. 11–1). In a broad sense, environment includes all surrounding factors that affect the animal such as management, nutrition, and disease. It also includes climatic factors such as temperature, humidity, and ventilation, which must be properly managed, or modified if practical, for efficient animal performance.

This chapter focuses on the more specific factors of the animal's environment such as the (1) surrounding (ambient) temperature, humidity, air velocity, radiation, and light, and their effects upon the animal. Also discussed are (2) thermoregulation (heat conservation, production, or dissipation) by the animal in maintaining its rather constant internal temperature and (3) environmental and management modifications that enable the animal to be relatively comfortable and productive.

Considerable research has been conducted to determine the optimum climate and environment for maximum production among animals and to ascertain the degree of impairment that results from climate and other environmental factors that are not optimum.

For many years, producers and scientists have worked to develop management systems that maximize average daily gain, feed utilization, milk production, eggs per hen, and other production measures; maximum animal performance continues to be very important to a successful enterprise. Today, however, with increased energy, labor,

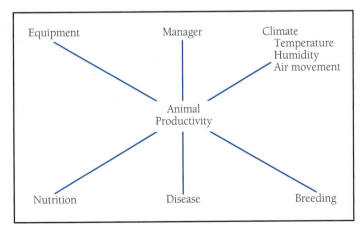

Figure 11–1
Animal productivity is influenced by genetics and by many environmental factors.

and other costs, interest has shifted somewhat from *maximum* to *more economical* levels of production. Cost-benefit calculations disclose the relative profitability of providing shade, insulation, supplemental heat, cooling, fans, sprinklers, and related environmental control devices.

Upon completion of this chapter, the reader should be able to

1. Define key terms related to animal environments, such as radiation, conduction, convection, critical temperature, wind chill index, poikilotherms, and homeotherms.
2. Explain how animals physiologically adapt to high or low temperatures.
3. Contrast summer and winter weather in terms of typical effects on livestock and poultry.
4. Differentiate among shade, evaporative cooling, refrigerated cooling, and sprinklers in terms of how they achieve cooling and the species and circumstances for which each might be more or less desirable.
5. Explain the influence of extreme heat or cold on feed and water intake of animals, and ration modifications that may be made to accommodate the heat or cold.
6. Explain and provide illustrations regarding how space per animal and nature of floor surface may influence profitability of poultry or swine production.
7. List reasons why ventilation is important and helpful in confinement facilities for livestock and poultry.

11.1   HEAT LOSS AND HEAT PRODUCTION

> **Poikilotherms**
> Animals with little or no ability to maintain an even body temperature, so temperature is influenced largely by the environment. Referred to as cold-blooded animals.

Animals may be grouped into two general classifications: cold-blooded (poikilotherms) and warm-blooded (homeotherms). **Poikilotherms** do not maintain constant body temperature; internal body temperature changes in response to the environment. This group includes fish, aquatic animals, and to a minor extent, poultry during the first 10 days of life.

Homeotherms, which include all farm mammals and birds, maintain a relatively constant body temperature. If the animal's body core temperature changes more than a few degrees from normal, death will occur. Among farm animals, normal body temperatures range from 101°F in horses to about 106°F in poultry (see Table 12–6). Some daily (diurnal) body temperature variation occurs; it varies several degrees, usually lowest in early morning and highest in late afternoon.

> **Homeotherms**
> Animals that utilize or dissipate energy to maintain body temperature, usually in the range of 95° to 105°F. Usually called warm-blooded animals.

To maintain a constant body temperature, heat produced and received must equal heat lost. Animal heat may come from several sources, such as routine body processes of digestion, muscle activity, rumen fermentation, exercise, and cellular respiration, as well as from solar radiation and other external sources. To maintain homeothermy the animal must dissipate excess heat produced by internal processes or received from external sources.

> **Radiation**
> The exchange of heat between two objects that are not touching; emission of heat rays, from the warmer to cooler objects.

The animal has four basic avenues of heat loss or gain: (1) radiation, (2) conduction, (3) convection, and (4) evaporation (Fig. 11–2).

Radiation is the exchange of heat between two objects that are not touching. Heat flows from warm to cold objects. For example an animal (or person) standing in the sun on a winter day receives solar heat from radiation. Radiant energy moves via electromagnetic waves through space. It is transformed to thermal energy upon contact

Figure 11–2
Routes by which heat may be gained or lost by an animal.

with the animal. Thus, an animal can feel warmth on a bright winter day, especially when in an area protected from the wind. An animal can also lose heat by radiation to the walls and ceiling when their temperature becomes lower than the animal's. However, the reverse occurs when the walls and ceiling exceed the body temperature of the animal, and the animal must dissipate heat to maintain homeostasis.

      Conduction is the flow of heat from warm to cold objects that are touching. The warm molecules, through their motion and contact with the colder molecules, transfer kinetic energy to the colder object. The amount of conductive heat transferred depends on the surface temperatures of the objects touching and their thermal conductivity. A pig lying on cold concrete slats will lose heat by conduction to the slats; an animal lying on a floor heated with hot water will gain heat by conduction.

> **Conduction**
> Transfer of heat through a medium; warm molecules impart kinetic energy upon contact with cooler molecules.

Convection
Flow of heat through air or water.

Convection is the flow of heat from a warm area to a cooler area through air or water movement. Warm air blowing over an animal may provide convective heat gain to the animal. An animal standing in a 20 mph wind at 0°C is exposed to "effective temperature" of −12°C (the "wind chill index") and experiences convective heat loss.

Evaporative cooling
A means of heat loss from the body (sweating, panting); warmer liquid molecules escape from the body via evaporation, resulting in a cooler body mass.

Evaporative cooling causes heat loss from the animal. As environmental temperature approaches body temperature of the animal, heat losses by radiation, conduction, and convection become less and the animal must dissipate heat via moisture evaporation. The amount of heat loss depends on the amount of evaporation, which is influenced by the temperature of the skin, the speed of air movement, and the temperature and relative humidity of that air. Heat is also lost by evaporation of moisture from the lungs and respiratory tract. Ability to perspire and lose heat by evaporation varies among species. (See Section 11.3.) Most of the common farm animals are limited in their ability to perspire. In humid areas or days of high humidity, evaporation contributes less to heat loss. A summary of how animals adapt to ambient temperature is provided in Box 11–1.

BOX 11–1   Farm Animals Adapt to Their Environment

Can you imagine the horse in Fig. 11–2 coping with high humidity and ambient temperatures exceeding 100°F during the hottest days of summer and enduring the coldest winter days well below 0°F? Of all the farm animals, the horse is probably most capable of adapting to weather and temperature extremes. However, remember that all animals managed in animal agriculture are homeotherms, except for the cold-blooded species in aquaculture. The following summarizes how warm-blooded animals adapt to increasing and decreasing temperatures:

- *Decreasing temperatures:*
  1. Reduce heat losses by
     (a) vasoconstriction of peripheral blood vessels
     (b) increasing body insulation (increasing adipose fat, increasing haircoat thickness [longer and greater density], and piloerection of hair)
     (c) seeking shelter from wind, snow, and rain (e.g., leeward side of hills)
     (d) reducing surface area (humping up and/or grouping up)
  2. Increase heat production by
     (a) increasing food intake (more energy intake and warmth from heat increment of digestion)
     (b) increasing physical activity (e.g., signs of friskiness) and involuntary shivering in extreme cold stress
     (c) seeking exposure to solar radiation

- *Increasing temperatures:*
  1. Increase heat losses by
     (a) vasodilation of peripheral blood vessels
     (b) decreasing body insulation (shedding of hair coat)
     (c) increasing body surface area (e.g., resting in stretched-out position)
     (d) increased evaporative cooling by perspiring and/or panting
        As the environmental temperature approaches body temperature, perspiring and panting become the main methods of heat dissipation. Radiation, conduction, and convection can become ineffective; in fact, these methods of heat exchange may actually contribute to an increased heat load to the body.
     (e) avoiding exposure to solar radiation (e.g., seeking shade)
  2. Reduce heat production by
     (a) reducing feed intake (lowered thyroxine and lowered metabolic rate)
     (b) reducing activity

It is the responsibility of animal managers to keep their animals as comfortable and productive as practical. They must provide special attention and care in extremely hot and cold weather.

Source: Compiled by the authors.

11.2   THE COMFORT ZONE AND CRITICAL TEMPERATURE

For animals there is a range of temperatures at which performance is maximized. This range is termed the **comfort zone** or *thermoneutral zone*. The lower end of the zone is called the lower **critical temperature** and is that air temperature below which heat-producing mechanisms in the body must be increased to keep the animal at a constant body temperature. The critical temperature varies among species and depends on the age of the animal, degree of fatness, hair coat, weight, and level of feeding. Any factor that alters the heat production or heat loss abilities of the animal will also affect critical temperature.

At temperatures below the critical temperature, caloric requirements for body maintenance are higher due to increased heat loss (Fig. 11–3). The animal responds to the lower temperature by increasing feed intake, by shivering, or by nonshivering thermogenesis (heat production). If the animal continues to be cold or below the critical temperature zone, blood will begin to shunt (shift) away from the limbs and toward the intern organs. This is what leads to "frostbite" commonly seen in some animals not protected from severe weather.

The upper limit of the comfort zone, or upper critical temperature, is more difficult to define and varies greatly among species, and to some extent within species. Those more tolerant to heat stress generally have greater ability to perspire and more skin surface area per unit weight. Zebu cattle (*Bos indicus*), such as the Brahman, are more tolerant of high temperature than are European breeds (*Bos taurus*). Heat stress

> **Comfort zone**
> The temperature range in which an animal will be comfortable and produce at a maximum level.

> **Critical temperature**
> The temperature at which an animal must increase or decrease the oxidation of energy sources in order to maintain body temperature.

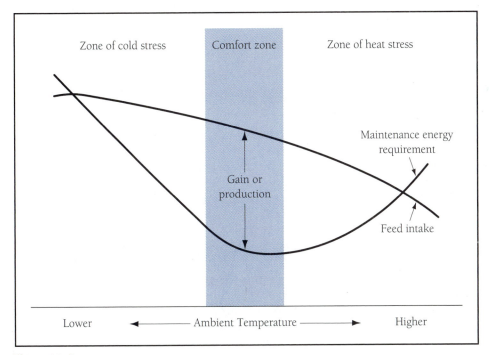

Figure 11–3
Feed intake generally decreases as temperature increases because the body is warm. At either temperature extreme, energy required for maintenance increases, so energy available for animal gain or production decreases.
(*Source:* Iowa State University)

may result in panting (more rapid or laborious breathing), which increases caloric requirements for maintenance. Generally, animals that are exposed to heat stress eat less, thereby reducing energy available for gain, as shown in Fig. 11–3.

11.3   PHYSIOLOGICAL ACCLIMATIZATION TO TEMPERATURE STRESS

Animals will partially adjust or acclimatize[1] to temperature when exposed over a period of time. You may have noticed how much colder a 30-degree temperature seems in April than it does in January. **Acclimatization** or adjustment to the environment involves both hormonal and physiological thermoregulatory changes.

> **Acclimatization**
> Adaptation by an animal to several environmental factors over several days or weeks.

During temperature stress, hormones of the adrenal and thyroid glands are secreted to help the body adjust. During cold temperatures, thyroxine secretion is increased, causing increased metabolic rate and heat production by the animal. Corticoid secretion from the adrenal gland also will increase as a result of cold stress, further contributing to increased metabolic rate.

During heat stress, the thyroid gland secretes less thyroxine and the adrenal gland secretes lower levels of corticoids, permitting lower heat production.

Adaptation of a species over successive generations can result from selection and genetic change in the population, permitting the species or a group within a species to cope with its environment. Among bovines, for example, Brahman cattle, which developed in India, have large sheaths and briskets and a loose, light-colored skin, and so are more adapted to high temperatures than are most breeds developed in Europe, such as Angus, Hereford, and Charolais. The latter tend to have a thicker, denser hair coat and less skin surface per unit weight, and so are better able to withstand cold stress.

The remaining sections of this chapter discuss environmental management practices that can be made to keep animals more comfortable and productive.

11.4   CONTROLLING HEAT STRESS

In the United States, the summer months are more limiting on livestock and poultry performance than are the winter months. Over 50 percent of the cattle in this country are raised in areas where the average summer temperature is above 75°F. A majority of the hogs are yet grown in the upper Midwest, but swine production has increased in the southeastern region of the United States. There has been some increase, too, in the number of dairy cattle, sheep, and beef in the Southeast. A majority of the poultry is produced in this region.

Productivity per animal is markedly lower in most of the hotter regions of the earth. Though a major reason is probably the direct effect of "hot climates" on the animals, other factors are parasites, tick-borne diseases, low-protein and high-fiber content of deceptively lush-looking forage, management emphasis on cereal food crops in regions where total food supply is limited, and cultural traditions that place more prestige on herd or flock size than on volume of product produced.

---

[1] *Acclimatization* refers to adjustments an animal makes to environmental changes or factors over a period of weeks or months. *Acclimation* is a term sometimes used to imply a more immediate response to a single factor, or stressor.

Most farm animals, with the exception of the horse, cannot dissipate excess body heat as easily and rapidly as humans. Pigs possess sweat glands that respond poorly to heat stress, partially due to the fat layer beneath their skin. Sheep have sweat glands that provide some evaporative cooling on the skin surface, but their wool limits rapid heat removal from the body, although it does provide excellent insulation from the cold. Chickens have no sweat glands and with their feather covering are particularly susceptible to heat stress. If the heat doesn't kill the chicken, then the production (i.e. eggs) will be reduced. More specifically, when egg laying hens eat less, egg production goes down and that's why egg production is better during fall and spring, when the weather is mild.

A typical cow has sweat glands, but they are not well distributed on the skin surface, and though she "perspires" some, she has a maximum evaporative cooling rate equal to about one-eighth that of a human per unit of surface area. A lactating cow has some difficulty dissipating the heat she produces when temperatures exceed 75°F, especially when humidity is 50 percent or higher.

Heat stress has a significant negative effect on reproduction efficiency of both the male and female. These effects are discussed and illustrated in Section 15.4.

Animals apparently have some ability to adapt to hotter weather. Dairy cattle, for example, may reduce hair thickness by shedding, may increase reflectivity by slight changes in coat color and increases in skin secretions, and may reduce thyroid secretion to slow body metabolism and lower heat production. During severe heat stress, animals also may increase their respiration rate and replace warm air from their lungs with drier inhaled air. This "panting" form of evaporative cooling is especially helpful to birds because of their inability to perspire.

Shade

Shade is the simplest, and a relatively inexpensive, tool for reducing heat gain from solar radiation. Shade trees scattered in the pasture and artificial shades can offer considerable protection from solar radiation.

Did you ever watch cattle lie in the shade? If the weather is extremely hot, they pick shade on top of a hill where a breeze blows. They lie on the north side and away from the tree, so the radiant heat they emit will not reflect back from the lower branches.

Properly designed shades will reduce radiant heat gain by cattle, hogs, and sheep up to 50 percent. Research indicates that shades should be 10 to 12 feet high to allow for proper air movement. If possible, they should be on top of a hill and open on all sides. If the long dimension of the shade is east–west, more radiation is prevented and temperatures beneath the shade will be cooler; but if the long dimension is north–south, sun may cover the entire area part of each day, keeping the ground dryer and improving sanitation. If the roof slopes, it should be higher on the north side than the south. This allows greater protection from the warmer southern sky.

If metal or wood shades are used, the top should be white or shiny and the underside should be dull and dark to avoid reflecting the animal heat it receives. For maximum benefit, wooden shades should have little space between boards—enough for hot air to rise out but not enough to let significant sunlight through.

Commercially available mesh (netting) shades are now being used by livestock owners in open areas where tree or building shade is not available. With care, mesh shades can be expected to last about 10 years and are easier to construct and move than most metal, wood, or straw shades. A 20-by-50-foot mesh shade costs about $250, excluding poles and labor. Sixty square feet of shade per animal is apparently adequate for cattle. Hogs and sheep need 20 to 25 square feet each.

TABLE 11–1	Evaporative Cooling Effects on Swine		
	Control	Fogged	Sprinkled
Average daily gain (lb)	1.15	1.28	1.54
Daily intake (lb)	4.31	4.89	5.22
Feed-to-gain ratio	3.75	3.84	3.39

Source: *Kansas State University Agr. Exp. Sta. Report 371.*

Evaporative Cooling

One of the most effective means of controlling heat stress is the use of *evaporative cooling*. Sprinklers, foggers, and sometimes evaporative coolers pay big dividends, especially with poultry and swine. Underground water is already cool and evaporation occurs when natural air movement exists.

With increased confinement-rearing of swine, cooling systems have become a necessity. Because swine have little ability to perspire and are inefficient panters, they must rely on water (foggers, sprinklers, etc.) to provide evaporative or conductive cooling during heat stress. For maximum evaporative cooling, it is important that animals be permitted to dry. Ideally, hogs should be alternately wetted and given time to dry. This is also true for poultry. Studies at Kansas State University demonstrated the value of foggers and intermittent sprinklers in reducing heat stress (Table 11–1). Both foggers and sprinklers improved average daily gain and average daily feed intake, compared with controls. Intermittent sprinklers also improved feed efficiency, compared to both foggers and controls.

In Texas experiments, wallows, sprinklers, and foggers increased gains by 10 percent when temperatures exceeded 85°F. Similar findings were observed by Georgia researchers.

The use of wallows, a practice popular for many years to inexpensively provide both conductive and evaporative cooling, has declined. With increased confinement of swine, sprinklers are much more desirable. Wallows are difficult to construct in confinement systems and also permit more disease problems than do other cooling systems.

Water cooling is naturally more effective in arid climates where evaporation is more rapid. In humid areas, sprinklers can actually increase the humidity in the immediate vicinity of the animal and lower the evaporation rate. With cattle, in humid areas, a fine spray that wets only the outer hair coat serves as an insulator, preventing heat dissipation. To achieve cooling, cattle need to be drenched with cool water or alternately wetted and permitted to dry.

Wetting the outer wool of sheep would also be a hindrance. Because the water would be some distance from the skin, evaporation would do little to cool the body. Water cooling is seldom used in poultry units because it contributes to the moisture problem and ventilation load in the unit.

Commercially constructed evaporative coolers, designed for homes or offices, also can be used in regions where humidity is relatively low. Because these are relatively compact and enclosed, a powered fan is necessary to force air movement necessary for substantial evaporation and cooling.

Evaporation of moisture from grass in grazing systems likewise gives a cooling effect. To experience this effect, stand at the leeward end of a dry, cultivated field on a hot summer day, then move to the leeward end of a clover or alfalfa field. You will find the breeze blowing off the forage to be considerably cooler than that from the cultivated field. So, if practical, in pasture systems, why not put your brood sows under some shade on a slight knoll north of the clover field (assuming traditionally southern summer breezes)?

Refrigerated Cooling

Cooling with water is less effective in humid areas and where there is less air movement. It offers some sanitation problems, and water supply is sometimes limited. Also, because humidity and air movement fluctuate, it is difficult to maintain a relatively uniform cooled atmosphere. Refrigerated cooling is sometimes necessary.

An increasing number of swine raisers have insulated buildings for year-round farrowing and confinement rearing. Fans and a duct system are used for ventilation and heating. Refrigerated air conditioning is practical in some cases, but the cost of this system and of energy must be considered. The installation of **geothermal** *systems* is becoming more common and practical. These move atmospheric air through a series of tubes buried well below the earth's surface and cool it to the temperature of the soil, usually about 55 to 60°F.

Complete insulation of a building makes the cooling more effective and efficient. With lowered humidity, temperature does not need to be lowered so far. Profitability naturally depends on cost versus increased productivity and feed savings.

Cooling has particular application in farrowing houses for swine. By use of ducts and geothermal or air conditioner systems, cool air can be blown directly to the sow's head. This has the psychological effect of cooling the entire sow. With this system the sow remains cool and comfortable while the little pigs are kept warm. In extreme heat, the ducts also can be turned directly to the pigs.

Profitability of cooled air systems for cattle or poultry is less likely. A rare exception may be the cooling of bulls or rams in order to increase fertility and spermatogenesis in extremely hot weather.

> **Geothermal**
> A term indicating a system of building ventilation utilizing air drawn through ducts buried in the earth; provides a more uniform temperature throughout the year.

Feed and Water Needs during Heat Stress

Ample water has long been known to be essential for economical livestock and poultry production. Cattle drink three to four times as much water as they consume feed; in hot weather they need even more water. Laying hens consume about two times as much water as feed at 50°F and up to five times as much at 90° to 100°F.

Water intake varies with environmental temperature and feed intake. When temperatures rise and evaporative heat losses increase from the animal, water requirement increases drastically. It is important that animals exposed to heat have plenty of clean fresh water. The 1996 NRC recommendations for beef cattle indicate that water requirements for cattle increase significantly as environmental temperatures increase. For example, the recommended needs of 800-pound feedlot cattle at 50, 70, and 90°F are

Temperature (°F)	Estimated Daily Water Intake Needs (gallons)
50	7.9
70	10.7
90	17.4

TABLE 11–2	Estimated Water Intake of Sheep (Pounds of Water Per Pound of Dry Matter Consumed)					

| Environmental Temperature (°F) | Growing and Fattening Sheep | Ewes Carrying Single Lambs[a] | | | | |
		1st Mo. Gestation	2nd Mo. Gestation	3rd Mo. Gestation	4th Mo. Gestation	5th Mo. Gestation
Less than 59°	2.0	2.0	2.8	3.0	3.6	4.4
59°–68°	2.5	2.5	3.5	3.75	4.5	5.5
Over 68°	3.0	3.0	4.2	4.5	5.4	6.6

[a]*Total water intake of ewes carrying twins is about 20 percent greater in the third month of pregnancy, 25 percent greater in the fourth month, and 75 percent greater in the fifth month than for ewes carrying a single lamb.*

Source: *Adapted from* Nutrient Requirements of Farm Livestock, *Agricultural Research Council, London, 1965.*

Animals, such as swine, that eat large amounts of dry feed also need large volumes of water. Swine will consume approximately twice as many pounds of water as pounds of feed. Water intake will also increase during pregnancy, as illustrated with sheep in Table 11–2.

Cattle on the range tend to do most of their drinking in the forenoon and late afternoon and evening, as long as air temperature is below 80°F. As temperature increases, cattle drink more often, and at 90°F they will drink at least every 2 hours. This means that wells on the ranches of the hot Southwest need to be fairly close. There is naturally a practical limit, but many ranchers arrange wells so a cow is never over a mile from water, even though grazing areas are vast.

South Dakota research has shown that over 300 feet between feed and water will reduce pig gains in drylot. This may be important with lambs and cattle on feed, though probably less so because of the volume of water "stored" in the rumen.

Cooling the drinking water for feedlot cattle may increase animal comfort and performance. In California tests, daily gains were 0.19 to 0.50 pound higher when water was 65°F rather than 90°F. Due to high energy costs, mechanical cooling of water may be uneconomical, but several other techniques will help. Water from wells is colder than water from ponds or creeks. Water moved from the deeper portion of ponds or lakes by underground pipe to the water tank helps to ensure cooler water. Shade over a tank will keep water two or three degrees cooler. A shallow tank means fresher water (and thus cooler if from a well), and water in such a tank stays cooler because of surface evaporation.

Feed needs of the animal in hot weather also must be considered. With any species, high-fiber diets contribute to heat stress. Such feeds produce a high **heat increment** per **calorie** of intake, which means that a large amount of extra body heat develops in the process of digesting and utilizing such feeds. This extra heat must be dissipated by the animal, a difficult task in hot weather. Feeds with a high heat increment are a benefit to the animal during cold stress. In general, summer rations should contain considerable concentrate, and any roughage used should be highly digestible and have a low fiber content.

In general, animals decrease feed intake during heat stress. This lowered consumption, along with increased maintenance requirements during heat stress, leaves the animal with less energy to convert to useful product. Poultry producers sometimes use a "top-dress" feed to help maintain feed consumption in extremely hot weather. This

Heat increment
The energy used up in the consumption, digestion, and metabolism of a feed.

Calorie
The amount of heat required to raise 1 kg of water 1°C or 1 lb of water approximately 4°F.

top-dress feed usually contains high levels of trace minerals and B vitamins. Certain B vitamins stimulate appetite and increase feed consumption. This high-potency feed is periodically sprinkled on top of the regular ration in hot weather. Such a practice might also have value in swine feeding.

11.5   CONTROLLING COLD STRESS

When considering winter stresses on livestock, it is important to take into account the energy balance of the animal. To maintain homeothermy, the animal's heat loss must equal heat production. During winter cold stress, heat loss by convection and conduction may increase markedly. To assist the animal in coping, the producer can supply supplemental heat to the animal or reduce loss by use of such means as shelters and windbreaks.

Swine, especially small pigs, and poultry, due to their relatively sparse hair coat and large surface area per unit weight, have less tolerance to cold stress. These species require good shelter from cold temperatures in order to produce efficiently at high levels.

Cattle and sheep, however, have less need for shelter, especially when mature. Sheep have a heavy, insulating coat of wool and therefore are naturally adapted to cold temperatures. Cattle are much larger animals and therefore have less surface area per unit weight. This means that less body heat is lost, in relation to that produced. Because both of these species are ruminants, they eat roughages, which have a high heat increment, producing more heat inside their bodies relative to size.

Effective Temperature and Wind Chill Index

The degree of cold stress in animals is influenced by many factors, including temperature, wind, humidity, and shelter. *Effective temperature* is the sum of many environmental effects on the animal. It is defined as the total heating or cooling effect of the environment. Although the thermometer may indicate a temperature of 40°F, the effective temperature may be many degrees less. For example, wind and rain reduce the insulating ability of a steer's hair coat and increase convective heat loss, thereby decreasing the effective temperature. On the other hand, a steer standing in the sun on a cold day may feel many degrees warmer than what is indicated on the thermometer. An example of how effective temperature is determined is the **wind chill index**. The effect of wind chill is shown in Table 11–3.

> **Wind chill index**
> A measure of the net chilling effect on animals of temperature and wind.

TABLE 11–3	Wind Chill Index for Cattle with Winter Coats					
Wind Velocity (mph)	**Air Temperature**					
	30°F	*20°F*	*10°F*	*0°F*	*−10°F*	*−20°F*
0	30	20	10	0	−10	−20
5	27	16	7	−6	−15	−26
10	16	2	−9	−22	−31	−45
15	11	−6	−18	−33	−45	−60
20	3	−9	−24	−40	−52	−68
25	0	−15	−29	−45	−58	−75

Source: *Compiled from USDA studies.*

Windbreaks

The use of windbreaks can substantially reduce the degree of cold stress on the animal. Cold drafts and winter winds increase heat loss, promote stress, and restrict normal cattle production.

Windbreaks can reduce convective and some evaporative heat loss and improve animal performance, as well as help alleviate other winter problems. Windbreaks, if properly placed, can reduce snow drifting in livestock facilities. Windbreaks also reduce feed wastage by preventing finely ground feed from blowing out of the bunks.

Windbreaks are generally of two types: (1) natural, such as those provided by shrubs and trees, and (2) human-made or artificial. An example of an effective natural windbreak would be rows of trees planted near a farmstead to reduce wind velocity. Artificial windbreaks include snow fences, wooden partitions, and 80 percent solid fences. Natural windbreaks are less expensive to build and maintain but require years of planning ahead. Artificial windbreaks provide immediate protection.

When positioning windbreaks, it is important to consider prevailing winds. Cold, northerly winds should be diminished, but windbreaks should not interfere with cooling summer breezes. These breezes help keep livestock comfortable and also reduce dust and insect problems during the summer.

When wind speeds diminish, snow settles out and creates drifts. To prevent snow accumulation in livestock facilities, buildings and feedlots should be located within the area of wind protection, but not so close to the windbreak that drifts will accumulate in the lots.

Artificial windbreaks may be solid or partially slotted. Although solid windbreaks reduce wind speeds to a greater extent, they are more expensive to construct and may create a swirling effect on the leeward side. To reduce this swirling, it is recommended that 20 to 25 percent of the windbreak be open. Such openings usually increase the effective distance of the windbreak by reducing the swirling effect.

11.6   RATION ADJUSTMENTS DURING HEAT OR COLD STRESS

It is well established that animals vary their feed intake in response to temperature. During periods of cold stress, total intake increases as the animal consumes more energy to maintain body temperature. Increases in energy required to maintain body temperature cause decreased feed efficiency.

Animal rations are designed to meet both the energy and protein needs of the animal. During periods of cold stress, the energy needs of the animal increase, whereas protein requirements remain fairly constant. With this in mind, research has shown that the protein percentage of the ration can be reduced without reducing performance when the animal is exposed to cold stress. Because the animal consumes more pounds of feed, it receives the same amount of protein.

During heat stress the animal will decrease feed intake. Some poultry and swine producers have found that fat additions to the diet are helpful during periods of heat stress. In the case of laying hens, grams of protein needed per day for maintenance and egg production remains steady. Therefore, should feed consumption decline in hot weather, the percentage of protein in the ration should be increased proportionately. The use of a high-potency (vitamins, minerals, and protein) ration to stimulate feed intake in poultry was discussed in Section 11.4.

11.7   OTHER ENVIRONMENTAL FACTORS

Other factors, such as sound, light, and radiation, are parts of the animal's environment. Producers once were concerned about aircraft "sonic boom" effects on feedlot performance. Apparently sonic booms have little effect on performance after a brief period of acclimation. Several studies have compared the effect of intermittent versus continuous sound. Results indicate that intermittent sound has a large negative influence on animal performance.

The effects of light on performance often are attributed to temperature because the amount of daylight and temperatures are closely linked. Light factors can be broken down into three main categories: (1) quality (wavelength),(2) *intensity,* and (3) *duration* (photoperiod). By far the most important of these factors is photoperiod. Day length or duration of daylight is associated with reproduction in several of the animal species, and light is also involved with hair growth and shedding in cattle and horses. Light appears to have little effect on growth and efficiency of most species, but reduced intensity has been shown to improve growth rate in poultry, perhaps due in part to reduced agonistic behavior in dim light.[2]

Radiation effects on animals can include listlessness, loss of appetite, buckling of joints, diarrhea, bloody scours, and muscle weakness. There are three types of **radiation**—alpha, beta, and gamma—and the effects are influenced by the penetration of tissues and nature of building materials. Beta and gamma radiation are the most damaging to livestock and affect the most rapidly growing cells of the body. Factors that affect the severity of exposure to radiation include distance from the source, length of exposure, and degree of shielding from the source.

> *Radiation*
> The exchange of heat between two objects that are not touching; emission of heat rays, from the warmer to cooler object.

11.8   VENTILATION

Air movement, primarily to hold down high humidity, is essential for confined livestock and poultry, regardless of temperature. Note the example of the daily moisture production that might be expected from the cow and pig (Fig. 11–4) and heat and moisture production of broilers (Table 11–4).

The quantity of air that needs to be moved depends on the number and weights of animals confined, the temperature inside and outside, and the humidity of the outside air. Ventilation should be designed according to the maximum needs anticipated.

Well-designed ventilation systems, as illustrated in Fig. 11–5, are especially effective in removing moisture during winter because of the principle that the water-holding capacity of air is doubled with each 20°F increase in temperature. Outside air at 30°F and 90 percent relative humidity contains 0.25 pound water per 1,000 cubic feet. When warmed to 50°F, it will hold 0.54 pound (at 90 percent relative humidity); at 70°F it will hold 1.04 pounds. Theoretically, then, a ventilation system in a broiler house could remove 790 pounds (almost 100 gallons) of water by pulling in a million cubic feet of air (30°F, 90 percent relative humidity), letting it warm as the result of bird heat, and returning it to the outside. Most animals beyond a few weeks of age produce enough heat that such a housing system can be completely effective without supplemental heat. Also, a real concern is electrical power loss because in confined buildings, when the

---

[2] Agonistic behavior means a tendency to be aggressive and active, as in fighting, or defensive, as in fleeing or submitting to other animals of the same group or species.

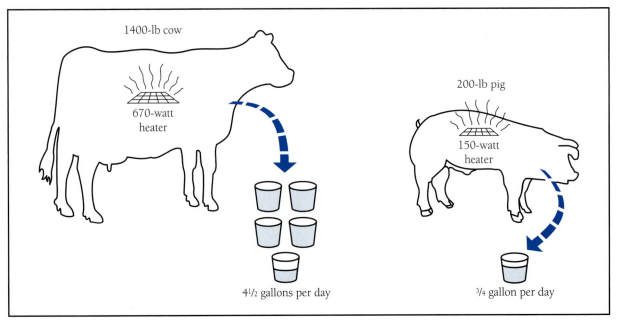

Figure 11–4
Heat and moisture production from a 1,400 lb dairy cow and a 200 lb pig.
(*Source:* University of Wisconsin Ext. Pub. No. 2812.)

| TABLE 11–4 | Average Heat and Moisture Production per 1,000 Broilers at 68° to 70°F | | | |

| | | Heat Output (BTU/hr) | | Water Output[a] (lb/day) |
Age	Average Body Weight (lb)	Per lb, Live Weight	Total	
2 weeks	0.45	26	11,700	130
5 weeks	1.75	14	24,500	350
8 weeks	3.75	10	37,500	500

[a]*Forty percent from respiration, 60 percent from droppings. Water wasted from waterers not included.*
Source: *Adapted from Lampman, C. E., et al.* Idaho Agr. Exp. Sta. Bull. 456.

power goes off the curtains fail to operate and the temperature races upward. In nearly all large confined buildings, a backup generator is used.

Preventing a buildup of ammonia is another objective of the ventilation system. Ammonia forms quickly because of high nitrogen and moisture content of the droppings. It can cause watery eyes due to irritation, and at high concentrations over long periods it can increase susceptibility to respiratory disease. Frequent removal of droppings will help hold down ammonia levels.

Ventilation is very important for animals in confinement because humidity builds up quickly in closed buildings. Then moisture condenses on walls, the bedding becomes damp, and the animals can develop colds or other respiratory ailments. If confined to such shelter, their body coats may become damp and they chill when exposed to cold air.

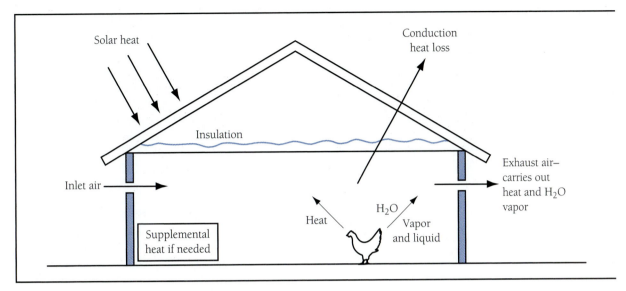

Figure 11–5

A simplified schematic of a poultry house ventilation system showing heat and moisture production and dissipation. (*Source:* North Central Regional Ext. Pub. No. 183)

Calves naturally need more shelter than older cattle, but even here ventilation must be adequate and a shelter, such as calf hutches, closed on three sides is ample.

Ventilation needs can be held down by using less water in cleaning, providing sufficient floor slope and floor drains so that most urine and water escapes as liquid, and using minimum bedding. In the case of poultry, where droppings contain much water, frequent removal of droppings also reduces ventilation needs.

Because ventilation is so critical, older buildings that have considerable "natural ventilation" are often better for cattle, sheep, and hogs than newer tightly closed buildings. Unless animals are very young, they can withstand a lot of cold as long as they are dry and protected from drafts, strong winds, sleet, and snow.

Most modern automated confinement buildings are designed and equipped to provide a healthy and comfortable environment (Fig. 11–6). The criteria for effective ventilation systems can be summarized as follows:

- capacity to provide sufficient ventilation as needed throughout the year
- proper mixing and draft-free distribution of fresh outside air throughout the building
- automatic adjustment of ventilation rate with changing outdoor conditions
- control (dilution and removal) of moisture, odors, noxious fumes, and other air contaminants
- reduction in the amount of manual supervision and labor needed to maintain the desired building temperature[3]

---

[3] Building control systems are being installed in the larger animal units with the capability to monitor, from a central control center, environmental conditions and control items such as alarms, curtains, fans and air inlets, feeding, flush valves, heaters and heat exchangers, lights, spray-cooling, and watering.

Figure 11–6
This modern dairy freestall barn provides excellent fresh air ventilation through the continuous open ridge vent across the top of the roof and the adjustable panels on each side of the structure. (Courtesy of Purdue University.)

11.9   FLOORING OR LOT SURFACE

Regardless of season, mud is a problem both to the producer and to the animal housed outdoors. Feedlot performance is severely depressed when cattle must walk through deep mud. It is important that lots have proper drainage and slope to minimize mud problems. Concrete feeding slabs can reduce feeding problems during wet weather.

Increased use of confinement rearing of swine has decreased exposure to many of the weather elements. Hogs in confinement are raised on solid, concrete floors with bedding or on partially or totally slotted floors. Slotted floors allow wastes to fall through to a storage pit or gutter for removal. This system helps keep the pig drier and cleaner and improves labor efficiency.

Materials used for floor slats include concrete, wood, stainless steel, steel, plastic, fiberglass, and expanded metal. All of these materials work well in certain phases of a swine operation, and choice of slat material depends mostly on cost, followed by durability, warmth, ease of cleaning, and the size of the pig housed on the slats.

11.10   SPACE REQUIREMENTS

With climate control, mechanical feed-handling and cleaning equipment, and close operating margins, maximum livestock or poultry numbers per dollar invested in facilities are desired. An example is laying hens (Table 11–5). As bird density increases, production drops slightly. But feed consumed per egg mass plateaus or declines, and profit may be higher.

Where layers are maintained on the floor, 1 to 1¼ square feet per bird is recommended for birds weighing about 4 pounds.

TABLE 11–5	**Effect of Bird Density on Egg Production**				
Birds per Cage[a]	**Eggs per Day per 100 Hens**	**Mortality (%)**	**Body Weight (lb)**	**Lb Feed per lb Eggs**	
4	78.3	7.6	3.65	3.03	
5	77.2	9.1	3.43	2.84	
6	75.1	6.4	3.30	2.88	

[a]*Cages 50.8 cm (20 in.) deep and containing 1,935 cm². *

Source: *Cunningham, D. L., and Ostrander, C. E.,* J. Poultry Sci. *60: 2010, 1981.*

TABLE 11–6	**Space Requirements for Pigs (Square Feet)**			
			Confined in Building	
Animals	**Open Concrete Lot**	*Concrete*	*Slatted Dunging Alley*	*Slatted*
Sow and litter	—	35	40	40
10- to 40-lb pig	4	4	2	2
80- to 120-lb pig	6	6	4.5	4
160- to 200-lb pig	10	10	6	5.5

Source: *Norris, W. H. M. and Nygaard, A.,* Purdue Univ. Agri. Exp. Sta. Res. Bull. 762, *1963.*

There is similar economic pressure to intensify livestock feeding units, especially swine, decreasing the space allotted per animal (Table 11–6). Slatted floors, which permit waste to fall into a pit below, are used in many cases to permit higher concentration, to maintain a reasonable sanitation level, and to hold down labor costs. Slatted floors are most effective, in fact, with a high animal concentration—solid waste is worked off the top of the slats by the animals' feet (Fig. 11–7).

11.11 ANIMAL WASTE MANAGEMENT

One of the major problems facing livestock producers, as discussed in Chapter 1, is management of animal waste. With large confinement units having a high density of animals, manure handling becomes very important. The handling, storing, and application as fertilizer of animal waste must be economically and environmentally sound.

Animal flocks and herds produce large volumes of manure (Table 11–7). Wastes may be handled as a solid, slurry, or as a liquid. Manure handled as a solid is loaded on spreaders and spread on farmland. If handled as a liquid or slurry, it may be stored in holding tanks or pits until it can be hauled out and spread, or knifed into the soil. In some cases it is pumped out via irrigation systems on land. Another system of liquid manure handling involves the use of lagoons, in which biological decomposition of waste

Figure 11–7
Modern hog rearing enclosures have been engineered to provide excellent air exchange, automatic feeders and waterers, and slotted floors.
(Courtesy of Lester Building Systems.)

TABLE 11–7	Annual Manure Production of Animals	
Species	**Animal Weight (lb)**	**Total Manure Production (tons/yr)**
Dairy cow	1,400	19–21
Beef cow	1,000	11–12
Beef feeder	600	6–8
Swine	200	3–4
Sheep	100	.75–1.0
Poultry broiler	4	.05
Horse	1,000	8–9

Source: *Adapted from* Livestock Waste Facilities Handbook, *2nd ed., 1985.*

occurs. Lagoons are a relatively low-cost means of manure handling where large volumes of waste are involved, as in large confinement operations, and where nutrients cannot be used economically for crop production. Also, where two or more lagoons are utilized, recycled water can be utilized in some flush systems. Any manure handling facility has risk, but today if a large spillage occurs, killing wildlife or fish, this could be the end of production for that farmer through legal action—as observed in a recent case in Indiana.

It is emphasized that the manure weights in Table 11–7 pertain to the form in which it is excreted and do not include rainwater, cleaning water, or bedding that may be present in the total waste that drains from or is removed from animal quarters.

In addition to the factors mentioned above, nutrient contents of animal waste also vary because of variations in digestibility of feeds and species differences. However, estimates of nutrient content and amounts applied to the soil can be made, and the amount of commercial fertilizer purchased and applied can be reduced accordingly. Also, legislation in animal waste management must be considered and proper application rates and methods must be followed.

QUESTIONS FOR STUDY AND DISCUSSION

1. Explain how temperature regulation differs for homeotherms and poikilotherms.
2. Define and provide an example of heat loss or gain via (a) radiation, (b) conduction, and (c) convection.
3. What is meant by diurnal body temperature variation?
4. Name two methods available to most animals for losing heat via evaporative cooling.
5. What is meant by the term *comfort zone?*
6. What changes in metabolic rate of animals can be caused when animals are exposed to steadily decreasing winter temperatures? What occurs in summer? What principal hormones are involved?
7. What is meant by the terms *acclimatization* and *acclimation?*
8. Explain why unusually high temperatures have an adverse effect on the productivity of animals.
9. Explain how "panting" helps in cooling an animal.
10. What materials and design would you use in constructing a shade for livestock? How would you orient it?
11. Give examples of providing effective supplemental cooling via evaporative cooling.
12. What is the principle of geothermal cooling?
13. What are the objectives of a properly designed ventilation system?
14. What is the relationship between feed intake and water intake? Give examples of how water can best be provided in hot weather.
15. What is meant by "heat increment" of feeds? Compare roughages to concentrates.
16. Compare the wind chill effect when an animal is exposed to an air temperature of 20°F and a wind velocity of 5 and 15 miles per hour, respectively.
17. What changes in the percentages of nutrients would you anticipate in calculating rations during very hot weather?

12
Animal Health

Animal health is a broader topic than the presence or absence of an infectious disease. It refers to the physiological well-being of an animal. Animal health is certainly affected by a variety of infectious diseases, but it is also affected by noninfectious diseases, animal density, stress, boredom, nutrition, availability of water, temperature, cleanliness, attention and care, shelter, and other factors. Disease control of livestock and poultry must be based on planned programs of prevention and controlled management when disease problems occur.

The reader with limited animal background experience should become especially observant of the appearance and behavior of healthy animals. For example, the normal, healthy animal is content, alert, and productive. A laying hen will have bright eyes, an attentive and alert manner, the head held high, and absence of skin, head, or leg blemishes. Healthy cattle on pasture will graze contentedly and intermittently. They will have a smooth hair coat, walk with ease, and generally will be with or close to others of the herd. A healthy young lamb will nurse aggressively, play, run, and jump with other lambs, and alternately stretch out in the sun to rest. A riding horse in good health will hold the head high, be inquisitive, carry a smooth hair coat, seek affection, be lean and trim, and walk or run with a free and clean stride. For new animal caregivers, it is important to observe animals to learn signs of good health, so unhealthy animals can be recognized immediately.

Alert managers, as well as veterinarians, must know the signs of good health and regularly observe these traits, and quickly note deviations in appearance or behavior—listlessness, rough hair coat, drooping head, dull eyes, arched back, slow movement, separation from the herd or flock, or unusually soft or firm feces—that suggest health is impaired.

Subclinical illnesses or low levels of parasitism that cause decreased feed intake and gain may not manifest themselves in an obvious manner, however, unless feed consumption and weight gains are recorded.

Upon completion of this chapter and related sections in other chapters, the reader should be able to

1. List a series of factors that influence animal health.
2. Differentiate among them in regard to the intensity of effect that negative factors may have on animal comfort and productivity.
3. Define the word *disease* and differentiate between infectious and noninfectious diseases, and provide examples of each.
4. Describe the immunization process and explain why vaccination is practiced.

5. Describe conditions in livestock and poultry that disclose to a herd or flock manager that health is impaired.

6. Better understand the importance of herd or flock health programs and explain how a specific program would be implemented.

7. Explain the role of the veterinarian.

12.1   ECONOMIC AND HUMAN SAFETY CONCERNS

Losses caused by animal diseases in the United States annually have been estimated to be as much as 15 percent of the potential gross income from animals and animal products. Mortality, however, is only a part of the monetary cost. Other costs are vaccines and chemicals used to prevent or control diseases, veterinary expense, reduced productivity of diseased animals, cost of federal and state disease-eradication programs, and value of product condemned at processing plants. The USDA estimates the losses due to mastitis (inflammation of the mammary gland) to the dairy producer to be in excess of $190 per cow per year in costs of treatment and lost milk, yet the disease does not cause mortality.

Routine vaccination is deemed necessary to prevent diseases such as Newcastle disease and infectious bronchitis in chickens, erysipelas in swine and turkeys, rabies in dogs, shipping fever in cattle, and enterotoxemia in lambs, at considerable cost to producers. Table 12–1 lists types of losses attributable to a single disease and benefits derived by disease prevention measures.

Antibiotics and other chemicals are fed to most young monogastric animals to prevent certain bacterial diseases. *Anthelmintics* are used to eliminate or control intestinal worms, and *coccidiostats* are fed to young chickens to prevent the protozoan disease coccidiosis. The necessity of using these products increases production cost. At the time of this revision, antibiotics are permitted in many livestock and poultry; however, there is a good chance the United States will follow Europe where antibiotics are not used.

> **Antibiotic**
> A compound that inhibits life. Each has some specificity, inhibiting only certain species or strains, such as certain kinds of bacteria.

TABLE 12–1	Benefits Derived from Control of Marek's Disease Through Vaccination	

Source of Benefits	Annual Amount ($ millions)
Reduced leukosis broiler condemnations	27.3
Reduced other broiler condemnations	6.3
Reduced broiler mortality	5.6
Reduced broiler breeder mortality	5.0
Reduced egg-type chicken mortality	15.4
Reduced condemnation of egg-type chickens	0.1
Improved feed utilization	3.2
Increased egg production	105.5
Total	168.4

Source: *Purchase, H. G., and Shultz, E. F., Jr., "The Economics of Marek's Disease Control in the United States,"* World's Poultry Sci. J., 34: 198–204, 1978.

Because of concerns that heavy antibiotic use in both animal rations and human medicine may result in evolution of antibiotic-resistant disease organisms, antibiotic use in animal rations may, in time, be limited to therapeutic use at certain stages of production.

Some diseases are given special consideration because they affect both domestic animals and humans. These diseases, called **zoonoses**, include bovine tuberculosis, brucellosis (undulant fever in humans), psittacosis (parrot fever), rabies, trichinosis, Q fever, salmonellosis, and others. Certain of these have been brought under control through federal testing programs. (See Section 12.11.) Others are controlled by sanitation, preventive medication, and thorough cooking of food products. For more detailed discussion of possible foodborne illnesses, see Box 25–1. Revenue lost due to consumers' fear (usually unwarranted) of contracting disease through animal products is not measurable.

> **Zoonoses**
> Diseases that can be transmitted between humans and animals.

12.2   *DISEASE* DEFINED AND CLASSIFIED

The term *disease* may be defined in livestock and poultry production as an impairment to normal body health, an illness, or a malady.

Infectious diseases are caused by specific, pathogenic agents. These are

- viruses
- bacteria
- fungi
- protozoa
- internal parasites
- external parasites

Some infectious diseases are contagious—they can be transmitted to other animals directly or indirectly. Examples include coccidiosis, anthrax, and leptospirosis.

Noninfectious (or noncontagious) diseases may include certain genetic abnormalities, nutritional deficiencies (such as anemia), metabolic diseases (such as pregnancy toxemia), poisoning (such as lead), and other maladies.

The term *disease* is not defined in the same way by all. The above definition may be considered broad. Some restrict the term to maladies that are infectious or to those that are contagious and would list the other types of impairments mentioned in this chapter simply as parasite infestations, genetic anomalies, nutritional deficiencies, stress syndrome, and so forth. Because of intense confinement space occupied by large numbers and density of animals, the threat of disease is significantly increased.

The purpose of this chapter and related sections of other chapters, however, is to discuss the total topic of animal health and the groups of factors in livestock and poultry production that may impair health and, therefore, animal comfort and productivity.

12.3   PREDATORS AND INJURIES

Confinement of animals to feedlots, swine production units, drylot dairies, and broiler and layer houses minimizes risk from predators. Wildlife predators remain a problem, however, to the sheep rancher and, to a lesser extent, the cattle rancher. In fact, predators may be a more serious problem to ranchers in the early years of the new century than before there were imposed constraints on predator control.

The coyote is the most significant predator in ranching areas,[1] but the bobcat, bear, and cougar have also been known to cause problems. Bands of domestic dogs may be predators as well, both in ranching areas and where ewe flocks or lambs are maintained near residential areas. Losses in sheep flocks are significant, resulting in an economic loss each year approximating $100 million.

Among the more common injuries are wire cuts in horses, especially among those kept in small pens, and teat injuries in cows from frostbite, from being cut by coarse grass or weeds, or from being stepped on where cows are bedded in open sheds. Cannibalism in poultry and tail and ear biting in swine are discussed in Section 13.13.

12.4 NUTRITIONAL DEFICIENCIES

When many swine were routinely produced on pasture and when layers and poultry for meat ranged over a farmstead and could consume grass and insects, nutritional deficiencies were less common or, because of less advanced knowledge at that time, less observed. Today, however, most concern is with deficiencies in individual vitamins, minerals, or amino acids.

As confinement of nonruminant animals became more common, nutrients were available only through provided rations, and as genetic selection and other techniques were used to increase productivity per animal unit, more nutritional deficiencies were observed and identified.

Deficiencies may result from several causes: (1) a low level of the nutrient, such as an essential mineral in the soil, resulting in forage or grain being especially low in that nutrient; (2) a prepared ration that is too low in a key nutrient; (3) impaired absorption of a nutrient; (4) incomplete or blocked metabolism of a nutrient in the body cells; (5) abnormal requirement for a nutrient, due perhaps to incomplete metabolism but possibly to some environmental factor; or (6) failure of the animal to synthesize a nutrient that is normally synthesized in the digestive tract or body cells (and thus considered nonessential in the diet).

Where feed supplies are short, as in prolonged drought, or when good animal management is lacking, energy or protein deficiencies may cause general unthriftiness, low prolificacy, and susceptibility to infectious diseases. Some of the more common nutritional deficiencies observed in livestock and poultry are listed in Table 12–2.

Nutritional deficiencies generally are avoided by providing nutritionally complete **rations**, formulated according to available research information. Where soils may be deficient in certain minerals, a trace mineral mixture is made available or mixed with concentrate. Where locally grown and lower cost feeds are the major ration ingredients but may be deficient in a nutrient, the ration is designed to include a rich source of that nutrient, even though that ingredient may be more costly. An example is corn-soybean meal rations for swine and poultry in the Midwest. Such a ration is usually low in the amino acids lysine and methionine, so food sources such as blood meal, fish meal, or commercial sources of lysine and methionine can be included in the ration. Manufactured or mined nutrients, usually vitamins and minerals, are often added.

> **Ration**
> A combination of feeds, perhaps mixed together, that are fed to livestock to meet nutrient requirements. May vary in quantity.

---

[1] Terrill, C. E. "Trends of predator losses of sheep and lambs from USDA mortality statistics." *Third Symposium Proceedings on Test Methods for Vertebrate Pest Control and Management,* Fresno, CA: American Society for Testing and Materials, March 7, 1980.

TABLE 12–2	Common Deficiency Diseases in Domestic Animals	
Element or Nutrient	**Species Affected**	**Disease Condition**
Vitamin A	Primarily cattle and sheep	Weak calves at birth
	All species to some degree	Night blindness
		Keratitis
		Reproduction problems
		Congenital defects
Vitamin D	All species	Rickets
		Joint disease
		Osteomalacia
Vitamin K	All species	Hemorrhagic condition
Calcium[a]	All species	"Milk fever" in dairy cows
		Osteoporosis
		Rickets
Phosphorus	All species	Pica
		Poor growth
		Infertility
Selenium and Vitamin E	All species	Muscular dystrophy
		Retained placenta
		Encephalomalasia in poultry
Magnesium	Primarily cattle	Grass tetany, hyperexcitability
		Transportation tetany
Zinc	Primarily swine	Parakeratosis
Manganese	Primarily cattle	Infertility
		Skeletal deformities
	Poultry	Perosis or slipped tendons
		Skeletal deformities
Cobalt	Primarily cattle and sheep	Unthriftiness
		Emaciation
Iron	All species	Anemia
Copper	Sheep	Enzootic ataxia, anemia
	Swine	Anemia
Iodine	All species	Goiter, hairlessness
		Partial or complete alopecia

[a]*Calcium deficiency is closely associated with calcium/phosphorus ratios and availability of Vitamin D. High levels of calcium during the nonlactating period of the dairy cow may contribute to incidence of milk fever.*

Source: Adapted from National Research Council publications: *Nutritional Requirements of Sheep, 1986; Nutritional Requirements of Dairy Cattle, 1988; and Nutritional Requirements of Beef Cattle, 1996.*

It is important to recognize that there may be two levels of deficiency. An acute or severe deficiency would likely cause the visible symptoms described in Table 12–2. Today, however, such deficiencies are rare in well-managed herds and flocks. Deficiencies that are borderline and that are difficult to perceive or identify may simply prevent health and production from being at the *optimum* level.

Considerable research is directed toward the latter, and producers in many cases will provide a "safety factor," an extra amount of an apparently critical nutrient in the ration. This is especially common where knowledge appears incomplete, where the

TABLE 12–3	Common Metabolic Diseases of Domestic Animals	
Species	**Disease**	**Causative factor**
Horse	Lactation tetany of mares	Calcium deficiency (early lactation)
	Azoturia	Sudden increase in muscular work or high carbohydrate diet
Cattle	Rumen acidosis	Abrupt increase in ingestion of highly fermentable carbohydrates
	Ketosis	Hypoglycemia
	"Fat cow syndrome"	Excessive conditioning, especially during dry period
	"Milk fever"	Hypocalcemia
	Transportation tetany	Calcium-magnesium deficiency
	Grass tetany, hypomagnesemia	Magnesium deficiency
	"High mountain" or "brisket disease"	Chronic hypoxia related to high elevations
	Urinary calculi in males	Calcium-phosphorus imbalance; unknown factors
	Postparturient hemoglobinuria	Phosphorus deficiency?
	Whole-milk tetany in calves	Magnesium deficiency
	Photosensitization	Liver pathology—may be related to plant toxins
Swine	Hypoglycemia (piglets)	Postpartum deprivation of milk
	PSS (porcine stress syndrome)	Heritable defect
Sheep	Pregnancy toxemia (ketosis)	Impaired metabolism of carbohydrates or advanced pregnancy stress

Source: *Adapted from National Research Council publications: Nutritional Requirements of Sheep, 1986; Nutritional Requirements of Dairy Cattle, 1988; and Nutritional Requirements of Beef Cattle, 1996.*

animals may be under considerable stress, or where the nutrient content of feed ingredients may vary. At the same time, excessive levels should not be added. Extremely high levels of vitamin A will cause epithelial problems, for example, and excess levels of certain minerals will interfere with absorption or utilization of other nutrients. In the case of calcium and phosphorus, the ratio of the two is considered as important as the actual levels. Today, there are very few nutritional deficiencies observed as a result of feeding livestock and poultry ratios.

12.5   METABOLIC DISORDERS

Metabolic disorders are often difficult to separate from nutritional deficiencies. The blockage of a metabolic pathway may cause a deficiency at the body cell, even though the ration is entirely adequate. The fact that increasing the nutrient content of the diet may eliminate the symptom (by offsetting the metabolic blockage) often adds to the belief that a dietary deficiency exists.

Common metabolic disorders of domestic animals are summarized in Table 12–3.

12.6   TOXINS AND POISONS

Toxin
A harmful chemical of animal or bacterial origin.

For this discussion, **toxins** are defined as those harmful chemicals produced by animal cells or microorganisms. *Poisons* are defined as those chemicals harmful to animals where the source is a growing or harvested plant or a material that has been chemically or biologically manufactured. In practice the terms are used interchangeably, so we will discuss them as a group.

Illnesses initiated by toxins are called **toxemias**. Toxins generally can be divided into two categories: (1) antigenic toxins, produced by bacteria or other microbiological life, and (2) metabolic toxins, produced by animal body cells.

Antigenic toxins are usually proteins and trigger an antibody reaction in the animal. Body temperature goes up and clear symptoms of illness result. Enterotoxemia, caused by *Clostridium perfringens* type D in feedlot lambs, is an illness caused by an antigenic toxin.

Ketosis in the cow is caused by metabolic toxins. Improper carbohydrate metabolism, or lack of sufficient carbohydrate (low blood sugars), results in massive catabolism of stored fats to supply energy. End products of this catabolism—ketones—accumulate and are toxic to the animal.

High levels of nitrates in rapidly growing plants, plants whose growth may be interrupted by an early frost, or plant materials harvested during a rapid growth stage may cause poisoning. Nitrate is an intermediate step in protein manufacture, and large quantities would be present at a given time in a rapidly growing plant. Frost or harvesting prevents the plant from converting nitrate to protein. Toxicity occurs when nitrates are converted to nitrites in the rumen more rapidly than the microbes can effectively incorporate the nitrites into microbial protein.

A number of inorganic and organic compounds have been incriminated in livestock toxicity or poisonings, including the following:

Inorganic	Organic
Lead	Hydrocyanic (prussic) acid
Mercury	Nitrate or nitrite
Arsenic	Oxalate
Copper	Polychlorinated biphenol (PCB)
Molybdenum	
Fluorine	
Sodium chloride	
Selenium	
Zinc	
Sulfur	

Especially high levels of selenium are contained in the soils of parts of south-central South Dakota and in the grasses and grains that grow there, resulting in malformed hooves and horns on cattle and reduced prolificacy. Cases of lead poisoning in animals that lick lead-based paints are less common today because of the use of latex paints.

Fluorine has been a serious problem where it has settled on grass leeward from aluminum manufacturing plants or plants where rock phosphate is converted to the feed ingredient dicalcium phosphate.

Anthelmintics (worm-controlling agents) such as carbon tetrachloride, hexachlorethane, phenothiazine, nicotine, and cadmium are safe when administered under optimum dosage and conditions but may be poisonous at high doses or when administered to debilitated animals.

Some animal or plant insecticides, such as chlorinated hydrocarbons, organophosphatic compounds, and rotenone, as well as herbicides, rodenticides, and wood preservatives, have been incriminated as animal poisons.

Toxemia
An illness or malady caused by a toxin.

Ketones
Potentially toxic substances created by only partial oxidation of fatty acids; can result in ketosis in cows, does, or ewes.

12.7   PARASITES AND PROTOZOA

Livestock and poultry are subject to infestation by both internal and external parasites. Here are the common internal parasites, by host species:

Cattle

Stomach worms
Intestinal worms:
   Roundworms
   Tapeworms
Lungworms
Liver fluke
Grubs

Sheep and Goats

Stomach worms
Intestinal worms:
   Roundworms
   Tapeworms
Lungworms
Nose bots

Horses

Stomach worms (bots)
Intestinal worms:
   Strongles
   Pinworms
   Roundworms
   Tapeworms

Swine

Stomach worms
Intestinal worms
Lungworms
Swine kidney worms
Trichina

Poultry

Coccidia
Intestinal worms:
   Roundworms
   Tapeworms
Capillary worms
Grapeworms

The principal intestinal parasites affecting domestic animals are roundworms and tapeworms. Adult roundworms produce eggs that pass out with feces. If temperature and moisture are adequate, the eggs develop to larvae. After ingestion by a suitable host, the larvae (immature worms) burrow into the intestinal mucosa, causing irritation and damaging the absorptive surfaces of the intestine. Eventually mature roundworms emerge into the intestinal lumen and produce eggs, completing the life cycle. In addition to damage caused by the immature worm, adult worms may become so numerous that they cause nutritional deficiencies in the host or even cause blockage of the intestines.

Tapeworms, which are more common in sheep, cattle, and poultry, are flat, segmented worms with indirect life cycles—part of the life cycle occurs within one or more specific *secondary* hosts (insects, arthropods, slugs, etc.).

Within the digestive tract of the animal, the head and neck sections of the immature worm are embedded permanently in the intestinal lining. Segments grow out of the neck section in a chainlike fashion, and the worm may reach several feet in length. Each segment has both a male and female gonad and, as the segments mature, thousands of eggs are produced. As the egg-filled segments break off, they pass out of the animal's body with the feces. Some are ingested by a secondary host and the eggs undergo changes that make them infectious for another secondary host or the primary host animal.

The fact that the head section of the tapeworm remains permanently embedded in the intestinal lining means that eliminating tapeworms is usually more difficult than eliminating roundworms. It also suggests that the appearance of tapeworm segments in the feces of treated animals does not ensure that the head and neck sections have been removed.

External parasites (arthropods) include the following:

- lice
- mites
- flies
- sometimes fleas and ticks

Lice may damage their host by biting or blood sucking, depending upon the species. Some mites, including most poultry mites, live on the skin surface, feeding at night and leaving the host during daylight. However, the tiny northern fowl mite, the most common external parasite in chickens, spends its entire life on the bird's body. Other mites burrow into hair follicles, skin glands, or skin tissue per se. Mange in dogs, swine, and other mammals is caused by the burrowing type of mite.

Heel flies are the adult stage of the cattle grub larvae and cause much economic loss to cattle producers. Although the adult flies neither bite nor sting, they greatly annoy cattle. These large flies attach eggs to hairs on the heels or legs of cattle. Developing larvae penetrate the animal's skin, migrate through the animal's body, and after 5 or 6 months appear under the skin of the back. Each larva, called a grub, makes a breathing hole through the skin. Eventually the grub emerges, drops to the ground, and pupates to form young adult flies. Hides of infected animals have reduced value; annual losses caused by this parasite are estimated at well over $100 million.

Horn flies and stable flies are annoying blood-sucking insects that cause reduced growth rate in beef cattle and reduced milk production in dairy cattle. Other external parasites of importance include fleas, ticks, horseflies, and mosquitoes.

Blood-sucking external parasites may also be vectors of disease (mosquitoes, for example, transmit equine encephalitis virus, and ticks and some flies spread anaplasmosis in ruminants). Others may neither bite nor suck blood but transport microorganisms from one host to another.

Microscopic examination of skin scrapings and fecal material is often necessary to achieve a complete and accurate diagnosis of internal or external parasite or protozoan infestations.

Though some chemicals are used as preventatives, animal management systems that include effective waste management, periodic and rigid cleaning of facilities, good nutrition and care, rapid treatment of injuries, and rotation of animals among different pastures or lots are generally effective in reducing parasite burdens. Each of these contributes to breaking the life cycle of the parasite or increasing the resistance of the animals to infestation.

In the case of some parasites, prevention systems are more specific. Horn flies and face flies are credited with the spread of eye infections in cattle, especially the bacterial disease pinkeye. While cattle are on pasture, subjected to eye stresses such as wind, sunlight, pollen, weeds, and tall grass, pinkeye infections are common. Insecticide-impregnated plastic tags attached to the ears of cattle have effectively reduced the presence around the head of horn flies and face flies. Their use has reduced the incidence of pinkeye and resulted in heavier weaning weights of beef calves.[2]

Treatments are generally internally or externally administered chemicals that serve as poisons to the parasite or protozoan. Obviously such chemicals should be used according to established recommendations, and the manager should be especially alert to state or federal restrictions on use in the case of laying hens or dairy cows, or where affected animals may soon be slaughtered for meat.

12.8 TYPES OF INFECTIOUS AGENTS

> **Infectious**
> The ability to be transmitted by disease; denoting a disease caused by a microorganism.

The terms **infectious** and **contagious** are often used loosely and interchangeably. For this discussion, an *infectious agent* is defined as one that is capable of causing an infection in the body. A *contagious agent* is an infectious agent that may be readily transmitted from one body to another. The term *communicable* is considered a synonym for the term *contagious*. Localized abscesses of animals are often infections caused by the bacterium *Corynebacterium pyogenes*. However, that bacterium does not readily spread from the affected animal to another animal. By contrast, anthrax, caused by the bacterium *Bacillus anthracis*, is both infectious and highly contagious; it is transmitted readily from one animal to another.

> **Contagious**
> The characteristic of a disease that permits it to be readily transmitted from one animal to another.

Disease-producing agents are sometimes referred to as *primary* or *secondary* agents. For example, "lumpy jaw" in cattle may be initiated by a plant awn breaking the mucosa within the mouth and then the bacterium *Actinomyces bovis*, a common inhabitant of the bovine mouth, establishing itself and causing the infection termed lumpy jaw. The plant awn would be the primary agent, which predisposes the animal to *Actinomyces bovis*, the secondary agent.

Infectious agents range in size from multicelled fungi or large single-celled protozoa to viruses so small that they can be observed only with an electron microscope.

Viruses

> **Virus**
> A large and complex protein material that is capable of causing disease and that reproduces only inside a host cell.

Viruses must reproduce inside living cells of the host. They lack cytoplasm and utilize the metabolic system of the host cell to replicate. The host cell loses its ability to divide, eventually degenerates, and the virus particles which are then released invade other cells.

Many viruses are host-specific and may be tissue-specific. For example, hog cholera, formerly one of the most serious swine diseases, occurs naturally only in swine. Eastern equine encephalitis virus, pathogenic to both humans and horses and capable of infecting birds, primarily affects the host's central nervous system tissue. Many respiratory diseases such as equine rhinopneumonitis, canine distemper, virus pneumonia of pigs, and infectious bronchitis in chickens are caused by viruses.

Some viral diseases are transmitted from host to host only by specific vectors (e.g., equine encephalitis by mosquitos), whereas others such as hog cholera may be spread by direct contact between animals or by indirect contact (feed, water, boots, vehicles). Control measures include isolation, strict sanitation, elimination of vectors, exclusion of visitors from production units, and vaccination programs.

In the case of widely disseminated viral diseases, vaccination of susceptible animals is routine. See Table 12–4 for a sample vaccination program for replacement pullets. A more detailed discussion of vaccination programs is provided later in Box 12–1.

---

[2] Francis, E. N. Kansas Cooperative Extension Service, private correspondence, November 9, 1981; Williams, R. E., et al. *J. Animal Sci.* 53 (1981): 1159.

Table 12–4	Sample Vaccination Schedule for Replacement Pullets

Week of Vaccination	Type of Vaccination
1 day old	Marek's
15 days (1/2 dose)	Infectious bursal
20 days (1/2 dose)	Infectious bursal
25 days	Bronchitis, Newcastle, infectious bursal (typical brand name Combo Vac 30)
30 days	Bronchitis, Newcastle, infectious bursal (typical brand name Combo Vac 30)
49 days	Bronchitis, Newcastle, infectious bursal (typical brand name Combo Vac 30)
10 weeks	Fowl pox and laryngotracheitis (commonly referred to as LT)
12 weeks	Combo Vac 30
13 weeks	Avian encephalomyelitis (commonly referred to as AE)
16 weeks	Newcastle

Source: *Commercial Egg Production and Processing, Purdue University Extension Publication, 2000, AS-545-W.*

Bacteria

Bacteria are microscopic, single-celled organisms that reproduce by binary fission and that vary greatly in size, shape, pathogenicity, and host specificity. The zoonotic diseases tuberculosis, brucellosis, and salmonellosis are caused by bacteria belonging to the genera *Mycobacterium, Brucella,* and *Salmonella,* respectively.

> **Bacteria**
> A class of single-celled organisms.

Bacteria damage their hosts' tissues primarily by the production of toxins. Some bacteria (e.g., staphylococci and streptococci) produce and release toxic substances as byproducts of their metabolism. These toxins are referred to as **exotoxins**, and symptoms appear in the animal soon after infection occurs. Other bacteria, such as salmonellae, produce *endotoxins*—cellular components that are released to cause damage to the host only when the bacterial cells lyse or disintegrate. Symptoms, therefore, appear in the animal later than do symptoms caused by exotoxins.

> **Exotoxin**
> A toxin produced by an organism other than the host animal, such as organisms of the intestinal tract.

Bacterial infections are reduced by use of antibiotics or other chemicals, vaccination, and testing programs that provide for elimination of infected individuals.

Protozoa

Protozoa are single-celled animals, microscopic in size but larger than bacteria. The most common protozoan diseases of domestic animals are in a group referred to as "coccidiosis" and caused by protozoan parasites called coccidia. Typically coccidia invade the epithelial cells of the digestive tract and exist as intracellular parasites throughout most of their life cycle. The coccidia often replicate asexually within the cells of the host's intestinal tract, cause rupture of the host cells, and invade other cells for continued replication. After a number of asexual generations, male and female gametes are formed and unite within the host's cells, forming oocysts. The oocysts are eliminated with feces and, given adequate conditions of temperature and moisture, sporulate (divide) outside the animal's body. The sporulated oocysts can remain viable and infectious for several days or weeks. If oocysts are ingested by a new host during that time, the disease may spread.

> **Protozoa**
> Single-celled organisms that reproduce by fission. Many may cause animal diseases.

BOX 12–1

History and Development of Vaccination Programs

In the case of certain diseases, the causative organism or virus is rare enough that the young animal doesn't normally encounter it (and therefore doesn't develop a natural immunity) and/or is so virulent that when encountered it often causes serious illness and death. For such cases, scientists have developed systems to create immunity in the animals that may later be exposed. The following paragraphs describe the basis for these systems, which began about a century ago.

The French chemist and biologist Louis Pasteur developed in the 1870s and 1880s a serum that would provide resistance to anthrax in an animal that had received the serum. The serum had been separated from the blood of an animal that had been infected with the disease, had developed an antibody resistance to it, and had recovered. Test sheep were first injected with serum, then were injected with an infectious culture of the anthrax organisms. Those sheep that had first received the serum remained well. A second group, which had not received the serum, became ill and died when injected with the culture of anthrax organisms.

Anthrax is a virulent (quick-acting and strong) disease. Apparently the serum provided temporary and passive immunity until the infected animals could develop their own antibodies (active immunity) to an extent sufficient to prevent the disease.

In the ensuing century, a series of discoveries and developments resulted in a variety of vaccination techniques for all species and for a variety of infectious diseases. Most were directed toward triggering the production of immunoglobulins or other disease-fighting agents in the susceptible animal. During the earlier years, the technique that Pasteur used—(1) inject serum from a recovered animal, (2) inject a culture of the disease-causing organism, and (3) depend on the animal to achieve permanent resistance—was followed in most cases. Species that were not as susceptible to the disease and/or that would yield large volumes of serum were kept and used in serum manufacturing plants. Horses and swine were especially useful. Cultures of the live agent—usually called "bacterin" or "virus"—were prepared in manufacturing laboratories.

In some instances, the disease was so virulent that there was much risk in using a live culture of disease-causing organisms or viruses on farms and ranches. A broken bottle or an accidental thrust of the syringe plunger could contaminate the soil or facilities for decades, or directly infect animals that had not received serum. Eventually, forms of attenuated (reduced virulence) or dead cultures that would trigger the antibody reaction but not cause severe illness were developed. This and other developments resulted, in some cases, in the injection of serum being unnecessary.

After this refinement was achieved, it took many years, of course, before production units or areas where live bacteria or live virus had been used could be rid of all apparent live material. In the case of hog cholera, special educational programs urged use of attenuated or dead virus. As a state or area approached a low incidence of the disease, laws forbidding the use of a live virus were passed in order to eventually achieve a cholera-free designation. This designation enhanced the market for breeding stock and feeder pigs, and suggested a safer environment for swine production.

Especially for poultry, where individual injection would be laborious and costly, vaccines were developed that could be administered via the water or as an aerosol.

Similar developments with other diseases and other species could be cited.

Source: USDA.

Coccidia are very host-specific. Thirteen species of coccidia affect cattle and 11 affect swine, but none of the cattle species are infectious for swine; chickens are affected by 9 species of coccidia belonging to a single genus *(Eimeria)*, none of which affect turkeys or other animals.

Coccidiosis causes extensive damage to the lining of the digestive tract, impaired digestion and absorption, diarrhea, and possibly death. The disease is prevented by including drugs called coccidiostats in the ration. When outbreaks of the disease occur, anticoccidial drugs are usually administered in the drinking water.

Fungi

Fungi (fun'ji) are single-celled or multicellular plants of a low order of development. Best known among filamentous fungi are species of *Tricophyton* which cause the external infection "ringworm" in pets as well as farm animals. Sometimes seen in humans, it can be acquired by handling infected pets.

Aspergillosis in birds is caused by a filamentous fungus. This internal disease results from inhalation of the fungal spores and their germination and growth in the respiratory system. The disease in young birds is often called "brooder pneumonia" because of the pneumonialike symptoms and its occurrence during the early life of the bird.

Histoplasmosis, a systemic fungal disease of canines and humans, is an infectious, noncontagious disease. The organism is usually inhaled with dust particles and often causes infection of the lungs.

Fungal diseases are difficult to treat and are best controlled by preventing exposure of susceptible animals. The feeding of moldy feedstuffs and use of moldy litter or bedding should be avoided.

> *Fungus*
> A group of plants without chlorophyll that reproduce by spores. Includes molds and can cause animal health problems.

Rickettsiae and Mycoplasmas

Rickettsiae are organisms that have properties similar to viruses but are larger in size and seldom affect livestock and poultry. *Mycoplasmas* are smaller than rickettsiae but larger than viruses, and have been shown to cause a serious form of mastitis, as well as respiratory problems in livestock and poultry.

12.9 IMMUNITY AND VACCINATION

Immunity is a state of resistance in the animal to a disease-producing agent. It may be classified as passive or active acquired (Table 12–5).

TABLE 12–5	Types of Acquired Immunity
Passive Immunity	**Active Immunity**
Acquired by introduction of antibodies	Acquired by production of antibodies
Natural species and/or genetic resistance in specific lines	Natural exposure and recovery
Colostrum	Administered vaccines and bacterins
Serum or gamma globulin	

Even though it has been observed for decades that some animal species are not bothered by certain infectious diseases, the scientific study of inherited resistance to livestock and poultry diseases has been pursued intensely only in recent decades. There is much evidence that tendency toward or resistance to mastitis and cancer eye in cattle is inherited. (See the heritability estimates in Table 18–1.) Many commercial poultry breeding programs seek to achieve genetic immunity to certain diseases, such as Marek's disease. The opportunities in other species may be high, but because of the lower prolificacy and longer generation time in those species, less has been learned, and what may be learned will have less rapid implementation.

In practice, most long-lasting resistance is acquired, either by exposure of the young animal to the existing disease or by some form of artificial exposure—vaccination. A newly born or hatched animal is immediately exposed, in the natural environment, to a multitude of microorganisms and viruses—in the bedding or litter, in the soil, on the teat or tongue of the dam, in the water, and so forth. Many of these are capable of causing infection, especially should the young animal be susceptible due to being in a weak condition or in a stressful situation.

But most newly born or hatched animals have naturally provided disease-fighting mechanisms. All have in the blood a relatively high circulation level of immunoglobulins, nonspecific disease-fighting proteins.

Preformed antibodies are passed across the placenta from the mother to the fetus in humans, dogs, cats, and guinea pigs, but not in pigs, cattle, sheep, and horses. In the latter mammals, the milk produced the first few days after birth of the offspring, colostrum, is especially high in immunoglobulins, other protein (used in production of other specific antibodies), and fat. (Composition of colostrum was shown in Table 6–5.) Thus, the young animal will likely acquire passive immunity to specific agents. It is very important that these newborn animals consume colostrum within a few hours after birth, because the ability of the intestine to absorb the large immunoglobulins conferring passive immunity rapidly diminishes. The newly hatched chick or poult is nourished by the egg yolk, which is also high in immunoglobulins.

Development of immunity by exposure of the young animal to an existing disease explains several phenomena in livestock and poultry production. Animals born or hatched and kept during their early weeks in especially hygienic conditions or in totally new facilities may become ill when moved to more normally managed, previously used facilities where there is a normal population of potential disease-producing agents. The same often occurs when stock is moved from one country to another, or one latitude to another, and a markedly new and different population of organisms is encountered.

12.10   REPRODUCTIVE AND RESPIRATORY PROBLEMS

Reproductive and respiratory problems are major concerns in many livestock and poultry production units in which a variety of infectious agents, nutritional deficiencies, and other factors may be involved. They illustrate well the present complexity of animal health problems when animals are under considerable production stress, animal density may be high, and movement of people, animals, and feed is wide ranging and rapid.

Infectious reproductive diseases among cattle include brucellosis, vibriosis, and leptospirosis, all of which can be carried and transmitted by sires. These diseases can be effectively prevented by vaccines. But vitamin A or phosphorus deficiencies, low energy or protein levels, high environmental temperatures, and other factors can also diminish prolificacy in either the dam or sire. The number of possible interactions is high, and the difficulty of sorting out likely causes of reproductive problems is apparent.

Respiratory problems in all species are complex. Though specific infectious agents are often identified in nasal cultures, a mixture of infectious agents often exists. Research has demonstrated that the apparent virulence of an agent may be influenced by the presence or absence of others, by dust, by the ration, and by stress. In fact, even though there are specific respiratory diseases, such as infectious bovine rhinotracheitis (IBR), many respiratory infections are often referred to simply as a "respiratory disease complex."

Horses may encounter equine viral rhinopneumonitis (EVR), swine often have influenza, and airsacculitis is common in poultry. It is emphasized that stress on the individual animal is usually a major contributing factor to the establishment or the consequences of these diseases.

12.11   ERADICATION AND REPOPULATION PROGRAMS

Gradual elimination of hog cholera by elimination of the use of live virus, continued vaccination with modified vaccines, and eventual absence of cases was described in Section 12.9. Other diseases have been eradicated or reduced in incidence to the point that they are no longer of major economic importance in the United States.

Foot and mouth disease and contagious pleuropneumonia are examples. The incidence of tuberculosis and brucellosis has been significantly reduced. In the case of tuberculosis, the major technique used was routine testing at state or federal expense of all breeding cattle and slaughter of all infected animals. Though some indemnity was paid the owner, there was often severe economic loss. In a similar brucellosis program, USDA indemnity payments were supplemented in most states by state or county indemnity payments.

After lower incidence levels are attained, most of the cases are detected in cattle at slaughter plants. These cattle are tagged, the owner is identified, and the total herd from which they came is tested.

When increased numbers in an area test positive, inspection and surveillance programs are increased. For example, animals as well as herds may be inspected, held in quarantine for 30 days, and inspected again. Carriers of the disease go directly to slaughter.

Eradication programs require complete cooperation by producers and others, both within and between states, plus inspection and quarantine provisions at national boundaries.

Where outbreaks of specific and especially virulent or economically significant diseases such as foot and mouth disease occur, elimination and proper disposal (by burning, burial, and/or cooking to produce tankage) of all susceptible animals within the area are essential. Following cleanup and disinfection, "sentry" animals are usually placed in the area. If, after a period of time and thorough examination, no evidence of the disease exists in the sentries, repopulation proceeds.

TABLE 12–6	Suggested Guidelines for a Swine Herd Health Calendar[a]	
Time (age)	**Vaccines and Parasite Control**	**Management and Breeding**
	Gilts/Sows	
6½ months	Deworm; treat for lice and mange; feed fresh manure from boars and sows. Repeat in one week. Commingle with cull sows, and initiate fence line contact with boars. Vaccinate for lepto, erysipelas, parvovirus, and PRV.[b]	Choose gilts according to established genetic selection criteria and program. Isolate purchased gilts for 60 days. Blood-sample purchased replacements for important disease(s) not already present in herd.
7½ months	Repeat vaccinations.	
8 months		Breed on 2nd or 3rd heat period (at least two matings per service).
3 weeks post-breeding		Pregnancy-check nonreturns to heat.
9 months		Pregnancy-check (35–60 days post-breeding).
6 weeks prior to farrowing	*Clostridium* toxoid.	
4–6 weeks prior to farrowing	*E. coli* bacterin, *Pasteurella* (AR), *Mycoplasma*, TGE, PRV. Treat for lice and mange.	
2 weeks prior to farrowing	*E. coli* bacterin, *Clostridium*, *Mycoplasma*, TGE, AR.	May include feed additives through lactation to prevent *Clostridium*.
7–10 days prior to farrowing	Treat for lice and mange and deworm.	May include feed additives to prevent constipation. Wash sows thoroughly with detergent before entering farrowing house.
Farrowing		Record litter and sow information. Pig environment—90–95°F. Sow environment—65–70°F.
3–5 weeks post-farrowing	Lepto, parvovirus, and erysipelas, PRV for sows. Treat for lice and mange.	Wean pigs. Provide comfort, sanitation, and adequate diet.
	Boars	
4–6 months		Select and bring to farm at least 60 days prior to breeding. (Boars are ready for limited use at 8 months of age.) Isolate purchased boars for 60 days. Blood-sample for important diseases not already present in herd.
1st 30 days following purchase in isolation	Test for brucellosis, lepto, PRRS, parvovirus, *Actinobacillus*, TGE, and PRV. Treat for lice and mange and deworm.	Feed unmedicated feed, and observe for diarrhea, lameness, pneumonia, and ulcers.
2nd 30 days following purchase in isolation	Vaccinate for erysipelas, lepto, and parvovirus.	Feed manure from other boars and sows. Commingle with cull gilts, and observe desire and ability to breed. Provide fence line contact with gilts and sows to be bred.
Every 6 months	Revaccinate for PRV, lepto, erysipelas, and parvovirus, then deworm. Treat for lice and mange.	

TABLE 12–6	(continued)	

Time (age)	Vaccines and Parasite Control	Management and Breeding
	Pigs	
1 day	*Clostridium* antitoxin.	
1–3 days	Iron injection.	Clip needle teeth. Dock tails. Ear notch. Castrate.
3–7 days	Vaccinate for AR and TGE.	Start oral iron, fed in tray.
10–14 days	Iron (injection or oral).	Start creep feed with oral iron mixed in; wean if SEW.
3–4 weeks	Vaccinate for AR, PRRS, *Mycoplasma,* and *Salmonella choleraesuis*	Expose to pre-starter feed. Wean.
Weaning + 10 days	Treat for lice and mange and deworm.	
Weaning + 20 days	Vaccinate with erysipelas and *Actinobacillus pleuropneumoniae* bacterins.	
10–12 weeks	Vaccinate for PRV and revaccinate with erysipelas and *A. pleuropneumoniae* bacterins.	Fecal exam.
5–6 months	Correctly follow all feed medication and vaccination withdrawal times prior to slaughter.	Health-check 20% to 30% hogs from a market group.

[a]*Shows most of the recommended vaccinations and parasite control measures used by the producers in consultation with their veterinarian.*

[b]pseudo-rabies virus

Source: *Pork Industry Handbook, PIH 68,* revised January 1997.

12.12   TYPICAL HERD HEALTH PROGRAMS

Most herd or flock managers follow a specific and planned health plan, based on the life cycle of the animals and/or based on the calendar. An example of a recommended swine herd health calendar, based on time and age, is shown in Table 12–6. Also, a detailed example of a health management plan for beef cattle, based on month of the year, is shown in Box 12–2.

All health programs depend heavily on complete records on individuals or on groups handled together.

Scheduled visitation and consultation by a veterinarian, perhaps under a contract that calls for payment per animal or per hour, is often an integral part of a herd health plan. A dairy herd contract may include vaccination, udder health, and foot care plans, plus periodic visits for pregnancy testing, vaccinating, and assessing mastitis, foot infection, calf disease, and breeding problems. An abbreviated example of a dairy herd health program is shown in Fig. 12–1. The objective of all herd health programs is prevention of disease and maintenance of normal health and high production in the herd or flock.

BOX 12–2	**Beef Cattle Health Management Calendar**

Month	Health Practices
January	Look for mange.
	Watch for abortions.
	Treat for lice.
February	Watch for foot rot.
	Consider *E. coli*–type vaccines for control of calf scours. These are given to cow before calving.
March	Provide clean, dry calving facility. Have heat lamps on hand. Know how to feed weak calves with stomach tube. Have serums on hand for calf that can't nurse.
April	Provide grass tetany prevention with MgO (0.5 oz per cow per day) in mineral mixture.
	Deworm all cows before turning out to pasture. Repeat in 3 weeks. Where trichomoniasis is a problem use vaccine before the breeding season. Use two injections 4 weeks before the breeding season.
	Castrate all bull calves and implant.
	Dehorn before fly season.
May	Fertility-check bulls.
	Vaccinate open cows for vibriosis, IBR, BVD, leptospirosis (5 serotypes) prior to breeding.
	Vaccinate all calves and young stock for pinkeye.
June	Get ready to use AI.
	If using natural breeding, one bull (fertile) is necessary for 25 cows.
	Implant calves with growth promotant implant specifically manufactured for calves.
	Check on use of estrus synchronization products. Cows should be cycling in order to settle. Estrus synchronization will work if management factors are used as recommended.
July	Vaccinate all calves more than 3 months old for pinkeye, clostridial group (Blackleg, etc.), Pasteurella, haemophilus, and BRSV.
	Vaccinate heifer replacements with strain 19 at 2–6 months.
August	Check on pinkeye and treat with antibiotics.
September	Watch for grass tetany.
	Check for foot rot and cancer eye.
	Check fly control.
October	Pregnancy-check cows.
	Precondition calves while pregnancy-checking cows.
	Deworm both cows and calves and treat for external parasites.
November	Precondition calves and wean.
December	Check cows for abortion and heat periods.

Source: Adapted from Herrick, J. B. "Cattle disease guide—Beef and dairy." *Feedstuffs* 68, no. 30 (1996): Reference Issue.

Dairy Herd Health Program

Newborn calves:
  Hand-feed first milking of colostrum.
  Treat navel with iodine disinfectant.
  ID calf and place in individual pen or stall.

Vaccination:
  2–6 months of age:
    Strain 19 for brucellosis, in some states.
  4–6 months of age:
    • IBR, parainfluenza, clostridial group, *H. somnus,* and BRSV. Do not administer simultaneously.
    • BVD, two weeks after those above, unless it is killed-type.
  11–13 months of age, or two months before breeding
    • Leptospirosis (5 serotypes) followed by annual vaccination.
    • repeat IBR, BVD, and *H. somnus.*
  Mature cow:
    • One month after breeding, administer booster vaccinations for Leptospirosis, IBR, BVD, and *E. coli.*

Reproduction:
  Herd examination by veterinarian every third Tuesday, to include:
    • Pregnancy exam of cows 30–45 days post-breeding.
    • Exam of cows 0–15 days postpartum and about 30 days postpartum.
    • Exam of cows that have not conceived after 3 or 4 services.
    • Exam of cows not observed in heat by 60 days postpartum.

Parasite control:
    • Worm heifers during first grazing season—3 and 6 weeks after going on pasture.
    • Use approved insecticide on all stock continuously for control of lice.

Udder health:
    • Use recommended pre-milking and post-milking practices.
    • Monitor with CMT at cow-side and Somatic Cell Count reports from DHI records.
    • Carefully ID all cows treated for mastitis; withhold milk for the prescribed period.
    • Use dry cow treatment program as prescribed by veterinarian.

Emergencies:
    • Dystocia: Call veterinarian on very large and/or abnormal presentations.
    • Sick or lame animals: Call veterinarian or treat with veterinarian's advice.
    • Aborted fetuses and dead animals: Arrange for prompt necropsy by veterinarian.
    • Displaced abomasum: replacement by surgery or manipulation.

Figure 12–1
An example of a dairy herd health program. Each herd should have a specific program designed by the herd manager and the veterinarian.
(*Sources:* Adapted from data from Ohio State University; "Cattle disease guide—Dairy." *Feedstuffs* 1999 Reference Issue.)

Figure 12–2
A veterinary field hospital in the Kansas Flint Hills, top. Note cattle working and sorting pens. Bottom photo shows cattle-restraining chute inside hospital.
(Courtesy of Kansas State University.)

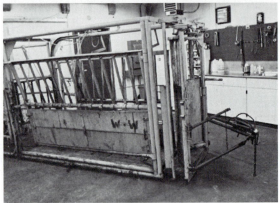

12.13   ROLE OF THE VETERINARIAN

The veterinarian functions in livestock and poultry production systems as a member of the flock or herd health team. That team includes the flock or herd manager, all workers, nutritionists, and perhaps other specialists. The veterinarian has concentrated training in animal diseases, medicine, surgery, and diagnosis (Figs. 12–2 and 12–3).

In pursuit of specific diagnosis, a veterinarian may depend on microbiologists, parasitologists, metabolic or reproduction physiologists, pathologists, biochemists, and scientists in other disciplines. All who work with animals should be familiar with the characteristics of the normal animal, including body temperature, pulse, and respirations. These are listed in Table 12–7. Normal temperatures will vary 0.5 to 1.0°C in 24 hours, with the minimum in the morning and the maximum in the afternoon.

High environmental temperatures will often cause elevated body temperature; so will exercise. Infections commonly cause a fever (elevated temperature). Toxemias or shock will often cause body temperature drop.

Pulse and respiration rates generally increase as the body temperature increases or as a result of excitement or exercise. The pulse rate may be determined by palpating the ventral side of the tail or facial arteries of the cow, the facial arteries of the horse, and the inner thigh of sheep and goats. It is difficult to obtain an accurate pulse in the pig by palpation.

Figure 12-3
Large animal-restraining table with 360° rotation
potential, top, and tilting surgical table, bottom, in a
college of veterinary medicine. Clinical cases referred
by practicing veterinarians provide training
opportunities for veterinary students.
(Courtesy of Kansas State University.)

TABLE 12-7	Average Temperature and Pulse and Respiration Rates of Common Domestic Animals		
Species	**Temperature (rectal)**	**Pulse Rate (per min)**	**Respirations (per min)**
Cow	101.5°F (38.6°C)	60	20
Horse	100.0°F (37.8°C)	35	12
Pig	102.5°F (39.2°C)	60–100	8–18
Sheep	102.5°F (39.2°C)	60–90	12–20
Goat	102.5°F (39.2°C)	60–90	12–20
Poultry	106.0°F (41.1°C)	250–350	17–27

Source: *Adapted from The Merck Veterinary Manual, 4th ed. Rahway, NJ: Merck & Co. Inc., 1973.*

Although temperature and pulse and respiration rates are important indicators of health, deviations from the normal behavior, such as listlessness decreased feed intake, slow movement, or unusual posture, are clues to impaired health. Physical traits can best be observed from a reasonable distance and in the animal's natural environment. The following are examples of signs that may help a veterinarian and the flock or herd manager diagnose an illness, the essential step before treatment can be initiated:

- Examination of individual animals to determine the nature of the illness and possible causes usually begins by examination of the eyes, nostrils, and ears. Excessive discharge from the eyes or nostrils suggests respiratory problems. Excess salivation may suggest immobility of the tongue, central nervous disturbances, dental problems, or blockage of the esophagus. The attitude (alignment) of the head and neck is often altered with difficult breathing; an animal with pneumonia often will extend the head and neck.

- Alterations from the normal respiration rate, rhythm, and depth suggest a respiratory disease such as pneumonia. Thorax and brisket edema (swelling with fluid) is commonly associated with a circulatory disturbance and respiratory distress.

- An arched back is commonly associated with pain in the ventral thorax such as that caused by traumatic reticuloperitonitis (hardware disease) where a foreign body (nail, wire, etc.) has penetrated the reticulum.

- The abdomen and its symmetry and fullness often provide clues. Bloat will initiate marked swelling of the dorsal, left abdomen; gastrointestinal blockage, on the other hand, will usually cause a general abdominal distention. Gauntness indicates that the cow likely has not eaten for at least several days. A combination of gauntness and tucking up of the abdomen might indicate abdominal soreness.

- Infections of the vulva, penis, and scrotum may cause swelling of varying degrees. Blockage of the urinary tract in the male, such as with urinary calculi, may be detected by checking the penis for the presence of calculi or stones. Vaginal discharges, other than those associated with the estrous cycle, may indicate infection of the genital tract.

- Swelling of the mammary gland often indicates mastitis, whereas a flaccid udder in a producing cow suggests a systemic infection or metabolic disturbance.

- The feces should be observed closely. The lack of fecal passage or constipation may indicate a digestive disturbance or may be the result of a systemic disease accompanied by high fever. Diarrhea may suggest inflammation of the intestine or stomach. Bright red blood in the feces suggests coccidiosis, or rectal injury. Blood discharged from gastric ulcers and/or the anterior of the intestine would cause very dark feces. Excess mucus glistens in the feces and may result from intestinal inflammation.

- Examination of the limbs for abnormalities is important, especially in the horse. Watching a horse walk or trot will disclose impairments. Founder (or laminitis) will slow and change the rhythm of the gait. So will fractures and dislocations of bones, joint strains, and abscesses on the feet. Soreness of the thorax (chest cavity) causes a changed gait and abnormal positioning of the elbows when at rest.

12.14 DIAGNOSTIC SERVICES

The veterinarian who services livestock and poultry producers and others concerned with animal health are supported in their work by one or more diagnostic laboratories administered by the state university college of veterinary medicine or department of veterinary science in most states, or in some by the executive branch of the state government. The laboratory subjects tissues, live animals, or dead animals to routine testing or to detailed diagnostic scrutiny to ascertain or confirm causes of illness or death, and issues timely reports by phone or otherwise so that treatment may begin or the management regime may be adjusted.

Colleges of veterinary medicine also operate a clinic or hospital, which provides learning opportunities for veterinary students and handles routine and emergency cases in the geographic area of the college and special or unusual cases from throughout its own and neighboring states. College, diagnostic laboratory, clinic, and extension veterinary faculty present conferences and short courses for veterinarians and producers, and publish bulletins and other materials on animal health.

Many producers of livestock and poultry also have diagnostic services provided by some of the larger commercial firms, especially in vertically integrated enterprises where responsibilities and returns are shared by both parties.

12.15 BIOSECURITY

A successful herd or flock health program requires a rigid "biosecurity" plan to reduce the risk of outside diseases being introduced into the home farm and/or production unit from outside sources. This means that all animals should be protected from the risk of diseases possibly spread by people, equipment, or vehicles. Some practices to be considered in implementing a biosecurity program include the following:

- Allow only farm personnel in animal units.
- Require personnel caring for animals in isolation or at other multisite facilities to follow an approved procedure before entering other animal units on the farm.
- Require requested or invited visitors to wear durable plastic boots.
- Follow approved procedure upon infrequent visits to other producers' facilities and return to home farm.
- Thoroughly wash and disinfect equipment from isolation units, other multisite units, or outside sources before and after use.
- Instruct all personnel responsible for service truck deliveries on biosecurity procedures upon entering the farm.
- Provide isolation unit for newly purchased livestock. Biosecurity procedures should be as rigid as those on the other home facilities. Any waste from the isolation unit should not be flushed into a lagoon where water will be recycled into other facilities.
- Birdproof confined structures and maintain an effective rodent control program.

Any biosecurity program should be practical, cost effective, and understood by all farm personnel.

QUESTIONS FOR STUDY AND DISCUSSION

1. What are several signs of a healthy animal?
2. What is meant by the term *zoonoses*? Name four such diseases.
3. What is the difference between infectious and noninfectious diseases? List two examples of each.
4. Which element is most likely deficient when each of the following disease conditions is observed? (a) rickets (b) parakeratosis (c) anemia (d) goiter
5. What causative factors are usually associated with (a) rumen acidosis, (b) grass tetany, and (c) urinary calculi?
6. What are the consequences of heavy infestations of internal parasites, such as intestinal roundworms or tapeworms?
7. Explain how the cattle grub is perpetuated.
8. What is a disease vector? Give an example.
9. Are all infectious diseases contagious? Explain.
10. Name two characteristics of viruses. How do their characteristics differ from bacteria?
11. How would you define *protozoa*? Give an example of a protozoan disease.
12. Name the type of organism that causes the following diseases: (a) brucellosis, (b) hog cholera, (c) coccidiosis, (d) salmonellosis, (e) equine encephalitis.
13. What are mycoplasmas?
14. Give examples of how an animal would acquire (a) passive immunity and (b) active immunity.
15. Explain how the "test and slaughter" program helped to eradicate brucellosis in cattle.
16. Explain how a herd health program might work.
17. What are examples of visual signs of illness the herd manager might note prior to communication with the veterinarian?
18. Where are diagnostic services available to the livestock and poultry producer?

13
Animal Behavior

Animals, like people interact, communicate, develop friendships or attachments, are dominant over or submissive to others, have some need for privacy or "territory," and are affected by social relationships.

A thorough understanding of animal behavior—how they function individually and in groups—plus thoughtful observation of animals can benefit livestock and poultry production enterprises. Behavioral patterns influence the manner in which cattle or sheep utilize grass on open ranges. Observation of these patterns may influence decisions on size and shape of pastures, numbers of animals to run per group, placement of watering facilities or salt and mineral mixtures, or the use of rotation versus season-long grazing. Similarly, understanding and observation may help determine the optimum size of swine finishing groups, design of equipment, the amount of feeding space needed for laying hens, or how soon and under what conditions ewes and young lambs might be penned with others.

Awareness of behavioral patterns and how they might be learned or conditioned permits use of devices or systems that can save operator time. Dairy cows have been conditioned to leave the pasture for the milking parlor at the ringing of a bell,[1] or to begin the milk let-down process for speedier milking at the sound of a radio in the milking parlor. The tendency of cows to enter the parlor in a fairly regular sequence (Figure 13–1) can make record keeping and milking management easier in herds without an electronic cow identification system.

This chapter is devoted largely to animal behavior as a science, components of behavior, animal communication, normal behavioral traits and how these can aid in managing production and certain behavioral problems. Domestic animal behavior can be viewed in two ways: (1) their behavior within a confined condition, or (2) their behavior in a more natural environment. In either case, scientists are trying to understand how these animals interact and function in the best welfare and economic manner.

Upon completion of this chapter, the reader should be able to

1. Define *ethology, social order,* and other key terms.
2. Differentiate among instinct, imprinting, and the learning that results from training.
3. List the sensory systems by which an animal receives stimuli.
4. Differentiate between seeing and perceiving.

---

[1] Kiley-Worthington, M., and P. Savage. *Applied Animal Ethology* 4 (1978): 119–24.

Figure 13–1
Cows enter a modern rotating milking parlor in a regular and particular sequence.
(Courtesy of Purdue University.)

5. Describe how behavioral traits of a species would influence the design of facilities for that species.

6. List several behavioral problems common in livestock and poultry enterprises.

7. Explain why it is important for an animal manager to imagine being in the animal's place as a member of that species when evaluating the appropriateness of animal facilities or animal handling systems.

13.1   ANIMAL BEHAVIOR AS A SCIENCE

Ethology
The scientific study of animal behavior in their natural or typical environments.

Ethology is the study of animal behavior in the animal's natural habitat. For nondomesticated species, this would be interpreted as being "in the wild." For domesticated species, it would be interpreted as the habitat in which the species is normally maintained, such as a pasture, poultry house, feedlot, or other facility. Scientists who study animal behavior with the goal of improving animal management are called *applied animal ethologists*. Ethology is a relatively new discipline within animal science, though some of its principles have been used in animal production for years.[2]

Three ethologists shared the 1973 Nobel Prize for physiology and medicine. Karl von Frisch discovered how bees transmit information, such as direction and distance of a nectar source, to other hive members. Prior to his discovery, most people believed that animals, especially insects, were not capable of communicating meaningful information to each other.

---

[2] Much of the early knowledge on animal behavior was obtained from nutritional and reproductive studies. Today, most agricultural institutions have researchers trained or specializing in ethology.

Konrad Lorenz, generally considered to be the founder of ethology, discovered **imprinting**, an especially rapid and relatively irreversible learning process that usually occurs within hours or a few days after the birth of birds and mammals. As a basic concept, imprinting includes an animal learning who its mother is and to what species it belongs. The concept has been expanded to include other types of knowledge, such as the imprinting of salmon to the stream in which they were hatched.

Nikolass Timbergen, the third recipient, elucidated survival behavior of animals. For example, he observed that in many species of birds, as eggs in the nest hatch, the parents carry the eggshells some distance away, reducing the attraction of predators to the nest.

> *Imprinting*
> The rapid and relatively irreversible learning that occurs within a few hours or days of birth.

13.2 DOMESTICATION AND GENETIC INFLUENCES ON BEHAVIOR

Just as there are inherited differences in physical traits among species, breeds, and individuals, there are also inherited differences in behavioral traits.

Genes provide the "blueprints" not only for the brain, nervous, and hormonal systems, but also for the biochemical, anatomical, and physiological processes which, in turn, govern such phenomena as perception and reproductive cycles.

The domestication process amounted to selecting, over time, those species that had behavioral traits permitting control and management of the animals by people. These traits, though influenced by environment, are largely inherited.

Several inherited behavioral traits found in most domestic animals permit their use in commercial agricultural enterprises. To an extent, these traits became fixed during evolution of the species and perhaps permitted domestication.

1. *Gregariousness.* Farm animals may be combined into and generally are content in herds or flocks. Some animals, such as the bear, live most of their lives in a solitary existence. Others generally live in small families or bands and would not flourish in a large group. Certain breeds of sheep, such as the Merino, tend to flock together; gregariousness is less evident in other sheep breeds.

2. *Social organization.* Members of herds or flocks organize by social dominance or "pecking order." This is discussed in more detail in Section 13.9.

3. *Promiscuous matings.* If farm animals mated for life, as occurs in many wild species, it would be necessary to have one bull for each cow, for example. Although preferences for certain members of the opposite sex have been demonstrated in most domestic species, it is a sufficiently mild preference that producers can use one sire to breed many females and can also make specific mating decisions to achieve herd improvement goals. (Animal behavior during estrus is discussed in Section 13.12.)

4. *Precocial young.* Foals, lambs, calves, piglets, and kids are born with their eyes open and can stand and follow their mothers within an hour or two of birth. This permits lower cost group management. They do not require as much maternal care as do human infants or puppies.

5. *Adaptability.* Domestic animals will adapt to a wide range of environments, including management systems and feeds.

6. *Limited agility and docile temperament.*   Cattle and most other commercial species can be contained with relatively simple, inexpensive fencing that deer, for example, would readily jump over or go through. There are breed differences. Hereford cattle are an especially docile breed, whereas Brahman cattle are more responsive and usually require stronger and taller fences.

Most of the above traits would be called *instinctive*—behaviors an animal exhibits in the absence of any opportunity to learn them. For those traits that are instinctive, the animal appears to be "preprogrammed" for its central nervous system to respond to a circumstance or a specific stimulus in an established way. A common example is the ability of animals to swim. A dog that previously has not been near a body of water usually can swim easily, whereas humans generally lack that instinct.

The tendency of a chick to break through the eggshell after it reaches a certain stage of development during incubation or of a newly born pig to work itself around to the sow's udder for nursing are other examples.

Maternal behavior in farm species appears to be largely instinctive. A heifer calf may be taken from its dam at birth and not experience or see the nursing process, but it will likely show normal maternal behavior when it matures and calves. Some other species of primates, such as chimpanzees and gorillas, must *learn* maternal behavior by observing their own mothers and other mothers in their troop. Mating behavior is largely instinctive; most domestic animals will copulate successfully after puberty without having observed others, although some males, because of youth and inexperience, have difficulty in initial matings.

Some instinctive behaviors occur even in the apparent absence of an appropriate stimulus or circumstance. Hens on open range or in the wild dust themselves from time to time, apparently to inhibit insect infestations. A hen maintained in a cage or on dry concrete may go through the motions of dusting itself even though there is no dust.

Because many behavioral traits are influenced by heredity, they can be changed by selection. Over the generations since domestication, there has been *natural* selection for desired behavioral characteristics—animals that were comfortable with confinement rearing grew well, remained healthy, and reproduced; animals that were not adapted did not or at least were less efficient. As a result, a higher proportion of the animals in subsequent generations were adapted to a human-controlled environment.

Beyond this, there has been the additional selection pressure of human decisions as to which female or male to breed, based consciously or subconsciously on observation of the animal's behavioral traits. For example, unruliness in a mare, cow, or ewe, or in a male may be the specific reason for not mating or for culling the animal; or the unruliness may exert a subconscious influence on selection or breeding decisions ostensibly made for other reasons.

The development of the egg incubator eliminated the need for broodiness (interrupting laying for "setting" on a clutch of 10 to 12 eggs) in laying hens, so selection to eliminate the trait was practiced. Poultry breeders increased egg production by reducing the incidence of this trait from over 90 percent of the birds to less than 20 percent in only five generations of selection.[3]

---

[3] Goodale, H. D., R. Sanborn, and D. White. *Bull. Mass. Agr. Exp. Sta!,* 1920.

Heritability estimates ranging from 0 to 0.57 have been calculated for such traits as social dominance or aggressiveness in livestock and poultry.[4] Sows that demonstrate good mothering abilities often produce gilts that demonstrate the same characteristics when they reproduce. A calm foal may be from a sire and dam with similar characteristics.

13.3 ENVIRONMENTAL INFLUENCES ON BEHAVIOR

Environmental effects on behavior appear even more complex and varied than genetic influences. The genetics of an animal is established at conception, whereas the environment has influence from that time forward and will change throughout an animal's lifetime.

Environment during fetal development can affect an animal's behavior after birth. It has been demonstrated in laboratory animals that psychological stress during pregnancy, especially during periods of rapid growth of brain and nervous tissue in the embryo, can cause permanent abnormal behavior when that embryo has become an adult.[5] Alcohol, caffeine, and other drugs, physical injury, and illness of the mother during pregnancy have been demonstrated to alter behavior of the offspring.

Imprinting, described in Section 13.1, as well as the learning during handling, feeding, and grazing, result from postnatal environmental influences.

Daylight is an environmental influence. **Photoperiod**, the amount of daylight exposure in a 24-hour period, influences egg laying in chickens as well as the breeding season in goats, sheep, and horses, and can be manipulated. Increasing day length stimulates egg production and decreasing day length suppresses it. An artificial photoperiod is also used to manipulate the breeding season in chickens, horses, and goats. (See also Section 33.4.)

> **Photoperiod**
> The length of daylight or artificial light provided.

The social group within which an animal lives is an important environmental influence. Farm animals tend to reach puberty earlier, for example, when kept with their own species, especially if they are kept near members of the opposite sex. Group size and the age, sex, and dominance ranks of other group members can all affect an animal's behavior. (See also Section 13.9.)

13.4 PHYSIOLOGY

The physiology of an animal is sometimes considered the link between heredity and environment and may be defined as the physical and chemical pathways by which the body functions. Physiology provides the means by which environmental stimuli are *perceived* and a body *reaction* occurs, within the potential established by **heredity**.

> **Heredity**
> A study or description of genes passed from one generation to the next through sperm and ova. The heredity of an individual would be the genes received from the sire and dam via the sperm and ovum.

Most actions of the nervous system result from sensory experience, received through sensory receptors—visual, auditory (hearing), tactile (touch), taste, olfactory (smell), and others. The body can initiate a reaction to stimuli at three points or levels: the spinal cord, the lower brain (hypothalamus and brainstem), and the brain cortex.

---

[4] Craig, James V. *Domestic Animal Behavior.* Upper Saddle River, NJ: Prentice Hall, 1981.
[5] Dahlott, L., E. Hard, and K. Larsson. *Anim. Behav.* 25 (1977): 958.

When the reaction is initiated at the spinal cord, it is called a *reflex*. For example, when a sensory nerve on the finger touches something hot, a message to the spinal cord stimulates a motor nerve to contract the appropriate muscles and withdraw the finger.

The hypothalamus and brainstem control *subconscious* activities necessary to maintain life, such as blood pressure, heart and respiration rates, body temperature, and coordination of movement. The hypothalamus is also involved in anger, excitement, sexual activities, pleasure, and hunger, though at times these may also be under conscious control.

Storage of information, learning, thought, and conscious control of movement are controlled at the brain cortex. The cortex is packed with neural cells that achieve these functions.

The endocrine system, including glands that produce and release hormones into the blood, regulates body metabolism. For example, estrogen and progesterone control reproduction cycles and estrous behavior of the female; prolactin causes broodiness in hens. The endocrine system is influenced by neural activity, especially through the hypothalamus, resulting from both internal and external stimuli.

Psychological stress, such as results from isolating a ewe from its flock or a cow from its herd, or moving animals long distances to a new environment, can, through endocrine function, reduce receptivity to a male, conception rate, or resistance to disease, and otherwise affect body functions.

Behavioral differences among species, breeds, or lines may be related to physiological or anatomical differences. Animals in outside lots or pastures devote a considerable amount of their activity to thermoregulation—moving to shade, seeking water or a breeze, or seeking shelter. Brahman cattle have about five times more sweat glands per unit of body surface than do Hereford or Angus cattle, and the Brahman glands produce more perspiration per gland (Figure 13–2). Swine have neither effec-

Figure 13–2
Hereford cattle resting in the shade while a herd of Brahman cattle in the next pasture graze during a hot afternoon in Texas.
(Courtesy of Texas A&M University.)

tive sweating nor a panting mechanism (as do dogs), so they tend to wallow in water or seek out a sprinkler or shade to maintain a normal body temperature during extreme heat.

13.5   PERCEPTION AND THE SENSORY SYSTEMS

Each species no doubt has a characteristic and unique perception of "reality." Further, because of varied genetic makeup and experiences, individuals may vary in their perceptions. Applied animal ethologists study animal perception because principles learned help explain why animals behave as they do. Figure 13–3 describes how perception occurs. It's also likely that certain animals are more aware or rely on specific sensory systems as compared to using them all equally. That is, some animals most likely use sight or hearing much more than touch (and vice versa for other animals).

Vision

The sensory systems of livestock and poultry are similar to those of humans, but the capabilities of the systems differ markedly. Consider the difference in size of the visual field and other items related to vision. Eyes placed on the side of the head in grazing animals provide an almost 360° field of vision (Figure 13–4). The slotted or rectangular pupils (Figure 13–5) provide a wider visual field than is provided by pupils of predators. Humans, eagles, and other predators, however, have more depth perception because the frontal placement of the eyes permits a larger field of *binocular* vision.

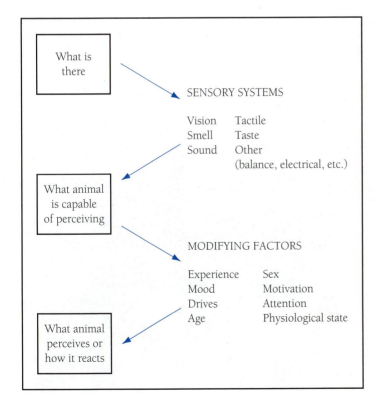

Figure 13–3
Perception in an animal results from the animal sensing what is there, then modifying, according to experience or other factors, what is sensed.
(*Source*: Texas A&M University.)

Figure 13–4
The approximate field of vision of a steer. Binocular vision aids in depth perception due to the stereo effect of two eyes. (*Source:* Texas A&M University.)

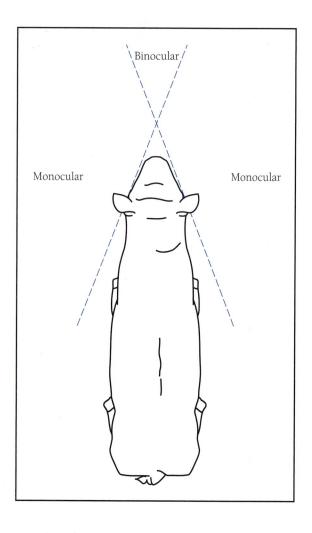

Figure 13–5
The shape of the pupils of grazing animals and the placement of the eyes on the sides of the head affords them an almost 360° visual field.
(Courtesy of Texas A&M University.)

Depth perception is necessary for judging distance. Grazing animals are prey species (subject to attack by predators) and their vision is adapted for watching a maximum area around them. They also apparently have a unique method of focusing that permits them to simultaneously focus on both near and distant objects. But because they lack a *tapetum* (area of concentrated light-sensing cells in the retina found in humans and other predators), their vision is thought to be not much sharper than the peripheral vision of humans.

There are also differences in the wavelength of light that can be perceived. In a typical classroom, electromagnetic waves range from about 10^{-7} to 10^{17} nanometers (nm) in length.[6] The average human can detect as visible light only that portion from 390 to 770 nm. Bees, however, can perceive light in the ultraviolet range (well under 390 nm) and therefore discriminate among flower markings and designs that are invisible to the human eye. Many researchers believe cattle, sheep, goats, and horses cannot see red, whereas swine and poultry are presumed to see all colors. However, there is little research to support this belief.

Smell

The sense of smell is so much more acute in livestock than in humans that it is difficult for humans to even imagine the environmental information livestock receive by smell. All livestock have an accessory olfactory organ called the vomeronasal organ (VNO) located between the mouth and nasal cavities. This structure enables animals to make ultrafine discriminations among odors humans cannot even detect. When a bull, ram, or stallion performs *flehmen* (smelling the urine of a female to determine if she is in estrus), he is inspiring molecules of odor into the VNO for identification (Figure 13–6).

Hearing

Livestock and poultry have a sense of hearing believed to be within the frequency range of humans, whereas bats and dogs can perceive sounds of much higher frequency than can humans. Those species with relatively large ears that can be directed independently,

Figure 13–6
Flehmen or "lip-curling" in a stallion after smelling urine from an estrous mare. (Courtesy of Texas A&M University.)

---

[6] A nanometer is one billionth of a meter.

Figure 13–7
Mules possess larger ears than horses which enable them to more easily detect and locate the source of most sounds.
(Courtesy of Texas A&M University.)

such as mules, can easily detect and locate the source of sounds. They may also listen simultaneously to sounds from opposite directions (Figure 13–7).

Touch

The tactile sense (touch) is well developed in farm animals. Cattle and horses easily detect an insect on their skin and can selectively shake skin areas to dislodge most insects. All species show considerable awareness of their bodies and will rub on posts or trees, scratch themselves with hooves or horns, and groom themselves.

Touch is important in farm animal communication. In general, gentle touch gives reassurance and helps establish and maintain social bonds, as is true in humans. Animals lick and scratch each other, swat flies from one another (observed in horses standing head to tail), and in other ways groom each other.

Pigs like to be physically against one another to the extent that much of their skin touches the skin of another pig. Brahman cattle have closer contact with one another when lying down than do European breeds of cattle. Mature horses do not lie against each other.

Observing where animals like to touch or scratch each other, and where it is difficult for them to scratch themselves, will suggest where an animal might enjoy being petted or scratched. Also, observing their preferences for physical contact can influence the decision to turn them out rather than stable them alone.

Taste

Livestock readily distinguish among the four tastes—bitter, sweet, sour, and salty—as well as among intensities of these flavors and flavor combinations. For example, it has been demonstrated that cattle can discriminate between a 4 and 5.5 percent sugar solution and between a 0.05 and 0.09 percent salt solution.[7] Feeding cows silage (sour taste) will decrease taste sensitivity to other sour feeds and increase sensitivity to sweets, similar to the effect with humans. Even though livestock rely significantly on taste in dis-

---

[7] Kudryavtzev, A. A. *Proceedings of the XVI International Dairy Congress,* Copenhagen, 1962, p. 565.

criminating among feeds, poultry are believed to have a relatively poor sense of taste and to rely more on sight. Coloring poultry feed can significantly influence intake, whereas feed flavor and odor have more influence on feed intake by cattle, horses, or swine.

Other Factors Affecting Perception

As with humans, everything a sensory system detects is not necessarily *perceived* by the animal. As sensed information feeds into the central nervous system, much is ignored and some is processed, together with information from memory. The sorting depends on experience or learning, age, physiological state, mood, attention, motivation, and other factors.

For example, consider a range cow that has never seen a feed trough and a dairy cow accustomed to eating from one, both placed in the same pen with an empty trough. Though both animals would *see* the trough, each would *perceive* it differently and behave differently. The dairy cow would likely associate it with feed and approach it; the range cow might perceive it as a strange object and show fear.

There are individual as well as breed differences in the sensitivity of animals to touch, sudden movements, and sounds. For example, Colorado State University studies suggest that Holstein dairy cattle are more sound and touch sensitive than beef cattle when subjected to sudden movements or intermittent sounds as observed in livestock auctions. Their studies also showed that beef steers and heifers are more motion sensitive than older bulls and cows.

A cow that is normally docile and easily approached may, after her calf is born, threaten a person who approaches closely. The cow's sight is no different after birth of her calf, but her perception is, and so too is her behavior.

An uneasy animal or one in new and strange surroundings may become frightened by something that normally would not cause fright. A good animal handler is aware of and alert to the factors that affect an animal's perception of stimuli and will anticipate the resulting behavior.

13.6   LEARNING AND TRAINING

Livestock and poultry constantly learn in response to changes in their environment. Animals learn to know their group members, location of water and good grazing or foraging areas, indicators of predators, and other items. A horse or pony learns to respond to cues from the rider. Below are four concepts basic to learning and training:

1. If reward for a behavior is to increase the occurrence of that behavior, or if punishment is to decrease its occurrence, the reward or punishment must occur simultaneously with or immediately after the behavior. Confusion will result if too much time elapses. For example, should a horse misbehave in a show ring, prompt and proper punishment should be administered. Punishing the horse after it leaves the ring causes confusion and fear.

2. Reward or punishment in response to a specific behavior usually needs to be repeated several times to achieve learning.

3. Occasional retraining or reinforcement is necessary. An animal forgets, and repeated training provides the basis for memory.

4. The amount of emotional arousal (excitement) can influence how rapidly and what an animal learns. If an animal is punished, to the extent of becoming fearful, for a behavior it cannot help, the animal may not learn to avoid the behavior but may rapidly learn fear of the handler.

Animals can learn from each other. Not all members of a herd of cattle have to be shocked by an electric fence to learn to avoid it. When one gets shocked, jumps, and runs, the whole herd will likely run and, in time, will learn to avoid the fence. If a herd is confined by an electric fence, only about 30 percent of the herd will ever be shocked.

A specialized type of learning is the development of an aversion (dislike). If an animal eats something that makes it very sick, the animal will likely develop a strong aversion for that feed, even if it had formerly been a preferred feed.

13.7   ANIMAL COMMUNICATION

Animals regularly communicate via several sensory systems with members of their own species. There is also some communication across species.

Poultry and livestock vocalizations have distinct meanings. Chickens, for example, have a feeding call that will attract other chickens to feed. In fact, a very special call occurs between the hen and her chicks during and after feeding.[8] An experienced poultry handler can detect distress among hens by listening to vocalizations. Four to seven distinct vocalizations have been described in the horse. A snort likely means danger; a whinny or nicker may be a distress call.[9] Additional meanings can be conveyed by variations in intensity and by repetition.

People have attempted to teach other species to use the human language but with little success. Other species lack the anatomy to generate the wide spectrum of sounds present in the human voice. People can, however, effectively communicate to other species with the human voice patterns. A calming effect on animals usually results from low and soft tones of long duration. Short, high-pitched tones will often excite animals.

Poultry and livestock rely heavily on visual cues as a means of communication. Many are so subtle they are overlooked by humans. A slight change in posture, eye contact, or position of the ears or tail may convey considerable meaning. Repeated or exaggerated behavior—rapid tail swishing or prancing—may convey meaning. Horses are so expressive that an observing human can "read" degrees of threat, submission, excitement, and playfulness (Figure 13–8).

Touch, discussed in Section 13.5, is the least studied of animal communication modes, but it may be one of the most important. This is suggested by the degree of mutual touching and grooming that occurs among animals. There is variation among species. Horses do considerable mutual grooming; swine rarely groom each other, yet are commonly observed lying in contact with one another. Humans generally can communicate reassurance and can calm livestock by long, soft strokes, whereas short, hard pats may generate fear or cause arousal.

---

[8] Woodcorse et al. SPSS Abstract, Atlanta, GA, 2002.
[9] Feist, J. D., and D. R. Z. McCullough. *Tierpsychol.* 41 (1976): 337.

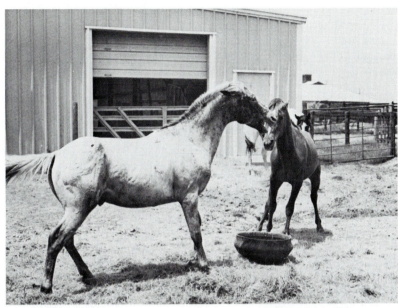

Figure 13–8
The stallion (left) has made a very severe threat (ears flat back, teeth bared, head and tail up, neck somewhat arched) to the horse that was eating at the feeder, which is now backing away while making a protest or bluff (ears slightly back, mouth closed and teeth covered, head and neck lower than that of the aggressor).
(Courtesy of Texas A&M University.)

13.8   UNDERSTANDING ANIMAL BEHAVIOR

Most successful operators understand their livestock or poultry to the extent that they anticipate their reactions to many stimuli. Such understanding is helpful when handling animals, designing facilities, and managing grazing, breeding, or other components of production systems. Recognition of what may frighten or startle an animal and cause it to bolt or kick may help prevent both animal and human injuries.

Many people treat animals as though they were humans, and are surprised that an animal does not react to a stimulus in the same way as a human. For example, most animals do not appreciate a pat on the head or scratching the forehead. The act of reaching over an animal's head to pat it is perceived as a threat in many species, and scratching a bovine between the eyes is perceived as a challenge, an invitation to start butting. Animals also tend to treat people like others of their own species.

Two members of the same species rarely maintain eye contact or stare at each other for very long. Similarly, animals become uneasy when humans stare at them. When a person stares into a dog's eyes, it will usually turn away or growl. The stare is perceived as a threat.

Many problems in handling livestock or poultry, and in designing facilities for them, can be avoided if the handler and designer have a high level of empathy for the animal—can imagine themselves in the animal's place as a member of that species. This is difficult, except among the most observing, attentive, and thoughtful people. It requires a knowledge of both the animal's behavior repertoire and how it perceives things.

Figure 13–9

The open walls of this cattle chute allow animals to see all that is occurring in front as well as all around them. The frightened cattle will balk, and attempt to back up or turn around. The time and labor required to work them will increase, as will the stress to the cattle.

(Courtesy of Texas A&M University.)

Figure 13–10

A well-designed cattle handling facility with hydraulic squeeze chute and dipping vat. Cattle readily go through the curved lead-up chute because they sense they are heading back to where they came from. Also, from a steer's point of view, right, they can see an escape route without being distracted by what is going on around them.

(Courtesy of Texas A&M University.)

To illustrate, you might pretend to be a steer walking through a cattle chute in order to determine what may cause cattle to balk. You would need to imagine almost 360° vision, little depth perception, that the head is close to the ground, and that there is hypersensitivity to stimuli due to fear. Note the differences in design of cattle chutes shown in Figures 13–9 and 13–10.

You might imagine yourself as a horse backing out of a trailer. The horse can see, through its peripheral vision, the dropoff at the back of the trailer, but because it cannot turn around and look with binocular vision, it lacks a perception of distance. A horse that has been trailered frequently may trust that the dropoff is not large, but a less experienced animal may hesitate, testing the distance carefully with a hind leg, or even balk. A person with empathy who can effectively imagine being in the animal's place would interpret the reluctance to back out as normal, not as misbehavior. Interpreting the reluctance as normal behavior will likely result in patience on the part of the handler, giving the animal time to test. Interpreting the reluctance as misbehavior and administering punishment may cause fright, increase distrust, make it more difficult to trailer the horse in the future, and perhaps cause injury to the horse or handler.

Those who are unable to effectively imagine themselves in the animal's place as a member of that species may ask when they see a sow in a farrowing crate or chickens in a cage, "How would I feel confined like that?" rather than, "How would I feel if I were *that animal?*" A sow, which spends as much as 80 percent of her day asleep, may be content when routinely confined. In contrast, a horse requires more exercise and is stressed by close confinement for long periods. Constructive research in animal welfare is being achieved by those with good understanding of animal ethology.

In general, the more control people have over animal activity and environment, the more beneficial their knowledge of animal behavior. Behavioral traits should be considered in the design of housing and feeding facilities, and these traits will suggest limits of the animal's adaptability. For example, domesticated species vary in their elimination patterns or habits. Stallions are deliberate in their selection of a location to defecate. Hogs, too, prefer to defecate where it is wet, downhill from their sleeping area, and away from their feed. Slotted floor sections that allow feces and urine to fall through should be provided, floors should be sloped, and feeders and waterers should be arranged to take advantage of this preference. Swine will more likely establish good dunging patterns in a rectangular pen rather than in a square pen. In contrast to horses and swine, sheep and cattle do not appear to have patterns of elimination.

The following sections provide more detailed discussions of behavioral traits that are significant in the controlled conditions under which many commercial animals are maintained.

13.9   GROUP BEHAVIOR AND SOCIAL ORDER

Farm animals naturally form herds or flocks, which permit them to be managed in large groups with reasonable efficiency of feeding and management. In fact, they have such a strong group instinct that isolating an individual from its group can be very stressful. Isolated individuals usually become nervous and their behavior is more difficult to predict than that of a group.

Free-ranging herds usually restrict their activity to a given area called a **home range.** Home ranges frequently overlap, and the occupants of a range generally do not attempt to defend it from other herds. Some breeds of poultry, on the other hand, are very territorial. The flock will claim a geographical area that it defends against invasion by members of other flocks. In such a case, feed and water should be located so that members of one flock are not forced to enter the territory of another. In some species these tendencies are called pack or territorial instincts. Dogs are used to protect people's homes and bands of sheep partly because they have strong pack and territorial instincts (Figure 13–11). For more discussion of the breeding and behavior of guardian dogs in the protection of herds or flocks from predators, see Box 13–1.

Members of flocks, herds, or packs are organized by social dominance or pecking order. The term "pecking order" is often used because it was first described in relation to how hens peck each other. Generally, if 10 unacquainted birds are placed in a pen, they will begin a pecking process that will establish social superiority or dominance of individual birds. That bird dominant over all others will be free to peck the other 9 but generally will not be pecked by them. The second-ranking bird will peck all but the

> **Home range**
> The territory an animal or group of animals normally occupies and in which it feels comfortable.

Figure 13–11

This female Komondor (a breed of guard dog) is being used to defend the flock of sheep under the tree from stray dogs and coyotes. The dog perceives the sheep as its pack members and the pasture as its territory. The dog would not allow the photographer to enter the pasture.
(Courtesy of Texas A&M University.)

BOX 13–1   Livestock Guardian Dogs in Control of Predators

The guardian dog shown in Figure 13–11 is a Komondor breed originating in France. Several other livestock guardian breeds have been introduced into the United States in recent years. Most are from well-established, white-colored breeds originating in European countries. In addition to the Komondor, other popular white-colored breeds are

● Akbash from Turkey
● Great Pyrenees from France
● Kuvasz from Hungary
● Maremma from Italy

All have been described as being tall, white, and courageous. They generally weigh from 85 to 120 pounds when grown and range from 25 to 34 inches tall at the withers. Unlike herding dogs (e.g., the border collie) that work closely with their owners in herding animals with their quick movements and intimidating eyes, guardian dogs were bred to work alone in protection of animals and property. This accounts for their independent nature and ability to care for themselves.

Bonding, or the acceptance of the herd or flock as its social group, is essential for the guardian dog to perform well. It is introduced to the species of livestock to be guarded within 8 to 12 weeks of age. In a short period, it bonds with the animals, accepting them as its equal and, in some cases, as dominant to itself. It seldom if ever becomes confrontational with the guarded animals. Once trained, it stays with and receives its food with the guarded animals.

The guardian dog spends much of its time at a location overlooking the animals and surrounding area. Should an intruder approach, most guardian dogs assume their preventive, defensive role and take a position between the intruder and the herd or flock. With most dogs, an intruder will not be attacked until it has ignored the guardian dog's warnings and does not withdraw. Training and management of a guardian dog requires considerable patience and understanding from its owner, especially of the independent nature that has been bred into the dog through many generations.

Source: Adapted from Taylor, Tamara. "Livestock guardian dogs—one answer to predators." *Dairy Goat Journal,* January/February 1996.

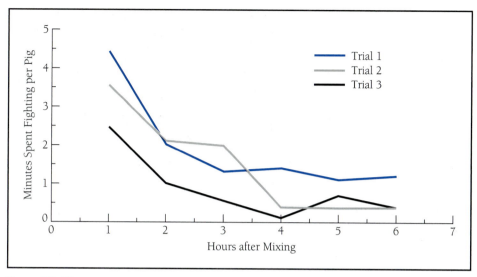

Figure 13–12
Amount of time spent fighting per pig during the first 6 hours after mixing.
(*Source:* Stookey, J. M., and H. W. Gonyou. *J. Anim. Sci.* 72 [1994]: 2804.)

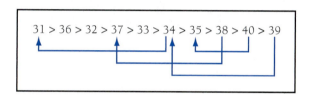

Figure 13–13
Dominance hierarchy in a herd of 10 horses. Number 31 dominates all except 34. Number 34 is subordinate to all listed above it except 31.
(*Source:* Houpt, K. A., et al. *Applied Animal Ethology* 4 [1978] 273–83.)

top bird, and so forth, down to the bottom bird which may be pecked by all others but will not peck any of them back. Once the order is established, the birds may well live in contentment as long as there is ample feed, water, and space. Should these be limited, the dominant birds at the top of the order will tend to have and demonstrate priority. Those at the bottom of the order may starve. Social order serves, then, as a system that gives group members a priority of access to something, such as feed.

When a group of unacquainted animals are first put together, a large number of fights usually occur while they establish the dominance order, which takes about 3 to 6 hours in small groups of pigs and perhaps a day or more in larger groups. Stookey and Gonyou[10] concluded that regrouping pigs 2 weeks before marketing will negatively affect daily gains without sufficient time for compensatory gain or recovery (Figure 13–12).

Dominance orders are usually linear (A–B–C–D) in small groups but become more complicated in large groups of animals. In one study of 11 herds of domestic horses and ponies, linear hierarchies tended to be formed in herds of 3 to 5, but in larger herds, triangular relationships were also observed (Figure 13–13). Dominance

---

[10] Stookey, J. M., and H. W. Gonyou. *J. Anim. Sci.* 72 (1994): 2804.

was determined by use of a paired feeding test—2 horses and 1 bucket of grain. Biting, kicking, threats to bite or kick, and other forms of aggression were demonstrated, and the horse that controlled the feed bucket for the longest period was considered dominant.

In most species, social dominance is expressed rather strongly among males or among females at breeding time. When a young bull is to be used in a herd where older bulls are also used, it is usually wise to put the younger bull and a portion of the cows to be bred in a separate pasture. Section 19.7 includes some discussion of aggressiveness versus fertility.

Stress may be greater and growth or production less among animals while social order is being established, and this stress may well continue, especially for those at the lower end of the order. Among laying hens, the use of cages, with 3 to 6 birds per cage, diminishes that stress. In swine farrowing operations, managers build weaned litters of pigs into larger groups. In some less-intensive operations, two sows and their litters may be put together, then several litters combined at weaning time.

13.10   GRAZING AND FEEDING BEHAVIOR

Horses graze closely to the ground surface, by use of upper and lower incisors. Cattle and sheep have only lower incisors, and cattle cannot graze close to the ground surface because of the thickness of the lips and their tendency to grasp grass with the tongue, then pull or cut it off by use of the incisors on the lower jaw and the upper dental pad. Sheep, however, being smaller and with thinner lips, are able to graze close to the ground surface.

The horse will cover considerable area while grazing, taking a step or two with every few bites. Sheep or cattle will travel less and also will intermittently lie down to rest and ruminate (Figure 13–14). (See Section 4.6.) Times of more active grazing tend to be near sunrise and dusk, plus periods during the morning and afternoon. There are breed and herd variations in grazing patterns that may also be influenced by temperature, wind, and quality of forage.

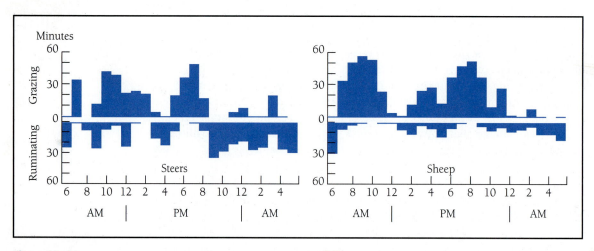

Figure 13–14
Time devoted by sheep and cattle in a 24-hour period to grazing trefoil pasture and to ruminating. (*Source:* Lofgreen, G. P. et al., *J. Animal Sci.* 16 [1957]: 773.)

Animals will tend to avoid areas of recent urination or defecation (especially of their own species). Because animals often urinate or defecate in specific areas—near gates, buildings, and shade trees—these areas are well fertilized and grass is lush, but animals won't graze there. Especially in high-rainfall areas where grass is lush and animal density on the pasture is high, and to some extent in other grazing areas, rotation grazing is often practiced. The herd or flock is moved periodically among several pastures. A pasture is grazed down, then during the weeks of regrowth, rain and time diminish the effect of urine or feces on grazing.

Ranchers in the Kansas Flint Hills observed that cattle tend to face and move into the wind as they graze. Because summer breezes are usually from the south, rangeland might be overgrazed on the south part of the pasture whereas grass in the north part goes unused. To prevent this, the ranchers routinely provide water—wells or ponds—and salt in the north part of the pasture.

Cattle, sheep, and horses tend to drink one to four times per day, depending on temperature, feed intake, and type of feed. Pasture forage and silage are relatively high in water content, so animals consuming these feeds will drink less water. Also, the volume of the digestive tract of ruminants is such that more water can be consumed at one time by these species.

Swine and chickens generally are fed in confinement, are fed dry feeds, and eat and drink many times a day. Pigs like wet feeds, though, and can effectively use *limited* quantities of highly nutritious forage such as alfalfa. Because swine and poultry are nonruminants, they generally consume high-energy, highly digestible feeds. High feed intake is desired, so they are often "self-fed," and the amount of feeder space per bird or per pig is important. They may eat or drink anytime of the day or night, though specific patterns often develop.

13.11   SLEEPING AND RESTING BEHAVIOR

Horses apparently sleep a few hours in each 24-hour period, and most of this sleep occurs with the horse standing. Duration of sleep varies; it may be short and irregular.

All species of domesticated livestock and birds have been observed to experience deep sleep, where there is muscle relaxation, lower heart and respiration rates, and even evidence of dreaming—brain wave patterns similar to those of humans who are dreaming, rapid eye movements under closed lids, and facial or limb muscle movement that suggests response to dream stimuli.

Deep sleep is extremely rare, though, in mature cattle and sheep, in part because the almost perpetual movement and contractions of the rumen and reticulum require that the sternum be held in place. Deep sleep might result in the animal lying on its side or in relaxation of the sternum. This may explain why some mature cattle or sheep are observed in relatively deep sleep only while lying against a feed bunk, wagon wheel, or fence, which would help keep the sternum in a normal position relative to the rumen and reticulum. Baby calves or lambs, whose rumen and reticulum do not yet function, are often observed in deep sleep.

Pigs of all ages tend to sleep soundly and for longer periods and more hours daily than is the case with cattle, sheep, or horses. In their sleep and rest characteristics, they tend to be very similar to dogs and humans.

Birds appear to sleep while standing or perched on a roost. This suggests retention of muscle tone and balance, with a higher level of consciousness than in sleeping swine. Their ability to sleep in this position also can be explained by the fact that, when the bird's hock is flexed, as is usually the case in a perched bird, the toes are pulled downward by tendons, thereby closing the toes around a perch.

Except for swine, it appears that most of the "rest" experienced by domestic animals is in the form of a drowsy state or a light sleep. It is common for cattle and sheep to devote 4 to 10 hours in each 24 period to ruminating. This may be broken into 15 or 20 periods during the 24 hours, and considerable rest can occur during these times.

13.12   SEXUAL AND REPRODUCTIVE BEHAVIOR

Chapters 14 and 15 include discussions of the reproductive cycles, hormonal influences, and management systems related to reproductive behavior and efficiency. This section is devoted to the visible behavioral traits of animals as they relate to reproduction.

Livestock and poultry (except for geese) tend to be *polygamous* (free breeding and without specific mates). Early in the breeding season, mutual grooming or courtship is observed, especially among animals on pasture. But the presence of estrus or "visible heat" among most female livestock will tend to attract breeding males almost at random. A cow coming into estrus usually will bawl or vocalize more, will lick other animals, may attempt to mount another cow, and once into estrus will stand for others to mount. A sow in estrus is usually restless, there may be some riding, and the sow sometimes emits soft, rhythmic grunts. Also, if provided the opportunity, she will spend more time near a boar than a castrated male. A ewe usually will seek out the ram a few hours before estrus, spend time with him, and perhaps pursue a quiet courtship.

A mare in estrus demonstrates the most vivid behavioral traits including an extended stance, with rear legs extended, the tail arched, splashes of urination, and rhythmic contractions of the vulva. In all species, as estrus approaches, the vulva generally becomes turgid, and there may be mucous discharges.

Libido
Male sex drive.

Male sex drive, or **libido**, varies among species, breeds, and individuals. It tends to be higher in seasonally breeding species such as sheep. Males with reasonable levels of libido respond quickly and aggressively to females in estrus. Courtship behavior generally includes nosing the area of the vulva, nudging the female from the rear or in the flank areas, flehmen (see Section 13.5), bellowing or other vocal displays, stamping and pawing the ground, or challenging another male.

In the case of most poultry, the male tends to initiate the courting and mating process. The male chicken or turkey moves among the hens and approaches one appearing more receptive. The receptive female may crouch. The male may place one foot on the back of the hen or circle around the hen while fluttering his feathers. A step-by-step mutual stimulation seems to occur. Each response by the female encourages the male to take the next step, and vice versa. In a short time, if **copulation** is to occur, the male stands on the back or the outstretched wings of the crouched hen, the female raises her tail, the rooster lowers his (while grasping the feathers at the back of the hen's head in his beak for balance), both cloaca evert, the vents meet, and ejaculation occurs.

Copulation
The act of sexual intercourse, where semen of the male is deposited in the reproductive tract of the female.

The mating process in livestock involves the male mounting from the rear, insertion of the erect penis into the vagina, and ejaculation. Courtship time varies among species, as does the time of penetration. Penetration normally ranges from one thrust in the case of ruminants to 4 to 6 minutes (up to 10 or 20 minutes in some cases) of penetration in the case of swine. It is common for males to breed or serve the same female several times during one estrus period, 3 or more times in the case of sheep and up to 10 or 11 times in the case of horses and swine, especially if there are no other females in heat.

At the end of pregnancy and as parturition approaches, the ewe, cow, mare, or gilt develops a noticeably distended udder, the vulva becomes pinkish and swollen, and discharges may appear. The animal begins to show restlessness and sometimes goes through the motions of building or finding a "nest."

At birth, the young usually emerge head first, with the head between the forelegs. The newborn breaks through the amniotic sac and begins breathing, the mother nudges or licks away the remains of the amniotic sac and tissues, and the newborn struggles to find the udder to begin nursing. This is the point, of course, where imprinting (see Section 13.1) is most intense.

13.13 BEHAVIORAL PROBLEMS AND ABNORMALITIES

Thorough acquaintance with the normal behavior of livestock and poultry permits an alert manager or veterinarian to detect abnormalities and use these as diagnostic clues to illness, stress, inadequate nutrition, or other problems. The gait of a horse—directness, balance, smoothness—will disclose any lameness, leg injury, or nerve problem that would affect leg movement. A drooping head, abnormal nervousness, or evidence of discomfort or pain may be a clue to any number of disorders.

Cannibalism is a behavioral problem sometimes observed in chickens. Though some may consider it an inordinate perpetuation of the pecking that establishes social order, it tends to resemble feeding behavior rather than aggression, and in hens is more common among high-producing birds. In rare instances it can be severe, to the extent of feather loss, bleeding, and death. Low light intensity—one foot-candle or less—and debeaking of chicks usually prevents the problem. (See also Section 33.4.)

Tethered animals or those closely confined may, due to boredom or irritation with constraint, excessively lick or chew on stalls, pen partitions, mangers, or troughs. In the case of swine, tail or ear biting can result from mixing two or more groups of hogs midway during the feeding period, or from overcrowding or boredom. Some producers hang tires or other "toys" in the pens to reduce boredom, and most producers dock pigs' tails at an early age as a preventive measure.

Rejection of the newborn by cows or ewes is a fairly common behavioral problem faced by managers of breeding herds and flocks. This sometimes results from an especially difficult birth, temporary separation of the calf or lamb from its mother immediately after birth, a storm, herd or flock migration, or the birth occurring in a pen crowded with other animals. In instances such as these, maternal bonding to progeny is hindered, and by the time the mother and young are reunited, the familiar odor may be lost or diluted with the odor of other animals. Many techniques are used to achieve bonding, including penning the mother and her young together, separate from the herd or flock, or capitalizing on the sense of smell by rubbing the newborn with any remaining amniotic fluid, or with tissue, milk, urine, or feces of the mother.

Not all behavioral problems exist in the production units. In some terminal markets and at several major livestock slaughtering plants, it is difficult to move animals, especially cattle, toward and into the plant. They appear to dislike the smell of blood. Forced ventilation may be used to direct odors away, and recently constructed plants have been designed so that animals approach the plant and the killing area with the wind, minimizing this problem.

QUESTIONS FOR STUDY AND DISCUSSION

1. What is the scientific term for the study of animal behavior?

2. When does "imprinting" usually occur?

3. Of the fixed or inherited behavioral traits, give examples of the following types of behavior in domesticated farm animals: (a) gregarious, (b) social, (c) sexual.

4. How can the farm manager take advantage of the instinctive eliminative behavior of some domesticated species?

5. What three levels of the nervous system can initiate a reaction to a stimulus?

6. Explain how the field of vision and depth perception of livestock and humans differ.

7. Can animals distinguish color? If so, which species?

8. Compare the sense of smell in livestock to that of humans.

9. Which species of animals often prefer to lie in contact with each other?

10. What are good guidelines for animal reward and punishment?

11. Which is most likely to be effective in communication with animals, the human voice or visual clues?

12. To best utilize visual perception of cattle, how should handling facilities such as chutes be designed?

13. Which is the best strategy when moving animals in with other animals, one at a time or several? Explain.

14. Members of which domesticated species experience very little deep sleep?

15. Is most farm animal sexual behavior polygamous or monogamous?

16. Name situations where the environmental surroundings as in confinement housing might increase the incidence of abnormal behavior in animals. What are some abnormal behaviors?

14

Physiology of Reproduction

Increased frequency and number of offspring from the breeding herd or flock results from excellence in reproductive management. For the breeder of superior or elite animals, successful reproduction with the production of a desired offspring is especially gratifying. For the commercial producer, reproductive efficiency is an important key in keeping animal productivity high and production costs low.

Reproduction in farm livestock and poultry, as in other mammals and birds, is accomplished by union of the male sex cell, the sperm, with the female sex cell, the ovum, in the female reproductive tract and by gradual development in the reproductive tract of the zygote thus formed. This chapter will describe the importance of reproductive performance and provide details about the sperm and where and how it is produced, transported, and united with the ovum. It will similarly describe the ovum and the cycles of events that are peculiar to female reproductive physiology.

Egg laying by chickens, turkeys, and other fowl is biologically a part of the reproduction process, but also yields a highly nutritious product that has become a staple in the human diet. The physiology of egg production is discussed in this chapter, and effects of nutrition and other environmental factors on egg production are discussed in Section 15.6. Physical characteristics of the egg are described and discussed in early sections of Chapter 34.

Upon completion of this chapter, the reader should be able to

1. List advantages of high reproductive efficiency in herds and flocks.
2. List, locate, and describe each of the structures in the male and female reproductive system.
3. Describe specifically where oogenesis and spermatogenesis occur.
4. Describe the endocrine control of the gonads and their endocrine role(s).
5. Identify the structures involved in thermoregulatory control of the testicles.
6. List the male accessory sex glands and the functions of fluids produced.
7. Discuss why most meat animals are castrated rather than left intact.
8. Compare the differences in appearance of the ovaries, uterus, and cervix in the cow, sow, and mare.
9. Describe the differences in appearance of the Graafian follicle and corpus luteum.
10. Describe ovarian function in relation to the different phases of the estrous cycle.
11. List the main parts of the placenta.
12. List hormonal changes preceding parturition, and the stages of parturition.

13. List the major differences in the reproductive tracts of poultry and mammals.

14. Describe the steps in production of an egg in the fowl.

15. List the reasons for using artificial insemination, estrus synchronization, superovulation, and embryo transfer.

14.1   REPRODUCTIVE PERFORMANCE AND PROFITABILITY

High reproductive performance in the breeding herd or flock is essential to profitability of the enterprise. Managers who know the proper anatomy and function of the reproductive systems, and who place a high priority on improving reproductive performance, are generally more successful, for several reasons:

- **More total offspring will be born.** More may be marketed or saved to return to the breeding herd or flock. The latter allows more intensive selection.

- **More females will conceive within the scheduled breeding period.** Therefore, the offspring will be born in a shorter time span, will be more uniform in age, and so will be easier to manage as a group.

- **Feed, labor, and breeding costs should be less.** When the female doesn't conceive on schedule, extra feed, time, and labor are required until the next estrus (heat) period. Also, in artificial insemination, the costs for semen and labor involved in breeding and checking for pregnancy are higher per breeding animal or per animal born.

- **Fewer genetically superior animals will be culled because they didn't conceive or give birth.** Selection pressure can be devoted to other production or quality traits. Also, animals marketed for lack of reproduction bring fewer dollars than they would as breeding animals.

- **A dairy herd will have a shorter average calving interval, more days in lactation, and a higher proportion of productive life near the peak of lactation.** This means higher milk production per herd. Long calving intervals result in less milk per day of life because more days of production are during late lactation when milk yields are lower or the herd has more nonlactating (dry) days.

14.2   THE MALE REPRODUCTIVE SYSTEM

The role of the male in reproduction is less complex than that of the female. Whether boar, bull, ram, stallion, tom, or rooster, all males have the same basic roles: (1) to produce ample amounts of viable **sperm,** and (2) to be willing and able to deliver semen (containing sperm) into the reproductive tract of the female in natural service or into a proper receptacle for use in artificial insemination.

The male reproductive system consists of the

> **Sperm**
> The male sex cell, produced by the testicle and carrying a sample half of the genes carried by the male in which it was produced.

- testicles, which produce sperm
- duct system, which delivers sperm and semen
- accessory glands, which produce fluid to lubricate and to carry the sperm
- penis, or male organ of copulation, which deposits the semen in the female

The reproductive organs of the bull and stallion are diagrammed in Figure 14–1.

Once puberty is reached in the male, the production of sperm and male sex hormones (*androgens*) is essentially continuous. The gonads of both the male and female originate in the region behind the kidneys. Normally, the testes of the fetal male descend to the scrotum by birth or usually within the first day. In all except poultry, the testicles normally descend outside the abdominal area into the scrotum (a cutaneous skin sac) near or soon after birth.

The **scrotum** aids sperm production and survival by keeping the testicles 4 to 8°F cooler than deep body temperature. This is necessary because production and survival of spermatozoa cannot occur at normal body temperatures in farm animals other than poultry. Further regulation of temperature of the testicles is attained by three systems working together. These are

- The external **Cremaster muscle**, which contracts during cold weather to hold the testicle closer to the abdominal wall and relaxes during warm weather to allow the testicles to be more remote and therefore cooler.

- The smooth muscle fibers known as the **tunica dartos** within the skin of the scrotum, which also contract to reduce the total surface area and hold the scrotum closer to the body during cool weather and relax during warm weather.

- A network of blood vessels above the testicle known as the **pampiniform plexus**. In this system, the cooler venous blood cools the warmer arterial blood as it enters the area of the testicles. This heat-exchange mechanism is especially important in animals such as the stallion and boar, where the scrotum is less pendulous.

Cold ambient temperatures are less harmful to the production and survival of sperm than hot temperatures. During extended hot weather, breeding males such as the boar should have access to excellent shade and water sprinklers. Breeding males exposed to high temperatures may have lowered sperm production (number of viable sperm) and fertility.

Evaluation of semen of all breeding animals should be done just before the breeding season (or routine use), after an illness causing high fever, after extreme and extended exposure to cold weather, and especially after exposure to heat stress.

Testicles

The testicles of all species perform the same role of spermatogenesis and male sex hormone (androgen) production, although their location and size vary. Generally oval-shaped, their internal structure is composed of lobes of connective tissue (Figure 14–2). Within the lobes are **seminiferous tubules** that represent about 90 percent of the total mass of the testicle (Figure 14–3). These tubules produce spermatozoa cells from primary sex cells located within their walls. The initial step in spermatogenesis is the mitotic division of the primary sex cells to increase their numbers. The second step is the process of meiosis that results in spermatids that contain the haploid number of chromosomes. The spermatids are nourished and guided toward the lumen (hollow center) of the tubule by *Sertoli's cells,* or *nurse cells,* also located within the tubules. During this migration toward the lumen, each spermatid is transformed into a spermatozoon with developed head, midpiece, and tail.

Scrotum
The saclike extension from the abdominal cavity of a male that contains the testicles.

Cremaster muscle
The muscle that suspends the testicle and contracts or relaxes to enhance temperature regulation.

Tunica dartos
The elastic tissue and muscle fibers within the scrotal wall that contract and relax to aid in temperature regulation of the testicles.

Pampiniform plexus
An arrangement of blood vessels above the testicle that aid in temperature regulation of the testicles.

Seminiferous tubules
Tubular structures within the testicles where spermatozoa are produced.

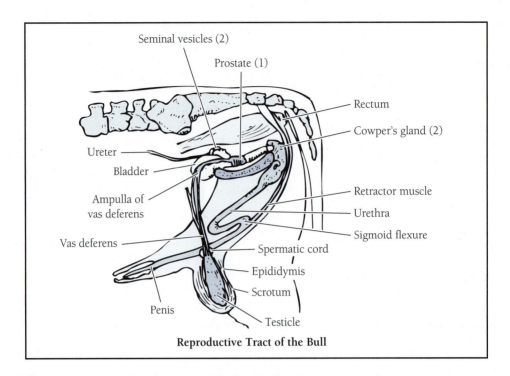

Reproductive Tract of the Bull

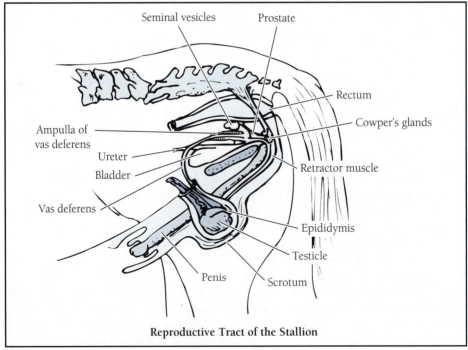

Reproductive Tract of the Stallion

Figure 14–1
The reproductive tracts of the bull and stallion, with the individual organs or parts identified. The testicles, epididymis, and vas deferens are, of course, paired, and the latter unite where they join the urethra. Also note the accessory glands, which include the ampulla, seminal vesicles, prostate, and Cowper's glands; all are paired except the prostate gland. Also note the sigmoid flexure present in the bull but not in the stallion; it is also present in the boar and ram.
(*Sources:* Iowa State University and Purdue University.)

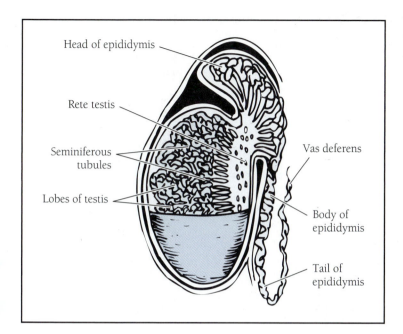

Figure 14–2
The testicle and epididymis.
(*Source:* Purdue University.)

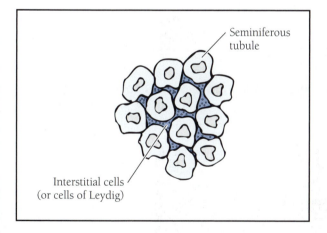

Figure 14–3
Seminiferous tubules and cells of Leydig.
(*Source:* Purdue University.)

It is interesting to note that normal development and function of the seminiferous tubules, and thus spermatogenesis, is regulated by a pituitary hormone, **follicle-stimulating hormone (FSH)**. This is the same hormone that stimulates growth of the ovarian follicles in females.

A new sperm cell (Figure 14–4) is not immediately capable of fertilization. It must first undergo a series of maturation processes in the testicle as well as in the **epididymis** where the sperm are stored. For example, bull sperm require approximately 8–11 days to mature, so management must be sensitive so as not to overwork the animal. Sperm are minutely small, as little as 1/20,000, the volume of an ovum, even though they are equivalent in hereditary significance. Unlike ova, which are produced in small numbers in the female, sperm are produced continuously and in extremely large numbers.

> *Follicle-stimulating hormone (FSH)*
> A hormone produced by the pituitary gland that promotes growth of ovarian follicles in the female and sperm in the male.

> *Epididymis*
> A tortuous tube leading from the testicle. A site of sperm storage and maturation.

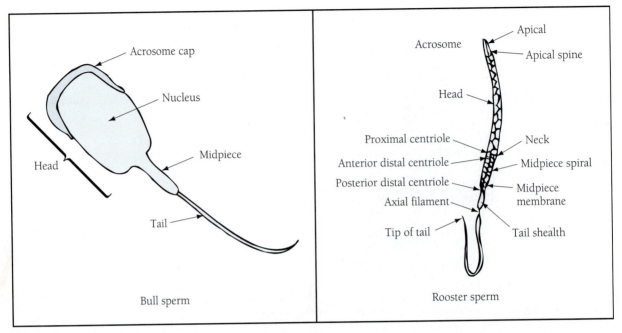

Figure 14–4
Diagrammatic sketch of a bull sperm magnified about 2,400 times and of a rooster sperm magnified about 1,000 times. (*Sources:* Bull sperm drawing from Iowa State University; rooster sperm drawing adapted from Storkie, P. D. *Avian Physiology,* 3rd ed. New York: Springer-Verlag, 1976.)

TABLE 14–1	Approximate Sperm Production in Several Species		
Species	**Average Volume (mL)**	**Millions of Sperm per mL**	**Potential Number of Matings per Ejaculate**
Bull (cattle)	2–10	300–2,000	100–600
Ram (sheep)	0.7–2.0	2,000–5,000	40–100
Buck (goat)	0.6–1.0	2,000–3,500	15–40
Stallion (horse)	30–300	30–800	8–12
Boar (swine)	150–500	25–300	15–20
Cock (chicken)	0.2–1.5	0.5–60	8–12
Tom (turkey)	0.2–0.8	0.7	30

Source: *Nalbandov, A. V., Reproductive Physiology of Mammals and Birds, 3rd ed., W. H. Freeman and Company, San Francisco, 1976; and other sources.*

A typical ejaculate of bull semen may contain over 5 billion sperm, sufficient to inseminate at least 500 cows. Numbers of sperm per ejaculate are also high for other species (Table 14–1). Number of sperm produced per week is highly correlated with testes size, as measured by the circumference of the scrotum, according to research on young bulls at Cornell University.

The second important function of the testicles is production of male sex hormones (androgens). Testosterone is the principal androgen and is produced by specialized cells located in the spaces between the seminiferous tubules. Known as **cells of Leydig,** or *interstitial cells,* their production of testosterone is controlled by a hormone of the anterior pituitary, luteinizing hormone (LH), and sometimes referred to as interstitial-cell stimulating hormone (ICSH). The production of testosterone is essential for typical male development and behavior, such as development of secondary sex characteristics (e.g., heavier muscular development and masculine body conformation), the accessory sex glands, and sex drive, or libido.

> **Cells of Leydig**
> Interstitial cells of the testicle that produce the male sex hormones, androgens.

The many segments of seminiferous tubules converge into a network of central connecting tubules, the *rete testis.* Because of continual production, sperm are moved through the tubules into the epididymis. Adhered tightly to the outer surface of the testicle, the initial portion of the epididymis consists of several tubules leading from the rete testis, then becomes a long, single convoluted tubule. It is composed of a head, body, and tail, the latter portion connecting to the vas deferens which extends into the abdominal cavity.

The epididymis provides both a passageway and storage site for the immature sperm. Its fluids nourish sperm and help bring about their maturation. Movement of the sperm through the epididymis is due to peristaltic action of smooth muscles in its walls. During ejaculation, peristaltic action of the vas deferens moves spermatozoa from the tail of the epididymis through the vas deferens and into the urethra.

Large numbers of sperm may be stored in the epididymis for several days or weeks in an inactive male, and viability diminishes. Especially if heat stress has occurred, ample time should be allowed for the production of new viable sperm and perhaps semen tests made prior to use of the male.

The tissues above the testicles, which consist largely of blood vessels, nerves, the **vas deferens,** and the external Cremaster muscle, are collectively referred to as the **spermatic cord.** It is this portion that is clamped, or severed, when nonsurgical castration is performed.

> **Vas deferens**
> The passageway for sperm from the epididymis to the urethra.

> **Spermatic cord**
> The area between the testicles and the body that includes the vas deferens and the blood, lymphatic, and nerve supply to the testicles.

The vas deferens leads from each epididymis into the abdominal cavity, and the two converge near the neck of the urinary bladder to connect with the urethra.

Pathway for sperm

Seminiferous tubules → Rete testis → Epididymis →
Vas deferens → Urethra

> **Ampulla**
> An enlargement of a passageway, such as the innermost portion of the vas deferens in some male species, which is one of the accessory sex glands.

Accessory Glands

Prior to leaving the body during ejaculation, the sperm are combined with secretions from the accessory sex glands. The first glands to add fluid to the ejaculate, the **ampulla** (not present in swine), are actually enlargements of the vas deferens just prior to where they converge with the urethra (Figure 14–1). The **seminal vesicles** are paired, lobulated, or sponge-shaped hollow structures and are also located near the neck of the urinary bladder. The single **prostate gland,** somewhat nodular in shape, is located at the junction of the vas deferens and urethra in some species such as the horse. In others, it is embedded in the wall of the urethra for a length of several inches. Enlargement of the prostate, as sometimes occurs with age (especially in humans) or with infection, can

> **Seminal vesicles**
> Paired glands attached to the urethra, near the bladder, that produce fluids to carry and nourish the sperm.

> **Prostate gland**
> A gland that surrounds the beginning of the urethra in the male.

> **Cowper's glands**
> The paired accessory male sex glands posterior to the prostate and on each side of the urethra.

result in difficulty in urination. The **Cowper's glands**, or *bulbourethral glands,* also paired, are the outermost of the accessory glands; proportionally larger in the boar, they secrete a more gelatinous fluid.

Sperm + Fluids from Accessory Glands = Semen

- Ampulla (if present)
- Seminal vesicles
- Prostate gland
- Cowper's glands

The sperm, combined with the accessory gland fluids, make up the semen. The fluids add volume to the ejaculate and enhance survival and movement of the suspended sperm by producing nutrients and electrolytes. The relatively nonmotile sperm become activated upon contact with the fluids. The fluids contain sodium chloride, potassium chloride, nitrogen, citric acid, fructose, and several vitamins, which may be utilized by the sperm cells for energy and metabolic processes, as well as certain enzymes. Certain accessory fluids also serve as lubricants during the mating process. Survival is also enhanced by the neutralization of urinary acid residues in the urethra as accessory gland fluid moves through this passageway.

Urethra

> **Urethra**
> The portion of the urinary tract that carries urine from the bladder; in males, also provides passage for semen ejaculate.

The **urethra** is a long tubular structure that extends from the opening of the urinary bladder through the entire length of the penis. It serves as a passageway for both semen and urine. Although physiology of reproduction and most anatomical parts are similar in male animals, the structure and functioning of the penis during copulation differs among species. For example, the erection of the penis of stallions prior to copulation is largely due to erective tissue and increased blood volume. However, the penis of the ram, bull, and boar contains less erective tissue, but possesses an S-shaped structure, known as the *sigmoid flexure,* which lengthens upon erection to extend the penis.

Factors Affecting Sperm Production

After sexual maturity of the male, sperm production is continuous under most conditions. There is much individual variation in age for sexual maturity and initiation of sperm production.

Many environmental factors, as well as heredity, influence rate of sperm formation. Environmental effects are usually only temporary. Extreme underfeeding, especially after animals have been well fed, will decrease the formation. Also, a major decrease in protein content will directly reduce the number of viable sperm in the male.

Vitamin A and protein seem to be the critical nutrients involved in sperm formation in most areas of the country. Lack of these nutrients is most common in dry seasons.

As mentioned earlier, extremely high temperatures will decrease sperm formation and also cause a higher proportion of sperm that are incapable of fertilization. Low conception rate among ewes bred during late summer is common, and some sheep raisers keep their rams in a cooled stall and let them out with the ewes only at night. High body temperatures that result from illness or infection will also impair sperm production.

Other factors may also limit or prevent sperm production, including heredity, limited energy intake at puberty, and cryptorchidism (discussed later in this section).

Castration

Castration of male meat animals—removal of the testicles or rendering them inactive by crushing of the spermatic cord—is routine. One reason for castration is that castrates are easier to manage than intact males; without sex libido they are calmer in the feedlot. Also, castration helps ensure that the meat from males will have no sex-related odor, especially in the case of swine. Historically meat processors have paid less for young boars, lambs, or bulls, but considerable data suggest equal or higher value of young intact males, especially lambs and bulls, because of a higher proportion of lean and more efficient production, and because animals are marketed before risk of strong flavors.

> **Castration**
> Removal or permanent alteration of the testicles of a male animal.

   Castration should occur as early as possible because stress is usually less at a young age. Some horse owners prefer to castrate the young stallion as late as 2 years of age, after it has developed some secondary sex characteristics, especially in the neck and shoulder areas. Others may castrate as soon as the testicles have descended into the scrotum, which in the young stallion may be several months of age. As with castrated meat animals, the gelding is easier to manage and there is less fighting and risk of injury when the animal is kept with other geldings.

Cryptorchids

When one or both testicles fail to descend into the scrotum, the animal is referred to as a **cryptorchid** (or sometimes, *ridgeling* in horses). When both testicles are retained in the abdominal cavity, the cryptorchid usually will be sterile because the higher body temperature prevents spermatogenesis. The cryptorchid will, however, develop typical secondary sex characteristics and the behavior of an intact male because the retained testicles still produce testosterone.

> **Cryptorchid**
> A male animal in which one or both of the testicles remained in the body cavity and did not descend into the scrotum during fetal development.

   Surgical castration of cryptorchids is, of course, difficult. Cryptorchidism is inheritable and those that are not sterile should not be used for breeding purposes.

14.3   THE FEMALE REPRODUCTIVE SYSTEM

Female reproduction systems and functions are more complex than those of males. As with males, the system must be anatomically sound and the animal must have the physiological desire to mate. The female sex cell, the ovum, must mature and be released at the appropriate time in the estrous cycle, conception must occur, and the developing embryo and fetus must be nourished and protected. In addition, the female must successfully give birth and accept and nourish its newborn. Finally, to justify continuation in the herd or flock, the female must again conceive within a relatively short period after parturition.

   Some variation exists in size and shape of the female reproductive system of the common farm females (cow, sheep, sow, mare), but the functions and locations are similar. All parts except the vulva are located within the body (Figure 14–5). The structures (from innermost to the exterior of the body) are

- ovaries, left and right
- oviducts (or fallopian tubes), left and right
- horns of the uterus, left and right
- body of the uterus
- cervix
- vagina
- vestibule
- vulva

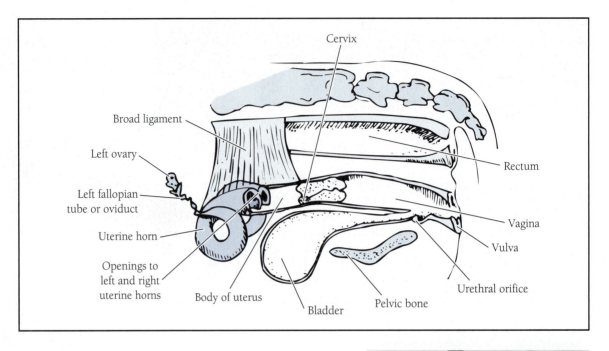

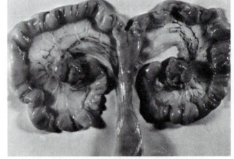

Figure 14–5
In the drawing, the female reproductive tract of the cow. Uterine horns in sheep, cattle, and horses are less elongated than in swine. In the photo, a gilt reproductive tract. Note the longer uterine horns and the ovaries, with developing follicles at the innermost portion of each horn.
(Drawing from Purdue University; photo courtesy of Iowa State University.)

> **Ovary**
> The female sex organs, which produce ova after sexual maturity.

> **Ovum**
> The female sex cell, produced on the ovary and carrying a sample half of the genes carried by the female in which it was produced. Pl., Ova.

Ovaries and Ovarian Function

The **ovary**, the primary organ of the female, produces the female gamete (**ovum**) and female sex hormones (estrogens). As is true with the testicles, the ovaries develop embryonically in the lumbar region of the abdominal cavity, behind the left and right kidneys. Whereas the testicles descend into the scrotum, the ovaries remain in this region, suspended from the broad ligament, with the outer surface of each ovary exposed to the adjacent fallopian tube. The ovary is well served by a blood supply that provides oxygen and nutrients. The blood also carries hormones from other glands to the ovaries to affect their functioning, as well as carrying hormones produced by the ovaries to other target body tissues.

The ovaries of the cow are somewhat oval- or almond-shaped and measure about 3/4 inch by 1 1/2 inches (Figure 14–6). In the mare, they are more bean-shaped and measure about 1 1/2 inches by 3 inches. The sow possesses ovaries that are lobulated, almost grapelike, and approximate 1 inch by 1 1/2 inches in size. The size of

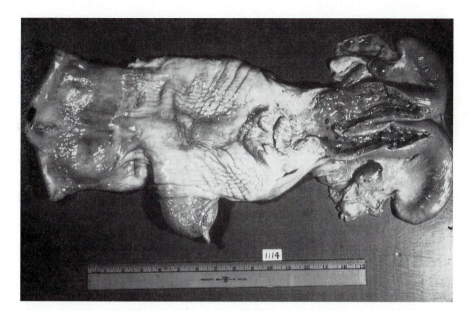

Figure 14–6
A reproductive tract from a pregnant cow. Note the well-developed corpus luteum (CL) on the right (lower) ovary. (Courtesy of Purdue University.)

ovaries will vary considerably, depending upon the development stage of the follicles or the presence of ovarian cysts.

Each ovary consists of an inner **medulla** and an outer **cortex**. The medulla receives the blood and nerve supply, and the cortex contains the ova. It is in the cortex where the Graafian follicles develop, estrogen is produced, and, after rupture of the follicles and release of the ova, the corpus luteum (or lutea) develops and produces progesterone.

Oviducts (or Fallopian Tubes)

Each **oviduct** connects the tip of the uterine horn with the respective ovary. Thus, oviducts provide a passageway for both the sperm en route to the ovum and the resulting fertilized ovum or zygote as it travels to the uterus. The ovarian end of the oviduct, shaped somewhat like a funnel, is called the **infundibulum**. It is attached at one point to the ovary, and with its fimbriated border, almost surrounds the surface of the ovary at the time of ovulation to successfully direct the ovum into the infundibulum. If viable sperm are present, fertilization normally occurs in the infundibulum within about 12 hours after ovulation. The cilia-lined lumen and wall muscles of the oviduct enhance movement of both the sperm and the **zygote**. The zygote undergoes mitotic division as it travels toward the uterus during a period of several days.

Uterus

The uterus with its two horns provides the site for implantation of the embryo and placentation (development of the placenta), essential for the survival and development of the fetus. The uterus and its horns are attached by the broad ligament and suspended in the abdominal and pelvic cavities. The shape and size of the uterus differ with species. In the sow, the body of the uterus is quite small, but the horns are of considerable length to accommodate as many as 8 to 10 fetal pigs in each horn. The uterine body of the cow extends only a few inches beyond the tip of the cervix, then joins the two well-developed

---

Medulla
The inner portion of a body organ.

Cortex
The outer portion of an organ, such as ovary or kidney.

Oviduct
The tube leading from each horn of the uterus to the corresponding ovary.

Infundibulum
A funnel-shaped passageway such as the innermost portion of the oviduct of the hen.

Zygote
The diploid cell resulting from the union of haploid female and male gametes.

Myometrium
The muscular wall of the uterus.

Endometrium
The inner lining of the uterine horns and body.

Cervix
The constricted necklike structure located between the vagina and uterine body.

horns (Figure 14–5). On the other hand, the mare has somewhat blunted (rather short) uterine horns but a well-developed uterine body.

A cross section of the uterus or uterine horn reveals three main layers. These consist of (1) the outer covering, (2) the intermediate smooth layer, called the **myometrium,** which provides uterine contractions, and (3) the innermost lining, called the **endometrium,** which is highly glandular. The health and proper functioning of the endometrium is especially important for the many reproductive processes (e.g., implantation, placentation).

Cervix

The **cervix** is sometimes called the gateway to the uterus because it is located between the vagina and body of the uterus and restricts entry. It is a thick, elongated, smooth muscle sphincter of 1 to 2 inches in diameter and several inches in length. The opening of the cervix, known as the *os cervix,* protrudes into the vaginal cavity. In artificial insemination, care must be taken to ensure that the inseminating rod is guided through the os cervix rather than against the adjoining depression of the vaginal wall.

The cervix of the mare consists of numerous folds of muscle layers that, although restrictive, are rather easy to manipulate during insemination. The inner surfaces of the cow or ewe consist of several ridges, or interlocking rings, which during insemination must be grasped through the rectal wall in order to guide the inseminating rod through it. (See Section 14.6.) The cervix of the sow is less difficult to penetrate, although its corkscrew-like muscle arrangement is somewhat restrictive. In natural service, the penis of the boar actually penetrates into the opening of the cervix.

In all species, the cervix secretes considerable amounts of thin mucus during its dilated condition at estrus. A thicker mucus is produced upon pregnancy and provides a hardened seal or plug that blocks access to the uterus. At parturition, the cervix must dilate and stretch considerably to allow delivery of the newborn.

Vagina, Vestibule, and Vulva

The vagina is considered to be the female organ of copulation because it receives the penis during natural service. It also provides a passageway for birth of the fetus. The vagina and the vestibule are divided at the junction of the ventrally positioned external urethral orifice, the opening from the urinary bladder. Just forward of this junction is the hymen, a ridge of tissue that sometimes, because of its membrane, causes the initial act of copulation to be difficult.

The vestibule extends from the urethral orifice to the external genitalia of the vulva. The vulva consists of thickened folds of skin that cover the clitoris, a rudimentary structure homologous to the penis and of the same embryonic origin. The clitoris is located on the inner ventral wall of the vestibule. Visible swelling of the vulva occurs during estrus and aids in estrus detection.

Estrous Cycle

Ovarian functions occur in a cyclic manner following puberty, and this is referred to as the estrous cycle. Oogenesis, the production of the female sex cell, through the process of meiosis results in the female sex cells or ova containing one-half the chromosome number of the somatic (body) cells. Each ovum is surrounded by a fluid-filled or blisterlike follicle that enlarges on the surface of the ovary. At the same time, estrogens are produced by specialized cells in the walls of the developing follicle(s). Ovulation occurs at the end of heat (estrus), or shortly after in the cow. At this time, usually one follicle

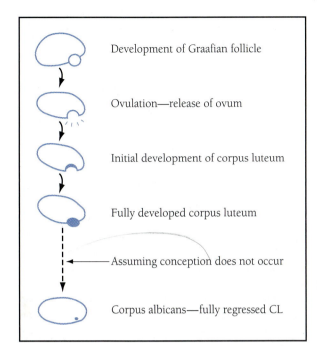

Figure 14–7
Ovulation and corpus luteum
development (sequence of events
at the same location on ovary).
(*Source:* Purdue University.)

Development of Graafian follicle

Ovulation—release of ovum

Initial development of corpus luteum

Fully developed corpus luteum

Assuming conception does not occur

Corpus albicans—fully regressed CL

in single-birth animals (several in multibirth animals) will rupture, with the release of the ovum into the corresponding oviduct.

Upon ovulation (Figure 14–7), cells in the wall of each ruptured follicle multiply to form the **corpus luteum (CL)**. The CL usually protrudes beyond the surface of the ovary as a firm, turgid structure within a few days after ovulation. If pregnancy occurs, the CL is retained (as shown in Figure 14–6), and it performs as an endocrine gland in the production of progesterone, a hormone necessary for maintenance of pregnancy.

If pregnancy doesn't occur, prostaglandin causes the CL to regress (become smaller), production diminishes, and within a few days the next heat (estrus) and subsequent ovulation occur. The CL regresses to a tiny scar on the surface of the ovary and is known as the **corpus albicans.**

The sequence of ovarian activities and their relationship to other events occurring within the female reproductive system is better understood by reviewing the major hormones involved. A brief summary follows:

1. *Gonadotropin-releasing hormone (GnRH)* is produced in the hypothalamus of the brain and controls the release of two gonadotrophic hormones from the pituitary gland. These are the follicle-stimulating hormone (FSH) and luteinizing hormone (LH).
2. *Follicle-stimulating hormone (FSH)* stimulates follicle development and estrogen production.
3. *Luteinizing hormone (LH)* triggers ovulation and is necessary for development and maintenance of the corpus luteum. Although FSH and LH have specific functions, they are synergistic in their effects on the reproductive system.

Corpus luteum (CL)
Active tissue that develops on the ovary at the site where an ovum has been shed. If conception does not occur, the tissue gradually regresses. If conception does occur, the tissue becomes functional, producing progesterone.

Corpus albicans
The term to denote tiny scars that remain after complete regression of earlier lutea.

4. *Estrogen,* produced by the ovarian follicles, causes sexual excitability, increases fluid production and muscular contractions of the reproductive tract, and triggers increased release of LH from the pituitary body.

5. *Progesterone,* produced by the functional corpus luteum, prevents further estrus during pregnancy and thus maintains pregnancy. It causes increased vascularity of the uterine horn and body to enhance survival of the embryo during implantation, and increases development of the alveolar structures in the mammary gland which produce milk.

6. *Prostaglandin* ($PGF_2\alpha$) is a hormone or hormonelike substance produced in the uterine wall that causes regression of the corpus luteum when pregnancy doesn't occur.

The controlled release of hormones into the bloodstream and their effects upon the target sites are more logically understood by reviewing the functioning of these structures in relation to the estrous cycle. The **estrous cycle** is defined as the interval from the beginning of one heat period (estrus) to the beginning of the next estrus. Estrous cycles are approximately 21 days in the cow, sow, and mare, and about 16 to 17 days in the ewe (Table 14–2). The estrous cycle consists of four phases in the cycling nonpregnant female: proestrus, estrus, metestrus, and diestrus. The phases within an estrous cycle can be summarized as follows:

> **Estrous cycle**
> Physiological events occurring in the reproductive system of a female between one estrus period and another. About 17 days in sheep; 21 days in cattle, swine, and horses.

- *Proestrus:*   The period between regression of the CL and estrus, when follicular development is occurring and estrogen production is increasing; about the 18th to 20th days of the bovine cycle.

- *Estrus:*   The period when the female is most sexually receptive, due to high levels of estrogen. Depending upon the species, length of estrus ranges from about 12 hours to several days. Estrogen levels bring about a surge of LH and FSH, the increased LH triggering ovulation toward the end or shortly after estrus; day 1 in the bovine cycle.

- *Metestrus:*   The phase following estrus when the CL forms and begins to produce progesterone; about days 2 to 5 in the bovine cycle.

- *Diestrus:*   The phase when the CL is highly active in its production of progesterone. If pregnancy occurs, the CL is maintained and further estrus

TABLE 14–2	Length of Estrous Cycle, Duration of Heat, and Time of Ovulation			
Species	**Days in Cycle**	**Duration of Heat (hr)**	**Time of Ovulation**	**Suggested Time of Mating**
Cow	21	12–18	12–15 hr after estrus	4–8 hr before end of estrus
Sow	20–21	48–72	18–40 hr after start of estrus	24 hr after start of estrus
Ewe	16–17	24–36	18–26 hr after start of estrus	12–18 hr after start of estrus
Goat	19–20	34–39	9–19 hr after start of estrus	Alternate days during estrus
Mare	19–23	90–170	1 day before to 1 day after estrus	Alternate days during estrus

Sources: *Adapted from Nalbandov, A. V.,* Reproductive Physiology of Mammals and Birds, *3rd ed., W. H. Freeman and Company, San Francisco, 1976; and Bearden, H. J., and J. W. Fuguay,* Applied Animal Reproduction, *Reston Publishing Company, Reston, VA, 1980.*

is inhibited. If conception doesn't occur, release of prostaglandin from the uterine walls causes regression of the CL, and then proestrus begins again. Diestrus occurs from about days 5 to 18 in the bovine cycle.

Obviously, if pregnancy occurs, the estrous cycle is interrupted due to the production of progesterone by the corpus luteum. In some cases, cystic ovaries result in a persistent corpus luteum, one which does not regress. In this case, although not pregnant, the female displays no signs of estrus and is referred to as being in *anestrus*. Females of species that are seasonal breeders are also anestrous between the breeding or mating seasons. On the other hand, frequent or almost continuous estrus may also be caused by cystic ovaries which continue to produce high levels of estrogen, with the animal showing almost continuous signs of heat.

Females of species that have no specific mating or breeding season, such as the cow and sow, are called continuous breeders, or are **polyestrous**. Others, such as the mare and most breeds of sheep, may mate only during certain seasons of the year. Because the latter have several estrous periods during the breeding season, they are termed *seasonally polyestrous*. With shortening of daylight, as in autumn, the estrous cycle of the seasonally polyestrous breeds of sheep is initiated. Most nonpregnant and nonlactating mares become sexually active in early spring when the length of day is longest. However, increasing the mare's exposure to light to about 16 hours per day in the winter months may initiate their estrous cycle earlier in the year.

> **Polyestrous**
> Refers to animals that have several estrous cycles per year or breeding season.

14.4   PREGNANCY

The period of time from conception to the birth of the offspring, or pregnancy, is also called the **gestation period**. The approximate lengths of the gestation period of different species are given in Table 14–3. Gestation length varies considerably among animals within a species and is influenced by age, breed, number of fetal developments, as well as environmental factors.

The optimum time for mating livestock, to achieve the highest conception rate, is influenced by the time of ovulation in relation to the heat period. In cattle, more will conceive at first service if bred about 12 hours after the beginning of heat because ovulation does not occur until 12 to 15 hours after heat. The cow must be receptive to the bull, but the sperm should not be deposited longer than necessary before ovulation. In ewes and sows, where ovulation occurs during but near the end of heat, it is easier to synchronize mating with ovulation.

> **Gestation period**
> The duration of pregnancy.

TABLE 14–3	Length of Pregnancy in Farm Animals
Cow	279–290 days
Sow	112–115 days
Ewe and doe	144–151 days
Mare, light	330–337 days
Donkey	365 days

Figure 14–8
Identical twins result when an ovum divides after fertilization. They have, therefore, identical genotypes.
(Courtesy of Iowa State University.)

Section 13.12 provides a thorough discussion of sexual and reproductive behavior, including behavior of the male and female during courtship and copulation.

The exact, detailed process of sperm and ovum combination is not completely known. In some species the surface of the sperm head attaches itself to the ovum. The tail of the sperm, giving some motility, may also help the sperm penetrate the membranes around the ovum. Enzymes of the sperm apparently continue to play a role as the head, and then the tail, of the sperm penetrate the ovum and become engulfed in the ovum cytoplasm. Membranes surrounding the ovum then change to prevent entrance of other sperm.

Inside the fertilized egg, the nucleus of the sperm and the nucleus of the ovum come into close contact, nuclear membranes disappear, and the complete chromosome pairs become evident. Then begins the first cell division and growth of the zygote. In rare cases, the division of a single zygote will occur, resulting in two genetically identical twins (Figure 14–8).

In the cow, ewe, mare, or sow, the fertilized ovum or zygote moves slowly down the oviduct to the uterine horn where, if it survives, it attaches to the uterine wall. The zygote may include several cells by this time, as the result of cell division. After implantation, capillaries develop at the site to nourish the developing embryo during pregnancy.

Implantation
After fertilization and passage through the oviduct, the embryo enters the uterus at about the 16-cell stage, known as the *morula*. Once there, it locates against the endometrial lining where its enzymes digest tissue to form a small depression. At this location, it implants and continues to develop, nourished by secretions of the endometrium.

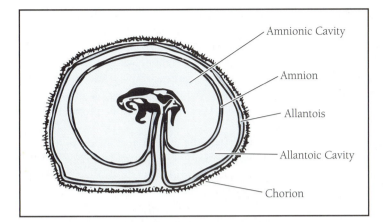

Figure 14–9
The placenta.
(*Source:* Purdue University.)

Amnionic Cavity

Amnion

Allantois

Allantoic Cavity

Chorion

Figure 14–10
Note the prominent cotyledons on the outer surface of the chorion and the size of the fetal calf within the placenta at 80 days of pregnancy.
(Courtesy of Purdue University.)

Placentation

Placentation consists of development of extraembryonic, highly vascular membranes (Figure 14–9) that provide movement of oxygen and nutrients from the endometrium of the dam to the fetus, and metabolic waste from the fetus to the dam's circulatory system. The structures of the **placenta**, are briefly described as follows:

> **Placenta**
> A structure in mammals that provides the exchange of nutrients and waste materials between the maternal and fetal system.

1. *Chorion:* The outermost membrane in contact with the endometrial lining of the uterus of the dam (Figures 14–9 and 14–10).

2. *Allantois:* A second membrane that lines the inside of the chorion and joins the amnion to form the fluid-filled allantoic cavity. Membranes of the allantois also lead to the fetal bladder to provide for waste removal.

3. *Amnion:* The innermost membrane that surrounds the fetus and forms the second fluid-filled sac or cavity. Its fluids cushion the fetus from shock.

The umbilical cord of the fetus consists of arteries that carry oxygen and nutrients from the dam to the fetal placenta and the veins that return metabolic wastes from the fetal bladder to the dam. There is no direct exchange of blood between the maternal and fetal system; rather, nutrients, oxygen, and metabolic waste pass via placental membranes. Also, the placental membranes are quite complex and prevent the absorption of antibodies from the dam by the fetus. For these reasons, the intake of antibodies from colostrum as soon as possible after birth is essential for providing early passive immunity.

Farm animals differ in the manner that the placenta attaches to the endometrium of the dam. In ruminants, **cotyledons** (cuplike structures) on the surface of the chorion (Figure 14–10) attach to corresponding **caruncles** (protuberances) of the endometrium to form very secure connections, called *placentomes.* The placental attachment in the horse is more diffuse; that is, the attaching villi cover a greater area of the chorion but are not as strongly attached. The placental attachments in the pig are even more diffuse than in the horse. "Retained placenta," a more common problem in cattle after parturition, likely is related to the type of placental attachment.

Parturition

Toward the latter days of gestation, increased estrogen production by the placenta stimulates production of prostaglandin from the uterine wall. This brings about regression of the CL and lowered progesterone levels: The hormone *relaxin,* from the corpus luteum, "relaxes" the pelvic muscles and ligaments in the pelvic area, and the fetus is moved into the birth position. **Prolactin** levels increase to stimulate milk synthesis, and *oxytocin* brings about steadily increasing uterine contractions.

The birth process involves three stages. Stage 1 consists of contractions moving in a posterior direction and entry of the fetus into the dilated cervix. Stage 2 is the expulsion of the allantochorionic sac through the cervix, causing the membrane to rupture. Generally, at this time, the dam will be lying down and complete expulsion of the fetus will be brought about by more intense uterine and abdominal contractions. Stage 3 is the expulsion of the afterbirth. Parturition in the sow is generally less complicated than in the cow, ewe, or mare. The fetal pigs may be born either in an anterior or posterior position, and because in most cases the placenta isn't retained, the pig is surrounded by the placental membranes when born.

14.5   THE REPRODUCTIVE SYSTEM IN DOMESTICATED FOWL

The Male Reproductive System

Many differences exist between the reproductive system of the bird and that of mammals. The reproductive tract of the **cock** (rooster) is shown in Figure 14–11. The first notable difference is the location of the two testicles, high inside the abdominal cavity, along the backbone and near the front end of the kidneys. The body temperature (about 104°F) doesn't inhibit spermatogenesis as it does in most mammals. Also, because the sex chromosomes in the male are identical (ZZ), unlike in male mammals, it is the female with two different sex chromosomes (WZ) that determines the sex of the chick.

In the bird, production and maturition of large numbers of spermatozoa occur rapidly within the *seminiferous tubules.* The *epididymis* on the side of each testicle is quite small and provides limited storage. Thus, the *vas deferens,* which lead from the testicles to the *cloaca,* are the main storage site for sperm.

Cotyledon
Cup-shaped structure; areas of the fetal chorion that connect with the uterine caruncles of the maternal host.

Caruncles
Disklike areas of the uterine endometrial lining which with the fetal cotyledons form a strong connection known as the placentome.

Prolactin
An anterior pituitary hormone needed for synthesis of milk.

Cock
Most commonly refers to a sexually mature, intact (uncastrated) male chicken.

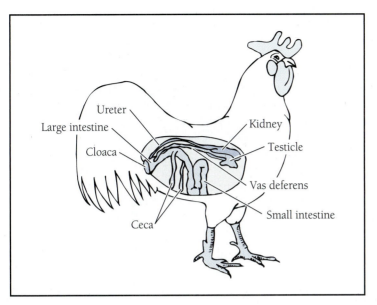

Figure 14–11
The reproductive tract of the male chicken is shown in this diagrammatic cross section. The vas deferens is longer and more convoluted than shown. The testicles, vas deferens, and kidneys are, of course, paired.
(*Source:* Purdue University.)

Another major difference is the lack of accessory sex glands in the bird. As sperm leave the testicle in the bird, they are carried in a *seminal fluid* produced by the testicle. Also, a transparent fluid is ejaculated with the sperm, it being derived from a portion of the cloacal wall. *Papillae,* or small projections on the cloacal wall, serve as the intromittent or copulatory organ. During the act of copulation, the vent of the female relaxes to expose the interior of the cloaca which receives the ejaculate from the male's cloaca.

Castration, or caponization, of the male broiler-type bird at 2 to 4 weeks can be accomplished by a small incision on one side of the back that allows removal of both testicles. Capons are grown to heavier weights but still retain the tenderness and juiciness of younger broilers. Most are grown to about 12 pounds at about 20 to 24 weeks of age and sold as a specialty product.

The Female Reproductive System

The female reproductive system in birds is also quite different from mammals in both anatomy and production of offspring. As in mammals, during embryonic development of the female, both left and right ovaries and their respective oviducts are present. In birds, however, the right side atrophies, leaving only the left ovary and its oviduct to function in the production of the ovum within the yolk. The supporting nutrients are added to form the complete egg in the oviduct (Figure 14–12).

The eggs produced by birds are much larger than the ova produced by cattle or other domesticated mammals. Because the avian embryo develops inside the egg and without continued nourishment from the blood of the dam, all nutrients for embryonic development until hatch, or birth, must be provided within the egg.

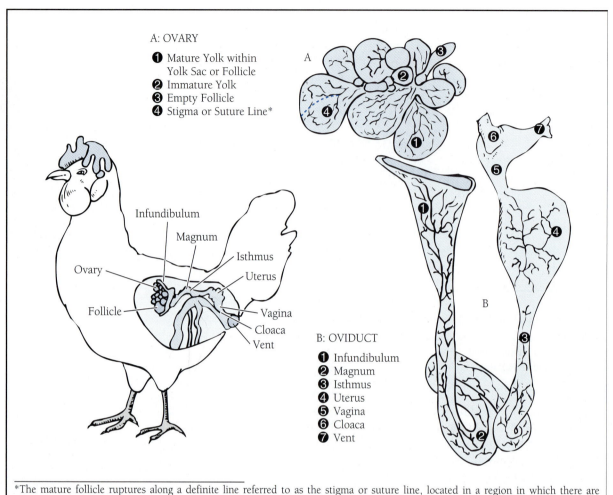

A: OVARY

❶ Mature Yolk within Yolk Sac or Follicle
❷ Immature Yolk
❸ Empty Follicle
❹ Stigma or Suture Line*

Infundibulum
Magnum
Isthmus
Ovary
Uterus
Follicle
Vagina
Cloaca
Vent

B: OVIDUCT

❶ Infundibulum
❷ Magnum
❸ Isthmus
❹ Uterus
❺ Vagina
❻ Cloaca
❼ Vent

*The mature follicle ruptures along a definite line referred to as the stigma or suture line, located in a region in which there are normally no blood vessels.

Figure 14–12
The reproductive tract of the hen, showing the functional left ovary and developing follicles on the ovary. The funnel, magnum, isthmus, uterus, and vagina are collectively called the oviduct. The funnel is also called the infundibulum. The left drawing shows relative location in the hen; the right is enlarged and shows more detail.
(*Source:* Left drawing, Purdue University; right drawing is from *Ohio State University Coop. Ext. Serv. Pub. MM-207.*)

The estrous cycle is not a characteristic of poultry. Rather, a pullet (young female) lays her first egg when sexual maturity is reached, usually at about 20 to 21 weeks of age, and generally will continue on a daily basis. In nature, unselected or wild birds would lay only in the spring and would lay about 13 eggs, at which time gonad-stimulating hormones would decrease and egg laying would cease. Prolactin production, which stimulates the mothering instinct, would increase and the hen would "set" on the eggs until they hatched. After hatching, the hormone production shift would reverse and egg production would resume. Selection for egg production in chickens and turkeys, plus daily gathering of eggs and regulation of lighting, has resulted in almost uninterrupted daily laying of eggs for up to 300 or 350 days.

Egg Formation in the Fowl

The right drawing in Figure 14–12 shows an enlargement of the left ovary and oviduct of a hen. Microscopic examination has disclosed that a fowl ovary may contain as many as 12,000 reproductive cells, which represent a theoretical total egg-laying potential for the bird. (One reproductive cell may be observed as a small white germinal disc on the surface of each yolk.) Under natural light, ovulation of the egg to be laid the following day usually occurs early in the morning and, under either natural or artificial light, about 30 minutes later than the egg of the previous day. This is induced by *gonadotropin-releasing-hormone* (GnRH) from the hypothalamus and release of **luteinizing hormone** (LH) from the anterior pituitary which, in turn, are stimulated in part by light length and intensity.

Before ovulation, the ovum, which is in fact the yolk of the developing egg, will have grown in size with deposition of yellow and white yolk granules, high in fat and containing protein and other nutrients. This will have been stimulated by *follicle-stimulating hormone.*

When the yolk is fully formed, increased blood levels of luteinizing hormone triggers follicle rupture, and the ovum (yolk) moves into the funnel or infundibulum of the oviduct (Table 14–4).

Should the hen have been inseminated, by natural service or artificial insemination, fertilization will likely occur in the infundibulum. Fertilization, though, has no effect on the later processes of egg formation by the oviduct, such as albumen secretion and shell formation.

During about 3 hours in the *magnum* of the uterus, high protein and viscous albumen (thick white) is secreted from magnum glands and deposited around the yolk. Precursors of albumen are provided, from the blood, largely by a filtration process in the magnum glands.

The developing egg will move through the **isthmus** in slightly over an hour, and during that time two thin membranes will be secreted to surround and contain the albumen. These are sometimes called shell membranes because they lie just inside the shell of a finished egg.

The *uterus* in the bird is sometimes called the "shell gland." During a period extending as long as 20 hours, protein, calcium, and other materials, sometimes including pigments, are secreted to form the shell. Thin, watery albumen is added by diffusion

> **Luteinizing hormone (LH)**
> A gonadotrophic hormone produced by the pituitary gland that causes rupture of ovarian follicles in the female (ovulation), and secretion of testosterone in the male.

> **Isthmus**
> The relatively short section of the bird's reproductive tract where the shell membranes are produced, creating the shape of the egg.

TABLE 14–4	Formation of the Egg		
Area of Oviduct	**Structure Length (inches)**	**Occurrence**	**Time**
Infundibulum	3.5	Fertilization	15 min
Magnum	13.0	Albumen added	3 hr
Isthmus	4.0	Membranes added	1 hr
Uterus	4–4.7	Shell added	20 hr
Vagina	4.7	—	Few min
Cloaca	1.0	Egg rotates	50 min

Source: Adapted from Winter, A. R., and E. M. Funk, 1946. *Poultry Science and Practice.* J. B. Lippincott Company, Chicago, Philadelphia, New York.

through the shell membrane early in the egg's stay in the uterus. Eggshells are comprised largely of calcium carbonate ($CaCO_3$), but the very thin outer layer, sometimes called the cuticle, is largely protein.

There is no cervix in the female fowl reproductive system, but there is a sphincter between the uterus and vagina, and the vagina opens directly into the cloaca. The smaller, more pointed end of the egg is generally the leading end during the egg formation process, but before the egg is "laid," it is usually rotated 180° in a lateral manner.

The act of laying is stimulated by *oxytocin*, another hormone of the posterior pituitary, which causes vigorous muscle contraction of the uterus. The uterine wall, in fact, prolapses or everts, generally moving the egg through the vagina and the cloaca, without its touching the cloaca wall, and depositing the egg externally of the bird.

Incubation and Hatching

Normal number of days of incubation for various species of fowl are given in Table 14–5, and chicken embryo development is shown in Figure 14–13. Fertilized eggs may be stored several days while a sufficient number for an incubator accumulate. They should be stored a minimum time, no longer than 7 days, and at 55° to 60°F. Longer storage or holding at higher or lower temperature normally reduces hatchability. Storage in plastic bags has shown some promise, particularly with turkey eggs.

The head of the developing embryo generally is toward the larger end of the egg, where the air cell is located. Hatching eggs typically are stored, and placed in the incubator, with the large end up. The yolk tends to float, which could permit the embryo to move against the shell where contamination or dehydration could kill the embryo. The eggs are therefore automatically tilted back and forth through a 90° angle several times daily during incubation to keep the yolk centered in the albumen and away from the shell.

Incubators for all domestic fowl eggs are generally operated at about 100°F, with about 75 percent relative humidity and at least 21 percent oxygen (the amount in air at sea level). Fans provide circulation in the incubator to provide sufficient oxygen and to hold the carbon dioxide level below 0.5 percent. Significant deviations from the above will reduce hatchability. At high altitudes, supplemental oxygen is sometimes necessary.

TABLE 14–5	Incubation Time for Fowl Species

Species	Days of Incubation
Japanese quail (bobwhite quail)	17–18 (23–24)
Pigeon	16–18
Chicken and bantam	21
Pheasant	24
Guinea	26–28
Turkey	28
Duck (muscovy duck)	28 (35–37)
Goose (except Canada and Egyptian)	29–31
Ostrich	42
Emu	50

Source: *Poultry Digest*, Clemson University, and Purdue University.

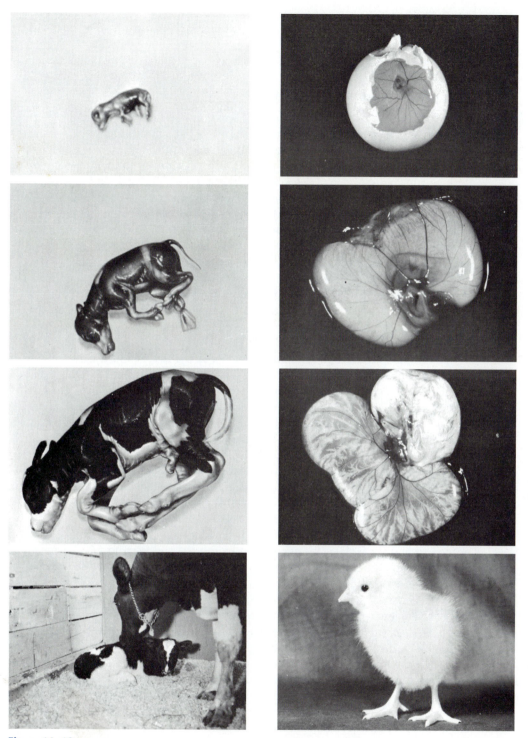

Figure 14–13
Calf and chick embryos at successive stages of development and after birth or hatch. Large amounts of nutrients are required by the embryo during the last third of pregnancy or incubation. (Courtesy of Purina Mills, Inc.)

Eggs are transferred from incubators to hatcher units 2 or 3 days before hatching. In these units, temperature is slightly lower and humidity slightly higher than in the incubator.

14.6   ARTIFICIAL INSEMINATION

Artificial insemination (AI) provides several opportunities for breeders and producers to increase the performance and potential profitability of their herds or flocks. Its principal advantages are (1) more rapid genetic improvement through use of superior sires; (2) reduction or elimination of cost and risk of maintaining a sire for a small herd or flock; and (3) reduced risk of spreading of certain reproductive diseases.

Artificial insemination was reported to be used as early as 1322 in horses. It was first used commercially in the United States in 1938, primarily in dairy cattle. Initially, its use generally involved an AI technician who traveled from farm to farm on call. Some 60 years later, over two-thirds of the dairy cows and heifers are being bred artificially, and almost two-thirds of these are by owners or herd managers.

Because of the development of broad-breasted and heavy turkeys and their difficulties in natural mating, AI was introduced in the mid-1970s to improve fertility rates. Now almost all U.S. turkey breeders use AI. On the other hand, herd and flock management systems and other factors have limited AI to about 5 to 6 percent of the beef cattle, sheep, and horses in the United States.

AI has become a more common practice in the swine industry. Although used only by purebred swine breeders for many years, it now has become a well-established practice in many commercial swine operations. Researchers have been unable to develop technology that gives acceptable conception rates and litter sizes with frozen boar semen. Therefore, most swine producers using AI collect and process fresh semen on their own farms, or receive fresh semen from commercial sources.

Improvements in delivery services and semen extenders that enhance semen viability have increased the use of commercially available boar semen. Although only a few commercial AI firms serve the swine industry, boar semen has been sold on a limited basis since the early 1980s. Today, several companies provide specialized AI services, from the collection of semen to the sale of semen, or even breeding boars.

Because a single ejaculate of semen contains many times the number of sperm needed to breed a female, the semen can be diluted and used in smaller portions. Table 14–1 gives the number of females that might be inseminated per ejaculate of semen in several species.

Semen collection from bulls is commonly done with the use of an *artificial vagina* and a dummy steer or cow, or, in cases where mounting of a dummy is difficult, the *elector-ejaculator* may be used. The latter instrument, inserted rectally, causes ejaculation over a 15- to 20-minute period by means of electric impulse. Collection from stallions and rams is typically done using only the artificial vagina. Boars are most commonly trained to mount a dummy structure, with semen collected using a *gloved hand* technique that applies continued and appropriate pressure to the penis.

Poultry semen is generally collected by a stroking and milking technique. Stroking of the pelvic arch down to the pubic bones will generally cause the male to raise its tail and evert its cloaca. Semen stored in an enlarged section of the vas deferens near its entrance to the cloaca is squeezed out into a container by pressure of the thumb and forefinger.

The semen is checked for sperm concentration and motility, then diluted to the desired concentration. If it is to be used fresh, it is usually maintained at exactly 40°F, or under carbon dioxide at room temperature. For storage and later use, diluted semen is usually placed in a small plastic straw, chilled slowly to about 0°C, then rapidly frozen in liquid nitrogen vapor. Freezing stops metabolism.

Because the sperm are single, living cells with little environmental protection, they must be handled with care. The dilutents include such materials as glycerol, egg yolk, milk, and chemical buffers, which help keep the sperm alive. Glycerol dehydrates the sperm, preventing damage from ice crystals during freezing. Bovine semen is usually held below −196°C and at that temperature can be stored successfully for many years.

For use, the frozen semen is thawed by placing the straw in a 32° to 35°C water bath for 30 seconds. It is then deposited by means of an insemination syringe or "gun," which holds the plastic straw, into the reproductive tract of the female (Figure 14–14). In mammals the end of the syringe is usually inserted into the cervix and the semen is deposited deep in the cervix or just in the body of the uterus. If any vitality of the sperm has been lost during storage, depositing the semen in the uterus may be of benefit. Poultry semen does not stand freezing and thawing well so is generally used within 2 hours after collection. Extended boar semen provides optimal conception rates for 72 hours.[1]

In poultry, about 0.05 to 0.1 cc of semen is usually deposited by means of a syringe or small tube into the vagina. Most turkey breeders use artificial insemination exclusively. Whereas natural service results in 50 to 60 percent conception, artificial insemination results in conception rates of 85 percent or higher. In addition, fewer toms need be maintained and there are fewer bruises on the hens when they are eventually sold for slaughter.

In some herds, poor insemination techniques contribute to low conception rates. Skilled AI technicians pay careful attention to details of estrus detection and thawing, protection, and proper placement of the semen.

14.7 ESTRUS SYNCHRONIZATION

Practicability of artificial insemination has encouraged efforts to achieve control of the estrous cycle so that all or a specific portion of a herd or flock may be bred at one time. This would reduce labor and permit a more organized and efficient production system. For example, enough gilts could be mated at a predetermined time to fully utilize, at farrowing, a set of farrowing crates. All pigs would be essentially the same age and could be handled as a unified group until marketing.

With range cattle, the need for daily detection of cows in heat and sorting them out for insemination theoretically would be eliminated. If all came in heat at the same time, all could be inseminated while handling the group only once.

During the 1970s, research in several laboratories established that a natural substance, **prostaglandin**, or chemically produced analogs, had the ability to regress the corpus luteum and to cause estrus. In late 1979, the U.S. Food and Drug Administration authorized the sale of a commercial product called Lutalyse, a chemical analog of

> *Prostaglandin*
> Certain organic acids of the body; involved in bringing about regression of the corpus luteum.

---

[1] Extended semen is semen that contains additional materials to increase the volume of the original ejaculate and aid in the preservation and longevity of the spermatozoa.

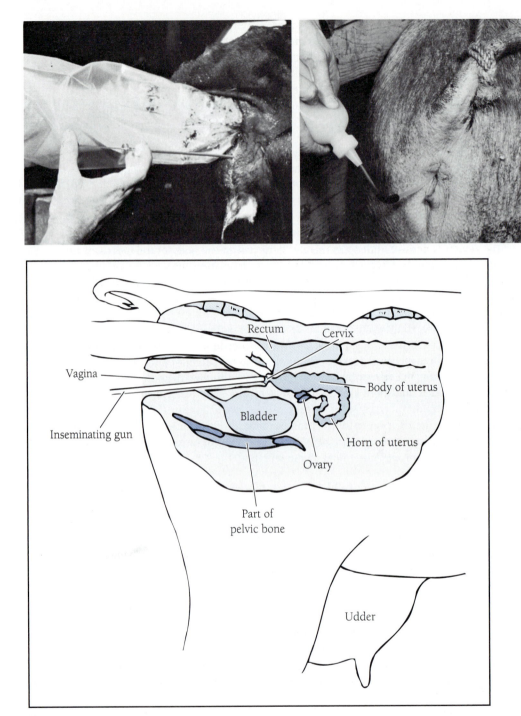

Figure 14–14

For inseminating cows, the left hand is inserted into the rectum and guides the gun holding the plastic straw through the cervix. This is not possible with sows and gilts; instead, a spiral end catheter is placed against the cervical opening and gently rotated counterclockwise to aid entry and passage through the spiral-shaped cervix.

(Photos courtesy of Kansas State University; drawing from Iowa State University.)

prostaglandin, for estrus synchronization of cattle. Because it is effective only when a corpus luteum exists, it will be effective on most cows from day 5 through day 18 of their 21-day estrous cycle. One recommended practice is to use two injections of Lutalyse 11 days apart. The first injection regresses the corpus luteum in those cows where it exists, and the cow shows estrus in 48 to 72 hours (day 2 or 3). Cows in which a corpus luteum did not exist at the time of the first injection would have one within 11 days, so that on day 11 a second injection could be given to all cows; 48 to 72 hours later all would show estrus and could be bred.

Some producers achieve the level of synchronization they seek with one injection. By this method, the single injection results in cows with functional CLs that regress, to come into estrus within 48 to 72 hours. Close observance should detect those cows that become synchronized. At the same time, cows that did not respond to the injection should be approaching estrus, so observance of these cows also should result in all cows being bred closely together.

Another method of estrus synchronization is to use one of several commercially available compounds termed *progestins*. Whereas use of prostaglandins is dependent upon the presence of a functional CL, progestins simply delay estrus and ovulation in cycling females. These compounds mimic the effects of **progesterone**, the natural hormone that blocks estrus and maintains pregnancy. If, by chance, a pregnant female should receive progestins, no harm should result.

> **Progesterone**
> A hormone produced by the corpus luteum on the ovary that aids in maintaining pregnancy.

Progestins can be administered by injections or implants, or orally in the feed. After 12 to 15 days, progestin treatment is stopped. Most females within the treated group should reinitiate their estrous cycles and be synchronized together.

The inducement of puberty in beef heifers also was found to occur when both progestin and prostaglandin were administered. In a multistate study[2] involving five groups of heifers, synthetic progesterone (melengestrol acetate, or MGA) was fed for 14 days and then followed by a prostaglandin (Lutalyse) injection 16 days later. This combination was effective in bringing more heifers into estrus for the first time and in synchronizing estrus in each group of heifers (Figure 14–15). Similar results have been obtained in sows and gilts treated with FSH and LH, or with compounds[3] that have similar actions (follicular development and ovulation).

Although estrus synchronization is not 100 percent effective, with good management it can be effectively implemented. In addition to earlier advantages mentioned, an overall benefit from concentration of breeding and parturition is that all animals can be managed more uniformly in all phases of management (e.g., nutrition, health, marketing).

14.8 MULTIPLE OVULATION

A 90+ percent calf crop—90 or more calves weaned per 100 cows—is considered by most beef cattle ranchers to be satisfactory. A 150 percent lamb crop is a reasonable goal for many sheep flocks, and 9.5 pigs weaned per litter is achieved most years by good swine producers. Economic incentive to increase output per breeding animal is high,

---

[2] Whittier, J. C., et al. *J. Anim. Sci.* 70 (1992): 2622; and NCR-87 project report.
[3] Reported in Britt, J. H., et al. *J. Anim. Sci.* 67 (1989): 1148. The study used P.G. 600, tradename for a compound that contains both pregnant mare serum gonadotrophin (PMSG) and human chorionic gonadotrophin (HCG), two hormones that mimic the effects of FSH and LH on follicular growth and ovulation.

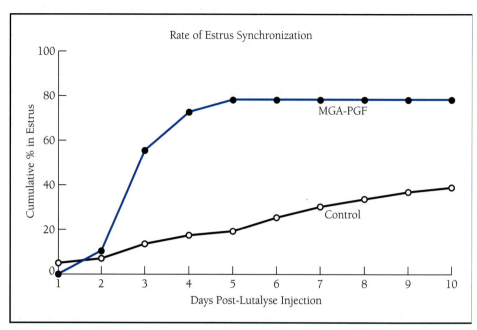

Figure 14–15

The cumulative percentage of heifers observed in estrus during the first 10 days of the breeding season. The upper line denotes the effects of MGA and prostaglandin in bringing 77 percent of heifers into estrus within 6 days. The lower line shows that only 25 percent of the control heifers were in estrus during this same period.
(*Sources:* Whittier, J. C., et al. *J. Anim. Sci.* 70 [1992]: 2622; and NCR-87 project report.)

however, and new knowledge in reproduction physiology coupled with intensive management of breeding herds and flocks permit higher reproduction rates.

Cows and ewes traditionally reproduce once each year; sows slightly more than twice. Estrus control, multiple ovulation, and selection of breeding stock (especially in sheep) that will mate and conceive any season, and adjustments in herd or flock management, such as early weaning, can permit five crops of lambs in 3 years or three crops in 2 years.

Today, multiple ovulation is brought about to provide sufficient ova for subsequent fertilization and embryo transfer to recipient animals. A successful procedure used by cattle breeders for inducing superovulation is to administer several injections of FSH to the donor cow between the 9th and 14th day of her estrous cycle. On the 3rd and 4th day of the schedule, the amount of FSH is reduced, and prostaglandin is injected to cause regression of the corpus luteum (or lutea). Within 36 to 60 hours after the first prostaglandin injection the cow usually shows signs of estrus. All recipient animals are given an injection of prostaglandin for estrus synchronization about 1 to 2 days before the donor receives prostaglandin.

In a cow, usually only one ovum matures during the estrous cycle and is released after estrus. In Oklahoma research[4] in the early 1990s, however, a combination of hor-

---

[4] Turman, E. J., et al. *J. Animal Sci.* 32 (1971): 962; and Turman, E. J., private communication, June 1, 1981.

mone treatments—two injections of PMS (pregnant mare serum, which causes development of more than one ovum during the cycle) and one injection of chorionic gonadotropin (which causes the several ova to be released)—resulted in a high proportion of multiple births. Of 134 cows treated, 77 conceived at the service immediately following hormone treatment, and these 77 dropped 1 set of quintuplets, 3 sets of quadruplets, 11 sets of triplets, 22 pairs of twins, and 40 singles. Thirty-four additional cows conceived at a subsequent estrus.

Livability and growth rates of twins were satisfactory, but the subsequent conception of cows nursing twins was delayed. There was high death loss in the triplets, quadruplets, and quintuplets, and approximately one-half of the multiple-birth cows had retained placentas (the placenta had to be removed by hand). Because of these problems, the more practical application of multiple ovulation in cattle appears to be in embryo transplanting. (See Section 14.9.)

The major application of multiple ovulation in cattle is to produce several ova in a high-quality cow for transplanting, after conception, to host cows. Embryo transplanting provides opportunity to spread more widely the benefits of a high-quality female, just as artificial insemination has permitted more genetic change in a herd or population by the wider use of a high-quality sire.

The practice of *flushing*, or increasing the daily feed allowance for about 2 weeks before breeding as a means of increasing ovulation rates and multiple births, is described in the next chapter. (See Section 15.3.)

14.9 EMBRYO TRANSPLANTING

The first recorded transplantation of fertilized ova in mammals was in 1890 at the University of Cambridge, when two ova were transferred from an Angora rabbit doe, bred 32 hours earlier, to the fallopian tube of a Belgian hare doe, which had been bred 3 hours earlier.[5] The Belgian hare doe gave birth to six young, two with Angora characteristics and four resembling the Belgian and her mate.

Experimentation continued in the twentieth century, and by the 1960s, laboratory transplantation was fairly common in many species. Introduction of exotic breeds of beef cattle to North America and their rapid growth potential and popularity, coupled with stringent quarantine regulations and restrictions on importations, provided incentive to speed reproductive rate by embryo transplantation.[6]

In 1979 about 17,000 bovine pregnancies were produced by superovulation and embryo transplantation (ET) in North America.[7] Since that time, ET has been widely used in beef and dairy cattle. It has been useful to the breeder in increasing the offspring from selected animals hopefully to be sold for breeding purposes. Through ET techniques, the chances of obtaining an outstanding sire from the superovulated host dam are somewhat enhanced. Its widespread use in the dairy industry is evidenced by more than 20,000 ET offspring registered each year with the Holstein-Friesian Association of America.

ET also has value in the development of a particular herd of animals, highly related and hopefully of desired type and production, such as those that might be desired in a herd for research purposes.

---

[5] Heape, W. *Proc. R. Soc., Lond.* 48 (1890): 457–58.
[6] Betteridge, K. T. *Embryo Transfer in Farm Animals.* Canadian Dept. of Agriculture Monograph 16, 1977.
[7] Seidel, George E., Jr. *Science* 211 (1981): 351.

| TABLE 14–6 | Developmental Stage and Quality of 1495 Transferable Embryos Recovered from Superovulated Dairy Cow Donors 7 Days after Estrus |

Embryo Quality Grade	Developmental Stage of Embryo					
	Morula	Compacted Morula	Young Blastocyst	Blastocyst	Expanded Blastocyst	**Total**
1 (excellent)	9	348	173	95	66	691
2 (good)	28	317	133	52	10	540
3 (fair)	35	147	30	7	2	221
4 (poor)	15	27	1	0	0	43
Total	87	839	337	154	78	1,495
Mean (SD)[a] quality grade per stage	2.6 ± 0.9	1.8 ± 0.8	1.6 ± 0.7	1.4 ± 0.6	1.2 ± 0.4	1.7 ± 0.8

[a]*Standard deviation.*

Source: *Callesen, H., et al., J. Anim. Sci. 73:1539, 1995.*

After the donor animal has been induced to superovulate and estrus occurs, she is normally artificially inseminated twice—at 12 hours and again at 24 hours after the first signs of standing heat. Then, fertilized ova usually are flushed from the reproductive tract of the donor female following conception but before implantation. They are evaluated microscopically for normal morphology and may be directly implanted in recipients or placed in a medium and stored as frozen embryos. Successful implantation can occur with frozen embryos; however, freezing and thawing will damage some embryos, and pregnancies are reduced about 50 percent, compared to the use of fresh embryos.

The many procedures to be carried out for successful embryo transfer make it a costly practice. These include the cost and administering of hormones to superovulate the donor and synchronize the recipients, and the specialized skills needed to harvest, sort, store, and implant embryos. These costs and the risk of low embryo yields and successful transfers should limit its use only to elite female animals.

In a 6-year study, Callesen et al.[8] evaluated embryos recovered from superovulated dairy cows 7 days after estrus. A total of 2,211 eggs were nonsurgically recovered from 217 donors. Of these, 1,495 were graded as transferable embryos based on morphological evaluation of development and quality (Table 14–6). Then, these acceptable embryos were nonsurgically transferred to recipients to produce 623 calves, for a 42 percent pregnancy rate. The two major factors associated with variation in embryo development were effects of the donor animal and the embryologist. Note in Table 14–6 the variation in the five stages of development. Because the quality grading was influenced by the embryologist, the researchers concluded that sexing of embryos by simple morphological evaluation is not possible.

Potentially, the discreet use of embryo transplantation could result in several benefits. These include

- More offspring from outstanding females (see Figure 14–16)
- Opportunity for genetic testing of males suspected to be carriers of an undesirable trait

---

[8] Callesen, H., et al. *J. Anim. Sci.* 73 (1995): 1539.

Figure 14–16
Shown is an outstanding Jersey cow with her 10 embryo transfer (ET) daughters—an excellent example of how daughters can be more rapidly produced from one female. Also, the use of ET techniques enables more rapid determination of the genetic potential of the donor cow, as well as the sire selected to provide semen for artificial insemination.
(Courtesy of Agri-Graphics, Ltd. Used with permission.)

- World movement of embryos (rather than cattle) made easier, less costly, and with less disease risk
- Increased diversity of germ plasm available from many sources
- Long-time storage provided by cryopreservation (freezing)

Successful cooling, packaging in straws, freezing, and thawing of embryos involves several critical steps but is similar to preservation of spermatozoa. Pregnancy rates for unsorted frozen embryos can be expected to be about 40 to 50 percent compared to 60 percent for nonfrozen embryos. With successful freezing, collected embryos need not be immediately transferred to recipient females but can be stored for future use.

The *splitting of embryos,* for production of identical twins or triplets, is possible with microsurgery and micromanipulation. Embryo splitting and transfer could more quickly increase a herd's or flock's productivity. Also, use of these genetically identical animals greatly reduces the numbers needed in animal experimentation.

It should be noted that processes of embryo transplantation and splitting do not result in offspring that have the same genotype as the donor animal. If so, they would be *clones.* When cloning techniques are perfected and made readily available to animal breeders, tremendous advances in genetics will have been made.

Sexing of embryos also has been achieved with about 85 percent of success in prediction of the correct sex. This advancement will enable breeders to select either male or female when purchasing embryos. Further success in the techniques of sexing of spermatozoa will provide even greater potential than embryo sexing and would enable the breeder through the usual AI procedures to inseminate sexed semen. To date, only limited success has been attained in techniques that separate sperm containing an X chromosome from sperm containing a Y chromosome. Other examples of technological advances in these areas are provided in Box 14–1.

BOX 14–1   Today's Reproductive Techniques—Tomorrow's Practices?

Predetermination of Sex

Sperm occur as either an *X* chromosome or a *Y* chromosome. Separation of these two types of sperm would enable the breeder to predetermine the sex of upcoming offspring. Female offspring should result from inseminating only *X* sperm and male offspring when only *Y* sperm are inseminated. The technique for separating *X* and *Y* sperm without great losses in viability has been developed by USDA researchers at Beltsville, Maryland.

These workers determined that the major difference between the *X* and *Y* chromosome is that the *X* chromosome is larger with slightly more DNA than the *Y* chromosome. By their sexing technique, the sperm are treated with a bluish fluorescent dye. The *X* sperm with more DNA absorb more dye than the *Y* sperm. This difference is determined by passing the dyed sperm through a laser beam of a flow cytometer. Optical detectors in this instrument measure the laser's light. The next step involves the coupling of the flow cytometer to a computer and "earmarking" the brighter *X* sperm with a positive charge and the dimmer *Y* sperm with a negative charge.

In the final step, sperm are directed single file through two electrically charged plates. Then, the negatively charged (*Y*) male sperm are collected in one container and the positively charged female (*X*) sperm are collected in another container.

14. Collection of Ova

Ova can be removed from the ovaries of a live animal through a technique known as transvaginal ultrasound-guided follicular aspiration. By this technique, a needle, guided by ultrasound images through the vaginal probe, is inserted into the ovaries. The eggs are aspirated out, collected, and sorted.

Combining These Techniques with in Vitro Fertilization and ET

With "presexed" sperm, the collected ova can be fertilized *in vitro* with either *X* or *Y* sperm. The fertilized egg is incubated for 7 days, and the embryo is then implanted into a recipient female.

Through USDA and English collaboration, the first calves resulting from predetermined sex were born in 1992. In later English work, the insemination of cows with presexed male sperm resulted in 36 out of 40 calves as males.

Sources: Adapted from Carnaham, W. E. *Hoard's Dairyman*, November 1994; and USDA communications.

QUESTIONS FOR STUDY AND DISCUSSION

1. List four reasons why high reproductive performance of farm animals should have a high priority.

2. Name two roles of the male in reproduction.

3. What mechanisms help ensure a lower temperature in the scrotum? Name the structures involved.

4. What type of environmental stress could render the dam infertile?

5. Name the specific structures that (a) produce sperm, (b) produce testosterone, and (c) provide a storage reservoir where sperm mature.

6. Name the structures that carry sperm from the testicles to the urethra upon ejaculation.

7. Name the accessory sex glands of the male and three functions they perform.
8. How does the reproductive structure of the stallion differ from that of the bull, boar, or ram?
9. Name two reasons for castration of male meat animals. At what age is castration typically performed?
10. What should a breeder do when numerous cryptorchids occur in the herd or flock?
11. Name the major structures of the female reproductive tract.
12. Describe the appearance of a Graafian follicle. How does this structure differ in appearance from the corpus luteum?
13. Name the functions of the fallopian tubes (oviducts).
14. Describe how the uterus and uterine horns differ in shape between sows, cows, and mares.
15. Name the innermost layer of the uterus.
16. Describe the appearance of the cervix. What are its functions?
17. What role(s) does the vagina perform?
18. Where are ova produced? What hormone causes their production? What hormone is most responsible for their release of time of ovulation?
19. What major hormone is produced by the cells of the Graafian follicle? What effects does this hormone have upon the female?
20. What is the principal hormone of the corpus luteum? What effect does it have on the female that is (a) nonpregnant and (b) pregnant?
21. What is the role of prostaglandin and where is it produced?
22. Name the phases of the estrous cycle.
23. Which of the domesticated farm animals are seasonally polyestrous?
24. Define the terms *implantation* and *placentation*. Name three principal structures of the placenta.
25. What are the main functions of the placenta? Which species have more pronounced cotyledon-caruncle attachments?
26. What principal hormones are involved in bringing about successful parturition? Name the stages of parturition.
27. Name two differences in the male reproductive system of the bird as compared to the mammal.
28. How does copulation occur in birds?
29. Name the parts of the oviduct in the hen. Where do the following occur: fertilization, addition of albumen, membrane formation, outer shell formation?
30. What are approximate incubation lengths for the chicken, duck, and turkey?
31. Why will artificial insemination of most farm animal species become increasingly popular?
32. What are advantages of multiple ovulation, estrus synchronization, embryo transplantation, and embryo sexing?

Management Regime and Reproductive Efficiency

15

Heritability of prolificacy (reproductive performance) is low[1]. This means that differences among animals in prolificacy, including rate of lay by hens, are not caused nearly as much by differences in genes carried by the animals as by differences in environmental factors such as nutrition, presence or absence of disease, management, and possible physical injury.

Diseases and infections of the reproductive tracts of the male or female that inhibit or prevent reproduction can usually be diagnosed by a competent veterinarian. Some may be treated and cured; others may be relatively incurable, so, once identified, such animals should be sold for slaughter.

Good management by the producer will help prevent physical injury. Properly textured or grooved floors around feeders and waterers ensure good footing, reducing injury by falls, which could cause fetal death and abortion. Cows on grooved concrete are less likely to become injured and move more freely, and the signs of estrus are more visible, making heat detection easier for the herd manager.

Temperature and season influence fertility. High body temperatures decrease sperm production, motility, and vigor in roosters, rams, bulls, and boars. These high temperatures may be caused by high environmental temperatures in summer or might result from infection or sickness. In either case, the partial or complete infertility is usually temporary.

Nutrition is a major influence on prolificacy, affecting both the potential sire and dam. Sperm production, health, and survivability are influenced by previous and present nutritional adequacy. Ovum production, conception rate, and embryonic survival are also subject to nutritional influences. Hatchability is a factor in poultry reproduction. Health of the offspring at birth or hatching and subsequent survival during the first few days can be affected by the nutritional status of the dam.

Upon completion of this chapter, the reader should be able to

1. List those nutrients that are especially critical, in common rations, for efficient reproduction.
2. Explain the effect of the level of energy intake by young female livestock on their reproduction efficiency at maturity.
3. Define *flushing* and explain why it is practiced.
4. List four or more factors other than heredity or level of nutrition that may influence conception rate in animals.

---

[1] Heritability estimates are discussed in Section 18.3 and shown in Table 18–1.

5. Describe how the ration should be adjusted from the beginning to the end of pregnancy.

6. Describe the influence of hen weight on egg production.

7. Define *colostrum* and explain why it is valuable to newborn livestock.

15.1   CRITICAL NUTRIENTS FOR REPRODUCTION

Certainly energy is needed for sperm and accessory fluid formation in males and ovum production in females. In fact, *flushing* (see Section 15.3) is sometimes practiced, or increasing the energy in the diet several weeks before breeding to encourage maximum ovulation and conception rate.

Energy level and good body condition are especially important at the time of rebreeding, after parturition, in ewes, cows, sows, and mares. The dam is using considerable energy for milk production, and lack of adequate dietary energy can result in failure of the estrous cycle to recur or, if it does, failure to conceive at first service. Although nutrient functions were discussed in Chapter 3, their more specific roles in animal reproduction are reviewed in the following paragraphs.

Protein is a critical nutrient because it may be inadequate in some rations. This is especially true in arid range country where breeding stock may receive only mature pasture or hay, both rather low in protein (Figure 15–1).

Sperm, ova, and the embryonic tissue that results after conception and develops in the uterus are composed primarily of protein. Epithelial cells that line the male and female reproductive tracts contain high proportions of protein.

Accessory viscous sex gland fluids, containing certain amino acids, carry the sperm; collectively the fluid and sperm make up the **semen**. These fluids provide temporary nourishment to the sperm, ensuring longer life. Fluids in the female reproductive tract also contain amino acids. Fluids help reproduction by lubricating the tract and promoting eventual union of sperm and ovum.

> **Semen**
> A mixture of sperm and accessory gland fluids produced by the testicle and accessory organs.

Figure 15–1
Crossbred calf and its F_1 Hereford-Angus dam in semiarid rangeland. Supplementing pasture with protein feeds is sometimes needed for beef cows nursing their calves. (Courtesy of *Hereford World*.)

Most hormones that control reproduction in both sexes of farm animals and poultry are composed of amino acids. These hormones regulate initiation of sperm formation, ovum production, and ovulation, and also help maintain pregnancy.

Vitamin A is critical because it also plays a role in epithelial health and integrity and because, like protein, it may be scarce if breeding animals receive only roughages, or grains low in carotene (the precursor of vitamin A). These epithelial tissues that line the reproductive tracts of both sexes may, in vitamin A deficiency, become dry and hard (keratinized), inhibiting fluid production and movement of sperm and ova and preventing efficient implantation of zygotes.

Vitamin E has been demonstrated as essential for fertility in cockerels, and for prevention of embryonic mortality in chicken eggs during hatching.

Calcium is especially critical in eggshell quality. The shell, largely calcium carbonate, is about one-third of the dry weight of the egg.

Phosphorus may be considered a critical nutrient in reproduction because of the soil deficiency in many grazing areas where cow herds and sheep flocks are kept. Delayed onset of puberty, delayed postpartum estrus, and poor conception rates can result when moderate to severe deficiencies of phosphorus occur.

All nutrients essential for mature animals are needed by developing embryos because the same general functions are being performed. Except in certain areas or situations, other specific nutrients are not considered critical because they are normally adequate in typical rations. In addition to those discussed, several other vitamin and mineral deficiencies have been reported to have adverse effects upon reproductive performance. A more complete summary is provided in Table 15–1.

TABLE 15–1	Effects of Some Vitamin and Mineral Deficiencies upon Reproductive Performance of Farm Animals[a]	

Nutrient	Effects upon Fertility	Effects upon Reproduction
Vitamin A	Delayed puberty	Weak, blind offspring
		Retained placenta
Vitamin D	Delayed puberty	Rickets in offspring
	Suppressed estrus	Milk fever when Ca and P also are marginal
Vitamin E	No effects	Retained placentas when Se also is deficient
Calcium	Lowered fertility in birds	Rickets in offspring
		Milk fever in cattle
		Lowered shell quality in birds
Phosphorus	Delayed puberty	Stillborn or weak offspring
	Delayed postpartum estrus	Rickets in offspring
Iodine	No effects except reduced thyroxin output	Premature birth, weak or dead offspring
		Goiter in offspring
Selenium	Reduced fertility	Retained placenta when vitamin E is marginal
Copper	Delayed estrus	Retained placenta
	Reduced conception rates	Weak lambs
Cobalt	Reduced conception rates	No effects
Manganese	Reduced fertility	Abortion
	Reduced conception rates	Reduced birth weight

[a]*Other vitamin and mineral effects likely exist, especially those on growth and development of the young animal and on the onset of puberty.*
Source: *Compiled by the authors from National Research Council publications.*

Figure 15–2
Individual feeding of pregnant sows and gilts permits the restricting of energy intake. The same could be accomplished by self-feeding a very bulky ration containing a high proportion of silage or chopped hay.
(Courtesy of Kansas State University.)

15.2   GROWTH AND DEVELOPMENT OF BREEDING STOCK

For best reproductive performance, breeding stock should be in moderate body condition, neither emaciated nor obese. Farm animals that are extremely fat tend to be less prolific.[2] Thus, controlled feeding is important (Figure 15–2). Though potential fatness may be inherited, the realization of this potential—getting the animals overconditioned—is caused by too heavy feeding. Gilts, ewes, or heifers that carry too much finish may release fewer ova during the reproductive cycle, and, if conception occurs, fewer of the embryos may survive during pregnancy. The reasons for decreased fertility are not completely clear.

Research by Oklahoma State University scientists[3] demonstrated that feeding young heifers relatively low energy levels—just enough to maintain weight during the first winter as calves—does not materially reduce skeletal size at breeding time, or subsequent calf crop percentage. Heifers fed at higher levels, in some cases gaining 150 pounds or more during the winter, conceived a bit more readily but had more difficulty in calving. Heifers raised at the extremely high levels gave less milk during lactation, perhaps due to fat deposition in udder secretory tissue during growth.

In more recent years, selection for muscling and leanness in cattle and use of feeding regimes that provide only the energy level needed for growth result, in some cases, in energy intake during winter months being too low for adequate reproductive performance. Research at Purdue University (Table 15–2) reported in 1980 demonstrated the benefit of supplemental winter feeding of heifers prior to and during the breeding season.

A 20-year study on dairy heifers at Cornell University yielded considerable information on longtime reproduction efficiency. (See Table 15–3.) Puberty (evidence of first heat) was delayed considerably among heifers fed from birth at the low-energy level; it was more a function of size than age.

---

[2] In the case of cattle and horses, which generally produce only one offspring at a time, the term *fertility* would be more appropriate. Because most discussions pertain to all species of farm animals, the term *prolificacy* is used.

[3] Pope, L. S., et al. *Oklahoma Agr. Exp. Sta. Misc. Pub. 55, 57,* and *64.*

TABLE 15-2	**Effects of Winter Energy Intake on Condition and Reproductive Performance of Heifers**

	Ear Corn Fed per Day, Winter		
	0 lb	*2.68 lb*	*5.39 lb*
Number of heifers	112	113	112
Initial weight (lb)	496.1	502.0	493.5
Winter avg. daily gain (lb)	0.07	0.48	0.77
Summer avg. daily gain (lb)	1.72	1.50	1.32
Condition score, end of summer[a]	3.07	3.44	3.72
Conception rate (%)[b]	69.2	73.9	83.5
Avg. days pregnant[c]	42.8	55.8	63.6

[a]*1 = very thin, 3 = average, 5 = very fleshy.*

[b]*As percent of heifers bred.*

[c]*Sixty-five days after breeding season ended; estimated by rectal palpation.*

Source: *Lemenager, R. P. et al., J. Animal Sci. 51:837, 1980.*

TABLE 15-3	**Influence of Energy Intake from Birth to First Calving on Reproduction Efficiency of Dairy Heifers**

	Feeding Level		
	Low	*Medium*	*High*
At puberty:			
Number of animals	33[a]	34	34
Avg. age (mo)	20.2	11.2	9.2
Avg. weight (lb)	636	582	612
Avg. heart girth (in.)	59.9	57.5	58.4
Avg. withers height (in.)	46.8	44.8	45.4
At first calving:			
Number of animals	31	34	34
Percent conceiving at first service	79.00	68.00	58.00
Services for first conception	1.55	1.41	1.48
At seventh calving:			
Number of animals	18	18	21
Services per conception	1.61	2.11	2.05
Cows culled and reason:			
Sterility	1	4	7
Mastitis	11	19	15
Milk fever	4	1	0

[a]*One heifer died before puberty.*

Source: *Reid, J. T., J. Dairy Sci. 43:103, 1960; Reid, J. T., et al., Cornell U. Agr. Exp. Sta. Bull. 987, 1964; and private communication, 1969.*

All heifers were allowed to mate at 18 months of age, although the low-energy group had not yet reached puberty and so, in fact, did not mate until several months later. The percentage conceiving at first service was higher for those on a low-energy ration, perhaps because they were older when bred. Average number of services for first conception was higher for the low-energy group, *only* because a *few* heifers required numerous services. Such data indicate that optimum level of energy may vary among animals. Similar research at several institutions has shown that heifers fed relatively high levels of energy from birth usually require more services per conception.

Results indicate more efficient net reproduction among heifers fed at a relatively low energy level from birth to first calf. Fewer cows were lost because of sterility, number of services per conception was less (even after removing sterile cows from data), and weights of calves were similar.

The previous illustration simply indicates that breeding stock being developed need to be fed *only for adequate growth, not for fattening.* Relatively low levels of energy intake may delay physical and sexual maturity without lowering, and perhaps *raising,* net reproductive efficiency.

The "low" level of energy in the Cornell study was described as "65 percent of standard nutrient allowances" (as listed in National Research Council publications then in use). Because there is much genetic variation among animals, as well as considerable environmental variation—quality of feed, management, stress conditions—among farms or ranches where heifers are raised, specific energy levels used in the studies presented should *not* be interpreted as recommendations for all farms and ranches. This and other research provide bases, however, for developing recommendations for specific herds and environmental situations. The most profitable strategy is to feed and manage heifers so that most attain desired skeletal size and weight by about 15 months of age and thus calve at around 24 months of age.

A similar effect of nutritional regime during growth on egg production has been demonstrated with chickens. Restricting energy intake to at least 20 percent below that which would normally be consumed with a high-energy ration offered ad libitum, by limiting the feed per day, by increasing fiber content of the ration, or by other means, usually delays sexual maturity (time that the first egg is laid). In much research where energy restriction during growth was such that sexual maturity was delayed 8 days or less, laying house performance was not affected. Delays of 8 to 21 days were associated with *improved* annual egg production and egg weight; where maturity was delayed over 21 days, egg production was improved in heavy breeds, but not in Leghorn-type breeds.

During the 1960s, the trend toward leaner hogs to provide leaner retail pork cuts caused selection of leaner breeding stock, and prolificacy in this species generally improved during that time.

Spitzer et al.[4] studied the effects of body condition score (BCS) of 240 beef cows at calving and their weight gain following calving. Using scores of 1 (emaciated) to 9 (obese), they found that when cows calved with scores of 4, 5, and 6, calf birth weights were progressively higher without an increase in calving difficulty. This moderate condition also resulted in more cows exhibiting estrus and becoming pregnant within the scheduled breeding season.

---

[4] Spitzer, J. C., D. G. Morrison, R. P. Wettemann, and L. C. Faulkner. "Reproductive responses and calf birth and weaning weights as affected by body condition at parturition and post paratum weight gain in primiparous beef cows." *J. Anim. Sci.* 73 (1995): 1251.

Well-grown but not fat male breeding stock is also essential. For example, there is a positive relationship between growth rate in young bulls and their sexual maturity, testicular development, and sperm production. If possible, examination of the reproductive tract and evaluation of semen of all breeding males should be made periodically, and especially before the breeding season.

There is also some effect of the social or physical environment in which animals exist. For example, puberty in gilts is usually advanced by exposure to a boar or by moving and mixing with strange pigs. Socially dominant or larger bulls among a group will usually breed a higher proportion of cows or heifers. This means that younger or smaller bulls may be more useful when used in smaller or one-bull herds. Artificial lighting has been demonstrated to initiate egg laying in pullets and to advance the breeding season of mares.[5]

15.3   FLUSHING AND OTHER NUTRITIONAL CONSIDERATIONS

A common practice among sheep raisers is to initiate grain feeding or increase the grain allowance about a pound per day for ewes about 2 weeks before breeding, so they are gaining weight before and during estrus. This elevation of the nutritional level is known as **flushing** and is often, though not always, beneficial in promotion of ovulation and increase in the number of multiple births. When benefit does occur, it is apparently due to the increased energy consumption. As would be expected, prolificacy is more often improved when thin ewes are "flushed" than when ewes in fat condition are fed additional grain. Flushing is also accomplished by turning the ewes onto an especially lush, growing pasture, where the forage is highly palatable and highly digestible.

> **Flushing**
> The practice of feeding a higher amount of nutrients prior to and during the breeding season to improve ovulation rate; most commonly done in sheep management.

Because a large percentage of ewes are maintained primarily on grass, most are in a rather thin condition before the breeding season begins. This probably accounts for the popularity and relative effectiveness of this practice in sheep flocks. The sharp increase in nutritional state is often the factor, apparently, that determines the conception, development, and birth of twins rather than a single. Also, twinning is common in sheep, if breeding, nutrition, and management are satisfactory.

In cattle, birth of twins is not common. However, breeding heifers receiving a higher grain allowance prior to breeding may conceive earlier in the season (see Table 15–1).

Flushing has become increasingly more popular in swine herds because of the resulting increased ovulation rate and litter size. With increased selection of lean breeding stock, and sometimes limited feeding of potential breeding animals before they reach breeding age, evidence indicates that flushing of gilts and sows is worthwhile, with the greatest response in gilts that were on restricted feeding prior to flushing.

15.4   OTHER INFLUENCES ON CONCEPTION

Several factors may, at times, have significant influence on reproductive efficiency of both males and females (such as temperature, feed additives, hormones, and feeds containing certain molds or toxic materials).

---

[5] Koistra, L. H., and O. J. Ginther. "Effect of photo period on reproductive activity and hair in mares." *American Journal of Veterinary Research* 36 (1975): 1413.

In hogs, decreased conception rates and litter size often result when environmental temperatures exceed 85°F. Swine have little ability to dissipate heat by perspiration, so cooling by fans or other means is often helpful.

Boars, like males of other livestock species, are more affected by heat stress than are females because of heat damage to the immature sperm. Table 15–4 shows that in an Oklahoma test, gilts bred to boars held in outside lots with shade had lower conception rates than gilts bred to boars held in lots with shade and sprinklers or in a room with temperature held below 80°F. The breeding season extended from mid-June to mid-September, when daytime temperatures were generally above 85°F. Breathing rates of boars in outside lots averaged more than 100 per minute during August; rates above 50 per minute indicate heat stress in swine. Temperature stresses are further discussed in Chapter 11.

Many feed additives such as *antibiotics* have caused marked increases in growth. This may mean that certain animals will be heavier when they reach sexual maturity. Gilts when fed such rations may reach 200 pounds 2 to 3 weeks early, though sexual maturity has not been speeded up similarly. With heifers, however, most research indicates that weight at breeding time is not materially influenced by the use of antibiotics. The calves may grow faster early in life, when the antibiotics help resist certain low-level infections, but any difference in size is essentially eliminated by breeding time.

Much research has shown that antibiotics in sow rations during gestation have little influence on litter size, birth weight, or survival. Very slight improvements have been noted in some cases, but the differences have not been considered significant. It is apparent that continuous feeding of antibiotics to successive generations has no long-term effect on reproduction efficiency.

Where low levels of natural and synthetic *hormones*—progesterone, estradiol, testosterone, and others—have been used in livestock feeding programs, noticeable effects on size and structure of certain reproductive organs have been observed, but drastic interference with reproduction has not been demonstrated. Use of such products with breeding animals is definitely not recommended.

When *moldy* corn is fed to gilts, their vulvae often appear pink and somewhat swollen, as if they were in heat. The condition is not accompanied by other signs of estrus and disappears when the moldy corn is removed from the ration. Estrogenic compounds in certain forages may affect pituitary function and adversely affect ovulation of grazing animals.

Rye is often attacked by a black fungus, *ergot*. In Montana experiments, when sows were fed grain with 1 percent ergot during pregnancy, many of the pigs were so

TABLE 15–4	Conception Among Gilts Bred to Boars Held in Different Environments			
			Gilts Pregnant at 30 Days	
Boar Treatment	Number of Boars	Gilts Bred	Number	Percent
Shade	6	34	15	44.1
Shade and sprinkler	6	36	23	63.9
Cool room, less than 80°F	6	31	21	67.8

Source: Wetterman, Bob, *Hog Farm Management, July 1979, p. 10.*

weak that they died soon after birth and the sows showed almost complete lack of udder development. The exact cause of this harmful effect of ergot is not yet known.

Other influences of feeding and management on reproduction in farm animals are continually being studied. The cost of maintaining breeding herds and flocks, and the need for high conception rates, provides much incentive for improving efficiency of reproduction.

15.5   NUTRITIONAL MANAGEMENT DURING PREGNANCY

Although a relatively high level of feeding just prior to breeding is important in all farm animals—to ensure the production and release of sufficient ova in the case of swine and sheep or to ensure the release of one or several ova in the case of horses and cattle—continued high-level feeding after conception may decrease embryonic survival.

Nutritive requirements for early fetal development in the sow, ewe, cow, and mare are primarily protein and mineral. During early pregnancy, embryos increase little in size, so energy needs are not high. Tissue produced is primarily skeletal or is composed mainly of protein. During the latter part of pregnancy, however, fetuses grow considerably and nutrient requirements increase markedly. Enough nutrients also must be supplied to allow the prospective mother to prepare for milk production after parturition.

Research at several experiment stations has demonstrated the value of restricting feed intake of sows and gilts early in pregnancy, using the feed saved for additional nutrients later in pregnancy.

Table 15–5 shows correlations from USDA research that demonstrate that weight gain during pregnancy is positively correlated with number and total weight of pigs born.

The level of feeding of beef cows during pregnancy has received considerable attention. Grass or harvested forage is the major energy source but is often supplemented with oil meal (for protein) or grain (for energy). This supplemental feeding is relatively expensive.

Results of a longtime Oklahoma study that began with weaned heifer calves are presented in Table 15–6. Because of the relatively moderate climate, animals could graze year-round and were given supplemental feed only during the winter, from November

TABLE 15–5	Correlations Between Sow Weight or Gestation Gain and Prolificacy		
		Sow Weight Change	
Litter Characteristics	Sow Weight at Breeding	Breeding to 109 Days	Breeding to 24 Hr after Farrowing
Total pigs born	0.004	0.141	−0.154
Litter weight at birth	0.100	0.257	−0.086
Pigs born alive	−0.006	0.126	−0.128
Total weight, live pigs born	0.093	0.237	−0.065
Litter size at 21 days	−0.068	0.064	−0.127
Total litter weight, 21 days	−0.017	0.120	−0.064

Source: *Bereskin, B., and L. T. Frobish,* J. Anim. Sci. *53:601, 1981.*

TABLE 15–6	Influence of Level of Wintering (During Pregnancy) on Reproduction Efficiency in Beef Cows		
	Feeding Level		
	Low	*Medium*	*High*
Supplemental feed per day:			
Cottonseed meal (lb)	1	2.5	2.5
Oats (lb)	—	—	3
Number of heifers started	15	15	15
Avg. weight (lb)	473	471	476
Cows remaining 11 years later	14	10	6
Avg. weight (lb)	1,066	1,149	1,099
Percent calf crop weaned[a]	92.3	89.2	88.6
Avg. calving date	March 14	March 8	March 8
Avg. calf birth weight (lb)	77.3	76.9	78.5
Avg. calf weaning weight (lb)	479	479	468
Cow cost per cwt calf weaned	$7.13	$9.66	$13.30

[a]Based on total number of cows bred to calve each year. Heifers first calved at two years of age.

Source: *Pinney, Don, et al.,* Oklahoma Agr. Exp. Sta. Misc. Pub. 57, *1960, and* Pub. 60, *1960.*

to April, when grass was drier, less palatable, and lower in protein and digestible energy. Cows fed at the lower level lost more weight during the winter but gained it back during the summer. Nutrient storage during the summer months was apparently great enough to provide nutrients for fetal development during the winter, even among heifers wintered at the low level.

Also, heifers on the low level of supplemental feeding evidently consumed more forage during the winter than cows on the medium or high level. In a companion study, where heifers calved first at 3 years of age, average calf crop percentage was reduced even more on the medium and high levels of winter feeding.

More recently, in a 79-day grazing trial, also conducted by Oklahoma State University researchers in late fall and winter,[6] pregnant beef cows received sufficient nutrients from bermuda grass pastures during the first 30 days of grazing. However, cows needed 2 pounds per head daily of a 25 percent protein supplement to minimize weight loss during the final 49 days of the study.

15.6   REPRODUCTIVE EFFICIENCY IN POULTRY

Many breeders specify the optimum average body weight at sexual maturity for their line of pullets. The closer this optimum is approached and the more uniform the weight, the higher the production of the laying flock. Unless ration or management is adjusted, pullets raised during cool months will be heavier than optimum weight at sexual maturity and those raised during warm months will be lighter than optimum. Egg production

---

[6] Wheeler, J. S., et al. *Animal Science Research Report,* July 1998.

TABLE 15–7	Body Weight Uniformity and Egg Production	
Percent of Birds Within 10% of Average Flock Weight		**Effect on Number of Eggs per Bird per Year**
79		0
76		−4
73		−8
70		−12
67		−16
64		−20
61		−24
58		−28
55		−32
52		−36

Source: *Adapted from North, M. O.*, Poultry Digest 37, No. 442:606, 1978.

generally will be reduced by two eggs per pullet during their period of lay for every 0.1 pound above or below the optimum weight for that line. Table 15–7 shows the effect of body weight uniformity.

In the case of turkey breeder hens, however, feeding for maximum growth has generally resulted in a maximum number of settable eggs. In a summary of 37 research trials, it appeared that modest energy reduction often caused a decrease in number of eggs but no effect on fertility or hatchability.[7]

QUESTIONS FOR STUDY AND DISCUSSION

1. Explain why differences in prolificacy are due largely to environmental effects rather than heredity. Give three good examples of adverse environmental effects.

2. What nutrients would you expect to be highly concentrated in the composition of sperm, semen, ova, accessory fluids, and embryonic tissues? When might these nutrients be especially critical in the diet of breeding animals?

3. Relate the effects of overfeeding and underfeeding of breeding animals to their age at puberty, skeletal size, conception rates, and possible dystocia.

4. What is the purpose of increased nutrient allowance prior to and during the time of breeding in some species? Explain.

5. How might such factors as heat stress and hormone-containing feeds affect the performance of breeding animals?

6. At what stage of pregnancy does nutrient demand rapidly increase?

7. What is the relationship between uniformity of pullet flock body weight and potential egg production?

---

[7] Nestor, Karl E. *Poultry Digest* 38, no. 444 (1979): 110.

16
Lactation

Milk is secreted by all mammalian female farm animals, so this chapter provides information and discussion pertaining to all these species. The most efficient of these species is dairy cattle that have been developed primarily to produce milk for human consumption. Therefore, more studies of milk synthesis, milk let-down, and factors affecting milk production have been conducted with the dairy cow. The internal anatomy and physiology of milk secretion of farm species are similar. A sound understanding of milk secretion is very important to the owners and caretakers of mammals, regardless of the species.

In addition to dairy cows, adequate milk production is also important in beef cows, sows, does (female goats), mares, and ewes because they must produce enough milk to nurse the young until they are ready to eat sufficient grass or dry feed (Table 16–1). Ranchers sometimes complain about cows that give more milk than the calves can take, or that fail to dry up after the calves are weaned. On the other hand, there are some animals in these species that produce too little milk, especially as prolificacy increases in herds and flocks. A few fail to "let down" milk after parturition so the young can nurse.

Upon completion of this chapter, the reader should be able to

1. Describe how milk production is regulated by hormones.
2. Explain how parturition is related to the initiation of milk production by female livestock.

TABLE 16–1	Approximate Milk-Producing Ability of Farm Livestock[a]	
	Avg. Pounds per Day during Lactation	Percentage of Body Weight per Day
Dairy cow, Holstein	60–65	4.4
Beef cow	18	1.6
Sow	12.8	3.4
Ewe	3.3	2.2
Goat, dairy	9.0	7.0
Mare, pleasure (first 16 weeks)	30	3.2

[a]Comparisons cannot be made between species because of significant differences in composition. See also Figs. 16–2 and 16–3 and Table 16–2.

Source: Compiled by the authors and colleagues.

3. Chart a typical lactation curve.

4. Sketch the udder and identify major parts, including the milk ducts and cisterns and the alveoli.

5. Describe the physiology of milk secretion within the udder.

6. Explain the milk let-down process and list how it may be initiated.

7. List examples of how milk composition, flavor, or odor may be affected.

8. Discuss the importance of udder health, particularly in dairy herds.

Even though principles discussed in this chapter do apply to all species, most examples and illustrations pertain to milk secretion in dairy cattle.

Adequate intake of colostrum very soon after birth and an adequate quantity of milk until weaning are essential for healthy, thrifty, growing young farm animals. A strong nutritional start is especially important for young animals that are to be weaned from their mother at an early age, such as a baby dairy calf that is to be placed on a milk substitute diet, or pigs weaned to a dry ration as early as 3 weeks of age. The importance in weaning quickly means the dam can be bred back sooner, so the producer will have more animal units on an annual basis.

Heavier weaning weights of beef calves, lambs, or pigs are usually due, in part, to higher than average milk production of their dams, and result in more profit to their producer, especially when the animals are sold as young feeders.

All producers of brood stock, whether mares, cows, ewes, does, or sows, benefit from understanding the synthesis of milk and its let-down at nursing time. Familiarity with differences in milk composition enables owners of orphan animals to best utilize milk from other species (Table 16–2). For example, cow's milk can be altered for use by an orphan foal after the colostrum period by adding water to dilute the content of fat and protein and increasing the sugar content. Meeting nutritional requirements for lactation not only helps ensure heavier weaning weights of offspring but also helps to maintain the dam in a more desired state of body condition, especially important at the time of rebreeding for her next reproductive year.

Several factors besides genetics and nutrition influence milk production and its let-down. Extremely high temperatures, especially in association with high humidity, lower milk yields and alter milk composition. Age of the dam and number of nursing

TABLE 16–2	Approximate Composition of Milk During Mid-Lactation (percent)				
	Cow	**Ewe**	**Goat**	**Sow**	**Mare**
Water	87.5	81.5	87.5	82	90
Dry matter (total solids)	12.5	18.5	12.5	18	10
Protein	3.4	5.2	3.4	6.4	2.2
Fat	3.6	7.8	3.8	5.6	1.5
Lactose	4.8	4.7	4.6	5.3	5.8
Ash	0.7	0.8	0.7	0.7	0.5

Source: *Compiled by the authors from numerous scientific publications.*

offspring also have an effect upon the level of milk production. For example, dairy cows generally have their high-yielding lactation during the fifth or sixth lactation. Also, ewes nursing twins produce more milk than when nursing a single lamb.

All managers of farm animals and their offspring, and persons in charge of milking dairy cows, must understand the importance of milk let-down and the need for contented animals at nursing or milking time for effective harvesting of the milk. Also, they must be aware of the need to minimize mastitis, inflammation of the mammary gland caused by microorganism infections.

High-producing dairy herds almost always have a short calving interval, usually between 12.5 and 13 months. A short average calving interval means more cows are in the months of higher lactation at any given time.

16.1   IMPORTANCE OF MILK PRODUCTION IN VARIOUS SPECIES

The dairy cow usually attains peak milk production by 45 days after parturition. High-producing dairy cows often exceed 100 to 120 pounds of milk per day during peak lactation. After the peak, daily yields decline. Persistency—prolonged and steady production after the peak—is necessary for high production during the total lactation. (See Figure 16–1.) Persistency is affected by many factors, such as genetics, nutrition, disease, frequency and completeness of milking, as well as time of conception during lactation. Most of these factors, except for the effects of frequency and completeness of milking and pregnancy, are discussed elsewhere.

Dairy breeds differ in their average yield and composition of milk. Holsteins, which represent about 90 percent of all dairy cows in the United States, are known for their higher yields of milk. Many Holstein herds will average over 22,000 pounds of milk per cow per lactation, although the average for all Holsteins on official Dairy Herd Improvement (DHI) records is about 19,000 pounds.

The averages for the smaller breeds of Jersey and Guernsey on a DHI test range from 11,500 to 13,000 pounds. Although lower in total milk yield, their average percentage of milk solids is considerably higher than that of the Holstein. Because the price received is higher for milk containing more solids, many of the more productive herds of the smaller breeds are competitive with the Holstein. The Ayrshire and Brown Swiss breeds are typically intermediate in milk yield and composition compared to the Holstein and the Jersey and Guernsey breeds.

The daily milk production for some beef crosses shown in Figure 16–1 probably includes some yield response due to heterosis (hybrid vigor). An estimate of the lactation average per cow to weaning at 196 days in this study is about 3,000 to 3,500 pounds. Milk production of some of the continental (largely European) beef breeds introduced into the United States, such as the Limousin and Simmental purebreds, would be expected to be higher than that of the traditional English beef breeds (Angus, Hereford, Shorthorn) because they have been bred both for dairy products and meat production.

The sheep and goat provide milk for more people in the world than does the cow. A properly fed and well-managed milking doe (goat) can be expected to have a lactation of about 305 days in length and produce 1,800 to 1,900 pounds of milk. The milk is quite similar in composition to cow's milk. Demand for goat's milk is high enough in the United States that the more populated urban areas have access to modern goat dairies that produce and sell Grade A quality milk. (See Section 30.8.)

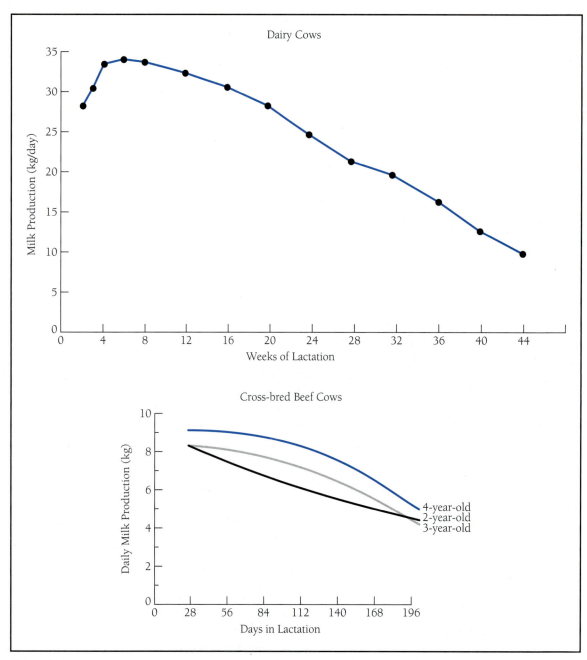

Figure 16–1

Lactation curves for dairy cows, top, and crossbred beef cows, bottom. Although many dairy cows produce more milk than shown, the above lactation curve is typical. The higher the lactation curve at peak lactation and persistency of lactation during midlactation and late lactation, the greater the yield of milk for the total lactation.

(*Sources:* Easkins, Charles T., and D. Craig Anderson. *J. Animal Sci.* 50 [1980]: 828; and Satter, L. D., and R. E. Roffler. *J. Dairy Sci.* 58 [1975]: 1219.)

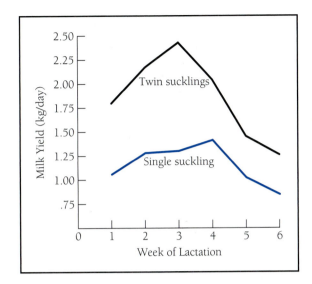

Figure 16–2
Lactation curve of the ewe.
(*Source:* Iowa State University.)

Top individuals in the five major dairy goat breeds in the United States have produced from about 4,800 to 6,650 pounds during a 305-day lactation, quite remarkable when considering that the average doe weighs about 120 to 150 pounds.[1]

In many parts of the world, the ewe provides milk for family use. However, in several European countries, milk is used primarily for the processing of cheese and yogurt. The milk of the ewe is considerably richer in fat and protein than cow's milk (see Table 16–2). Some European breeds have been developed for milk production and have produced as much as 3,400 to 3,500 pounds during their lactation.

The lactation curve of the ewe, shown in Figure 16–2, reaches a peak at about 3 to 4 weeks.[2] The effects of twin sucklings should be noted because the production of a ewe suckling twins was 59 percent greater than a ewe suckling a single lamb in this experiment.

Compared to cow's milk, the milk of the sow is richer in protein and fat. Peak daily yield of a sow suckling a typical litter would be expected to be as high as 15 pounds daily, followed by a steady decline until the pigs are weaned (Figure 16–3). Early weaned pigs (14 to 21 days) take maximum advantage of milk produced during peak lactation.

Mares produce milk somewhat lower in protein and fat but higher in lactose than cow's milk (see Table 16–2). Mares of the pleasure horse breeds as shown in Figure 16–4, may produce as much as 35 pounds daily during peak lactation, which occurs about 40 days after foaling. Heavier draft breed mares can be expected to produce as much as 40 to 50 pounds daily during peak production.

---

[1] The five major dairy goat breeds in the United States are the American LaMancha, French Alpine, Nubian, Saanen, and Toggenburg.
[2] Baylan, W. J. "Milk production in the ewe." *National Wool Grower*, April 1984.

Figure 16–3

Lactation curve of the sow.
(*Source:* University of Georgia.)

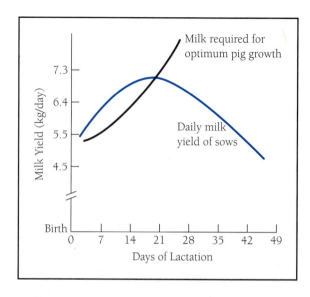

Figure 16–4

An Appaloosa mare and her colt. Lactating mares
should receive supplemental feeds to provide
additional nutrients (especially carbohydrates for
energy and protein) while the foal is nursing.
(Courtesy of Appaloosa Horse Club.)

16.2   HORMONAL INFLUENCES ON MAMMARY DEVELOPMENT
AND LACTATION

Embryonic development of the mammary gland, as well as its growth and development
from birth to puberty and from puberty to parturition, are similar among the common
farm mammals; the major variant is speed of development. The following discussion of
mammary gland development deals primarily with the bovine.

Mammary development is identifiable in the embryo as early as day 25, when a
single layer of cells has differentiated into a band known as the mammary streak. These
specialized mammary cells develop until about day 90, when a teat, complete with teat
cistern and a potential gland cistern, is grossly evident. At this time, close inspection might

reveal initial stages of a duct system, the ducts referred to as primary sprouts. By day 180, the primitive duct system has branched to form secondary and tertiary sprouts. There is little additional development in the mammary tissue from this time until birth, except for vascularization, appearance of suspensory connective tissue, and an increase in size.

During this *prenatal* time, other endocrines are making a contribution. The growth hormone STH is necessary for the increase in gland size; insulin causes cellular division, which is also a feature of growth; and the glucocorticoids from the adrenal cortex enhance the progesterone effects.

From *birth to puberty,* growth of the mammary glands is isometric, increasing at about the same rate as other parts of the body. Considerable development of the duct system occurs, primarily controlled by growth hormone.

With the *onset of puberty and each recurring estrous cycle,* female sex hormones stimulate pronounced growth of the ducts and cisterns of the mammary glands. Development now becomes allometric or more rapid than the rest of the body. The estrogen *estradiol* from the ovarian follicles predominates during the follicular phase preceding estrus and during estrus. Progesterone, produced by the corpus luteum, predominates following estrus, during what is termed the luteal phase of the estrous cycle.

After conception, progesterone produced by the corpus luteum dominates for most of the gestation period. This results in almost complete lobule-alveolar development necessary for milk synthesis. In addition to the development of secretory tissue, other changes are evident, such as extension of the vascular bed and organization of connective tissue into distinct layers and structures: the median and lateral suspensory ligaments.

As *parturition* approaches, there is a marked rise in placental estrogen and prolactin. The increased estrogen stimulates uterine release of prostaglandin and consequent reduction in progesterone secretion. Prolactin, necessary for the initiation and maintenance of lactation, is released from the anterior pituitary. With these hormones and probably others, the alveoli begin to secrete colostrum.

At parturition, a simultaneous peak of adrenal glucocorticoids, oxytocin, and prolactin bring the mammary glands into full lactational state, providing colostrum for the first few days, then increasing amounts of normal composition milk.

16.3   LOCATION, NUMBER, AND APPEARANCE OF MAMMARY GLANDS

In contrast to humans, other primates, and elephants, which have mammary glands in the pectoral (breast) region, farm mammals, except the sow, have mammary glands in the pelvic, or inguinal, region. In the sow, mammary glands develop in two parallel rows, one on each side of the medial ventral line.

Note the structures of the gland and udder illustrated in Figures 16–5 and 16–6, particularly the streak canal, the teat cistern, and the gland cistern. Numbers of glands vary among species, and in certain species, the number of visible external glands differs from the number of internal gland developments. Glands on the two sides have separate duct systems, and milk produced within each gland does not mix with milk from other glands.

The cow has four teats, each serving one of the four functional glands. The gland cistern receives the milk from the alveolar system of each quarter.

In comparison, the ewe and doe each have an udder consisting of two mammary glands, with internal gland anatomy similar to that of the cow. The mare also has two mammary glands as viewed externally, but internally each gland is served with two streak canals, two teat cisterns, and two gland cisterns.

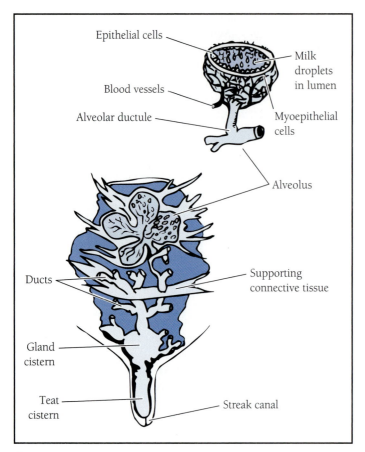

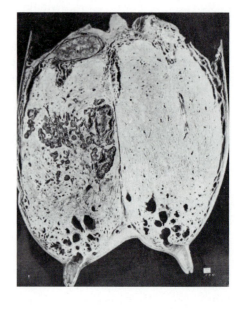

Figure 16–5

A cross section of a quarter, left. Note the horizontal strands of connective tissue that support the ducts and secretory tissue within the quarter. In the photo above, the connective tissue is not easily differentiated, but the cisterns and cross sections of the ducts are evident.
(Drawing from WestfaliaSurge, Inc.; photo courtesy of Iowa State University.)

The sow has two rows of mammary glands, ideally totaling 14 in number, in order to nurse large litters. Each teat in the sow is served by two streak canals and two teat cisterns. As in the case of the mare, each gland is served by two gland cisterns.

Extra nonfunctional teats or nipples sometimes are present. In the young heifer, such nonfunctional teats usually are located posterior to the rear teats. If left to develop, these may detract from the appearance of the animal or, more importantly, may become infected with mastitis-causing organisms. Thus, it is a sound practice to surgically remove these extra teats at an early age. Replacement gilts sometimes have *inverted* nonfunctional nipples; these tend to be inherited so should not be counted when appraising gilts for selection in a breeding herd. The front teats of the sow usually are spaced wider than the rear teats and are slightly more productive.

The teats of dairy cattle vary in shape and size. Because teat and udder size and structure are moderately heritable, appraisal systems for dairy animals should include soundness and suitability for machine milking, with shorter teats being preferred over long teats. Occasionally a beef cow or ewe will have an udder of such poor conforma-

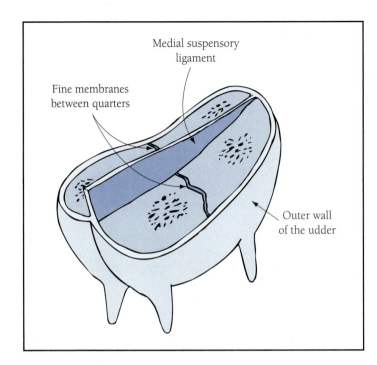

Fine membranes
between quarters

Medial suspensory
ligament

Outer wall
of the udder

Figure 16–6
A top cutaway of a cow's udder. Note the thick center membrane (medial suspensory ligament), which provides most of the support. (*Source:* WestfaliaSurge, Inc.)

tion (pendulous or with poorly spaced or sized teats) that the newborn has difficulty nursing. The teat of the mare is broader than that of the cow and tends to have a flat or blunt end.

16.4   ANATOMY OF THE UDDER

Note in Figure 16–6 that the two sides of the udder of the cow are separated by a thick layer of connective tissue known as the **medial suspensory ligament**. This strong, elastic tissue provides the main support for the udder and extends from the abdominal wall to the base of the udder and the intermammary groove. A well-defined intermammary groove on a full, distended udder indicates a strong medial suspensory ligament. Lateral suspensory ligaments extend inward from each side of the udder to the medial ligament, and the two ligament systems provide a cradlelike support for the udder.

> *Medial suspensory ligament*
> A primary support of the mammary system; heavy elastic tissue dividing the udder into halves; extends from the abdominal wall and to the base of the udder.

The combined weight of the udder tissue and milk produced between milkings for a high-producing dairy cow may exceed 100 pounds, so strong support of the udder is essential.

All quarters of the udder are not necessarily the same size. The rear quarters of the cow are usually slightly larger. Together, they produce about 60 percent of the milk, whereas the forequarters produce about 40 percent.[3] If one quarter of an udder is lost by injury or infection, the remaining quarters usually compensate slightly. The total production in such cases usually is not equal to the production before loss of the quarter.

---

[3] Rothschild, M. F., et al. *J. Dairy Sci.* 63 [1980]: 1138.

The teat or nipple of each gland is relatively hollow and consists of two main internal structures, the streak canal and the teat cistern. The *streak canal* in the end of the teat provides an opening to the outside. It is surrounded by a circular *sphincter muscle* that keeps the canal closed. Pressure at milking or nursing forces the streak canal to open, allowing milk to be removed. How rapidly a cow milks out, or how slowly, is largely dependent upon the strength of this sphincter muscle. Cows with weak sphincters tend to milk out rapidly but may also be more prone to microbial invasion of the gland and hence more susceptible to mastitis.

The small cavity above the streak canal is the *teat cistern* which is lined with rather delicate membranes. This teat cistern may hold small amounts of milk between milkings. Because this milk usually contains some bacteria, when milk is to be used for human consumption it should be milked out by hand and discarded prior to attachment of the machine milker.

The *gland cistern,* at the base of each gland near the junction with the teat, receives milk from the ducts and alveoli at milk let-down time. It should be noted that the capacity of the gland cistern is small, perhaps 500 mL in the cow. Therefore very little of the milk obtained at nursing (or milking) time has been held in this cistern; almost all comes through the cistern directly from the alveolar system where the milk is synthesized and held until let-down.

Note in Figure 16–5 that each *duct* is slightly constricted at the point where it joins a larger duct. These constrictions are opened or closed to control passage of the milk to the milk cistern. Each of the ducts acts as a storage cistern for milk until milking begins.

The *alveolar system* of the mammary gland is a structure of lobes and of lobules containing millions of milk-producing centers known as the *alveoli,* plus connective tissue that provides internal support for the alveoli (Figure 16–5). Each microscopic **alveolus** is a spherical structure with a layer of single *epithelial cells* comprising its wall and a hollow center, or lumen. The epithelial cells produce tiny droplets of milk continuously between milking, and these are discharged into the lumen. Each alveolus is served by a tiny duct that provides for milk removal and by a network of blood capillaries that provide the constituents for milk synthesis.

Another very important part of the microscopic structure within the mammary gland are the *myoepithelial cells* which comprise the cells or fibers that completely surround each alveolus. These fibers stretch as the alveolus fills with milk and contract to force out the milk when stimulated by the milk let-down hormone, oxytocin.

> **Alveolus**
> A hollow follicle of cells. In the case of mammary tissue, a spherical structure with cells that manufacture milk and secrete it into the lumen of the alveolus. Pl., *alveoli*.

16.5   MILK SYNTHESIS—COMPONENTS OF MILK

Circulating fluids—blood and, indirectly, lymph—are the immediate source of milk constituents and their precursors. It has been estimated that 300 to 500 pounds of blood passes through the udder for each pound of milk produced.

Milk is produced in the epithelial cells of the alveoli of each gland. Some of the milk components, such as vitamins, minerals, or water, apparently pass directly into the cells from circulating fluids through the wall of the capillary and the cell wall. Because milk is rather consistent in mineral content, some regulatory mechanism is likely involved. Other milk constituents, such as casein, lactose, and milk fat, are manufactured in the epithelial cells, being synthesized from certain precursors supplied by the circulating fluids.

The average composition of milk produced by different species is given in Table 16–2. Water, which makes up most of the weight of milk, apparently enters the mammary tissue by simple filtration. Sources of other nutrients that milk contains are discussed in the following paragraphs.

Lactose, a disaccharide that is the major carbohydrate in milk, is composed of glucose and galactose. These two compounds are monosaccharides, rather similar in chemical structure. Because both blood and milk contain glucose, it might be assumed that glucose moves into the cells of the alveoli by simple filtration. This apparently occurs to a great degree, but research indicates that some of the glucose is also formed in the cells from other small compounds, such as lactic, butyric, and propionic acids. Certain of these small free fatty acids are the major products of carbohydrate digestion in the ruminant, so the blood does contain significant quantities.

Galactose is apparently formed by a rather direct chemical conversion from glucose. This conversion is accomplished in the secretory cells with the aid of enzymes. One unit each of glucose and galactose are then chemically combined to form lactose.

Most of the protein in milk is in the form of *casein,* but some appears as lactalbumin and globulin. Each of these proteins contains different proportions of amino acids linked in unique patterns. It is apparent that the cells of the alveoli have the capability to build protein that is unique to milk.

Research indicates that large proportions of the amino acids used for casein, lactalbumin, and globulin come directly from the blood. They were present in the blood as free, circulating amino acids. There is also evidence, however, that blood proteins and glycoproteins (compounds made of both carbohydrate and protein) can be absorbed into the mammary gland and used as a source of amino acids. In addition, it has been demonstrated that mammary tissue is capable of *synthesizing* certain dietary nonessential amino acids for milk protein formulation.

A similar situation exists in the formation of *milk fat.* Many of the long-chain fatty acids that form a part of the milk fat apparently are absorbed directly from the circulating fluids. At least part of the short-chained fatty acids, however, are synthesized in mammary tissue from acetate (a small, 2-carbon unit) derived from absorbed acetate in ruminants, or from glucose in nonruminants.

Minerals in milk are absorbed directly into the alveolar cells from blood. Some are present as free elements; others are in a combined form. It may be that some type of selective absorption occurs to control the amounts and proportions of the minerals in milk in view of the fact that calcium and phosphorus are more concentrated in milk than in the blood, whereas sodium and chlorine are less concentrated.

Vitamins, or their precursors, come directly from the blood. Because vitamin intake influences vitamin level in the blood, it also has an effect on the vitamin level in milk.

16.6 RATE OF MILK SYNTHESIS

As tiny droplets of milk are formed within the epithelial cells of the alveolus (Figure 16–5), they are released into its lumen and then into the duct system. As milk synthesis continues during the interval between nursings or milkings, pressure builds up within the alveolus (Figure 16–7). As pressure increases, milk synthesis slows and is practically stopped as pressure within the lumen approaches that of the blood capillaries.

Because intramammary pressure influences rate of milk synthesis, some dairy operators practice three-times-daily milking rather than twice-daily. Although management, feed, and labor requirements are greater, three-times-daily milking usually will result in about 15 to 20 percent more milk production.[4] Probably half of the increase is

---

[4]Today, some dairy operations have "robot" milking systems which enable a cow to be milked more than three times daily.

Figure 16–7

Alveoli, full before let-down, and contracted during let-down of milk.
(*Source:* WestfaliaSurge, Inc.)

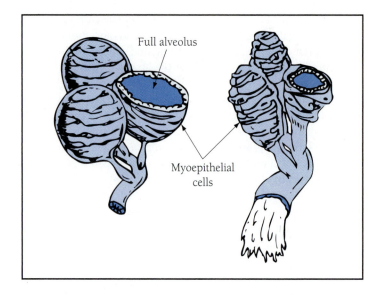

directly due to decreased pressure that allows increased synthesis between milkings. On the other hand, some dairy owners would like to milk only once daily, but research indicates that cows will decrease milk production by as much as 40 to 50 percent; again, this reduction can be explained, in part, by the slowed rate of synthesis due to increased pressure in the udder.

Udder pressure may also explain why the last milk drawn from a cow at a milking often contains a higher proportion of fat than the first milk removed. The large fat particles are more likely than the smaller particles or compounds, such as water, minerals, and carbohydrates, to be retained in the individual secretory cells of the alveoli when the udder pressure is high. As the cow is milked, udder pressure decreases and the fat globules leave the cells and flow down through the duct system.

Cows vary in the degree to which fat is held in the secretory cells, but whereas the strippings of some cows contain about 2.5 times as much fat as the average of all the milk, it pays to milk cows out completely.

16.7   MILK LET-DOWN

Most of the milk secreted is held within the alveolar system until milking or nursing occurs. Milk release results only when favorable stimuli bring about release of the milk let-down hormone, oxytocin. This is referred to as the *neurohormonal reflex* of milk let-down. The animal must receive a favorable stimulus associated with milking or nursing, such as nudging of the udder by the nursing offspring, sounds from the milking parlor and milking equipment, the warmth of the water as the teats are washed before milking, or the massage effect of drying the teats. The favorable stimuli are transmitted from the nervous system to the brain, then the posterior pituitary releases oxytocin.

The second portion of the neurohormonal reflex of milk let-down is due to the effect of oxytocin. Once released into the bloodstream, it is carried to the mammary gland. Once there, it causes the myoepithelial fibers that surround each alveolus to contract (Figure 16–7). Because some myoepithelial fibers surround some of the ducts, they

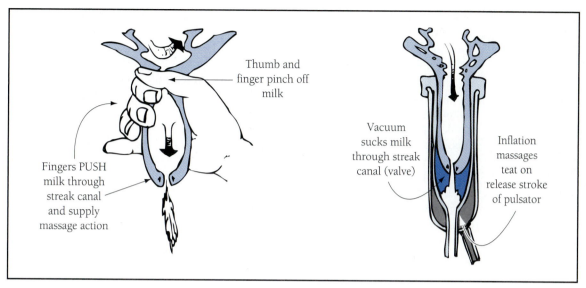

Figure 16–8

The milking process. Milk is forced through the teat canal by positive or negative pressure. (*Source:* WestfaliaSurge, Inc.)

also straighten to aid in milk removal from the alveoli. Because of an increase in pressure in the mammary gland, milk is easily removed by the negative pressure created by nursing or by the milking machine unit (Figure 16–8). The action of oxytocin lasts only for a short time, perhaps 4 to 6 minutes; therefore, the milking process should be completed within this period.

Animals should not be disturbed at the time of milking or nursing because the normal let-down process may be interrupted. Even minor changes in the milking pattern, or strangers entering the milking area, farrowing house, or nursing area for other species can cause an interruption of milk let-down. Irritations caused by rough teat-cup liners or the sensation caused by stray voltage in a machine milker may also interfere with normal let-down.

When the animal becomes frightened or nervous, the adrenal medulla hormones *epinephrine* and *norepinephrine* are released into the bloodstream. Once released, these hormones cause the blood around the myoepithelial fibers, which contains oxytocin, to be shunted toward the extremities of the body. Without adequate oxytocin, the fibers relax and milk let-down ceases. It is also possible that, through nerve effects, the hypothalamus fails to cause adequate release of oxytocin from the posterior pituitary or the myoepithelial cells fail to respond to oxytocin. Because less milk will be obtained when animals are disturbed, good milking management with dairy animals, as well as common sense with all lactating animals, is essential.

16.8   NUTRITION AND MILK COMPOSITION

Under normal feeding and management programs, the percentages of carbohydrate, protein, and fat in milk are not greatly influenced by the composition of the ration. Restriction of one of these macronutrients in the ration may cause a sharp reduction in volume of milk produced but will not alter composition appreciably. In fact, lack of energy (mostly supplied as carbohydrates in grains and roughages) is the most common limiting factor in milk production.

Numerous experiments have demonstrated that the kind of fat present in milk can be influenced, to a degree, by the kind of fat present in the ration. Increased levels of unsaturated fatty acids in the ration, for example, may cause the fat in milk to be slightly soft. Similar influences on protein and carbohydrate structure are not as common, apparently because of the way the precursors of these milk components are metabolized in the body and in the secretory tissue.

The method of feeding or the mechanical form of the ration may influence milk composition. Rapid changes in the dairy herd ration may cause a temporary drop in production or butterfat percentage. If the entire ration is very low in fiber or is finely ground or ground and pelleted, the butterfat percentage will likely be very low. Metabolic reasons for these effects apparently involve a change in the ratio of acetic to propionic acid. It may be that the mechanical form of the feed influences the proportions of fatty acids produced from carbohydrate digestion in the rumen and that this, in turn, influences utilization or production of fatty acids by the secretory tissue.

Milk composition normally changes after the first few days when colostrum is produced and as lactation progresses. The protein and fat percentages are *negatively correlated* with volume of milk produced; content of these nutrients increases during the latter part of lactation. The lactose percentage is positively associated with volume produced. Reasons for these changes are not apparent.

There is little evidence that the level of minerals in the ration (within reasonable limits) appreciably influences the mineral content of the milk. The percentages of calcium and phosphorus, present in large quantities in milk, usually are not influenced by the level in the ration. If the dairy cow's ration is low in calcium or phosphorus, the cow's blood levels will decline and she will use a considerable amount of these minerals stored in her skeleton before milk production declines.

This does not suggest, of course, that dairy producers need not be concerned with the mineral level of a cow's ration. Low blood calcium level and/or the inability of the cow to assimilate calcium at an adequate rate to meet calcium needs in the milk can cause *hypocalcemia,* or "milk fever." Any decrease in milk production is costly. Minerals removed from the skeleton for milk production must be replaced. Minerals perform other important functions in digestion and metabolism as well.

The influence of ration on vitamin content of milk varies according to the vitamin. In general, those vitamins that are dietary essentials for the cow—A, D, E, and K—are present in milk in somewhat the same proportions they are present in the feed. In the case of vitamin A, of course, some of the potency may be present as the precursor, carotene. Vitamin D content in milk would vary according to sunlight exposure or content in the ration. The B vitamins in the milk of the sow would be related to ration content, but they should be amply supplied via microbial synthesis in ruminants and the mare.

16.9   COLOSTRUM

As discussed in Section 6.5, colostrum, the product of the mammary gland during the last few days of gestation and the first few days postpartum, is especially high in protein, fat, minerals, and vitamins A and C. Nutrient content of colostrum is not necessarily a direct function of the ration, but adequate levels of certain of these nutrients must be present. These high levels of critical nutrients help the newborn get off to a good start.

Some of the extra protein in colostrum is in the form of gamma globulin, which is usually present at levels up to 300 times the level during midlactation and seems to function as a nonspecific disease fighter. It makes the newborn animal more resistant to diseases and infections, providing passive immunity to protect the nursing young.

Some research has indicated that specific antibodies might be produced in the cow and secreted in milk by exposing the cow to certain *human disease organisms*. The theory is that milk might provide a simple and painless way of causing immunity to certain diseases in people.

With the use of numerous nonnutrient feed additives in livestock production, there is some concern over the movement of such compounds through the body and, perhaps, through the mammary gland into the milk. Traces of certain fed or injected compounds such as antibiotics or ingested polychlorinated biphenols (PCBs) have been detected in milk, and, in these rare cases, the milk kept out of the milk lines and bulk tank cooler milk.

A common practice in dairy herds is to "dry-treat" cows. This involves the infusion of antibiotics into each quarter of the udder to provide protection against mastitis-causing organisms during the nonlactating period. Therefore, at the beginning of the next lactation, it is imperative that each treated cow's milk be tested and found free from antibiotics. Only then should it be allowed to enter the milk lines and commingle with the bulk tank milk. Federal and state regulations are designed to prevent sale of milk containing residues of many such additives.

16.10 FLAVORS AND ODORS

Most of the flavors in cow's milk are either lipid in nature or fat soluble—carried in the fat portion of the milk. Whole milk has more flavor than skim milk. Butter, made from milk fat, is used in cooking of food to enhance the flavor of cooked vegetables. Some variation in milk flavor may exist without distaste on the part of the consumer. Also, different people prefer different kinds or intensities of flavor.

The most common off-flavors in milk are "feed" flavors, due to ingested or inhaled materials. These flavors are probably volatile fatty acids that become part of the milk fat. Flavors that result from processing milk would not necessarily be due to the fat. Overheating of protein or carbohydrate, or other factors, may contribute to an off-flavor. "Oxidized" flavor, however, which results when milk is exposed too long to light and/or air, is probably mostly due to oxidation of short-chained fatty acids.

Long lists of feeds that will alter the flavor and odor of milk have been compiled. Most such feeds are not available to dairy herds, especially as more cows are fed in confinement. There are, however, certain common feeds used in these current management systems that can impart distinct flavors to milk.

Early in the spring when herds may be turned out on lush, rapidly growing pasture, milk sometimes has a "grassy" flavor. This flavor disappears after a few weeks as the grasses mature.

Rape, a member of the cabbage family, or wild onions will impart a distinct flavor to milk. Sweet clover, rye, and some ensiled grasses and legumes also can produce a characteristic strong flavor in milk. Bromegrass, which has become very popular as a forage for dairy cattle, influences the flavor of milk in certain seasons.

Feed flavors in milk are most evident if the cow grazed or consumed the feed 1 or 2 hours prior to milking. Some flavors reach the mammary gland in a few minutes, but the greatest concentration is reached after an hour or two elapses. After this, the concentration of the flavor in the milk decreases as it diffuses back into the blood for detoxification and elimination or is degraded by mammary tissue. It is therefore recommended to feed cows immediately *after* milking, or at least 3 hours before milking, if the feeds are suspected of potentially influencing milk flavor, and to bring cows in from early spring pastures 2 or 3 hours before milking.

Feed and other flavors might reach the milk by routes other than the digestive tract. Because most noticeable flavors are highly volatile, they can be inhaled and reach the circulatory system via the lungs. This emphasizes the importance of barn ventilation, especially where silages with strong aromas are fed inside the barn.

Certain processing techniques are rather effective in removing the volatile flavors in milk. Most, as would be expected, involve the principles of heating and/or subjecting the milk to a vacuum. Both conditions make evaporation easier, so the volatile compounds are removed.

16.11   APPEARANCE AND OTHER MILK QUALITY FACTORS

The most noticeable difference in milk appearance is the color variation associated with the breed. Guernsey and Jersey milk appears richer because of its more distinct yellow color. Milk of most other breeds is nearly white. The yellow color is caused primarily by carotene, which is a precursor of vitamin A, and other yellow pigments contained in grass, hay, silage, yellow corn, and certain other feeds. Most Holsteins, Ayrshires, and Brown Swiss do a very efficient job of converting carotene to *colorless* vitamin A, primarily in the wall of the small intestine and in the liver, so their milk is not so yellow. Guernseys and Jerseys, however, convert less carotene to vitamin A, so the yellow pigment passes directly to the milk. There is little difference in vitamin A *potency* of the milk from the different breeds as far as the human who consumes the milk is concerned. If the cow does not convert carotene to vitamin A, the job is done in the body of the consumer.

Season and kind of feed may influence milk color. In early spring, when cows may consume lush grass with high pigment content, milk will be richer in color. The effect is striking in breeds that are inefficient in carotene conversion, but also noticeable in other breeds because they do not convert all yellow pigments. As the season progresses, Guernsey and Jersey milk remains yellow longer, probably because these breeds draw from the yellow pigments stored in body fat.

If cows were fed weathered, bleached roughage, grain sorghum, and soybean meal as the main feeds, the milk would be almost pure white. Vitamin A, if supplied in pure form, would not contribute to the color of the milk. Because most good dairy rations include top-quality forage, considerable pigmentation is automatically supplied.

16.12   UDDER HEALTH, MASTITIS, AND MAMMARY GLAND INJURIES

The most common concern in udder health is mastitis—inflammation of the mammary gland. Though it is more common in dairy cattle, it occurs in all domestic animals, especially swine, and in humans.

Clinical mastitis is easy to detect. Flakes or clots appear in the milk (Figure 16–9) and the affected quarter is usually swollen, hot, and sensitive to the touch. *Subclinical mastitis*—a persistent low level of infection—is much more common in dairy herds, however, and significantly reduces both production and quality of milk in herds where

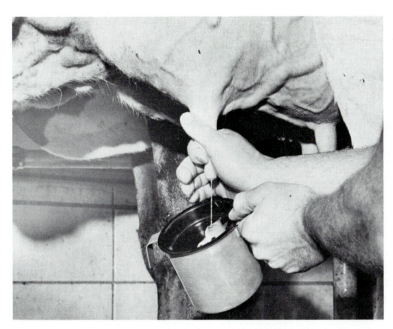

Figure 16–9
Drawing foremilk into strip cup to detect clinical mastitis infection.
(*Source:* WestfaliaSurge, Inc.)

it is prevalent. These low levels of infection are usually not evident to the milker-operator, but they can be detected by certain laboratory tests. When such an infection exists, milk contains higher than normal levels of leucocytes (produced in response to injury or infection) and epithelial cells that have eroded or died as a result of the injury or infection.

When a portion of milk is mixed with a reagent in the California Mastitis Test (CMT), a high concentration of leucocytes and/or epithelial cells will cause formation of a gel. Should there be some or considerable gelling, further laboratory tests can be used to identify the causative organisms, and treatment or culling can be implemented. The somatic cell count (SCC) is routinely determined in most milk testing laboratories to provide an indication of milk quality and severity of subclinical mastitis. Most dairy processing plants pay a premium for low somatic cell milk.

Good dairy managers effectively control the incidence and severity of mastitis by sound control measures. These include

1. ensuring that the milking system is properly cleaned and sanitized
2. ensuring that equipment functions properly
3. using correct milking procedures (udder examination and preparation, predipping, proper timing and application of teat cups, postdipping)
4. providing fresh feed for cows as they leave the parlor[5]

---

[5] Cows leaving the parlor have extremely dilated teat ends. By providing fresh feed, the cow remains standing for some period instead of lying down in a free-stall. While standing, the teat end returns to its constricted state and becomes less subject to bacterial invasion from environment sources (e.g., bedding).

5. monitoring somatic cell counts regularly and determining cause(s) if high

6. treating clinical cases promptly

7. implementing a "dry-treatment" program

8. eliminating cows with chronic problems that do not respond to treatment

Tendency toward injuries may be inherited to a degree because injuries are highly associated with pendulous udders (closeness of the teats and udder to the ground). Also, if the medial suspensory ligament is weak, the teats point out, and may be bruised as the cow walks.

Tendency toward udder infections also might be inherited and associated with conformation or structure of the udder. Teats that have weak sphincter muscles are more likely to become infected.

Infection of or injury to the udder of a young heifer may limit her milk production as a cow. If scar tissue develops in place of some of the secretory tissue, less milk can be produced and less milk can be stored in the udder between milkings. Overfeeding of heifers during growth may decrease the milk-producing capacity of the udder by causing formation of nonsecretory tissue.

QUESTIONS FOR STUDY AND DISCUSSION

1. Why should owners of animals other than dairy cows know the composition of both cow's milk and that of the species they own or manage?

2. How soon after calving should most cows peak in milk production?

3. What does persistency of milk production mean?

4. What is a typical amount of milk produced per lactation by Holstein dairy cows? How does this compare to the amount produced by the world-record cow?

5. Which dairy breeds produce milk of higher solids?

6. What are the benefits of good milk production by beef cows?

7. Compare the richness of milk of the sow and ewe to milk of the cow.

8. How would you alter cow's milk to feed an orphan foal?

9. Name the principal hormone(s) responsible for mammary development (a) from birth to puberty, (b) from puberty to pregnancy, (c) during pregnancy, and (d) to initiate lactation.

10. As viewed externally, how many functional glands are typical for sows, cows, ewes, does, and mares?

11. Of the above, which have two streak canals, two teat cisterns, and two gland cisterns per gland?

12. How can extra teats create problems with dairy heifers?

13. What are some abnormal nipple conditions in the gilt or sow?

14. What is the principal support of the bovine udder?

15. What is the purpose of the sphincter that surrounds the streak canal? What are the effects of extra strong or weak sphincters?

16. Upon secretion, where is most of the milk located within the mammary system?

17. Describe the network of milk ducts. What structures do they connect?

18. Describe an alveolus and its epithelial cells. What is the function of these cells?

19. Name the constituents of milk. Name the principal protein and carbohydrate.

20. What slows the rate of milk synthesis from milking to milking, or from nursing to nursing?

21. Explain the effect of proper stimuli upon oxytocin release. What is the target of oxytocin when released into the blood?

22. Explain what happens to oxytocin during milking or nursing when the animal becomes frightened.

23. What are the major differences in colostrum and normal milk?

24. Give an example of an undesirable feed flavor in cow's milk.

25. Explain why good managers take steps to ensure a low incidence of mastitis.

17
Genetics

The principal methods of improving productive efficiency and animal product quality are through the understanding and application of genetics, nutrition, reproduction, lactation, physiology, disease control, management, and marketing. All animal breeders have genetic improvement as their principal goal.

The study of modern biology involves more than how genes are transmitted from one generation to the next and their effects upon the traits of the next generation. Other areas of genetic studies involve population genetics, cellular and chromosomal genetics, molecular genetics, and gene mapping, as discussed earlier in Section 2.6. *Population genetics* views genes collectively and involves large numbers of observations. In contrast, *molecular genetics* involves the study of genetic materials (DNA and RNA), their structure, and how they control metabolic processes within cells. The next four chapters focus on a better understanding of the transmission of genes, their effects upon important traits of farm animals, and rules for improvement through selection and mating.

How are characteristics inherited—passed from one generation to the next? The only physical link between generations is the pair of reproductive cells, the sperm from the sire and the ovum from the dam, which unite at conception. The answer lies, then, in these cells, what they contain, how they develop in the testicle of the sire or ovary of the dam, how the zygote divides after conception, and just how the contents of these cells exert their effects.

Upon completion of this chapter, the reader should be able to

1. Define *gene*, *chromosome*, *DNA*, *nucleotide*, *genotype*, *phenotype*, *dominance*, *recessiveness*, and other key terms.
2. Explain how genes are transmitted from one generation to the next.
3. Differentiate between single and multiple gene effects.
4. Explain what is meant by sex chromosomes and sex-linked traits.
5. Differentiate between genetic and cytoplasmic inheritance.
6. Explain when, in the life cycle of an animal, environment begins to have an effect.
7. List the four rules for achieving improvement of livestock and poultry by selection.
8. Explain why genetic variation in a breeding herd or flock is good.
9. Explain how prolificacy and average generation time of a species influences the speed with which genetic change can be achieved.

17.1   GENES AND CHROMOSOMES

> **Deoxyribonucleic acid (DNA)**
> Double-stranded nucleic acid which is a component of chromosomes and which contains coded genetic information within its nucleotides.

Deoxyribonucleic acid (DNA) is recognized as the genetic material for all living organisms. DNA, with its double helical structure and nucleotides, forms long, threadlike strands of nucleoprotein (Fig. 17–1). These strands are generally not visible with an ordinary light microscope, but during nuclear division the nucleoprotein strands contract to become visible with such a microscope.

The strands of nucleoprotein are called *chromosomes.* Segments of the strand, called *genes,* are considered to be the simplest functional units of inheritance.

DNA contains many units of nucleotides, each composed of the sugar deoxyribose, phosphoric acid, and a nitrogen-containing base. These units might be linked in a variety of patterns, so various genes (segments of the nucleoprotein strand) differ in their chemical structure.

Figure 17–1
Diagram of a DNA structure arranged to illustrate its helix structure and showing key components. The nucleotide circled contains cytosine; adjacent nucleotides contain thymine and adenosine. A gene contains several hundred nucleotides. The sequence of components determines the code or "message" that is carried by another nucleic acid, ribonucleic acid (RNA), to the sites of protein synthesis and causes physiological and physical characteristics.
(*Source: Missouri Agr. Exp. Sta. Bull. 558.*)

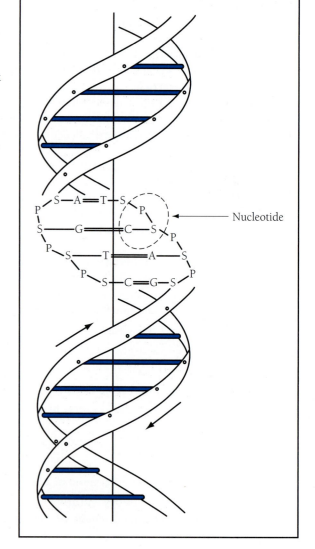

Nucleotide

PLATE A. DAIRY PRODUCTION

1. Young dairy heifers are necessary to replace cows culled from the milking herd. Individual attention is necessary until the calf is weaned at about 5 to 6 weeks of age. Colostrum, then milk, or milk replacer, grain starter and good quality hay make up the calf's diet.

2. Dairy heifers are group-reared from weaning until about 22 months of age. Most heifers are bred at 13 to 15 months, preferably to AI-available sires, in order to calve at 22 to 24 months of age. A few weeks before giving birth, heifers are moved into the maternity group and adapted to the herd ration and milking parlor routine.

3. The dairy cow normally peaks in lactation within 30 days after calving and should produce milk for at least 305 days, or until 60 days before the next calving. The cow should be bred and pregnant within about 100 days to achieve a 12.6 to 13.0 calving interval. Most cows are fed a "complete ration," and the high-producing group consumes about 4 percent of their body weight of a forage-concentrate ration (on a dry matter basis).

4. Most dairy herds are milked twice daily in approved Grade A milking parlors. Cows adapt quickly to milking procedures and very few are fed in the parlor. After proper udder preparation the milking unit is attached, milk is rapidly removed, then the milking units are removed automatically. Each teat is post-dipped or sprayed with an approved disinfectant to help reduce risk of infection between milkings.

Courtesy of Germania Dairy Automation, Inc.

5 & 6. Milk moves through properly sanitized stainless-steel or glass pipelines to the milk cooler that quickly reduces and holds the milk at about 38°F until picked up by the milk hauler.

7, 8 & 9. All milk is properly pasteurized at the processing plant. About 40 percent of all Grade A milk is utilized in the fluid form; the remainder is processed into popular cheeses, ice creams and other dairy products.

10. A goal of every breeder of livestock is to obtain an outstanding animal such as this world champion Holstein cow. This 6-year-old cow, ROBTHOM SUZET PADDY classified "Excellent-93-2E" and completed a 365 day DHIR record of 59,300 pounds of milk, 2,297 pounds of milkfat, and 2,038 pounds of protein. She is bred and owned by the Robert Thomson, Jr., Family, Springfield, MO.

All photos except #4 courtesy of Milk Promotion Services of Indiana, Inc., and Agri-Graphics.

PLATE B. SHEEP PRODUCTION

1. Sheep, as ruminants, make excellent use of vegetation that otherwise would go to waste. Most sheep breeds are seasonal breeders with the breeding season in the fall months as the photoperiod decreases. Average gestation length is 147 days; thus ewes bred in mid-October should lamb in mid-March. Twin lambs occur about 45 percent of the time. Proper nutrition of the ewe is essential for strong lambs and sufficient milk.

2. Lambs can be successfully weaned at 6 to 8 weeks of age if they are consuming 1.5 to 2.0 pounds of feed per head daily. Many farm flock lambs continue from weaning to market age on a high-concentrate–low-roughage ration. Typical market weights for the larger breeds are 105 to 110 pounds within about 130 to 140 days of age. Some lambs, especially those in range flocks, are allowed to graze throughout the summer and fall months. They then are placed in feedlots as "feeders" to be further grown and fattened for about 60 days before being sent to market.

3. Replacement ewe lambs should receive high quality pasture, close observation and treatment for parasites, if necessary, and protection from predators.
Courtesy of Purdue University.

4. Most well-bred and properly fed lambs will grade USDA Choice or Prime when marketed. Yield grades from 1 to 5 have been established. A Yield Grade No. 1 lamb should have a cutability of 47.3 to 49.0 percent of the boneless retail cuts, consisting of the leg, loin, rack (rib) and shoulder.

Courtesy of National Live Stock and Meat Board (now defunct).

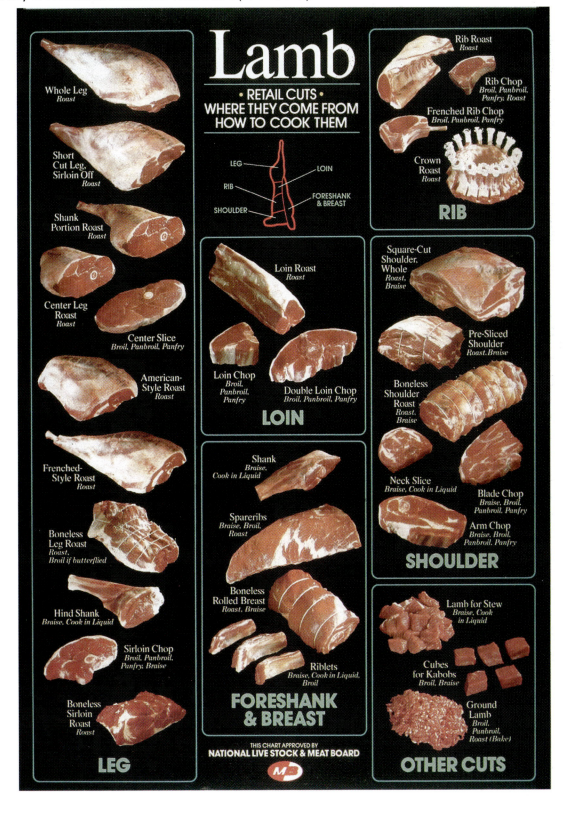

5. Lamb meat provides an excellent source of protein. Lamb chops and legs of lamb are most popular, but numerous recipes can be used to ensure that the lower priced cuts are prepared in a nutritious and delicious manner.

6. Wool provides a supplemental income for the sheep enterprise, with typical raw wool yield of 8 to 10 pounds per sheep. Wool is sold on the basis of yield (amount of clean scoured wool), grade (degree of fineness), and class (length of wool fiber). Wools of longer, finer, and uniform fibers that are free from contaminants are used for making fabrics such as gabardine and crepe. Shorter and coarser wools are better suited for tweeds, blanket wool and melton, a smooth fabric with a soft surface.

All photos except #3 and #4 courtesy of American Lamb Council.

PLATE C. BEEF CATTLE PRODUCTION

1. An important goal of beef cow-calf operations is to obtain a strong, healthy calf from each cow per year.

2. The beef calf normally suckles its mother until weaned at about 205 days of age. Sufficient milk should be produced by its dam so that the larger breed calf will weigh 550 to 600 pounds at weaning. Once a purebred herd has been established and improved, use of crossbreeding to provide heterosis is practiced by most herd owners. Generally a 60- to 80-day breeding period is allowed to ensure that a high percentage of the cows is successfully bred.

3. After a calf crop has been weaned and some heifer calves saved for herd replacements, the remaining steers and heifers are usually grown and "backgrounded" on a high forage ration through the following winter, spring and summer months. This system provides economical body weight gains prior to entering the feedlot for finishing for market.

4. Upon entering the feedlot, groups of 700- to 900-pound "feeders," sorted by sex, are fed a high energy ration until marketed at 1100 to 1300 pounds.

5 & 6. An increasing number of beef market animals are being marketed on a "grade and yield" basis. The higher value cuts are derived from the round, sirloin, and short loin wholesale cuts. Most carcasses should grade USDA Choice quality grade. Yield grades consider the percentage lean versus fat so well-muscled, low-fat carcasses are preferred.

Courtesy of National Live Stock and Meat Board (now defunct).

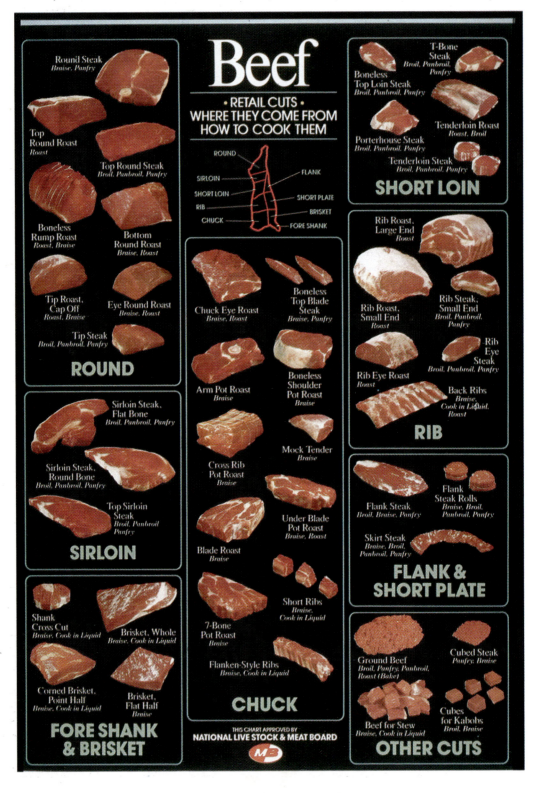

6.

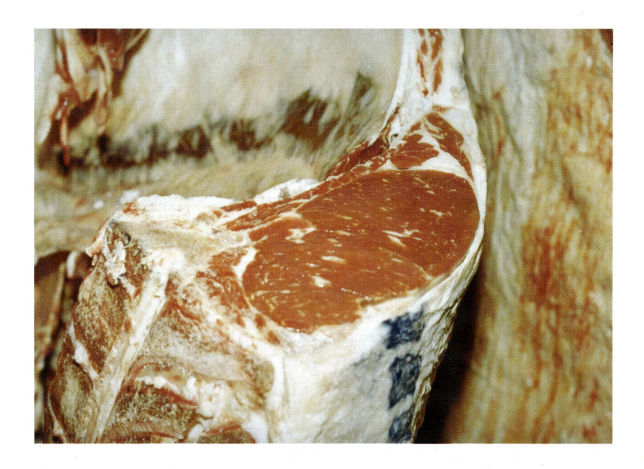

7. High-quality cuts of beef are very popular and a moderate meat portion can provide high-quality protein needed in the human diet in a delicious and palatable manner.

All photos except #5 courtesy of American Chianina Association, and Purdue University.

PLATE D. SWINE PRODUCTION

1. After a 114-day gestation period, the sow or gilt gives birth to about 9 to 13 piglets in the farrowing unit. Most successful operations average 2.1 litters per sow per year, and 9 to 11 pigs weaned per litter. Today, pigs may be weaned as early as 10 to 14 days in a "segregated early weaning" program.

2. Pigs from several litters are managed in groups in the nursery unit. Here special attention is given to the health and growth of the pig as it adapts to special formulated "early weaner" rations. After reaching around 25 pounds each at about 30 days of age, this group graduates to a "starter" ration for an additional 20 to 25 days.

3. Once most pigs reach 45 to 50 pounds they remain in groups in an "all-in all-out" system and receive a "grower" ration until 120 pounds each at about 100 days of age. At that time they are placed on a "finisher" ration. Some operations separate gilts from barrows, the gilts receiving a ration slightly higher in protein and amino acid percentage. Most hogs are marketed at 230 to 260 pounds at about 170 to 180 days of age.

4. The U.S. Yield Grade No. 1 pork carcass must yield 60.4 percent or more in the four more valuable (primal) lean cuts (ham, loin, picnic shoulder and Boston shoulder). Most large processors provide "boxed" pork according to the wholesale cuts desired by the retailer.

Courtesy of National Live Stock and Meat Board (now defunct).

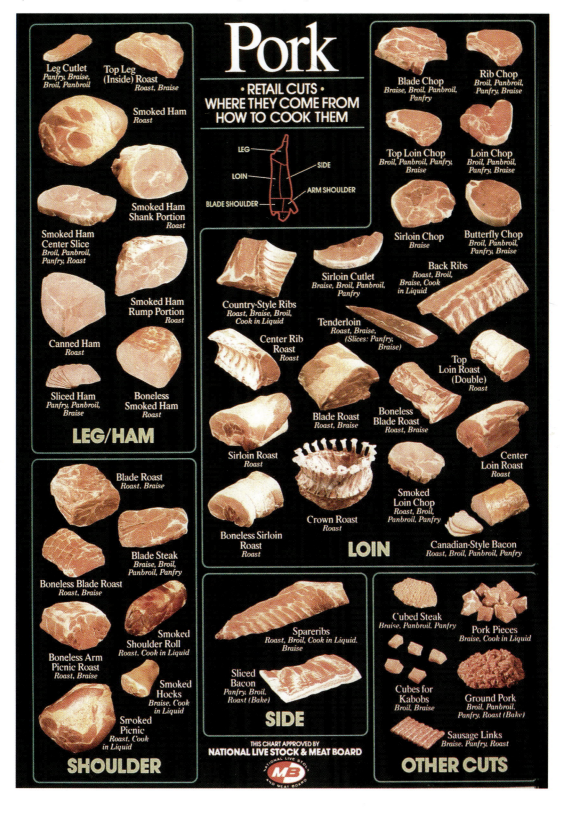

Pork

• RETAIL CUTS •
WHERE THEY COME FROM HOW TO COOK THEM

LEG/HAM

Leg Cutlet
Panfry, Braise, Broil, Panbroil

Top Leg (Inside) Roast
Roast, Braise

Smoked Ham
Roast

Smoked Ham Shank Portion
Roast

Smoked Ham Center Slice
Broil, Panbroil, Panfry, Roast

Smoked Ham Rump Portion
Roast

Canned Ham
Roast

Sliced Ham
Panfry, Panbroil, Braise

Boneless Smoked Ham
Roast

SHOULDER

Blade Roast
Roast, Braise

Blade Steak
Braise, Broil, Panbroil, Panfry

Boneless Blade Roast
Roast, Braise

Boneless Arm Picnic Roast
Roast, Braise

Smoked Shoulder Roll
Roast, Cook in Liquid

Smoked Hocks
Braise, Cook in Liquid

Smoked Picnic
Roast, Cook in Liquid

LOIN

Blade Chop
Braise, Broil, Panbroil, Panfry

Rib Chop
Broil, Panbroil, Panfry, Braise

Top Loin Chop
Broil, Panbroil, Panfry, Braise

Loin Chop
Broil, Panbroil, Panfry, Braise

Sirloin Chop
Braise

Butterfly Chop
Broil, Panbroil, Panfry, Braise

Country-Style Ribs
Roast, Braise, Broil, Cook in Liquid

Sirloin Cutlet
Braise, Broil, Panbroil, Panfry

Back Ribs
Roast, Broil, Braise, Cook in Liquid

Center Rib Roast
Roast

Tenderloin
Roast, Braise, (Slices: Panfry, Braise)

Top Loin Roast (Double)
Roast

Blade Roast
Roast, Braise

Boneless Blade Roast
Roast, Braise

Center Loin Roast
Roast

Sirloin Roast
Roast

Crown Roast
Roast

Smoked Loin Chop
Roast, Broil, Panbroil, Panfry

Boneless Sirloin Roast
Roast

Canadian-Style Bacon
Roast, Broil, Panbroil, Panfry

SIDE

Spareribs
Roast, Broil, Cook in Liquid, Braise

Sliced Bacon
Panfry, Broil, Roast (Bake)

OTHER CUTS

Cubed Steak
Braise, Panbroil, Panfry

Pork Pieces
Braise, Cook in Liquid

Cubes for Kabobs
Broil, Braise

Ground Pork
Broil, Panbroil, Panfry, Roast (Bake)

Sausage Links
Braise, Panfry, Roast

THIS CHART APPROVED BY
NATIONAL LIVE STOCK & MEAT BOARD
M3

5. An increasing number of market hogs are sold on a "grade and yield" basis, which provides an incentive to produce pigs yielding a well-muscled carcass with a high percentage of lean and a low percentage of fat.

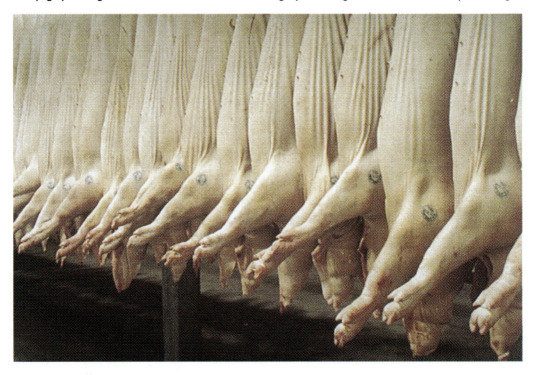

6. Pork has been successfully promoted as "the other white meat." It almost always is the main breakfast meat as bacon, sausage or ham. In addition, it often provides an attractive and tasteful meat dish and contributes to a healthy, balanced diet.

All photos except #4 courtesy of the National Pork Producers Council.

You might think of a chromosome as "carrying many genes in a row." Each gene occupies a specific location (locus) on a certain chromosome. Chromosomes occur in pairs in most body (somatic) cells. So, then, do genes. One member of each pair of chromosomes is inherited from the male parent and the other member of the pair from the female parent.

Fig. 17–1 illustrates DNA in diagrammatic form. Note the nucleotide; each contains the sugar deoxyribose (S), a phosphate (P), and a nitrogen-containing unit. These nitrogen-containing units are of four types: adenine (A), thymine (T), guanine (G), and cytosine (C). As many as 600 nucleotides may exist for each gene.

The number of chromosomes in body cells (or the diploid number—see Section 17.2) of some animal species is given in Table 17–1. The total number of genes on certain chromosome pairs is not known.

Although much genetic study is directed toward individual genes and how they work, most economically important traits of livestock and poultry are influenced or affected by several to many different genes. Transmittal of single genes or a very small number of genes can be easily studied, and a few examples are given in later sections. But when many genes are involved, inheritance patterns are not so simple, and different techniques are used for studying them. (See Sections 17.5, 17.8, 17.9, and 17.10.)

17.2   TRANSMISSION OF GENES

Every cell in the body except sperm and ova (and certain circulating blood cells) has a full set or **diploid** number (sometimes designated as $2n$) of chromosomes, half contributed from the dam and half from the sire. Thus, in the case of cattle, there are 30 pairs of chromosomes. However, each sperm and ovum carries only one member of each chromosome pair, or the **haploid** number (n).[1] This is accomplished by "reduction division" (**meiosis**) during the development of the sperm cell in the testicle (**spermatogenesis**)

TABLE 17–1	Chromosome Number in Body Cells of Several Species	
	Cattle, bison, goats	60
	Horses	64
	Donkeys	62
	Pigs	38
	Sheep	54
	Chickens	78
	Turkeys	80
	Rabbits	44
	Dogs	78
	Cats	38
	Humans	46

Source: *Charles E. Stufflebeam. 1989. Genetics of Domestic Animals. Prentice Hall, Inc., Englewood Cliffs, New Jersey.*

Diploid
A cell having two full sets of chromosomes in its nucleus.

Haploid
Having only one complete set of chromosomes within the nucleus of a cell, such as the sex cells.

Meiosis
Cell division early in the reproductive process, and in the formation of sperm and ova in the testicles and ovaries. Each pair of chromosomes in the cell being divided separates, and one member of each pair goes to each of the two newly formed cells.

Spermatogenesis
Production of spermatozoa within the seminiferous tubules of the testicle.

---

[1] Sperm and ova are said to contain a "haploid" number, or only one from each pair of the chromosomes in the somatic or body cell. The somatic cells that contain both members of the chromosome pairs contain the "diploid" number of chromosomes.

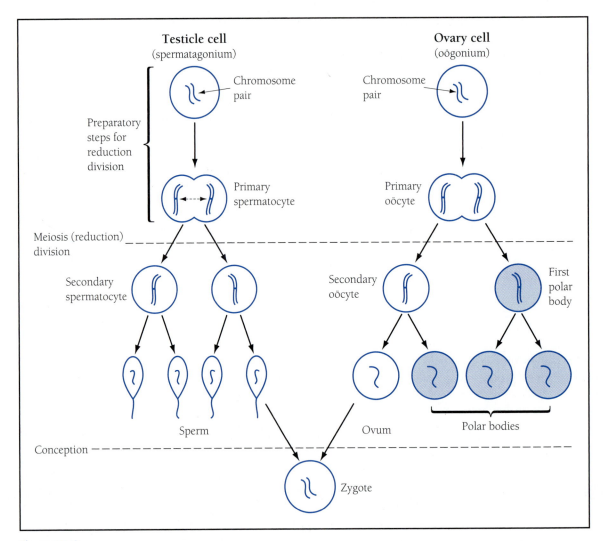

Figure 17–2
Sequence of events in meiosis to form sperm and ova, then their combination at conception to form a zygote. The illustration of spermatogenesis and oogenesis shows only a single pair of chromosomes. The same thing happens simultaneously to every pair of chromosomes within a species.
Source: (Iowa State University.)

Oogenesis
The process of formation and development of ova.

or the ovum in the ovary (**oogenesis**). It is important to note that genes segregate in the gametes during the process of spermatogenesis and oogenesis. They then *recombine* when the gametes unite to form the zygote (Fig. 17–2).

Note that in formation of the sperm cell in the testicle, a normal cell, called a *spermatogonium*, divides after one member of each chromosome pair migrates to each end of the cell and duplicates. Four sperm are formed from this one cell. These sperm develop, as described in Chapter 14, and in time are capable of combining with an ovum produced by the ovary of the female.

The process of meiosis in the ovary of the female is similar but not identical. In the case of the female, divisions result in one ovum and three "polar bodies," which are inert in terms of fertilization. Each ovum formed and shed from the ovary contains one member of each pair of chromosomes that was present in the original cell.

Meiosis differs from **mitosis,** which is replication of body cells during tissue maintenance and growth. In mitosis everything inside the cell, including each member of the chromosome pairs, divides so that the new cells formed contain a full complement of chromosome pairs, and the resulting new cells are exactly like the original cells.

It is emphasized that the sperm and the ovum each contain one member of each chromosome pair. When a sperm and ovum unite, then, the resulting fertilized cell (zygote) contains a full set of homologous chromosome pairs. Chromosome number 1 present in the sperm matches up with chromosome number 1 present in the ovum. The other chromosomes pair up similarly.

Because each gene is positioned at a specific site (or **locus**) on chromosomes, things that happen to chromosomes also happen to genes. During meiosis, one member of each pair of genes migrates *as a part* of the chromosome to opposite ends of the cell. When the cell divides, each new sperm or ovum carries one member of each gene pair. Then, at conception, when the members of the chromosomes match up, the genes automatically do likewise.

17.3   SINGLE GENE EFFECTS

All livestock and poultry traits, visible or measurable, are influenced by one or more gene pairs. In most cases the economically important traits are influenced by several gene pairs.

First let us review qualitative genetics where traits are influenced by just one pair of genes. These examples will also illustrate why offspring may be similar or dissimilar from their parents in certain characteristics.

The first illustration is the polled (absence of horns) characteristic in cattle. *P* designates the *dominant* gene, for the polled trait; *p* designates the corresponding *recessive* gene, for the horned trait. If the *genotype* (genetic makeup of the cell) is pure for the polled trait (*PP*), the animal is polled. If the genotype is pure for horns (*pp*), the animal is horned. If the animal received a *p* from one parent and a *P* from the other parent, the genotype is *Pp* and the animal is polled. The polled gene (*P*) is *dominant* over the horned gene (*p*) and will mask the expression of the gene for horns. Geneticists use capital letters to denote dominance and small letters to denote recessiveness. The term **homozygous** is used where both genes of a pair are identical (*PP* or *pp*) and the term **heterozygous,** where they are different (*Pp*). The term *alleles* is sometimes used to denote genes that can occupy the same loci on homologous chromosomes but have different effects.

In the preceding paragraph, a term was introduced that should be explained in more detail. **Genotype** refers to the genes and gene combinations on the chromosomes in the cells of the individual. **Phenotype** is the expressed (usually visible or measurable) characteristic of an individual such as the presence or absence of horns.

In the above example, if the genotype is *pp,* the phenotype is *horns.* If the genotype is *PP,* the phenotype is *polled.* The same phenotype (polled) would be observed with the genotype *Pp.* We see, therefore, that the phenotype or visible characteristics of an animal do not always disclose the genotype.

Mitosis
Cell division during normal growth of tissue. Each chromosome divides such that resulting new cells each have a full complement of chromosome pairs.

Locus
A specific site; commonly used to denote the area on a chromosome where a gene is located. Pl., *loci.*

Homozygote
An animal whose genotype for a particular trait (or pair of genes) consists of like genes.

Heterozygote
An animal whose genotype for a particular trait or pair of genes consists of unlike genes.

Genotype
The genetic makeup of an animal. A listing of genes carried by the animal, for one or several traits.

Phenotype
The characteristics that an animal shows or demonstrates. Includes both appearance and performance characteristics.

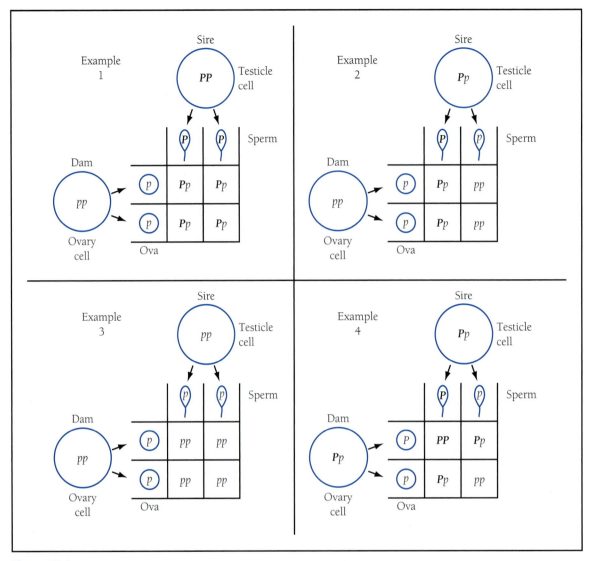

Figure 17–3

Four of the six fundamental types of mating, the examples involving the gene for the polled trait. Note the proportions of genotypes that occur in the offspring, and also that an offspring may differ in genotype from either parent.

Illustrations in Fig. 17–3 provide a basis for discussing inheritance of the polled characteristic. In Example 1, the sire has the genotype *PP*. The dam, or cow, has the genotype *pp*. After reduction division in the testicle, one *P* is present in each of the sperm that develop. After reduction division in the ovary of the dam, one *p* is present in each ovum produced. Regardless of which sperm unites with which ovum at conception, the resulting zygote will contain the *Pp* genotype and the offspring will be polled, but will be heterozygous for the trait and a *carrier* of the gene for horns.

In Example 2, where the sire is also polled but has a different genotype, *Pp,* two kinds of sperm will be produced. Half will contain the *P* gene; the other half, *p*. All ova produced by the dam will contain the *p* gene. So if by chance a sperm containing the

P gene unites with an ovum, the resulting zygote will have a *Pp* genotype and the calf will be polled. If, however, a *p*-containing sperm unites with the ovum, a horned calf with the *pp* genotype will be born.

What determines which type of sperm will unite with a particular ovum? *Only chance.* In this case, half of the sperm carry the *p* gene and half carry the *P* gene, so there is a 50:50 chance that either will unite with a particular ovum.

In Example 3, both the sire and dam are horned (genotype *pp*). Obviously any zygote formed will carry the pure (homozygous) *pp* genotype; these are the only genes that can be transmitted from the parents.

In Example 4, both the sire and the dam are polled, but their genotype is *Pp*. Half of the sperm will carry *P* (for polledness) and half will carry *p* (for horns). The same will be true of the ova. Chances are 25 percent that a calf born will have a *PP* genotype, 50 percent that it will have a *Pp* genotype, and 25 percent that it will carry a *pp* genotype; thus, the genotypic ratio will be 1*PP*:2*Pp*:1*pp*. If a large number of such matings are made, then statistically 75 percent of the calves will be polled and 25 percent will be horned; thus, the phenotypic ratio will be three polled to one horned. Two-thirds of the polled calves will carry a recessive gene *p* for horns. It is obviously impossible to look at a polled animal and tell whether or not it carries the recessive *p* gene. (Small hornlike scurs, more often seen on male calves in breeds or families that are generally polled, are apparently caused by a different set of genes.)

The two remaining types of matings (not shown) are homozygous dominant mated to homozygous dominant (*PP* × *PP*), and homozygous dominant mated to heterozygous (*PP* × *Pp*). In summary, when a trait is influenced by only two alleles, such as *P* and *p*, three genotypes (*PP*, *Pp*, and *pp*) are possible. These can result in six basic crosses, as follows:

1. *PP* × *PP*   4. *Pp* × *PP*
2. *PP* × *pp*   5. *Pp* × *pp*
3. *pp* × *pp*   6. *Pp* × *Pp*

The above series of illustrations shows that much of inheritance involves the laws of chance.

Another trait that may be influenced by just one pair of genes is color. For example, in Angus cattle, *B* represents black and is dominant over *b*, which represents red. Most Angus cattle carry the *BB* genotype. A few carry the *Bb* genotype. These, of course, are not apparent because they are black. A few individuals carry the *bb* genotype and are red.

Red Angus apparently originated from matings of two black cattle that both carried the red recessive gene. It is apparent that Red Angus cattle will breed true for color when mated within the breed. Because they carry only the *b* gene for color, this is the only gene they can transmit to their offspring.

Skin color in poultry, white versus yellow, is apparently influenced by a single gene pair. *W* represents white and is dominant over *w*, which represents yellow. A hen with the *WW* genotype will produce only white-skinned offspring, regardless of the genotype of the male to which she is mated.

A fourth characteristic apparently influenced by only one pair of genes is the "snorter dwarf" characteristic in beef cattle. Dwarf calves often die at birth or soon after. Those that live grow slowly and inefficiently, and are an economic loss to the producer. *N* represents normal body size; *n* represents dwarf. Because *N* is dominant over *n*, it is difficult to pick out the normal-sized animals that carry the dwarf gene. Only when such animals produce a dwarf calf are they positively identified. Because n is recessive, *both* parents are then known to carry this gene.

In this situation, though N is dominant and Nn animals appear essentially normal, there are a *few* telltale signs that expose some dwarf gene carriers. Often the dwarf carrier bulls or cows have shorter front legs and a shorter, wider head, and identifiable ridges on the ventral side of the lumbar vertebrae (detectable by X-ray). Other examples of traits with which an animal may be heterozygous and a carrier are *mulefootedness* in cattle and little or no wool or hair on newborn lambs.

The previous illustrations show why some offspring are like their parents in a certain characteristic and some offspring are not like their parents. One can also understand why a pig or calf or lamb may resemble its dam in certain traits and the sire in certain other characteristics.

Not all cases of simple inheritance involve complete dominance and recessiveness. There may be partial dominance, or two members of a gene pair may be equal in power, neither being dominant over the other. Such genes are said to be *codominant*. Color of shorthorn cattle is an example. R designates red; W designates white. An animal with an RR genotype is red, one with a WW genotype is white, and one with an RW genotype is usually roan or spotted. Some of the more common examples of dominant and recessive traits in farm animals, including those above, are shown in Table 17–2.

TABLE 17–2	Examples of Dominant and Recessive Traits in Farm Animals	
Species	**Dominant Trait**	**Recessive Trait**
Cattle	Black hair coat	Red hair coat
	Cloven hooves	Mulefoot
	Normal muscling	Double muscling
	Normal size	Dwarfism
	Polled (no horns)	Horns
	Red	Yellow
Horses	Bay	Non-bay (black)
	Black haircoat	Chestnut or sorrel
	Normal hair	Curly hair
Poultry	Barred plumage	Non-barred plumage
	Broodiness	Non-broodiness
	Feathered shanks	Clean shanks
	White skin	Yellow skin
Sheep	Wooly fleece	Hairy fleece
	White wool (except the Karakul and Black Welsh Mountain breeds)	Black wool
Swine	Black hair (Hampshires)	Red hair
	Cloven hooves	Mulefoot
	Erect ears	Drooping ears
	White belt (Hampshires)	No belt

Source: *Compiled by the authors and colleagues.*

17.4   INHERITANCE OF SEX

What determines the sex of a zygote? Again, chance. Each normal body cell contains a pair of chromosomes called the sex chromosomes. Among mammals, both members of this chromosome pair are similar in the female and are called X. Hence, the genotype for the female sex characteristic is XX and is homogametic (alike). One chromosome of the male is like that of the female and is called X; the other has a different content of sex genes and is designated Y. The male genotype, therefore, is XY and is heterogametic (unlike).

Just how the X and Y chromosomes differ is not fully known. It appears that the Y chromosome doesn't possess genes (though its presence appears related to sex determination).

In reduction division in the testicle, the X chromosome from the sperm mother cell[2] becomes a part of one sperm; the Y chromosome inhabits the other. Half of the sperm carried in semen, therefore, carry the X chromosome and half carry Y. Ova shed in the female, however, carry only the X chromosome.

At conception, there is a 50 percent chance that an X-carrying sperm will unite with an ovum to produce a zygote with an XX genotype (a female). There is also a 50 percent chance that a Y-carrying sperm will unite with an ovum and a male zygote (XY) will develop. Hence, we say that sex is determined by chance. Small variations from the 50 percent proportion of males and females born are not unusual. These may be due to chance, to difference in survival or activity of sperm carrying different sex chromosomes, or difference in survival of embryos according to sex. There is no known scientific basis for the common belief that certain matings will produce a high percentage of one particular sex, although there is evidence suggesting that more males may be conceived than females and higher prenatal death occurs among them.

For poultry, the reverse designations, XX for the male and XY for the female, could be used. Poultry geneticists, however, generally use the designations ZZ for the male and ZW for the female. The W denotes the lack of one of the "sex chromosomes," or the presence of an inert chromosome. Because the avian female carries both ZW chromosomes, the sex is completely determined by the female as opposed to the mammalian counterpart. Hence, this adds tremendous cost and effort to the layer industry because one-half of all egg-laying-type chickens hatched are males.

The above may explain certain *sex-linked traits*. Those traits influenced by genes on the sex chromosomes are called sex-linked traits, and their expression is in part dependent on whether one or two of the genes are present. In the case of poultry, two would be present in the male and one in the female; the reverse would be true in livestock. It is believed that sex-linked traits are primarily controlled by the alleles located on certain loci of the X chromosome in mammals, or the Z chromosome in birds.

17.5   MULTIPLE GENE EFFECTS

In livestock and poultry improvement, it is important to recognize that most economically important traits such as rate of gain, body type and conformation, egg production, milk production, and carcass merit are influenced by many gene pairs. It becomes very complicated to trace inheritance of a trait when two, three, or more pairs of genes are involved. The number of possible gene combinations in individuals becomes tremendous.

---

[2] The cell that divides to form two sperm cells. (See lower part of Figure 17–2.)

Animal improvement by selection therefore involves the selection of animals with *many good genes*. This is accomplished by putting all candidates for herd or flock sire, or all potential replacement heifers, gilts, ewes, fillies, or pullets, in the same or similar environment and then selecting those that perform best in terms of the characteristic being selected for—such as gain, finish, feed efficiency, egg production, or wool production (Fig. 17–4). The ones that perform the best in a certain trait when the environment is held constant are probably the ones with the most good genes for that trait.

This concept may not be as glamorous or as striking as the discussion on horns or dwarfism, but it is the whole foundation of practical poultry and livestock improvement. Though the improvement possible under these situations is achieved slowly, ultimate possibilities for improvement are practically unlimited.

As stated above, most of the economically important traits are affected by many genes. We know, for example, that milk production in livestock is influenced by at least 10 pairs of genes. The percentage of fat in milk also is influenced by at least 10 pairs. It is reasonable to assume (based on genetic studies) that other economically important characteristics are influenced by relatively large numbers of gene pairs.

Because each pair of genes occupies a particular location on the chromosome, the goal in livestock and poultry selection is simply to make sure that the genes that inhabit each location are on the "positive" side. That is, we want these genes to be the ones that will cause faster gains, more efficient feed utilization, higher quality carcasses, and higher milk production, rather than genes that might work in the opposite direction. We almost forget the concept of dominance and recessiveness, though we recognize it does continue to play a role, especially in undesirable traits related to the recessive gene and in the masking of recessive genes by dominant genes resulting in increased hybrid vigor, such as occurs in crossbreeding.

Figure 17–4
A central bull testing station (left) provides a relatively uniform environment, so genetic merit of bulls from different herds can be compared. The right photo shows on-the-farm boar testing facilities of a purebred swine breeder. In both cases, all animals receive the same ration, management, and care.
(Courtesy of Kansas State University.)

Evidence indicates that in economically important characteristics, dominance works in our favor. It is apparent that, more often than not, dominant genes are the good genes—at least in the characteristics of fast growth, efficiency, litter size, and a few other characteristics. (See Section 20.4.)

17.6 CYTOPLASMIC INHERITANCE

Chromosomes are contained in the cell nucleus. Surrounding the nucleus is the cell **cytoplasm,** which may contain materials or structures that affect certain animal traits.

Cytoplasmic inheritance exists when DNA in the cytoplasm, apart from nuclear chromosomes, self-replicates and independently transmits traits to daughter cells. There are several examples of traits that are cytoplasmically inherited in rats, wild mice, and humans, as well as in plants.[3] The two most widely demonstrated traits are the characteristics of mitochondria, involved in respiration in plant and animal cells, and chloroplasts, which contain double-stranded, mutable DNA molecules usually circular in form.

Maternal inheritance through the significant amount of cytoplasm of the egg in higher organisms would result in differences between reciprocal crosses, with the offspring more closely resembling the mother. Offspring of crossed lines A and B, where the mother is of line A, would more closely resemble A; offspring where the mother is of line B and the sire of line A would more closely resemble B.

The sperm contains very little cytoplasm; thus, paternal cytoplasmic inheritance is less likely. Future research may indicate an economic impact of cytoplasmic ladder inheritance in the animal industry and whether "maternal inheritance" is due to cytoplasmic inheritance or other maternal effects during reproduction.

Instances in livestock and poultry where offspring more closely resemble the mother in economically important traits generally have been demonstrated to be due to the maternal or mothering effect, which is *environmental.*

> **Chromosomes**
> Body cell structures within the nucleus that contain hereditary materials (genes) and are microscopically visible only during cell division.

> **Cytoplasm**
> The nonnuclear portion of a cell.

17.7 GENOTYPE AND PHENOTYPE

As described in Section 17.3, the *genotype,* or genetic makeup, of an animal is a listing of the specific genes carried on the chromosomes. The *phenotype* is a summary of the visible or measurable characteristics. A Black Angus bull may have the genotype *PPBBNN,* indicating that he is polled, black, and normal in size. If this were true, the genotype and phenotype would correspond. Because *P, B,* and *N* are dominant, it is apparent that the phenotype of an animal discloses some information about the genotype. Any Angus that is polled, black, and normal in size must carry at least one *P* gene, one *B* gene, and one *N* gene.

Breeders want to know as much as possible about the genotype of their breeding stock. They know a normal-sized bull carries one *N* gene, but what do they know about the second member of the pair? A bull calf born without horns is known to carry one *P* gene. How can you know what the other gene is?

In many traits the phenotype doesn't fully disclose the genotype. Though a ewe lamb may carry genes favorable for rapid growth, heavy wool production, and high

---

[3] Giles, Richard E., et al. *Somatic Cell Genetics* 6 (1980): 543.

Figure 17–5
The heredity or genetic makeup of an individual is determined at conception when the sperm and ovum unite to form a zygote. Anything that later influences or affects the individual is considered environmental.

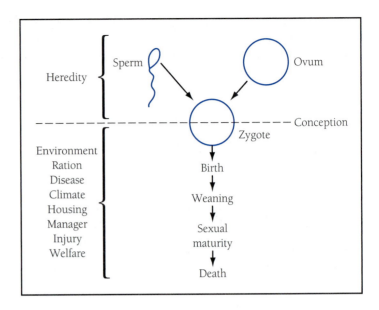

prolificacy at maturity, environment will determine the degree to which the genes are demonstrated. If a ewe lamb's mother should die soon after birth, leaving her without proper nourishment, she probably will grow slowly. If she receives a ration that is low in protein, her wool growth will be impaired. After maturity, if she is mated to a sterile ram, her genetic prolificacy will not be demonstrated.

Although a chicken may carry the genotype for yellow skin, a ration composed largely of grain sorghum and lacking yellow corn or alfalfa meal may not provide sufficient quantities of yellow pigment to make the skin yellow.

Environment often greatly masks genotype, or hereditary potential. This topic is discussed in more detail in Chapter 18.

It is also important to realize when during the life cycle of an individual heredity and environment exert their influences. This is illustrated in Fig. 17–5. *Heredity* of an individual is determined at conception when the sperm, the reproductive cell from the male parent and the ovum, the reproductive cell from the female parent unite to form a zygote. The details of this fertilization are discussed in detail in Chapter 14.

Any influence that is exerted upon the individual following conception is considered **environmental**. This includes the time when the developing embryo and fetus are contained inside the uterus of the dam or the embryo of poultry within the eggshell. For example, nutritional deficiencies that may occur during pregnancy can seriously impair the development and later health of the offspring. One specific example involves vitamin A. During extended droughts, many cows receive very little carotene or pro-vitamin A during pregnancy. Calves are often weak at birth; some are blind or have poor vision. Therefore, such a defect at birth may have been due to an environmental cause.

The presence or absence of disease may influence whether or not the embryo lives long enough to be born. Leptospirosis or brucellosis can exert a direct effect upon the embryo and can kill it before birth. In such a case, the death of the embryo is caused by an environmental factor (disease), not by heredity.

> **Environment**
> All of the conditions to which an animal is exposed after conception.

It should be emphasized that the genetic makeup (heredity) of an individual does not change once it is established.[4] Environmental factors that might influence performance or visual characteristics of an individual do not change the genetic makeup. For example, a cow with a "blind" quarter (one teat not functional) because of injury or mastitis infection will not genetically transmit this trait to her calves. A sow stunted in size because of an inadequate ration will not genetically transmit this characteristic to her offspring. Nor will a debeaked hen produce chicks with a short beak.

The above examples illustrate some of the environmental influences that may alter the phenotype of animals. The degree to which the phenotype represents the genotype is an important factor in determining the rate of genetic improvement through selection.

17.8   RULES FOR IMPROVEMENT BY SELECTION

Four rules for achieving improvement of livestock and poultry by selection are listed below. Following these rules closely allows progressive breeders to make relatively rapid progress. It is assumed, of course, that selection is based on *economically* important traits. The rules are as follows:

Rule 1:   *Have maximum genetic variation* (see Section 17.9).

Rule 2:   *Spend selection efforts on traits largely influenced by heredity.* This means the traits for which phenotype is a good indicator of genotype. (Chapter 18 is devoted to the different traits in this respect.)

Rule 3:   *Observe or measure accurately the traits carried by a prospective breeding animal* (see Chapter 19).

Rule 4:   *Use the selected animal or animals* (carrying the traits you want) *most effectively.* (Effective programs are discussed in Chapter 20.)

17.9   GENETIC VARIATION

Poultry and livestock breeders or students of animal science may assume that uniformity is desirable in a breeding herd or flock. For effective and rapid improvement by selection, however, it is *not* desirable.

Breeders naturally take much pride in a herd or flock that is uniformly good. This is certainly justifiable. But the emphasis should be on merit rather than on uniformity per se.

Cattle and lamb feeders like to buy uniform groups of animals for feeding. Processors are usually impressed by groups of slaughter hogs, lambs, or cattle that are uniform in size, conformation, and finish. They often bid higher on such a group than on a group of the same average quality but with more variation. Carcasses from a uniform group might all be sold to the same retailer, and there is less chance of an extremely low-quality animal being in the group.

---

[4] Rare changes in genetic makeup are caused by mutation, but these are not of common occurrence and will not be discussed in this text.

Egg producers want uniformly sized eggs. Broiler producers want broilers that will all reach market weight at the same time. This justifiable preference for uniformity is well embedded in the livestock and poultry industry. Prospective buyers of breeding stock are often noticeably impressed by extreme uniformity in the herd or flock in which the breeding stock was raised. But selected breeding animals—cocks, bulls, rams, stallions, boars, and selected females—don't transmit "uniformity" or "variation" to other offspring. They transmit genes.

Also realize that a breeding herd or flock does not exist just to be looked at and to be a source of pride. It exists to be improved and to serve as a source of progressively better breeding stock for commercial producers. Little improvement can be made within a herd or flock that is extremely uniform.

For example, the groups of gilts owned by Jones and Smith, illustrated in Fig. 17–6, average the same in total merit score, or index, though Jones's gilts are much more *uniform*. Both breeders plan to keep the five best gilts in their herd for breeding and to sell the remainder for slaughter. Which breeder will make the most improvement in quality?

Smith should make more progress because the difference in performance (phenotype) of the selected gilts as compared to the average of the herd was largely due to their genotype. The *greater variation* within the Smith herd should result in a greater "selection differential" (discussed in Section 18.6) and potentially greater genetic progress in the next generation. Note that greater uniformity, as illustrated in Jones's herd, resulted in selected animals with performance closer to the average of the herd.

Figure 17–6

A scale of merit, on which are plotted gilts belonging to Jones and Smith.

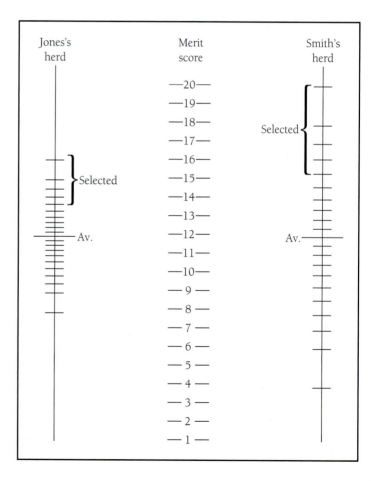

TABLE 17–3	Average Performance Data of Spring-Season Pigs at Iowa Swine Testing Station, 5-Year Intervals						
	Pen Average		**Boar**		**Barrow**[a]		
Season	Gain (lb/day)	Feed Efficiency (lb feed/ 100 lb gain)	Backfat Probe (in.)	Index (in.)	Carcass Length (in.)	Percent Ham and Loin	Loin Eye Area (in.²)
1956	1.89	285	1.46	101	29.1	32.3	3.22
1961	1.80	284	1.17	125	29.3	35.6	3.87
1966	2.16	259	1.01	175	29.4	40.4	4.29
1971	2.11	249	0.86	188	30.1	43.8	5.15
1976	2.13	254	0.79	191	30.8	45.2	5.20
1981	2.01	244	0.78	189	31.5	54.8	4.94

[a]*Through 1970, barrows were slaughtered near 200 pounds. From 1971 through 1976 they were slaughtered near, and data were adjusted to, 220 pounds. Since 1978, they have been slaughtered near, and data adjusted to, 230 pounds. Data not available after 1981.*

Source: *Stevermer, Emmett, Spring 1981 Summary, Iowa Swine Testing Station, Ames, IA*

17.10 EXAMPLES OF GENETIC IMPROVEMENT

Livestock populations have changed in their characteristics. Examples in hogs are easily found because (1) changes can be rapid, as previously discussed, and (2) goals have changed, several times in some countries.

In early U.S. history, products such as salted jowl and fatback were easily preserved, high-energy foods needed by a laboring people. Pork fat was used for soap, candle, and munitions manufacture. Now, however, such markets for pork fat no longer exist, and people want lean meat.

Table 17–3 shows the progressive improvement among pigs entered in the Iowa Swine Testing Station during its use in the spring of 1956 through 1981. Note that improvement in leanness has been great, as indicated by lower backfat probe values, higher percentage of ham and loin, and larger loin eye area. This change does not necessarily mean that the total hog population of Iowa, or even of the breeders involved, has changed so rapidly. Even though the station has permitted identification of muscular boars that have been used effectively to improve leanness in many herds, some of the increased leanness seen in hogs at the station, especially in earlier years, probably was due to improved ability of breeders to choose lean pigs to bring to the station for testing. They naturally want to bring pigs that will perform relatively well.

During the 1960s, rate of gain probably received more emphasis in the breeders' selection of testing-station entries. Considerable improvement in leanness had been achieved and probably the breeders were more able, with the aid of on-farm records, to identify pigs that were lean and had rapid gaining ability.

Improvement of beef cattle or sheep by selection is more difficult to demonstrate because of the tremendous *seasonal effect* on performance. Hogs are usually fed a concentrated ration in confinement so environment does not vary much from season to season. But weaning weights, type score, and other traits of cattle and lambs are influenced greatly by rainfall, which controls pasture growth, as well as other environmental factors.

Changes in egg production and growth rate in poultry can be relatively rapid because of (1) high heritabilities—35 percent or higher; (2) the large number of offspring

TABLE 17–4	Growth and Egg Production Changes in Turkey Hens, Caused by Selection		

| | | | Birds Selected for | |
		Control Birds	*8-week Weight*	*Egg Production*
After four generations of selection for growth:				
8-week body weight (g)		1,503	1,889	
24-week body weight (g)		5,997	6,905	
24-week breast width (cm)		5.0	5.3	
After five generations of selection for egg production:				
Days to first egg[a]		19.5		17.8
Eggs produced in 180 days		57		116
Egg weight (g)		84		78
Broody periods per hen, first 84 days		1.47		0.25
Length of broody periods (days)		18.4		7.4
Offspring per hen, first 84 days		27		40
8-week body weight (g)		1,471		1,475
Body weight at 50% production (g)		7,810		7,400
24-week breast width (cm)		4.8		4.8

[a]*Measured from the first day of lighting (February 1).*

Source: *McCartney, M. G., et al.,* Poultry Sci. 47:981, 1968.

per bird, permitting more rigid selection; and (3) the fact that birds are incubator-hatched and usually reared in a closed building with a relatively high degree of environmental control, so inherited traits are not significantly masked by maternal effects or other environmental factors.

Research by scientists at the Ohio Agricultural Research and Development Center has shown changes in turkeys as a result of selection for growth or egg production. Because there can be a "year" effect (environment may vary from one year to the next), a nonselected control flock was maintained through the total experiment, and data in Table 17–4 are from the control and selected flocks the last year. In the flock selected for growth, 14 to 21 percent of the females each year were chosen as parents for the next generation; for egg production, 17 to 38 percent of the dams were selected as parents. Note that egg production, however, more than doubled in five generations. Growth rate, as the result of selection, increased about 26 percent in four generations.

Why the more rapid progress in egg production than in growth? Data in Table 17–4 suggest a sharp decrease in broodiness (periods of nesting time with no eggs laid) as the result of selection. In other words, in turkeys, broodiness is a big factor in limiting egg production. The selected birds also came into production slightly earlier.

17.11   INFLUENCE OF PROLIFICACY AND GENERATION INTERVAL ON IMPROVEMENT RATE

Plant breeders are envied by livestock breeders in their efforts to achieve genetic improvement. The characteristics of a corn or small-grain population can be changed rapidly. Two parent kernels can produce 700 to 2000 offspring (kernels) each generation.

	Weight	Feed/Pound	Age Marketed
Year	**(lb)**	**(lb)**	**(weeks)**

TABLE 17–5 **Progress in Broiler Performance**

Year	Weight (lb)	Feed/Pound (lb)	Age Marketed (weeks)
1935	2.80	4.4	16.0
1950	3.00	3.5	11.0
1975	3.75	2.0	8.0
1994	4.65	1.9	6.5
2003	4.5	2.0	6.0

Source: Poultry Tribune, *September 1995 (1935–1994)*. *Purdue Poultry Extension Specialist (2003)*.

Two crops (generations) can be grown in one season, by taking advantage of semitropical climates. Think how fast selected traits can be reproduced! Assuming that only one ear of corn with 800 kernels was selected for a new "line" or "strain" and that succeeding generations produced only one such ear per stalk, two years' reproduction could result in 327,680 billion descendants available for planting on farms. The opportunity for *rigid selection* and *rapid duplication* of selected genes is tremendous, compared to livestock.

The *generation interval*—the average period of time between the birth of one generation and the birth of the next—varies greatly between farm animal species and affects the rate of genetic progress. For example, rapid progress can be made in poultry. Egg-type pullets begin laying eggs at 20 to 21 weeks of age. Each breeding hen is capable of producing over 250 chicks per year, half of which would be females. Each rooster can sire thousands. A complete generation passes every year, or perhaps in only 7 to 8 months. This explains the rapid development of specialized strains of layers and broilers in the chicken industry and the equally rapid use of these strains to replace traditional breeds. Selection can be for only the best, and when the best are found, they can be multiplied rapidly.

Combined improvements in genetic selection and management have resulted in remarkable progress in broiler performance, as evidenced by increased growth rate and feed conversion (Table 17–5). The time required to produce a broiler that weighs 4.65 pounds, rather than 3.0, has been reduced by almost 40 percent since 1950, with 46 percent less feed.

Hogs are the most prolific of large animals on farms. Sows normally farrow first at 1 year of age, thereby completing one generation, and well-managed operations exceed an average of two litters per year. Today, a selected gilt will produce about nine pigs by 1 year of age and another nine or more by 18 months of age.

The potential offspring of a selected boar might also be calculated. Two years after a boar is selected, assuming normal use and natural service, he might have from several hundred to several thousand offspring available for breeding. But remember, half are boars and half are gilts, whereas in corn every kernel can be both a male and female "parent."

In many flocks, ewes do not lamb until 2 years of age, and twins generally occur less than 50 percent of the time, even in well-managed flocks. So a ewe selected now will not have much effect on average flock quality for several generations, or many years.

How about cattle? Most heifers calve at 2 to 3 years of age, and the calf crop is usually under 100 percent. Though 2 to 3 years might be considered the theoretical generation time, 5 years is more realistic as an average because of some nonbreeders or abortions and because some of the calves are not born until the cow is 4, 5, or even 10 years of age. So here, relative improvement by selection is *very slow,* although increased use of multiple ovulation and embryo transfer, and sexed semen can greatly reduce the generation interval and reduce the time necessary to adequately progeny test potential breeding animals.

Although prolificacy and generation time have significant effect on the rate of genetic change in a herd or flock, and in some species may be considered "bottlenecks" to improvement, much improvement has been made in all species over the past century. Also, recent developments may speed improvements, by increasing prolificacy (see discussions of artificial insemination, multiple ovulation, and embryo transplanting in Chapter 14).

QUESTIONS FOR STUDY AND DISCUSSION

1. Define *deoxyribonucleic acid.* How is it related to the gene(s) and chromosome(s)?

2. What are the three basic components of a nucleotide?

3. What is the established number of chromosomes in the body cells of pigs, sheep, cattle, horses, and chickens?

4. Define *mitosis* and *meiosis.* How do they differ? Which occurs only in specialized cells in the ovaries and testicles?

5. Does each sperm and ovum contain a diploid or haploid number of chromosomes? Which number exists in the newly formed zygote?

6. What is the major difference in the processes of spermatogenesis and oogenesis (Figure 17–2)?

7. Review the following terms:

alleles	homologous
chromosome	homozygous
dominant	locus (loci)
genotype	phenotype
heterozygous	recessive

8. What is the expected distribution of the genotype for haircoat color of the progeny resulting from mating a heterozygous (black) individual (*Bb*) and a homozygous recessive (red) individual? What is the expected phenotypic distribution?

9. Name at least three traits each affected by one pair of recessive genes.

10. What is meant by "sex-linked" traits?

11. Why is it practically impossible to predict accurate genotypes and phenotypes of most economically important traits?

12. Give several examples of "environmental" effects that may limit optimal performance of an animal and full expression of its genotype.

13. Explain why considerable variation in performance within a population is needed for maximum genetic progress. Give an example.

14. List four rules for effective improvement by selection.

15. Compare the generation interval of common domesticated species: chickens, swine, sheep, cattle, and horses. What effect does the generation interval have upon the rate of genetic improvement?

Heritability and Genetic Improvement 18

Every experienced animal breeder should be able to accurately observe or measure differences in animal performance. However, their precision becomes much less in their selection of breeding stock that produces superior offspring. Why do we sometimes see disappointing results in the next generation? The principal reason is our inability to accurately separate the differences in animal performance into two proportions—those due to genetics and those due to the environment. This chapter is devoted to a better understanding of heritability and the amount of genetic progress that can be expected through selection.

Heritability may be defined as the *relative* importance of heredity in influencing poultry and livestock traits. All other influences are considered to be environmental. Heritability of traits may range from 0 to 100 percent. Turn to Section 17.7 and Figure 17–5 for a quick review of where in the life cycle of farm animals heredity and environment exert their effects.

Upon completion of this chapter, the reader should be able to

> *Heritability*
> The degree to which heredity influences a particular trait.

1. Define *heritability*.
2. Explain why heritability applies to the trait, not the animal.
3. List those groups of livestock and poultry traits for which heritability is generally high, those generally intermediate, and those generally low, and explain why their heritability differs.
4. Explain how the effective heritability of a trait may be raised or lowered within a herd or flock, or a population of animals.
5. Define and calculate *selection differential*.
6. Calculate expected improvement in a trait for one generation, given selection differential and heritability.

18.1 HEREDITY VERSUS ENVIRONMENTAL INFLUENCES

Because the genotype (heredity) of an animal is established at conception, it can be said that the genetic potential of an animal is determined at that time. If genes for prolificacy, rapid gaining, efficient feed utilization, high carcass merit, and other desired traits are provided by the uniting sperm and ovum, the embryo that develops and is eventually born or hatched has a high genetic potential. The degree to which the potential is reached during the animal's lifetime is determined by environment—ration, housing, climate, incidence of disease, management, and related factors. If environment is excellent, the high potential may be approached or even reached. The best rations, shelter, sanitation, and management will allow rapid gains, efficient feed utilization, efficient reproduction, and top carcass quality.

If the genotype of an animal provides a low potential, however, it is apparent that even the best rations, shelter, sanitation, and management will not give top production. Good environment will help, though, and will allow the animal to demonstrate the limit of its productive potential.

Environment may influence some traits more than others, after the potential (genotype) has been established at conception. The polled trait in beef cattle is established by heredity (genotype *PP*). Coat color is established by heredity. An Angus with genotype *BB* is black; one with *bb* is red. White-feathered birds will produce white-feathered chicks. True-breeding Duroc hogs transmit only the red hair coat. Suffolk sheep transmit the distinctly black color markings on the head and shanks. These traits are high in heritability; heredity is the major influence and environment exerts little effect.

But how about *prolificacy?* Think of all the environmental factors that can encourage or limit litter size in hogs, egg production in hens, the percent calf crop in a cow herd, or the incidence of twins in a sheep flock. Fertility of the sire or dam, time of breeding, ration amount and quality, incidence of disease, possibility of mechanical injury, and other factors are all important. Even though the genetic potential has been established, many different environmental factors can exert much influence on whether that potential is reached. Heritability of such a trait is usually low.

It is apparent that both heredity and environment influence all livestock traits, but in varying degrees. Figure 18–1 illustrates one technique used to demonstrate effects of both heredity and environment on rate of gain in cattle.

Two sets of identical twin steers were used, one member of each set being fed in pen 1 and the other member of each set in pen 2. The two pens were in different locations, there was a contrast in shelter provided, and rations fed in the two pens were different. Steers A and A′ were not related to steers B and B′.

The difference of 0.6 pound gain per day between the steers in pen 1 and those in pen 2 indicates a sizable environmental influence. It is not apparent which of the contrasting environmental influences exerted the largest effect—ration, location, or shelter.

Averages in the right column twin indicate a genetic difference in gaining ability. Because one member of each twin pair was subjected to the same environment, this difference of 0.2 pound per day is apparently hereditary.

In the above case, environmental factors had more influence on rate of gain than did the differences in heredity. This is not always true. If the two pens had been

	Pen 1		Pen 2		Average
	Animal	Daily Gain	Animal	Daily Gain	
Twins A & A'	A	1.20	A'	1.80	1.50
Twins B & B'	B	1.40	B'	2.00	1.70
Average		1.30		1.90	1.60

Environmental difference = 0.60

Figure 18–1
Differences in livestock performance can be caused by both heredity (genes) and environment. In this case the difference in environment had a greater influence on rate of gain than did the difference in heredity.

adjacent, with similar rations and facilities, the difference caused by environmental factors would have been smaller.

It should be noted that geneticists throughout the world study the differences in trait performance, as in Figure 18–1. As more heritability estimates for a trait are made they are added to the compilation of other estimates. Then, the overall average for each heritability estimate is made available to the breeder, as shown later in Table 18–1.

18.2   THE MEANING OF HERITABILITY

Heritability was previously defined as the relative importance of heredity in influencing certain poultry and livestock traits. Heritability might be defined another way. There are many differences among animals in each characteristic. The total variation that exists among animals in a certain trait is caused by heredity and environment. The proportion of the variation or differences caused by variations in heredity (ancestry or parentage) is the heritability. Thus, heritability is the proportion of variation that can be expected to be transmitted to the next generation.

The symbol for heritability is h^2. Most estimates of heritability consider the effects of only the additive genes, and it may be defined as the additive genetic variation as a proportion or percent of the total variance:

$$h^2 \frac{\sigma^2 a}{\sigma^2 a + \sigma^2 n + \sigma^2 e}$$

where h^2 equals heritability, σ^2 (or sigma squared) equals variance, a equals additive gene effects, n equals nonadditive gene effects (e.g., gene dominance, interactions, or epistatic deviations), and e equals environmental effects. Heritability, as expressed by this formula, gives an indication of the progress possible through animal selection.

Heritability means other things too. It means the *degree of relationship* between genotype and phenotype in a certain trait. For example, if heritability of leanness (muscling) in hogs is high, a boar's degree of leanness should disclose fairly closely his genotype for leanness. (Remember that most traits, such as leanness, are influenced by many genes and the individual genes are not named or designated. But in this case it would be apparent whether or not the boar carried many genes that promoted leanness.)

Heritability means the degree of relationship between phenotype and breeding value. If heritability for leanness is high, a lean boar has "lean" genes, and these are the genes he will transmit to his offspring.

At Miles City, Montana, several years ago a "Line 10" bull ranked top in rate of gain among seven bulls tested on a standard ration. A year and a half later, steers sired by each bull were compared in a similar feeding test. Test conditions, including ration, were similar for steers of different sires. The group of steers sired by the Line 10 bull ranked first. In no case were the progeny (offspring) better or worse than their sires by more than one rank. This illustrates that rate of gain is significantly heritable—influenced considerably by heredity.

18.3   HERITABILITY OF INDIVIDUAL TRAITS

The *trait* is heritable, not the *animal*. Though breeders must select or reject a potential breeding animal as a unit, they select because of the traits it carries. Heredity establishes the potential of all traits, but environment influences some much more than it influences others. Therefore, heritability of different traits varies within a species (Table 18–1). Heritability estimates commonly are expressed as percentages and categorized as

- Zero to 20 percent—Low
- 20 to 40 percent—Moderate
- Above 40 percent—High

Note that, regardless of species, those traits related to body or carcass structure and physical composition, especially those observed or measured before weaning, are more highly inherited than "performance" traits. This simply means that environmental factors on farms and ranches or other production units—factors such as ration, housing, disease control, and management—can be a greater benefit or detriment to prolificacy, preweaning growth rate, or survivability before weaning than they can be to carcass traits. It is worthwhile to discuss the influences of environment on certain traits.

Why Some Traits Are Low in Heritability

Prolificacy generally is low in heritability. Sows that were members of large litters don't necessarily produce large litters. A number of experiment stations have demonstrated that ewes born as singles are almost as likely to give birth to twins as ewes that were born as twins. In cattle, the interval between calvings is used as a measure of prolificacy, and the heritability of this, too, is generally low. Note that the heritability of services required per conception is low among dairy cattle. Environmental factors, such as disease or injury, may be the main reasons cows do not conceive at first service.

Some swine breeders consistently identify and save for breeding only gilts that are from large litters. The value of this practice might be questioned. The large litter demonstrates that the dam had a genetic potential for high prolificacy and that environment was apparently favorable for realization of this potential. Presumably, some of the genes for prolificacy would be transmitted to the gilts in question. But there is no guarantee that environment will be equally favorable for these gilts.

Similarly, there is no guarantee that gilts of a small litter will inherit a low genetic potential for prolificacy. The fertility of the boar used, condition and feeding level at breeding, time during estrus when the dam was bred, feeding level during pregnancy, incidence of disease, or possibility of mechanical injury—any or all of these factors could severely limit litter size. Low heritability suggests that in a large proportion of the cases, environmental factors such as these are major causes of variation in litter size.

Why Some Traits Are Moderate in Heritability

Heritability estimates for conformation score for several species, shown in Table 18–1, range from about 15 to 25 percent. These low to moderate heritability estimates may be surprising, especially because much past selection of breeding stock has been done visually and there has been considerable emphasis on "type" in the showring. Remember that heritability values are calculated, usually from data which show how much more closely relatives appear or perform than do nonrelatives. Data may be actual weights, carcass measurements, or other objective measures, but for type-scoring the scores are subjective.

An objective method of measurement, such as a scale, will measure weight with a consistent degree of precision. But type scores are applied by people and might be influenced by factors such as (1) mental attitude that particular day, (2) other animals seen since the last group or previous generation was scored, (3) angle from which the animal is

TABLE
18–1

Approximate Heritability of Certain Poultry and Livestock Traits[a]

	Dairy Cattle	Hogs	Beef Cattle	Sheep	Poultry	Low (L) Moderate (M) High (H)
Type and conformation:						
Conformation score	25	30	25	15	—	M
Amount of spotting (Holsteins)	95	—	—	—	—	H
Face covering	—	—	—	55	—	H
Skin folds	—	—	—	40	—	M,H
Number of functional nipples	—	50	—	20	—	M,H
Shank length	—	—	—	—	25	H
Body depth	—	—	—	—	45	H
Reproduction efficiency:						
Services per conception	5	—	10	—	—	L
Reproduction interval	0	—	10	—	—	L
Number born	—	10	5	10	—	L
Birth weight	60	5	40	30	—	L,M
Hatchability of fertile eggs	—	—	—	—	10	L
Production:						
Weight at weaning	—	10	25	30	—	L,M
Growth rate to 12 weeks	—	—	—	10	40	L,M
Postweaning pasture gains	—	—	30	—	—	M
Postweaning feedlot gains	—	25	50	60	—	M,H
Mature weight	60	—	—	40	50	H
Milk produced	25	—	—	—	—	M
Percent butterfat	50	—	—	—	—	H
Feed efficiency	—	30	40	20	—	M
Staple length	—	—	—	50	—	H
Fleece weight	—	—	—	38	—	M
Egg production	—	—	—	—	35	M
Egg weight	—	—	—	—	60	H
Health and soundness:						
Survivability	—	—	—	—	10	L
Heat tolerance	20	—	—	—	—	L,M
Mastitis resistance	25	—	—	—	—	M
Longevity	10	—	—	—	—	L
Cancer eye susceptibility	—	—	30	—	—	M
Carcass traits:						
Length	—	60	—	—	—	H
Loin eye area	—	50	55	53	—	H
Thickness of fat covering	—	50	40	50	—	H
Percent ham	—	55	—	—	—	H
Percent lean or lean cuts	—	48	40	35	—	M,H
Tenderness	—	—	60	—	—	H
Carcass grade	—	—	35	—	—	M
Marbling score	—	—	50	—	—	H

[a]See Chapter 32 for heritability estimates of horses.

Source: *Adapted from Craft, W. A., J. Animal Sci. 17:960, 1958; Shelton, Maurice, and Carpenter, O. L., Texas Agr. Exp. Sta. Lamb Report 21, 1957; Warwick, E. J., J. Animal Sci. 17:922, 1958; Minyard, J. A., and Dinkel, C. A., J. Animal Sci. 24:1072, 1965; Kinney, T. B., Jr., et al., Poultry Sci. 47:113, 1968: Carson, J. R., Purdue Agr. Exp. Sta. Bull. 754, 1962; Warwick, E. J., and Legates, J. E., Breeding and Improvement of Farm Animals, New York: McGraw-Hill, 1979; and other sources.*

viewed, (4) hair coat, (5) finish, or (6) temperament of the animal. These factors and others constitute a source of "error" and invariably lower the heritability estimate calculated.

Feeding performance of livestock and poultry is apparently inherited to a rather high degree. Values given in Table 18–1 range from about 30 to 60 percent for gain and efficiency. For animals confined to a feedlot or growing house, an important part of the environment is uniform and contributes less to performance variation, so most of the variation among animals would be due to their genotype. In general, animals confined to feedlots or growing houses are all fed high-energy rations designed for fast gains and relatively quick marketing. Note that heritability of postweaning pasture gains of beef cattle is only about three-fifths as high as heritability of feedlot gains.

Milk production, milk fat percentage, and related factors are influenced significantly by heredity. However, differences in environment—feeding, age at first calving, method of milking—also contribute considerably to the variations in production that exist. Heritability is high enough to indicate that it pays to select heifer replacements from the best-producing cows and from sires that rank high in predicted transmitting ability (PTA) for milk production.

Why Some Traits Are High in Heritability

Variations among carcass traits of meat animals are largely influenced by heredity. Vertebrae number in swine, for example, is established early during embryonic development, giving little time for environmental effects to play a role. The same thing generally would be true for traits such as carcass length and leg length, which are primarily a function of skeletal proportions.

The ratio of lean to fat is highly inherited, and this is economically very important. The relatively high heritability of beef tenderness suggests that feeding high-energy rations to beef cattle may affect tenderness relatively little. (See Section 26.7.) Perhaps tenderness can be achieved by selection.

Determination of heritabilities of horse traits has been limited. Herd sizes are small and animals are often managed and trained differently within and between generations. Some estimates are available, however, and are shown in Table 18–2.

TABLE 18–2	Approximate Heritability Percent of Traits of Horses
Traits	**Heritability (%)**
Pulling power	20–30
Walking speed	40–45
Trotting speed	35–45
Running speed	35–40
Movement	40–50
Height at withers	45–50
Heart girth	20–25
Body weight	25–30
Temperament	25–30
Intelligence	30–40

Source: *Warwick, E. J., and J. E. Legates,* Breeding and Improvement of Farm Animals, *7th ed. New York: McGraw-Hill, 1979 (used with permission); and other sources.*

18.4   LEVEL OF HERITABILITY AND SELECTION EFFORT

Remember the four rules for improvement by selection. (See Section 17.8.) The first rule, "Have maximum genetic variation," was described in Section 17.9. This section describes the second rule:

Rule 2: *Spend selection efforts on traits largely influenced by heredity.*

If a trait is highly heritable, selection can be effective, for several reasons. Consider, for example, leanness in hogs. Because leanness is known to have a high heritability, breeders can be fairly confident that boars and gilts that are lean have genes for leanness. Because the phenotype corresponds well with the genotype, breeding stock can be selected with confidence. Presumably these "lean" genes will be transmitted to the offspring and will be a major influence in causing the offspring to be lean. The phenotype of the offspring also will correspond well to the genotype. In other words, if heritability of a trait is high, potential breeding animals with that trait will be good breeders for that trait—they'll tend to put that trait in their offspring. Moreover, whereas selection is very effective, you can make rapid progress by selection. This is especially true in hogs or poultry where reproductive rate is high.

We'll go one step further. If the trait you are selecting for is highly heritable and economically important, you can afford to pay well for a sire that has a high index for the trait. He is more likely to bring a quick and high return on investment if he is used effectively.

If, on the other hand, a trait is low in heritability, environment often masks the genotype. This makes it difficult to select herd or flock replacements with confidence. Consider, for example, prolificacy in ewes, a trait known to be low in heritability. One ewe born a single may nevertheless have genes for high prolificacy, whereas another ewe born a single may lack genes for high prolificacy. How is a breeder to distinguish between the two?

Moreover, once good breeding animals are selected and mated, individual offspring will not perform so closely to their genotype. Environmental factors will mask the genotype here, too. So, in effect, when heritability of a desired trait is low, selection is less effective and improvement by selection is much slower.

When the heritability of an economically important trait is low, the breeder should *spend less time, effort, and money on sire selection but spend more time, effort, and money trying to improve the environment.* For high prolificacy, the breeder should follow rigid sanitation practices, check sires for fertility, feed for top reproduction (see Chapter 15), provide adequate facilities, and be there at farrowing, calving, foaling, or lambing time.

It still may be difficult to appreciate that prolificacy in farm animals is controlled primarily by environment and little influenced by heredity. Most farmers can cite numerous cases where gilts or boars from large litters begot large litters, where selection for prolificacy was apparently effective. But realize that the typical livestock producer who selects rigidly and carefully is also a top-notch manager and caretaker.

Breeders who have ear-notched thousands of pigs and consistently selected for litter size might disbelieve research demonstrating that much of the improvement that has occurred in litter size was automatic and would have happened without conscious selection. On any farm where there might be an equal number of large and small litters, there would be automatic selection for prolificacy. With absolutely no conscious selection, and gilts being saved for the breeding herd at random, more gilts from large litters would be saved simply because there are more of them! The same would be true for other species—dairy and beef cattle, sheep, or poultry.

Ceasing to select for prolificacy in farm livestock is not being suggested here. If a cow calves every 12 months, you know she has the genetic potential (genotype) for relatively high prolificacy. If a ewe gives birth to twins or triplets, she too has a high genetic potential for prolificacy. So does a sow that farrows 15 pigs. These genes can likely be transmitted to the offspring. So if environment can be made and kept optimum, selection for prolificacy can be worthwhile. Although genetic improvement in traits with low heritability will be relatively slow, the *important ones should be included in the long-range goals of the breeder.*

18.5   METHODS TO IMPROVE SELECTION ACCURACY

A livestock or poultry breeder wants selection to be effective (animals with the best genotypes chosen for mating) to achieve the most rapid improvement. Breeders can make phenotype (rate of gain, leanness, or other visible or measured traits) a more accurate indicator of genotype by reducing environmental variations. Measured or observed differences in performance or traits will then be due more largely to heredity.

Selection within a herd or flock, where all prospective breeding animals receive the same feed, housing, and care, tends to be more accurate; variation in performance caused by environmental influences is minimized. But when the task is the purchase of a sire from one of several breeders, in different areas, using different feeding programs, with varying facilities, and where other unknown variations exist, larger environmental differences can be expected and it is almost impossible to know whether the differences noted among animals are due primarily to inheritance or to environment.

A relatively effective solution has been the development of swine, beef, and sheep testing stations. In these stations, all pens and facilities are identical. All animals are fed the same rations, from standard beginning weights. Every effort is made to minimize environmental variations that might affect animals. Prospective purchasers (and owners) can be relatively sure that the animals performing the best in certain traits have the most good genes for these traits and can transmit them to their offspring. A more detailed description is provided in Box 18–1.

Where it is not possible to bring prospective breeding animals to a central testing station, efforts are sometimes made to *standardize the environment* among farms where they are raised. Such efforts may be referred to as "on-the-farm" testing programs. All breeders in a specific program agree to a standard feeding trial. They all use similar rations, start animals on feed at the same age or weight, conclude the test at the same age or weight, and make the same kinds of measurements and observations. Some environmental variations still exist, but they are reduced and more of the variation in performance is due to heredity, so effectiveness of selection is improved.

In herds on Dairy Herd Improvement (DHI) official testing, it is common to mathematically adjust production for certain environmental variations known to have influence. Age at calving, season of calving, times milked per day, length of lactation, and other factors affect production to varying degrees; detailed records have disclosed how much. (See Chapter 30.) Because these are environmental factors, minimizing their efforts on records by mathematical adjustments makes selection for production more effective.

BOX 18-1

Example of Rules for Performance Testing of Beef Bulls
Excerpts from Rules for 1996 Summer Performance
Test Indiana Beef Evaluation Program, Inc.

Purposes

1. To promote performance testing and carcass evaluation of beef cattle and serve as an educational tool to acquaint producers with their value; (2) to complement on-farm performance testing; (3) to provide a common environment for evaluating the performance of young bulls; (4) to assist breeders in identifying sires whose progeny excel in growth rate, feed conversion and carcass value; and (5) to aid beef producers in obtaining superior, performance tested bulls which have been evaluated for breeding soundness, pelvic area, rib eye area and rib fat depth.

Eligibility

1. Any beef producer is welcome to enter bulls in IBEP performance tests.
2. Each individual or firm must be a current member of the Indiana Beef Cattle Association, or Junior IBCA.
3. Entrants are encouraged to have their herds enrolled, and be currently participating in an official on-farm performance testing program (breed or state) recognized by the Beef Improvement Federation (BIF). IN ORDER FOR BULLS TO BE ELIGIBLE FOR IBEP AUCTIONS, ENTRANTS MUST BE PARTICIPATING IN AN OFFICIAL ON-FARM HERD PERFORMANCE TESTING PROGRAM, AND WEANING DATA ON THE MAJORITY OF CALVES IN THE HERD MUST BE REPORTED TO AND PROCESSED BY THEIR PERFORMANCE ORGANIZATION.

Bulls

1. Must have been born May 1 through October 31, 1995.
2. Actual birth weights are mandatory for entry in IBEP Performance Tests.
3. Purebred and percentage bulls eligible for registry in a national or international breed association are eligible for testing. A copy of the registration paper, or application for recording, must be submitted upon delivery of bull at Station. Nonrecorded bulls will be accepted provided that: a) all other eligibility requirements are met, b) the sire is registered/recorded in his respective breed association, and c) the breed composition of the bull(s) entered is specified.

4. Bulls must be polled, or dehorned prior to delivery; horn cavities should be healed and free of infection at time of delivery or bulls may be rejected.
5. Must have a minimum weight of 450 pounds, and a minimum weight per day of age (including birth weight) of at least 2.25 lbs. on delivery at the Station.
6. Bulls will be inspected at time of delivery and will not be accepted if sick, wild, or have a physical defect that may impair future reproductive performance such as: cryptorchidism, testicular hypoplasia, stifled, foundered, "double muscled," etc. Scrotal palpation of prospective entries is recommended at time vaccinations are given.
7. When the test station is full, the number of bulls tested per owner is limited to five units. A unit is defined as three bulls from one sire (Get-of-Sire), or one bull from a different sire. Ownership must be consistent with how the animal is/will be registered in his breed association.
8. Please prioritize entries (gets and/or individuals) beyond five (5) units. If the test station is not full, a 6th unit will be accepted from each owner in the same order as requests for pen space were received. If space remains, a 7th, 8th, etc., unit will be accepted from owners who requested such until the test station is full. Normally, the test station is not completely full in the summer.

Health Requirements

1. Bulls must be tested and found negative to brucellosis within 30 days of delivery to the Station, or originate from accredited and certified herds, or they will be rejected at delivery. Animals from states not accredited TB-free also need a negative tuberculosis test. Bulls also must originate from herds free of Johne's Disease (paratuberculosis).
2. Animals must be free of active ringworm lesions and warts.
3. Animals must be accompanied by an officially approved Certificate of Veterinary Inspection and have been vaccinated according to manufacturers' recommendations with the following vaccines and bacterins: Bovine Rhinotracheitis (IBR); Bovine Virus Diarrhea (BVD); Parainfluenza$_3$ (PI$_3$); Haemophilus Somnus; Leptospira-five strains; Clostridial organisms-7 way; Bovine Respiratory Syncytial Virus (BRSV); and Pasteurella.

(continued)

Another technique is to calculate the degree to which a potential breeding animal deviates in a trait from the mean of its herd or flock (Figure 18–2). The environment in which a herd or flock exists tends to pull each individual's performance toward the herd or flock mean. Should an individual perform in a manner sharply superior to its herd or flock average, you can be more confident that individual has more good genes for the trait.

The example shown in Figure 18–2 also suggests that Smith may be a better manager, providing a more balanced ration, better facilities, and more care. If that is in fact true, the performance of animal B is augmented more by environment, less by genes. The performance of animal A is augmented less by environment, more by genes.

Because selection of parent stock in the poultry industry is done primarily within a particular company, environmental influences are more easily and routinely minimized. The foundation breeder will compare sire candidates within a single building, on a standard ration, and under a single caretaker.

Also, because a single mating can produce many closely related birds (full brothers and sisters), and because production of breeding stock is concentrated in a small number of companies, emphasis is on breeding "lines" rather than on individuals to fill the sire and dam roles in poultry improvement.

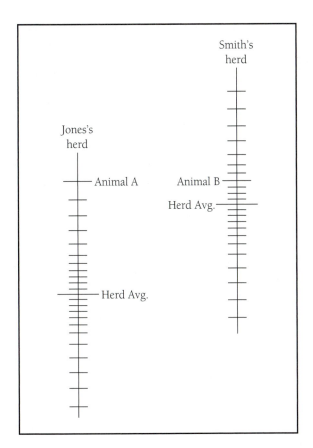

Figure 18–2
If animals of equal performance (equal phenotypic trait) are in herds (or flocks) with different averages, the animal that deviates most, in the desired direction, from the herd or flock average likely has the most good genes for that trait. Animal A is likely to be genetically superior to animal B.

18.6   PREDICTING RATE OF IMPROVEMENT

High heritability of selected traits, maximum selection differential (sometimes called "reach"), and accurate appraisal of prospective breeding animals all contribute to rapid improvement. The effects of heritability on the speed of improvement by selection were discussed in Section 18.4. As shown in Figure 18–3, predicted improvement (in progress) is determined by calculating selection differential and multiplying it times the heritability percentage for that trait.

   Selection differential can be defined as the difference between animals selected to be parents and the average of all animals in the herd or flock available for selection for a particular trait. The selection differential for that trait can be maximized if there is wide variation within the herd or flock (see Section 17.9) and if the animals are selected with care and accuracy (see Chapter 19).

   Rate of improvement per generation can be predicted with reasonable accuracy by simple calculation. After breeding animals have been selected with the greatest accuracy and objectivity possible, the selection differential is determined and multiplied times the heritability of the trait. Figure 18–3 provides several hypothetical examples. Heritability values were taken from Table 18–1.

> *Selection differential*
> The difference in a trait between the average merit of the population (usually a herd, flock, or segment thereof) and the average merit of those animals selected to be parents of the next generation.

A. *Weaning weight of beef cattle:*
   Avg. of calves selected for breeding herd   =   580 lb
   Avg. of all calves   =   500 lb

                                                          Selection differential   =   80 lb
                                                                   Heritability   =   .20
                                                          Predicted improvement   =   16 lb

   Predicted avg. weaning weight of offspring of
      selected calves = 500 + 16 = 516 lb

B. *Backfat thickness of pigs at 230 pounds:*
   Avg. of pigs selected for breeding herd   =   0.8 in.
   Avg. of all pigs   =   1.4 in.

                                                          Selection differential   =   −0.6 in.
                                                                   Heritability   =   .50
                                                          Predicted improvement   =   −0.3 in.

   Predicted avg. backfat thickness of offspring
      (at 230 lb) of selected pigs = 1.4 − 0.3 = 1.1 in.

C. *Milk production of dairy cattle:*
   Avg. of animals selected to be parents of herd replacements   =   22,000 lb
   Avg. of the herd   =   18,000 lb

                                                          Selection differential   =   4,000 lb
                                                                   Heritability   =   .25
                                                          Predicted improvement   =   1,000 lb

   Predicted avg. milk production of herd
      replacements for the herd = 18,000 + 1,000 = 19,000 lb

Figure 18–3
Predicted Improvement = Selection Differential × Heritability. Raising either speeds improvement.

In Examples A and B, calculations were relatively simple and direct because the traits—weaning weight of beef cattle and backfat thickness of pigs—are observed directly in the potential breeding animals. Averages have, of course, been adjusted for sex, age of dam, or other environmental factors, and selected males were counted as heavily as selected females in calculation of the averages. Note that when predicting performance of the next generation, the predicted improvement is added to (or decline is subtracted from) the average of all animals in the herd or flock (e.g., +16 pounds weaning weight, or −0.3 inch backfat).

If selection of only females was made in Examples A and B, and they were mated to sires that represented herd average in merit, the selection differential would be half as great and so would expected improvement. Where selection differential is greater for one parent than for the other, you can calculate selection differential as in Figure 18–4.

Because few sires, in relation to the number of females, are needed in a breeding herd or flock, sire selection can be more intense. The sire or sires chosen will likely be farther from the mean than will the average female chosen. Therefore, the selection differential for the male parent(s) will likely be larger than for the female parents chosen. This is especially the case where artificial insemination is used and is the reason why most of the genetic improvement in livestock and poultry herds and flocks is credited to the sires.

In dairy herds, all cows are bred and should calve, but only certain heifer calves are saved for the herd. Those cows that are selected to be *parents* of herd replacements (Example C in Figure 18–3) need not be designated, in most cases, until they have com-

Figure 18–4

Calculation of average selection differential.

> x = Average of selected males
> y = Average of selected females
> z = Herd average
>
> Selection differential $= \dfrac{(x - z) + (y - z)}{2}$

pleted at least one lactation. Though they would have calved prior to that first lactation, any heifer calves born would not yet be placed in the herd.

The breeding value of the bull that sires a replacement heifer must, of course, be estimated (preferably from production of his daughters compared to their sisters or other herdmates). If the sires, as well as the dams, of the replacement heifers in Example C averaged 4,000 pounds higher in "breeding value" than the herd average, then the selection differential given is correct. This is certainly feasible where artificial insemination is used or in large herds where several bulls, ranging considerably in breeding value, are used. In herds where only one bull is used, however, selection is made only on the basis of the dam's production, and selection differential would be half as great. The replacement heifers get only half their genes from the selected dams. To calculate expected improvement per year, divide the product of the previous calculations by the number of years that elapse per generation.

Livestock breeders also must realize that the improvement achieved by one set of selection decisions does not cause a corresponding improvement in total herd or flock average. The dams of animals not saved for breeding are still in the herd or flock, with unchanged genetic merit, and contribute to the herd or flock average. A dairy cow may produce at a 17,000-pound rate (below herd average in Example C), so her heifer calves are not saved for replacements. But she may remain.

This emphasizes the fact that the average generation interval, for example, in sheep flocks and cow herds is considerably longer than the time it takes for an animal to produce one lamb or one calf.

These examples of predicting improvement rate are given only to show that it can be done. Genetic progress of a trait will be determined by the *size of the selection differential* and *size of the heritability estimate*. The breeder can control selection differential size through number of sires and number of females chosen relative to the size of the herd or flock. Remember that certain circumstances can, in effect, raise or lower heritability estimates and slow or speed up improvement.

18.7 MULTIPLE-TRAIT SELECTION AND RATE OF IMPROVEMENT

In the preceding sections, generation interval, heritability estimate, and size of the selection differential have been discussed as major factors in determining the rate of genetic improvement. The breeder also must be aware of slowed progress in multiple-trait selection.

A general formula that indicates the effects of including additional traits in the selection program is as follows:

$$P = 1\sqrt{N}$$

where P equals percentage of progress of any one trait, and N equals number of traits selected for simultaneously, with similar heritability estimates and traits inherited independently.

With the addition of a second trait, progress of the first trait is reduced to 71 percent of the original value. If three traits are in the selection program, progress of the first trait is reduced to 50 percent. This adverse effect of multiple-trait selection can be expected. Several animals within a herd, flock, or population may excel in one economically important trait. However, as more traits are included in the selection process, you would expect the number of animals superior in all traits to decrease.

Methods used in multiple-trait selection include the

- tandem method
- independent culling method
- selection index

Some breeders use the *tandem method* to select for genetic progress in only one trait at a time. When the one trait reaches a desired level of performance, it is followed by the second most important trait in the selection program. This method is not recommended because it requires several generations to make progress in more than one trait. This method is also not recommended when a negative genetic correlation exists between two traits.

In cases where a negative correlation exists between traits, selection for one results in slowed or lost progress for the second trait. However, progress in both traits occurs when a positive correlation exists.

For example, selection for milk yield also results in an improvement in both milk fat and protein yield. However, a negative correlation exists between milk yield and the percentage of milk fat or protein. Therefore, overall progress in these two traits in dairy herds has been reduced because economics dictated selection for greater milk yield.

When selecting for two traits, the breeder may choose to use the *independent culling method*. This method establishes minimum values for both traits. Animals that attain below the minimum value in one trait are not chosen to be in the breeding herd or flock regardless of their performance in the other trait(s).

The method of choice in multiple-trait selection is the *selection index*, which results in a single numerical measure of overall merit performance for each animal. Use of an index allows progress to be made in several traits at the same time. Indexes are discussed in more detail in Section 19.10.

QUESTIONS FOR STUDY AND DISCUSSION

1. Define the term *heritability.* How is it quantitatively expressed?
2. Give several examples of economically important traits with low heritability estimates. Where should emphasis be placed if high performance in these traits is to be expected? Explain.
3. Why would you expect traits such as services per conception, hatchability, weight at weaning, and survivability to have relatively low heritability estimates?
4. Name four traits that have high heritability. Why are they high?
5. How are heritability estimates derived?
6. Define the term *selection differential.* Give an example of its calculation.
7. Study the examples of predicting improvement by use of selection differentials and heritability estimates in Figure 18–3.

Note: When calculating predicted performance, always add the expected improvement to (or subtract the decline from) the average performance of the herd or flock.

Evaluation of Breeding Animals

19

Two of the four rules for effective improvement of livestock and poultry by selection have been described in previous sections. We are now ready to focus on Rule 3:

Rule 3: *Observe or measure accurately the traits carried by a prospective breeding animal.*

This chapter describes how the important traits can be appraised visually and by means of certain measuring techniques. Some traits can be accurately evaluated by the eye; others cannot.

While reading this chapter, keep in mind an important point emphasized in the previous chapter: If environment is uniform, phenotype will be a better indicator of genotype, and most of the variation in the animals available for selection will be due to heredity. In other words, if all prospective breeding animals are fed and handled as nearly alike as possible, it will be easier to identify the animal or animals with the best traits. Both the eye and measuring techniques can be more effectively and accurately used for selecting the best herd or flock replacements (Figure 19–1).

Although some of the points discussed in this chapter are also applicable to poultry and dairy cattle, most emphasis is on swine, beef cattle, and sheep. Section 30.6 discusses appraisal of dairy cattle for milk production. Sources of poultry breeding stock are discussed in Sections 33.1, 33.7, and 35.3.

Upon completion of this chapter, the reader should be able to

1. List traits that are very important in breeding animals.
2. List factors that influence *which* traits should attract attention by a livestock or poultry breeder in a selection and breeding program.
3. Describe a simple and useful animal identification system.
4. Explain how age of livestock can be estimated, should birth records not be available.
5. Explain why central testing stations or adjustments of weight to a standard age are used in breeding animal appraisal.
6. Describe two systems for measuring meatiness in live animals.
7. List ways of assessing breeding efficiency of sires *before* they are placed in service.

Figure 19–1
Some heifers in this herd will produce calves that are heavier at weaning and gain faster in the feedlot. If they can be identified and used and if others are culled before breeding, the herd should be more profitable.
(Courtesy of Beefmaster Breeders United. Used with permission.)

8. List traits that generally cannot be assessed with reasonable precision by the eye, and traits that can, and give the reasons.

9. Define *index* and *selection threshold* and differentiate between them as devices used in selection decisions.

10. Calculate repeatability of a trait, given the data, and explain why repeatability provides a clue as to the wisdom of culling a breeding animal after one breeding season.

19.1   DESIRABLE TRAITS

Health is obviously a primary consideration in selection of breeding animals and in profitability of livestock and poultry enterprises. Poor health, whether resulting from injury, contagious disease, or malnutrition, will impair productive and reproductive efficiency. All new breeding animals should be blood-tested or otherwise certified free of reproductive diseases. These diseases can be spread very rapidly, especially when animals are bred by natural service. A newly purchased animal should be isolated for at least 3 weeks to permit latent symptoms of disease to appear.

Any addition to a flock or herd should be *prolific*. Females should come in heat regularly, shed a sufficient number of ova, and conceive upon first service. Bulls, boars, and rams should be masculine and aggressive. They should produce a sufficient volume of semen with a high concentration of normal, healthy, long-living sperm and should be test-mated to evaluate their reproductive performance prior to the breeding season.

In most cases a potentially long life is important. For example, after a ewe reaches two years of age, she is of little value for slaughter. (Compare current prices of fat lambs and mature ewes for slaughter.) Yet considerable investment has been made to grow her to breeding age. For maximum profit to the flock owner, then, the best ewes must possess longevity and must produce a large number of lambs before they die. Ten-year-old ewes are not uncommon in well-managed midwestern flocks. In range flocks, few ewes are kept after they reach seven or eight years of age.

DEKALB® DELTA

PLATE E. EGG PRODUCTION

1. Most commercial egg laying birds are lines or strains of the White Leghorn breed. This highly efficient hen is capable of producing about 300 eggs in her first year of production.

2. During flock production, special attention must be given to floor-reared chicks to assure that they have a warm environment and that they adapt rapidly to feed and water. Pullets are grown and developed with special attention to nutrition and lighting so that egg laying is initiated at 20 to 21 weeks of age.

3. Most egg-laying flocks, some with houses of 60,000 to 100,000 layers, are maintained in elevated cages, each containing about 4 to 6 birds. Carefully formulated rations, water, lighting, and ventilation are provided to help assure comfort and productivity. After 52 weeks of production, the flock manager may replace all hens with new pullets, or retain the hens into the second year following a period of "forced molting."

4 & 5. Highly automated housing systems assure that fresh, clean eggs are collected and graded to provide a high-quality cartoned product. White-shelled eggs prevail throughout the U.S. except on the East Coast.

4.

5.

6. Eggs provide very high quality protein in the diet and contribute to the attractiveness and palatability of many prepared dishes. The yolk portion of eggs contains a higher-than-normal level of cholesterol and may be restricted in cases where an individual's blood cholesterol is elevated. However even in these cases, the egg white portion can be consumed freely.

Courtesy of DeKalb Poultry Research, Inc., The American Egg Board, and Purdue University.

PLATE F. BROILER AND TURKEY PRODUCTION

1. The U.S. broiler industry is highly efficient in providing the consumer with a high-quality meat product in the desired cuts at an economical cost. Broiler production begins with genetically superior chicks as future breeding stock.

2. A routine and well perfected procedure known as "vent sexing" is performed with day-old chicks to separate male from female breeding birds.

3. Both breeder stock chicks and meat type broilers are placed on a health program where appropriate vaccines are used to produce immunity from potentially dangerous diseases.

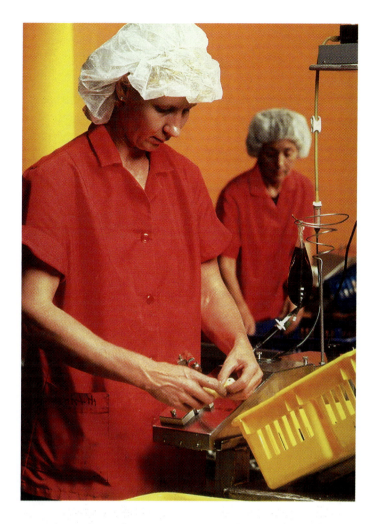

4. Broiler chicks are hatched from eggs produced by the breeding flocks. Strains and lines of the Cornish-White Rock breeds have been most common in commercial broiler production. The highly integrated broiler industry provides proper breeding, housing, nutrition and management of large numbers of floor-reared broilers. Generally broilers are scheduled to go to market within 48 to 50 days at a live weight per bird of 4.3 to 4.7 pounds.

5. Per capita consumption of broilers has steadily increased. Their popularity can be credited to efficient production of a high-quality product available to the consumer as a whole bird or individual portions. Attractive and nutritious poultry dishes can be prepared at a very economical consumer cost.
   Courtesy of Ross Breeders, Inc.

6, 7, & 8. Commercial turkey production also depends upon breeding stock with superior genetics for rate of growth, feed efficiency and high quality carcasses. As in broiler production, systems consist of flocks of several thousand birds on a scheduled growing and management program. Per capita consumption of turkeys continues to increase, although it is somewhat more seasonal than broilers.

6.

7.

Courtesy of Nicholas Turkey Breeding Farms.

8.

A similar situation exists in beef herds. Heifers are usually 2 to 3 years of age before they calve the first time. Because the slaughter value of a mature animal is usually lower and the expense of raising the heifer to breeding age is large, cost per calf is tremendous if the heifer produces only three or four calves during her lifetime. Longevity permits the genetically superior cows to remain in the herd and contribute to total herd productivity.

In swine, longevity is not so important. Gilts usually farrow first before 1 year of age and again at 17 or 18 months. The relative cost of raising the gilt to breeding age is less because she is sexually mature earlier. Second, she produces litters—usually nine or more offspring—rather than singles or twins. A third important factor is that the value of the carcass from a 2-year-old gilt (many are sold for slaughter after weaning the second litter) is only a little less, per pound, than the value of a carcass from a 6-month-old gilt.

Efficient growth is important. Feed represents 60 to 90 percent of the cost in livestock enterprises. Animals that utilize feed more efficiently are more profitable.

Quality of product is another major consideration. For a rancher, the product may be feeder lambs or feeder calves—though the ultimate product for all meat animals is a carcass. Because most aspects of carcass quality are so highly heritable (see Table 18–1), a breeder should select sires or dams that would themselves yield high-quality carcasses. Well-muscled animals that would yield a high proportion of high-priced cuts should be selected. A detailed discussion of desirable carcass traits is given in Chapter 23.

Wool sales contribute income for the sheep raiser. Amount and quality of wool are determined primarily by heredity, largely through the selection and breed of the ram and the mating system used with the ewe flock.

To which of the above traits should the livestock or poultry breeder devote most attention in a selection and breeding program? The answer is, the trait that will contribute most to long-term profit; but that trait may not be the same for all situations. Here is why:

1. Traits desired in a sire may be influenced by traits already present in the female herd or flock. A farmer with sows that are prolific, lean, and meaty, but which gain slowly, probably would emphasize growth rate when selecting a boar, especially in the terminal cross (last mating) utilized in the production of slaughter barrow and gilts.

2. Some traits are desired in the animal being selected, as well as in its **offspring**. Prolificacy, longevity, and feed efficiency are examples. Other traits need to be demonstrated only in the offspring—rate of gain and carcass quality, for example.

> **Offspring**
> Animals born to a parent; descendants, either the first or a later generation.

3. For traits where demonstration of the characteristic in the offspring is important, heritability of the trait should be considered. If the trait is poorly inherited and influenced largely by environment (and its demonstration in the animal being selected does not contribute directly to profit), then it is not an important trait to consider in selection. If, however, the trait is highly heritable, it would be very important in potential breeding animals.

4. How many offspring the selected breeding animal will have should be considered. A selected heifer, ewe, or gilt generally transmits genes to relatively few offspring in her lifetime. A bull, ram, or boar, however, may sire hundreds or thousands of offspring. This doesn't influence *which* trait is most important, but it does influence *how much more important* a trait, desired and carried by the sire, is.

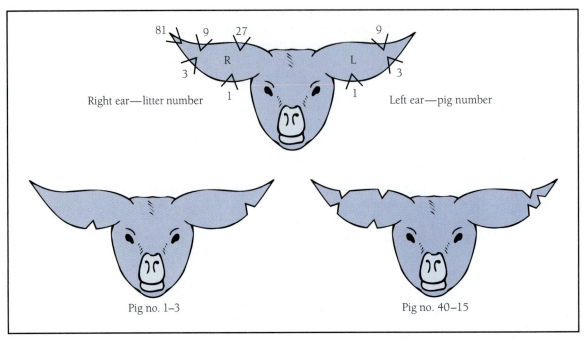

Right ear—litter number

Left ear—pig number

Pig no. 1–3

Pig no. 40–15

Figure 19–2

Pigs can be numbered by notching the ears soon after birth. The system shown above has been adopted as official by several purebred swine associations. The pig's right ear is used for the litter number and the left ear is used for its number. The system permits numbering up to 80 litters without notching the tip of the right ear. By using the 81 notch at that location, one can identify 161 litters, and modifications could permit numbering more.

19.2   IDENTIFICATION SYSTEMS

Proper identification of all animals is the basis of any complete record-keeping and livestock selection program. The breeder may be familiar with every animal, but lack of a positive identification system makes any record-keeping program worthless if he or she should no longer be available.

Most sheep and cattle breed registry organizations require that an animal's registry number be tattooed inside one ear with permanent ink. Some, especially for sheep, also provide small ear tags. Because registration numbers are usually long, most breeders also use some type of private system. Such numbering systems are also used on grade (not purebred) herds or flocks. A common system used in swine herds is illustrated in Figure 19–2.

For easy identification on pasture or range, or in drylot, large, easily read paint brands are used on sheep (Figure 19–3), and ear tags are often used on cattle. Various brands may range considerably in durability and retention. In a 23-month Kansas cow herd trial,[1] percent retention of four brands of ear tags ranged from 3 to 98 percent. At 5 months the retention ranged from 73 to 98 percent. Large numbers can be applied with special, long-lasting paint. Freeze-branding is also used on solid-colored cattle. Supercold branding irons cause new hair growth to lack pigment, so white numbers eventually appear on the red or black background (Figure 19–4). Hot brands for cattle have numerous disadvantages but are still used in large herds where proof of ownership is the primary purpose.

---

[1] Sprott, L. R., and L. R. Corah. *Progress Report 394.* Kansas Agr. Exp. Sta., 1981.

Figure 19–3
For numbering sheep, special paint can be used that will not smear in rain, yet will wash out when clipped wool is scoured. (Courtesy of Iowa State University.)

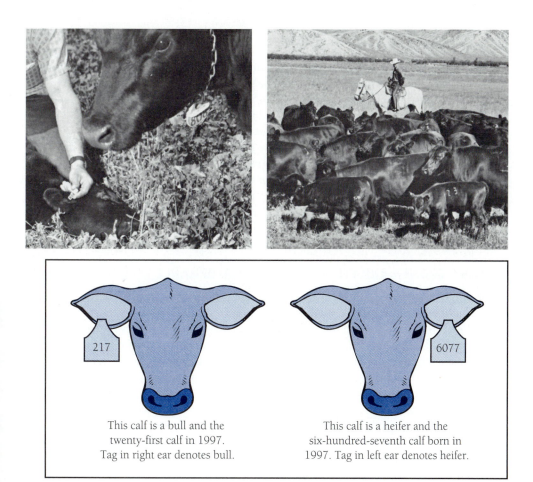

This calf is a bull and the twenty-first calf in 1997. Tag in right ear denotes bull.

This calf is a heifer and the six-hundred-seventh calf born in 1997. Tag in left ear denotes heifer.

Figure 19–4
Calves should be ear-tagged very soon after birth (left photo). A practical numbering system is shown in the lower sketch. The upper-right photo shows cows and calves that have been hot-branded. (Photos Courtesy of Nation's Agriculture and Union Pacific Railroad; drawing from University of Wisconsin.)

Many dairy herds utilize electronic identification of their cows. For example, each cow is equipped with a neckworn transponder that identifies her to the computer system as she enters the milking parlor. Then all milk produced by each cow is electronically recorded and credited to that individual. Poultry can be identified by numbered or coded leg bands.

19.3   RECORDING BIRTH DATES

Records of birth dates are, of course, basic to any later appraisal of growth, and also are helpful in estimating time at first estrus or for other management purposes. However, in the infrequent times when records are not available, age of cattle, horses, and sheep can be closely estimated by looking at the teeth (Figure 19–5). Animals within these species tend to lose their temporary incisors at rather uniform ages. Age can also be estimated, by an experienced livestock raiser, by noting the size of head, the length of ears and tail, and length of legs. Remember that the skeleton has priority over lean and fat tissue in receiving nutrients for growth, so it tends to grow at a rather consistent and uniform rate.

19.4   IMPORTANCE OF RECORDS

Write it down! Birth dates, litter size, and number born dead, as well as environmental stresses or injury that might later influence selection decisions, can be recorded on the spot in a small notebook computer. Such information can be transferred later to permanent record sheets, books, or computerized records.

Breeders benefit from records showing dates females came in heat and were bred. Late and inconsistent breeders can be detected more easily, and calving, lambing, or farrowing dates can be anticipated with more accuracy.

Birth weight is often related to the ability of the newborn animal to withstand the stresses to which it will be subjected. Animals that are heavier at birth usually gain faster. Experienced livestock raisers can estimate birth weight with considerable accuracy, but a scale weight provides precise information to use later in selection decisions.

19.5   GROWTH AND FEED EFFICIENCY PERFORMANCE RECORDS

Weaning weight is important to livestock breeders for two reasons. It indicates the relative *milk production of the dam,* especially in beef herds and sheep flocks where creep feeding is not practiced and the calf or lamb depends on milk as the major source of nutrients. It also discloses, to some degree, the young animal's inherent *ability to gain.* If management, milk supply, and other factors are similar, the animal that gains more rapidly up to weaning time probably will continue to gain more rapidly and will tend to transmit this trait to its offspring.

Where hogs are raised in confinement, all pigs are weaned at the same age so weaning weights can be compared directly. With beef herds and sheep flocks this is not practical so adjustments of weights to a standard 205-day age can be quickly calculated using a hand-held computer, or weights recorded, then processed electronically using an appropriate software program. Also, adjustment tables or charts have been developed that allow weighing many calves within a certain age range at the same time and adjusting their weights to a *standard 205-day age* or, in the case of lambs, adjusting on the basis of 60- or 90-day weaning weights. Growth rate in swine is typically expressed as

Lamb: All temporary incisors

Yearling: One pair of permanent incisors

Two-Year-Old: Two pairs of permanent incisors

Three-Year-Old: Three pairs of permanent incisors

Figure 19–5
Teeth provide a close estimate of the age of a ewe or ram up to 4 years of age. (Courtesy of Iowa State University.)

days to 230 pounds. Also, you may want to review the rate of growth measurements and formula for adjusted weaning weight previously discussed in Section 9.2.

It is also important to adjust the weaning weight of calves or lambs for *age of dam,* because very young and very old cows and ewes give less milk. Single lambs may weigh as much as 10 or 11 pounds more than twins at weaning time in drier range areas where vegetation is sparse. In midwestern farm flocks, however, where feed is usually plentiful and lambs are usually creep-fed, the difference may be insignificant.

If productivity of dams is being compared to decide if some should be culled, the offspring's weaning weight should also be *adjusted for sex.* Bulls will be about 15 pounds heavier than steers, and steers about 20 pounds heavier than heifers. This, too, varies among areas.

Weaning weight, as a measure of the dam's producing ability, is better than weights recorded later. After weaning, the animal's growth is influenced more by feed supply and less by the dam's milking ability.

There is a relatively high positive correlation between weaning weights of calves, lambs, or pigs, and gains during subsequent periods. This is especially beneficial in sheep flocks and beef herds where breeders like to select their replacement females early and sell the remainder as feeders.

Most breeders and commercial breeding organizations want more precise information, especially in evaluating prospective sires. Because a sire may transmit genes to hundreds of offspring in a year, it is worthwhile spending more time, money, and effort for precise appraisal of the sire's productive merit.

Swine, bull, and ram testing stations available for use by many breeders have been developed in many sections of the United States. (See Section 18.5.) Some breeders, in cooperation with universities or breed associations, operate their own testing units. Such units are designed to measure rate of gain and feed efficiency for a standard period after weaning and before breeding age. Potential sires from different farms and ranches are more fairly compared at these stations because environment is standard.

Some stations operated by livestock organizations require that all animals meet certain growth, feed efficiency, and, perhaps, meatiness standards in order to be sold or taken home for use. Those that do not meet the standards are castrated and sold for slaughter.

All feed consumed is carefully weighed so that *feed efficiency values* can be calculated. Individual feeding is necessary if feed efficiency values are to pertain to a single animal.

Selection for gaining ability (and/or feed efficiency) does not preclude selection for desirable conformation. Gaining ability and meatiness, or other conformation traits, can be improved by selection at the same time. They are not genetically antagonistic. Obviously, though, less rapid progress probably will be made in one trait if some of the "selection opportunities" are being used on a second trait.

19.6 MEASURING MEATINESS OR MUSCLING

Carcass data from relatives could be used as a guide to meatiness of potential breeding animals, but is sometimes not available and is of limited value because the relatives are genetically different, to a degree, from the animal in question. Other techniques have been developed to measure directly the meatiness of animals in some species.

Backfat thickness in pigs can be measured with either a metal ruler or, more commonly today, ultrasonic measurement, sometimes called an *ultrasonic probe.* Backfat thickness as measured by the probe is highly negatively correlated (-0.90 or lower) with the percentage of lean in the carcass. Gilts or boars are probed at about 230 pounds. With the metal ruler, small skin incisions are made and the ruler inserted at three points, about 2 1/2 inches off the midline of the pig (Figures 19–6 and 19–7). The ruler penetrates the fat layer easily but stops at the layer of connective tissue covering the eye muscle. If pigs are probed at some other weight, values can be adjusted to a 230-pound basis.

The ultrasonic probe—an instrument that emits high-frequency sound waves and records the time it takes for them to bounce back from a junction between two

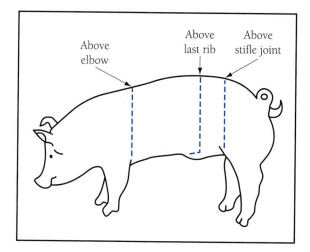

Figure 19–6
Locations for measuring backfat thickness.
(*Source:* Purdue University.)

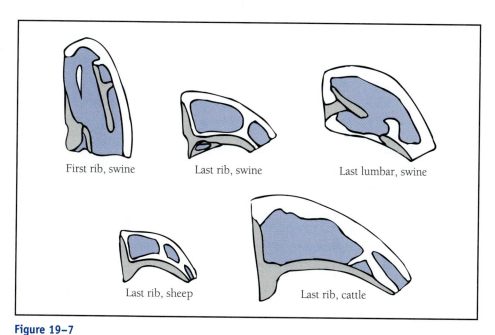

Figure 19–7
Muscle patterns at probe or sonoscope measuring sites in swine and comparable cross sections of lamb and beef at last rib.
(*Source:* USDA.)

tissues—is also used to effectively appraise meatiness in beef cattle and sheep, in addition to hogs. The instrument was originally used as a flaw detector for metal parts and ceramics. High-frequency sound waves emitted by the machine pass through tissue at speeds that vary according to density of the tissue. When they reach a junction with a different tissue, some of the waves bounce back. These "echoes" may be displayed on a cathode ray tube or printed, then measured by hand or automatically measured, to determine depth of fat or lean. The electronic procedure in pigs is similar to that of the metal ruler procedure with measurements at the same locations (Figure 19–6), but without skin penetration.

One reason the probe is such an excellent indicator of percentage of lean is because fat thickness is measured near the middle of the eye muscle. A pig with a shallow, kidney-shaped muscle will have a thicker layer of fat at the measuring site, whereas one with a bulging muscle will have a thinner fat covering.

The probe is not as accurate an indicator of percentage of lean in beef as it is in pork. The irregular shape of the eye muscle makes it difficult to obtain consistent measurements.

Where it is necessary to use carcass data on relatives to evaluate meatiness of a possible breeding animal, percentage of lean cuts and area of the loin eye cross-section are probably the most useful measurements. Electromagnetic scanning[2] for total body electrical conductivity to determine total carcass lean may provide an accurate, rapid evaluation system in the future. Also, Box 23–1 provides more details.

Carcass grade is a rough approximation, not a precise measure, of merit, so is of limited value. Meatiness can be estimated in breeding animals by visual appraisal of width of shoulders, bulginess of muscles, and thickness of ham, round, or leg. An animal can show good muscling even when it carries little finish. In fact, too much finish often fills in the slack spots and disguises the degree of muscling.

19.7 REPRODUCTION EFFICIENCY

The breeding potential of a bull, boar, or ram is influenced by the fertilizing potential of the semen—volume, sperm concentration, and sperm viability—and also by the "serving capacity" (the number of females bred in a given period). Serving capacity includes **libido**, which is the willingness or eagerness to mount and service, as well as the physical ability to complete the service.

Libido
Male sex drive.

Of 10,940 bulls checked by Colorado State University scientists, 20 percent (2,266) were found to be questionable or unsatisfactory for breeding (Table 19–1). The figures in the table total more than 2,266; some had two or more negative characteristics.

Criteria established by the Society for Theriogenology for evaluating breeding soundness include sperm motility and morphology, as well as scrotal circumference. The latter is positively correlated with testicle size and semen production, especially in young bulls. Sperm concentration and percent of live sperm are not used because these traits have been found to be low in repeatability under range or pasture conditions.

---

[2] Electromagnetic scanning studies conducted by D. L. Wishmeyer, et al., of Utah State University used procedures that restrained lambs on a carrier table that passed them through the TOBEC™. In their initial studies, they did not find the scanner to be a reliable tool for predicting body composition of live lambs. However, they did find it to be reliable in predicting chemical composition of swine and lamb carcass. (*J. Anim. Sci.* 74 [1996]: 1864.)

TABLE 19–1	Reasons for Classifying 2,266 of 10,940 Bulls as Questionable or Unsatisfactory for Breeding	
Reasons	**Number of Bulls**	**Percent of Total**
Semen quality questionable or unsatisfactory	1,832	16.7
Defects of testes	1,931	17.6
Defects of penis	388	3.5
Defects of epididymis or vas deferens	112	1.0
Defects of limbs and feet	597	5.5
Abnormalities found by rectal palpation	547	5.0

Source: *Elmore, R. G., C. J. Bierschwal, C. E. Martin, and R. S. Youngquist,* J. Theriogenology 3:209–18, 1975.

TABLE 19–2	Variation in Breeding Performance of Beef Bulls During a 90-Day Breeding System				
Bull I.D.	Number of Cows Bred	Number of Cows Calved	Percent Calving in		Calf Crop (%)
			First 21 Days	*First 42 Days*	
5219	20	20	65	85	100
1116	19	19	37	63	100
0200	21	20	62	86	95
0310	19	18	16	53	95
2010	21	19	71	90	90
5238	21	19	38	66	90
2906 ('67)	18	7	6	28	39
2906 ('68)	22	9	23	32	41
2710 ('67)	20	7	26	36	36
2710 ('68)	22	21	45	81	95
6290	23	0	0	0	0

Source: *Rich, T. D.,* Effects of Management on Bull Reproductive Ability, *Great Plains Beef Cow Handbook, GPE8451, Oklahoma Cooperative Extension Service, 1975.*

There is considerable range in libido among males, and there is some evidence that the trait is reasonably heritable.

Table 19–2 illustrates wide variation among bulls in total breeding performance. The percentage of cows that calved in the first 21 days after the initial cow calved is indicative of the closeness in days that cows conceive from a particular bull, and is important in keeping the average calving interval close to 365 days, with a more uniform calf crop as well.

Reproductive efficiency in the female is less easily appraised. As discussed in Section 19.1, reproductive performance is repeatable and some culling decisions can be made with confidence after the first breeding season.

Temperament of a heifer, ewe, or gilt may affect net reproduction rate. If the dam is restless, offspring may be unable to nurse readily, and in swine, an extremely nervous gilt is more likely to lie on some of her pigs.

19.8   CONFORMATION TRAITS

When selecting herd or flock replacements, the main criterion should be performance, but structural soundness and correctness should also be evaluated. Conformation traits are relatively subjective but nevertheless important and must be appraised in prospective breeding animals with all the accuracy that can be mustered.

Some breeders use a committee of two or three experienced livestock workers who score conformation of heifers or ewe lambs by means of some numerical system. This can be done at weaning time, as they leave the working chute or scale. Scoring systems can provide as many quality groups as desired. A verbal description for each group is usually developed, and appraisers try to "set their sights" together.

A committee adds validity. Conformation scores can be averaged and committee members serve as a check on each other. Emphasis should be on visual characteristics that relate to performance, such as correct angle of joints and bones, frame size, distribution of muscle, and correctness of reproductive and mammary systems.

Most breeders prefer animals that have a spacious thoracic cavity, giving plenty of room for the heart, lungs, and other organs to operate. Hence, they look for animals with a wide chest and with the ribs wide apart—a lot of "spring of rib."

Number of teats is easily determined by counting, but location and prominence is a relative thing. Teats should be far enough apart and large enough so nursing can be done without difficulty. Size and prominence of teats are highly heritable, so should be noted when appraising breeding stock, both female *and* male. Inverted nipples, often nonfunctional or difficult to suckle, are rather common in hogs.

Animals that will be in a herd or flock for many years—females that will carry the load of pregnancy many times and sires that will breed often for many years—need straight, strong, and well-placed legs and feet. Dairy breeders also place considerable emphasis on soundness of feet and legs when selecting AI sires with hopes that future herd replacements will not be eliminated from the herd with feet and leg problems. A straight leg will support more than one that slants and for a longer time. There will be less strain on the joints. If legs are placed "out on the corners," less of the body hangs over to create leverage.

Extremely straight or extremely crooked legs are distinguished by anyone, but those in between sometimes vary in appearance. Straightness of legs may be influenced by the current weight of the animal and whether it is standing on soil, bedding, or concrete.

After conformation scores are assigned and weights recorded and adjusted, a breeder probably will do the final selections. Naturally those animals that are chronically ill or that have certain traits definitely not wanted (such as color markings in a purebred herd or flock that will not allow registry) will already have been eliminated. Selecting from hand- or computer-tabulated data prevents the breeder from allowing personal, subjective bias to influence his or her decision. To some, this may seem absurd, but it happens. For example, the breeder may hate to eliminate a particularly affectionate calf, lamb, or gilt, and be reluctant to keep one that is a bit independent or aggressive.

19.9   OTHER PRODUCTION TRAITS

Points of conformation have a certain aesthetic value to livestock raisers, especially the purebred breeder. In other words, such things as depth, straightness of legs and topline, and features of the head often have "beauty" to the breeder, sometimes beyond their true economic importance. Some traits can be appraised more objectively. Wool quality, teat placement and prominence, and size of udder, for example, apparently have little aesthetic value, and breeders can scrutinize them with extreme objectivity.

A breeder can check quickly to see if the teats are spaced so offspring can nurse with ease and that no teats are inverted. An experienced sheep raiser becomes skilled at appraising wool quality. Samples of wool grades can be compared with the wool on the lamb. Wool density and length, which influence amount of wool produced per year, and wool quality can be compared among animals considered for the flock. Amount of wool covering the face is also apparent.

19.10   USING SELECTION INDEXES

How can the *net merit* of prospective breeding animals be established? Because they vary in all traits, how can the candidates for herd or flock sire or replacement female be fairly compared?

Indexes, which are actually a net merit score, can be calculated considering the relative economic importance of the traits as well as their heritability. Separate indexes may be calculated for different herds or flocks, or different years, as the relative importance of the traits may vary.

The index for boars that completed the performance testing program at the Iowa Swine Testing Station (Fall 1984) emphasizes growth rate, feed efficiency, and backfat probe equally and is calculated as follows:

$$\text{Index} = 500 + (\text{Daily gain} \times 100) - (\text{Backfat} \times 100) - \text{Feed efficiency}$$

This index example takes into account feed costs, market differential for lean versus fat hogs, and the heritability of the various traits (see Table 18–1). The factor 500 is included so that a good-quality boar will index above 100. A boar that gains 2.21 pounds per day, on 2.57 pounds of feed per pound of gain, and probes 0.80 inch at 230 pounds will index 19.2.

Where a litter mate barrow is slaughtered at the end of a performance test, information on carcass length, backfat, percent ham and loin, and loin eye area is also available to aid in evaluation of the full-sibling boar.

Obviously, a different index that emphasizes leanness can be used if there is a larger price advantage for producing lean pork carcasses or if feed and labor are cheap enough that rate and efficiency of gain are less important.

Also recognize that a breeder who already has a group of lean sows may ignore the index and base his or her selection mainly on growth rate and feed efficiency.

A more recent evaluation program for genetic improvement in swine is referred to as STAGES (Swine Testing and Genetic Evaluation System).[3] Developed at Purdue

---

[3] STAGES is a cooperative project involving Purdue University, the USDA Agricultural Research Service and Extension Service, the National Association of Swine Records, the National Pork Producers Council, and the individual purebred swine associations.

University, it consists of a series of computer programs that analyzes performance data provided by participating purebred swine producers. Both reproduction and postweaning data are analyzed, then genetic evaluations for these traits are combined into three alternative indexes—maternal, general, and terminal sire.

Data on beef bulls performance-tested by the Indiana Beef Evaluation Program (Summer 1996) included measurements of backfat, scrotal circumference, and pelvic area. In addition, each bull was indexed within each breed as follows:

$$\text{Performance index} = 60 \left[\frac{\text{Bull avg. daily gain}}{\text{Breed group ADG}}\right] + 40 \left[\frac{\text{Bull wt. per day age}}{\text{Breed group WPDA}}\right]$$

| Weight | Conformation grade | | | | | | | | |
| | Large frame | | | Medium frame | | | Small frame | | |
	Thick muscle	Medium muscle	Thin muscle	Thick muscle	Medium muscle	Thin muscle	Thick muscle	Medium muscle	Thin muscle
570+		23		31		40			
560–570	24		47		20		16		
550–560		13 42	30 49	45					
540–550				22		48 35		46	
530–540	19				41 15				
520–530		34	12					27	
510–520	43			44		38	14		
500–510		11			32				36
490–500	25		21						
480–490		50		39					
470–480	Possible					29			
460–470	culling line	26	18						
450–460									
440–450					17				
430–440			33						
420–430								28	
410–420									
400–410									
390–400									
380–390									
370–380									
360–370									
350–360									

Figure 19–8

A selection sheet for heifer calves at weaning time. After weights are adjusted and as conformation appraisals are made, all potential herd replacements are entered, by number, on this sheet. A single selection line can be drawn or a line can be drawn within each column, according to the breeder's goals and number of needed replacements.

Rather than use indexes, breeders may employ other techniques to arrive at a selection decision. A sheep rancher who wants to save 25 out of 100 ewe lambs may first select the top half of conformation, then save only the half of these that gained fastest.

A breeder also might use a selection sheet like that in Figure 19–8. Because about 40 percent of the heifers must be selected in order to maintain herd size, there is a limit to the severity of selection.

19.11   USING RELATIVES TO APPRAISE MERIT—PROGENY TESTING

Performance of offspring is considered by many to be the best indicator of the breeding merit of an animal. This is termed **progeny testing**. In poultry, the young male can be mated to a large sample of females and the performance of offspring—broilers or pullets—measured within a year. The sire, if demonstrated to be of high merit, can then be used extensively.

> *Progeny testing*
> Testing of offspring to determine which parents should be chosen for more extensive use in breeding programs.

Where artificial insemination is commonly used, permitting the potential value of an extremely superior sire to be extremely high, as in dairy cattle, it is often considered worthwhile to use progeny testing. Young bulls in an artificial insemination "stud" are mated to a random sample of heifers over the country, and the sire owner and/or breeder service organization then awaits the performance records of the heifer offspring. The wait will, of course, be close to 4 years. In the meantime, however, semen may be collected and frozen for possible use with thousands of cows if the offspring excel.

In the case of dairy cattle, a large proportion of the heifers that result from artificial insemination are in herds where DHI records are maintained. These data, plus a record of the heifers' sires and dams, are processed by computer and stored in a multistate cooperative program. If needed, all the records of the offspring of a particular sire can be retrieved in minutes. USDA sire summaries that rank bulls according to **predicted transmitting ability (PTA)** within breeds are updated twice yearly. Also, these summaries and other information such as type-production indexes and linear-type evaluations are made available by the different dairy breed associations.

> *Predicted transmitting ability (PTA)*
> A term that has replaced *predicted difference and cow index;* values predicting the genetic merit of dairy animals.

With increased use of artificial insemination and computer processing of performance data in other species, use of progeny testing in these species is likely to increase.

Brothers and sisters (sibs) may also be used to help appraise the genetic merit of a potential breeding animal. This is especially helpful where carcass information is needed. In the case of swine, litter mates to a boar or gilt considered for use in a breeding program could be slaughtered and carcass traits measured. In the case of beef cattle or sheep, half sibs (with a single parent in common, usually the sire) might be used to gather carcass data.

Full sibs, with the same sire and dam (which could be expedited by embryo transfer techniques), are as closely related to the animal being appraised as are offspring. An offspring receives 50 percent of its genes from each parent, so it is sometimes said that an animal and its offspring are "50 percent related." The 50 percent that are transmitted theoretically occur at random. In the case of full sibs, this random sorting that occurs in ova and sperm formation suggests that half the genes received by sib A from the sire would be the same as those received by sib B and that half the genes received by sib A from the dam would be the same as those received by sib B. The consequence is that sibs A and B are 25 percent related through the sire and 25 percent related through the dam, a total of 50 percent relationship. Half sibs, with one parent in common, would be "25 percent related."

It is recognized that the above explanation doesn't take into account homozygosity in gene pairs or the fact that a total population may be essentially homozygous for certain genetic traits. When this is taken into account, you would say the percentages above pertain to the degree the relationship between parent and offspring or between sibs exceeds the average genetic relationship of the population.

19.12   CULLING FROM THE BREEDING HERD OR FLOCK

Appraisal of breeding animals usually continues even after they have entered the herd or flock. An animal must continue to justify its place in the herd or flock. Considerable and continuous culling occurs with livestock, where the breeder has sufficient information to justify culling. When heifers, ewes, or gilts are selected for breeding at a rather young age, some mistakes in selection will be made. The true productive ability is not known at that age and the guides or clues used for selecting are sometimes misleading.

In the case of poultry, however, the high level of quality and uniformity in egg production that has been achieved by years of intense selection results in little culling by commercial producers after birds are placed in the laying houses as the flock is often replaced after a 52-week laying period and it is not economical to invest time in such a practice.

> **Repeatability**
> The tendency of animals to repeat themselves in certain performance traits in successive seasons, pregnancies, or lactations.

Can a breeder cull with confidence at the end of the first season? Is the first season's production of a heifer, gilt, or ewe a good indication of her later production? This depends on the trait. The approximate repeatabilities for various traits are given in Table 19–3. **Repeatability** is the degree to which an animal repeats itself in productivity of the same trait in successive years or seasons. Because the repeatability varies for different traits, the speed with which a breeder can cull with confidence also varies according to the trait being considered in culling.

The true meaning of repeatability can be illustrated by showing how a repeatability figure of 0.46 for weaning weights of beef cattle is calculated. A herd of cows is divided equally into two groups, based on weaning weight of their first calf (Figure 19–9). Then the weights of all subsequent calves of each group are averaged.

If the subsequent calves from the high group had averaged exactly 89 pounds heavier than the subsequent calves from the low group, repeatability would have been 1.00. This means the breeder could have culled down to the number desired to keep after the first calf crop was weaned and be sure correct selection decisions were made. If both subsequent groups of calves had averaged the same, repeatability would be 0; the breeder could not have culled on the basis of weaning weight of the first calf with any degree of confidence. The repeatability value of 0.46 indicates that the extremely low-

TABLE 19–3	Approximate Repeatability of Productive Traits in Livestock		
Trait	**Beef Cattle**	**Sheep**	**Swine**
Prolificacy		0.10	
Birth weight of offspring	0.30	0.30	
Reproduction efficiency	0.10	0.20	0.10–0.16
Weaning weight of offspring	0.45	0.40	
Grade of offspring at weaning	0.22		
Annual wool production		0.50–0.60	

Source: *Compiled by the authors and colleagues from several sources.*

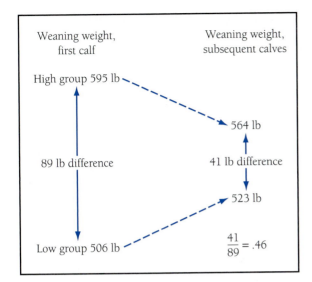

Figure 19–9
How repeatability of weaning weight is calculated.

weight heifers can be culled after one season, but the breeder should wait until the second year for those which might be "borderline."

Repeatabilities of reproducing interval, tendency to settle at first service, tendency for heifers to calve once every 12 months, and related traits usually are high enough that when a heifer, ewe, or gilt does not settle the first breeding season, she probably should be sold for slaughter. The cost of maintaining a heifer or ewe for a year is too great and the chances that she will conceive later are too poor. With multiple or continuous farrowing, gilts needn't be maintained a year, but a gilt that won't settle after two or three services to a fertile boar probably should be culled.

Repeatabilities generally are calculated in such a way that all effects due to permanent differences among cows (or ewes or sows) are included. Because some of the permanent differences among females may be environmental in origin, repeatability will be as high as or higher than heritability.

QUESTIONS FOR STUDY AND DISCUSSION

1. List four desirable traits in breeding animals that contribute to increased flock efficiency and productivity.

2. Why might a special emphasis be placed on certain traits in the selection of the herd or flock sire that are quite different than those traits in the present breeding flock or herd?

3. List the more common methods of animal identification that are suitable for use in record keeping, culling, and selection.

4. Explain the ear-notching system for swine. Which ear denotes the litter number? What do the notches in the opposite ear signify? Sketch an outline of pig number 7 in litter number 12.

5. Select either sheep, cattle, or swine and make a list of the more important information that should be recorded and transferred to permanent herd or flock records.

6. Briefly explain how an ultrasonic instrument provides a means of measuring backfat and loin area in the live animal.

7. In the selection of sires, what is meant by (a) fertilizing potential of semen and (b) serving capacity?

8. Where should emphasis be placed on conformation traits when selecting breeding animals?

9. What traits might be included in an index to be used in selecting beef bulls, rams, or boars for breeding stock?

10. Compare the methods of production, performance, and progeny testing. Which should be the best indicator of an animal's breeding value? Why? What is the greatest limitation in the use of progeny testing?

11. What does the term *repeatability* mean?

20
Mating Systems

The three preceding chapters discussed the fundamentals of gene inheritance and their effects and emphasized three of four rules for improvement by selection. These three rules are:

Rule 1: *Have maximum genetic variation.*

Rule 2: *Spend selection efforts on traits largely influenced by heredity.*

Rule 3: *Observe or measure accurately the traits carried by a prospective breeding animal.*

This chapter focuses on the fourth and final rule:

Rule 4: *Use the selected animal or animals most effectively.*

Genetic variation among animals permits the breeder to select the top-performance animals for parents of the next generation. Once the breeding animals are selected, the mating system or breeding program should make maximum use of their traits, or desirable genes. The old adage "it's in the genes" is certainly true. In the twenty-first century, the most useful tool to predict or understand genes and function is through the use of molecular biology. Molecular tools allow the scientist to more accurately determine the overall expression level of a specific gene(s) and its influence on other related genes.

Mating systems followed by successful livestock and poultry breeders (random mating, inbreeding, outbreeding, crossbreeding) vary depending on species, the breeders' goals, and other factors. Various types of livestock mating systems are described in this chapter, and the situations in which each might be used effectively are listed. Poultry mating systems are based on the same principles but are covered in more detail in Chapters 33 and 35.

In studying each mating system, the reader should think about the following:

1. What is the purpose(s) for using this mating system?
2. Does it increase homozygosity or heterozygosity of the genes?
3. Does it increase or decrease hybrid vigor—heterosis?
4. When is it best used by the breeder?

Upon completion of this chapter, the reader should be able to

1. Define *random mating, inbreeding, outbreeding,* and *heterosis.*

2. Explain why inbreeding is used in the development of breeding lines or strains and not generally used in commercial livestock and poultry production.

3. List those groups of traits that tend to show high heterosis when breeds are crossed and those that tend to show low heterosis.

4. Describe or chart a rotational crossbreeding system and explain why such is commonly used in commercial swine herds.

5. Describe or chart a three-breed terminal cross system and explain why such may be used in some commercial cattle herds.

6. Contrast using heterosis versus only intense selection and random mating of selected animals as systems for achieving improvement in a commercial species.

20.1   RANDOM MATING

> **Random mating**
> Allowing selected animals to mate at random.

The term **random mating** doesn't imply mating without selection. It refers to the way that a selected number of males are mated to a selected number or group of females within a herd or flock.

Random mating means that mating of selected breeding stock is not controlled. Selected rams and ewes are kept together and mate at random. All bulls run with all the heifers, boars mate with gilts at random, or all roosters and hens are combined, so that each male has an equal opportunity to mate with each female. This is practical if the breeder has no knowledge that certain controlled matings might produce better offspring. Less labor is required and an entire herd or flock can be handled as a single unit during the breeding season.

Obviously, random mating cannot be used in purebred herds and flocks with more than one sire where breed registration is required, because both the sire and the dam of animals must be known and recorded. Relationship of animals cannot be determined when random mating is practiced; thus, it is commonly used in commercial herds.

Random mating would not be followed if it were apparent that much could be gained by specific matings. A commercial Angus raiser in Missouri, for example, may raise feeder calves for Corn Belt feedlots, but also will select and keep a certain number of the heifers for replacements. The goal is to improve herd quality, thus the breeder would probably breed the best bull to his or her best cows to get top-notch replacement heifers. Resulting heifers probably would be noticeably better than heifers that could have been selected under random mating. Some of the bull calves born from these matings might be good enough to sell to another breeder or, if castrated, bring top price as feeder steers.

20.2   INBREEDING

> **Inbreeding**
> Mating animals that are related. Varies in degree, depending on degree of relationship.

Inbreeding is defined as mating relatives within a breed. The more closely the sire and dam are related, the greater the degree of inbreeding. Many species of plants can be self-pollinated, giving the highest degree of inbreeding possible. Inbreeding cannot be that extreme in farm animals; maximum inbreeding in a herd or flock would be accomplished by continued brother-sister matings in successive generations. Even this is seldom practiced, however.

Nearly all purebred animals are inbred to some degree. Because matings are confined to registered animals within a breed that usually have some common ancestry, any two animals mated would be more closely related than two animals selected at random within the species. A purebred Duroc boar, for example, has some ancestors in common with any purebred Duroc sow to which he might be mated. If the common ancestors were only one or two generations earlier, the degree of inbreeding would be relatively high. If there were no common ancestors closer than four to six generations, the degree of inbreeding would be insignificant.

From a genetic standpoint, relatives are more likely to have the same genes (or more homozygous pairs) than animals selected at random within the species. Herein lies the explanation of the benefits and also the disadvantages or risks of inbreeding.

Livestock raisers usually inbreed to concentrate the good genes known to be present in a superior animal or family into certain future offspring. Such offspring are therefore more likely to *be* top-quality and also to beget top-quality offspring. They carry many good genes, so they probably will transmit many good genes.

The *theoretical* goal of inbreeding is to finally develop a herd or flock of animals that carry only the good genes and breed true for all traits in subsequent generations. A more practical and realistic goal is the development of inbred lines or strains that will breed true for certain traits. Two lines, or strains, might be developed for different traits, and then animals from each line mated to give offspring with both sets of traits. (See Section 20.4.)

Inbreeding is not recommended for most farms and ranches, particularly in initial herd or flock establishment because of certain risks and disadvantages. Just as mating relatives tends to concentrate good genes in individual offspring, it also tends to concentrate bad genes that might be present in the line (Figure 20–1)

Such bad genes may not all be as detrimental as the dwarf gene, but highly inbred animals are *less vigorous* and *more difficult to raise.* They are often smaller in size and more susceptible to diseases and other environmental stresses. Also, there are

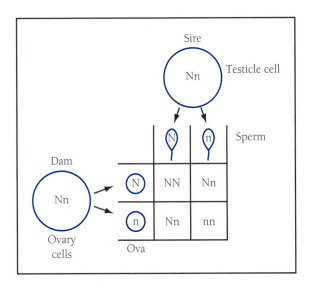

Figure 20–1
Genotype of possible offspring that might result from mating two cattle that carry the dwarf gene (*n*). If the cow and bull are related, they are more likely to have the same genotype. They could be brother and sister. Note that such mating concentrates the good genes (*NN*) in one-fourth of the offspring and the bad genes (*nn*) in one-fourth.

specific recessive genes that can cause death or abnormalities when one gene is inherited from each normal-appearing parent. Rigid culling and selection is essential in minimizing or preventing these adverse effects of inbreeding.

Because of the risk involved and the difficulty of raising animals of inbred lines, few producers can profitably develop inbred lines for use on their own farms and ranches or in poultry production units. Only those breeders who can afford the risk, because of larger size or probability of selling many breeding animals from a relatively pure line, find it practical to inbreed. In corn and poultry, individual companies have developed many inbred lines; a few companies have developed inbred lines of hogs. With cattle and sheep, however, development is so slow because of low prolificacy that most inbred lines are developed by state agricultural experiment stations and/or the USDA.

It is apparent that inbreeding can be worthwhile with constant selection when the goal is eventual development of an inbred line. Assume that the bull and four *Nn* cows (Figure 20–1) are considered the original stock for development of an inbred line. Half of the original genes are *n*. One generation later, after the dwarf is culled, one-third of the genes are *n*. Continued mating within this line and continued culling of dwarf calves would further reduce the proportion of *n* genes, and the line eventually would be relatively true breeding for the *N* gene, having very few *n* genes present.

It is also apparent that inbreeding can result in more genetic variation within a group of animals. So culling, and selection, can be easier and more effective.

A sheep raiser with 25 ewes, for example, may have 5 excellent ewe lambs selected for use as replacements in the flock. All were sired by an excellent ram. The ewe lambs inherited half of their genes from the ram, so their genotypes are probably relatively similar. Because the ram and ewe lambs show excellent traits—gains, wool growth, meaty conformation—they probably both carry many genes that promote these traits.

If the raiser keeps the same ram and mates him to his own daughters, chances are high that the resulting offspring will be very high-quality animals. There is considerable evidence that the ram and his first daughters both had excellent genotype, thus chances are very high that, if mated, their offspring will inherit many of these good genes.

Once a sire is proved to be desirable, continued use of him and/or his close relatives tends to either maintain or concentrate his genes into the line of breeding.

> **Linebreeding**
> A form of inbreeding to concentrate the genes of a particular individual into the pedigree.

Linebreeding is a form of inbreeding in which animals are mated so their offspring retain a high relationship to an outstanding individual. It increases homozygosity but through less intense inbreeding. Rather than sire-to-daughter mating as mentioned above, the mating of grandsire to granddaughter would be more common in linebreeding.

20.3   OUTBREEDING

> **Outbreeding**
> Mating animals relatively unrelated, usually with diverse type or production traits. Varies in degree, depending on the degree of divergence in type or production traits.

The opposite of inbreeding is **outbreeding**. Broadly speaking, it could include "crossbreeding"—the mating of animals of different breeds—as well as "species crossing"—the mating of animals of different species. However, the more common form of outbreeding is "outcrossing."

In outcrossing, animals to be mated are less related than the average of the breed, generally unrelated for four to six generations. This system typically is used by a breeder for the sole purpose of introducing a desired trait (actually genes for the trait) into the line or herd. Outcrossing may be a temporary interruption in an inbreeding program, practiced when the breeder feels the line (or strain) being developed lacks in a certain trait and that an outcross with another line would provide the desired trait.

A Hereford breeder, for example, who has developed a line of cattle with good traits except that they are below average in a particular trait, may seek a bull from another line of Herefords, noted for strength in that specific trait. Thus, outcrossing can be used to correct trait weaknesses intensified through inbreeding.

Outcrossing also provides an opportunity for a moderate amount of hybrid vigor within a breed; however, it also results in heterozygosity, or gene impurity, and tends to conceal recessive genes. Therefore, once the herd or flock has attained a desired trait level, the breeder should initiate an inbreeding program, which should increase gene purity.

20.4   CROSSBREEDING

Crossbreeding is defined as the mating of animals of two different breeds. As practiced in the United States, it usually includes the breeding of selected purebred sires to females of another breed or crossbred females of two or more breeds.

> **Crossbreeding**
> Mating animals of different breeds. A distinct type of outbreeding.

Crossbred animals—poultry, pigs, sheep, or cattle—tend to be more vigorous at and soon after birth and more resistant to environmental stresses, so they tend to gain faster, and are more prolific when it is their time to reproduce, than the average of their parents. These benefits of crossbreeding are often referred to as **heterosis**. In general, this type of breeding program would be the preferred choice of large integrated livestock and poultry complexes.

> **Heterosis**
> The tendency of the offspring of a cross to perform better than the average of their parents.

Swine research at Iowa State several decades ago provides an excellent example of the beneficial effects of crossbreeding (Table 20–1). Some gilts of each breed were mated to boars of each of the breeds shown. Data for the offspring of these matings are summarized after "cross." "Pure" pigs had sires and dams of the single breed.

Why are crossbred animals usually more vigorous, faster gainers, more prolific? The discipline of genetics may provide some explanation.

TABLE 20–1	Influence of Crossbreeding on Weight at Eight Weeks and at Five Months				

| | Litter Size | | Pure or | Weight (lb) | |
Breed	Birth	Weaned	Cross	8 Weeks	5 Months
Berkshire	8.1	6.1	pure	29.9	124
			cross	34.8	157
Duroc	10.3	6.7	pure	36.6	159
			cross	35.8	159
Hampshire	8.7	6.6	pure	31.1	144
			cross	39.6	166
Landrace	8.2	6.3	pure	33.2	150
			cross	38.0	176
Poland	8.0	6.3	pure	37.3	165
			cross	38.4	173
Tanworth	8.9	7.0	pure	31.8	134
			cross	36.5	155
Yorkshire	11.9	10.5	pure	35.2	143
			cross	38.2	150

Source: *Hazel, L. N., Iowa State University, 1961.*

	Breed A	Breed B	Crossbred Offspring
Number of pigs born per litter	16 (AA)	4 (aa)	16 (Aa)
Percent surviving after 10 days	25% (bb)	100% (BB)	100% (Bb)
Number of live pigs at 10 days	4	4	16

Figure 20–2

An extreme and oversimplified illustration of the genetic basis for heterosis in hogs. Genotypes are given in parentheses.

It was previously mentioned that dominance tends to help livestock and poultry production. Good genes are more likely to be dominant genes; bad genes are more likely to be recessive genes. Figure 20–2 shows heterosis that might result when two hypothetical breeds of hogs are crossed. The example is extremely oversimplified, compared to a true situation. The two traits—number of pigs born per litter and percentage surviving at 10 days—for the purpose of illustration, are each presumed to be influenced by one pair of genes. The two breeds have been developed, over many generations, by selection for different traits. Breed A has been selected for number of pigs born per litter and is now carrying only the dominant good genes (*AA*) for this trait. No selection has been exerted for survival in breed A, and assume, for now, that the genes carried are recessive. In breed B, the exact opposite is true. Selection for survival (*BB*) has been successful, but number of pigs born per litter is small. Take a look at the genotype and phenotype of the offspring.

In this example, the offspring not only performed better than the average of the parents, they performed much better than either parent. It is emphasized that this is an extremely oversimplified example. First, many pairs of genes influence each of the traits. Second, variations in environment do not allow performance to be completely controlled by genotype, especially in prolificacy and survivability. Third, it is unlikely that either breed could have been inbred and developed to the point of carrying only good, dominant genes. The development of inbred lines of livestock relatively pure in certain genes is slow, especially with cattle and sheep. The more gene pairs that are involved, the slower it would be. Fourth, it is improbable that each breed would carry only bad, recessive genes influencing the trait that was not selected.

It is important to note that crossbreeding, through heterosis, usually results in improvement of traits of low heritability, such as increased survivability, faster gains, and increased fertility. *Remember that new genes are not created, only manipulated in mating systems, and crossbreeding must be continued to maintain the benefits of heterosis.* (See Sections 20.6 and 20.7.) Figure 20–3 also illustrates a typical heterosis effect.

Research by USDA scientists in Nebraska, Montana, and Virginia, and by scientists at many state agricultural experiment stations, has shown effects of heterosis in crosses involving many breeds of beef cattle. Table 20–2 demonstrates heterosis effects among crossbred calves, from the time of conception through the first calving of the heifers born. In this study, the Hereford and Angus cows were purchased from commercial breeders and were bred by artificial insemination to bulls selected for performance in progeny testing programs.

The crossbred animals include about half carried by Hereford cows and sired by Angus bulls, and half carried by Angus cows and sired by Hereford bulls. The last row of the table, "Heterosis," shows how the performance of the crossbred animals differed from the average of the straightbred animals of the two breeds.

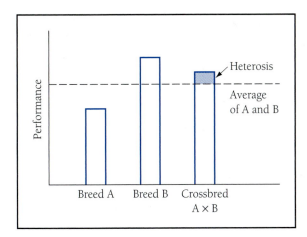

Figure 20–3
Crossbred animals tend to perform better, in certain traits, than the average of their parents. Heterosis is that amount by which performance in the crossbreed exceeds the parental average.

Note that crossbred calves, on the average (1) were heavier at birth, (2) were born with less difficulty, (3) had a lower mortality rate, (4) and were heavier at weaning. Puberty in the heifer calves was determined by detection of estrus. Note that the crossbred heifers (5) reached puberty at a significantly earlier age, and a significantly larger proportion produced a calf. Also note again that all of these improved traits are low in heritability.

Male calves were fed as steers for slaughter. Carcass, meat quality, and palatability data were gathered, and showed little heterosis. As described elsewhere in this book, those traits associated with growth and vigor exhibit higher heterosis; those associated with structure and body composition exhibit little.

Crossbreeding research with dairy cattle in the United States and Britain has shown less death loss and faster growth in crossbred calves, and a shorter calving interval (better conception) for crossbred cows. Some crosses have shown a heterotic effect on milk production, at least in the first lactation. There is limited use of crossbreeding in dairy enterprises, however, largely because of the predominance of the Holstein breed.

Heterosis has been repeatedly demonstrated in sheep. Crossbred lambs are (1) usually larger and (2) more vigorous at birth, and (3) gain more rapidly. Note, in Table 20–3, that in all three traits compared, crossbred lambs performed better than purebred lambs in four of five cases. Crossbred ewes also usually perform at a level higher than the average of their parents' lines in wool per ewe, as well as in survival and weaning weight of offspring.

Crossbreeding may also be done to introduce new traits from some other breed into a commercial herd or flock. An example is the typical ewe flock in the Rocky Mountain area. Most ewe flocks originally were of relatively pure Rambouillet or Merino ancestry. Both are "fine wool" breeds and the Rambouillet was developed primarily from the Merino, so they are similar. Being fine-wool breeds, they have been developed with strong selection for heavy production of high-quality, fine-diameter wool. The high value of wool justified intense selection for wool. Because the rangy Rambouillet ewes were more adapted to the rough terrain, most flocks had little evidence of "meat-type" conformation.

In time, wool prices declined. Lambs not needed as flock replacements, formerly considered a byproduct of wool production, became relatively more important. Because lamb feeders prefer blockier lambs that will produce a higher proportion of meaty cuts than will typical fine-wool sheep, it became apparent that such lambs would bring higher prices if they carried some meat-type breeding.

TABLE 20–2

Performance of Straightbred and Crossbred Animals from Conception Through Weaning and Reproduction of Heifers[a]

Breed	Conception to Weaning							Heifers			
	Number	Gestation Length (days)	Birth Weight (lb)	Calving Difficulty (%)	Preweaning Mortality (%)	Weaning Weight (lb)	Number	Age at Puberty (days)	550-Day Weight (lb)	Percent that Calved
Hereford	141	285.5	76.5	18	8.6	472	62	415	673	78.2
Angus	166	281.6	68.1	12	8.6	496	64	366	673	82.3
Average of Hereford and Angus		283.55	72.3	15	8.6	484		390.5	673	80.25
Crossbred	375	282.9	74.3	11	3.8	507	132	371	715	93.0
Heterosis		−0.6	+2.0	−4	−5.8	+22		−19.5	+43	+12.75

[a]These data are from an extensive study involving the U.S. Meat Animal Research Center at Clay Center, Nebraska, the University of Nebraska, and Kansas State University.

Source: Adapted form USDA, ARS, Evaluating Germ Plasm for Beef Production, Cycle 1, Progress Report No. 3, U.S. Meat Animal Research Center, Clay Center, Neb., 1976.

TABLE 20–3	**Purebred and Crossbred Lambs Compared**					
	Lambs Weaned per 100 Ewes Bred		Birth Weight of Lambs (kg)		Average Daily Gains of Lambs (kg)	
Breed	Purebred Matings	Crossbred Matings, by Sire Breed	Purebred Lambs	Crossbred Lambs, by Sire Breed	Purebred Lambs	Crossbred Lambs, by Sire Breed
Hampshire	87.8	98.3	4.75	4.89	0.259	0.272
Suffolk	79.1	97.0	5.30	4.97	0.297	0.285
Dorset	55.2	78.2	3.88	4.19	0.223	0.251
Targhee	102.2	91.4	4.82	4.85	0.231	0.252
CSC[a]	99.7	104.9	4.12	4.49	0.219	0.254

[a]CSC is a strain that resulted from Columbia, Southdown, and Corriedale stock.

Source: Sidwell, G. M., and Miller, L. R., J. Anim. Sci. 32:1084, 1090, 1971.

Flock owners gradually began using more muscular rams—Hampshire, Suffolk, and others—on many of their ewes. The fine-wool ewes still produced top-quality fleeces, but also produced feeder lambs more desirable for feeding. Enough ewes were bred to fine-wool rams to provide needed flock replacements. The genes for meatiness and muscling simply were not present in the fine-wool breeds to the degree needed.

Selective crossbreeding can result in the two breeds complementing each other to fill the void in each breed. This is referred to as *complementarity*. In the above example, the crossing of a fine-wool breed with a meat-type breed (e.g., Rambouillet × Hampshire) resulted in an improvement of these traits in the offspring.

A Midwest farmer with a herd of beef cows occasionally may use a sire of a breed developed for both meat and milk to provide genes for increased milk-producing ability in the future heifer replacements.

Hog producers might use a Duroc boar to increase prolificacy in later generations, a Landrace boar to increase length, or a Spotted boar to increase muscling. Other breeds might also be used to provide these or other desired traits.

Crossbreeding between or among breeds naturally causes mixed and nonuniform color patterns among offspring. This is of no consequence for meat, milk, wool, or egg production.

Remember that crossbreeding improves many of the traits that have low heritability. However, both good management and inclusion of high-quality animals of all breeds in the crossbreeding system are essential to be successful.

20.5   CROSSING LINES OR STRAINS

Most broilers and egg-laying stock in commercial poultry units result from crosses of two or more inbred lines or strains, the lines or strains being maintained by one of a small number of breeder hatcheries. The birds in the commercial production unit are usually a *line cross* or a *strain cross* and are identified by a copyrighted name and/or number. They are extremely uniform in color, growth rate, mature size, egg production, and/or carcass quality. This uniformity is achieved by using as parents those inbred lines or strains that consistently have produced uniform and productive birds in pretests. Because poultry are highly prolific, such pretesting is feasible and worthwhile.

> **Line**
> A group of animals descended from or related to a specific animal or source of genetic stock.

The term **line** usually refers to an inbred line as described in earlier sections. **Strain** generally refers to a family within a breed, perhaps a flock that is *relatively less related* to the rest of the breed. They may have a modest degree of inbreeding. The terms are sometimes used interchangeably.

> **Strain**
> A group of animals within a breed with characteristics that distinguish them from others in the breed.

Primary poultry breeders generally maintain many lines or strains and, in their quest for crosses that will be popular with franchise dealers and producers, make many test crosses to determine which crosses perform best. Some lines are found to combine well with many other lines, so are said to have general combining ability. Others may have specific combining ability—combine well with only one or two other lines.

In such testing programs, *reciprocal crosses* usually are made. Lines A and B are crossed two ways, the first with line A as the male and the second with line B as the male. Performance of offspring from the two kinds of crosses may vary considerably, especially in the females (Figure 20–4).

This may be explained by the sex chromosome and the genes it carries. (See Section 17.4.) Females of an A × B cross (male is A) would carry a functional sex chromosome only from line A. Females of a B × A cross would carry a functional sex chromosome only from line B.

In Figure 20–4, lines of the White Leghorn (WL) and New Hampshire Red (NHR) were used in crossbreeding. The numbers indicate the eggs per 100 hens during the laying season of the test. Note that heterotic effect is greater where the White Leghorn line was used as the male parent.

20.6   ROTATIONAL CROSSBREEDING PROGRAMS

In commercial production of "single-cross" hybrid chickens, inbred lines are maintained for *both* parents. This is practical because an inbred hen can beget 250 to 275 chicks per year. In addition, it costs relatively little to produce and maintain a pullet, relative to the cost of producing and maintaining a gilt or cow.

But a heifer of breeding age is expensive. She can produce only one calf per year. Gilts and ewes are less expensive to raise and more prolific, but still strikingly different from chickens.

It is obvious that it is financially impractical to maintain inbred lines of livestock as a source of female breeding stock on farms and ranches. Much time and money

		Female Parent	
		NHR-53.1	WL-61.4
Male Parent	NHR-53.1		62.5
	WL-61.4	66.5	

Figure 20–4
Percent egg production in parent lines and reciprocal crosses.
(*Source:* Hyre, H. M., et al. *W. Va. Agr. Exp. Sta. Bull.* 479.)

are involved raising them. Only half of the offspring in the line would be females. Remember also that inbred livestock may be more difficult to raise.

In livestock production, much of the heterosis of crossbreeding can be achieved more economically by rotational crossbreeding systems. In livestock production, females are raised in the herd. Only sires are purchased. Because more crossbreeding is done in swine, they will be used as an example (Figure 20–5).

The first boar purchased for a rotational crossbreeding program should be a different breed, and should carry additional good genes for certain traits that need to be improved in the herd. (See Section 20.4.) The first crop of crossbred pigs will tend to carry the good traits of both breeds and will be more vigorous, if the cross is successful in terms of heterosis. More will probably survive and they will gain faster. These pigs, called the F_1 *generation,* will have a consistent, though relatively small, advantage over purebred pigs. When the crossbred gilts reach sexual maturity, they will be more prolific and will give more milk. This is the big advantage to rotational crossbreeding as compared to maintaining inbred lines for female stock.

In the rotational breeding program shown in the Figure 20–5 example, the crossbred gilts (one-half Yorkshire and one-half Duroc) are then mated to a purchased boar unrelated to them. This boar (a Spotted) is selected not only to cause heterosis but also to supply some new or additional good genes for certain traits in which the gilts are

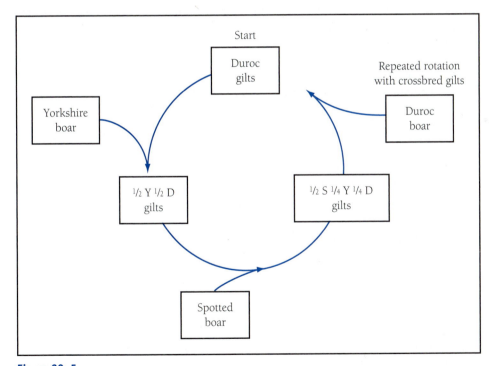

Figure 20–5
An example of a rotational breeding program in a swine herd, starting with gilts of relatively pure breeding (Duroc), then rotating boars (Yorkshire first, Spotted second, and Duroc third) on selected crossbred females in successive generations.
(*Source:* Iowa State University.)

lacking. The offspring of these crossbred gilts and the second boar breed (Spotted) generally will have about a 30 percent advantage over purebred pigs in those traits that show heterosis effect. Female offspring of this cross, to be saved for subsequent mating back to a Duroc boar, are one-half Spotted, one-fourth Duroc, and one-fourth Yorkshire.

Gilts mated to the Duroc boar contain only one-fourth Duroc breeding, so heterosis would still be expected. A three-breed rotational crossbreeding system maintains heterosis at about 86 percent of maximum, 20 percent higher than is possible by rotating only two breeds.

Most who follow such a rotational breeding program in their swine herds employ only three breeds. The gain in heterosis by using a fourth breed is small, about 7 percentage points over the three-breed cross or 93 percent of the maximum, and breeders often find it difficult to locate good boars, with the traits they want, of four different breeds in their area. It is better to use three good boars in a rotational program and lose a trace of heterosis than to use three good boars and one poor boar.

A program that is becoming more popular in commercial beef production involves the combination of a two-breed *rotation* and a *terminal cross* system. In a two-breed terminal cross, all offspring benefit from maximum heterosis but all go to market. The disadvantage of this system is that all F_1's are marketed, and replacement females must be purchased.

The more popular *rotaterminal* cross system combines the benefits and eliminates the drawbacks of the rotational and terminal systems. The younger one-half of the cow herd (15 through 36 months of age) is mated in a two-breed rotation to produce replacement heifers. The older cows are mated to a terminal breed to capitalize on growth and carcass merit in the three-breed cross calves. This system is practical only in large herds because it requires skilled management in the use of three breeds of sires and several separate pastures with natural service.

Although most swine growers utilize the three-breed rotational cross, some have adopted a terminal or rotaterminal system. In the latter system, most use three white breeds to produce replacement gilts on the farm. Then each crossbred female is mated to the least-related white breed of boar to maintain high heterosis. About 20 percent of the herd is used in the gilt production program and 80 percent in the terminal market hog program (Figure 20–6).

In addition to maintaining heterosis, this rotaterminal system takes advantage of the sow productivity (prolificacy and milking ability) of the white breeds (e.g., Landrace, Chester, Yorkshire, or Large White). Then crossbred females are crossed with a breed that promotes growth, muscling, and leanness on the sire side (e.g., Hampshire, Duroc, or Spotted) in the production of market pigs.

Crossbreeding for heterosis has been routine since the early 1950s in market hog production (Figure 20–7) and rotational crossbreeding has long been practiced in sheep flocks (Figure 20–8) for both heterosis and production of better feeder lambs. It is also a common practice in beef cattle (Figure 20–9), though implementation came later in beef cattle.

Table 20–4 illustrates the benefit of the two-breed and three-breed rotations, using data from a Nebraska study involving Hereford, Shorthorn, and Angus, used in nearly all combinations.

20.7   THE FUTURE—HETEROSIS VERSUS SELECTION

Heterosis, though influencing the whole animal, is expressed in a limited number of traits, mostly those related to prolificacy, survivability, and rate of gain.

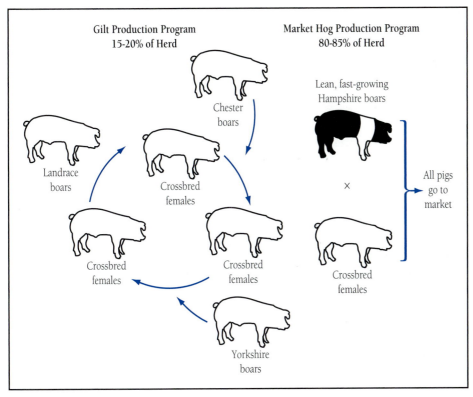

Figure 20–6
A sketch of a rotaterminal crossbreeding system, utilizing three white breeds.
(*Source:* Adapted from *Pork Industry Handbook,* PIH 39 and PIH 106.)

Figure 20–7
Most groups of feedlot hogs are crossbred and so contain a variety of color patterns.
(Courtesy of Cargill, Inc.)

Figure 20–8
Crossbred yearling ewes resulting from purebred Suffolk ewes sired by a purebred Montadale ram.
(Courtesy of Montadale Sheep Breeders' Association, Inc.)

Figure 20–9
A Chianina bull and Angus cows in a crossbreeding program on a Texas ranch.
(Courtesy of American Chianina Association.)

TABLE 20–4	**Effects of Heterosis in Beef Cattle Rotational Crossbreeding Systems**		
	Control	**Two-Breed Rotation**	**Three-Breed Rotation**
Cow	Straightbred	F_1 cross	F_1 cross
Calf	Straightbred	Backcross	3-breed cross
Number of matings	431	410	211
Calf crop weaned (%)	75	79	83
200-day weight (kg)	196	216	221
Weight weaned per cow (kg)[a]	147	171	183
Increase in weight weaned per cow (%)[a]	—	16	24

[a]*Per cow exposed to bull for breeding.*

Source: *Gregory, K. E. and Cundiff, L. V., J. Anim. Sci. 51:1224, 1980.*

Breeders continue to select for prolificacy, survivability, and rate of gain. Cross-breeding is employed only to gain the greatest possible use of selected animals. If livestock and poultry breeders could ever reach, by selection, the optimum level of prolificacy, survivability, and other traits enhanced by heterosis, then the benefit of crossbreeding for the purpose of attaining heterosis would drop.

Some sheep ranchers in the dry range country do not appreciate twins because ewes are not able to graze enough forage most years to provide sufficient milk. A single lamb has a distinct advantage. So crossbreeding for the sake of increasing prolificacy is not always advantageous.

Perhaps improved shelter and management—environmental control—will raise prolificacy, survivability, and rate of gain to an optimum point. Perhaps some of our herds and flocks now have the genetic potential for optimum production if environment were optimum.

It is unlikely that selection will be effective enough in the near future to eliminate the benefits of crossbreeding in livestock. Poultry breeders still are crossbreeding (and linecrossing), and they continue to be ahead of livestock breeders in selection achievement. For sheep, swine, and beef cattle, the hope of reaching an optimum level of production by selection alone seems more remote. However, this statement doesn't minimize the importance of selection. Rigid selection is essential for genetic improvement.

20.8   THE ANIMAL IMPROVEMENT INDUSTRY—LIVESTOCK AND POULTRY BREEDERS

A livestock breeder is one who is in the business of raising and selling breeding stock, male or female—even hatching eggs in the case of poultry—to commercial flock and herd owners. A breeder is a specialist in animal selection. The successful breeder has top-quality breeding stock as the foundation of the herd or flock and practices continued and rigid selection to maintain and improve production traits in his or her animals. A good breeder also makes carefully planned matings to achieve maximum value from the good sire and dam.

Whether breeders operate a 220-acre farm in southern Indiana, a 10,000-acre ranch in Montana, or a breeder hatchery in California, they are analogous to a supplier of "ingredients" in the feed industry, because they provide breeding stock for the commercial grower, feeder, egg producer, or dairy producer. That is, they supply the "seed stock" for commercial herds and flocks, just as hybrid corn companies supply seed for commercial corn growing.

Most livestock breeders raise and sell registered purebreds—animals that are registered by one of the established breed organizations. A few suppliers of breeding stock, especially of hogs, have developed inbred lines, each developed for certain recommended mating systems. As an example, white lines have been developed from the white swine breeds—Chester White, Large White, Landrace, and Yorkshire—that are used more commonly in providing maternal lines. These companies usually sell hybrids (crosses) that have been developed by crossing two or more lines, or lines that will be crossed to produce hybrids.

Livestock and poultry breeders exert great influence on the quality of future animals. Traits passed from one generation to the next by inheritance are considered permanent. They can be eliminated only by selection: culling the animals that carry the traits. Breeders are obliged, therefore, to exert much effort to fix desired traits in their breed or line.

Many livestock breeders are extremely loyal to their breed, though more are now providing seed stock from several breeds. Many are active members of their state and national breed associations and participate in state, regional, and national shows, competing for prizes, advertising their herds or flocks, and promoting the breed.

Most breed organizations have almost identical goals. Secretaries and field personnel are employed to promote the breed, handle registration of animals, and help members with production problems, sales, and purchases. The main function of breed organizations, however, is to improve quality and production traits within the breed. Within species, breed organizations often work together in developing uniform standards for use in selection and improvement programs. Effective sire evaluation programs have been developed and included in numerous breed improvement programs throughout the United States.

Most U.S. poultry breeding stock is developed by less than a dozen very specialized breeders. Broilers, turkey poults, and pullets generally are not identified as "breeds," but rather carry the identification of the breeder, a copyrighted name, and/or a number. Poultry breeds, per se, lost their significance in the commercial poultry industry with development of specialized breeder hatcheries during the 1940s.

QUESTIONS FOR STUDY AND DISCUSSION

1. Do mating systems create new genes or manipulate existing genes of the selected animals? Explain.

2. Under what conditions might you see "random mating" being practiced?

3. Define and compare inbreeding and outbreeding systems within a breed. Which should result in the most (a) gene purity, (b) heterosis, and (c) expression of traits influenced by recessive gene pairs?

4. When might intense inbreeding be recommended? What risks are involved?

5. When might a breeder who is utilizing an inbreeding program benefit from temporarily switching to an outbreeding program?

6. Define *crossbreeding* and list its principal advantages in the production of meat animals.

7. What traits can be greatly improved through crossbreeding? Do these traits have low or high heritability?

8. Define *heterosis* and provide a simple explanation of how it occurs.

9. How have the poultry breeders used mating systems to develop desired lines or strains?

10. Sketch a rotational crossbreeding program that will maintain adequate heterosis in the breeding herd or flock, and in meat animals intended for market.

21
Breeds

A **breed** of livestock or poultry is a group of animals that results from breeding and selection and that have distinguishable characteristics. In most cases, selection has been practiced for many generations in order to fix certain characteristics in the breed.

Some breeds that once existed are, for all practical purposes, extinct today. They may have reached a certain stage of development, then were crowded out by new and better animals. Or, as is known in many cases, several former breeds may have supplied foundation stock for development of a new breed.

Today, most of the red meat in the American diet is derived from high-quality crossbred slaughter steers and heifers, lambs, and hogs.[1] As discussed in the previous chapter, best results are obtained from crossbreeding only when the animals of each breed are genetically superior. Therefore, maintenance and improvement within breeds are very important. Genetic diversity of the numerous breeds provides greater opportunity for improvement in the economically important traits. Dedicated purebred breeders are essential in maintaining the purity and identity of the breeds. They deserve much of the credit for the efficient production of high-quality meat, milk, and eggs.

Only since the middle and late 1800s has breed registry been closed to animals whose parents were not registered. Before registry was closed, genes from other breeds or strains often were introduced into a particular breed in this or other countries by certain matings, in an attempt to improve certain traits of the breed. This was a practical thing to do because livestock raisers were more concerned with maximum and efficient production than with extremely uniform color markings.

Upon completion of this chapter, the reader should be able to

1. Explain why there are different breeds within species.
2. Locate on a chronology chart the century or approximate decade his or her favorite breeds of sheep, cattle, and swine achieved identity.
3. Identify in which decades most livestock breeds that originated elsewhere were first imported to the United States.
4. Explain what is meant by an "exotic" breed.
5. List five major breeds of swine, beef cattle, and sheep (breeds of horses and ponies are listed, pictured, and discussed in Chapter 32).
6. Explain the key points in breed selection.

---

[1] Crossbred animals are the offspring from the mating of animals from different breeds. (See Section 20.4.)

21.1   WHY BREEDS HAVE DEVELOPED

Primarily, breeds have been developed to provide increased production of meat, milk, eggs, and wool within a particular environmental area.

Livestock and poultry producers through the ages have observed that some animals were more adapted to certain areas or production situations than others. A sheep raiser in the rough, mountainous land of Scotland in the sixteenth or seventeenth century, for example, may have observed that certain sheep in his flock were more adapted to the rocky terrain. They were more surefooted than others, would graze faster and closer, and therefore could produce more wool. He and other sheep raisers of the area naturally saved offspring of these particular sheep and not of the others, so in time there may have developed a sheep breed with excellent wool production in that particular environment.

At the same time, a sheep raiser in some fertile valley in England, where nutritious grass and root crops grow well, may have observed that certain of her rams and ewes consistently produced lambs that fattened and were ready for market sooner than lambs from others. They apparently were able to make better use of the feeds produced in this area, to produce meat and at the same time produce a good fleece. She naturally developed her flock by selecting replacements from her best ewes and rams, and eventually she and her neighbors, or their descendants, may have developed a breed consistently excellent in meat production under those particular conditions.

Some breeds of chickens, especially in early England, were developed for game purposes. In Italy, Spain, and other Mediterranean areas, eggs were considered important dietary components, so breeds with good egg production evolved. Most early French breeds had good meat-producing qualities.

The above examples have been repeated in some manner hundreds of times, among all livestock and poultry species, and in nearly every country on the globe. In brief, most breeds have developed from animals that were particularly well adapted to a specific environment, producing in that environment the kind of product desired.

It is certainly possible that two or more breeds may have been developed simultaneously for adaptation and top production in the same kind of environment. This was certainly feasible in the seventeenth and eighteenth centuries, when most of the world's present breeds were being developed. With little or no communication, livestock and poultry raisers in one section of the country or world could have been totally unaware of developments occurring in other regions. This may explain similarities in production traits observed among two or more breeds today. A breed of dairy cattle, selected and developed for roughage utilization and maximum milk production in southern Sweden, for example, may have similar production traits and be adapted to the same kind of environment as a breed developed for the same traits in southern Chile.

There is considerable pride associated with breed development. Those who have been responsible for development and maintenance of a breed strive for an ideal in breed type, those characteristics that help make the breed unique, at least in color markings and other obvious traits, as well as in production characteristics. This pride is passed on, through generations of breeders, and has persisted for centuries.

In the first paragraph of this chapter, a breed was defined as a group of animals with distinguishable characteristics. These distinguishable characteristics—usually color markings, but also sometimes wool or haircoat-cover patterns, size, and shape—soon become an effective and well-known "trademark," though they may have little economic significance in terms of meat, milk, egg, and wool production.

21.2 CHRONOLOGY OF BREED DEVELOPMENT

Breed development may have begun soon after animals were first domesticated. Even though time of domestication is not known for all species, reports indicate that swine were domesticated in Eastern Asia about 2900 B.C. and in Europe about 1500 B.C. Ancestors of today's chickens were domesticated in China and India before 1000 B.C. The turkey was first domesticated in Central America, but the date is unknown. Cattle and sheep were domesticated in prehistoric times. The Bible contains many references to use of the various livestock species and hints that some attention had been paid to differences in producing ability among animals. Table 21–1 presents a sketchy chronology of the development of most present breeds.

Several significant items are apparent in this table. The first is the early development of distinguishable breeds of cattle for milk production. This may have been influenced by the recognition of milk as a valuable human food efficiently produced from forage. Development of specific dairy breeds in particular areas also may have been encouraged by local customs of cheese- or butter-making. Because milk was a highly perishable product that had to be produced in the immediate area, cows that were the best producers were easily identified. Local pride in high production and uniformity, and the proximity of the cattle raisers, may have allowed and encouraged exchange of breeding stock and development of a specific breed within an area.

Holstein-Friesian cattle, now referred to as Holstein, are known to have originated in western Europe more than 2000 years ago, in the area now known as the Netherlands. Although the name Holstein was not attached to them until 1864, they were highly selected and relatively uniform in traits for many years before. Brown Swiss, Jersey, Guernsey, and Ayrshire also developed within rather limited geographical areas. Brown Swiss were developed in accord with the cheese-making industry of Switzerland; Jersey were especially well adapted for producing milk and butterfat on Jersey Island's lime-deficient soil. Guernsey milk produced excellent butter, a major product of the island of Guernsey, which has an area of only about 24 square miles. The Ayrshire breed combined good milk production with economical beef production, desired by the livestock raisers in the low-income area of southwestern Scotland where the breed developed.

In contrast to early developments of our presently popular dairy breeds, beef breeds, and some of the sheep breeds, many of today's swine breeds are of recent differentiation. This does not mean that hog raisers centuries earlier did not have breeds. History indicates they did. It probably does mean that the higher prolificacy of swine has allowed swine raisers to develop new breeds more rapidly, to more adequately fit the changing environmental conditions. As the United States was settled, breeding animals were brought from Europe and used as foundation stock for breeds adapted specifically to this country. At the same time, European breeders used some of the same, or other, breeding stock to develop new breeds for their needs.

Several chicken breeds, especially for meat production, were developed in the New England states during the first half of the twentieth century. Closeness to market was probably the main factor in this development. Also, rather sizable business volumes could be achieved by producers on the small New England farms. Broilers were raised in other regions of course, but not as intensively.

Small farm flocks for egg production were common throughout the country before 1950, though most common in the Midwest, where feed grains were plentiful and the laying flock served as part of a diversified farming operation. Most large

TABLE 21–1

Chronology of Breed Development and Importation into the United States[a]

Year	Cattle		Swine	Sheep	Poultry
	Dairy and Dual-Purpose	*Beef*			
1000 B.C.	About 100 B.C.—Holstein-Friesian—Europe	Before 700 B.C.—Chianina—Italy		About 1500 B.C.—Angora goat—Asia	
A.D. 1000	Guernsey—Guernsey Island			Merino—Spain	
1500		Angus—Scotland Limousin—France Shorthorn—England			
1700		Hereford—England	Berkshire—England	Lincoln—England Cheviot—England and Scotland	
1750	Jersey—Jersey Island			Leicester—England Rambouillet—France Southdown—England	
1775		•Shorthorn			
1800	Ayrshire—Scotland Brown Swiss—Switzerland •Ayrshire	Simmental—Switzerland Charolais—France •Hereford •Brahman	Yorkshire, Tamworth—England Hampshire—U.S. •Berkshire	•Tunis •Merino •Southdown Romney—England	Leghorn—Italy
1825	•Guernsey •Jersey		Chester White—Pennsylvania	Hampshire, Dorset—England Cotswold, Oxford—England •Cheviot	•Leghorn
1850	•Holstein •Brown Swiss	•Angus	Poland China—Ohio	•Rambouillet, Hampshire Angora goat, Oxford Shropshire, Suffolk—England •Shropshire	Cornish—England White Holland—U.S.

Year	Cattle	Swine	Sheep	Chickens	Turkeys
1875	•Milking Shorthorn	Duroc—U.S. •Tamworth •Yorkshire Landrace—Denmark	Corriedale—New Zealand •Dorset, Suffolk	Barred Plymouth Rock—U.S. Plym. Rock, Wyandotte—U.S. Orpington—England •Cornish •Orpington Rhode Island Red—U.S.	
1900	Beefmaster—U.S. Senepols—Virgin Islands	Spotted—Indiana Hereford—Missouri	•Romney, Karakul Columbia—U.S. Finnish Landrace—Finland		
1925	Santa Gertrudis—U.S. •Charolais Brangus—U.S. Charbray—U.S. •Limousin •Simmental Chianina[c]	•Landrace Minn. 1, 2; Mont. 1; Md. 1; Palouse—U.S.	Targhee—U.S. Romeldale—U.S.		Broad Breasted Bronze Turkey—U.S.[b] New Hampshire—U.S.[b] Beltsville Small White Turkey—U.S.[c]
1950			Debouillet, Monadale—U.S.		
1960–70			•Finnish Landrace Coopworth—New Zealand Perendale—New Zealand Polypay—U.S.		
1970–80	•Gelbvieh •Pinzgauer •Salers •Marchigiana •Maine Anjou •Blond D'Aquitaine		•St. Croix		

Note:

[a] *Breed names not preceded by a black dot are placed to show the approximate time the breed attained identity as a breed (though the name may have been different), according to information available, and the area where it was developed or became identified as a breed. Breed names that are preceded by a black dot are placed to show the approximate time animals of the breed were first imported into the United States, according to information available.*

[b] *Turkeys listed are considered to be varieties, not breeds.*

[c] *Not imported; rather, semen imported and used on U.S. cattle, to produce American Chianina.*

Sources: Information was gathered from many sources, including Jull, M. A., National Geographic 51 (1927): 379 (with permission); Briggs, H. M., Modern Breeds of Livestock, 3rd ed., New York: Macmillan, 1969; American Poultry Association, Standard of Perfection, 1958; and Miller, E. H., Mt. Vernon, Ohio, private correspondence, 1969, 1996.

Red and White cow

Brown Swiss cow

Guernsey cow

Holstein cow (ideal model)

Jersey cow

Figure 21–1
Some of the major breeds of dairy cattle raised in the United States.
(Courtesy of respective breed associations.)

Year	Dairy Cattle	Beef Cattle	Swine	Sheep	Poultry
1875	•Milking Shorthorn		Duroc—U.S. •Tamworth •Yorkshire Landrace—Denmark	Corriedale—New Zealand •Dorset, Suffolk	Barred Plymouth Rock—U.S. Plym. Rock, Wyandotte—U.S. Orpington—England •Cornish •Orpington Rhode Island Red—U.S.
1900		Beefmaster—U.S. Senepols—Virgin Islands	Spotted—Indiana Hereford—Missouri	•Romney, Karakul Columbia—U.S. Finnish Landrace—Finland	
1925		Santa Gertrudis—U.S. •Charolais	•Landrace Minn. 1, 2; Mont. 1; Md. 1; Palouse—U.S.	Targhee—U.S. Romeldale—U.S.	Broad Breasted Bronze Turkey—U.S.[b] New Hampshire—U.S.[b] Beltsville Small White Turkey—U.S.[c]
1950		Brangus—U.S. Charbray—U.S. •Limousin •Simmental Chianina[c]		Debouillet, Monadale—U.S.	
1960–70				•Finnish Landrace Coopworth—New Zealand Perendale—New Zealand Polypay—U.S. •St. Croix	
1970–80		•Gelbvieh •Pinzgauer •Salers •Marchigiana •Maine Anjou •Blond D'Aquitaine			

Note:

[a]Breed names not preceded by a black dot are placed to show the approximate time the breed attained identity as a breed (though the name may have been different), according to information available, and the area where it was developed or became identified as a breed. Breed names that are preceded by a black dot are placed to show the approximate time animals of the breed were first imported into the United States, according to information available.

[b]Turkeys listed are considered to be varieties, not breeds.

[c]Not imported; rather, semen imported and used on U.S. cattle, to produce American Chianina.

Sources: Information was gathered from many sources, including Jull, M. A., National Geographic 51 (1927): 379 (with permission); Briggs, H. M., Modern Breeds of Livestock, 3rd ed., New York: Macmillan, 1969; American Poultry Association, Standard of Perfection, 1958; and Miller, E. H., Mt. Vernon, Ohio, private correspondence, 1969, 1996.

egg-production units were near larger cities, however, and breeds were developed for these units. Those adaptable to farm flock conditions also became popular throughout the Midwest.

During the 1950s, *lines* and *strains* were commercially developed from one or more breeds, for use as parents in the increasing specialized broiler- and egg-producing industries.

It is apparent in Table 21–1 that many of today's breeds were developed in Europe, especially in England or Scotland. Certainly the livestock and poultry raisers in these areas were industrious, observant, and good selectors of breeding stock, but livestock and poultry raisers in other areas were making progress in breed development as well.

Remember that most of the early American settlers came from western Europe, thus much of our trade and communication has been with this area because of closeness, ancestry, and similar habits and customs. Recognize, too, that the climate and topography of much of the United States is similar to that of western Europe. Livestock and poultry that were well adapted to conditions in western Europe were likely to be raised successfully in the United States.

Other countries of the world have been very active in breed formation and have successfully developed breeds of livestock and poultry specially adapted to their conditions. Only in the past 40 to 50 years, however, have many of these other countries been tapped as sources of breeding stock for livestock improvement in the United States.

Of special note are several "Continental beef breeds" imported from Europe.[2] Among these is the Chianina, introduced into the United States from semen imported in 1971. Known as the largest size and oldest of all cattle breeds (see Table 21–1) the Chianina was developed for both work and beef production in the Chianina Valley in Italy.

Animal breeders often refer to a newly introduced breed from another country as an *exotic breed.* Only a limited number of animals, or frozen embryos, or amount of sire's semen of a new breed is available; thus, describing the breed as exotic is appropriate.

21.3   VARIATION WITHIN BREEDS

Though breeds each have certain distinguishable characteristics, there is also much variation within each breed, especially in economically important production characteristics. Some of the variation observed may be due to environmental causes, because breeds are raised under widely varied conditions, but much of the variation is known to be genetic. It is not uncommon for beef calves within one herd to vary 75 to 100 pounds in weight at a standard weaning time. Dairy cows within a breed and herd may vary considerably in fat and protein percentage, and more in milk production. Similar variation exists within breeds of other species.

Most breeders have extreme loyalty to their breed, and considerable ire is raised when someone ranks a breed lower in a particular trait than another breed. Therefore, when asked a question such as "Which is the best breed for milk production?" an educator is often thankful for the variation and may say, "There is more variation within the breed than among the averages of breeds."

This statement is appropriate when discussing swine and beef breeds. It is less appropriate, and in fact would be misleading, in discussions of sheep or poultry breeds.

---

[2] Continental beef breeds include the Blood D'Aquitaine, Charolais, Chianina, Gelbvieh, Limousin, Maine Anjou, Marchigiana, Pinzgauer, Salers, Simmental, and others originating in Europe.

Breeds of the latter species were developed for divergent purposes; breeds of swine and beef cattle were generally developed for single purposes. Discussions in later sections and photos in Figures 21–1 through 21–6 illustrate this point.

That genetic variation exists within breeds simply indicates that the breeds are not genetically "pure." Even though essentially all animals in the breed may carry the same genes for color markings or other noticeable traits, there is considerable range in genes that influence other traits. As long as cross-fertilization of male and female is required for animal conception, breeds or lines of animals totally pure in gene composition will be impossible to achieve except by chance. Even if such occurred by chance and a male and female of identical gene makeup were mated, the breeder probably would not realize it.

Remember, also, that genetic variation is essential in order to make progress by selection. (See Section 17.8.)

21.4 CATTLE BREEDS

For purposes of adaptation to geographical areas and products desired, cattle breeds are differentiated into two general groups—dairy and beef. Some of the more common breeds in the United States are listed below.

Dairy	Beef	
Ayrshire	Angus	Limousin
Brown Swiss	Brahman and Brahman related:	Maine Anjou
Guernsey	Brangus	Polled Hereford
Holstein	Simbray	Polled Shorthorn
Jersey	Charbray	Red Angus
Milking Shorthorn	Charolais	Santa Gertrudis
Red and White	Chianina	Shorthorn, horned
	Gelbvieh	Salers
	Hereford, horned	Simmental

Typical dairy and beef breeds are shown in Figures 21–1 and 21–2. The classification given is not completely descriptive. An estimated 10 to 20 percent of the beef consumed in the United States, for example, comes from cattle that are primarily of dairy breeding. Similar breeds in some other countries are the major source of beef. During much of its early development, the Ayrshire was considered a dual-purpose animal, adapted to both dairy and beef purposes, and the Brown Swiss is considered such by some breeders today. Though the Milking Shorthorn is a distinct part of the Shorthorn breed, excellent dairy herds of Milking Shorthorns have been developed. Some milk consumed by Americans comes from cows with considerable beef breeding.

In modern livestock raising, one pays less attention to the "textbook classification" of livestock and more attention to the specific traits—rate of gain, pounds of milk produced, percentage of lean in the carcass—of the breed or of animals within the breed.

Although all traits of all breeds cannot be described in these pages, it should be emphasized that significant data are available. DHI records, described in Chapter 30, provide sufficient national data to characterize the major dairy cattle breeds in terms of cow size, average milk or butterfat production, and other traits. Some data are presented in Table 30–1. In the case of beef breeds, organized record programs are provided through most of the breed associations established within the states.

Red and White cow

Brown Swiss cow

Guernsey cow

Holstein cow (ideal model)

Jersey cow

Figure 21–1
Some of the major breeds of dairy cattle raised in the United States.
(Courtesy of respective breed associations.)

Angus cow and calf

Brahman cow and calf (ideal model)

Brangus bull

American Herford bull and calf

Charolais bull

Shorthorn cow and calf

Figure 21–2
Some breeds of beef cattle raised in the United States.
(Courtesy of respective breed associations.)

(continued)

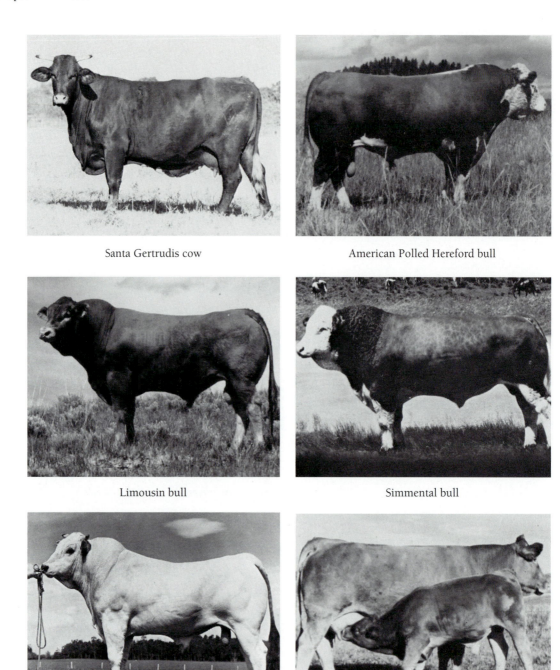

Santa Gertrudis cow

American Polled Hereford bull

Limousin bull

Simmental bull

Chianina bull

Gelbvieh cow and calf

Figure 21–2
Continued

(continued)

Pinzgauer cow and calf

American Tarentaise cow

Beefmaster cow

American Salers young cow

Barzona cow and calf

Maine Anjou cow and calf

Figure 21–2
Continued

21.5   SWINE BREEDS

Swine are raised to produce lean, high-quality pork. All breeds have this common purpose, though there are differences in the degree to which the goal is achieved. At one time, breeds of hogs grown in this country were classed as "lard type" and "bacon type." The individual breeds continue, but this classification has disappeared and hog raisers, regardless of breed or breeds used, are all aiming for the same kind of product, the "meat-type" hog. Note the extreme differences in conformation between the lard-type hog produced in the mid-1940s (Figure 21–3) and the modern meat-type hogs shown in Figure 21–4.

Because crossbreeding is almost universal in commercial hog production, breeds may be selected and used for crossing that would not be raised as a pure breed for commercial production. When the Landrace breed was first imported into this country, for example, boars of that breed were generally more valuable for crossing with sows of other breeds such as Duroc, Poland, or Hampshire than they were for mating with Landrace sows. At that time the Landrace were highly inbred, difficult to raise, and not efficient meat producers on the average American farm. But when crossed with another breed, the offspring were vigorous and fast-gaining, and the carcasses produced were leaner than those then being produced normally by the other breed.

Common breeds of hogs raised in this country (Figure 21–4) are listed below:

Berkshire	Spotted
Chester White*	Tamworth
Duroc	Yorkshire*
Hampshire	Landrace*
Poland China	

Between the mid-1930s and mid-1950s, scientists at state agricultural experiment stations and the USDA developed several breeds—Minnesota Nos. 1, 2, and 3;

---

* The Chester White, Yorkshire, Landrace, and Large White breeds are often referred to as the "white breeds"; also see Section 20.6 on their use in rotational crossbreeding programs.

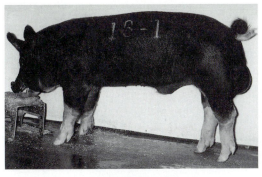

Berkshire boar

Chester White boar

Duroc boar

Chinese extreme

Hereford boar

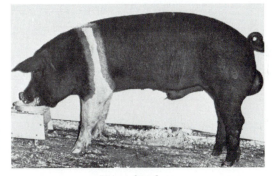

Hampshire boar

Figure 21–4

Some breeds of hogs raised in the United States. Note: The Chinese extreme is not commonly raised in the United States.

(Courtesy of respective breed associations.)   *(continued)*

Montana No. 1; Maryland No. 1; Beltsville Nos. 1 and 2; and Palouse. They provided stock for some commercial crossbreeding operations and contributed to the genetic base of some commercial lines of swine breeding stock. They were developed as and could be called *inbreds* because they were developed by rather intense inbreeding, after being started by the crossing of two or more breeds. The more established breeds were started similarly, but developed over a longer span of time and with much less intensive inbreeding.

Landrace boar

Poland China boar

Yorkshire boar

Tamworth gilt

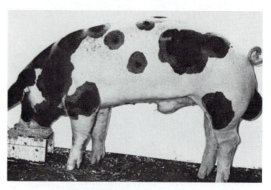

Figure 21–4
Continued

Spotted boar

Several commercial companies now provide seed stock to swine producers from selected maternal and paternal lines and from crossbred hybrid lines. Also, some companies have included breeds and crosses not common to the United States such as the Camborough, Chinese breed, Large White, and Saddleback in their testing and selection program in an effort to provide highly unrelated animals and maximum heterosis.

Though space does not allow consideration here, more is known about the productive merit of swine breeds and individuals or families within the breeds than is true for sheep or beef cattle. Consumer demand for high-quality pork, high feed costs, and narrow profit margins has encouraged early development of swine testing stations and on-the-farm testing programs to identify potential breeding animals with top genetic merit. Such testing programs have allowed accumulation of much data on breed performance.

21.6 SHEEP BREEDS

The major breeds of sheep raised in the United States are listed below:

Fine-Wool Breeds	Meat Breeds	
Merino	Cheviot	Oxford
Rambouillet	Columbia	Shropshire
	Corriedale	Southdown
	Dorset	Suffolk
	Hampshire	

The Merino and Rambouillet were developed for high production of fine, high-quality wool, but much lamb and mutton comes from animals that carry Merino or Rambouillet breeding. With the major sheep breeds (see Figure 21–5), wool represents a small but important proportion of total flock income.

Merino ram

Rambouillet ram

Cheviot ram

Columbia ram

Figure 21–5
Some of the major breeds of sheep raised in the United States.
(Courtesy of *Sheep Breeder* and *Sheepman*.)

(continued)

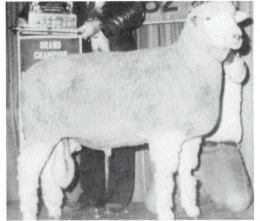

Corriedale ram

Dorset ram

Hampshire ram

Oxford ewe

Shropshire ram

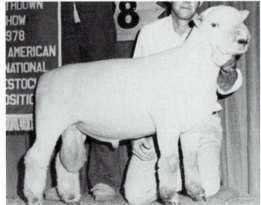

Southdown ram

Figure 21–5
Continued

(continued)

Finn ewe

St. Croix ram

Tunis ram lamb

Suffolk sheep

Figure 21–5
Continued

Other meat breeds developed in this country include Targhee, Panama, Romeldale, Debouillet, and Montadale. More recently developed sheep breeds include the Coopworth and Perendale from New Zealand, Boorda Merino from Australia, Cormo from Tasmania, and Polypay from the United States. The Finnish-Landrace (or Finnsheep) breed was also introduced into the United States in the late 1960s and is used primarily in crossing with the meat-type breeds to provide more prolificacy in the crossbred progeny.

The opinion is sometimes expressed that there are more breeds of sheep raised in the United States than necessary. Nevertheless, each breed has enthusiastic supporters and breeders. Also, certain large sheep raisers have, in effect, developed their own "breed" or family of sheep specially adapted to their area and needs.

Relative fineness of wool from the various breeds is shown in Table 27–2 in Chapter 27 and indicates considerable range. Also, some breeds (Barbados, St. Croix) possess hairlike or furlike body coverings, such as the Karakul breed (Box 21–1). There is much range in other production characteristics, too. Southdown and Shropshire sheep, for example, are relatively small and compact. Though they produce excellent carcasses, other breeds such as the Hampshire, Suffolk, and Columbia are more adapted to commercial lamb production because of their larger size and more rapid gains.

There are several breeds of long-wool sheep that are also classed in the meat group because their meat production is considered most important and their wool is relatively low in quality. These breeds, including the Leicester, Lincoln, Cotswold, and Romney, are not popular in this country, partly because their low-quality wool is not protected from foreign competition by import tariff.

Sheep breeders also classify sheep breeds according to their commercial use (particularly in crossbreeding) as *ewe breeds* and *ram breeds*. Ewe breeds are strong in traits such as prolificacy, mothering and milking ability, body size, and wool production. Ram breeds have strong characteristics in growth rate, muscling, and carcass quality. The dual-purpose breeds have traits that enable them to be used either as a ewe or ram breed. The types of ewe, ram, and dual-purpose breeds are as follows:

Ewe Breeds	Ram Breeds	Dual-Purpose
Border Leicester	Cheviot	Columbia
Corriedale	Hampshire	Dorset
Debouillet	Montadale	Lincoln
Delaine Merino	Oxford	Romney
Finnish-Landrace	Shropshire	
Rambouillet	Suffolk	
Targhee	Southdown	

Research at several locations has compared breeds of sheep in reproductive, growth, and carcass characteristics. Research reports are summarized in Tables 21–2 and 21–3. These data do not present information on more recent breed performance and should not be interpreted as characterizations of the total breed in the United States. However, the research workers did gather breeding stock for the research at that time from as many breeders and geographic locations as feasible.

BOX 21–1

The Karakul Sheep and Their Wool

The twin black lambs and the adult long-fleeced sheep shown in Figure 21–6 are a Karakul breed of sheep. The Karakul may be the oldest breed of sheep, evidenced by a distinct Karakul type found on ancient Babylonian temples. Native to Central Asia, Karakuls were bred to adapt to a region of high altitude with scant desert vegetation and limited water supply.

Karakuls became known as the "fur" sheep because the young lambs possess a lustrous coat of fur instead of wool covering their bodies. This fur coat on the newborn Karakul lamb gave the markets of the world the highly prized and popular fur known as Persian lamb and Broadtail, famous for its beauty and durability.

The breed was introduced to the United States between 1908 and 1929 for pelt production of the young lambs. As evidenced by Figure 21–6, Karakuls differ radically in conformation from many other breeds. One major difference is that they are of the fat-tailed type of sheep. In their native land they were able to store huge amounts of energy in the form of fat in their tails and use it when vegetation became scant.

Its colored fleece, due to a dominant black gene, also distinguishes the Karakul. Most lambs are born coal black with lustrous wavy curls, with the face, ears, and legs usually showing smooth, sleek hair-like fibers. As the lamb grows, the curls open and lose their pattern, the color generally turns brownish or bluish gray, then becomes grayer with age.

The Karakul is considered a rare breed in the United States with a current population of about 1,300 animals. However, large flocks are still found in Central Asia and South Africa where it successfully survives the harsh environment and provides a source of milk, meat, tallow, pelts, and strong-fibered long wool.

Source: Adapted from material provided by the American Karakul Sheep Registry and Letty Klein, breeder of Karakul sheep.

Figure 21–6
Twin Karakul newborn lambs with their characteristic black fleece (left); adult Karakul rams with their long, lustrous fleece (right).
(Courtesy of American Karakul Sheep Registry and Letty Klein.)

| TABLE 21–2 | Reproductive Performance of Four Breeds and the CSC Developed Strain of Sheep |

Breed[a]	Reproduction		Lamb Performance			
	Number of Ewes Bred	Lambs Weaned per 100 Ewes Bred	Number of Lambs	Birth Weight (kg)	Weaning Weight (kg)	Average Daily Gain (kg)
Hampshire	156	87.8	93	4.75	27.0	0.259
Suffolk	176	79.1	139	5.30	30.6	0.297
Dorset	167	55.2	92	3.88	23.9	0.223
Targhee	290	102.2	299	4.82	24.6	0.231
CSC	205	99.7	203	4.12	23.6	0.219

[a]Stock for each of the first three breeds was obtained from nine different breeders across the United States. The Targhee source was the USDA Sheep Experiment Station, Dubois, Idaho, where the breed was developed. CSC refers to a strain that resulted from Columbia, Southdown, and Corriedale stock developed at the Dubois station and at Middlebury, Vermont.

Source: Sidwell, G. M., and Miller, L. R., J. Anim. Sci. 32:1084–1090, 1971.

| TABLE 21–3 | Growth and Carcass Characteristics of Ram Lambs from Six Breeds |

Breed[a]	Live Weight at 26 Weeks (kg)	Dressing Percent	Quality Grade[b]	Rib Eye Area (cm²)	Percent Boneless Lean Cuts[c]	Yield Grade[d]
Suffolk	63.6	53.2	6.23	16.5	46.2	2.61
Hampshire	56.6	53.5	6.92	13.9	45.7	2.89
Dorset	48.5	53.4	6.27	13.3	46.4	2.46
Rambouillet	56.1	48.9	4.77	13.0	45.6	2.92
Targhee	56.5	50.8	4.71	12.5	45.2	3.13
Corriedale	52.8	50.1	4.54	11.2	44.7	3.41

[a]Source flocks ranged from one to eight for the ewes in each breed and from three to nine for the rams in each breed.

[b]4 = Low Choice, 5 = Choice, 6 = High Choice, 7 = Low Prime.

[c]Leg, loin, rack, and shoulder.

[d]The lower the number, the better the yield grade.

Source: Dickerson, et al., J. Anim. Sci. 34:940, 1972.

21.7   GOAT BREEDS

Interest in goat breeds has increased in recent years, in part because of the increased popularity of goats for 4-H projects and for alternative family milk supply in suburban areas. There also are commercial goat dairies in most states; DHIA state summaries often include records from several goat dairies.

The French Alpine, Nubian, Saanen, and Toggenburg are the more popular breeds in the United States. They and the American La Mancha are pictured in Figure 21–7.

American La Mancha doe

French Alpine doe

Nubian doe

Saanen doe

Toggenburg doe

Figure 21–7
Breeds of dairy goats raised in the United States.
(Courtesy of *Dairy Goat Journal*.)

These breeds have their ancestral roots largely in India, Egypt, France, Switzerland, and England. Each has been developed by selection for high milk production and, in the case of the Nubian, for high butterfat percentage.

Although there are distinct breed characteristics—pendulous ears and convex nose on the Nubian, no ears or very short ears on the American La Mancha, and white stripes down the face of the Toggenburg—there are also variations in color patterns or shades within most breeds.

Some goat breeds and their crosses are slowly becoming established in the United States for the production of goat meat (**chevon**). Most of the dairy breeds lack

> **Chevon**
> Meat from the goat.

the desired conformation and carcass traits. However, the Nubian, originally a dual-purpose breed, has been crossed with native "Spanish" goats of the Southwest. This cross combines the carcass quality and milk yield of the Nubian with the ability of the native range goat to graze and survive well under range conditions. Researchers also are evaluating the Boer goat, a compact, short-legged, meat-type breed from South Africa, for commercial use in the United States.

21.8   POULTRY BREEDS AND VARIETIES

Private breeders and primary poultry breeders maintain pure groups of the major breeds as a continuing source of genes for line or strain development. The lines or strains are developed by crossbreeding, inbreeding, and selection. In broiler or egg production units, therefore, it is the commercial name or identification number of the egg-producing or broiler stock that has significance, rather than the breed, per se. Selected birds are shown in Figure 21–8.

H & N p.g./two, a white egg layer of Leghorn stock

Hubbard Golden Comet, layer of brown eggs

Male parent (New Hampshire stock) and female parents (synthetic stock) of Hubbard Golden Comet

Barred Plymouth Rock Hen

H & N Meat Nick broiler breeding stock

White Holland male turkey

Figure 21–8
Selected chicken breeds, commercial lines of broiler or egg-producing stock, and turkey varieties in common use in the United States. (Courtesy of USDA, Hubbard Farms, Hy-Line Indian River Company, Pfizer H & N Inc., and DeKalb Ag Research.)

(continued)

Broad-Breasted Bronze turkey hen

Beltsville Small White male turkey

New Hampshire Red hen

Leghorn hen

Hi-Line W-36 layer

DeKalb DK layer

Rhode Island Red hen

Cornish male

Figure 21–8
Continued

The White Plymouth Rock breed has been the source of much of the breeding stock for today's broilers. The breed is meaty, and dressed carcasses are more appealing because they lack the small dark "pinfeathers" of some other breeds. Many of the male lines used in the broiler industry carry a high proportion of Cornish ancestry.

The Leghorn is considered to be the standard of commercial egg-producing stock. Its small size, early sexual maturity, and high egg production incorporated into lines or strains permit production that is efficient in terms of feed, time, and space (housing). Also, the eggshells are white and are preferred by most consumers.

Breeders maintain chicken flocks of many breeds, and selected individuals are employed in the continued development of breeding strains or lines, which in turn are used to produce stock for efficient meat and egg production (Figure 21–9).

The term *breed* has seldom been used in turkeys, though the Bronze, White Holland, and Beltsville Small White varieties are differentiated in much the same way as are breeds of other species. The Broad-Breasted Bronze, a developed selection from the Standard Bronze, attained popularity for meat production during the 1940s and 1950s. In recent years, consumer preference for carcasses free of dark pinfeathers prompted development of the Broad-Breasted or Large Broad White, from the white varieties back-

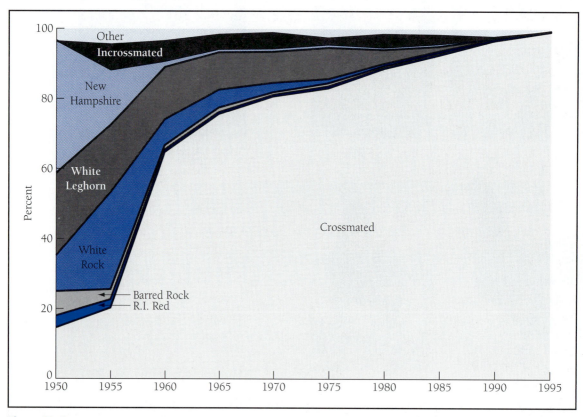

Figure 21–9
Breed distribution of flocks participating in the National Poultry Improvement Program, 1950–1997. Commercial varieties have predominated in recent years; presently about 98 percent of all birds are crossbreeds.
(*Source:* USDA.)

crossed to the Broad-Breasted Bronze. The Beltsville Small White variety was developed by the USDA to accommodate the market demand for 6- to 8-pound carcasses. This development aided acceptance of the turkey as a year-round food item. The Large Broad White can be fed for slaughter at varying ages or weights.

21.9 KEY POINTS IN BREED SELECTION

As discussed in Section 21.3, there is considerable variation within breeds. Obviously, there also are major differences between breeds. The prospective beef breeder can choose from 30 or more breeds of beef cattle. However, very few choices are available in the poultry breeds that are considered to be competitive in meat and egg production. Regardless of the species chosen, the prospective breeder should carefully consider the consensus of geneticists regarding the following key points:

- All breeds have both strong and weak points.
- No one breed is best for all important characteristics under all conditions.
- Much hereditary variation exists in all breeds.

The prospective breeder should

- determine the prevailing and most productive breed(s) in his or her geographical area
- study market demand for animals of different breeds and relate these to potential productivity and profitability
- consider other breeds that may be stronger in certain economically important traits than the prevailing breeds in the area

Regardless of the breed, the serious breeder should select superior animals and concentrate upon further improvement through use of sound genetic principles and breeding practices.

QUESTIONS FOR STUDY AND DISCUSSION

1. Why might some breeds within a species be more adaptable to certain areas of the United States?
2. What is probably the oldest and largest size beef breed existing today?
3. How would you define today's "exotic" beef breeds in the United States? Are these breeds also presently exotic in the country where they were developed?
4. Name three additional beef breeds introduced into the United States since 1970.
5. What are the main benefits of Brahman crosses in the southern portion of the United States?
6. Name three recently developed breeds of sheep. What is the major benefit in using the Finnish-Landrace?
7. What are the "white" breeds of swine? Name at least two breeds of swine being used in some breeding herds today that are not commonly raised in the United States.
8. Choose your favorite breed within a species of livestock or poultry. Why do you prefer this breed? Can you agree that "there is usually more variation within the breed than between breeds of that species"?

22

Marketing Meat Animals

This chapter focuses on the marketing of meat animals—animals that are ready for slaughter or livestock that may be en route from a rearing facility or ranch to a feeding enterprise. The ultimate goal of producers is profitable marketing of their animals.

Markets are the link between animal producer and processor, as well as a link among producers. This chapter does not include the marketing of processed meat by the processor to the retailer or consumer; that topic is fully discussed in Chapter 26. Chapter 25 includes discussion of the processing steps that prepare meat for sale to the consumer. Egg marketing is covered in Chapter 34, and there are further discussions of poultry marketing in Chapter 35.

A market is where buyer and seller interact; it may be a physical facility or it may be by phone, contract, or other means of communication. The buyer may be looking for animals for slaughter or feeding.

Few breeding or dairy animals are sold through established markets—terminal markets or weekly auctions; most is sold by private treaty or at special auctions.

Most slaughter animals are sold directly to meat processors at the end of the feeding period. A few are contracted prior to the end of the feeding period. **Direct marketing,** as a term, also includes sales to or through dealers and order buyers, who either take title to the animals or receive a commission from the buyer for purchases handled.

Auctions increased in significance during the last half of the 1900s and until the mid-1990s. They continue to handle large numbers of slaughter livestock and a relatively high proportion of all feeder animals marketed.

Terminal markets, or public stockyards, which handled most of the feeder and slaughter animals in the first half of the 1900s, now handle less than 10 percent of the slaughter livestock and relatively few feeder animals.

Since 1990 the number of cattle and calves marketed through auctions and terminal markets has been fairly steady. However, the volume of hogs marketed through firms selling on commission has decreased more than 50 percent as most producers now sell their hogs through direct marketing to processors.

Discussion of livestock and poultry marketing also includes supply and price patterns. The relationships between supply and demand determine price, and there are many factors that normally influence supply and demand and, hence, price.

This chapter, therefore, includes discussions of long-term trends and cycles in livestock and poultry numbers, or supply, as well as seasonal patterns that tend to repeat themselves year after year.

In general, as livestock and poultry feeding have become concentrated in fewer and larger units, facilities and production management talent have been increasingly utilized year-round. Thus, supplies of livestock and birds for slaughter have become more uniform through the months of the year.

Direct Marketing
Direct transaction between the livestock producer and the meat-processing establishment in the sale of animals.

Auction
A sale where successive bids are received and the animal is sold to the highest bidder.

Terminal Market
A market where animals are gathered and sold by commission agents on behalf of the owner.

Upon completion of this chapter, the reader should be able to

1. Explain differences among auctions, direct markets, and terminal markets.
2. Explain why price is usually a function of supply and demand.
3. Describe a long-term cycle and the events within a cycle.
4. Explain why long-term cycles in swine numbers are shorter than in cattle numbers.
5. Chart a 12-month supply pattern for the meat animal species and explain why it generally occurs.
6. Explain why the supply of broilers for slaughter is relatively uniform over the 12-month year.
7. Differentiate among auctions, terminal markets, and direct marketing in terms of trends in numbers of animals marketed.
8. Describe what function an order buyer or dealer performs in animal marketing.
9. Identify the federal agencies involved in supervising and reporting the marketing function.

22.1 SUPPLY, DEMAND, AND PRICE

Price is a function of supply and demand. Supplies of livestock and poultry tend to fluctuate in two rather distinct patterns: cyclic and seasonal. Long-term cycles and the factors that cause them will be discussed in Section 22.2. Seasonal patterns, which tend to repeat annually, vary among species and between feeder and slaughter animals.

Factors that constitute demand differ between feeder and slaughter animals. Also, because the demand for feeders tends to follow more of a seasonal pattern than does the demand for slaughter livestock, the effects on price are markedly different. Demand is important in helping to establish price, consequently producers, processors, and retailers should be alert to factors that influence demand and should be aware of any changes in demand. (See Chapter 26.)

Increases in contract sales agreements between producers and processors, especially with poultry and swine, have tended to dampen cyclic patterns. These agreements, in effect, apply the demand pressure early in the production cycle, perhaps before birds or livestock are placed in the broiler house or feeding unit, to some extent preventing production excesses or deficits.

Other fluctuations in livestock and poultry supply and prices, unrelated to the cycles and patterns mentioned in the first paragraph, certainly do occur. There are reasons, however, for their occurrence. Perhaps an understanding of the causes of cyclic patterns will permit understanding, or even anticipation of, these other fluctuations.

22.2 LONG-TERM CYCLES IN LIVESTOCK NUMBERS

Livestock numbers, meat supplies, and prices tend to be cyclical in the long term. Note the rather pronounced peaks in numbers of breeding sows and gilts and ewes shown in Figure 22–1. The cycles for hogs and sheep are generally somewhat shorter than for cattle.[1] Generally, 6 to 8 years of increase in cattle numbers in the United States are followed by 4 to 8 years of decline, a total of 10 to 16 years per cycle.

---

[1] Since 1975, market hog prices have peaked every 3 to 4 years: 1975, 1978, 1982, 1986–1987, 1990, 1993, and 1996.

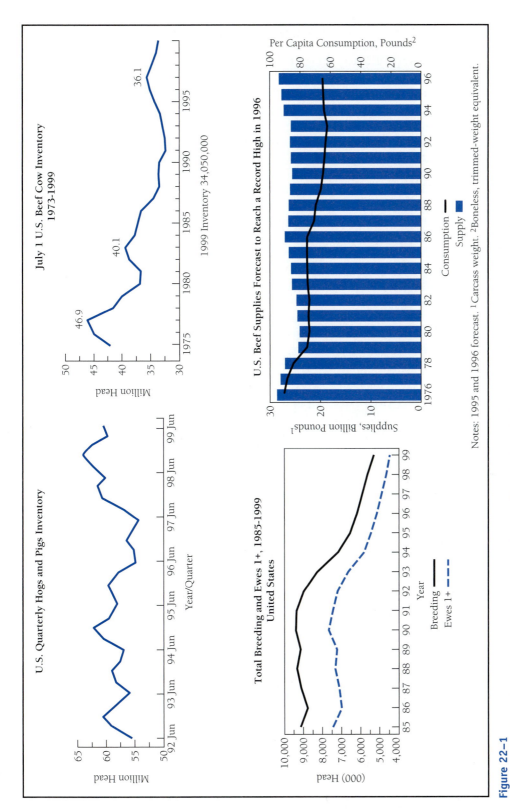

Figure 22–1

Long-term cycles in hog, sheep, and beef cattle production. The graph on the lower right shows the increased supplies of beef in recent years as per capita consumption remains relatively steady.
(*Source:* USDA.)

From about 1990 to 1996 a continued upward trend in beef cow and U.S. beef supplies occurred. This increase was due primarily to continued population growth and, until recent years, as noted in Figure 22–1, increased consumption of beef.

Why do these cycles occur? Let us discuss any one of the cycles, beginning at a low point. For example, a low "supply" of beef cattle (e.g., in 1980) meant that beef production was relatively low—lower per capita than in previous years. Hence, beef prices were relatively high in relation to the previous years or to the trend line. (Per capita demand was generally stable or increasing during these years.) Cattle raising and feeding were relatively profitable. More people started raising cattle; those who already had cow herds kept more heifers and culled fewer mature cows to increase herd size.

However, it takes several years for larger cow herds to result in an increase in the amount of beef on the consumer's table. The heifers have to mature, be bred, and calve, and the calves then have to grow and be fed for slaughter. During this time beef prices continue high and the beef business continues to appear attractive. So cattle producers continue expansion and still more enter the business.

Finally, in 6, 7, or 8 years, supply usually equals or surpasses demand. Beef production has expanded more rapidly than population. Beef prices drop. Raising beef cattle becomes less profitable in relation to previous years or in relation to other enterprises.

Raisers would like to decrease numbers. But there is a snag. Reducing the herd size means culling and selling more old cows for slaughter, or keeping fewer replacement heifers and sending more, directly or indirectly, to slaughter. This puts more beef on the market, decreasing prices even more. Note the depressing effects of continued high beef supplies on 1995–1996 steer prices and the consequent decline in beef cow numbers (Figure 22–2).

Depressed prices have an inhibiting effect on further sales, and therefore, reductions in cow herd size tend to occur less rapidly than desired. But reductions do continue as long as beef prices appear unfavorable and until, in time, marketings decline relative to demand and prices or price expectations increase.

Then the cycle repeats. The pendulum swings—profit incentive creates momentum in cattle expansion; then the burden of low prices creates momentum in decreasing cattle numbers.

Certainly other factors influence the regularity, length, and smoothness of the cycles. Because each species represents a meat product, supplies and prices of one influence supplies and prices of the others. Wars, depressions, prolonged drought, feed supplies, and other factors also exert obvious influence. Though meat supplies may be high and prices low, extremely abundant supplies of cheap feed may encourage livestock feeding, increasing total meat supply and causing still lower livestock prices.

Sheep numbers are drastically influenced by certain economic factors. Historically, sheep numbers seem to have been inversely proportional to the farmer's or rancher's economic plight. When prices have been high and livestock raising profitable, sheep numbers have dropped. When prices have been low and farming and ranching relatively unprofitable, sheep numbers have risen. In the United States, sheep numbers declined in the 40 years prior to 1986 and then slowly increased to peak in 1990 at 11.36 million head. Since that time, sheep numbers declined to about 9.24 million in 1999. In 2002, there were approximately 6.4 million sheep in the United States. The largest volume of sheep are found in Texas followed by California, Wyoming, South Dakcota, and Colorado.

Sheep compete with cattle for a place on farms or ranches—they both eat forage. Cattle will yield more dollar volume and sometimes more profit per acre, per

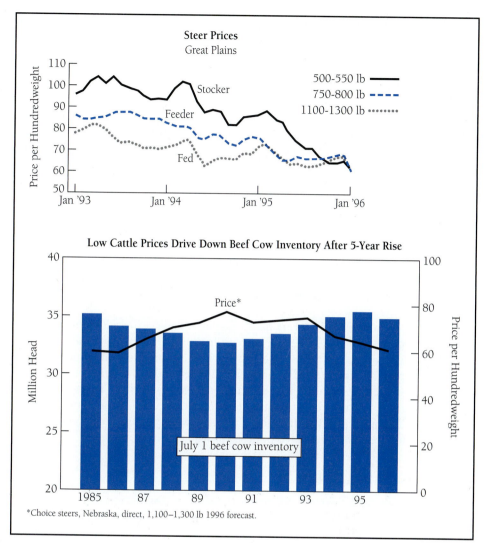

Figure 22–2

Declining stocker, feeder, and slaughter steer prices occurred as a result of high cow numbers and high beef supplies, shown in Figure 22–1. High feed prices also contributed to lower cattle prices. Also note in the second graph that these lower cattle prices caused the first decline in beef cow numbers in 6 years. This decline in mid-1996 indicated the beginning of the contraction phase of the most recent long-time cattle cycle.
(*Source:* USDA, ERS.)

dollar invested, or per hour of labor. The trends to larger units and higher cost labor have made cattle more competitive.

Sheep and hog cycles, as noted earlier, are shorter than cattle cycles. Why? This is caused mainly by differences in prolificacy—age at sexual maturity, gestation period, and number of offspring. Salvage value of culled or surplus breeding stock is also a factor.

The earlier discussion of a cattle cycle explained that it took several years for an extra heifer saved for herd expansion to contribute to increased beef supply at the market. Heifers are usually bred at 13 to 15 months of age, have approximately a 9-month pregnancy, and usually give birth to one calf that probably will be slaughtered at 15 to 20 months of age. Contrast this with a ewe, which is bred younger, has a 5-month pregnancy, and has one or two lambs that may be slaughtered in 4 to 6 months. Or compare it to a gilt, which is bred at 8 months, farrows nine or more pigs at about a year of age, and the pigs can be slaughtered in 4½ to 6 months.

Another factor is that swine herds can expand faster. Each gilt may farrow five boars and five gilts; the five gilts can go into the breeding herd. A heifer has but one calf, and chances are only 50 percent that it will be a heifer for the breeding herd. If it is, the increase in marketable beef is further delayed.

When the time comes to decrease numbers, the hog producer can do it faster and usually with less sacrifice. Surplus sows often will sell for slaughter at prices only slightly under those for barrows and gilts. A ewe that has lambed, however, is worth much less for slaughter. The price differential per pound for mature cows is also greater than for mature sows, and a cow is a large animal. Cattle raisers have more money invested in their breeding herd than hog raisers, in relation to production, and so can lose more by selling off surplus breeding stock at low prices. Hence, cattle raisers are more hesitant to sell, tending to wait for better prices, so the decline in cattle numbers is delayed.

Long-term poultry cycles are hardly distinguishable, especially in recent years. There are several reasons: (1) The recent trend has been sharply *increased* production and consumption of poultry meats; (2) contracts between producer and processor have tended to keep production in steadier harmony with demand; (3) the short life cycle (75 days or less between the time eggs are set in an incubator and chickens are processed) and high prolificacy permit almost immediate response to overproduction or underproduction; and (4) most large poultry processors can switch the type of product offered (i.e., make cutup parts on demand as opposed to whole birds).

22.3 SEASONAL PATTERNS

Seasonal patterns tend to repeat themselves year after year, except as influenced by movement of the long-term cycles or other factors. They usually are caused by *seasonal weather variations,* which affect feed supply as well as the animals, and by *reproductive cycles.* In some cases seasonal weather extremes dictate that certain livestock production schedules be followed. Though modern housing and equipment may counter these weather extremes, the same schedules often are followed year after year by producers because of habit or tradition.

Factors that influence the supplies and prices of feeder cattle and lambs are similar. Heavy movement of feeder animals occurs in the fall. Note in Figure 22–3 that placement each year is higher in the fall months.

Why are many feeder calves placed in feedlots in the fall? Most calves and lambs are ready to be weaned then. Most are born in the spring, in the range areas, and they are 5 or 6 months of age by September. Advantages of calving or lambing in the spring in these areas are obvious—warm weather and grass for the calf or lamb, and lush grass for the cow or ewe for heavy milk production.

Most cow-calf operators do not want to sell calves or lambs before September because they are the market outlet for grass during the entire growing season, and the higher market weight at the end of the grazing season will contribute to larger dollar volume. Yet the rancher likes to have them sold and away from the ranch by the middle of November, before snow flies, to avoid feeding more expensive harvested feed.

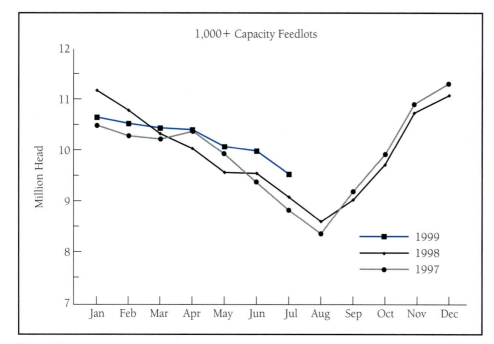

Figure 22–3
Monthly cattle placement into commercial feedlots for 1997 through July 1999.
(*Source:* USDA.)

An exception is those in the cattle industry who harvest considerable forage. They probably will winter the calves on hay, let them graze the following summer, and sell them, or place them in feedlots, as yearlings the second fall. Carryover of lambs is not practiced because their carcass value is sharply reduced after they pass a year of age. Feeders want only young lambs so they can be finished and marketed by their first birthday. Supply of feeder cattle and lambs is, to a considerable extent, a function of convenient and economical production.

There is trading in feeder cattle and lambs almost continually through the year. Fall-born calves and lambs raised in the South are marketed in different seasons. Intensified feeding operations provide a constant demand for feeders throughout the year. But there continues to be a strong seasonal pattern. Weather extremes in the northern United States and Canada, especially in range areas, continue to encourage fall peaks in feeder supply.

Because many who raise and sell feeder pigs have facilities for farrowing in any season and keep these facilities busy, there is a rather steady supply of weanling feeder pigs year-round. Accurate data are not available on seasonal movement of feeder pigs because few go through established markets.

In heavy grain-growing areas, demand for feeder pigs of any weight is noticeably heavier at harvest time, especially in years when crops are good. Higher feeder pig prices may encourage other hog raisers to sell, and movement increases.

Because "long-term" hog supply cycles are rather short and move up and down sharply, they and the factors that cause them are probably the major influences in establishing feeder pig prices. Local conditions also exert an effect.

Factors that influence supply of slaughter animals and prices received for such livestock and poultry are the same for all species. The degree of influence varies among

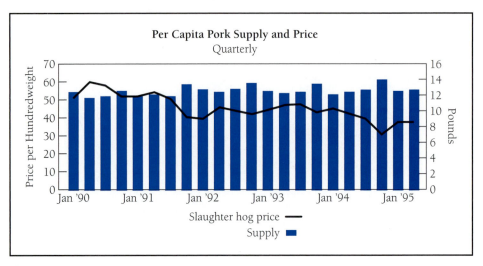

Figure 22–4
Prices of slaughter animals are the inverse of supply.
(*Source:* USDA.)

species. The first factor is that *total demand for meat does not change much from season to season*. There is some variation in consumption, especially of individual meat items. More stews and roasts are used in the winter; hot dogs and steaks are more popular in the summertime. Turkeys are a holiday favorite. The total demand for meat, however, remains about the same throughout the year.

The second factor is *the basic economic law that price is a function of supply and demand*. If demand doesn't change, price change becomes a function of supply change. In the case of meat, it is sometimes a drastic function of supply change; a slight change in supply of meat often results in a drastic price change, in the opposite direction.

Note in Figure 22–4 that in January 1995, pork supply per person attained a new high. During this same period, slaughter hog prices were much lower than in any of the previous months shown.

The third factor is that *meat consumption does not react sharply to slight price changes*. Demand is rather constant. If supply increases a bit, prices drop. Consumption usually increases, but typically there is also an accumulation of surpluses for a period during which prices remain low.

The fourth consideration is that *supply is, in some cases, largely a function of convenient and economical production*. This was pointed out previously in the discussion of feeder cattle and lambs. It is likewise true for slaughter hogs and for some fed cattle and lambs. Confinement animal systems increasingly accommodate the steady year-long consumer demand for product.

Although many hog producers farrow year-round, some farrow only in early spring and in the fall. Routine feeding and management make hogs ready for slaughter at 5 to 6 months of age. The result, then, is more slaughter hogs on the market during October through December, from farrowings that reach a peak in April through June (Figure 22–5). A marketing peak in early spring is caused by hogs farrowed in the fall. Seasonal patterns tend to repeat themselves year after year.[2]

---

[2] Since the mid-1970s, the lowest monthly prices for market hogs have occurred in April and November.

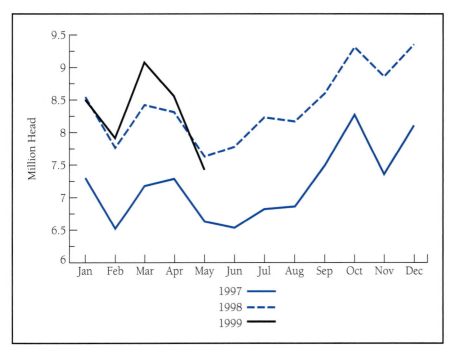

Figure 22–5
Monthly hog slaughter for 1997 through May 1998. Generally fewer hogs are slaughtered
during the summer months as a result of fewer sows being bred in the previous winter months.
(*Source:* USDA.)

Trends in recent years have been toward continuous farrowing and rapid finishing of hogs. This has caused a noticeable leveling of the seasonal supply pattern. This
trend will no doubt continue, though there still will be an advantage for seasonal farrowing on farms with limited housing and equipment and with a large field crop enterprise that demands summer labor.

Figure 22–6 shows seasonal production and slaughter patterns for sheep and
lambs over several years. Note the slightly lower production levels during the summer
grazing season. A sizable proportion of lambs, as well as cull ewes, are sold for slaughter directly off grass, which accounts for higher slaughter in the fall season. Peaks are
caused largely by lambs from feedlots.

There are two major groups of slaughter lambs—grass lambs and fed lambs.
Lambs from farm flocks of the Midwest and South are usually finished on grass, though
a few may have been creep-fed grain. Some years many western lambs also are finished
enough when they come off grass in the fall.

There are exceptions. Some midwestern flock owners lamb early, creep-feed
their lambs, and market early. In the South, many lambs are born in the winter and raised
on temporary pasture for spring slaughter. These lambs are the usual source of "spring
lamb" that appears in retail outlets.

The price pattern for slaughter lambs tends to fluctuate less, in relation to supply, than that for hogs or cattle. This may be because lamb and mutton represent a small
proportion of total meat supply and slaughter lamb prices therefore are influenced by
supplies and prices of beef and pork, or even of chicken and turkey.

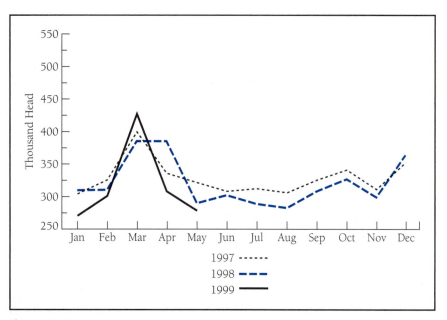

Figure 22–6

Seasonal marketing patterns for sheep and lambs from 1997 through June 1999. More lambs are generally marketed during March and April after lambs have been finished in the feedlots during the previous late fall and winter months.
(*Source:* USDA.)

Supply patterns of slaughter sheep and lambs will likely remain relatively stable. Commercial lamb feedlots, which tend to handle lambs year-round, are feeding a higher proportion of the lambs, and this tends to hold supply steady. Also, production in different regions tends to have a leveling effect (Figures 22–6 and 22–7). Lambs are most economically produced on grass, however, where seasonal weather extremes control production patterns. This contrasts with hogs and poultry, which are adapted to harvested concentrates and are adaptable to confinement rearing.

Seasonality in cattle slaughter is shown in Figure 22–8. The relatively steady rate is due partially to the high proportion of cattle fed in commercial and custom lots that operate year-round, and also to the fact that various classes slaughtered at different periods of the year tend to balance each other. Most beef cows culled from herds are sold at the end of the grazing season, after calves are weaned, and contribute to the peak slaughter number in late summer. Dairy cows and those beef cows culled because they have not conceived are culled in all seasons of the year.

In sections of the country where average-quality, lightweight beef is preferred, some calves are slaughtered at or soon after weaning. Most are of mixed or dual-purpose breeding, and their dams are good milk producers, so they carry sufficient finish at weaning time. Because such calves may not have the desirable conformation to demand high prices as feeders, they sell better for slaughter. Some are fed for short periods by the producer, to reach 600 to 700 pounds. The season of slaughter varies with geographic region.

Most steers and heifers in the United States and Canada are sold for slaughter after some feeding in drylot. Even though length of feeding period varies greatly, most are fed about 120 to 150 days and slaughtered when they reach the Choice grade.

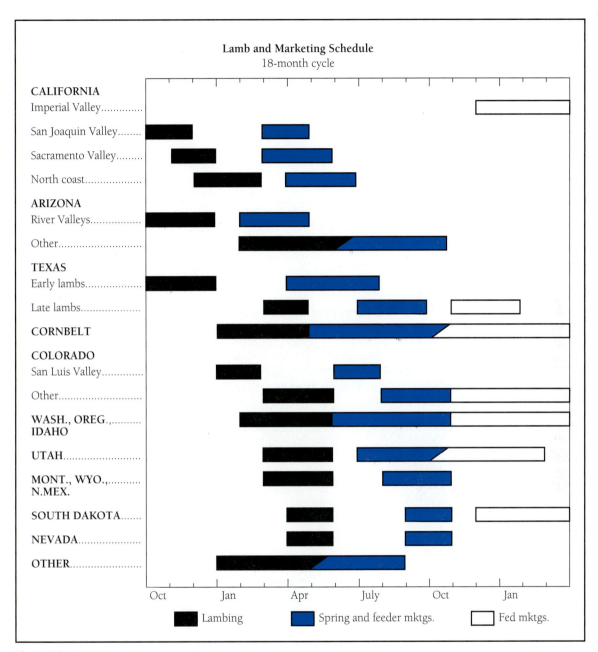

Figure 22–7
Supply is a function of convenient and economical production, largely dictated by climate.
(*Source:* USDA.)

Table 22–1 summarizes the quarterly production of broilers and turkeys in the United States for 1998. Production of both is relatively steady. Most years there is a slight dip from January through April resulting from (1) lack of winter production in a few units due to weather extremes, and (2) interruptions in hatching and filling broiler houses during the holiday season.

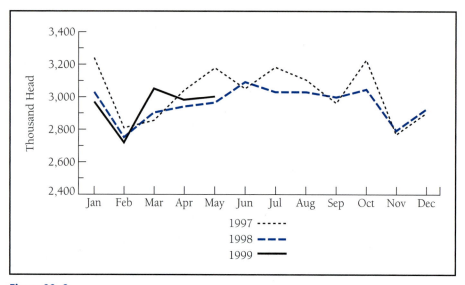

Figure 22–8
The monthly cattle slaughter from 1997 through May 1999 reflects some seasonality.
(*Source:* USDA.)

TABLE 22–1	Broiler and Turkey Supply, Utilization, and Price, by Quarter, 1998					
	Broilers			**Turkeys**		
Quarter	Production *(million lb)*	Per Capita *(lb)*	Price *(cents/lb)*	Production *(million lb)*	Per Capita *(lb)*	Price *(cents/lb)*
I	6,846	20.3	56.39	1,286	3.9	54.96
II	6,990	21.0	61.05	1,326	3.9	54.90
III	6,942	21.4	70.40	1,302	4.2	64.51
IV	7,085	21.8	64.21	1,367	6.0	71.33
Total	27,863	84.5	63.01 (avg.)	5,281		62.18 (avg.)

Source: *Adapted from USDA*, Livestock, Dairy, and Poultry Situation and Outlook, *May 25, 1999.*

Retail price summaries for broilers and turkeys in Table 22–1 indicate that price is largely a function of demand rather than being the inverse of supply. Note that both consumption and price received for turkeys during the fourth quarter reflects seasonal consumer demand.

22.4   DIRECT MARKETING

Today, direct marketing is the method used for marketing about 90 percent of all U.S. market hogs, mostly through *carcass merit pricing* (grade and yield). The main buyers of hogs by this method are packing (processing) companies and packer-operated buying stations.

Whereas direct marketing implies no intermediate agent, that is not always the case. Transactions often are made directly between the buyer and seller or their agents, such as an order buyer. Direct marketing of both feeders and slaughter animals has increased in recent years.

An accurate estimate of the percentage or number of feeder cattle or lambs sold direct is not available because such animals do not regularly pass through established points. Direct marketing of hogs represents a major portion of all hogs marketed (see Table 22–2) and an increasing number of hogs are produced and marketed under contract.

Most broilers and turkeys are produced under contract, so they are not discussed in this section. Contract growing and moving birds to the processor are discussed in Sections 35.4 and 35.5.

As livestock feeders and ranchers have become more specialized and larger, and as communication and transportation have improved markedly, it has become easier for buyer and seller to negotiate directly. Ranchers tend to know current prices and the value of their feeder calves, lambs, or yearlings. Feedlot operators, who buy in large groups and on a continuous basis, similarly are more current on values and prices.

Certain associations of growers and feeders assist direct marketing. The Nebraska Sandhills Cattle Association and the South Dakota Stockgrowers Association, for example, send direct-mail advertising to midwestern feeders listing age, breed, and sex of feeders available on certain ranches. Some county cattle feeder associations in Iowa and Illinois organize trips into feeder-producing areas to acquaint feeders with ranching practices and allow them to look at available cattle. Rancher groups reciprocate by visiting feedlot areas.

Increased direct marketing of slaughter livestock has been due to several factors:

- Animals being produced in fewer but larger units
- Movement of processing plants from cities into producing areas
- Excellence in trucking and highway systems for animal shipment
- Establishment of buying stations and producer alliances for larger animal grouping and marketing
- Modern communication systems that keep producers abreast of market situation and more confident in dealing directly with the buyer
- Increased opportunity to sell on *grade and yield basis*

The movement of processing plants to producing and feeding areas has made direct selling of fed cattle, lambs, and hogs more convenient. A number of surveys have shown that decisions on selection of a market often are based significantly on convenience or location. Also, transportation costs are lower because of shorter distances. There is no yardage fee or commission charge to pay. Though direct marketing may or may not mean more net return to the seller, lack of these obvious marketing costs makes it attractive to many producers.

Because the feeder may obtain a bid by phone or by a buyer visiting the farm or feedlot, the price per hundredweight can be established before the animals leave the premises. This makes the feeder feel more secure than when animals are hauled to a distant market or auction. However, with the increase in large-scale commercial cattle feedlots and swine production units, the relatively small farm feeder is at somewhat of a disadvantage

TABLE 22-2

Numbers of Livestock Sold at Selected Public Stockyards and Direct Marketing of Hogs (Thousands)

Sheep and Lambs: Receipts at Selected Public Stockyards[a]

Year	Denver	Fort Worth	Kansas City	Omaha	National Stock Yards	South St. Joseph	South St. Paul	Sioux City	All Others Reporting	Total Markets Reporting[b,c]
1972	9	17	43	111	60	83	228	107	1,895	2,553
1982			21	33	21	18	137	24	1,176	1,430
1986			8	14	15	9	87	10	738	881
1994				11		8	64	82	1,238	1,403
1997				2		0.8	47	53	997	1,060

[a]Total rail and truck receipts unloaded at public stockyards.
[b]Rounded totals of complete figures.
[c]The number of stockyards reporting varies from 41 to 68.

Cattle and Calves: Receipts at Selected Public Stockyards[a]

Cattle

Year	Denver	Fort Worth	Kansas City	Omaha	National Stock Yards	South St. Joseph	South St. Paul	Sioux City	All Others Reporting	Total Markets Reporting[b,c]
1972	16	122	437	1,165	360	373	916	748	9,141	13,278
1982			245	612	241	185	549	425	6,707	8,964
1986			161	320	153	140	391	230	5,430	6,825
1994				79		127	188	10	4,169	4,573
1997				54		121	180	48	4,549	4,953

[a]Total rail and truck receipts unloaded at public stockyards.
[b]Rounded totals of complete figures.
[c]The number of stockyards reporting varies from 41 to 68.

TABLE 22–2 (continued)

Calves

Year	Denver	Fort Worth	Kansas City	Omaha	National Stock Yards	South St. Joseph	South St. Paul	Sioux City	All Others Reporting	Total Markets Reporting
1972	3	96	61	12	15	30	126	27	571	941
1982					3	10	53		67	133
1986					1	13	27		31	72
1994							36		12	48
1997							3		92	96

[a] Total rail and truck receipts unloaded at public stockyards.
[b] Rounded totals of complete figures.
[c] The number of stockyards reporting varies from 23 to 46.

Hogs: Receipts at Selected Public Stockyards and Direct Receipts at Interior Markets[a]

	Receipts at Selected Public Stockyards										Direct Receipts in Interior Iowa and Southern Minnesota[d]
Year	Denver	Fort Worth	Kansas City	Omaha	National Stock Yards	South St. Joseph	South St. Paul	Sioux City	All Others Reporting	Total Markets Reporting[b,c]	
1972	355	74	1,087	1,514	1,452	1,747	2,056	1,832	8,181	18,298	20,506
1982			246	821	1,141	740	1,125	1,406	4,671	10,150	21,499
1986			160	667	967	560	738	940	3,330	7,332	21,983
1994				352		451	406	468	1,368	3,045	28,669
1997				139		192	258	155	734	1,479	28,624

[a] Total rail and truck receipts.
[b] Rounded totals of complete figures.
[c] The number of stockyards reporting varies from 41 to 68.
[d] Covers receipts at 14 packing plants and 30 concentration yards.
Source: Agricultural Marketing Service, compiled from reports received from stockyard companies.

due to reluctance of some packer buyers to trade for a few head at a time. Also, many small feeders do not have experience or qualifications for direct marketing (Figure 22–9).

Direct marketing of slaughter livestock affords opportunity for selling animals on the basis of carcass grade and yield, eliminating the necessity of appraising live slaughter animals. Pricing inaccuracies exist when slaughter animals are appraised alive because of errors in estimating yield, grade, meatiness, and other factors. It should be emphasized, however, that pricing inaccuracies are not necessarily eliminated when slaughter livestock is sold on the basis of grade and yield. Carcass grade, whether applied by the processor or a USDA employee, does not always indicate the true value of the carcass. This is because the carcass grade assigned results, in part, from a subjective appraisal and may be applied by persons who vary in their ability to grade carcasses according to specifications.

Procedural regulations for purchasing livestock on a "grade and yield" basis have been established by the Packers and Stockyards Administration of the USDA. They require processors to (1) pay sellers on the basis of hot carcass weight (without a cooler shrink factor); (2) use uniform equipment—hooks, rollers, and so forth—in tare weights; (3) provide sellers with detailed information concerning factors such as terms of contract and method of grading; (4) maintain identity of each seller's livestock; and (5) make payment on the basis of carcass price. **Cutability** or yield grades may be used to establish price, but if they are private grades, the processor must provide the seller written specifications of these grades.

> **Cutability**
> The proportion of the weight of the carcass that is salable product after the original carcass has been trimmed or processed for sale.

In general, producers and feeders have found it more advantageous to sell high-quality, high-yielding animals on a carcass grade and/or yield basis, and less advantageous to sell low-quality, low-yielding animals on that basis. Although there are other factors, buyers who appraise live animals tend to underestimate the value of the best and overestimate the value of the poorest.

Grade standards are independent of animal color or geographic origin; however, animals of certain origin or of dairy breeding may be discounted by buyers when

Figure 22–9
A packer buyer at the feedlot, left, bids on fed cattle that will be delivered directly to the processing plant. The company's head buyer, right, keeps in touch with and provides guidance to all buyers by radio via satellite. This is an example of direct marketing.
(Courtesy of Iowa Beef Processors.)

they appraise the animals on a live basis. Where such practices may occur, and where these cattle produce relatively high-quality carcasses, it may be advantageous to sell on a carcass grade and yield basis.

The feeder who is familiar with the quality of carcasses normally produced can do a better job of deciding the selling method to be used. If breeding stock is of consistent quality and feeding and management practices are uniform, the feeder or producer can predict, with relative accuracy, dressing percent and carcass grades of animals ready for slaughter.

22.5 AUCTIONS

Livestock auctions are facilities operated by one firm where animals are sold by public bidding to the buyer who offers the highest price per hundredweight (100 pounds) or per head (Figure 22–10). Livestock auctions are popular. The number in this country rose from about 200 in 1930 to over 2,300 in the 1960s, then declined to about 1,300 in 2000.

Auctions have flourished in feeder cattle-producing areas, such as Missouri, Texas, and the Sandhills of Nebraska. (See Figure 22–11.) Many auctions are also located in the intensive cattle feeding areas. Auctions also are used heavily for sale of feeder pigs.

Auctions differ from terminal markets in that all prospective buyers view and bid on the animals at the same time. At terminal markets commission salespeople deal with each buyer individually in private negotiations and usually accept or reject bids before that buyer leaves the alley and another buyer comes to bid. A few auction firms also conduct private negotiation selling, usually on a 1-day-per-week basis.

Sales in an auction are made with a nod of the head, displaying a number, or by spoken word. Traditionally, the auctioneer starts livestock at a reasonable price, asks for and receives successively higher bids, and sells animals in a few minutes to the highest bidder. Commission rates may be a flat charge per head or based on value of animals sold. There may also be a yardage fee or other charges.

Figure 22–10
Auctions have remained popular in the marketing of livestock.
(Courtesy of Kansas State University.)

Figure 22–11
A modern livestock trading center in southwestern Missouri. This facility serves as a regional stockyard and also provides a modern auction ring for the displaying and selling of livestock.
(Courtesy of Purdue University.)

Traditionally, animals are in the auction ring during bidding and prospective buyers are in bleachers, surrounding the ring on three sides. During the early 1980s, however, tele-auctions and computer networks were implemented.

Tele-auctions usually are organized by a cooperative and require preauction evaluation and marking of animals on the farm by qualified graders. Animals that meet market standards are then consigned to the auction by the seller and grouped with other animals of similar size and grade. When well organized, animals can be gathered at several locations but sold through only one auction, making more animals available to the buyers at one time. Information on weight, quality, and other characteristics is transmitted to prospective buyers by computer network and bids are conveyed by phone. Auctions generally are not used for marketing poultry.

22.6 TERMINAL MARKETS

Terminal markets are also called public stockyards or central public markets. They are livestock trading centers with complete facilities for receiving, caring for, handling, and selling livestock on a private treaty basis. All buyers and sellers are privileged to use the facilities. Essentially all livestock is sold by commission firms, who act as agent for the seller.

Terminal markets are less important now in volume of animals marketed. Numbers and sizes in the United States continued their decline in the 1970s to the late 1990s as direct marketing increased and auction volume held steady (Table 22–2). For example, the number of cattle sold through Omaha stockyards decreased more than 93 percent, from 1,165,000 head in 1972 to only 54,000 head in 1997. Today, less than 2 percent of all slaughter hogs are sold through public stockyards.

For 1994, the Packers and Stockyards Administration reported 21 terminal markets under its supervision, with only 4 reporting significant sales of all three species—cattle, hogs, and sheep. These are located in Omaha, St. Joseph, St. Paul, and Sioux City.

Several agencies at a terminal market assume rather specific functions. These usually include the stockyards company, commission firms, livestock exchange, dealers and traders, and a market news service. They sometimes include banks and insurance and transportation companies.

The stockyards company provides the facilities—pens, alleys, scales, unloading docks, feed, water, and sorting equipment. Cattle pens are usually in the open; sheep and hog facilities typically are sheltered. The stockyards company normally provides the central office building at the market. In effect, it operates a "livestock hotel." Its employees help unload or load livestock, move the animals to and from pens and scales, distribute feed, and operate the scales.

Commission firms act as agent for the seller, in most cases. For this service they receive a set commission per head, just as a real estate agent receives a commission for selling property. Commission firms operate within certain areas on the market. The stockyards company may assign certain cattle, sheep, and hog pens to each commission firm according to their normal volume of business.

As agents for the seller, personnel of the commission firm sort and pen the livestock to make them appear most attractive to buyers, and strive to obtain the highest possible prices. Commission rates are usually per head, not a percentage of sale price, and are controlled by government regulations, although the degree of rate control has been relaxed in recent years. Commission firms assume responsibility for obtaining payment for livestock sold; after deducting commission fees, feed cost, yardage fees for the stockyard company, insurance, and other expenses, they forward the remainder to the seller or the seller's bank. The yardage fee and feed sales income are the major source of income to the stockyards company; commission firms usually do not pay rent on the pens in which they operate.

Some terminal markets have a livestock exchange, an organization of commission firms that operate on the market. The livestock exchange promotes and publicizes the market, encourages fair dealing, settles disputes that might arise, and encourages all member commission firms to charge standard commission rates and provide good service. Most livestock exchanges employ an executive secretary to handle much of this work.

Dealers or traders operate on many markets, especially dealing in feeder cattle. They rent pens from the stockyards company, buy and take title to livestock, then sort and group for resale, hoping to make a profit. At some markets a certain group of the cattle pens is designated as the "dealer and trader division."

A variety of livestock buyers operate on a terminal market. Buyers for processors adjacent to the market may purchase slaughter cattle, hogs, and sheep of various weights and grades. They may be competing with buyers for other processors, or order buyers, each of whom might buy for several distant processors.

Farmers and feeders may be on hand to purchase feeder cattle or lambs, though most buy through feeder order buyers or auctions. Few feeder pigs are sold at terminal markets.

Terminal livestock markets have certain specific characteristics. A nod of the head or a spoken word takes the place of a written contract. All livestock is sold on a live basis. Payment is made on the basis of weights taken soon after a sale is made. For slaughter animals, the buyer's judgment of dressing percentage and carcass quality is pitted against that of the seller or commission firm representative.

22.7   DEALERS AND ORDER BUYERS

Some livestock dealers own or rent specific "yards" where they actively buy and sell feeders or slaughter livestock. Independent dealers operate buying stations for fed hogs, cattle, and lambs in the Corn Belt and in other major feeding areas. Animals purchased are then sold to close or distant processors, the dealer hoping to make a profit on price per hundredweight or on weight.

Many dealers in or adjacent to heavy grain-producing areas buy large numbers of feeder lambs, calves, or pigs for resale. In some cases, feeder cattle may be purchased for the dealer in the producing area by an order buyer. The same may be true for lambs and pigs.

Dealers in many parts of the country handle dairy herd replacements, breeding animals, and a variety of other livestock. Most dealers also have some other business, such as an auction or a truck line. A few operate feedlots. Because dealers often buy odd groups to hold for an expected price increase, feedlots or a farm to handle these animals is helpful.

Even though many dealers operate at a specific establishment and do a relatively routine business buying and selling feeders or slaughter animals with steady clients, some are much more speculative. They buy odd lots of livestock at the market, at country auctions, or from other dealers, then sort and regroup for later sale at the same or other markets.

The term *order buyer* was mentioned earlier as one who buys fed cattle, lambs, or hogs from a producer or at a terminal market or auction for some distant processor. The order buyer does not take title, but receives a commission payment for this service, and handles all sorting, loading, and insuring of purchased animals for the client.

An order buyer may also buy feeder animals at a terminal market, at an auction, or on the ranch or farm where they have been raised. Order buyers are not restricted to place of operation. Processors and feeders often are willing to pay for the services of an order buyer because (1) the buyer is where the stock is available, and (2) they depend on the order buyer's judgment and ability to buy what they want at the lowest possible prices.

Terminal markets, auctions, dealers, and order buyers are separated here for discussion purposes, but actually there is much overlap. Many who own or manage auctions also act as dealers or order buyers. So may commission salespeople who work primarily on terminal markets. Depending on the size of the market involved, many such arrangements come under scrutiny of federal regulations. (See Section 22.10.)

22.8   MARKET POOLS AND GROUP MARKETING

In some situations, the medium- and small-sized producers pool their animals with those of other producers for the purpose of attracting more or different buyers, exercising some group control over supply and thereby strengthening price, or saving on marketing costs. Though volume marketed through pools has not been high, pools have been used in many geographic areas for all species, especially in the marketing of breeding ewes and rams and feeder pigs. For example, approximately 300 hog producers in Iowa and neighboring states are producing "natural pork" as members of the California-based Niman Ranch group. Most are processed by Sioux-Preme Pork at Sioux Center under a toll arrangement, and the destination is the white tablecloth market on the east and west coasts. Producers are inspected about once a year for compliance to deep litter production and no antibotic use.

Group marketing (sometimes referred to as "alliances") has significantly increased in the swine industry in recent years, especially with groups who can provide a large volume of hogs with high-lean, uniform carcasses and command a higher price. Some hog producer groups pool together primarily to reduce trucking costs and to receive higher prices through access to additional markets. Other groups are formed for the purpose of negotiating market agreements.[3]

22.9 MARKETING COSTS

A number of expenses are incurred in livestock marketing, varying according to type of market, distance livestock are moved, and other factors. These may include commission charges, yardage fees, insurance in transit and at the market, transportation, and shrinkage of or injury to the livestock. Deductions from sale price also may be made at the market or processing plant for use in meat promotion. (See Section 26.5.) This is especially true for slaughter livestock.

Commission rates and yardage normally charged at terminal markets are relatively small—each under 1 percent of the value of the animals sold. Where fees are on a per-head basis, it may cost more to sell a poor-quality animal on the market in relation to its value.

Commission and other charges for auctions are not markedly different from those at terminal markets. Other charges, such as insurance and transportation, vary so greatly that a thorough discussion is not worthwhile.

An advantage often given for direct marketing is that there is no out-of-pocket marketing cost charged to the producer or feeder. However, there are still transportation, transit insurance, and shrinkage costs. When processor-owned buying stations are used, there is some cost in maintaining them as well. Because there is no yardage fee, this cost eventually must be reflected in livestock prices. Also, mileage cost for buyers who visit farms or feedlots for purchasing must be paid, and may well influence livestock prices. Inexperienced producers may also sustain a hidden cost due to inability to judge the market value of their livestock (for example, weight, quality grade, yield grade, dressing yield), inability to bargain on a par with a professional buyer, or inadequate knowledge of current market prices.

Weight loss—shrinkage—may be a large expense in livestock marketing, depending on when the shrinkage occurs and when the animals are weighed to establish sale weight. If weight loss results only from loss of gastrointestinal contents or urine, then the shrinkage is only superficial and there is no real loss of tissue. If this is the case and slaughter animals are sold on a carcass weight basis, there is no financial loss to the seller. If, however, animals are sold on a live weight basis with the price already determined, financial loss to the seller and benefit to the buyer may be substantial.

Seldom does much tissue shrinkage, usually due to loss of tissue moisture, occur in normal, short-haul movement and marketing of slaughter animals. Extremely hot weather, long hauls, and poor watering facilities at the market would, however, tend to allow greater tissue shrinkage.

Most of the shrinkage due to gastrointestinal content or urine loss occurs during loading and the first few hours of a trip. Note in Table 22–3 that percentage of shrinkage in fed cattle doesn't change much in trips of 2 to 9 hours.

---

[3] *Pork Industry Handbook,* PIH-12, 1996.

TABLE 22–3	Shrinkage of Fed Cattle in Transit		
Hours in Transit	Number of Shipments	Total Number of Cattle	Shrinkage (%)
1	7	615	1.70
2	24	1,138	4.24
3	42	1,415	4.98
4–6	24	1,001	5.42
7–9	50	2,132	5.06
10–17	852	29,769	6.20
18–35	97	5,531	9.63
36–59	85	3,610	7.53
60–83	39	2,470	8.60
84 +	22	1,078	10.81

Source: *USDA, SEA,* Guidelines for Establishing Beefpacking Plants in Rural Areas, *Agriculture Handbook 513, 1978.*

It is apparent that the time and conditions under which animals are weighed are important to determine who, if anybody, loses because of transportation shrinkage. Although shrinkage may be less when slaughter animals are sold direct to a local processor, the amount depends on when they are weighed—at arrival or later—and whether they have access to feed or water after arrival and before weighing. Because of greater distance, on the average, to terminal or auction markets, shrinkage may be more, but animals usually have access to feed and water before they are sold and the pay weight established.

Tissue shrinkage is more likely to occur in movement of feeder cattle and lambs because longer distances are often involved. Unless animals are in poor condition due to shipment stress, such shrinkage might be quickly made up once the animals arrive and are given ample feed and water.

It is common in certain market situations for a "pencil shrink" (calculated shrinkage value) to be imposed. This is especially common where weight is determined at or near the feedlot or production unit, and avoids both the cost and disagreements regarding feed and water given to the animals before slaughter or after arrival at a feedlot.

Most livestock goes to market in trucks and trailers. Interstate highways, special truck routes in cities where markets are located, and much emphasis on driver education to prevent crippling and bruising of stock have greatly decreased losses in truck transportation.

Most transportation losses—bruising, crippling, and death—are the fault of a person: the one who loads, drives, or handles the stock. Overcrowding causes bruising, especially of thin cows and steers that lack the protective layer of fat. Mixing species or different weight groups is another common cause of injury. Extremely hot, humid weather is a main cause of death in hogs and sheep in transit. There are other causes of death, crippling, and bruises; good judgment can avoid most.

22.10   REGULATORY AGENCIES

The Packers and Stockyards Act, enacted in 1921 and amended several times since, provides for governmental scrutiny of public livestock markets, dealers, and processors who engage in interstate trade, as well as processors who engage in livestock feeding. The regulation covers terminal markets and auctions, as well as meat processors. Markets and meat processors who operate completely within state boundaries usually are subject to the laws of that state.

The act, administered and enforced by the USDA, provides that there be no unfair competition, monopoly, price controlling, or restraint of trade. Because all such markets are considered public markets, the "public" is to be protected. Regulations require that records of all transactions be kept and made available for routine inspection. Processors are prohibited from giving any undue or unreasonable preference or advantage to any person or any locality.

Enforcement offices of the Grain Inspection, Packers and Stockyards Administration (GIPSA), part of the USDA, are maintained at major markets to provide routine checking of operations and to receive complaints of alleged infractions. GIPSA officials check scales for accuracy and monitor weighing procedures. Those guilty of scale manipulation or of causing false weights are subject to provisions of the Packers and Stockyards Act, including heavy fines and imprisonment.

22.11   MARKET NEWS

Most commercial feedlots, many swine, broiler, and turkey producers, and several other farmers and ranchers subscribe to some form of electronic market data service that provides immediate price and related information on a video screen in the manager's or producer's office. Services are commercially available that provide futures prices on live cattle, feeder cattle, live hogs, pork bellies, corn, and soybean meal; option prices for meat and grain; cash prices for livestock, broilers, turkeys, meat, grains, and hay at many locations; USDA and private market commentary; historic futures charts for many commodities; and related information—from a few dollars to about $30 per month.

Processors and traders would have the same or more complete electronic services, and such video screens are often provided in the lobbies of terminal markets, auctions, and other centers of market activity. The input to these services includes both the Market News Service of the USDA and private sources. The Market News Service, often in cooperation with state agencies, collects information at major markets, buying stations, auctions, and processors in person or by phone throughout the day, and electronically dispenses it to media and other points.

Today, online service provides rapid electronic mail communication and access to the Internet's "information superhighway" to many American farmers. Its worldwide popularity is evidenced by a doubling in its use in 1 year to 26 million people worldwide in 1996. Its number of users is estimated to exceed 50 million people in the early 2000s.

Radio, television, and newspapers provide market news information, but generally with less detail and with a time lag.

Private market and consulting services and some producer associations provide newsletters for a subscription fee, generally on a weekly basis, and many provide subscribers or members a call-in number for more current analyses and sales advice; the taped message is updated several times daily. The Nebraska Cattlemen's Association and Texas Cattle Feeders Association provide such services, including example sales for different classes and weights of animals.

QUESTIONS FOR STUDY AND DISCUSSION

1. Describe the typical U.S. cattle and swine cycle. Relate the price of livestock to livestock numbers and per capita consumption.

2. Why are poultry cycles less obvious?

3. What are some seasonal factors that affect price and availability of livestock?

4. What effect does increased confinement and continuous production of animals such as broilers, turkeys, veal, pigs, and market cattle have upon seasonal patterns?

5. Define *direct marketing*. Is this method of marketing increasing or decreasing? Why? Also define *group* or *alliance marketing*.

6. What are two possible advantages of direct marketing to the seller of livestock?

7. How are animals sold on a "grade and yield" basis? Which seller benefits the most, the one with high- or low-quality animals?

8. Why might a farmer prefer to sell through a local auction rather than through a terminal market?

9. Review the typical procedures of selling by auction.

10. Why has the number of terminal markets steadily declined? What are the functions of (a) the stockyards company and (b) commission firms?

11. Describe what an order buyer does.

12. What is meant by "shrinkage" in market animals? Which method of marketing usually would result in higher shrinkage and transportation costs?

13. What are the functions of the Packers and Stockyards Act?

14. Review the current methods that provide the livestock producer up-to-date market news.

Evaluating 23 Slaughter Animals

All modern meat-processing plants have adopted new technology for measuring carcass composition and establishing carcass value. (See Chapter 25.) The breeders and feeders of livestock respond to the need for change in carcass quality as demanded by processors and consumers (both domestic and international). Breeders select seedstock that produce leaner carcasses. Feeders carefully evaluate feeder animals that best conform to industry standards.

As some animals continue to be sold on a "live" basis, buyers must be able to visually appraise their value with a high degree of accuracy. In the evaluation of live slaughter animals, buyers for meat processors estimate dressing percent, carcass grade, cutability, and other quality characteristics, then offer a bid per 100 pounds of live weight (Figure 23–1). Accurate live animal evaluation depends on the evaluator's knowledge and ability to determine how various factors—meatiness, finish, age, weight, bruises or injuries, and health—influence carcass merit. Evaluators (buyers or sellers) should also consider the value of byproducts such as wool, hide, and viscera.

Animal evaluation research has disclosed the degree to which certain live animal traits are correlated with carcass merit. These are discussed in this chapter. Also, several surveys that have measured the accuracy with which buyers estimate dressing percent and carcass merit are summarized.

Accurate and fair evaluation of slaughter animals is not always accomplished. For this reason, more livestock are being sold on a carcass grade and yield basis. This method of selling eliminates the necessity for appraising carcass traits on the live animal, or of estimating carcass value.

Problems yet remain in carcass appraisal. Processors, and also their customers, are continually searching for quicker and simpler techniques for accurately appraising carcass traits, such as percentage of lean, tenderness, and flavor.

Live animal and carcass grades, used as standards for comparing and evaluating as well as for grouping, are listed, illustrated, and discussed in Chapter 24.

In the case of broilers and turkeys, age, weight, feeding regime, and genetic background are generally so standard within a geographic region that appraisal is not a difficult task. Discussion of carcass value of slaughter poultry will be restricted, therefore, to brief treatment in Chapter 35.

Upon completion of this chapter, the reader should be able to

1. List significant factors that influence the value, per pound, of animals ready for slaughter.
2. Differentiate between fed steers and heifers of the same age in finish, dressing percent, and cutability.

Figure 23–1
When livestock are not sold on a grade and yield basis, both seller and packer buyer must be able to appraise slaughter animals accurately.
(*Source:* USDA.)

3. List the visual clues and also the measuring systems that are used to differentiate among animals in terms of the proportion of lean their carcasses will yield.

4. Define *cutability* and list the factors used to determine cutability.

5. List injury locations and illnesses that are most common in slaughter livestock.

23.1   AGE AND WEIGHT

It generally is assumed that meat from older animals is less tender. That is true when you compare fed cattle or calves with aged cows or 6-month-old pigs with mature sows. But within the age brackets that most finishing occurs, most research has shown a positive correlation between age and *tenderness,* likely due to the length of the finishing period. The correlation is usually not high; 5 to 6 percent of the variation in tenderness has been shown to be due to age. Inheritance is also a significant factor in tenderness. Heritability of tenderness in cattle is about 60 percent, which is quite high (Table 18–1).

Meat from older animals—cow beef, mutton, and pork from older hogs—usually has a more intense and distinct flavor. Veal, young spring lamb, and roasting pigs have a relatively mild flavor.

Amount of flavor preferred by consumers varies greatly and is usually not a basis for discriminating against older or younger animals in slaughter livestock appraisal, except in the case of sheep and lambs. The flavor of lamb is unique and slightly stronger than beef or pork. Therefore, many consumers purchase only young lamb because they do not care for the stronger "mutton" flavor that comes with advanced age in yearlings and ewes. There is a rather severe price discrimination on yearlings and ewes, compared to lambs.

Older animals carry more *finish,* unless they have been on a limited ration, so the meat is usually juicier and more palatable. As the percentage of fat in carcasses increases, within a given quality grade, the percentages of lean, bone, protein, and moisture decrease. In general, these also are the heavier carcasses.

TABLE 23–1	Effect of Slaughter Weight on Carcass Composition of Swine					

	Live Slaughter Weight[a]					
	73 kg	*85 kg*	*100 kg*	*112 kg*	*126 kg*	*137 kg*
Carcass length (cm)	72.1	74.4	77.3	81.6	83.0	86.0
Ham (%)	25.5	25.4	25.1	24.4	24.6	24.4
Loin (%)	21.64	21.59	22.52	22.60	23.30	23.40
Lean in ham and loin (%)	56.90	56.20	55.45	56.35	54.70	53.55
Fat trim from ham and loin (%)	28.45	30.35	32.00	31.30	33.75	35.30

[a]*Converted to pounds the corresponding weights are close to 33, 39, 45, 51, 57, and 62 pounds, respectively.*
Source: Martin, A. H., et al., J. Anim. Sci. *50:699, 1980.*

The USDA grade specifications favor younger animals. Older animals grade lower when finish and other quality traits are similar. Lighter and younger animals grade higher with less finish and also yield a lower ratio of carcass to live weight. Carcasses from older animals have larger, harder, whiter bones and usually a darker colored lean. The connective tissue in meat becomes tougher as animals advance in age.

Hogs generally are full-fed a growing-finishing ration and tend to fatten early in life. After hogs attain 50 pounds, the rate of increase in fat deposition is faster than the rate of increase in muscle growth (Table 23–1). Therefore weight, in conjunction with other quality measurements, is helpful to a processor in appraising carcass merit of hogs.

Because sheep and cattle are raised and finished on varied feeding programs using rations differing in energy level, weight of these animals is not as highly correlated with meatiness and carcass merit.

Wyoming studies[1] suggest that heavier carcasses with lean lamb characteristics can be produced from later-maturing breeds such as the Columbia. The studies imply that the commercial practice of discounting the value of heavy carcasses is inappropriate and punishes the acceptable carcasses from breeds of larger mature size.

Even with sheep and cattle, however, weight is important to processors because of consumer desire for cuts in certain weight groups. Shoppers generally prefer moderate-sized cuts of beef, pork, and lamb. Restaurants, hotels, and institutions, on the other hand, often prefer heavier cuts. They serve large groups, and heavier cuts usually shrink less during cooking because of less surface per unit weight, yielding more servings per pound of meat purchased.

23.2 INFLUENCE OF SEX AND SEX CONDITION

Sex, or gender, is of little importance in appraising slaughter lambs or hogs, except as it might, in the case of pregnancy, influence dressing percent. (See Section 23.5.) Hogs normally are slaughtered at 5 to 6 months of age, when there is little chance of gilts having been bred. Also, all male pigs intended for market are castrated at an early age to avoid processor discounts for potentially strong meat odors.

---

[1] Snowder, G. D., et al. *J. Anim. Sci.* 72 (1994): 932.

When gilts and barrows are fed as a group[2] and are being appraised for slaughter, normally the group will contain a consistent 40 to 45 percent gilts and 55 to 60 percent barrows. Gilts, as compared to barrows, normally carry about 0.1 inch less backfat, larger loin eye muscles, and a higher percentage of lean cuts such as the ham, shoulder, and loin. Gilts also grow more slowly than barrows, but are more efficient in feed conversion. Because gilts are leaner and have less backfat, they can be grown to higher market weights with less risk of a discount. Today most sows are contracted or sold directly to companies that specialize in sausage (e.g., Jimmy Dean, Johnsonville, etc). Thus, much of the sausage comes from sows. Depending on the market, some of the primal cuts (loins, bacons, etc.) are sold directly to specialized groups and helps support some lower end markets.

A similar situation exists for slaughter lambs, except that market groups normally have relatively fewer ewes. (More must be saved for breeding than in the case of hogs.) Wether lambs have slightly less finish than ewe lambs. Ram lambs are often discounted $1.00 or more per hundredweight at slaughter time, though substantial reasons for this in terms of carcass quality are not apparent when such rams are sold at 4 to 6 months of age. Carcasses from older rams and ewes, of course, have a stronger, less desirable flavor, so are heavily discounted. Such carcasses typically are not sold through regular retail channels, but rather are boned out and mixed with other meats for use in "processed" meat items.

Slaughter heifers almost always sell for $1.50 to $3.00 less per hundredweight than steers of similar weight and finish. The price differential is usually larger in older and heavier animals, where heifers are more likely to be pregnant and dressing percentage is consequently lower. The practice of pregnancy checking of feedlot heifers and aborting those in early pregnancy has reduced the number of pregnant slaughter heifers.

At the same age, heifers tend to be slightly fatter and therefore attain a similar degree of finish earlier than steers. They also yield a smaller proportion of trimmed lean cuts, due largely to their smaller carcass size and tendency to be overfed (Table 23–2).

TABLE 23–2	Influence of Sex and Breed on Carcass Composition of Cattle		
	Heifers	**Steers**	**Bulls**
Holstein:			
Avg. slaughter weight (kg)	281.6	331.0	328.3
Muscle (%)	58.55	58.14	61.93
Fat (%)	18.95	17.68	14.52
Bone and tendon (%)	21.00	22.26	22.08
Angus:			
Avg. slaughter weight (kg)	300.6	357.6	324.0
Muscle (%)	54.12	56.66	61.68
Fat (%)	27.74	25.13	18.58
Bone and tendon (%)	15.86	16.64	17.61

Source: *Fortin, A., et al.,* J. Anim. Sci. *50:81, 1980.*

---

[2] *Split-sex feeding* is practiced in many hog operations as a method for increasing efficiency and returns. Gilts and barrows are penned, managed, and fed different rations separately.

Steers normally are marketed separately from heifers because of their different rates of growth and finishing, even though they may have been fed together. Therefore, a price differential can be made more easily.

Young bulls, even when well finished, generally are discounted in price compared to steers because they are assigned a lower quality grade at the same age.[3] However, they convert feed to gain more efficiently and yield higher percentages of retail cuts than steers or heifers.

23.3   MEATINESS

The goal in meat animal production is to produce a carcass that contains a high proportion of high-quality lean, or muscle. It is certainly desirable to have a moderate degree of marbling (visible intramuscular fat) within the muscle, but excess quantities of external fat or heavy seams of fat between muscle groups (intermuscular fat) are definitely not desired. Therefore, a major task in evaluating slaughter animals is to appraise meatiness—the proportion of lean meat (Figure 23–2). It has been demonstrated widely and repeatedly that there is much range in meatiness within every species.

There are three major kinds of tissue in animal carcasses—lean, fat, and bone. If there is a higher percentage of lean, there must be less fat and/or bone. Studies have indicated that the proportions of lean and bone are correlated, though the correlation is not high. In other words, animals with a higher proportion of lean usually have a somewhat heavier skeleton and relatively less fat.

Ultrasonic measurements (see Section 19.6) generally are more accurate in predicting the percentage of lean cuts or total lean in pork carcasses than any other measurement that can be taken readily on a live hog (Table 23–3). It is certainly more accurate than any visual scoring or appraisal system that has been used, even though considerable skill in estimating backfat thickness or percentage of lean cuts can be developed.

The backfat probe was first used to appraise fatness in slaughter hogs and because of its accuracy is also used for appraising fatness of prospective breeding gilts and boars. The probe measurement of backfat thickness on live hogs is as accurate as, or more accurate than, backfat measurements taken on a carcass, as an indicator of the percentage of lean. Correlations of live probe and other measurements with certain carcass traits are given in Table 23–3.

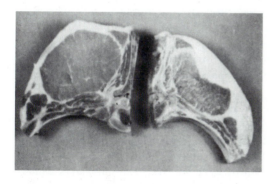

Figure 23–2
A contrast in proportion of lean in two pork rib chops.
(Courtesy of Iowa State University.)

---

[3] Stress before slaughtering of bulls must be kept to a minimum to reduce the incidence of "dark cutting beef." Darker colored lean occurs when glycogen content of the lean is reduced at or just preceding the time of slaughter. This darker color adversely affects acceptability and value.

TABLE 23–3	Approximate Correlations Between Certain Live Animal or Carcass Measurements and Carcass Traits			
Measurement	**Carcass Trait**	**Hogs**	**Cattle**	**Sheep**
Fat thickness, live	Percent lean cuts	−0.70		
	Percent primal cuts	−0.60		−0.55
	Fat trim	0.65	0.65	
	Carcass fat thickness	0.80		0.75
	Percent muscle		−0.65	
Depth of rib eye muscle, live	Depth of rib eye muscle		0.80	0.45
Width of leg, live	Percent lean cuts	0.65		
Fat thickness, carcass	Total lean	−0.60		−0.60
	Percent trimmed primal cuts	−0.60	0.40	−0.65
	Fat trim	0.70	0.70	0.65
	Percent muscle		0.60	
Area of rib eye, carcass	Total lean	0.70		0.40
	Percent trimmed primal cuts	0.70	0.30	0.40
Percent kidney fat	Percent lean		−0.55	−0.60
Percent brisket fat	Percent lean		−0.70	

Source: *Compiled by the authors and colleagues.*

A review of how meatiness is measured is provided in Section 19.6, and Figure 19–8 illustrates why the probe is not so valuable as an indicator of meatiness or total lean in cattle and sheep. The irregularities of the rib eye muscle circumference prevent accurate measurements.

Ultrasonic measurements are more adapted to assembly-line use in appraising carcasses on a "carcass merit" pricing basis, or hogs in an alleyway, than the carcass backfat measurement or live probe, and is much more accurate than visual grading. For more details on electronic sensors for assessment of carcass composition and value, see Box 23–1.

Variations in relative meatiness among beef carcasses of the same quality grade are significant. Some retailers personally select their own carcasses; they are able to choose those which will "cut out" the highest proportion of trimmed retail cuts.

Although the probe and other measuring techniques are emphasized as being extremely accurate in appraising meatiness, people also can develop skill and accuracy if they are observant, gain much experience, and compare their appraisals on a periodic basis with some objective and accurate measurements, such as live probe, carcass backfat thickness, or percentage of lean cuts.

Visual appraisal of leanness in hogs should include observation of the jowl, underline, ham, tailhead, and other points (see Appendix A). In lean hogs the jowl is trim and solid, not heavy, soft, and loose. The underline and side are trim and smooth; heavy creases and wrinkles do not appear as the animal walks. The ham is usually somewhat tapered at the base, not necessarily deep and full. The tailhead is usually prominent on a lean hog, similar to that on a heavily muscled horse. The tailhead of an extremely fat hog usually appears countersunk, like the stem of an apple, surrounded by a roll of fat.

BOX 23-1

Advancing Technology in Pork Quality Measurements

More and more producers of high-quality meat carcasses choose to have their animals marketed on the grade and yield basis. Most processors agree, but want a method that is accurate and rapid enough that it does not slow down slaughtering and processing time. Researchers in educational institutions and private industries are very much aware of both producer and processor needs. To date, they have made several technological advancements for assessing carcass quality on-line. The methods described below have been tested primarily with pork carcasses, although their principles should apply to red meat species as well.

● *Electromagnetic Scanning.* Scanners consist of an electromagnetic coil large enough to pass whole carcasses or primal cuts through the core. The energy of the weak magnetic field generated in the coil is differentially absorbed by the highly conductive lean of the carcass or cut and the less conductive fat tissue, which absorbs little or no energy. Automated positioning of the carcass upon entering the scanner eliminates operator error from this technology. However, a disadvantage is that an operator is required to return carcasses to the rail. Once this is overcome, this system could provide an accurate and rapid method of evaluation. To date, most researchers have worked with an electromagnetic instrument (TOBEC™) to measure total body electrical conductivity.

● *Ultrasound.* Two forms of ultrasound application are being used or tested on-line. One U.S. plant uses ultrasound-generated images to measure fat depth and lean depth over the loin, off midline at multiple points between the tenth and last ribs (counting from anterior to posterior). Average multiple readings appear to provide greater precision compared to single-point measurements. This instrument appears to be accurate and have very few maintenance problems.

Another ultrasound scanning instrument being tested measures fat depths as the carcass is pulled over ultrasound sensors mounted on a curved bar. This instrument would have the advantage of fully automated operation, reducing the chance for operator error.

● *Optical Fat/Lean Probe.* This method has been the most prevalent technology for measuring carcass composition on-line. Developed in New Zealand and Europe, it has been adopted as an official technique for several years. It has had widespread use in the United States since about 1991.

This device consists of a needle probe with two light diodes near the end. One diode emits light and the other detects the reflectance of emitted light. The higher reflection of light from fat versus lean tissue allows the device to measure both fat and lean depth. The fat depth is the main predictor of composition with this instrument. The main disadvantage associated with this technology is that operators must be highly trained and consistent in probing at the proper location. Correct probe placement is 7 cm off the carcass midline between the third- and fourth-last ribs. Both proper probe placement and angle of probe are important. Research has shown that an inherent bias in the single-point fat/lean depth measurement may result in underestimation of lean in some of the leaner pigs.

● *Ruler.* The simplest method is to use a small ruler to measure the thickness of backfat along the midline of the split carcass directly opposite the last rib. Its biggest drawback is that accuracy is highly dependent upon the accuracy of the carcass split. Also, it is difficult on high-speed lines to discern the exact line of demarcation between fat and lean because the connective tissue that covers the muscle may be exposed and appear as fat. A third drawback is that the measurement has an inherent bias that overestimates the lean in fatter hogs. These limitations have prompted researchers to pursue the previously described methods.

Source: Adapted from a progress report; courtesy of Dr. John C. Forrest, Purdue University.

It is apparent that the loin of a hog must be wide enough for a large, bulging eye muscle on either side of the backbone, yet without excess finish (Figure 23–3).

Other observations help in predicting meatiness, regardless of species. Especially meaty (muscular) animals are not uniform in width. Their shoulders and hind legs (round, leg, or ham) are *bulging* and muscular. They are wider in the shoulder than

Figure 23–3
Diagrammatic profiles of lean and fat market hogs, cattle, or sheep.

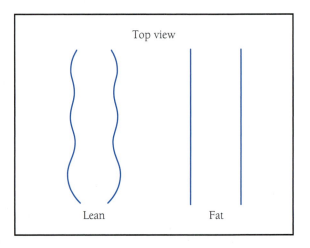

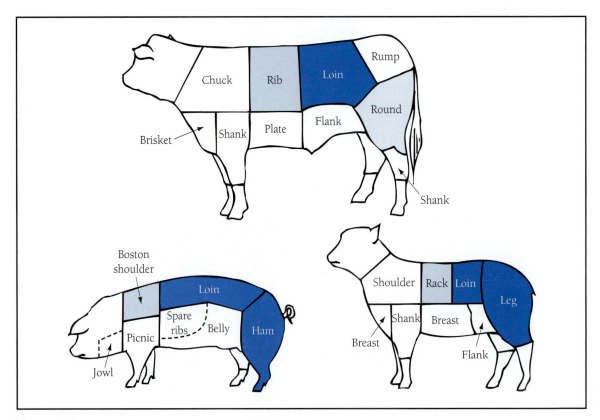

Figure 23–4
Normal wholesale cuts of beef, pork, and lamb. Darker shading indicates higher total value (weight × price). (*Source:* Iowa State University.)

immediately behind the shoulder, and wider at the hind leg than immediately in front of the leg. It is worthwhile to review typical muscle patterns in an animal anatomy book. Such a review demonstrates that animals extremely uniform in width are usually not meaty; they just have surplus fat in the areas lacking in muscling.

Muscling in lean animals is especially noticeable as the animal walks. Muscle movements are not hidden by thick layers of fat if the animal is meaty. The hindquarter and the rib area, yielding mostly steaks, chops, and the better roasts, are the most valuable in carcasses of all red meat species. Figure 23–4 indicates the approximate relative value of the wholesale cuts (see also Table 25–1).

It is apparent that high-priced cuts are those with a higher proportion of thick muscles that run in the same general direction. This makes cooking, carving, and eating easier and simpler. These cuts also carry muscles that normally are more tender because they are largely structural muscles and used less for locomotion. In wholesale and retail outlets, the hindquarter normally sells for about 20 percent more per pound than the forequarter.

23.4 FINISH

During the finishing phase of feeding, fat is deposited outside the muscle groups— below the hide and on the carcass exterior (subcutaneous fat), in the abdominal cavity, and around individual muscle groups—and also within muscle groups. Rate of deposit is discussed in Chapter 9. Much of the fat deposited outside muscle groups (intermuscular fat) is trimmed away during processing, in preparation for cooking, or at the table, and so contributes little to the pleasure of eating meat. That fat deposited within muscle groups, when finely dispersed through the meat, is called *marbling* (Figure 23–5) and contributes to palatability and desirability—juiciness, flavor, appearance, and perceived tenderness.

Juiciness is probably the single most important quality factor contributed by marbling. It apparently makes meat more desirable by providing lubrication for chewing and, by being in liquid form in meat that is hot, carrying the flavors quickly to

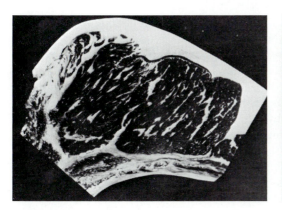

Figure 23–5
A contrast in marbling. The steak on the left has much more marbling, what the USDA describes as moderately abundant. This gives more juiciness to every bite. The steak on the right has "slight" marbling. Despite this marbling, customers may choose one over the other and that preference could easily change as the desires of a consumer change.
(Courtesy of USDA.)

sensitive taste buds. Some fat on the *outside* of a meat cut helps prevent weight loss during cooking. Extreme quantities of external or internal fat, though, are undesirable. Over 200 meat flavor components have been identified, and many of them are carried in the meat fat.

Appearance of meat cuts is influenced by degree of finish. Those with little marbling, especially pork, are soft and watery. They do not "hold their shape" in display coolers and are generally not appealing to the consumer. After cooking they are dry and, in the case of roasts, tend to fall apart, making carving difficult. Cuts with too much fat are likewise undesirable to the consumer.

Although the direct effect of marbling on meat tenderness, as measured by mechanical devices, may be slight, those who eat meat with considerable marbling think it is more tender. This is probably because it is somewhat juicier and generally more palatable.

The amount of marbling in meat is not easily appraised by looking at a carcass or evaluating the finish on a live animal. Generally, but not always, animals with more external fat carry more marbling within the meat. Correlations between external finish and marbling are usually between 0.40 and 0.60.

Even so, until more accurate techniques are developed and tested, visual appraisal or other methods of measuring external fat of live slaughter animals must be used as an indicator of marbling. Finish over the back, loin, and rump, as well as over the ribs, can be felt. Because animals normally fatten first over the back and fore ribs, feeling over the loin, rump, and rear ribs often will disclose the degree of fattening that has occurred.

Well-finished lambs are thick in the dock (the tailhead region). In steers the brisket gets bulgy and that portion of the scrotum remaining after castration fills with fat. Cattle and lambs with considerable finish walk with some difficulty, swinging a bit from side to side as fat rolls in front of the flanks, rather than with a free-and-easy stride. Amount of fat is inversely related to amount of lean; techniques for measuring relative quantities of the two kinds of tissues are presented in Section 23.3.

23.5   DRESSING PERCENT AND CUTABILITY

Dressing percent (chilled carcass weight divided by live weight × 100) of animals has no influence on the merit of the carcass but is important in appraisal of live slaughter animals. Packer buyers of live animals are familiar with market value of carcasses, but need to calculate the value of the live animal per hundredweight in order to make a bid. If a buyer overestimates dressing percent of a 1,200-pound steer by one percentage unit, the animal will yield 12 pounds less carcass than had been presumed. If the carcass is worth $133 per hundredweight, the animal yields $16 less in carcass value. Obviously, then, accurate estimation of dressing percent is important. Approximate dressing percent values for various classes and grades of slaughter animals are given in Table 23–4.

What factors influence dressing percent? Conformation and finish of the carcass have some effect. Fatter animals dress high and usually are thick and blocky in shape. Thin, long-legged, long-necked, angular animals, such as utility cows, dress very low.

Another major influence on dressing percent is the volume of feed and/or water in the digestive tract—the "fill." Excessive fill lowers dressing percent markedly because the carcass then represents a smaller proportion of live weight. Thickness and weight of cattle or calf hides or sheep pelts influence dressing percent. Also, a heavy growth of wool means a lower dressing percent.

TABLE 23–4	**Approximate Dressing Percents of Livestock and Poultry**				

Hogs		Cattle		Sheep (Shorn)	
Average	70	Average	60	Average	50
Barrows and gilts:		USDA:		USDA:	
		Steers and heifers:		Prime lambs	54
U.S. No. 1	70	Choice and Prime	63	Choice lambs	53
U.S. No. 2	71	Standard and Select	59	Good lambs	50
U.S. No. 3	72	Commercial	57	Fat ewes, utility	49
U.S. No. 4	73			Thin ewes	44
U.S. Utility	69	Cows: USDA utility,			
Sows	68–72	cutter, and canner	45–53		

Chickens		Other Fowl	
7-wk broilers:		Ducks, 5.6 lb	71.4
Male, 3.8 lb	76.3	Geese, 12.7 lb	68.5
Female, 3.0 lb	76.6	Turkeys:	
12-wk roasters:		Male, 5.5 lb	80.0
Male, 8.2 lb	73.2	Female, 13.8 lb	79.0
40-wk Leghorn hen	66.4		

Source: *Adapted from USDA, AMS,* Livestock Marketing Handbook, *AMS 556, June 1978; Snyder, E. S., and Orr, H. L.,* Poultry Meat-Processing, Quality Factors, Yields, *Ontario Dept. of Agriculture, Publication 9; and Jensen, L. S., "Turkey Growth Standards,"* Turkey World *56:10, 1981.*

In visual appraisal, extremely low-dressing animals or groups generally were estimated too high and high-dressing animals or groups were estimated too low. Also, there was a tendency to underestimate dressing percent more often than to overestimate. This might be an expected, cautious thing for a packer buyer to do.

Dressing percent is important in appraisal of slaughter hogs, but not as important as in cattle or sheep. The major reason is that there is less variation caused by differences in content of the digestive tract because the pig is a nonruminant.

As earlier indicated, carcass value is not necessarily determined by a high dressing percentage. To the contrary, excessively fat pigs tend to dress higher but yield a lower proportion of lean cuts than lean pigs.

A better indication of carcass value is *cutability,* that proportion of the carcass that is actually salable to the consumer. A high-cutability carcass has a high proportion of its weight in trim, heavily muscled, preferred cuts. Cutability is influenced by (1) the amount of excess finish and (2) the degree of muscle development. These are estimated by measuring external fat, kidney and pelvic fat, and rib eye area in beef, or leg conformation in lambs. These factors are used in an equation to provide an overall cutability value or yield grade (see Section 25.4).

23.6   GRADE

USDA carcass grades and specifications, as well as private grading systems, are discussed in Chapter 24. *Grades* are recognized as grouping of carcasses according to cutability—proportions of lean or valuable cuts—and relative quality of the meat. The purpose of this section is to discuss the accuracy with which carcass grades of live animals can be estimated.

TABLE 23-5	Accuracy of Estimating Carcass Grades on Live Hogs					
	Hogs Graded Individually		Hogs Graded in Groups		Difference in Percentage	Years Buying
Buyer	Number	Percentage Accuracy	Number of Groups	Percentage Accuracy	Accuracy	Experience
A	50	76.0	2	98.0	22.0	9.0
C	100	62.0	4	84.0	22.0	3.5
D	149	51.6	6	77.4	25.8	6.0
E	171	55.2	5	93.1	37.9	5.0
G	53	60.4	6	81.1	20.7	16.0
P	123	49.0	5	83.7	34.7	0.6
S	151	55.3	6	76.6	21.3	0.5
T	159	58.2	6	76.4	18.2	3.0
U	100	46.9	4	75.5	28.6	2.0
V	151	34.5	6	52.7	18.2	20.0
Average	128	54.0	4.8	80.1	26.1	9.9

Source: *Adapted from* Purdue Univ. Agr. Exp. Sta. Bull. SB 650, *1957.*

As with dressing percentage, buyers more accurately predict the carcass grade of a group of animals than of individuals. Individual errors in appraisal tend to cancel out each other, to some degree, in group appraisal.

There is a tendency for buyers to overestimate the grade of lower quality and underestimate the grade of higher quality animals. There seems to be a basic human tendency to avoid making estimates that are extremely low or high. This means that if animals are priced according to grade estimates, sellers of very low-quality or low-cutability animals profit but sellers of high-quality and high-cutability animals lose, by live animal appraisal.

The accuracy of 16 packer buyers in estimating carcass grade of individual market hogs and groups of hogs was determined, and results are shown in Table 23–5 for the five highest and five lowest. Note the large range in estimating ability. The next-to-last column of the table shows improvement in appraisal accuracy achieved when grades are estimated on a group rather than on individual hogs.

The correlation between years of experience in buying live hogs and accuracy of estimating carcass grade was low. However, those buyers who had spent considerable time in appraising carcasses (data not shown) tended to do a better job estimating carcass grades on the live hogs.

It is apparent that there is considerable range in the ability of buyers to appraise carcass grade on live hogs. Consequently, pricing accuracy may be poor when slaughter animals are sold on a "live" basis. This discussion illustrates that live animals can be appraised somewhat accurately by some buyers.

23.7   BYPRODUCTS

Byproducts represent a rather large portion of the value of slaughter sheep, cattle, and calves (Table 23–6; also see Box 25–2). There is considerable variation in byproduct value from season to season or from year to year, but percentages given in the table for

TABLE 23–6	Approximate Relative Value of Carcasses and Byproducts, as Percent		
	Carcass (including Fat Trim)	**Hide or Pelt**	**Other Byproducts**
Lambs	80.2	14.0	5.8
Calves	86.8	7.8	5.4
Cattle	90.9	5.0	4.1
Hogs	92.9	—	7.1
Poultry	98.0	—	2.0

noncarcass byproducts are reasonable averages. Certain portions of the *carcass*, such as some bones, also may be defined as byproducts.

In the case of calves, cattle, and hogs, the value of byproducts does not vary greatly among animals or groups of animals, except where cattle hides might be damaged by brands or grubs. The value of wool carried by sheep and lambs may vary greatly, however, depending on age, wool growth, and wool quality. A further discussion of the variety of byproducts yielded in animal slaughter is given in Section 25.11.

23.8 INJURIES AND HEALTH

Most bruises cause a rupture of muscle cells and hemorrhaging of the small blood vessels in the muscle tissue, usually resulting in a dark and unappealing discoloration. Because these bruised areas can't be used for human food, they must be removed from the carcass by trimming. Thus, they represent a direct loss and may diminish the market value of a major section of the carcass.

Those who regularly appraise livestock are familiar with situations in loading, hauling, and unloading where bruising is common. Hauling mixed loads—for example, cattle with hogs or sheep—usually causes more bruising on the smaller animals. Most cattle bruises occur on the prominent sections of the animal, especially where bone lies not far below the skin surface. Cattle bruises, in order of occurrence, are on the shoulder, edge of the loin, and rib. Most hog bruises are on the ham, and small numbers occur on the belly, shoulder, and back. In lambs, the major areas of injuries are the leg and loin. Loading too tightly or giving a few animals too much room in a truck or railroad car with slick floors will contribute to injuries and bruises. In such cases, buyers may assume some loss due to bruises and lower their appraisal accordingly.

Cattle that are especially thin bruise more easily because they lack the cushioning layer of fat. There are usually more bruises on carcasses of cattle that carry horns or that are extremely nervous. Buyers note these items and also note those producers whose livestock often carry bruises. Bruises may be caused by irate producers or handlers of stock, those who lack proper facilities for restraining or loading for shipment, and those who forget that valuable cuts of meat lie just under the animal's skin.

Though federal and state inspection generally ensures that meat sold at retail is free of disease and fit for human consumption, appraisers and buyers of slaughter animals must be alert to the symptoms of serious illness. When a carcass is condemned in the processing plant because of disease or other reason, the loss is suffered by the processor, unless the animal was purchased "subject to inspection."

Major causes of condemnation for all species include inflammatory diseases such as nephritis, pericarditis, peritonitis, and pneumonia; arthritis; and infections or septic conditions, which include abscesses, septicemia, and toxemia.[4]

QUESTIONS FOR STUDY AND DISCUSSION

1. Relate to meat tenderness the effects of (a) age as associated with increased connective tissue, (b) age as associated with longer finishing period, and (c) heritability.
2. What effect would tenderness, flavor, and color of lean have upon consumer preference?
3. If of comparable age and management, which would possess the greater lean-to-fat ratio: (a) steers or heifers; (b) bulls or steers; (c) wether or ewe lambs; (d) barrows or gilts?
4. Explain the basis for discounted price per hundredweight of feeder or slaughter heifers compared to comparable-quality feeder or slaughter steers.
5. Name three or more areas or parts of the live animal evaluated by the buyer to determine degree of muscling and fatness.
6. Name the more valuable wholesale cuts of beef, pork, and lamb.
7. Relate the degree of marbling to flavor, appearance, and juiciness of retail cuts of meat.
8. Although dressing percentage is not a good indicator of carcass merit, explain factors buyers of livestock use in evaluation to ensure a good dressing percentage. What are average dressing percentages for hogs, cattle, and sheep?
9. Explain the differences in calculations of dressing percentage and cutability. Which is the best indicator of carcass value?
10. Which tendency do buyers have when visually estimating the grade of low-quality animals? of high-quality animals?
11. Give the carcass areas where most injuries occur as animals are moved to market.

---

[4] USDA, FSQS. *Statistical Summary, Federal Meat and Poultry Inspection for 1980.*

Live Animal Market 24 Classes and Grades

Classes and *grades* of livestock and poultry have been established so that (1) reports of sales and prices from different market locations can be more accurately interpreted (Figure 24–1); and (2) prices can be quoted and sales made by phone, wire, or mail. Grades applied to carcasses or meat cuts serve as a *quality* guide to the consumer and, in the case of beef, pork, or lamb, provide a guide to cutability or proportion of lean.

A **market class** is defined as a group of animals separated according to use. The major classes of cattle, sheep, and hogs are "slaughter" and "feeder." As the titles imply, animals in the slaughter class are intended for slaughter and feeders are being sold for further feeding. A third group, "stockers," is often differentiated as a separate class in cattle, though most people think of stockers collectively with feeders.

In our discussion, stockers refers to young animals, both steers and heifers, that are past weaning age but which lack sufficient growth and development to be placed into the feedlot. "Stocker heifers" can refer to similar animals intended for the breeding herd. In Chapter 8, we indicated that stocker cattle receive lower energy, higher fiber rations, whereas feeders are fed a higher energy, lower fiber ration. That chapter also discusses evaluation of and grades for feeder cattle, lambs, and pigs.

Subclasses within the major livestock classes, or market classes of poultry, are made according to age (Cornish game hen vs. broiler vs. roaster), sex, sex condition (steer vs. bull vs. stag), and weight.

Market grades are further subdivisions of the subclasses. Animals are grouped according to relative merit. Among feeders, the major criteria are frame size and thickness. Among slaughter livestock and poultry, both cutability (yield) and quality are considered.

Upon completion of this chapter, the reader should be able to

> **Market classes and grades**
> Standards established by the USDA that segregate animals, carcasses, and animal products into uniform groups, largely based on the best use of an animal and quality guidelines within that use.

1. Define and differentiate between *market class* and *market grade*.
2. List and define the major classes of each species.
3. List those major characteristics that are used to classify animals or carcasses into the classes.
4. List the factors that influence live animal or carcass grade.
5. List the live animal and carcass grades for each species.
6. Differentiate between private and federal meat grading.

TOPEKA
Livestock Auction

126 Pigs (actual weights) - 10:00 A.M.

30-40 lbs.	$23.00 to $34.00
40-50 lbs.	$36.00 to $51.00
50-60 lbs.	$54.00 to $57.50
55 Pigs	avg 31 lbs at $34.00
6 Pigs	avg 48 lbs at $51.00
11 Pigs	avg 58 lbs at $54.00
6 Pigs	avg 69 lbs at $57.50

212 Dairy Cows & Spr. Heifers - 11:00 A.M.

Hol. Spr. Hfr.	$1160.00
Hol. Spr. Hfr.	$1150.00
Hol. Spr. Hfr.	$1110.00
Hol. Spr. Hfr.	$1100.00
2 Hol. Spr. Hfrs.	$1085.00
5 Hol. Spr. Hfrs.	$1060.00
4 Hol. Spr. Hfrs.	$1050.00
7 Hol. Spr. Hfrs.	$1035.00
2 Hol. Spr. Hfrs.	$1025.00
3 Hol. Spr. Hfrs.	$1010.00
3 Hol. Spr. Hfrs.	$1000.00
47 Head	$800.00 to $985.00
Hol. Fresh Cow.	$1100.00
2 Hol. Fresh Cows	$1050.00
Hol. Fresh Cow.	$1025.00
Hol. Fresh Cow.	$1010.00
Hol. Fresh Cow.	$1000.00
11 Head	$800.00 to $985.00

13 Breeder Bulls - 3:00 P.M. ...top $600.00

190 Replacement Hfrs. & Stockers - 3:30 P.M.

2 Hol. Fdr. Hfrs.	238 lbs at $62.50
2 Hol. Fdr. Hfrs.	445 lbs at $57.00
5 Hol. Fdr. Hfrs.	539 lbs at $56.00
2 Hol. Fdr. Hfrs.	550 lbs at $55.00
3 Hol. Fdr. Hfrs.	650 lbs at $52.50
2 Hol. Fdr. Bulls	183 lbs at $41.00
3 Hol. Fdr. Strs.	497 lbs at $40.00
3 Hol. Fdr. Strs.	375 lbs at $38.00
3 Hol. Fdr. Strs.	503 lbs at $38.00
3 Hol. Fdr. Strs.	608 lbs at $37.00

82 Veal - 5:30 P.M.

Back to Farm Calves	$45.00 to $62.50
Light to Cull Calves	$10.00 to $40.00

76 Bu. Lambs - 2:30 P.M.

Bu. Lamb	85 lbs at $89.00
3 Bu. Lambs	133 lbs at $87.00
13 Bu. Lambs	117 lbs at $85.00
6 Bu. Lambs	110 lbs at $84.00
3 Bu. Lambs	105 lbs at $81.00
4 Bu. Lambs	100 lbs at $80.00
Bu. Ewes	$26.00 to $45.00
4 Bu. Goats	$40.00 to $45.00

298 Bu. Hogs - 3:00 P.M.

252 Direct Bu. Hogs ....$54.00 to $55.50

NEW HOLLAND LIVESTOCK AUCTION

CATTLE CALVES SHEEP GOATS TODAY 1295 696 187 342 LAST THURSDAY 1429 759 257 224 LAST YEAR 421 202 153 160 CATTLE...[USDA]...[Supply included 32% slaughter steers and 47% cows]...sl. steers weak to 1.00 lower, sl. heifers not fully tested, Breaking Utility and Commercial cows steady, Canner and Cutter 1.00 to 1.50 higher, bullocks scarce, bulls firm to 1.00 higher. SLAUGHTER STEERS: Choice 2-3 1100/1450 lbs. 72.75-76.75, Choice and Prime 2-3 1250/1500 lbs. 76.50-77.75, Select and Choice 2-3 1075/1450 lbs. 67.00-72.00, few Choice 2-3 950/1400 lbs. 56.75-65.00. HOLSTEINS: High Choice and Prime 2-3 1300/1525 lbs. 59.50-63.50, Choice 2-3 1275/1600 lbs. 55.75-60.00, Select 1-3 1200/1600 lbs. 51.50-54.50. HEIFERS: couple Choice 2-3 1180/1265 lbs. 69.00-70.50. COWS: Breaking Utility and Commercial 2-3 bulk 36.00-39.00, high dressing 39.00-41.50, Cutter and Boning Utility 1-3 low dressing 29.00-32.00, bulk 32.00-37.00, high dressing 37.00-39.00, Canner and Low Cutter 1-2 low dressing 22.00-26.00, bulk 26.00-29.50. BULLOCKS: few Select 1-2 1100/1400 lbs. 45.50-50.00. BULLS: Yield Grade No. 1 1300/2000 lbs. 42.00-48.00, couple 1390/1620 lbs. 50.50-54.00, couple 1375/1470 lbs. 59.00-61.00, No. 2 1200/2165 lbs. 34.50-39.50. CALVES...Vealers and slaughter calves steady. Demand good for Holstein bulls and heifers to return to farm. VEALERS: few Choice 170/365 lbs. 57.00-67.00, High Good and Low Choice 150/300 lbs. 32.00-50.00, 80/110 lbs. 15.00-28.00, Standard and Low Good 60/80 lbs. 10.00-17.00. SLAUGHTER CALVES: Good and Choice 270/500 lbs. 30.00-51.00. RETURNED TO FARM: bulk 90/125 lbs. Holstein bulls 75.00-100.00, several 100.00-110.00, 80/90 lbs. 55.00-80.00, small frame 90/110 lbs. 45.00-80.00, 75/95 lbs. 28.00-50.00; bulk 90/120 lbs. Holstein heifers 110.00-150.00, few 150.00-175.00, small frame 90/110 lbs. 70.00-120.00. SHEEP...Slaughter lambs fully steady, small supply slaughter ewes steady. Supply included an estimated 90% slaughter lambs with the balance slaughter ewes. SLAUGHTER LAMBS: Choice 1-2 40/55 lbs. 135.00-154.00, 55/75 lbs. 115.00-140.00, 80/105 lbs. 85.00-95.00. SLAUGHTER EWES: Utility and Good 1-3 20.00-40.00. GOATS...[All sold by the head]. BILLIES: couple Large 124.00-148.00, Medium 75.00-90.00. NANNIES: Medium and Large 45.00-70.00. KIDS: Choice 30.00-45.00.

USDA ESTIMATED COMPOSITE OF BOXED BEEF CUT-OUT VALUES

Based on FOB Omaha basis carlot prices of fabricated beef cuts and on industry yields as of 11:30 a.m. 139 Fab Lls CH 1-+ CH 1-3 SE 1-3 50s Lds 550/700 700/850 550/700 , 700/850 14 Grd Lds Values => 118.83 116.97 97.39 96.37 -0.79 -0.94 # -0.16 ` -0.27 Rib 194.92 189.81 146.74 145.51 Chuck 75/90 lbs. 14.00-20.00, Utility 85.07 80.38 75.59 73.74 Round 110.57 110.57 90.25 90.25 Loin ! 170.98 171.37 130.41 128.87 Brisket 71.73 71.73 67.28 67.28 Shrt Plate 67.63 67.52 67.84 67.67 Flank 77.46 77.46 77.31 77.31.

EGG PRICES (Lambright's Inc., LaGrange) Week of Dec. 2, Large 84¢/doz, Medium 62¢/doz, Small 36¢/doz. Last Week, Large 84¢/doz, Medium 62¢/doz, Small 36¢/doz.

INDIANA FARMERS LIVESTOCK MARKET

Cattle 215...[PDA]...Compared with last week's sale...steers, heifers & cows steady. STEERS: Choice 1100/1450 lbs. 67.25-74.50, Select 57.00-65.75. Standard 40.00-56.00. HOLSTEINS: few Standard 34.75-38.50. HEIFERS: Choice 1050/1400 lbs. 65.00-70.00, Select 52.50-63.25, Standard 35.00-46.50. COWS: Breaking Utility and Commercial 32.00-36.00, Cutter and Boning Utility 26.00-31.50, Canner and Low Cutter 18.25-26.50. Shells down to 14.00. BULLS: few Yield Grade No. 1 1250/1870 lbs. 33.50-35.00, No. 2 900/1550 lbs. 25.00-31.25. FEEDER CATTLE: STEERS: few Medium Frame No. 1 275/450 lbs. 43.00-54.00, Medium & Large Frame No. 2 300/950 lbs. 29.00-40.00; HEIFERS: Medium Frame No. 1 300/775 lbs. 35.50-45.00, Medium & Large Frame No. 2 325/800 lbs. 22.00-33.50; BULLS: few Medium & Large Frame No. 2 300/750 lbs. 20.00-27.50. CALVES 85...Farm calves steady VEALERS: few Choice 140/250 lbs. 45.00-55.50, Good 140/265 lbs. 24.00-42.50, Standard and Good 50/80 lbs. 6.00-13.00. FARM CALVES: No. 1 Holstein bulls 90/120 lbs. 55.00-77.50, No. 2 90/125 lbs. 24.00-52.50; few No. 2 Holstein heifers 80/100 lbs. 52.50-80.00. HOGS 64...Barrows and gilts steady to 4.00 higher. BARROWS AND GILTS: US 1-3 225/270 lbs. 54.25-58.00, US 2-3 250/285 lbs. 45.50-49.50, US 1-3 105/215 lbs. 44.00-47.50. SOWS: US 1-3 350/600 lbs. 47.50-56.00. FEEDER PIGS 13...US 1-3 20/50 lbs. 8.00-25.00. — per head. SHEEP 15...SLAUGHTER LAMBS: Choice 65/85 lbs. 90.00-93.00. SLAUGHTER SHEEP: 17.00-32.00. GOATS 1...One Medium @ 49.00— per head.

Figure 24-1

Partial market summaries for live animals, eggs, and futures contracts from several newspapers. Note the various classes and grades listed.
Source: Compiled by the authors from various news sources.

24.1 DEVELOPMENT OF CLASSES AND GRADES

The use of classes to differentiate and describe animals probably developed early in the history of livestock trading. There were also terms, no doubt, that were used rather uniformly to denote relative quality.

Market classes as known and used today have simply developed from common usage, but are surprisingly uniform across the country. The USDA grades in use today have been developed and standardized since 1916. Picture standards (See Appendix A) and verbal descriptions of grades are available from the USDA for nearly all classes of meat animals. They continue to exist to serve the industry and are revised as need and knowledge warrant. Their use is voluntary, and many producers, traders, and consumers depend on them as a standard of quality.

Anyone is free to apply "unofficial" class and grade designations to live animals. Continued confidence in those who use these designations, however, obviously depends on their experience, knowledge, and ability to discriminate. Application of "official" USDA grades to carcasses, however, must be done by a designated USDA grader. This person also has responsibility for checking the market class of the carcasses involved. If a "yearling" is found among a group of "lamb" carcasses, it must be designated as such.

Other carcass grades, besides those of the USDA, are used. Many processors have a private grading system, including grade names that may be copyrighted and that are applied by their own employees. Public confidence in these grades depends on discrimination exercised in applying the grades to carcasses or cuts.

24.2 CATTLE CLASSES

The major classes—slaughter and feeder, as well as stocker—were previously discussed. Other classification criteria will be discussed here, with reference made to Table 24–1.

Subclasses are further defined below, recognizing that there may be some discrepancy in interpretation among various sections of the United States.

- *Steer.* A male bovine castrated when young and which had not yet begun to develop the secondary physical characteristics of a bull.
- *Heifer.* An immature female bovine that has not yet developed the physical characteristics typical of cows.
- *Cow.* A female bovine that has developed, through reproduction or with age, the relatively prominent hips, large middle, and other physical characteristics typical of mature females.
- *Bull.* A mature uncastrated male bovine.
- *Bullock.* A young (uncastrated) male bovine that has developed or begun to develop the secondary physical characteristics of a bull, usually grown and fattened like feedlot steers and heifers.
- *Calf.* A bovine usually between 3 and 8 months of age.
- *Vealer.* A bovine under 16 weeks of age (intended for slaughter) and having subsisted largely on milk or a milk substitute. "Fancy" or "formula-fed" veal receive no grain, whereas "conventional" veal usually are fed milk and grain.
- *Heiferette.* A term sometimes used to describe a heifer that calved young, perhaps at 2 years of age, then was quickly fattened for slaughter. Carcass traits are between those of a heifer and a cow.

TABLE 24–1			Commonly Used Class Designations of Cattle			

Feeder			Slaughter			
Sex and Sex Condition	*Age*	*Weight*[a] *Range (lb)*	*Sex and Sex Condition*	*Age*	*Weight*[a] *Range (lb)*
Steer	Calf	300–500	Steer		1,050–1,300
	Yearling	500–800	Bullock		1,050–1,300
	2-year-old	800–1,050	Heifer		1,050–1,300
Heifer	Calf	300–500	Cow	Young	
	Yearling	500–750		Aged	
Cow	Young	750–1,200	Bull	Young	
	Aged			Aged	
Bull				Calves	Under 500
				Vealers	300–450

[a]*Weight groups are often used in market reports, such as lightweight cattle (less than 500 pounds), medium-weight cattle (500 to 800 pounds), and heavyweight cattle (above 800 pounds).*

Source: Adapted from the *USDA*.

Classes and subclasses listed under "Sex and Sex Condition" and "Age" are commonly used in market reports (Figure 24–1). Animals in certain of these classes may be absent, or present in small numbers, at some markets. Relatively few 2-year-old steers are sold, for example, and certain classes of cattle are relatively rare at some eastern markets. Age subclasses other than those listed in the table sometimes may be used.

Calves and vealers, for slaughter, often are not differentiated as steers, bulls, or heifers, because the secondary sex characteristics have not developed to the point of influencing carcass traits at the time of slaughter. Weight groups are usually divided into 200- or 300-pound increments.

Classes do not denote whether animals are of beef, dairy, or dual-purpose breeding, or whether they may be crossbreds. Many slaughter cows have been culled from dairy herds; most vealers are of dairy breeding. There are also many steers of dairy and dual-purpose breeding as only a small percentage of the males are kept as bulls.

24.3 SHEEP CLASSES

Table 24–2 summarizes the most common classes and subclasses of sheep. Nearly all feeder sheep are lambs. Drastic reduction in the slaughter value of sheep over 12 to 14 months of age almost dictates that lambs be sold for feeding before 8 to 9 months. A well-finished slaughter lamb weighing 100 to 120 pounds live weight is preferred.

Typically no distinction is made between ewes and wethers, unless the animals are 2 years of age or older. A ram lamb is often discounted in price when sold for feed-

TABLE 24–2	**Commonly Used Class Designations of Sheep**

	Feeder			**Slaughter**		
Sex and Sex Condition	*Age*	*Weight[a] Range (lb)*	*Sex and Sex Condition*	*Age*	*Weight[a] Range (lb)*
	Lambs	Lightweight 60 Heavyweight 90+		Lambs[a]	70–120
				Lambs (shorn)	70–120
				Yearlings	100–130
Ewes	Young		Ewes	Young	100–180
	Aged			Aged	
			Rams	Young	120–300
				Aged	

[a]*Predominantly wethers and ewes; some markets may include young ram lambs when well grown and finished to market weights of about 100 to 110 pounds.*

Source: Adapted from the USDA.

ing or slaughter. (See Section 23.2.) Weight groups given in market reports usually are divided into 10-pound increments.

Most yearlings intended for slaughter are just slightly over a year of age. In most cases it probably was not intended that they be so mature before being sold for slaughter. Ewes and rams intended for slaughter generally have been culled from flocks for various reasons. They may range considerably in age or weight, though most would be relatively old.

Some of the terms used to describe certain classes or groups of sheep are given below. Note the similarity of some definitions to those given for cattle.

- *Wether.* A male ovine (sheep) castrated when young, prior to developing the secondary sex characteristics of a ram.
- *Ewe.* A female ovine of any age.
- *Lamb.* An immature ovine, usually under 14 months of age, that has not cut its first pair of permanent incisor teeth.
- *Yearling.* An ovine, usually between 1 and 2 years of age, that has cut its first pair of permanent incisor teeth but not the second pair.
- *Ossification.* The formation of bone or of a bony substance.
- *Ram.* An uncastrated male ovine.
- *Sheep.* An ovine, usually over 24 months of age, that has cut its second pair of permanent incisor teeth.

Because age is so important, USDA graders check indicators of age on the carcass. Lamb carcasses have an exposed *break joint* (made possible by a soft cartilage layer near the lower end of the ankle bone on the front legs) that can be snapped because it hasn't calcified, narrow and reddish rib bones, a relatively narrow forequarter, and a light red, fine-textured lean.

Yearling mutton carcasses may show either break joints or spool joints (the end of the bone is exposed because the cartilage has calcified and the break joint did not snap) at the front pastern. They have moderately wide rib bones with only traces of red, a slightly wide forequarter, and slightly dark red, coarse lean. Mutton carcasses (2 years and over) show spool joints, wide and white rib bones, a wide forequarter, and dark red, coarse lean.

24.4   SWINE CLASSES

Because nearly all hogs are handled similarly—on growing-finishing rations for rapid gains—there are few age classes as compared to cattle. Common classes are charted in Table 24–3 and shown in Figure 24–2.

Feeder pigs are all young and generally are not sorted by sex. They usually are listed in market reports in weight increments of about 20 pounds. "Shoats" is a term used in some areas to describe relatively heavier and older feeder pigs.

Barrows and gilts can be sold together for slaughter. They sometimes are referred to collectively as "butchers" or "market hogs." Most slaughter groups contain about 40 percent gilts; there is little difference in carcass merit of barrows and gilts. Prices usually are given for 20-pound weight brackets. Where split-sex feeding and management is practiced, groups may be marketed as all gilts, or all barrows.

TABLE 24–3	**Commonly Used Class Designations of Hogs**				

Feeder			**Slaughter**		
Sex and Sex Condition	*Age*	*Weight Range (lb)*	*Sex and Sex Condition*	*Age*	*Weight Range (lb)*
	Pigs	30–180	Barrows and gilts		160–360 (230–260 preferred)
	Shoats	80–180	Sows	Young	250–600
				Aged	
			Boars	Young	
				Aged	
			Stags	Young	
				Aged	

Source: Adapted from the *USDA*.

Feeder pigs

Figure 24–2
Two classes of market hogs. (See Section 24.9.)
(Courtesy of Purdue University)

Feeder shoats (intermediate in age and size)

Many young sows that have farrowed only one or two litters are sold for slaughter. Often referred to as "lightweight sows," they sometimes bring almost as much per pound as immature gilts. Hog raisers like to sell them before they get too heavy. Large, more mature sows are too expensive to maintain in most sow herds.

Because boars are discounted heavily when sold for slaughter, some are castrated, then sold a month or so later as stags. Some boar carcasses carry a strong, offensive odor. This is less noticeable in bulls and rams.

For clarity and comparison, various class terms are briefly defined below.

- *Barrow.*   A male porcine (hog) castrated before secondary sex characteristics develop.
- *Gilt.*   A female hog under a year of age and before farrowing a litter of pigs. (Some refer to a sow that has farrowed once as a "one-litter" or "first-parity" gilt.)
- *Sow.*   A female hog over a year of age, or which has farrowed one or more times.
- *Boar.*   An uncastrated male hog.
- *Stag.*   A male hog castrated after it has developed or begun to develop secondary sex characteristics.

24.5 POULTRY CLASSES

Although a major proportion of poultry marketed for meat is in the form of 7-week-old broilers, there are other classes and species in poultry marketing channels. The classes of live, dressed, and ready-to-cook poultry listed below are in general use in all segments of the U.S. poultry industry, though the wording generally refers to the dressed carcass. The classes are listed within species.[1]

Chickens

- *Rock Cornish game hen or Cornish game hen.*[2]   A young, immature chicken (usually 5 to 6 weeks of age) weighing not more than 2 pounds ready-to-cook, which was prepared from a Cornish chicken or the progeny of a Cornish chicken crossed with another breed of chicken.

- *Broiler or fryer.*   A young chicken (usually under 13 weeks of age[3]), of either sex, that is tender-meated with soft, pliable, smooth-textured skin, and flexible breastbone cartilage.

- *Roaster.*   A young chicken (usually 3 to 5 months of age), of either sex, that is tender-meated with soft, pliable, smooth-textured skin and breastbone cartilage that may be somewhat less flexible than that of a broiler or fryer.

- *Capon.*   A surgically unsexed male chicken (usually under 8 months of age) that is tender-meated with soft, pliable, smooth-textured skin.

- *Hen, baking or stewing chicken, or fowl.*   A mature female chicken (usually more than 10 months of age) with meat less tender than that of a roaster and nonflexible breastbone tip.

- *Cock or rooster.*   A mature male chicken with coarse skin, toughened and darkened meat, and hardened breastbone tip.

Turkeys

- *Fryer-roaster turkey.*   A young, immature turkey (usually under 16 weeks of age), of either sex, that is tender-meated with soft, pliable, smooth-textured skin, and flexible breastbone cartilage.

- *Young turkey.*   A turkey (usually under 8 months of age) that is tender-meated with soft, pliable, smooth-textured skin, and breastbone cartilage that is somewhat less flexible than in a fryer-roaster turkey. Sex designation is optional.

- *Yearling turkey.*   A fully matured turkey (usually under 15 months of age) that is reasonably tender-meated and with reasonably smooth-textured skin. Sex designation is optional.

- *Mature turkey or old turkey (hen or tom).*   A mature or old turkey of either sex (usually over 15 months of age) with coarse skin and toughened flesh.

---

[1] Material adapted from the USDA Agricultural Marketing Service Publication (7CFR Part 70), May 1987.

[2] Though not part of the USDA specifications, some suggest substituting the term *squab broiler* for Cornish game hen.

[3] In practice, most are 50 to 52 days of age.

Ducks

- *Broiler duckling or fryer duckling.*   A young duck (usually under 8 weeks of age), of either sex, that is tender-meated and has a soft bill and soft windpipe.
- *Roaster duckling.*   A young duck (usually under 16 weeks of age), of either sex, that is tender-meated and has a bill that is not completely hardened and a windpipe that is easily dented.
- *Mature duck or old duck.*   A duck (usually over 6 months of age), of either sex, with toughened flesh, hardened bill, and hardened windpipe.

Geese

- *Young goose.*   A goose of either sex that is tender-meated and has a windpipe that is easily dented.
- *Mature goose or old goose.*   A goose of either sex that has toughened flesh and a hardened windpipe.

Guineas

- *Young guinea.*   A guinea of either sex that is tender-meated and has flexible breastbone cartilage.
- *Mature guinea or old guinea.*   A guinea of either sex that has toughened flesh and a hardened breastbone.

Pigeons

- *Squab.*   A young, immature pigeon of either sex that is extra tender-meated.
- *Pigeon.*   A mature pigeon of either sex, with coarse skin and toughened flesh.

24.6   USDA GRADES FOR SLAUGHTER ANIMALS AND CARCASSES

Even though official USDA grades are applied only on carcasses by a government grader, they also serve as a basis for grading live animals intended for slaughter. Verbal descriptions for live animal and carcass grades are printed in the *Federal Register* or Code of Federal Regulations, and are reprinted for distribution by the USDA. Standard grade pictures of animals and carcasses are likewise available (see Appendix A).

The ability of buyers and sellers to assign unofficial grades to live animals varies, and depends partly on experience. (See Section 23.6.) In general, buyers and sellers, as well as USDA market reporters, have a good concept of live animal slaughter grades and are able to make transactions and report sales with considerable accuracy. Note that live grades are used in the price quotations in Figure 24–1.

The procedure for applying grades on carcasses during or after processing is discussed in Section 25.4. It is emphasized that the use of the USDA carcass grades (and a USDA employee to apply them) is *voluntary* on the part of the processor. The service is available on an hourly or weekly contract basis. Because many consumers and retailers depend on USDA grades as a standard of quality, especially with beef and lamb, many processors feel it advantageous to employ this service on predominantly USDA Choice and Prime quality carcasses.

Current poultry marketing procedures, including extensive use of marketing contracts between raiser and processor that specify age and quality, preclude the need for live poultry grades. Grades have been established by the USDA, however, to correlate with grades for carcasses and parts and for use by producers in determining when birds are ready for market.

Standards are available for pork carcasses, but no official pork carcass grading is done by USDA graders. There is relatively less merit in grading pork carcasses, for two reasons: (1) Because nearly all hogs are fed for early marketing, there is less variation in age and weight and carcass quality traits influenced by age and weight; and (2) most pork is processed into "boxed pork" in the plant, cuts are trimmed to a uniform fat depth, and many cuts are cured, smoked, and otherwise processed. Recognizing quality differences that exist among various parts of carcasses, there may be more merit in grading pork cuts than pork carcasses.

USDA pork carcass grades have been used extensively, however, as a basis for developing private carcass grades in use by meat processors and for educating producers regarding variations in quality.

Among market lambs, there is likewise considerable uniformity in slaughter age and weight, so variation in carcass quality and carcass grading is not so common as in beef.

Among beef cattle, however, there may be considerable variation in slaughter age, weight, and finish because of the numerous feeding programs. Today, the higher quality carcasses of beef are graded and sold in uniform cuts and weights as fresh "boxed beef."

Changes are made periodically in grade names, specifications, or rules of application. Therefore, any discussion of a USDA grading system must be based on the *principles* of carcass grading. Grading specifications and standard pictures, as well as the names of the grades, must be considered subject to change.

The purpose of carcass grading is to *categorize carcasses according to merit*. Because a carcass is a large unit and various parts may differ considerably in value, carcass grades are based on two different considerations, both important. One is the *quantitative proportion of the various parts*, or expected yield of edible meat. Is there a high proportion of loin, rib, or round, the expensive cuts on a beef carcass, for example? Are these cuts "plump" or "bulging," with a high proportion of lean meat in relation to amount of bone and/or fat?

The other major consideration, probably most important to the consumer, is the *qualitative evaluation of the edible tissue*, believed to be associated with palatability. The latter would include texture of lean, intermixture of fat with lean, and other characteristics.

In the case of poultry, appearance and absence of such items as skin tears, cuts, pinfeathers, and discoloration of skin are considered. In poultry carcass grade specifications, most emphasis is on quality. Such items as "well-developed covering of flesh" or "plump and full legs" are mentioned, but no quantitative specifications are given and the final grade assigned is primarily on the basis of quality.

Cutability—proportion of trimmed retail cuts a carcass may yield—is given treatment equal to and separate from *quality* in USDA beef and lamb carcass grade specifications.

In pork carcass grades, relative emphasis is placed on *proportion of lean*. Ratio of lean to fat may be considered a quality factor, of course, but it is a quantitative factor to the processor or retailer whose customers demand *lean, trimmed* pork cuts.

Obviously the conformation or cutability characteristic is of largest concern to the processor, wholesaler, or retailer who breaks beef and lamb carcasses into cuts and assigns a different price to each cut. A beef carcass yielding a higher proportion of

TABLE 24–4	USDA Quality Grades for Beef and Veal	
Steer, Heifer, and Cow	**Bullock**	**Vealer and Calf**
Prime[a,b]	Prime	Prime
Choice[a]	Choice	Choice
Select[a,c]	Good	Good
Standard[a]	Standard	Standard
Commercial[d]	—	—
Utility	Utility	Utility
Cutter	—	—
Canner	—	—

Note: *Grades and their specifications are revised as deemed desirable by the USDA. Publications giving detailed specifications are available from the USDA, Ams. Quality grades for bull and stag have been eliminated.*

[a]*Maximum age for grade is 42 months.*

[b]*Cows are not eligible for Prime grade.*

[c]*Replaced the grade of "U.S. Good" in 1988.*

[d]*The Commercial grade is restricted to animals over 42 months of age.*

Source: *USDA.*

trimmed loin, rib, and round should, from that standpoint, grade higher. The consumer, however, who appraises a *single cut* of meat is interested primarily in the *quality* appraisal of that particular cut.

24.7   GRADES FOR SLAUGHTER CATTLE AND BEEF CARCASSES

Grade designations used by the USDA for slaughter cattle and beef and veal carcasses, along with the dates of the most recent revisions, are given in Table 24–4. Illustrations of **quality grades** and **yield grades** are shown in Appendix A. The approximate age limitation for Prime through Standard grades is 42 months.[4] Age is appraised on carcasses by meat texture, size of bone, and whiteness of bone.

Cows, stags, and bulls are not eligible for the Prime grade; few grade Choice unless they are exceptionally young and high in quality. Note that in 1988, the USDA Good grade was replaced by USDA Select for steers, heifers, and cows. The Standard grade is reserved for animals under 4 years of age and their carcasses. The Commercial grade is limited to cattle over 4 years of age and their carcasses, except for bulls and stags, where there is no age specification. There is no age limitation on other grades. There are no quality grades for bulls and stags.

Excerpts from the current specifications indicate that Choice grade slaughter steers, heifers, and cows "30 to 42 months of age have a fat covering over the crops, back, loin, rump, and ribs that tends to be moderately thick. Cattle under 30 months of age carry a slightly thick fat covering."

Conformation is no longer a factor in determining quality grades but will affect cutability of the beef carcass. Though verbal descriptions are also used for carcasses of

> **Quality grades**
> Official standards established by the USDA for evaluating palatability; indicate characteristics of the carcass of market animals.

> **Yield grade**
> A USDA grade, that indicates the amount of trimmed, boneless major retail cuts that can be derived from a carcass; cutability.

---

[4] Almost all USDA Prime and USDA Choice slaughter steers and heifers are less than 24 months of age.

each grade, Figure 24–3 best illustrates the traits needed for the Choice or other grades. The carcass quality grades are determined largely by (1) degree of maturity and (2) degree of marbling, but firmness of lean, color, and texture are also considered.

Five cutability or yield grades for beef are used. The factors used to establish Yield Grades 1–5 are rib eye area, fat thickness, percent kidney, heart and pelvic fat; and carcass weight.

Yield Grade 1 denotes highest cutability, and the live grade specifications indicate that "width through the shoulders and rounds is greater than through the back . . .

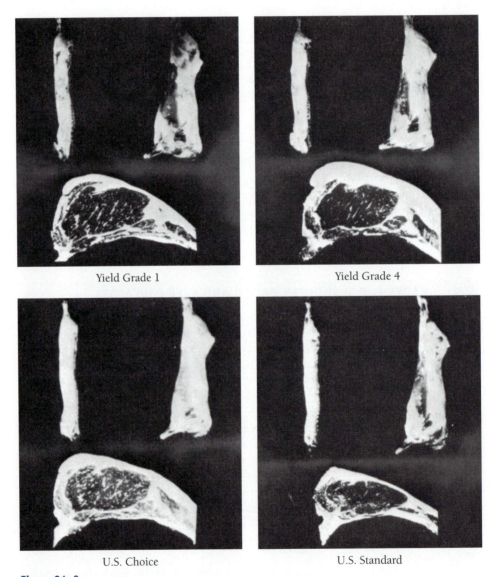

Yield Grade 1   Yield Grade 4

U.S. Choice   U.S. Standard

Figure 24–3
Carcass views and rib eye cross sections for two yield grades and two quality grades of beef carcasses.
(Courtesy of USDA, FSQS; from *Meat Evaluation Handbook,* National Livestock and Meat Board.)

top is well-rounded with no evidence of flatness . . . shoulders are slightly prominent." A 500-pound carcass in this grade should have about a 11.5-square-inch rib eye with about 0.3 inch of fat over it. There is usually "a thin layer of fat over the outside of the rounds. . . . Muscles are usually visible through the fat."

In Yield Grade (or cutability group) 4, descriptions for live animals indicate "low yields of boneless retail cuts . . . wider over the top than through the shoulders or rounds . . . back and loin are very thick and full, nearly flat, and break sharply into the sides." For carcasses, the description refers to carcasses being completely covered with fat, with the only muscles usually visible being on the shanks, plate, and flank, and large deposits of fat in the flanks and cod or udder.

Carcass cutability group specifications provided by the USDA also give calculation procedures, so that carcass measurements and weight can be used to compute the cutability designation of a carcass. The equation, with factors listed in order of importance, is

$$\begin{aligned}
\text{U.S. Beef Carcass Yield Grade} = {} & 2.50 \\
& + (2.50 \times \text{fat thickness, inches}) \\
& - (0.32 \times \text{area of rib eye, inches}) \\
& + (0.20 \times \text{percent kidney, pelvic, and heart fat}) \\
& + (0.0038 \times \text{hot carcass weight, pounds})
\end{aligned}$$

The range among grades, in carcass value or yield, is illustrated in Table 24–5.

It is impossible to develop ability and skill in grading slaughter animals or carcasses from a text, even with the aid of pictures. Much practice and experience are needed.

TABLE 24–5

What USDA Beef Yield Grades Mean to the Retailer

	Yield Grade				
	USDA 1	USDA 2	USDA 3	USDA 4	USDA 5
If a retailer sells 30,000 pounds of retail cuts and bought carcasses of these USDA yield grades:					
This many pounds of carcass beef would be needed:	36,585	38,760	41,210	43,990	47,170
If 600-pound carcasses were purchased, this many would be needed:	61	65	69	73	78
After preparing and trimming the retail cuts, this would be the percentage yields of					
Fat trim	7.5	12.6	17.7	22.8	27.9
Bone and shrink	10.5	10.0	9.5	9.0	8.5
Trimmed cuts	82.0	77.4	72.8	68.2	63.6

Source: *USDA, AMS.*

TABLE 24–6	USDA Grades for Lambs and Sheep

Lamb	Yearlings and Sheep[a]
Prime	—
Choice	Choice
Good	Good
Utility	Utility
Cull	Cull

[a]*Technically, according to USDA criteria, a sheep is an ovine, usually over 24 months of age, that has cut its second pair of permanent incisor teeth. "Mutton" refers to the carcass.*

Source: *USDA.*

24.8   GRADES FOR SLAUGHTER LAMBS AND SHEEP AND THEIR CARCASSES

Quality grade designations currently used by the USDA for slaughter lambs, yearlings, and sheep (called "mutton" in the carcass), as well as for their corresponding carcasses, are shown in Table 24–6.

Animals over 2 years of age, or their corresponding carcasses, are not eligible for the Prime grade. Though age of carcass is appraised by the USDA grader, it is a criterion for *classification*, not grading, and is discussed in Section 24.3. Note that "conformation" is considered when determining grades for lambs.

Prime lambs are described as "moderately low-set and blocky and thickly fleshed. . . . moderately wide over the back, loin, and rump." In contrast, cull lambs are described in USDA standards as "extremely rangy, angular, and thin-fleshed, and extremely narrow and shallow-bodied."

Prime carcasses must be "thickly muscled throughout, moderately wide and thick in relation to length . . . have moderately plump and full legs; moderately wide and thick backs; and moderately thick and full shoulders." In contrast, cull carcasses are "extremely angular . . . extremely thin-fleshed . . . legs are extremely thin and concave . . . flesh is soft and watery." In today's markets, practically all lambs sold will be in the Prime, Choice, or Good grades.

Five USDA yield grades for lamb and mutton were established in 1969. The principal factors that determine yield grade of lamb carcasses are fat thickness, percent kidney and pelvic fat, and leg conformation.

Yield Grade 1 lambs are described similarly to Yield Grade 1 beef cattle or carcasses and "might have 0.1 inch of fat over the rib eye, 1.5 percent kidney and pelvic fat, and an average leg conformation score of Prime."

Yield Grade 4 lambs are "extremely wide through the back and loin . . . shoulders and hips are smooth . . . flanks are moderately deep and full." The carcass might have 0.4 inch of fat over the rib eye, 4.5 percent of its weight in kidney and pelvic fat, and average Choice leg conformation score.

Yield Grade 1 lamb carcasses must have a cutability of 47.6 to 49.4 percent, based on percent trimmed boneless retail cuts of leg, loin, rack and shoulder. U.S. Yield Grade 4 would have a cutability of 42.2 to 43.6 percent.

24.9   GRADES FOR SLAUGHTER HOGS AND PORK CARCASSES

USDA grade designations for slaughter hogs and pork carcasses are U.S. Acceptable U.S. No. 1, U.S. No. 2, U.S. No. 3, U.S. No. 4, and U.S. Utility. Figure 24–4 illustrates how U.S. yield grades are determined, based upon backfat thickness and muscling score. Illustrations of live hogs are shown in Appendix A, and photos of carcasses are presented in Figure 24–5 on the following page.

These specifications serve primarily as a guide to producers and processors in evaluating relative carcass merit, as well as a guide to development of private grade designations by packers.

Grades U.S. Nos. 1, 2, 3, and 4 have acceptable lean quality and acceptable belly thickness. These grades are based entirely on carcass yields of the four lean cuts—ham, loin, picnic, and Boston shoulder. Expected yields of chilled carcasses in the four grades are shown in Table 24–7.

The U.S. Utility grade includes those carcasses that have "lesser development of lean quality . . . also . . . carcasses which do not have acceptable belly thickness . . . and all . . . which are soft and oily."

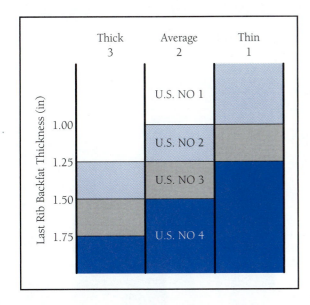

Figure 24–4
Backfat thickness and muscling score relationships used to determine U.S. pork carcass yield grades.
(*Source*: USDA.)

TABLE 24–7	Expected Yields of the Four Lean Pork Cuts Based on Chilled Carcass Weight, by Grade

Grade	Yield of Lean Cuts
U.S. No. 1	60.4% and over
U.S. No. 2	57.4 to 60.3%
U.S. No. 3	54.5 to 57.3%
U.S. No. 4	Less than 54.4%

Note: *These yields will be approximately 1% lower if based on hot carcass weight.*
Source: Adapted from the *USDA*.

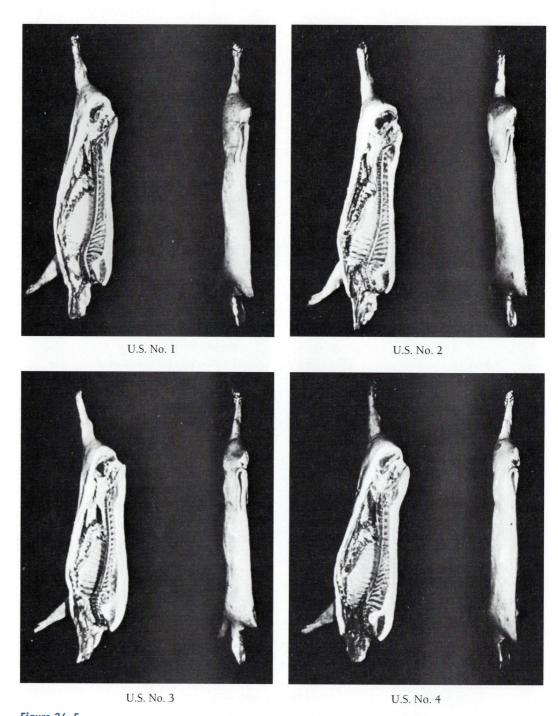

U.S. No. 1

U.S. No. 2

U.S. No. 3

U.S. No. 4

Figure 24–5
The top four grades of U.S. Acceptable pork carcasses. Note that U.S. No. 1 has a thinner layer of fat over the vertebrae, less internal body fat, and a more prominent ham and shoulder as viewed from the top. (Courtesy of USDA.)

Quantitative specifications were used in pork carcass grades more than 10 years before the yield or cutability grades were adopted for beef carcasses. This was possible because (1) slaughter hogs are more uniform in age and weight and (2) earlier research with pork carcasses had shown high correlations between certain measurements and percentage of lean cuts or other conformation and quality traits.

24.10   GRADES FOR POULTRY[5]

USDA poultry grades are written for live birds, carcasses, and parts thereof, including certain poultry meat products. (Specifications are summarized in Table 24–8.) Application of these standards is made to parts (quarters, legs, breasts, etc.), provided they are cut according to certain specifications. The specifications are determined by the buyer—for instance, Kentucky Fried Chicken will specify that chicken parts be of a specific size. By doing this, the customer ultimately will receive uniform pieces of meat. They are also used to designate quality of rolls or other deboned, ready-to-cook poultry products. For boneless products, the A-quality grade is dependent on the meat from young A-quality carcasses, as well as other quality attributes.

24.11   PRIVATE MEAT PRICING AND GRADING

As discussed earlier, the trend in marketing of high-quality animals is toward *grade and yield*. However, in addition, many pork processors utilize their own carcass value pricing systems in which they may provide premiums for leanness and discounts for fat. See Box 24–1 for a summarization of pork carcass pricing systems used by pork processors.

Also, many processors, as well as some retailers, employ their own system of grading meat products, including carcasses, cuts, or processed meats such as ham or turkey rolls.

Processors may have fewer or more grades, depending on the range in quality they handle and the degree of categorization they and their customers want. Some apparently use the USDA specifications and categories, but apply their own grade names. Review your local newspaper for examples of private grades, as well as USDA grades, used in retail advertising. You probably are familiar with many.

As for official USDA grades, accuracy and dependability of unofficial private grades depend on competence of the grader and the degree of discrimination employed. Many retailers and consumers prefer USDA grades because they were applied by a clearly "unbiased" person, not an employee of the processor. Companies that use a private system want to establish and maintain respect for their top grades, however, so they are likewise concerned that their employed graders be accurate and discriminating.

Private carcass grading may be less costly because a company employee does it. If not fully employed in grading, the grader may be used elsewhere. Also, where USDA graders are used, a company grader usually reviews the work or else precedes it to designate carcasses to be graded.

A private system may be more meaningful in a particular area or for a particular line of meat products than USDA grades, which are standard for the entire country. Large processors, with intensive advertising, may effectively establish one of their top

---

[5] USDA, FSQS. Adapted from *Agr. Handbook No. 31.*

TABLE 24-8

Summary of Specifications for Standards of Quality for Individual Carcasses and Parts of Ready-to-Cook Poultry (Not All-Inclusive) (Minimum Requirements and Maximum Defects Permitted)

Factor	A Quality	B Quality	C Quality
CONFORMATION:			
Breastbone	Normal / Slight curve or dent	Moderate deformities / Moderately dented, curved, or crooked	Abnormal / Seriously curved or crooked
Back	Normal (except slight curve)	Moderately crooked	Seriously crooked
Legs and wings	Normal	Moderately misshapen	Misshapen
FLESHING:	Well-fleshed, moderately long, deep and rounded breast	Moderately fleshed, considering kind, class, and part	Poorly fleshed
FAT COVERING:	Well covered—especially between heavy feather tracts on breast and considering kind, class, and part	Sufficient fat on breast and legs to prevent distinct appearance of flesh through the skin	Lacking in fat covering over all parts of carcass
DEFEATHERING:			
Nonprotruding pins and hair	Free	Few scattered	Scattering
Protruding pins	Free	Occasional	Occasional
EXPOSED FLESH:			

Carcass Weight (lb)	A Quality			B Quality			C Quality	
	Breast and Legs	Elsewhere[a]	Part	Breast and Legs[b]	Elsewhere[b]	Part	Elsewhere[b]	Part
0–2	None	1 in.	Slight trim on edge	½ of flesh exposed on each part on carcass provided meat yield not appreciably affected			½ of flesh exposed—meat yield not appreciably affected	No limit
Over 2–6	None	1½ in.						
Over 6–16	None	2 in.						
Over 16	None	3 in.						

DISCOLORATIONS:[c]

Carcass Weight (lb)	Breast and Legs	Elsewhere[a]	Part	Breast and Legs[b]	Elsewhere[b]	Part	
0–2	¾ in.	1¼ in.	¼ in.	1¼ in.	2¼ in.	½ in.	No limit[d]
Over 2–6	1 in.	2 in.	¼ in.	2 in.	3 in.	1 in.	
Over 6–16	1½ in.	2½ in.	½ in.	2½ in.	4 in.	1½ in.	
Over 16	2 in.	3 in.	½ in.	3 in.	5 in.	1½ in.	

DISJOINTED AND BROKEN BONES:	Carcass: 1 disjointed and no broken Parts: None	Carcass: 2 disjointed and no broken, or 1 disjointed and 1 nonprotruding broken Parts: May be disjointed, no broken	No limit
MISSING PARTS: (Whole carcass only)	Wing tips and tails[e]	Wing tips, 2nd wing joint, and tail Back area not wider than base of tail and extending halfway between base of tail and hip joints	Wing tips, wings, and tail Back area not wider than base of tail extending to area between hip joints
FREEZING DEFECTS: (Consumer packaged)	Slight darkening on back and drumstick. Overall bright appearances. Occasional pockmarks due to drying. Occasional small areas showing layer of clear or pinkish ice.	May lack brightness. Few pockmarks due to drying. Moderate areas showing layer of clear, pinkish, or reddish ice.	Numerous pockmarks and large dried areas

Note: Not all-inclusive; minimum requirements and maximum defects permitted.

[a]*Maximum aggregate area of all exposed flesh due to cuts, tears, and missing skin. In addition, carcass may have cuts or tears that do not expand or significantly expose flesh, provided the total aggregate length does not exceed the permitted tolerance for the weight range.*

[b]*For purposes of definition, the parts of the carcass shall be each wing, leg, entire back, and entire breast, with each permitted to have one-third of the flesh exposed by cuts, tears, and missing skin.*

[c]*Flesh bruises and discolorations such as "blue back" are not permitted on breast and legs of A-quality carcass or on these individual parts. No more than one-half of total aggregate area of discolorations may be due to flesh bruises or "blue back" (when permitted), and skin bruises in any combination.*

[d]*No limit on size and number of areas of discoloration and flesh bruises if such areas do not render any part of the carcass unfit for food.*

[e]*In ducks and geese, the parts of the wing beyond the second joint may be removed, if removed at the joint and both wings are so treated. Tail must be removed at base.*

Source: Adapted from the U.S. Government Printing Office: 1988–201–053/80013.

477

BOX 24–1   Features of Pork Carcass Value Pricing Systems by Processors

The following table is part of the results from a survey sent to most of the large pork processors (more than 1 million head slaughtered per year) in the midwestern and southeastern regions of the United States in 1995. As indicated by the authors, it shows the importance of swine producers having a clear understanding of the various premiums and discounts for fat, muscle, and weight when making production and marketing decisions.

Packer	Base Live Price	Base Carcass Yield Base Carcass Price	Backfat[a]	Loin Depth (LD)	Weight Sort
A	Average Ind./Ohio and Iowa/S. Minn. packer.	Average all hogs previous quarter. Assume 73.4+% BCP = BLP/0.734	Base =47.5–49.4%.	Variable in percent lean.	Base =170–210 lb. No premium <160 lb.
B	Average Ill. and Ind./Ohio direct top + $1.25 cwt.	Standard = 74.4% BCP = BLP/0.744	Report in inches. Base = 55.5–56.49%.	Variable in percent lean. Premium/discount based on wholesale market prices previous week.	Base = 161–200 lb. No premium <161 lb.
C	Average Iowa/S. Minn. packer.	Standard = 75.5% BCP = BLP/0.755	If < 27 mm, premium = 0.07 × BCP. If > 27 mm, discount = 0.014 × BCP.	If < 47 mm, discount count = 0.02 × BCP. If = 47–54 mm, no discount. If > 54 mm, premium = 0.02 × BCP.	Base = 167–208 lb. Base = 167–222 lb If BF < 25 mm.
D	Top Iowa/S. Minn. packer.	Standard = 73.5% BCP = BLP/0.735	Base = 46.5–48.4%.[b]	Variable in percent lean. If < 50 mm, no discount. If = 50–60.9 mm, premium = $0.50/cwt. If > 60.9 mm, premium = $1.00/cwt.	Base = 166–205 lb.
E	Local market price but ≥ average Iowa/S. Minn. or Ind./Ohio or ill. packer or Missouri direct.	No carcass price. Assume 73.4% for comparisons. Premium of 4 cents/cwt.for each tenth above 73.5%.	Base = 45.0–47.9%. Premium and discount are rolling average last four loads. Applied to current load's base price.[c]	Variable in percent lean.	Base = 162–201 lb. Discounts are $/live cwt.
F	No live price under carcass pricing system.	Assume 73.4% for comparisons. Ave. of low 47–48% lean and high 49–50% lean in USDA western Corn Belt.	Premium and discount based on backfat and carcass weight ranges. Base = 1.01–1.20 in.	No premium or discount.	Base = 172 –194 lb.
G	Top Iowa/S. Minn packer.	Standard = 73.5% BCP = BLP/0.735	Base = 47.1–49%.	Variable in percent lean.	Base = 163– 213 lb.
H	Market-determined.	Standard =74.0% BCP = BLP/0.740	Base = 49%.	Variable in percent lean.	Base = 170–221 lb.
I	Market-determined.	Carcass Weight Yield < 184 73.50 185–193 73.75 193–201 74.10 202–209 74.45 <210 74.80 BCP = BLP/yield	Base = 1.2–1.3 in.	No premium or discount.	Base = 162–192 lb.
J	Average Ill. and Ind./Ohio . direct prices. Some adjustment for competitiveness.	Base = 74.5% BCP = BLP/0.745	Base = 50%. Formula for percent lean includes BL, LD, and CW. (% lean –50) × 0.01 × 1.5 × BCP	Base = 50%. Formula for percent lean includes BF, LD, and CW (% lean –50) × 0.01 × 1.5 × BCP	Base = 170–209 lb. Add $1.00 BCP if in desired weight range.

[a]All packers using Fat-O-Meater on hot carcass measure at 10th rib except Packer F, which measures at last rib with a ruler.
[b]If load average lean = 43–44%, lower BCP by $1.00 cwt. If lean < 42%, lower BCP by $2.00 cwt.
[c]Program does not pay on the individual carcass but on the whole load.

Source: Kenyon, D., J. McKissick, and J. Lawrence. "Move to carcass value pricing requires understanding of premiums, discounts." *Feedstuffs*, June 3, 1996.

grades as a "trademark of dependable quality." It is to compete against this that small processors like to use USDA grades. They feel consumers also respect these as "trademarks of dependable quality."

A significant advantage of private grading of beef, lamb, or pork is that it can be and often is employed to grade *cuts* instead of carcasses. Because of the quality gradient that exists between or among parts of carcasses, or before and after trimming or processing, grading of cuts is often worthwhile. A loin from an extra-heavy barrow carcass (which would grade U.S. No. 3 or 4) might be trimmed of excess backfat and sold as chops labeled "Jones Meaty Select." The shoulder from the same carcass, however, probably would carry too much fat *between* muscles and effective trimming to the "Meaty Select" grade would not be possible.

QUESTIONS FOR STUDY AND DISCUSSION

1. What are the two principal market classes of cattle, sheep, and hogs?
2. How are USDA class and grade terms helpful to the buyer, feeder, and seller of livestock and poultry?
3. List common market subclasses of cattle. What is the definition of a bullock? of a heiferette?
4. Distinguish between the terms *lamb* and *mutton*.
5. What is the preferred market weight of slaughter barrows and gilts?
6. What is the difference in definition between a broiler and capon?
7. Are grades of live animals official or unofficial? Explain.
8. Why aren't pork carcasses officially graded for quality?
9. What are the two main factors that determine carcass value?
10. List the four top-quality grades for steers and heifers, bullocks, cows, vealers, and calves.
11. What grade has been replaced by U.S. Select?
12. Rank the U.S. yield grades of beef from highest (best) to lowest. What factors are used to establish these yield grades?
13. Name the four top U.S. quality grades for lambs. Are the yield grades similar to cattle? What factors are used to determine lamb yield grades?
14. Name the two quality grades of swine. What are the four U.S. yield grades for swine?
15. Explain how yield of lean cuts changes as yield grade changes.
16. Name the three U.S. quality grades of poultry. Name at least four factors used to determine poultry quality grades.

25

Processing Meat Animals

This chapter will cover meat processing, from live animals to the value-added finished product, as well as varied byproducts.

Upon completion of this chapter, the reader should be able to

1. Define *packer, processor, slaughterer, viscera,* and other key terms.
2. Describe the meat-processing industry, in terms of relative concentration and approximate number of processing plants.
3. Describe the slaughtering steps that yield a dressed carcass.
4. Describe the inspection and grading processes and identify the state or federal agencies involved.
5. Differentiate among the slaughtering procedures for beef intended for Jewish consumption.
6. Differentiate among wholesale, primal, and retail cuts.
7. Explain the significance of hot boning, boxed beef, and mechanically deboned poultry meat in the efficiency of meat-processing operations.
8. List three meat preservation techniques and explain why each prevents spoilage.
9. List specialty or processed meat items.
10. List nonmeat byproducts of meat processing.

Modern, high-volume plants make up the majority of today's meat-processing industry, resulting in an industry with the following characteristics:

- Companies compete for both purchase of materials and sale of finished product.
- Meat is perishable—it must be processed and sold rapidly.
- Because meat is perishable and sold rapidly, there is a quick turnover of product.
- Many byproducts contribute to total company sales. (See Section 25.11.)
- The industry is considered essential because it puts food in usable form.

The term *processing* is used in the above paragraphs to encompass steps in the manufacture of meat items. It should be pointed out that some companies and some plants restrict their operations to the slaughtering of animals and closely related functions. Some process only carcasses or parts of carcasses. Others process only byproducts. Some companies handle the product the entire route, within one plant, or by moving the

Packer
Refers to the buyer of animals to be slaughtered and processed for meat.

product among plants for different processing steps. The term **packer** was used during most of the nineteenth and twentieth centuries to describe companies in this business. The term was adapted from the early practice of packing cured hams, bellies, or other cuts under salt in barrels for storage or shipment. Though some use of the term continues, it is not currently descriptive.

In this chapter, the term *slaughterer* will refer to those units involved in slaughtering and related steps. The term *processor* will, in general, refer to units that process carcasses or parts of carcasses, or that handle all steps.

There is a large range in size of meat slaughterers and processors who operate in the United States and Canada. Most large slaughterers are located in livestock- or poultry-producing areas; some are near terminal markets. These companies buy most of their livestock direct from the producer, through order buyers, or through their own personnel at terminals or auctions.

As discussed in Section 2.5, the number of red meat and poultry slaughtering and processing plants is fewer but larger. Since 1986, the number of red meat slaughtering plants decreased 60 percent—from almost 5,200 to about 2,050. The poultry-processing industry also has undergone similar changes, with fewer but larger, more efficient plants. In Figure 25–1 note the locations of the 6,526 plants involved in slaugh-

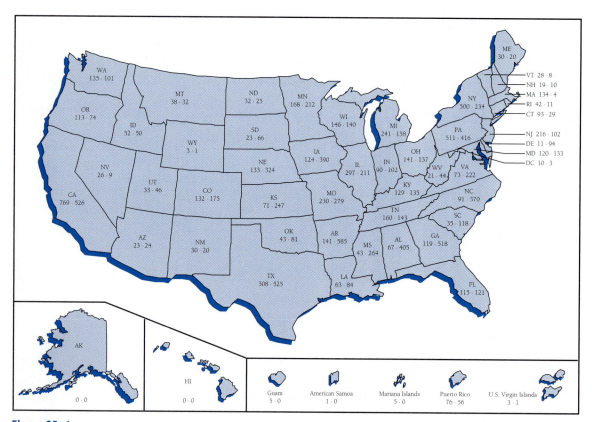

Figure 25–1

The location and number of plants and number of Food Safety and Inspection Service employees involved in federal meat and poultry inspection in fiscal year 1994. Total plant numbers in each state are shown, followed by number of employees. The overall number of plants at that time totaled 6,526; however, plant numbers continue to decrease each year.

TABLE 25–1	Leading Meat and Poultry Companies, 2003		
Rank	*Company*	*Number of Plants*	*Gross Revenue (billions of dollars)*
1	Tyson Foods, Inc.	128	23.0
2	Excel Corporation	27	12.0
3	Swift Company	10	8.4
4	Smithfield Foods	13	7.1
5	Farmland Foods	14	4.8
6	Hormel Foods	29	3.9
7	Sara Lee Foods	15	3.7
8	Oscar Mayer	70	3.6
9	Keystone Foods	8	2.7
10	Perdue Farms	8	2.7

Source: *Meat Processing, June 2003.*

ter and processing of both red meat and poultry in fiscal year 1994. This higher number of plants, all under federal inspection, included many very small companies providing products for their local towns and communities. Only federally inspected plants may sell their products interstate or in foreign commerce. Over 131 million head of livestock were inspected in fiscal year 1997.

The 10 leading companies in the meat and poultry industry in 2003, based on total sales, are shown in Table 25–1. The top three companies were Tyson Foods, Inc., Excel Corporation, and Swift Company. Tyson Foods, Inc. was the leading poultry processor with a weekly processing average of 148.8 million pounds. (See Chapter 35.)

25.1   SLAUGHTER PROCEDURES

Although procedures vary among species, animal slaughter generally includes rendering insensible, bleeding, removing the hair, feathers, or skin, and removing the viscera—trachea, lungs, esophagus, stomach, intestines, accessory organs, heart, and reproductive organs. The head is also removed, and the "carcass" remains.

To achieve *humane slaughter,* required in plants operating according to humane slaughtering laws, animals are rendered insensible by stunning. The initial procedure for cattle, hogs, and sheep will be described.

Cattle

Almost all cattle are *stunned* with an air-driven bolt gun, then are shackled and hoisted from an overhead rail. Calves are sometimes stunned electrically. In either case, bright lights are often used to blind the animals to the approach of the stunner's hand. The next step is *sticking,* insertion of a sharp knife in front of the brisket and severing the carotid arteries and jugular veins. Because the animals are only stunned, upon sticking, the heart forces out most of the blood.

Low-voltage carcass *stimulation,* a process that passes an electrical current through the carcass, is performed by some packers within 15 minutes after stunning. Electrical stimulation causes a series of muscle contractions and relaxations that speed

Figure 25–2
A specially designed electric saw used to split beef carcasses. The cut is made through the middle of the vertebrae, from the rump to the neck.
(Courtesy of Iowa Beef Processors.)

up the depletion of muscle energy and decrease in muscle pH. This process results in improved meat quality (color, marbling score, texture), improved bleedout, and easier hide removal. Some processors utilize a high-voltage stimulation method that can be performed later in the slaughtering process because it acts directly on the muscles rather than on the entire carcass.

All medium to large beef slaughter plants use "on-the-rail" dressing, in which the carcass is moved along a rail during the sequence of dressing steps. First, after head removal, one hind shank is skinned out and the foot removed, then the other. Beef skinners, frequently using electric- or air-powered skinning knives, open up the hide and skin out critical areas, such as the chest and flanks (Figure 25–2). Most plants use a hide puller to remove the hide from sides and back of the carcass. Viscera is removed, the carcass is split, washed, inspected, sometimes shrouded, and taken to the chill room. The hide of light calves and veal usually is not removed until after the carcass chills.

Dressing on the rail contrasts with the cradle system formerly used in large plants and still used in smaller and locker-type plants. In this system, the bled animal is lowered to and held on its back in a cradle placed on the floor. Feet are pointed upward. The shanks, chest, and flanks are skinned out, then the sides. The carcass is then raised back toward the rail so skinning of the back can be completed.

Beef intended for consumption by traditional Jews is slaughtered by the Kosher method.[1] Because a complete bleed is necessary, cattle are not stunned before bleeding. A rapid cut across the throat, which severs the trachea, esophagus, vagus nerves, carotid arteries, and jugular veins, is accomplished with a special knife (chalef) that is perfectly sharpened. The cut is painless and the animal loses consciousness within 3 seconds. Besides the manner of kill, the cattle must not be diseased, and the lungs must be completely healthy. Koshering must be performed under the supervision of a rabbi or a representative of the Rabbinical Board. Because of further restrictions on beef consumed by Kosher-observant Jews (see Section 26.4), a large percentage of healthy, high-quality beef cattle are slaughtered by the Kosher method.

---

[1] Donin Rabbi Hayim Halevy. *To Be a Jew, A Guide to Jewish Observance in Contemporary Life.* New York: Basic Books, 1972.

Hogs

Humane slaughter of hogs usually involves electrical stunning. After sticking and being hung to the rail, most hogs are scalded in a vat at about 143°F for 3 to 4 minutes. This loosens the hair follicles so most hair is easily removed by paddles (belt scrapers) of a dehairing machine that removes hair mechanically. Next, the carcasses are rehung by rail for remaining steps. The few small hairs not removed by machine are singed off as the carcasses pass through a gas flame. In some plants, hogs are skinned by a combination of hand and mechanical knives and skin pulling, much as cattle are skinned. The skin is later dehaired. Such skin has a higher value than scalded skin because it can be used for leather and for surgical treatment of burns for humans. The head is then removed after skinning and before evisceration.

Sheep

After rendering insensible—usually by compression bolt or electrical stunning—sheep and lambs are hoisted to a rail for bleeding, and the head may be removed at that time. The feet are removed at the "break joint," and the pelt is opened with a knife, but the knife usually is not used for removing the pelt. Sheep and lamb skin, as well as the natural fine membrane called the "fell" which remains on the carcass, are so tender that they are easily cut with a knife. Instead, fists are forced in between the skin and the fell to remove the pelt. Experienced workers can do this job rapidly without damaging the pelt or the appearance of the carcass. The fell helps prevent moisture loss and shrinkage of the carcass in the cooler and also gives the carcass a smoother appearance.

All Species

Evisceration of all species is essentially the same. The sternum and pelvic bone of cattle and hogs are split with a saw or knife for easier evisceration. In sheep only the sternum is split. Evisceration proceeds with the animals hanging from the rail by the rear shanks. Reproductive organs and intestines are first cut loose, and removal of the remainder of the viscera (stomach, esophagus, lungs, heart, liver, etc.) follows. Kidneys, and the fat that surrounds them, usually are left in beef and lamb carcasses but are removed from pork carcasses. By tradition, the spleen usually is left in lamb carcasses.

In plants that are under federal or state inspection, the head and viscera removed from an animal must be identified with the corresponding carcass. They usually move along the processing line together, the viscera on a tray beneath the carcass, until both are checked by the inspector. The liver and heart are then separated from the remainder of the viscera. (See Section 25.2.) Value of the viscera is further discussed in Section 25.12.

To facilitate rapid chilling of the meat, large carcasses are split down the backbone with a saw. Beef carcasses are split completely and sides are handled separately. Pork carcasses usually are split to the neck.

Carcasses are then thoroughly washed and prepared for the cooler. In a few plants, beef sides may be covered with a *shroud*. This heavy piece of cloth, previously soaked in salt water, is pulled tight around the outer portion of the side so that after chilling, when the shroud is removed, the side will present a smooth, neat appearance. The small amount of salt bleaches the outer layer of fat. The shroud also prevents some cooler shrinkage.

With increased fabrication of beef carcasses, shrouding has been discontinued by most processors in an effort to reduce costs and because shrinkage is not excessive when carcasses are processed rapidly. The fell on sheep carcasses serves the same function as the shroud; so does the skin on pork carcasses and the hide on light calves.

Figure 25–3
Turkey poults on the killing line (left) and carcasses after evisceration (right).
(Courtesy of Ralston Purina Company and Pfizer H & N.)

Poultry

Poultry are individually shackled and hung by the shanks to overhead tracks (Figure 25–3), immobilized, and then bled by severing the jugular vein just below the ear, or at the juncture of the head and neck. Birds usually are stunned by electrical shock before bleeding. Many plants have mechanical killing lines that utilize a whirling knife to sever the jugular vein and carotid artery. The heads of the birds are held in position on the killing line by bar guides.

Kosher slaughter of poultry requires that the jugular vein be severed in such a way that the windpipe "pops out." This ensures free and thorough bleeding.

After breathing has stopped, almost all birds (such as broilers) are scalded 20 to 30 seconds at 125° to 126°F. Hot scald temperatures up to 180°F sometimes are used for mature fowl, but for shorter periods. Feathers are removed mechanically by moving carcasses through rubber-fingered pickers on a rotating wheel, which, by friction, pull out the feathers under water spray. Vestigial (hairlike) feathers are usually removed by singeing or by other means, before evisceration; then the shanks, head, and oil glands are removed. Duck and geese carcasses are passed through molten wax that hardens rapidly on exposure to air. The wax is then stripped off, removing the vestigial feathers. In federally inspected plants, the inspector must be present at the time of evisceration for inspection of the viscera and the body cavity. Most slaughter steps, including evisceration, are automated in modern plants.

25.2   INSPECTION

In the United States, a circular stamp (Figure 25–4) on a cut of meat bearing the message "U.S. Inspected and Passed" means that the meat was processed in a federally approved plant, *is wholesome, and is safe for human consumption.* The stamp is applied by trained inspectors employed by the USDA. It appears on all U.S. meat and meat products, including poultry, that move in interstate or foreign commerce, plus meat from certain other plants. In Canada, employees of the Meat Hygiene Directorate perform the function.

Figure 25–4
The federal meat inspection stamps or marks applied to wholesale cuts or carcasses, packages of processed poultry parts, or packages of processed meat items such as chili or wieners. The number refers to the plant where the stamp was applied.
(*Source:* USDA.)

The Federal Meat Inspection Act of 1906 and the U.S. Wholesale Meat Act of 1967 provide that all plants that handle meat or meat products to be shipped interstate must be equipped to ensure sanitary preparation, handling, and storage. Included in the requirements are an ample supply of clean water, an adequate sewage system, abundance of natural or artificial light, adequate ventilation, and ample hot water for cleaning.

These laws also require the existence of an approved state meat inspection system for plants that do not ship interstate. If a state does not provide such a system (with assistance of federal funds), such plants must come under federal inspection.

The Poultry Products Inspection Act provides similar inspection of poultry-processing facilities and operations in plants that ship poultry products interstate.

Because meat inspection is mandatory in plants that come under the law, the government, through taxpayer dollars, hires, pays, and assigns the meat inspectors (Figure 25–5). Costs borne by the processor include equipping and maintaining the plant according to requirements and most losses resulting from condemnation of animals, carcasses, or products. (Where animals have been grown under contract, or sold on "grade and yield," the losses from condemnation may be borne by the producer.)

All animals are inspected before slaughter (antemortem) and their meat inspected postmortem; in processing operations, which may produce ham, chili, turkey rolls, or other products, there is continuous inspection. Animals that display symptoms of conditions that would render their meat unfit for food are tagged "U.S. Suspect" or "U.S. Condemned," or otherwise segregated, and do not enter the processing area. Suspect animals are slaughtered separately, then given a very thorough postmortem examination. Animals bought by the processor "subject to inspection" (Section 23.8) are also usually handled this way.

Inspectors, including trained veterinarians in charge and lay inspectors, are located at certain points along the dressing line to examine each *carcass, head,* and *viscera.* The lungs, liver, spleen, and heart, and also lymph glands on the carcass, are

Figure 25–5
USDA inspector (right) and plant production supervisor check federally inspected beef carcass. Note also the grade stamps along the back and shoulder.
(Courtesy of Oscar Mayer and Company.)

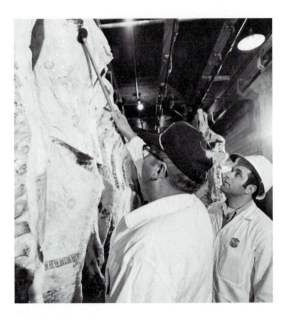

checked closely for evidence of disease. If evidence exists, each carcass or portion deemed unfit for food is marked "U.S. Inspected and Condemned," then consigned to sealed rendering tanks according to specific regulations. Portions of a carcass may be condemned because of bruises or localized infections and the remainder of the carcass passed.

In 2002, the Agricultural Marketing Act of 1946 was amended and now requires the Department of Agriculture (USDA) to issue regulations to implement a mandatory country of origin labeling (COOL) program not later than September 30, 2004. Commodities covered by the COOL rule include the following: a) muscle cuts of beef (including veal), b) lamb, c) pork, d) ground beef, e) ground lamb, f) ground pork, g) farm-raised fish, h) shellfish, i) wild fish and shellfish, j) perishable agricultural commodities (fresh and frozen fruits and vegetables); and k) peanuts.

25.3   ADDITIONAL FOOD SAFETY MEASURES

For most of the past century the principal program to ensure food safety and reduce the risk of foodborne illnesses through meat and poultry has been the mandatory meat inspection system. The Federal Meat Inspection Act and the U.S. Wholesale Meat Act required the visual inspection of the animal and carcass throughout the slaughtering and processing process, as well as inspection of the facilities and processing procedures.

Additional meat safety rules were implemented into the meat inspection system in 1996. These rules, designed primarily to reduce the risk of *microbial contamination* of meats, requires the testing for pathogenic organisms and implementation of procedures to reduce risk of contamination during slaughtering and processing. These changes in meat and poultry inspection were prompted by the outbreaks of foodborne illness from *Escherichia coli* contaminated foods, including ground beef, in 1993.

E. coli and other foodborne diseases (described in Box 25–1) have been estimated to cause as many as 5 million human illnesses (about 2 of every 100 people) each

BOX 25–1

Possible Foodborne Pathogens and Related Illnesses

This discussion is intended to acquaint the reader with potential pathogens in food and related illnesses. Hopefully, it also will create a greater appreciation for USDA Food Safety and Inspection Service efforts to educate and ensure that animal products used for human food are safe.

Contaminated food can cause infections when pathogens are consumed in food. Foods most likely to carry pathogens are high-protein, nonacid foods such as meat, poultry, seafood, dairy products, and eggs. These can become established in the body, usually multiplying within the intestinal tract. The effects can be irritation of the intestinal tract lining, or, in other cases, invasion of other tissues. Some pathogens are not toxic themselves but cause illness by production of harmful or deadly toxins. An abbreviated description of several foodborne pathogens is provided below.

- *Campylobacter jejuni.* Symptoms from infections range from a mild illness with diarrhea lasting a day, to severe diarrhea and fever for 2 to 10 days.

- *Clostridium perfringens.* Ingestion of food bearing large counts of this bacteria often results in a mild gastrointestinal distress, occurring within 6 to 24 hours after ingestion and lasting for about 1 day.

- *Escherichia coli.* The disease related to this organism is usually a mild intestinal illness that occurs 3 to 5 days after eating contaminated food. However, *Escherichia coli* 0157:H7 disease can re-

sult in illness requiring hospitalization. Symptoms include hemolytic uremic syndrome (HUS), characterized by severe abdominal cramps, little or no fever, and severe (often bloody) diarrhea. Only about 5 percent of cases develop HUS, but these can be life-threatening, especially in the very young and the elderly.

- *Listeria monocytogenes.* This bacterium can cause either mild or severe illness. Milder cases include sudden onset of fever, severe headache, vomiting, and other influenzalike symptoms. Severe cases can result in chronic illness and death.

- *Salmonella.* Illness usually appears as soon as 6 or as long as 70-plus hours after eating contaminated food and lasts for 1 to 2 days. Common symptoms are nausea, diarrhea, stomach pain, and sometimes vomiting. In rare cases it can cause death in very young or very old persons.

- *Staphylococcus aureus.* Illness usually occurs within 30 minutes to 6 hours following consumption of the toxins produced by the bacteria. Symptoms include severe nausea, vomiting, cramps, and diarrhea for 1 or 2 days. In severe cases, hospitalization is necessary.

- *Toxoplasma gondii.* This parasite causes acute or chronic illness when undercooked pork, mutton, and some other meats are consumed. The acute illness has mild flulike symptoms.

Source: Adapted from an article prepared by Jean C. Buzby and Tanya Roberts in *FoodReview* 18, no. 2 (May–August 1995):37.

year in the United States. Visual inspection of carcasses cannot detect the contamination of carcasses of these pathogenic organisms that are potential causes of foodborne illnesses.

In addition to pathogens that may be carried by the marketed animal, pathogens may be introduced to meat products from sources such as feeds, animal waste, or improper sanitation of personnel, equipment, and facility. Pathogens also can be introduced through improper meat-processing and food-handling practices (e.g., inadequate heating or cooling, poor hygiene) and result in their survival and multiplication, further increasing the risk of foodborne illness.

The 1996 food safety rules are based on prevention of meat contamination, whereas the old program relied more upon detection and elimination of meat deemed unfit for human consumption. This most recent program is overseen by the Food Safety and Inspection Service (FSIS) branch of USDA and is briefly described below.

- All processing plants must have a Hazard Analysis and Critical Control Points (HACCP) plan. A critical control point is a point, step, or procedure where controls can be used to prevent, reduce to an acceptable level, or eliminate food-safety hazards. Food-safety hazards include the presence of chemical residues, metal fragments, or other materials that may cause a food to be unsafe for human consumption. Once a plan is developed, the plant must then establish and monitor critical limits for each hazard at each critical control point.

- Each plant is required to develop written sanitation standard operating procedures to verify how they are meeting daily sanitation requirements. This rule is important to ensure a reduction of pathogens on meat brought about by unsanitary practices.

- The Food Safety and Inspection Service will test for *Salmonella* on raw meat and poultry products to verify that pathogen reduction standards for *Salmonella* are being met. *Salmonella* was chosen for testing because it has been the most common cause of foodborne illnesses.

- Slaughter plants are required to test for generic *E. coli* on carcasses to verify that they are preventing and removing fecal contamination.

The following studies provide further evidence of the concern for meat safety in the United States and prompt actions taken when foodborne illnesses are associated with products in the human food chain.

Texas A&M University, with financial support from the USDA, conducted studies to reduce fecal contamination on beef carcasses. Researchers' work showed that washing of carcasses, followed by organic acid treatment, performed better than trimming or washing alone on all surfaces except the inside round, where organic acid treatments and trimming performed equally well.

Also, a consortium of five land-grant universities, with support from the National Live Stock and Meat Board, found that trimming and spray washing greatly reduced fecal material placed on meat samples. Numerous other research grants were awarded through the USDA to develop rapid methods to detect and count certain bacterial pathogens in meat and poultry products. Rapid tests are a necessary tool in improving on-line meat and poultry inspection.

The new HACCP system has been estimated to cost the industry about $100 million annually, or about one-tenth of a cent per pound of product. However, it is estimated to result in savings of $1 to $4 billion in reduced costs related to foodborne illness. For more detailed discussion on possible foodborne illnesses, see Box 25–1.

25.4   COOLING

Carcasses traditionally have been cooled rapidly and thoroughly to ensure preservation. Poultry carcasses are chilled to 40°F or below in minimum time, in ice-water vats or by air chilling. Chilling in-line can be accomplished by carcasses being transported through

50- to 200-foot vats with slush ice at 32° to 33°F. Carcasses then are generally packed in boxes and covered with crushed ice for shipment or for short-time holding until further processing is done. Other poultry (such as turkeys) are vacuum packed, hard-chilled (−20°F), and kept frozen in heat-shrinkable plastic bags.

Coolers in beef, pork, and lamb packing plants are ideally held at 27°F, though temperature may rise above that when large numbers of hot carcasses enter the cooler. Large fans keep the cold air moving for quicker and more efficient cooling of the carcasses.

Pork carcasses usually are chilled 24 hours before cutting. This allows the inside temperature of the thicker portions, such as the hams and shoulders, to be completely chilled to 38°F.

Lamb and mutton carcasses are also cut, or loaded for shipment, after 18 to 24 hours in the cooler. Shipment, of course, must be in refrigerated rail cars or trucks.

Because beef carcasses are larger and thicker, complete cooling to a desired temperature of 33°F requires more time. Carcasses usually are not cut until 30 hours or more have elapsed. Top-quality beef is sometimes kept in the cooler longer for aging. (See Section 25.8.)

25.5 MEAT GRADING

Just as the federal inspection stamp (Figure 25–4) signifies that meat bearing the stamp[2] is safe for human food, the federal grading stamp (Figure 25–6) designates relative meat *quality*. Reasons for having and using grades were presented in Chapter 24.

Unlike meat inspection, federal meat grading is *not mandatory,* but any beef or lamb carcasses that are quality graded must also be yield graded. The service is available to processors who want it; the processors therefore pay the USDA for the service.

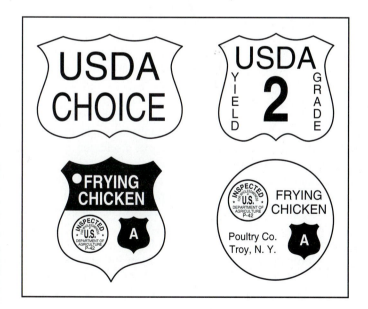

Figure 25–6
The federal meat quality grading stamp for beef or lamb, top left; beef yield grade stamp, top right. For poultry, both inspection and grade marks, bottom left and bottom right, may appear on a package or on a wing tag.
(*Source:* USDA.)

---

[2] The inks used in the application of the inspection and grading stamps are made from an approved vegetable dye considered safe for human consumption.

A yield grade on beef or lamb carcasses is an estimate of the carcass that is edible meat. It has a close relationship to carcass cutout value. Yield grades range as follows:

Yield Grade 1—very trim and muscular, high yield of edible portion

to

Yield Grade 5—very fat and wasty, low yield of edible portion

As discussed in Chapter 24, yield grades are based on fat cover over the rib eye; area of rib eye; percent kidney, pelvic, and heart fat; and carcass weight. The yield grade is stamped on the front and hind quarters.

Many retailers, especially large chain stores, demand USDA grades on the beef, lamb, or poultry they buy. Essentially no pork is "government graded," but many processors do their own carcass grading according to USDA standards, especially when hogs are bought on a grade and yield basis. Grading standards have been established for veal and calf carcasses, but grading is performed only on carcasses after the hide is removed.

The men and women who do government grading are hired, supervised, and paid by the USDA. Graders usually are assigned to a certain plant, or perhaps to several adjacent plants.

Where federal grading is employed, beef or lamb carcasses must be chilled before grading. Grade specifications are followed more easily after carcasses have chilled and the fat is firm. Beef carcasses are cut between the 12th and 13th ribs so the grader can appraise the quality of the lean in the eye muscle, as well as the carcass maturity. Most lamb carcasses are not ribbed for grading and leave the plant intact.

The federal grader determines the *quality* grade, then usually applies a temporary indicating mark. The grader or assistant then "rolls" the entire length of the carcass so the USDA quality grade stamp appears on every wholesale cut and on some of the retail cuts from that carcass.

Poultry grades may be applied to live birds, dressed or ready-to-cook poultry, or poultry food products. In many instances a random sample of a lot is individually graded, for the purpose of ensuring sale contract specifications. Poultry are graded on an individual basis following evisceration, so each bears a grade stamp. The grader often sorts the birds by grade into bins or onto different conveyors.

Where private or processor grades are used, they are usually applied in the same manner. Though specifications similar to those used in USDA grades may be employed, the private grade designation or stamp must be clearly different from those used by the USDA.

Chapter 24 described the USDA grades, and the above discussion described how people follow the written grade standards in subjectively grading carcasses. Research has been directed toward automated or semiautomated grading utilizing cameras, sensing devices, and computers that will rapidly quantify proportions of fat and lean (see Box 23–1). Such data available to the grader could increase consistency in grading by decreasing subjective judgments.

Quantitative measurements—to determine yield (see Figure 24–5)—have been used more in the grading of pork carcasses, but increased precision and grading consistency has been sought. The large volume of pork that moves among European Economic Community countries has caused these countries to do considerable research on automated measuring and grading devices.[3]

---

[3] Kempster, A. J. *Pig News and Information* 2 (1981): 145. Published by Commonwealth Agricultural Bureau, Slough, United Kingdom.

25.6 OTHER REGULATIONS OF MEAT PROCESSING

The Food Safety and Inspection Service (FSIS) of the USDA has a major responsibility in ensuring a wholesome meat supply through its handling of antemortem, postmortem, and continuing inspection. Continuing inspection covers compliance with requirements for sanitation, temperature control, ingredients, and labels. The FSIS must approve labels, formulations, and processing procedures to be sure they comply with regulations. It determines if products are adulterated (contain harmful or unexpected ingredients or amounts of ingredients) or misrepresented (with false or misleading labels or containers). Products must comply with standards of identity, especially with regard to minimum quantities of meat.

Also, in a program to control pathogens in meat and to educate food handlers, the FSIS now requires meat and poultry products that are not ready-to-eat to carry *safe handling instructions*. These instructions warn consumers that some food products may contain bacteria that could cause illness if the product is mishandled or cooked improperly.

The Food and Drug Administration (FDA) has responsibilities for foods, drugs, cosmetics, and related products and is especially concerned with additives and with harmful agents. The 1906 Food and Drug Act and the more comprehensive Food, Drug, and Cosmetic Act of 1938 provided the FDA the authority to remove adulterated and obviously poisonous foods from the market. The 1958 Food Additives Amendment and the 1960 Color Additives Amendments regulate food additives. The 1958 and 1960 amendments shifted the burden from the government (to prove that a food additive is unsafe) to the manufacturer (to prove it is safe). This is frequently a complicated and expensive process.

Today, some 2,800 substances are added intentionally by processors to foods to produce a desired effect. Additives may be used to maintain or improve nutritional value (not frequently done to meat products). They may also help maintain freshness. Examples would be use of antioxidants to delay onset of rancidity in lard or tallow, or of sodium nitrite to protect cured meats from *Clostridium botulinum* toxin development. Many additives may help in processing or preparation and others make food more appealing in appearance or flavor.

Two major categories of additives are exempt from the testing and approval process. The first is a group of about 700 substances "generally recognized as safe" (GRAS) by qualified experts. These have been used extensively in the past, but the inclusion of some substances on the GRAS list is under constant scrutiny because of public health concern. An example is the use of sodium nitrite in cured meat. Also exempt from testing are "prior sanctioned substances," those that had been approved for use by either the USDA or FDA prior to 1958. Some of these are also on the GRAS list. The Federal Trade Commission is involved in monitoring advertising, and its functions sometimes overlap those of the FDA and USDA.

25.7 FABRICATION—WHOLESALE AND RETAIL CUTS

Wholesale and retail cuts of lamb, beef, and pork and where they come from on the carcasses are presented in Color Plates B, C, and D. Even though most lamb leaves processing plants as intact carcasses, some carcasses are broken into wholesale cuts. This allows more effective merchandising. Heavy lamb legs or loins are preferred by many hotels and restaurants; other parts of the carcass are merchandised elsewhere.

Boxed beef (pork)
Precut portions of beef or pork prepared by the meat-processing plant for wholesale delivery.

Today, 90-plus percent of all beef and pork goes to wholesalers, retailers, chain store breaking plants, and restaurants as **boxed beef (pork)**—subprimal cuts that are vacuum packaged and transported in boxes containing about 60 to 80 pounds of "saw-ready" or "knife-ready" cuts (Figure 25–7).

Boxed pork also is available and preferred by the retailer. This method provides the wholesaler or retailer a more uniform product of the desired cuts. Benefits to the packer are savings in transportation costs and better use of the trimmings.

When subprimal cuts are sold by the processor, most bones, excess fat, and rough cuts are retained at the plant. This saves about 30 percent of shipping tonnage. The usual subprimal cuts include chuck backs, chuck arms, and boneless chucks; rib cuts, short loins, and sirloins; inside and outside rounds, round tips, and regular or semibone-less rounds. Processors also will sell coarse ground beef with an identified fat percentage and a variety of thin or special cuts. Customers have the option to buy beef carcasses as subprimal cuts in boxes, in which case they receive credit for thin cuts, fat, and bone but pay a standard labor charge. They can also purchase any amount of any subprimal cut.

Vacuum-packaged boxed cuts appear dark but "bloom" (brighten) upon unpackaging and cutting. They retain the ability to bloom up to at least 28 days. Boxed meat also allows each retailer to purchase more high-demand cuts and respond to the area's buying preferences.

Meat fabricators and retailers have much closer inventory control with boxed meat. This system also shifts more of the labor requirement to the fabrication plant from the retail store.

Some processors have gone further in merchandising of meat, providing family-sized, transparent, vacuum-sealed packages bearing the processor brand name. Retailers merely distribute the product and contribute little to the processing. Continued movement to central processing of retail cuts is likely.

More retailers are providing a combination of store-named and brand-name products. In these cases, their store name is placed on the product along with the name of the processor.[4] Benefits of merchandising "case-ready" products include more consistency, greater availability of value-added products, more safety and consumer confidence in branded products, and nutritional information on the packaging.

Processors also are receiving more assistance in marketing technology through such systems as the Computer Assisted Retail Decision Support (CARDS) computer program developed by Texas A&M and the value-based meat management (VBMM) system provided by the National Live Stock and Meat Board's Meat Marketing Technology Center. These systems enable the retailer to compare profits of traditional and close-trim items to know the cost of every item in the meatcase, and to develop sales forecasts.[5]

A significant proportion of beef and pork is ground, either at the primary processing plant or by specialized processors who form and freeze beef patties for the fast-food or pizza industry. Batches of the ground product are sampled and tested for lean content, then the batches are blended in the proportion needed to meet federal or state quality requirements and to achieve a uniform product. After blending and rapid chilling with CO_2, the product usually is ground again and formed into quarter-pound (or other weight) patties by machine. The patties then are quickly frozen, sometimes by

---

[4] *Meat & Poultry* 40, no. 7 (July 1995):52.
[5] Ibid.

Figure 25–7
In low-temperature rooms, beef carcasses are separated into wholesale cuts (upper left), and some bones are removed from the cuts (upper right). The cuts are then placed in vacuum-sealed bags and boxed for shipment (lower left).
(Courtesy of Iowa Beef Processors)

passing through a tunnel into which −320°F gaseous nitrogen is pumped, and are shipped and stored in the frozen state.

Quality control steps, in addition to sanitation checks, may include passing the formed patties through a metal detector to check for the rare presence of buckshot, metal product identification tags, or other metal.

Most pork carcasses are reduced to wholesale and retail cuts at the slaughtering plant because (1) hams, bellies, and sometimes other cuts are cured and smoked, requiring expensive equipment, and (2) the larger percentage of fat trim can be rendered directly into lard by the packer. Boxed pork is cut into desired cuts, vacuum-packed, and shipped to retail outlets in a manner similar to boxed beef.

Mechanization has reduced labor needs in cutting carcasses, though much skilled labor is still used. Because carcasses vary in size, shape, degree of fatness, and other characteristics, development of labor-saving equipment to handle carcasses has been slow.

Extremely low-quality carcasses—such as Cutter and Canner grades of beef, the Cull grade of lamb, yearling, or mutton, or extremely soft pork carcasses—are not merchandised as wholesale or retail cuts. The meat certainly is edible and usually is highly nutritious but presents a less desirable appearance to the consumer. Such carcasses, or parts of carcasses, may be completely "boned out" at the plant. Edible meat is removed from the bones and then sold as boneless meat or hamburger, or is used in one or more of a wide variety of prepared meat items such as sausages or canned meat products.

The term *lean cuts* is used to categorize certain pork wholesale cuts and includes the trimmed ham, loin, **Boston shoulder**,[6] and picnic. On good-quality lean carcasses these cuts contain about 65 or 70 percent lean, whereas the belly contains only about 35 percent lean. High-quality, U.S. No. 1 pork carcasses yield 60.4 percent or more of their weight as trimmed lean cuts. (See Section 24.9.)

The term **primal cuts** refers to the more expensive cuts on a carcass, whether of beef, pork, or lamb (Figure 25–7 and 25–8). In pork, the primal cuts are the ham, loin, shoulder cuts, and belly. Primal cuts of beef are the round, loin, rib, and chuck. Note that most of the weight of the hindquarter of beef is composed of primal cuts. In lamb, the leg, loin, rack, and shoulder are the primal cuts, essentially the same portions of the carcass as in beef. The leg of lamb usually represents about half the value of a carcass, even though it comprises only about one-third of the weight. Such terms generally are not used in discussing poultry cuts.

Poultry carcasses that are further processed usually are cut by hand in the initial processing plant before or after chilling, but sometimes are cut by the wholesaler or retailer. For retail, cuts usually are wrapped in transparent materials and sold fresh or frozen;[7] they may be sorted and packaged as breasts, backs, thighs, or other parts (Figure 25–8).

Chicken and turkey frames, backs, necks, and wings are often mechanically deboned and the recovered meat used for processed products. Turkey thigh muscle frequently is used in production of "turkey ham," and chicken breast sandwiches have become popular in some fast-food outlets.

> **Boston shoulder**
> The top part of the shoulder of a pork carcass. Sometimes called the Boston butt.

> **Primal cuts**
> The most valuable cuts on a beef, pork, or lamb carcass.

25.8 AGING AND TENDERIZING

Protein-digesting enzymes (proteases) present in fresh meat tissue will cause meat to become more tender in time. These enzymes cause chemical breakdown of certain meat components, such as connective tissue and muscle fibers. The enzymes are effective at the higher temperatures attained during cooking.

Meat usually is not aged by processors because of the cooler space required and inventory costs. Seven days of aging will achieve increased tenderness; holding to 14 or 21 days causes little additional benefit. Where aging of meat is done, only the highest quality meat for the most discriminating customers is aged. Choice and Prime beef and lamb carcasses, or primal cuts from these carcasses, are aged more often. Besides the special cooler and equipment, extra labor is involved in handling the carcasses or cuts, and there is interest cost on inventory. The meat loses weight during aging, and the sur-

---

[6] The term *Boston shoulder* is adopted by the National Live Stock and Meat Board and by many retailers to replace the term *Boston butt*. The latter term still is used in some market reports and sections of the industry. The term *shoulder* is used to include both the picnic and the Boston shoulder.

[7] A rule, published in January 1995, prohibits the term *fresh* on the labeling of poultry that has ever been chilled below 26°F. Poultry previously frozen between 0° and 26°F must be labeled "previously frozen."

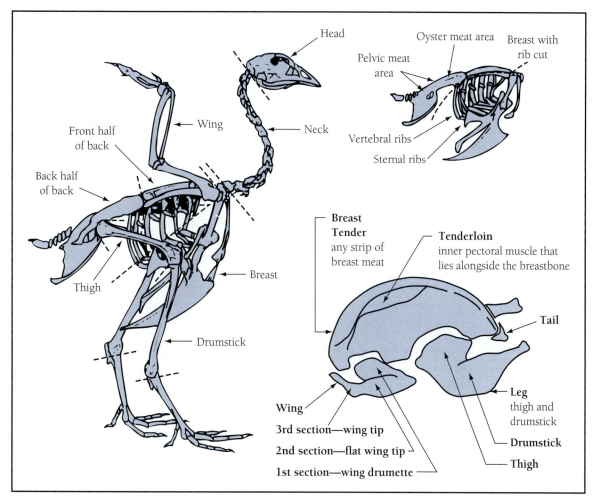

Figure 25–8

The broiler carcass and retail cuts. Outline of the skeleton (left and above) shows the bones that will be present in each cut, and also illustrates that most of the cutting of the carcass into retail cuts is at bone joints.
(*Sources:* USDA and Kansas State University.)

faces of the cuts usually need to be trimmed away afterward. Beef cuts in vacuum bags may be aged with little weight loss. Cuts comprised largely of heavily used muscles, such as the round or shoulder, rarely are aged. They are tenderized most efficiently by moist heat cooking.

Lower quality meat usually is not aged because the slight improvement in quality does not merit the expense, and such meat is used increasingly for ground and formed product. Also, beef, or lamb with less external finish, would shrink more.

The use of electrical stimulation of carcasses has become a widely developed practice in the beef-processing industry. As mentioned in Section 25.1, the use of low voltage, applied to the carcass within minutes after stunning, tends to accelerate the aging process by causing more rapid decline of muscle pH and release of the lysosomal enzymes (proteases). It also causes some stretching and tearing of muscle cells as the result of the sharp muscle contraction caused by the charge. This and connective tissue alteration also may contribute to some increased tenderness.

The degree to which tenderness is improved by electrical stimulation can be determined by measuring decreases in force required to shear a muscle or by subjective scoring. Results have ranged from 12 to 58 percent, depending on type of carcass and time and method of applying the charge.[8] Tenderness improvements of 25 to 30 percent are normal in commercial operations.

Because stimulation speeds normal postmortem changes, this development permits earlier "ribbing" and grading of beef carcasses in the plant, the boning of beef while carcasses are still hot,[9] and other processing modifications that reduce carcass holding time and inventory and storage costs.

Other benefits of electrical stimulation include improved flavor, perhaps associated with release of aromatic compounds due to muscle or cell disruption, and brighter colored lean, apparently due to more rapid postmortem glycolysis in the meat tissue.

Some processors tenderize beef by preslaughter injection of a purified enzyme solution, mostly *papain,* derived from certain tropical plants, into the bloodstream. The enzyme, which is a protein digestant similar to animal proteases, is quickly pumped to all tissues by the heart. Much remains in tissue capillaries, with circulatory fluid, even after the animal is bled.

The enzyme, or blends of several proteolytic enzymes, has little opportunity to cause tenderization until cuts are heated during initial stages of the cooking process because carcasses are chilled within a few minutes after slaughter. Also, the enzyme apparently is not released from the capillaries into the muscle tissue until cooking begins. This particular enzyme, at the concentration used, is most effective between 130° and 160°F. As cooking progresses and the temperature of the cut passes 160°F, in short-time, high-temperature cookery, action of the enzyme is essentially stopped, so there is little danger of "overtenderization."

Use is more common, of course, in those meat cuts such as outside round roasts that are less tender and that are therefore subjected to relatively long cooking time at low temperatures, 135° to 145°F. Enzyme concentration and length of cooking determine degree of tenderizing achieved.

Some processors use enzyme preparations in the form of a dip, spray, or injection on cuts prepared for institutional and restaurant trade. Tenderizers are also available in dry form for home use, often in combination with certain seasonings.

Meat tenderization by enzymes is centuries old, apparently first done in tropical countries where meat was wrapped with papaya leaves or boiled with unripe papaya fruit. Fruit and leaves of the papaya tree are the source of the proteolytic enzyme papain. Other sources of effective proteolytic enzymes currently used in commercial meat tenderizers are bromelin (from pineapple stems), ficin (from the fig tree), fungal and bacterial enzyme secretions, and trypsin (from hog pancreas). Others may also be used.

There is still much to be learned about the relative effectiveness of the various enzyme preparations. Most such materials are mixtures, and meat is composed of a mixture of different proteins. Each enzyme may be effective only on specific proteins. Essentially no pork or poultry is aged or treated with enzymes because tenderness usually is not a problem.

Some beef processors also use blade tenderization of boneless beef cuts to ensure tenderness. In the process, many small blades or cutting edges cut into the meat

---

[8] Smith, G. C. *The National Provisioner,* February 10, 1979, p. 50.

[9] Bowles, J. E., et al. *"Continuous versus intermittent electrical stimulation of beef carcasses and their effect on hot-boned muscle pH decline."* Progress Report 394. Kansas Agr. Exp. Sta., 1981.

structure. Sanitation and holding down microbial activity is especially critical because the blades significantly increase surface area. The technique is also used in processing of some pork and other meats. Though tenderization is achieved, the increased surface area can result in greater weight loss during cooking.

25.9   CURING AND SMOKING

Ham and bacon are the most popular cured meat products. Picnics, Boston shoulders, pork loins and tenderloins, beef briskets, chickens, turkeys, and poultry rolls are cured and smoked as specialty items.

Curing and smoking procedures vary among processors and according to the type and size of product. Curing is a preservation process; smoking adds a distinct flavor but also helps in preservation.

Salt, the main ingredient in curing solutions, promotes preservation by its mere presence in meat tissue, because many bacteria will not grow in high concentrations of salt. The salt also causes dehydration of tissue cells, further inhibiting bacterial growth. Sugar usually is included in a curing formula to help maintain desirable flavor. A small quantity of sodium nitrite also is used to promote a bright pink color, control *Clostridium botulinum* (producer of a toxin), and serve as an antioxidant.

Sodium isoascorbate (a form of vitamin C), or closely related compounds, accelerate development of and stabilize the bright pink color and also block formation of *nitrosamines,* potential cancer-inducing agents. Certain phosphate compounds are added to meat-curing mixtures to decrease moisture loss during smoking and cooking. Processors may add other distinctive flavors to their curing solution.

The curing ingredients are usually injected into the fresh cuts or birds in solution. Red meat cuts or poultry then may be immersed in curing solution. These methods allow uniform and rapid cure throughout the cut or bird.

Time of cure depends on size of cut or bird. Using present methods, bellies cure in about 2 days and hams in about 2 days per pound. Curing solution is injected into fresh pork bellies or hams by a multineedle injection machine.

Cured bellies to be smoked are hung on racks in smoke ovens and usually are smoked up to 18 hours at 125° to 135°F. Smoke to penetrate the cuts usually is generated from chips of some nonresinous wood, such as hickory, apple, oak, or maple. Hams and other pork cuts are heated to an internal temperature of 155°F, with the smoking oven temperature reaching as high as 180°F. Several meat processors use an assembly-line smoking procedure. Bellies are negatively charged, then exposed to positively charged smoke particles. Electrical attraction causes rather complete smoke penetration.

Many hams and some poultry are sold as "ready to eat." Such hams usually have been held in the smokehouse 18 to 24 hours, with the interior of the hams reaching and being held for at least 2 hours at 155°F. To produce ready-to-eat smoked poultry, an internal temperature of 160°F must be reached. Curing and smoking procedures, as well as other steps in meat processing, are closely checked by employees of the FSIS.

25.10   SPECIALTY MEAT ITEMS

The continued preparation of a large number of meat items in a typical processing plant would make a lengthy story—from chilling, slicing, and packaging cured and smoked bacon (bellies) to formulating and canning chili or chicken chop suey. Some processors specialize in certain prepared meat items such as luncheon meats, chili, wieners, unique sausages, or chicken and noodles, and may buy trimmings, poultry carcasses with skin tears, and other ingredients from other processors. Some companies produce only such prepared meat items.

Competition and management precision have led processors to fabricate meat items with predetermined and standard quality attributes. Examples of such items include "breakfast meats" with an established ratio of lean and fat and prepared in strips to simulate bacon, meat "cutlets" of blended lean and fat that have been flaked and restructured from raw material rather than chopped or ground so the consistency of the cutlet will more nearly approach the texture of natural cuts, and preformed chicken patties.

With higher proportions of U.S. meals served in restaurants, fast-food outlets, and group food services—such as university dining halls and plant cafeterias—uniformity and control of both size and quality attributes of meat servings has become increasingly important. The term "portion control" is used to describe this phenomenon. Servings of the size and shape of a standard pork chop, round steak, or turkey drumstick can be formulated, prepared, and marketed.

Many kinds of sausages are produced in meat-processing plants. Wieners and bologna are produced in the largest quantities. Wieners usually are made from beef and/or pork trimmings, finely chopped, and mixed with certain spices and flavors. This mixture is forced into long, artificial casings, then linked by machine, cooked, and smoked or sprayed with liquid smoke prior to cooking. Cooking solidifies the inside of the wiener. After chilling, another machine removes the artificial casing, yielding "skinless franks" (Figure 25–9). Meat inspectors check here, too, on processing equipment and methods. By chemical analysis of finished wieners, they make sure the wieners conform to limits for added fat or water.

Grinding meat automatically increases preservation problems because it increases surface area exposed to possible contamination by spoilage organisms.

Cereal products, milk products, and other nonessential materials are used in certain processed meats to increase water-holding capacity, reduce cost of production, or improve color, flavor, texture, or fat stability. Sausages produced in plants under federal inspection may contain up to 3.5 percent of these ingredients, except for high-protein soy additives, which are limited to 2 percent.

Figure 25–9
This equipment can stuff, link, cook, smoke, chill, strip, and package about 36,000 links per hour.
(Courtesy of Oscar Mayer and Company.)

Artificial casings are used most commonly for sausages, whereas natural casings sometimes are preferred and used for special types of sausages. Natural casings—intestines of lambs, hogs, or cattle—are rather expensive because they must be washed and trimmed carefully and thoroughly.

Tongue, considered a specialty meat item, may be sold fresh or smoked. Some also is used in prepared meat items. Heart and liver usually are sold fresh, though they too might be used in special sausages or other prepared meats.

Other merchandising techniques are used. One processor developed a process for canning a ham loaf so that it could be stored at room temperature for long periods. This has provided a significant outlet for trimmings that remain after boning hams, extra-large hams that retailers cannot sell effectively, or surplus hams available after marketing peaks.

Soups, stews, baby foods, and other canned products also use chicken, turkey, beef, pork, and lamb that may not be effectively merchandised as retail cuts.

25.11 MEAT DISTRIBUTION SYSTEMS

Meat wholesalers, processor "branch houses," and numerous retail food-chain warehouses serve as major links between processors and retailers, hotels, and restaurants.

There is some vertical integration in the meat business, where one company owns more than one type of unit in the feedlot–slaughter–processing–retail chain. The level of integration by species is estimated to be almost 100 percent for broilers, over 90 percent for turkeys, under 33 percent for pork, and under 20 percent for beef. Vertical integration allows a company to coordinate the capacity utilization of each stage of production, establish a single profit center, and control quality from beginning to end.[10]

Some foodstore chains own meat-slaughtering or -processing plants; a few also operate feedlots. Some meat processors own restaurants or retail stores.

The growth of independent wholesalers apparently is due to several factors. First is marked increase in meat consumption in institutions, plants, and restaurants served by such wholesalers. Second, the movement of processors to major production areas has caused processors to depend on wholesalers in meat-deficient areas to distribute their product. Also, some processors have tended to specialize more, and rely on wholesalers or brokers.

Most wholesalers employ salespersons who regularly call on and quote prices to retailers, institutions, and restaurants. There is much competition, and the product is perishable. Meat items are delivered by refrigerated truck.

25.12 BYPRODUCTS

Many consumer items, from pharmaceuticals to footballs, are byproducts of livestock and poultry processing. *Byproducts* are defined here as any products other than meat, prepared meats, or specialty meats. From 2 to 20 percent of the income received by processors comes from such byproducts, varying according to species, animal quality, and relative prices.

Swine usually yield a higher proportion of their total value as byproducts (see Box 25–2). In recent years, skin from hogs has been used to replace lost skin of burn victims, and hog heart valves are sometimes used to replace weakened human heart valves.

---

[10] Thornton, G. "Measuring the competition." *Broiler Industry.* September 1995.

BOX 25-2

Everything but the Oink!

Serving Essential Human Needs

No other animal provides society with a wider range of products than the hog.

Hogs are, of course, the source of high-quality animal protein in the form of the widest and most varied range of food products available from any animal.

Byproducts from hogs play a vital though less visible role in maintaining and improving the quality of human life. New and different byproducts from hogs are constantly being developed.

Insulin from hogs is used in the treatment of diabetes. Hog heart valves are used to replace damaged or diseased human heart valves. Skin from hogs is used to treat severe burn victims.

The amazing utility of the hog has motivated the saying, "We use everything but the oink."

A viable animal agriculture not only provides an abundant supply of vital nutrients found in meat but is also a ready source of essential and useful byproducts that humanity depends on so extensively.

Shown here are some of the important medical and industrial products we get from hogs.

Pharmaceutical Byproducts

Pharmaceuticals rank second only to meat itself in the important contributions hogs make to society. Rapidly advancing science and technology are continually adding to the list of life-supporting and life-saving products derived from the incredible hog.

Adrenal Glands
Corticosteroids
Cortisone
Epinephrine
Norepinephrine

Blood
Blood Fibrin
Fetal Pig Plasma
Plasmin

Brain
Cholesterol
Hypothalamus

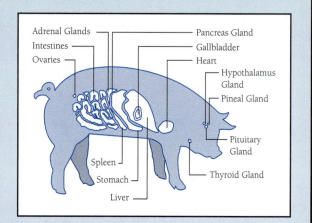

Gallbladder
Chenodeoxycholic acid

Heart
Heart Valves

Intestines
Enterogastrone
Heparin
Secretin

Liver
Desiccated Liver

Ovaries
Estrogens
Progesterone
Relaxin

Pancreas Gland
Insulin
Glucagon
Lipase
Pancreatin
Trypsin
Chymotrypsin

Skin
Porcine Burn Dressings
Gelatin

Spleen
Splenin Fluid

Stomach
Pepsin
Mucin
Intrinsic Factor

Thyroid Gland
Thyroxin
Calcitonin
Thyroglobulin

Pineal Gland
Melatonin

Pituitary Gland
ACTH—Adrenocorticotropic Hormone
ADH—Antidiuretic Hormone
Oxytocin
Prolactin
TSH—Thyroid Stimulating Hormone

Industrial Byproducts
Hogs also make a very significant contribution to the world of industrial and consumer products. Hog by-products are sources of chemicals used in the manufacture of a wide range of products that cannot be duplicated by syntheses. Of course, pigskin is used extensively as high-quality leather for clothing, shoes, handbags, sporting goods, upholstery—the list goes on and on.

Blood
Sticking Agent
Leather Treating Agents
Plywood Adhesive
Protein Source in Feeds
Fabric Printing and Dyeing

Brains
Cholesterol

Bones and Skin
Glue
Pigskin Garments, Gloves and Shoes

Dried Bones
Buttons
Bone China

BoneMeal
Mineral Source in Feed
Fertilizer
Porcelain Enamel
Glass
Water Filters

Gallstones
Ornaments

Hair
Artist Brushes
Insulation
Upholstery

Meat Scraps
Commercial Feeds
Feeds for Pets

Fatty Acids and Glycerine
Insecticides
Weed Killers
Lubricants
Oil Polishes
Rubber
Cosmetics
Antifreeze
Nitroglycerine
Plastics
Plasticizers
Printing Rollers
Cellophane
Floor Waxes
Waterproofing Agents
Cement
Fiber Softeners
Crayons
Chalk
Phonograph Records
Matches
Putty
Paper Sizing
Insulation
Linoleum

Source: Adapted from the National Pork Producers Council.

Cattle hides furnish most of the leather for shoes, luggage, purses, book bindings, garments, and athletic equipment. There are three types of leather used in these and other products: latigo, suede, and tooling. Hides vary in quality; those carrying brands or grub holes are less valuable because there is more waste in their use. The pelt is the most valuable single byproduct of sheep, the hide is the most important cattle byproduct, and feathers are a major poultry byproduct.

Beef cattle have thicker hides than do cattle of dairy breeds. Steer hides or hides from well-fed animals are thicker than hides from cows or poorly fed cattle. After slaughter, hides are cured by soaking in a salty brine for about 24 hours, then made into leather by tanning, so they will not be subject to bacterial action. Most tanners are specialists and buy cured hides from processors. The hides usually are cleaned for hair and any flesh by machine prior to shipment to the tanner. The hides are immersed for several weeks in tanning liquors made from certain wood bark, or for a few hours in a liquid containing chromic salts. Most heavy, top-quality hides are tanned by the former process to give maximum shock absorption, puncture resistance, and wear. Chrome tanning is used more for light cattle and calf hides.

Fats are a major byproduct of all species. Edible fats, kept fit for human consumption throughout processing, are rendered to make lard or tallow and eventually are used in spreads, pastries, candy, and other food items as shortening. The proportion of fat handled this way has increased with the trend toward boxed beef and pork—more of the fat is trimmed off at the processing plant. Inedible fats, generally gathered from dispersed retail and other outlets where trimming also occurs, are used for biodegradable detergents, flotation agents, candles, plastics, glycerines, other chemicals, lubricating oils, and hundreds of other materials. They also are used in livestock feeds, especially for poultry and swine, because of their extremely high energy value.

Blood meal, meat and bone scraps, mixed poultry byproduct meal, and bonemeal also are used in some livestock feeds and pet foods. Their names imply their origin, each product being dried and ground after excess fat is extracted. Tankage is simply a mixture of blood meal with meat and bone scraps. See Table 3–4 for nutrient content of these feeds.

Wool, pulled from the pelts after slaughter, is made into clothing, blankets, and other items, but also yields lanolin. This material is refined from wool grease and, because of its similarity to human skin secretions, has become popular as a base for many ointments and cosmetics.

Poultry feathers are used for millinery plumage, badminton shuttlecocks, arrows, decorations, bedding, brushes, and even garden mulch. The most extensive use of feathers, however, is hydrolyzed feather meal for livestock and poultry feeds. Carpet padding, insulation, gaskets, furnace and air-conditioner filter pads, plaster binder, small brushes, and dozens of other items are made from cattle hair. Most hog hair is used for cushion padding, but usually is mixed with longer cattle or horse hair. For effective use as padding, this hair first must be curled by machine.

Glue and other adhesives are mostly protein and are derived from collagen, a type of connective tissue extracted from hide or bone. Another type of animal adhesive is made from dried blood. Although gelatin is chemically and physically very similar to glue and is made from the same materials, it is used mostly in foods.

Dozens of drugs and pharmaceuticals are purified from glands and organs removed during and after livestock and poultry slaughter. ACTH, thyroid extract, and insulin are common to all. Others include epinephrine, used to relieve symptoms of hay fever and allergies; thrombin, which helps blood coagulation following injury or surgery; and heparin, which is an anticoagulant used to prevent undesired coagulation during surgery. Trypsin, a protein-digesting enzyme from the pancreas, is used to liquefy pus and debris in wounds. The liver yields an extract for treatment of anemia. Many other examples could be given.

QUESTIONS FOR STUDY AND DISCUSSION

1. What are four characteristics of the meat-processing industry, based on the larger volume companies?
2. What is the derivation of the term *packer*?
3. Where are the newer slaughtering plants primarily located? Why?
4. Name the top three companies in cattle slaughter.
5. What are the proper humane methods of rendering animals insensible for slaughter?
6. What is the purpose of stunning, not killing, prior to "sticking"? When is sticking done?
7. What is the value of electrical carcass stimulation?
8. Of cattle and hogs, which usually are skinned? scalded?
9. Why is the "fell" in lambs left intact?
10. Describe ante- and postmortem inspection. What three areas of the animal's carcass are inspected carefully during slaughter?
11. Who is authorized to be a meat inspector? Is meat inspection mandatory or voluntary? Who pays for it?
12. What are the qualifications of an official meat grader? Is this service mandatory or voluntary? Who pays for this service?
13. What is the desired temperature of meat coolers and pork, lamb, and cattle carcasses? What is the approximate time required for chilling of the carcasses?
14. Sketch and label (a) a federal meat inspection legend and (b) a federal meat quality legend.
15. What are the yield grades for cattle and lambs? What are the main factors in their determination?
16. What are the responsibilities of the FSIS of the USDA? What additional changes have been made in meat inspection to reduce risk of microbial contamination?
17. Why is more meat being fabricated (boxed)?
18. Name the wholesale cuts of (a) beef, (b) pork, and (c) lamb.
19. What methods have been (or are being) used to enhance tenderization of meat?
20. Name five inedible byproducts saved and processed by larger meat-processing companies that have value and add to their income.

26

Meat as a Food

Meat, the flesh of animals used for food, is the major product of the animal industry. Meat consumption trends, meat quality attributes, and the means by which those attributes are maintained or enhanced are very important topics for those in the industry. This chapter presents data and discussions on these topics.

Upon completion of this chapter, the reader should be able to

1. Describe trends in consumption of meat from several species.
2. Explain elasticity of demand as applied to meat or other foods.
3. List factors that influence meat consumption.
4. Describe meat promotional organizations and efforts in the industry.
5. Contrast major nutritional characteristics of meat and other foods.
6. List factors that may influence meat flavor, tenderness, color, and certain other attributes.

26.1 TRENDS IN MEAT CONSUMPTION

Per capita retail meat consumption in the United States (Figure 26–1) has risen slowly from about 166 pounds in 1970 to about 181 pounds in 2002, on a boneless, trimmed basis.[1] However, shifts from meats of one species to another have occurred during this period.

For example, beef consumption continues to be the American consumer's choice but has declined to about 65 pounds per person in recent years. Pork consumption peaked at about 59 pounds in 1986, but decreased, then stabilized at about 49 pounds per person in the late 1990s. In early 2000, the consumption of pork increased to 58 pounds per person. However, since 1970, the per capita consumption of poultry meat has risen from about 43 pounds to almost 67 pounds per person.

Several factors accounted for the rather dramatic increase in poultry meat consumption. First, both of the red meats have been more expensive than broilers and turkeys. Note that Americans continue to eat slightly more meat each year, but the trend has been toward purchasing the less expensive product. Second, much of society has become more health conscious and considers poultry meats to be lower in fat and cholesterol. More people are reducing each serving of red meat to a more "moderate" portion. A third factor may be that the consumers of poultry have and are purchasing more

---

[1] Some consumption values in the past have been reported based on retail weight in pounds and tend to be higher than current estimates that are based on meat that has been deboned and trimmed.

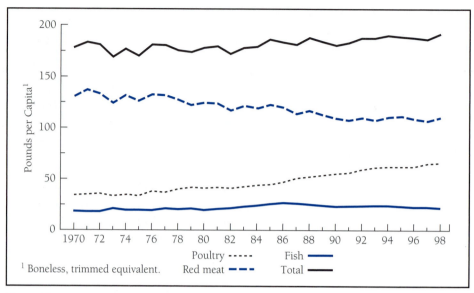

Figure 26–1

Total U.S. per capita meat consumption from 1970 to 1998. Total meat consumption per person for 2002 was about 30 pounds above the 1970–74 level—largely due to greater demand for poultry meat.
(*Source:* USDA, ERS.)

"brand name" and further-processed poultry products. The poultry industry has provided consumers with a good selection and consistent supply of high-quality products.

Lamb is also a "luxury" meat; higher income families eat more. But consumption has gone down during a time of increasing consumer purchasing power. Sheep population grows during economic depressions and recessions but constricts markedly in times of farm prosperity. Total sheep numbers dropped steadily from about 52 million head in 1942 to about 9 million head in 1999.

As lamb production declined, distribution problems arose. In the mid-1950s, for example, fewer than half of the retail stores in the United States handled lamb, and the percentage declined further after that. People patronizing stores not offering lamb lost the habit of eating lamb. Also, because lower volume means higher distribution and handling costs per unit, lamb had to be offered for sale at relatively high prices.

Trends in meat-buying habits other than the volume trends discussed above are apparent. Consumers prefer leaner cuts. They want cuts or prepared items ready for the pan, oven, or microwave. They shop more often and want more variety in their meals. An increase in single-parent families and in couples where both work outside the home means less time to prepare meals. A sharp increase in fast-food services and the volume of product moved through these, especially of ground beef, has markedly changed the demand for various grades and qualities of meat.

26.2   ELASTICITY OF DEMAND

The demand for food in the United States and other developed countries is relatively inelastic. This means that total food consumption doesn't change much in response to the consumer's ability to buy. In other words, the U.S. "stomach" does not change much in

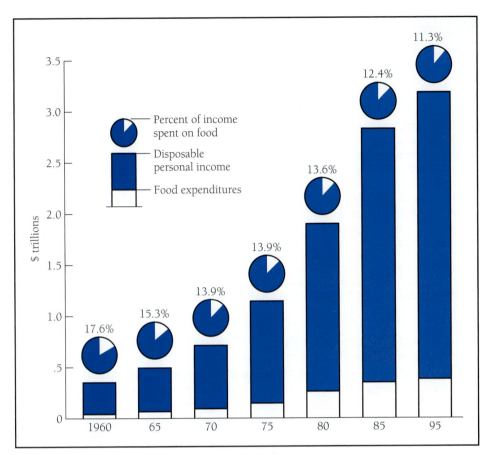

Figure 26–2
Expenditures for food have increased as disposable income has increased, but not as a percentage of income.
(*Sources: FoodReview,* May–August 1995; and USDA, ERS, November 1996.)

size when consumer income goes up or down. Although expenditures for food for the average American family have risen steadily, the percentage of income spent on food has declined (Figure 26–2). In 1997, only 10.7 percent of disposable income went to pay for food, with 6.6 percent for food at home and 4.1 percent for food away from home.

Data to support the above accepted principle, such as accurate estimates of the pounds of food dry matter consumed, are not available.

Few consumer items have a demand as inelastic as that of food. Though clothing and shelter are also considered essential, people buy more clothes and larger houses with more conveniences when they have more money to spend. When money is scarce, they cut back. Increases in disposable income cause people to buy more cars, more cameras, more sporting equipment, and more of other items that are generally classed as luxuries. The demand for these products is elastic.

This unique characteristic of food—relatively inelastic demand—is a basic principle for all persons involved in developed country agriculture to understand. Those who *produce* food, plus those who *service* the producers and *process* the food items, are subject to the principle. Let it be considered an "occupational hazard" that contributes

TABLE 26–1	U.S. Annual Average Food Consumption of Major Food Items				
	1970–74	**1980–84**	**1990–94**	**1998**	**2002**
Red meats, boneless wt. (lb)	126.2	121.4	111.3	115.6	113.6
Poultry meat, boneless wt. (lb)	34.1	42.3	60.4	65.0	66.9
Fish, edible wt. (lb)	12.1	13.0	14.9	15.1	22.0
Eggs, shell and processed (no.)	299.0	264.0	235.0	243.6	250.0
Milk, including beverage (lb)	770.1	665.1	629.5	582.5	521.5
Cheese (lb)	12.9	19.5	25.7	28.5	26.7
Other dairy products (lb)[a]	40.9	45.0	53.6	50.1	48.1
Fruits, fresh and processed (lb)	229.0	260.0	269.1	272.8	261.8
Vegetables, farm weight (lb)	335.6	339.0	394.2	363.1	358.4
Cereals and flour (lb)	135.1	147.0	191.6	175.5	181.2

[a]*Includes yogurt, cream, and frozen products.*

Sources: *Adapted from USDA, ERS,* Food Review *18 (2):6, May–August 1995,* Statistical Bulletin 965, *July 1999,* Agricultural Outlook, *July 1999, and* Food Review, *Winter 2002.*

to surpluses and low prices when too much is produced, or a "blessing" to people in such occupations when consumer purchasing power is low. Even then, people still have to eat.

People in developed countries tend to eat about the same quantity of food regardless of economic conditions, but there might be significant shifts among the kinds of food consumed. Some of these shifts have been prompted or influenced by health concerns that relate to consumption of fats, meats, eggs, and other items.

Table 26–1 shows some of the changes in consumption of various types of food during recent decades. It is emphasized that data in the table illustrate only trends for each group. Note that data for red meats are based on carcass weight, whereas for fish they are based on edible weight.

Surveys indicate that Americans received 33 percent of their calories from fat in 1994. This amount is down from 40 percent in the late 1970s and has been declining since 1993. However, the recommended limit for calories from fat is still above 30 percent. For most, energy needs are less, as they work fewer hours per week, use more machines, and do less strenuous physical labor. Most walk less and ride more. Furthermore, there has been strong and effective publicity on the benefits of reduced body weight. Insurance companies, educational institutions, and others have shown data indicating that people who are overweight die younger.

Some people are more conscious of their appearance and want to be trim. They have become more calorie-conscious in their selection of retail cuts of meat. They also have reduced the amount of fat in the food by fat trimming of certain meat cuts, and many have reduced their intake to more moderate portions of foods where a high proportion of food calories come from fat. The benefits of ample protein and less energy foods for continued good health, especially in older people, have been demonstrated widely. Note the suggested target levels recommended by the National Research Council in Table 26–2.

In the United States, the rate of obesity has nearly doubled over the past 10 years and today, an estimated 64 percent of Americans are classified as overweight or obese. Moreover, there has been a significant rise in the number of overweight children and adolescents. In fact, 15 percent of children and teens aged 6–19 are overweight, up from 5 percent in 1980. Not only will an obese condition lead to health-related complications

TABLE 26–2	Target Levels for Calories and Nutrients Based on Recommendations from Major National Health Organizations	

Item	Recommendation
Calories	Intake matched to individual needs and appropriate to achieve and maintain desirable body weight
Calories from	
Fat	30 percent or less for adults
Saturated fatty acids	10 percent or less for adults
Polyunsaturated fatty acids	10 percent or less for adults
Monounsaturated fatty acids	15 percent or less for adults
Cholesterol	300 milligrams or less per day for adults
Calcium	The recommended dietary allowance (RDA) for a person's age and gender
Iron	The RDA for a person's age and gender

Source: *Adapted from Glaser, L. K.*, National Food Review, *12(1989); 1.*

(increased risk for Type II diabetes, heart disease, and osteoarthritis) but it is also considered one of the primary factors affecting a person's self-esteem. The major contributors to the overweight problem in American are lack of exercise and eating away from home. As shown in the earlier chapters, the percentage of dollars spent on meals away from home is on the rise and, in many cases, these dishes included high-calorie items.

Consumers have more spendable income and, even with inflation, more purchasing power than in the first half of the century. They have been able to afford the more expensive high-protein foods, such as meat. Improved processing, preservation, and transportation of these more perishable foods also may have contributed to their increased consumption.

Although it is accepted that the total amount eaten per person does not change much, there is real competition among food groups, or even specific foods. If consumption of one food increases because of consumer income, health, fads, changes in relative food prices, effectiveness of merchandising, or promotion and advertising, then consumption of some other foods probably will decline.

Meat promotion, designed to increase consumption, is discussed in Section 26.5. Producers and processors of other food items—citrus, potatoes, wheat, milk, and others—also spend much money on product promotion. Although the net effect of these promotional programs is not apparent, certain programs have been very effective. It is doubtful, however, that total food consumption (dry matter) will increase as a result of the combined campaigns.

26.3   GEOGRAPHIC AND OTHER INFLUENCES ON MEAT CONSUMPTION

Per capita consumption of broilers tends to be higher in the southern states. Consequently, red meat consumption traditionally has been lower than in other regions of the country. Red meat consumption tends to be highest in the north central area, where cattle, sheep, and swine are more heavily concentrated. These consumption differences probably are due more to tradition, income, and other factors than to location, per se.

There is an apparent outlet for every weight and quality and yield grade of beef; however, there are significant geographical differences. Demand for high-quality, heavy-weight beef by wholesalers and retailers tends to be higher where income is higher. For example, business people, visitors, and tourists often choose beef in popular restaurants. Surveys have shown that people eat more meat, and often higher quality meat, when they spend money on food away from home.

Market lambs, especially from midwestern or eastern farm flocks, are quite uniform in size, condition and weight, and most grade USDA Choice or USDA Prime at slaughter. Therefore, there can be little geographical preference for quality of lamb. There may be a tendency for lamb consumed in some of the western states to carry less finish because more lambs marketed there come directly from the range and are fed less grain. This difference has diminished as more grain is being grown and more feedlots developed in the western states.

Geographic preferences for meat are not necessarily rigid. However, they usually do have basic causes, and shifts occur only gradually. Where geographical preferences exist, they probably are caused by an interaction of factors, including the following:

- Grain availability for finishing cattle and sheep to higher grades or for raising hogs—grain is less available in the Southeast and Southwest, as well as in the Rocky Mountain region.

- Cost of transportation for moving various grades of beef or lamb to all areas of the country.

- Current income level and quality standards.

- Inertia—habit and tradition—of meat processors, wholesalers, and retailers.

- Area of the country, rural versus urban location, income level, and other aspects of environment under which current residents of certain areas were raised—and the quality standards that resulted. Households in urban areas tend to spend more on food (including meat) than do those in rural areas, likely due to their higher incomes and more convenient access to food away from home.[2]

Household size also affects how dollars are spent on food. For example, larger households tend to be more frugal and spend a larger share of their at-home food dollar on basic ingredients and lower cost items (for example, ground beef vs. loin cuts).

Section 26.3 discussed the elasticity of demand for food and explained that it is rather low—people do not eat much more food when they have more purchasing power. Rather, they tend to eat fewer low-quality food items and more high-quality foods, such as meat, milk, and eggs. Consumers also direct a larger part of increased purchasing power to higher quality meat, or to species that have more esteem or luxury value, than to increased quantity.

In economic terms you might say that certain meats—lamb, turkey, and steak—have a greater elasticity of demand than pork, ground beef, or luncheon meats. Greater purchasing power may cause increased consumption of the former, but will have little effect or perhaps a decreasing effect on the latter.

---

[2] Smallwood, D. M., et al. *Food Review* 18, no. 2 (1995): 16.

Pork is less of a status meat; it is more of a stable dietary item for most families. Only in the west, where less pork is produced and pork on the table is more of a treat, is pork consumption correlated with income.

26.4 HOLIDAYS AND RELIGIOUS BELIEFS

A few relationships between meat consumption and holidays or religious beliefs are well known. Turkey has long been the traditional meat for Thanksgiving and Christmas, though ham is a favorite in some families.

Catholics are bound not to eat red meat and poultry on Ash Wednesday or Good Friday. Doctrine suggests that they abstain from meat on the Fridays during Lent, but this is not binding. The impact on meat consumption is much less now than formerly, when Catholics were required to "fast" through all of Lent, limiting themselves to one meat-containing meal per day.

The Kosher codes (Hebrew dietary laws) have some influence on meat consumed by the Jews in the United States. Orthodox and Conservative doctrine requires the strict observance of Hebrew dietary laws. The laws prohibit the consumption of pork and pork products. Beef and lamb can be eaten only after they have been slaughtered, inspected, and prepared according to stringent regulations.

The required procedures for slaughtering beef intended for Kosher consumption are discussed in Section 25.1. According to biblical command, the eating of blood, the sciatic nerve, or visceral fat is forbidden. Major veins and arteries must be removed from Kosher beef and lamb. Removal of the sciatic nerve and adjoining blood vessels from the hindquarter is a difficult and time-consuming task; therefore, hindquarters normally are sold with non-Kosher slaughtered animals. Traditional Jews utilize only forequarters of high-quality beef and lamb.

Much of the beef and lamb intended for the Kosher trade is slaughtered near the place of consumption. Fresh Kosher meat cannot be kept longer than 72 hours without a ritual washing. Even after such washing, aged meat is often difficult to sell. Kosher meat dealers prefer to buy their beef and lamb immediately after slaughter because it is easier to remove the blood vessels before the meat is chilled. Purchase of high-quality cattle and lambs by packers at market, for the Kosher trade, is timed accordingly.

Special Jewish holidays also have significance to the meat producer and processor. Because the lunar calendar of 354 days is followed, rather than the Gregorian calendar, the holidays are difficult to plot. Kosher slaughter is forbidden on Saturdays, and there are 13 Jewish holidays during the lunar year on which Kosher slaughter is forbidden. In two cases, 4 such days occur within a 10-day period, and in two other cases, 2 successive days are designated. These holidays, then, might have a marked influence on demand for high-quality, well-finished cattle and lambs at markets where animals normally are purchased for the Kosher trade, especially if they follow or precede a 3-day weekend.

26.5 PROMOTION OF MEAT AND MEAT PRODUCTS

Meat is a competitor of other high-protein foods of animal origin, such as cheese, eggs, and fish. Meat also competes, to a degree, with other kinds of food. The consumer has a choice—cereal versus bacon and eggs for breakfast, bean soup versus hamburgers for lunch, or a mixed casserole versus pork chops for dinner. Many Americans eat meat at least twice a day, but there is still competition with nonmeat food items.

Most promotion and advertising is aimed at urging the consumer to buy more of a commodity or to buy a higher quality product costing more money. This is certainly true for meat. Much promotion is based on the nutritional value of and need for plentiful quantities of meat in the diet, especially for youth and for older people. Some promotion is based on enjoyment—flavor, aroma, juiciness, tenderness—that comes from crisp fried chicken, roast beef, or cured ham. Though some advance the idea that there is prestige in eating more meat, there is less of this type of promotion for meat than for most consumer items.

Processors and retailers naturally promote meat. It is the processors' main business. Because retailers consider meat a relatively high-profit item, they spend considerable amounts on meat advertising. Promotional "specials" on meat products increase tonnage sold.

Many organizations sponsor meat promotion on a national scale. The former National Live Stock and Meat Board,[3] supported by producers, markets, processors, and retailers, and with a staff of many people, had an extensive promotional program. It provided professional talent, information, and visual aids for radio, television, cooking schools, meat exhibits, and cutting demonstrations; teaching materials for use in schools and colleges; and informational materials to food editors. The operations of this group were financed by contributions from growers and feeders, deducted by marketing agencies at marketing time, and matched by processors. The Board financed projects conducted by research teams at many universities.

The American Meat Institute, a national trade association for meat processors, responds to technical and political issues affecting the meat industry and sponsors educational conventions, seminars, and industry equipment shows. It provides technical information and helps members improve processing techniques.

Other meat trade associations include the American Association of Meat Processors (mostly small processors), the American Meat Institute (mostly large processors), the American Meat Association (formerly the National Independent Meat Packers Association), and the National Association of Meat Purveyors. All respond to technical and political issues and provide educational events for their members.

26.6 NUTRITIVE VALUE

To compare fairly the nutrient content of foods, such comparison should be made on the basis of comparable dry matter. Approximate nutrient content of certain meat cuts and other selected foods, on an "air-dry" basis, is given in Table 26–3.

Protein content of meat varies from 25 to 80 percent (grams per 100 grams) on an air-dry basis, depending primarily on degree of fatness. Cereal and vegetable products will range from 2 to 10 percent. Certain producers of cereal products have used the descriptive phrase "more protein than meat itself" in advertising. Some dry cereals may have more protein than certain cuts of *fresh* meat, but the instances are rare, and when figures are adjusted for the extra water meat contains, the protein content of meat is much higher.

Human dietary needs for essential amino acids (components of protein) are similar to those of other nonruminants discussed in Chapter 5. Human diets composed primarily of vegetable or plant material usually are lacking in the same critical nutrients

---

[3] The role of the National Live Stock and Meat Board has been assumed by the National Cattleman's Beef Association and the National Pork Producers Council Board.

| TABLE 26-3 | Nutrient Content per 100 Grams of the Edible Portion of Certain Meats and Other Foods, on an Air-Dry Basis |

Food	Energy (kcal)	Protein (g)	Fat (g)	Saturated (g)	Oleic (g)	Linoleic (g)	Vitamin A (IU)	Niacin (mg)
Bacon, broiled	554	26.1	52.2	16.3	2.40	4.6	0[a]	5.2
Ground beef, 21% fat	561	47.8	40.7	16.8	16.1	1.0	72	10.5
Steak, lean only, braised	282	45.5	8.7	3.7	3.3	0.4	22	8.9
Ham, roasted	563	41.4	43.7	15.6	18.2	3.9	0[a]	7.1
Broiler, broiled	422	73.9	12.3	3.9	4.4	2.3	281	27.3
Liver, fried	468	52.8	21.6	6.0	8.4	2.2	109,000	33.6
Potatoes, french fried	441	6.5	22.9	5.6	3.9	10.8	0[a]	5.2
Doughnuts, glazed	498	7.3	26.7	8.0	14.1	8.0	60	1.9
Oatmeal	56	2.2	0.9	0.2	0.3	0.4	0[a]	0.1
Whole milk	465	24.8	24.8	15.8	6.5	0.6	961	0.6
Skim milk	348	32.8	trace	1.2	0.4	trace	2,050	0.8
Cottage cheese, large curd	446	53.0	19.0	12.2	4.6	0.4	703	0.6
Eggs, poached	554	41.5	41.5	11.8	13.8	4.2	1,800	0.0[a]

Note: Presentation on an air-dry (90 percent dry matter) basis permits comparison of foods on a uniform basis and recognizes that water in foods serves the same nutritional function as water that is drunk.

[a]*Content is zero or negligible.*

Source: *Adapted from USDA, SEA,* Nutritive value of foods, *Home and Garden Bull. 72, September 1978.*

that all-plant swine rations lack. Meat provides the essential dietary amino acids, as well as other amino acids, in satisfactory proportions.

Note that, on an air-dry basis, there is not a large range in caloric content of the foods listed in the table except for oatmeal, which is composed largely of the fibrous portion of the oat kernel. There is considerable range among foods in content of total fat and saturated fatty acids, which some research has related to incidence of heart and circulatory disease in humans. Note especially the low levels of saturated fatty acids in broiler meat, skim milk, liver, and the lean of a steak.

Even though health of the heart and circulatory system are influenced significantly by heredity, amount of exercise, and amount of emotional stress, relationships between diet and health of the system have received considerable research attention.

Because of the high content of important amino acids needed for tissue building, meat is especially important in the diets of children, who are growing rapidly, and older people, whose amino acid needs for tissue repair may have accelerated. High-protein diets, including generous portions of meat, often are recommended following surgery or after accidents that cause major wounds and considerable loss of blood.

Chapter 3 explained that three classes of nutrients—carbohydrates, fats, and protein—supply energy. Meat contains only insignificant amounts of carbohydrates, so protein and fat are the major potential energy sources in meat.

Energy value of meat varies according to the cut, degree of finish, and proportion of the fat actually eaten. Many people trim excess fat and eat only the lean, or reduce the ounces eaten per day.

Many of the amino acids are used to build tissue or to make enzymes and hormones. Also, protein has a high "specific dynamic action"—meaning that much energy is used up while the body is metabolizing the protein. So, in effect, meat is not a high-energy food. In fact, lean meat is the major food in many reducing diets. It has a low "net" energy value, and because people on diets usually eat less total food, meat helps supply some of the vitamins and minerals they would otherwise get from normal intake.

Most meats compare favorably with other foods as sources of critical vitamins (Table 26–3) and minerals. It is an important source of B vitamins, especially B_{12}, and iron. Liver and certain other variety meats are especially high in these nutrients. The liver is the site of active metabolism in the animal body, where many vitamins and trace minerals serve as catalysts in important reactions, and is also a storehouse for such nutrients.

26.7   TENDERNESS

Tenderness of meat is distinctly important to the consumer, having much to do with the pleasure derived from eating meat. Research studies on meat tenderness often appear contradictory and even confusing. There are several reasons why, which include the following: (1) tenderness perceived by chewing may be entirely different from mechanical measures of tenderness, and both are used in research; (2) measurement of perceived tenderness is subjective and not especially precise; (3) many factors influence tenderness; and (4) perhaps not all factors that influence tenderness have been identified.

It is known and widely recognized that meats vary greatly in tenderness. There is variation among species, among animals within a species, and from one cut or muscle to another within a carcass. Variation among animals raised in the same environment and slaughtered at the same age, weight, and degree of finish suggests a genetic cause for some tenderness variation. Table 18–1 gave a heritability value of 60 percent for tenderness in beef, suggesting that heredity may be a major influence.

Some research has indicated that tenderness is associated with the size of the muscle fibers—the smaller the fibers and the finer the texture, the more tender the meat. As animals mature and the size of each muscle fiber increases, there would be an expected decrease in tenderness. Other changes during growth and maturity, however, tend to counteract this effect. Meat from beef animals fed a certain minimum time on a reasonably high-concentrate ration will have acceptably tender meat, and further feeding will have little effect on tenderness. Most research with poultry meat indicates that meat from older birds is less tender. Tenderness of broilers is rarely a problem because they are young when slaughtered.

Deposition of fat among the muscle fibers (marbling) as the animals grow and mature on a high-energy ration had been thought to improve tenderness somewhat. Correlations between degree of marbling and tenderness as measured by mechanical devices are lower than correlations between degree of marbling and tenderness as measured by a taste panel. Apparently the increased juiciness caused by the marbling makes taste panel members think the meat is more tender, whether or not it really is. In most studies degree of marbling has accounted for 3 to 11 percent of the variation in tenderness.

Collagen and other connective tissue materials that hold the muscle fibers together apparently have some influence on meat tenderness. Collagen and elastin are connective tissue fibers embedded in a third connective tissue substance, which varies

TABLE 26-4	**Average Shear Values for Turkey Thigh Muscle**			
	Carcass Chilling Temperature			
	61°F	*61°–46°–32°F*	*32°F*	*Average*
Toms[a]	8.83	8.84	9.58	9.08
Hens[a]	9.54	9.91	12.15	10.53
Average	9.19	9.37	10.86	9.81

[a]*Note the difference in tenderness between toms and hens.*

Source: *Welbourn, J. L, et al.,* Food Sci. *(1998); 450.*

from a fluid to gelatin consistency and is called "ground substance." Knowledge of the importance of these connective tissue components in influencing tenderness is limited due to the difficulty in quantitatively measuring the amount of each in meat. Cross-bonding within the collagen protein molecule, which increases with animal maturation, apparently reduces tenderness.[4]

Certain feeding programs are known to increase the proportion of connective tissue in meat.[5] Feeding may, therefore, have relatively direct influence on tenderness, in addition to the fattening effect.

Chilling turkey carcasses more slowly improved tenderness in research at Purdue University (Table 26–4). Some birds were cooked immediately, others after 3 hours of cooling, then shear tests were made on slices of thigh muscle. For effective preservation and maximum tenderness, chilling 45 minutes at 61°F, 45 minutes at 46°F, and 90 minutes at 32°F appeared best.

Cows and bulls may be less tender on average than heifers or steers simply because they are older when slaughtered. Frequently they may be thinly finished so that cold toughening occurs, reducing tenderness of steak and roast items. Such carcasses are suitable for ground and restructured meat products. They may also carry less marbling so seem less tender. USDA carcass-quality grades have limitations as indicators of tenderness because they are based on maturity and marbling. There is variation within all grades, and even lower grades include some carcasses that are tender.

Muscles that have been used more during the animal's life are called "exercise" muscles and usually are less tender. Muscles along the back, including the tenderloin (*psoas major*) and the eye muscle (*longissimus dorsi*) of beef, pork, and lamb are called "support" muscles and are relatively more tender. The eye muscle is the main muscle of the rib and loin and extends slightly into the chuck (or corresponding cut of lamb or pork). The tenderloin is the small muscle below the vertebrae of the loin.

There is much current interest in finding a live animal trait that could be used to predict tenderness of meat. Such a trait could be used not only for appraising slaughter animals but also in selection of feeders and breeding stock. Some scientists have suggested that a program of certifying length of time on feed would be more useful than carcass grading in ensuring tenderness. Others believe a minimum fat thickness is necessary to ensure tenderness.

---

[4] Dutson, Thayne R. *Proc. of Reciprocal Meat Conf.,* 1976, p. 336.

[5] McIntosh, E., Nelson, D. C. Acker, and E. A. Kline. *J. Agr. and Food Chem.* 9 (1961): 418.

Tenderness of meat can be increased by aging, mechanical blade action, high-temperature conditioning, freezing, electrical stimulation of the carcass immediately after slaughter, or treatment with enzymes. Several of these methods are discussed in Section 25.7. Cooking also influences tenderness, but the effect varies so much among cuts and among methods of cooking that space does not permit a thorough discussion here.

Freezing meat increases tenderness, apparently because fibers are ruptured by ice formation and connective tissue components are stretched and ruptured. Lowering temperature to about −10°F apparently causes a consistent increase in tenderness, but temperatures lower than this do not cause further tenderizing. Freezing at temperatures above −10°F, however, causes such large ice crystals that tissue is severely disrupted and there is much moisture loss during thawing.

26.8   FLAVOR AND AROMA

Flavor is more intense in meat from older animals and in muscles that are used most, such as those in the shank, shoulder, and flank. Marbling often is credited with contributing to flavor, though the semiliquid fatty deposits in hot meat may be more effective in carrying the flavor to the taste buds than actually contributing to flavor per se.

Many assume that meat flavor is due largely to the amino acids, especially glutamic acid. Purine bases, nucleotides, and nucleosides also contribute to meat flavor.

Some of the chemical factors contributing to beef flavor are nonvolatile, so apparently are different, in part, from those factors that contribute to aroma. Flavor factors apparently include certain sugars, fats, and small-molecular-weight proteins. The colored portion of beef extract (broth) contains many of these flavor components, as well as some of the aroma factors.

Research has identified at least 15 flavor-contributing components of chicken breast meat, these components also being present in the digestive tract and apparently due, at least in part, to specific microbial activity. Some studies have shown a stronger and more characteristic "chicken" flavor in meat from birds raised conventionally than in meat from birds raised under germ-free conditions and lacking a normal bacterial population in the digestive tract.

Much of the flavor of meat, especially the species-specific flavor, is carried by the fat. People generally are less able to differentiate among species of meat when fat is excluded from the meat serving.

Loss of flavor in fresh meat displayed unwrapped in open cases may be evidence that some flavor components are volatile. It may be, too, that the specific compounds are oxidized or otherwise chemically changed during display, causing the change in flavor.

Whereas it is difficult to distinguish aroma from flavor when eating meat, it is apparent that the chemical factors contributing to aroma are volatile. As in flavor, there is great range in intensity and type of aroma.

Because pork fat is chemically less stable than poultry, beef, or lamb fat, an oxidized, rancid aroma is observed more often. Oxidative rancidity of fat is promoted by air and salt. Cured ham, bacon, or sausage, poorly wrapped, may develop oxidative rancidity, especially in exposed surfaces of the cuts after several months' storage. It is recommended that pork be kept in a freezer not longer than 6 months, less time if cured.

A unique and undesirable odor sometimes observed in pork is often referred to as "sex odor" or "boar odor." Though often associated with carcasses of mature boars, it is not restricted to them. It has been reported in pork from gilts, barrows, and sows, as well as boars, and in every breed studied. It is not known if this odor is the same in every case reported, and the nature of the factors that contribute to the odor are not thoroughly understood, even though they are components of the fat.

26.9   MARBLING

Marbling, the dispersion of small fragments of fat among the muscle fibers and bundles, usually increases with length of time animals are fed high-energy diets. Marbling also is influenced by heredity.

Research has validated that marbling contributes to the pleasure a consumer derives from eating meat because it increases juiciness and contributes to both real and some perceived improvement in flavor and tenderness. Purchasers tend to select meat with some attention paid to marbling.

Meat technologists commonly appraise marbling in meat by visual scoring. This is subjective, of course, but some scientists have a high degree of repeatability—meaning they are very likely to score a specific cut the same on two different occasions—so the scoring technique has some value. Determining the defatted muscle (external fat removed) by use of ether extract measures total fat but not the fineness of dispersion, which is considered important. Other methods are (1) counting flakes of marbling in randomly marked squares on the cut surface, (2) measuring specific gravity of the defatted muscle, (3) measuring, with a sensitive light meter, the light reflected from the cut surface of meat, (4) electronic photo imaging, and (5) measuring by ultrasound.

26.10   COLOR

The most preferred color of beef is often described as a light red or a bright cherry red. Veal is expected to be grayish pink in color; lamb, dark pink to purplish red; fresh pork is normally grayish pink to pink. For poultry, a slight pigmentation to the skin is usually preferred over a "bleached" appearance.

Color is caused by muscle pigments and affected by both the concentration and their chemical state, which can be changed by certain environmental conditions—feeding, preslaughter treatment, storage, light, packaging, air, curing, and other factors.

There is considerable color variation within species and also among muscles within a carcass, especially in the case of pork. "Two-toned" hams, with darker lean near the pelvic bone and lighter lean near the outside, are rather common. Chickens and turkeys have dark, pinkish gray legs, thighs, and back muscles and light, translucent-appearing muscles on the breast and wings. Goose and duck carcasses have all dark-colored flesh.

Myoglobin is the major meat pigment. Its concentration in muscles, and the degree of conversion to its chemical derivatives, are the primary factors in meat color.

Reasons for variations in fresh meat color among animals within a species are not fully known but are related to muscle function. Concentration of myoglobin may be inherited to a degree. Concentration and the degree of conversion to related compounds may be related to feeding programs, treatment before slaughter, muscle pH drop after slaughter, and other factors. Cattle fed a high-energy ration right up to slaughter time tend to have a brighter colored beef. A combination of rapid chilling and lack of feed prior to slaughter increases the incidence of dark color in both beef and pork.

26.11   PRESERVATION

Most commonly used meat preservation techniques prevent spoilage by altering or eliminating one or more of the environmental factors spoilage organisms need—water, air, warm temperature, or other factors. Canning kills organisms and seals out oxygen, freezing inhibits or kills the organisms, drying eliminates most of the water, and curing (Section 25.8) increases the salt concentration so most organisms cannot function.

Certain preservation methods may cause undesirable flavor or color, or change other quality characteristics of meat. Such changes may occur at the time of preservation or may develop during later storage.

Section 26.10 mentioned that color changes occur when meat is cured. Sodium nitrite in the curing formula apparently fixes the cured color; salt and sugar contribute to the typical cured-meat flavor. These changes are considered desirable by people who like ham and bacon. But, as mentioned in Section 26.8, this high salt concentration may contribute to development of a rancid aroma and flavor after prolonged storage. Vacuum packaging will reduce the deterioration.

Freezing is probably the most satisfactory method of meat preservation at the present time. Meat is usually "sharp-frozen" at -10° to -30°F so ice crystals formed in the meat will be small. It is then stored at about 0°F. Freezing and storing at these temperatures kills or inactivates nearly all spoilage organisms and inactivates the enzymes meat contains, yet does not impair flavor, color, aroma, or other quality characteristics of meat during recommended storage times. If well wrapped, poultry, beef, or lamb can be stored in a freezer 6 months or longer. Six months is a practical limit for fresh pork, and 2 months for cured meat.

Though most turkeys, some broilers, and some processed red meat items are sold at retail in frozen state, the retailing of frozen meat has not become generally popular. Extensive research has demonstrated that frozen steaks and other cuts can retain customer appeal, that labor and space needed at the retail outlet is reduced, and that product uniformity is enhanced when meat is cut, packaged, and frozen at the processing plant.[6] But tradition and skepticism among both consumers and retailers, plus union contract constraints, have inhibited a significant shift to the handling of frozen meat cuts. Future increases in frozen meat sales likely will be largely in small convenience stores with limited shelf space and in those special cuts that move in smaller volume.

Dehydration of meat for preservation is effective and satisfactory, if freeze-dehydrated. If dried by vacuum or normal air-drying, undesirable flavor and texture changes occur. Freeze-dehydration is accomplished by drying the meat in the frozen state, usually under vacuum, permitting conversion of ice crystals directly to water vapor. Moisture can be reduced to 1 or 2 percent, and after dehydration such meat can be stored or shipped in packages at room temperature without spoilage. Rehydration can be accomplished in 15 minutes or less, before cooking. Such products are popular for backpackers, mountain climbers, and military personnel, and are also used in instant soup mixes.

Short-term preservation techniques, to provide sufficient "shelf life" to avoid product loss and provide a good product for the consumer, can be aided by several techniques. Kansas State University research showed that a 10 percent potassium sorbate

---

[6] Tuma, Harold J., et al. *Frozen Meat.* Kansas Agr. Exp. Sta. Research Pub. 166, 1975.

dip almost doubled the shelf life of ice-pack chicken, in terms of controlling odors and bacterial counts.

Radiation of food usually involves electron beams or gamma rays, emitted from cathode tubes or nuclei of radioactive atoms. These rays penetrate the food, killing certain or all organisms, depending on the dose of radiation given. High, sterilizing doses of radiation damage the flavor, aroma, and/or color of most foods, especially meat. The flavor change, apparently involving fat and sulfur-containing amino acids, is considered undesirable by most people and is more marked in beef than in pork. However, if meat is subjected to cryogenic freezing at -30°F or colder, the flavor problems are largely overcome. Radiation of fresh meat with high doses causes development of a brown color; a persistent pink to red color often develops when radiated meat is cooked.

Low doses of radiation have been demonstrated to be effective in lengthening storage time of meat, from 10 days to 60 days at 34°F in the case of chicken, while not causing a marked change in color or flavor. Low doses kill certain of the more common kinds of organisms. It is apparent that once meat is irradiated to kill organisms, it must be packaged to prevent entry of additional organisms.

Another limiting factor in the use of radiation for preservation of fresh meat is that radiation does not inactivate the enzymes in meat. Hence, if radiation were used to sterilize fresh meat for extremely long storage, the enzymes would continue to function and degrade the meat structure. Therefore, the main application of radiation may be for prolonged preservation of *cooked* meat, in which enzymatic activity has been destroyed. At the same time, prolonging storage time slightly by low radiation may allow some aging and tenderization of the meat cuts.

Subjecting cultures of some spoilage-type microorganisms to microwave treatment for 15 to 20 seconds has drastically reduced their numbers.[7] If similar results are shown for pathogens, this may extend refrigerator storage life for meat products in the home.

QUESTIONS FOR STUDY AND DISCUSSION

1. During the past 25 years, how has the per capita consumption of poultry changed in the United States? Describe the consumption pattern for beef, pork, and lamb during that same period.

2. List three factors that influence consumption of total meat or meat of individual species.

3. Give an example of how Kosher preparation differs in the processing of meat animals.

4. List three important nutritional characteristics of meat and tell why each is an important consideration in human diets and welfare.

5. Compare ham and oatmeal in nutritional characteristics.

6. What factors are known to influence meat tenderness? How can the tenderness of meat be increased?

7. What is the chemical source of meat flavor? of meat color?

8. List three methods of meat preservation and tell how and why each works.

---

[7] Cunningham, F. C. *J. Food Protection* 43 (1980): 651.

27
Wool and Mohair

Sheep are among the most versatile of all domesticated animals (see Color Plate B). As discussed in earlier chapters, they are very efficient ruminants and a major provider of milk to people throughout the world. Today's meat-type lambs provide tender, delicious meat, and many people savor well-prepared leg of lamb and lamb chops. This chapter focuses on another important role of sheep—as providers of a natural animal fiber, wool. It also discusses other fiber-producing animals.

Wool is the most valuable byproduct of meat animal production. It is by far the most used natural animal fiber for fabrics (other animal fibers include mohair, cashmere, and vicuna). To the consumer, it is a reliable and increasingly versatile fiber that, when woven into cloth, contributes to warmth and an attractive appearance.

Wool has been used for clothing and other fabrics for over 12,000 years. Today, essentially every wardrobe contains woolen garments, plus garments that contain wool along with some other natural and/or synthetic fiber.

The U.S. wool industry today is small compared to that of the leading countries of Australia and New Zealand. For example, Australia traditionally maintains a sheep population of more than 120 million sheep, with production of over a billion pounds of scoured wool[1] annually. The small country of New Zealand ranks second, with an annual production that approaches almost 50 million sheep and 470 million pounds of scoured wool.

In contrast, the number of sheep shorn in the United States in 1998 totaled about 6.4 million head and raw wool production amounted to about 49 million pounds, or on a scoured (clean) wool basis, about 28 million pounds. Today's U.S. wool production is about one-third of the total produced in 1970 (Figure 27–1). However, the American farmer and consumer prefer a meat-type lamb, and breeding animals have been selected primarily for meat rather than wool production.

Wool comprises a small but important proportion of all the fibers used by U.S. citizens; the amount of processed wool per person amounts to slightly more than one-half pound annually (Figure 27–2).

Currently, about 90 percent of the wool used by U.S. mills is used as apparel wool in the production of worsted and woolen fabrics. American sheep producers provide about 20 percent of the total, the balance being imported primarily from Australia and New Zealand.

---

[1] *Grease wool* refers to wool as it is shorn from the sheep, whereas the term *scoured* or *clean wool* refers to wool that has undergone a cleaning process that removes the grease and foreign material. Ten pounds of grease wool will yield about 5.8 pounds of scoured wool.

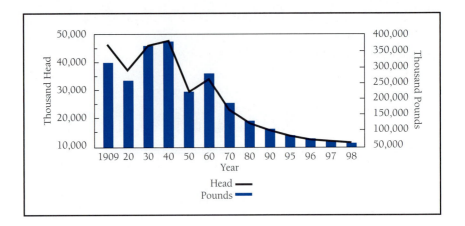

Figure 27–1
Total number of sheep and lambs shorn and wool produced from 1909 through 1998. Note the dramatic decrease in both sheep and wool production since 1940.
(*Source:* USDA.)

Figure 27–2
Per capita wool consumption in the United States, in proportion to other fibers, in recent years.
(*Source:* USDA.)

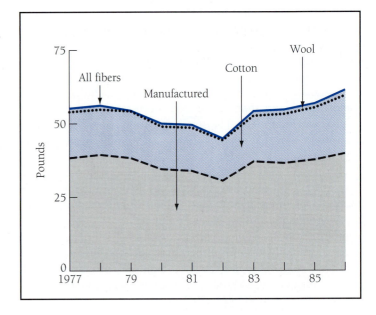

Upon completion of this chapter, the reader should be able to

1. List and describe the major characteristics of wool.
2. Differentiate among wool, mohair, and other fibers in major quality attributes or characteristics.
3. Describe the volume of wool used in the U.S. fiber industry, in comparison to cotton or chemically synthesized fibers.

4. Explain the wool grading systems, differentiating between the spinning count and blood systems.

5. List and explain the basic steps in wool processing.

6. Compare U.S. wool production volume with volume of other major wool-producing countries.

7. Describe how wool is removed from the animal and marketed.

8. List animals in addition to sheep and goats that yield commercially used fibers.

9. Define *grease wool, scoured wool, carpet wool, worsted yarn, woolen yarn, apparel wool,* and other key terms.

27.1   CHARACTERISTICS OF WOOL

A single wool fiber may be from 18 to 41 microns (or 1/1000 to 1/2000 of an inch) thick and 1½ to 5 or more inches long. Wool is composed primarily of amino acids, especially those containing sulfur (methionine and cystine). The manner in which the amino acids are linked chemically into peptide chains probably explains some of wool's unique characteristics.

Wool is *elastic*. It can be stretched 30 percent or crumpled tightly, and will recover its natural shape rapidly. This property becomes a built-in characteristic of fabric that has a high percentage of wool. It may be wrinkled, twisted, and stretched, but will regain its shape if allowed to hang overnight.

The billions of amino acid molecules in a wool fiber apparently are linked together in coiled chains that lie adjacent and that are chemically crosslinked. As a wool fiber is stretched these coils are distorted, and when tension is released they return to their natural position (Figure 27–3).

Wool has **crimp**. This natural wavy appearance adds to its effective elasticity, but also provides other advantages. Crimp prevents the individual fibers from lying close to each other in cloth. This produces a bulky effect with tremendous insulation value. Depending on texture and fineness of the fiber, from 60 to 80 percent of the volume of woolen fabric may be air.

> *Crimp*
> Natural waves in wool fiber that provide characteristics of elasticity and resilience.

Each wool fiber has an outer layer of flat, scalelike cells that overlap like shingles and that are covered with a thin membrane. Rain is repelled by the membrane, but water vapor can penetrate it. The protein cells in the center of the fiber absorb the moisture that may penetrate the membrane. This property allows water-soluble dye to react with the proteins so the color becomes an integral part of the fiber.

Crimp, the scales of the fiber, and the reactivity of certain of the spindle-shaped cells under the scales apparently allow the short wool fibers to adhere to each other as they are spun into a continuous strand of yarn.

Wool is *strong*. It is often said that a single wool fiber is stronger than steel of the same diameter.

27.2   WOOL QUALITY AND GRADES

The attributes of wool quality include fineness, length, crimp, color, strength, uniformity, and, in grease wool, percentage and kind of foreign material. *Fineness* is considered the most important.

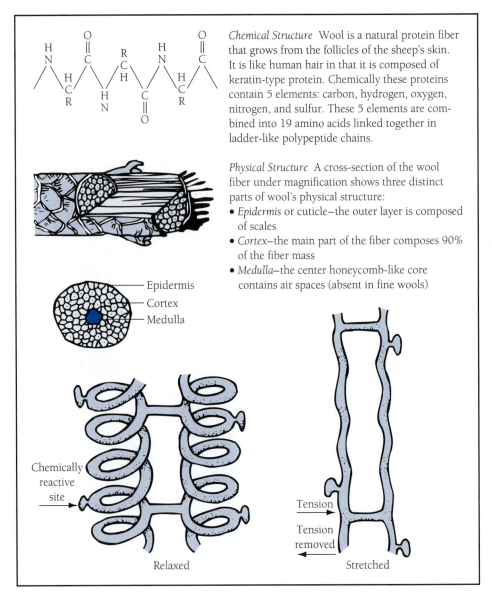

Chemical Structure Wool is a natural protein fiber that grows from the follicles of the sheep's skin. It is like human hair in that it is composed of keratin-type protein. Chemically these proteins contain 5 elements: carbon, hydrogen, oxygen, nitrogen, and sulfur. These 5 elements are combined into 19 amino acids linked together in ladder-like polypeptide chains.

Physical Structure A cross-section of the wool fiber under magnification shows three distinct parts of wool's physical structure:

- *Epidermis* or cuticle–the outer layer is composed of scales
- *Cortex*–the main part of the fiber composes 90% of the fiber mass
- *Medulla*–the center honeycomb-like core contains air spaces (absent in fine wools)

Figure 27–3
Characteristics of the wool fiber.
(*Source:* American Wool Council.)

The "spinning count" and "blood" systems used for grading wool according to fineness are shown in Table 27–1. The spinning count system is the basis of present USDA grades, even though the blood system is well known and used. The grade number in the spinning count system refers to the hanks (560 yards per hank) of yarn that theoretically can be woven from 1 pound of scoured wool. The USDA grades shown are for wool top (partially processed wool that exists as a continuous, untwisted strand of scoured wool fibers from which the shorter fibers have been removed by combing). The

TABLE 27–1	**Wool Grades and Specifications**		
USDA Grade (January 1969)[a]	**Avg. Fiber Diameter (microns)**	**Blood System Grade**	**Breeds According to Approximate Wool Grades[b]**
Finer than 80	18.1 or less		Merino
80	18.1–19.59		
70	19.6–21.09	Fine	
64	21.1–22.59		Rambouillet
62	22.6–24.09		Targhee
60	24.1–25.59	½ blood	Southdown
58	25.6–27.09		Montadale
56	27.1–28.59	⅜ blood	Shropshire
			Corriedale
			Hampshire
			Columbia
			Panama
54	28.6–30.09		Dorset
50	30.1–31.79	¼ blood	Suffolk
48	31.8–33.49		Cheviot
			Oxford
46	33.5–35.19	Low ¼ blood	Romney
			Leicester
44	35.2–37.09	Common	Cotswold
40	37.1–38.99		
36	39.0–41.29	Braid	Lincoln
Coarser than 36	41.3 or more		

[a]The date grade standards were last issued.

[b]This is not an attempt to rank breeds according to wool fineness. It is well known that there is much range within breeds and an average cannot be precisely determined.

Source: Wool grades are from USDA, C&MS, Marketing Bull. 53, 1971. Breed listing adapted from Montana Agr. Exp. Sta. Circ. 218, 1958, and Pohle, Elroy M., "New Wool Standards," National Wool Grower, November 1965.

USDA does not provide a grading service, but the standards are used as guidelines in virtually all buying and selling.

Originally, the blood system of grading wool was an indicator of the fraction of Merino breeding in the animal from which the wool came. The Merino breed produces fine, high-quality wool, and fineness is highly inherited. Since the development of other distinct breeds, the original blood system simply has been adapted to describe relative fineness.

Fineness is important because it allows the spinning of a finer yarn, tighter weaving of cloth, and production of lighter fabrics and garments. Fine wool often has more *crimp*, too, another attribute of wool quality. Crimp helps individual fibers cling

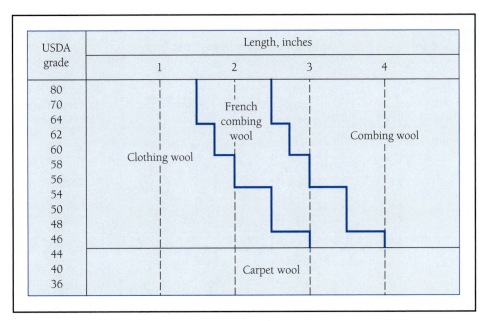

Figure 27–4
Length designations often used for wool, according to fineness.
(*Source:* USDA.)

together during spinning, so that a strong yarn can be woven with fewer fibers lying parallel. Crimp, then, also contributes to lighter weight garments and more efficient wool use. Fine wool often has 15 or more crimp per inch; coarser wool has less. Wool from the back usually has more crimp than wool from the thigh.

Length of fiber is a quality factor. Though fine fleeces usually have relatively short fibers, the longer the better. The reason for this is apparent, too, in spinning a strand of strong yarn.

Wool fiber length is measured easily in inches; however, a variety of special terms commonly are used in the wool industry to describe relative length, according to the way it can be processed. These are summarized in Figure 27–4.

"Combing" wool is long enough that regular combing machines can sort and straighten the fibers to make "worsted" yarn, which is smooth and used for light, high-quality cloth. This wool is also called "staple length." "French combing" wool is handled essentially the same as regular combing wool, except that special combing machines have to be used because the fibers are a bit short. "Clothing" wool is not long enough to be handled even by special combing machines. This wool, therefore, can only be carded and is destined for use in tweeds and other fuzzy fibers. Extremely short wool, from young lambs or extremely old ewes, is often called "scouring" wool. Because it cannot be effectively combed, it may be used for felt or similar materials.

It is apparent from Figure 27–4 that fine wool can be shorter and still be processed in regular combing machines.

Amount of foreign material and grease in the wool is considered in appraisal of grease wool. It is not uncommon for a 10-pound, fine-wool fleece to yield but 5.5 pounds of scoured wool, meaning it contained 4.5 pounds of grease and foreign material. In other words, the shrinkage on scouring was 45 percent.

Fine-wool fleeces normally carry more foreign material than medium or coarse fleeces. Many fine-wool sheep are raised in arid sections of the country where considerable dust blows and fine wool contains more grease, so shrinkage may be 60 percent or more. Medium wool from the Midwest normally shrinks 40 to 53 percent. *Skirting,* the practice of separating out inferior portions from the fleece at shearing, such as portions of the head, lower leg, and belly, and urine- and fecal-stained fibers, is performed by some growers to increase the value of their raw wool.

Traditionally, wool quality has been appraised for sale purposes by experienced buyers and sellers, who rather accurately estimate fineness, length, shrinkage, and other quality characteristics. In recent years the practice of "coring" bags of wool and subjecting core samples to objective measurements for appraisal purposes has increased. The standard instrument for measuring fiber diameter and distribution is a projection microscope with a wedge card scale for measurement of individual fibers. New techniques involving microcomputer-based digitization, or electrooptical laser instrumentation, although expensive, now can provide rapid, accurate measurements.[2]

27.3   WOOL PROCESSING

At the mill, each fleece is pulled apart and sections of the fleece are sorted according to fineness, length, and other characteristics. Fleeces are purchased by processors according to these factors, of course, but there is usually some variation within a fleece.

All wool is washed in several tanks of hot soapy water, rinsed, and dried. This is called *scouring* and removes grease and most other foreign matter. Lanolin is recovered from the washings, purified, and sold as a base for face creams and similar items.

Next the dried wool passes to carding machines where revolving cylinders covered with fine wire teeth remove burrs, straw, and similar matter and straighten and comb the wool into a thin veil or web.

From this point on, processing depends on diameter and length of wool fibers and whether "woolen" or "worsted" yarn is to be made. *Woolen* yarn is generally made of shorter and thicker fibers (clothing wool) that may lie in all directions, to produce thicker, fuzzier fabrics, such as tweeds. **Worsted fabrics**, such as gabardine, are made from yarns of longer, finer fibers (combing wool), so the fabrics will be lighter and have a harder, smoother finish. About half of the apparel wool in the United States is used in worsted fabrics. For more details, see Box 27–1.

> *Worsted fabric*
> High-quality woolen fabric comprised of long fibers that are small in diameter.

For the spinning of woolen yarn, the web from the carding machine is simply split into thin, soft strands called *roving*. For worsted yarn, however, the fine web must go through further sorting, combing, and straightening of fibers until a very thin, smooth strand of worsted roving can be obtained.

Spinning of roving into yarn is accomplished by mechanically twisting the roving. The machines then wind the spun yarn onto spools. From here on, yarn may be utilized in a variety of ways. Most, of course, is knitted or woven into fabric. Wool may be dyed in any stage of manufacturing—immediately after scouring, in the yarn stage, or as finished fabric.

---

[2] Lupton, D. J., et al. "Comparison of three methods of measuring wool fiber diameter." *Sheep and Goat, Wool and Mohair,* September 1988. Texas Agr. Exp. Sta., Texas A&M University System.

BOX 27–1

Woolen or Worsted—The Difference

Woolen and worsted are two major classifications for wool yarns and fabrics. Here are the differences:

Woolens	Worsteds
Processing	
Spun from *shorter* wool fibers.	Spun from *longer* wool fibers.
One to three inches in length.	Longer than three inches in length.
Spun from fibers of a *medium* or *coarse* diameter.	Spun from fibers of *fine* diameter.
Fibers are washed, scoured, and carded.	Fibers are washed, scoured, carded, combed, and drawn.
Yarn	
Bulky, uneven.	Fine, smooth, even.
Low to medium slack twist.	Tighter twist.
Tensile strength lower than worsted.	Higher tensile strength.
Fabric Appearance	
Soft.	Crisp.
Fuzzy.	Smooth, clear-faced.
Thick, heavier weight.	Lighter weight.
Characteristics	
More insulatory due to trapped air.	Less insulatory.
Not as durable as worsteds.	More durable than woolens.
Nap reduces shine.	May become shiny with use where abraded during wear.
Does not hold a crease well.	Holds creases and shape.
Uses	
Sweaters, carpets, tweeds.	Suits, dresses, gabardines, crepes.

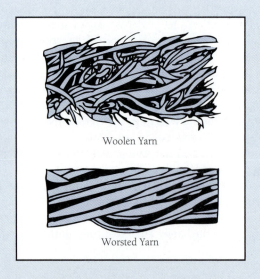

Woolen Yarn

Worsted Yarn

Source: *Fact Sheet #2 Processing Wool,* American Wool Council.

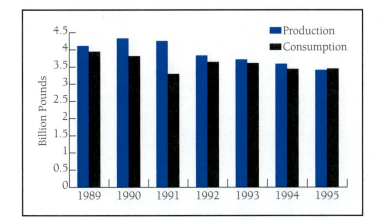

Figure 27–5
World production and consumption of scoured (clean) raw wool. (*Source:* USDA.)

27.4   GLOBAL AND U.S. PRODUCTION OF WOOL

Because one-fourth to one-third of the wool processed in the United States during the last several years has been imported, a consideration of global wool production is important (Figure 27–5). Currently, world production of clean wool is about 3.14 billion pounds (from about 5.6 billion pounds of grease wool), which amounts to slightly more than three-fourths of a pound per person. The leading countries in wool production traditionally are Australia, New Zealand, China, former Soviet Union, and Uruguay.

The five leading Southern Hemisphere countries (Australia, New Zealand, Argentina, South Africa, and Uruguay) represent about two-thirds of all world wool exports each year.

About 80 percent of the world production is apparel wool, used for clothing and similar fabric, and designated in Figure 27–4 as combing, French combing, or clothing wool. The remainder is carpet wool, used for rugs, carpet padding, and similar materials.

Because labor and other production costs are so much lower in most Southern Hemisphere countries, wool can be produced cheaply enough that exporters in those countries can pay the United States import duties and still compete in price with wool produced in the United States.

As mentioned earlier, wool production within the United States has steadily declined. Consequently, the needs of wool mills greatly exceed domestic supplies. For example, imports accounted for about 75 percent of all wool used by American mills in 1998. This trend will likely continue into the foreseeable future.

However, the value of U.S. wool produced annually amounts to over $29 million and wool sales are particularly significant in the leading states (Figure 27–6). The five leading wool-producing states (Texas, Wyoming, California, Montana, and Colorado) accounted for about 51 percent of all U.S. wool produced in 1998.

27.5   WOOL MARKETING

Most wool is shorn in the spring, and because shorn wool represents over 96 percent of total U.S. production, the supply is certainly seasonal. The remaining amount is pulled wool, pulled from pelts of slaughtered sheep and lambs.

Figure 27–6
The 10 leading states and their percentage of total U.S. wool production in 1998. All except Iowa are located in the western portion of the United States. (*Source:* USDA.)

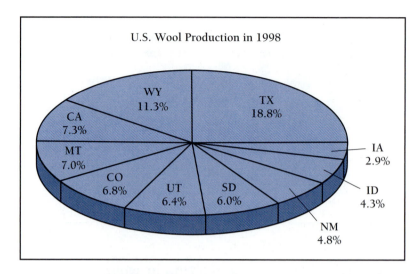

U.S. Wool Production in 1998

WY 11.3%

TX 18.8%

CA 7.3%

MT 7.0%

IA 2.9%

CO 6.8%

UT 6.4%

SD 6.0%

ID 4.3%

NM 4.8%

Figure 27–7
An expert sheep shearer is fast, makes no "second cuts," so wool fibers are long, and leaves the sheep free of remnants and skin cuts. Expert shearers are in strong demand during the spring. Shearers are paid on a per-head basis. In general, the price charged ranges from $2.00 to $2.50 per head.
(*Source:* Courtesy of Stewart Division, Sunbeam Corporation.)

Though production is seasonal, wool is not a perishable product if well handled, so processing mills can operate at a fairly uniform rate during the year. Also, processors can even out the supply by importing relatively more in the fall months, when wool is being shorn in the Southern Hemisphere.

The average fleece of grease wool in the United States weighs about 7.8 to 8.0 pounds. In certain western states, such as Montana and Wyoming, where wool production is a major enterprise and selection for heavy fleeces of good-quality wool has been more intensive and more effective, fleece weights often exceed 10 pounds.

Expert shearers, using power-operated clippers, can shear a ewe or lamb in a few minutes (Figure 27–7). Important steps in ensuring high-quality wool are to have sheep dry when shorn and to keep the wool dry in storage—as wet fleeces can become stained from mildew. The fleece is usually tied with paper twine, with the inside of the fleece out for better appearance, and packed in large bags.

Much of the wool produced in the United States is sold through producer-owned cooperatives, often called "wool pools." Such cooperatives have operated effectively for many years in plains and western states where wool production has long been a major enterprise. Some handle much of the production of several states.

The advantages of a wool marketing cooperative are apparent, especially in areas where wool production may be a sideline enterprise on many farms. Few producers are able to keep abreast of market prices, especially those who sell wool but once a year. Also, U.S. farm prices for shorn wool vary widely from year to year. For example, since 1977, farm prices for shorn wool have ranged from a high of 138 cents per pound in 1988, to a low of 51 cents per pound in 1993, and to 60 cents in 1988.

Few wool producers are excellent judges of grade or shrinkage so are not capable of dealing effectively with a buyer. Also, because few producers have storage facilities and know-how, if they were dependent on direct selling they would need to sell at shearing time when supply is high and prices are low.

A marketing cooperative allows employment of a wool grader who can appraise wool and keep abreast of supply and price trends. It allows provision of central storage facilities and sorting of wool into quality groups before it is offered for sale. It also creates a concentrated supply, attracting buyers and creating producer bargaining power.

Producers usually consign their wool to the cooperative, which makes partial payment and records the amount of each grade delivered. The wool is then stored; each grade is marketed at the time prices appear best. Producers are then paid the balance due them, minus the costs of operation. Some cooperatives also buy wool on a cash basis from producers, hoping, of course, to make a profit. Some wool is still marketed through a regular system of private dealers and brokers, who sell to larger brokers or directly to processors.

Boston is the center of wool marketing activity in the United States. Domestic and imported wool is sold at auction or by private treaty. Boston prices are often the basis of trading in most sections of the country. The Boston market probably developed because of the early settling of New England, development of the wool-processing industry there, and shipment of wool into the Boston harbor. In recent years, however, lower labor costs in the southern states and development of many southern ports have caused considerable shifting of mills to the South. The number of major domestic processors in the United States dropped to a low of six in 1996.

Until 1996, U.S. wool producers benefited from the National Wool Act of 1954. This act established an incentive payment plan to encourage wool production in the United States. While in effect, an average incentive, or support price, for grease wool was established each year by the USDA.

Producers sold their wool through regular channels, then the average sale price for the year was calculated. The percentage by which the support price exceeded the national average sale price was then used to calculate the payment to be paid each producer. In the 1993 marketing year preceding phase-out of the program, the support price was set at $2.04 per pound and the average price received by producers was 51 cents per pound,[3] or $1.52 less than the support price. Therefore, the 1993 government payment rate of 300 percent (1.52 divided by 0.51) brought the average price of all producers up to $2.04.

---

[3] The 51 cents per pound of shorn wool was the lowest price received by American wool producers since 1975, when the price averaged just under 45 cents per pound.

In 1996, the final year of the Wool Act phase-out, the average wool price was determined to be $1.04 per pound, or $1.08 less than the support price of $2.12. Prior to phase-out, the percent payment would have been about 104 percent. However, with a 50 percent reduction phase-out program, the producers received 52 percent.

Without the incentive payment from the National Wool Act, the average price paid for wool sold in 1998 was 60 cents per pound, down from 84 cents in 1997. Shorn wool production also was down an additional 8 percent during that period.

27.6   MERCHANDISING AND PROMOTION

In the mid-1990s, about 16.24 billion pounds of staple fibers were milled in the United States.[4] Of that, about 66 percent, or 10.7 billion, was chemically produced and 33 percent, or 5.2 billion, was cotton. Only 1 percent was derived from wool (80 million pounds) and flax and silk (120 million pounds). This accounts for the low demand for wool as compared to both chemically processed fibers and cotton. Today, wool can be considered more of a luxury fiber than a staple fiber, as it was several years ago.

In recent years there has been considerable use of wool in combination with other fibers. Because wool is a stable fiber of long-standing and good reputation, there is a natural tendency for apparel manufacturers to advertise garments containing any percentage of wool as "woolens." The Wool Products Labelling Act of 1939, however, requires that all products containing wool, except upholstery and floor coverings, bear a label showing the percentage of the total fiber weight that is wool and the percentage of each other fiber present. "Virgin" or "new" wool is used on labels to describe only wool that has not been formerly used and has not been reworked in any way.

Merits of wool as a fiber, alone in fabrics or in combination with other fibers, have caused garment manufacturers to promote wool and wool products aggressively. The Wool Bureau, Inc., of New York is an industry effort in wool and wool product promotion. It sponsors national advertising, helps individual companies with advertising of woolens, provides considerable educational material on wool, and contributes money to other wool promotion groups.

Producers, too, have contributed to the promotional efforts. Referendums of wool producers have voted to permit deductions from wool incentive payments, mentioned earlier, for the promotion of lamb and wool. Deductions have been as high as 1 cent per pound for shorn wool marketed and 5 cents per pound of lamb (unshorn) marketed for slaughter. The money was turned over to the American Sheep Producers Council, Inc., for promotional and educational efforts. Recently, leaders in the sheep industry have been working with the USDA to develop an industry check-off program that would benefit both the sheep and wool industry.

The American Sheep Producers Council annually sponsors the Miss Wool of America and "Make It Yourself With Wool" contests.

27.7   MOHAIR

Mohair is the fleece of the Angora goat. It is pure white, grows in ringlets, and the fibers after a year's growth may be as long as 10 inches (though most Angora goats in the southern United States and other warm climates are clipped twice yearly). The fibers are

---

[4] USDA, ERS. *Cotton and Wool,* November 1995.

smooth, and thus lack the felting property of wool. In diameter, the fibers are intermediate between fine and coarse wool.

Almost 90 percent of all mohair produced in the United States is derived from Texas flocks, the remaining percentages coming from New Mexico, Oklahoma, and Arizona. In 1998, the number of Angora goats clipped totaled about 705 thousand head. With a yield of about 7.2 pounds per animal, the annual yield totaled about 5.1 million pounds.

The traditionally leading countries in mohair production are South Africa, the United States, Turkey, Argentina, Lesotho, Australia, and New Zealand. The United States accounts for 42 to 44 percent of world production and exports several million pounds of clean mohair each year.

Mohair excels in luster, durability, and affinity for fast dyes, so it is used in goods of fine quality that are subjected to hard usage. Mohair as it comes from the animal usually contains 10 to 30 percent short undercoat fibers that are chalky and dull. For finer mohair fabrics, these fibers must first be removed by combing.

The National Wool Act of 1954 also provided an incentive for mohair production. In 1994, the incentive price for mohair was $4.74, but the average market price received was $2.56 per pound; thus, government payment represented about $1.98 of each dollar received by the producer. As discussed in Section 27.6, the Wool Act has been discontinued.

Much of the previous discussion on wool in this chapter, including marketing, processing, and incentive payments, also pertains to mohair. Of course, minor differences exist that cannot be discussed here.

27.8   OTHER FIBERS AND FURS

Fiber-producing animals other than the domestic sheep and the Angora goat are the llama, alpaca, vicuna, and guanaco. These animals are especially common in South America.

Alpacas thrive at higher altitudes. Alpaca wool is fine fibered, soft, and lustrous. It is drier and freer of grease than sheep wool. Annual wool production averages 2.5 to 3 pounds per animal. Alpaca wool grows in various tones of white, fawn, tan, cream, black, brown, and gray. Darker shades generally are finer fibered than lighter shades.

Llamas are larger animals and much better suited for lower altitude living. Their wool is coarser, harsher handling, and contains more kemp than alpaca.[5] Durability is one of the positive attributes of llama wool.

Cashmere is a very fine fiber (11- to 18-micron diameter) from the down of the goat. It is produced from some breeds of meat-type goats during fall and winter months. Although yield is low (0.2–0.3 pounds per fleece) when compared to wool or mohair, cashmere can bring as much as $45 per pound. Considered a luxury fiber, most processed cashmere is used for sweaters and other cashmere garments. The furlike coat of the Karakul lamb was discussed earlier in Box 21–1.

Leading countries in cashmere production are China, Iran, Afghanistan, and Outer Mongolia. Some breeds in Texas and breed crosses show potential in cashmere production because the environmental conditions are conducive to its production.[6]

---

[5] Kemp is an objectionable animal fiber characterized as being weak, straight, chalky white, and heavily medullated.

[6] Teh, T. H., et al. *J. Anim. Sci.* 68 (Suppl. 1, 1990): 497.

Vicuna wool is very fine and soft. Vicunas are small animals that yield as little as 0.5 pound of wool annually.

Guanacos resemble llamas in appearance and produce a soft coat fiber fawn to brownish in color.

QUESTIONS FOR STUDY AND DISCUSSION

1. List and describe five important characteristics of wool.
2. What are the major chemical components of wool? How are they linked to provide the elasticity of wool?
3. Name the two systems for grading wool and tell the basis for each.
4. Name two breeds of sheep that generally produce high-quality wool and two that generally produce lower quality wool.
5. List and describe the major steps in wool processing.
6. What is the major difference between "woolen" and "worsted" yarn or fabric? Which will be used for the higher quality and lighter weight garments?
7. How does mohair differ from wool?

The Business 28
of Producing Pork

Profitable swine production enterprises reflect the interest, education, dedication, and managerial ability of the operator. The successful producer is able to combine the art of husbandry and science of production with solid business methods to develop and maintain a profitable enterprise (see Color Plate D).

As in other meat-producing sectors, a high proportion of pork is produced by relatively few and large operations (Figure 28–1). In 2002, less than 10 percent of the pork operations had 75 percent of all hogs being produced in the United States. These larger operations can benefit through operational efficiencies and volume purchases and sales. Consequently, they tend to have higher profits in the industry's good years and less loss in the industry's poor years.

Most hogs are raised and sold under some type of "integrated" or contract arrangement. The party that feeds pigs from weaning to processing weight, usually from 30 to 40 pounds up to 260 to 280 pounds, may contract with a party that farrows the sows and gilts and raises the pigs until delivery to the feeder. The feeder may also contract with a processor to take the finished hogs. In each case, the contract will likely

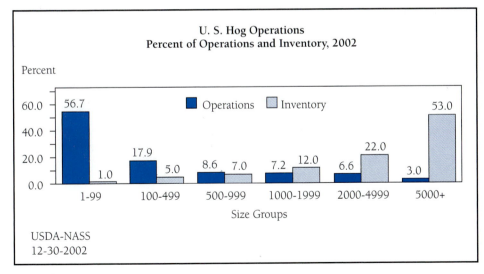

Figure 28–1
Most of the U.S. pork is produced in less than 3 percent of the operations—those with 5,000 or more animals on hand at a given time.
(*Source:* USDA, NASS.)

specify the number, weight, or weight range; the price or a formula by which the price will be determined; and, for the finished hogs, perhaps the time of day they are to be delivered to the processing plant. In some integrated systems, the hogs are owned and the ration is provided by a processor, feed company, or other party. The "producer" is paid for labor, facility use, and utilities on a per pig per day or per pound of weight gain basis. Today, the most integrated relationship is Smithfield Foods and Murphy Farms.

In pork production, there tends to be a "division of labor" between feeder pig production and the feeding of pigs from weaning to processing weight. Even in the case of a producer who owns and operates both farrowing and feeding, the two operations are usually in separate facilities, sometimes miles apart.

Swine production in the early twenty-first century, for most operations, is a year-round business, with industry focus on a steady supply of pork to serve a steady consumer demand. Processors, feeders, and those who farrow and raise the feeder pigs want to keep their facilities, workers, and management skills fully employed. Management focus is on pigs weaned per sow per year or weight of pigs marketed per unit of labor or per unit of feeding space, as well as return on equity investment.

This is in contrast to traditional pork production systems, practiced by a number of producers, and farrow-to-finish operations often based on seasonal advantages of spring and fall farrowing. Production costs can be reduced in these systems by use of pasture or timber lots for farrowing and less investment in facilities. Where a general farming operation is also involved, farrowing can be scheduled to avoid high-labor crop planting and harvesting times.

In the most recent decade consumer demand has developed for "natural pork" and organic pork, produced under more traditional systems, but with some refinements.

Because the elements of a feeder pig operation and a pig finishing operation are comparable to those segments of a farrow-to-finish business, discussion of the farrow-to-finish system precedes (in Section 28.3) sections on the specialized businesses of feeder pigs and pig finishing.

For any swine production system to return a reasonable profit for money and time invested, it must include a sound breeding program, a proper feeding program, high-quality personnel, and excellent management. Pigs are kept in a comfortable environment and the system is designed to minimize any odor or manure problems. Methods and procedures are reevaluated in light of new research information as it becomes available. Market trends and fluctuations are monitored, price risk is controlled and, where contracts are involved, terms are carefully negotiated.

This chapter describes these systems and refinements and components of the systems, as well as geographic distribution of pork production, common technologies and practices, investment, labor, feed, genetic sources, industry organizations, and related topics. Such terms as early weaning, split-sex feeding, phase feeding, multiple-site, and all-in-all-out are defined and discussed.

Upon completion of this chapter, the reader should be able to

1. Describe the geographic pattern of swine production.
2. List and describe the major types of swine enterprises.
3. List the significant financial factors to be considered in establishment of or involvement in a swine enterprise.
4. List major feeds used for swine production and differentiate among creep feeds, supplements, and complete rations.

5. Identify and describe the most common breeding system used in swine enterprises.

6. List the assets and liabilities of swine manure and its nutrients.

7. Explain the more recent trends in marketing hogs.

8. Identify the national organizations that promote pork and pork products.

Also, the reader with limited animal experience and background may want to become acquainted with the parts of the hog and their correct terminology, as illustrated and presented in Appendix B.

28.1 GEOGRAPHIC AND SEASONAL DISTRIBUTION

The upper Midwest continues to prevail in hogs produced and Iowa, the leading corn-producing state, continues to be the leading hog-producing state. However, North Carolina ranks number two among states because of its concentration of large-scale operations (Figure 28–2). The high volume in certain counties of western and southwestern states are due primarily to a few very large operations.

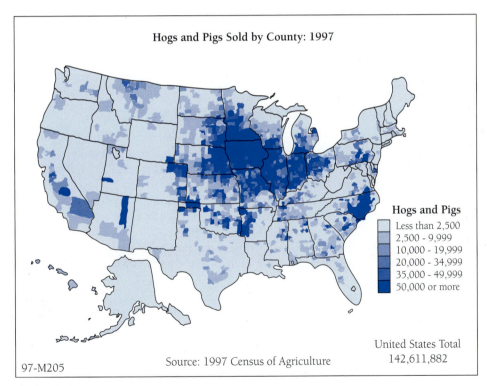

Hogs and Pigs Sold by County: 1997

Hogs and Pigs
Less than 2,500
2,500 - 9,999
10,000 - 19,999
20,000 - 34,999
35,000 - 49,999
50,000 or more

Source: 1997 Census of Agriculture

United States Total
142,611,882

97-M205

Figure 28–2
Although most hogs are raised in the Corn Belt, North Carolina has many intensive contract-production operations. Also, some very large production operations exist in less densely populated areas of the western United States.
(*Source:* USDA, NASS.)

TABLE 28–1	Ten Leading States in Hog and Pig Production	

Rank	State	12-1-02 Inventory (1,000 head)
1	IA	15,300
2	NC	9,600
3	MN	5,900
4	IL	4,050
5	IN	3,150
6	MO	2,950
7	NE	2,900
8	OK	2,490
9	KS	1,530
10	OH	1,440
Total 10 states		49,310

Source: *USDA, NASS.*

The 10 leading states in pig numbers in late 2002 are shown in Table 28–1. These states had more than 80 percent of the 59.8 million hogs in the United State at that time.

Because most hogs are processed at 5 to 6 months of age, December 1 or any monthly inventory data show only about half the year's hog production. With most swine production in year-round operations, there is little seasonality in hog marketing and processing, but slightly more go to processing in April through June and in September.

28.2   SWINE BUSINESS STRUCTURES AND OWNERSHIP

No sector of the animal industry has changed more in the late twentieth and early twenty-first centuries than the swine production sector. Large numbers of hog producers, especially those producing smaller numbers, quit the business (Figure 28–3). Large-volume producers rapidly increased their volume. The downfall of many small producers occurred in 1997–1998, when the price of pork dropped to $.10 per pound. Thereby, producers were only receiving approximately $26.00 for a 260-pound pig. The larger facilities sustained this low price and remain as opposed to the smaller producer.

Whereas nationwide hog production historically had been part of a multiple-enterprise farming operation, production units totally separate from crop or other animal production became commonplace. Ownership and management of these units focus solely on pork production; they generally purchase grain and other ingredients for on-site ration mixing or they contract for daily delivery of complete rations.

Ownership of these units varies, from a farm family living adjacent to the unit, to a cooperative or joint venture by farmers (fairly common for feeder pig production), to an individual or corporation that may own many units at diverse and distant locations. Some may be called part-time or "hobby" units, operated by a person with an off-farm job or profession.

Even where hog production remains part of a multienterprise farm operation—and there are many such well-managed operations—more hogs are now fed "on contract," with the farmer providing the facility, labor, and utilities, and another party, perhaps a feed supplier, providing the pigs, ration, and management supervision.

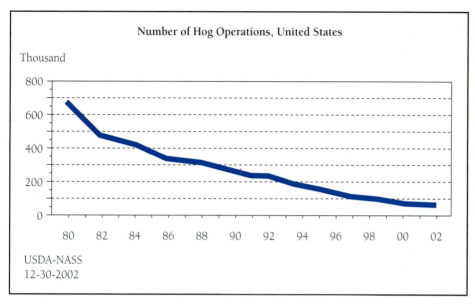

Figure 28–3

The decrease in numbers of hog operations reflects the specialization that has occurred in all of U.S. agriculture. Many former hog producers now focus entirely on crop production or another animal enterprise.
(*Source:* USDA, NASS.)

These and other industry changes have been made operationally feasible by recent technology, but they have been *driven* mostly by economic issues such as cash flow, return on equity, willingness or ability to carry financial risk, and leverage in negotiating prices for inputs or finished animals.

These changes, their magnitude, and their relative rapidity have resulted in a number of problems within and for the industry, especially environmental and sociological problems.

Manure smells, and it carries a high proportion of both phosphorus and solid matter. If surplus phosphorus reaches streams or lakes it prompts algae bloom. If massive amounts of manure solids (mostly nondigested carbohydrate from grain) reach streams or lakes, their decomposition uses up oxygen in the water and fish die. The larger the pig production unit, the more manure at the site and the more potential for odors offensive to the neighbors or a manure or lagoon pipe break that can result in serious damage to a stream or lake.

There are parallel sociological issues, perhaps as problematic to this changing industry as the risks. Some producers and other members of society object to this departure from the "traditional family farm" model. Some longtime producers feel "squeezed out" by the economies and competitiveness of the larger units.

There have been numerous lawsuits and large damage awards paid by unit owners to neighbors on the grounds that their residential property decreased in value or their health became impaired. Strict construction and operating regulations have been established in some states, and some producers have paid heavy fines for improper manure handling or discharging. In one case, a farm was forced by law to close operation following a manure spillage. Federal regulations now require permits for operations above a certain size and some states and counties require approval before new construction.

Consumer interest and, therefore, producer interest in the growing and processing of "natural" pork has developed, partially in response to these problems and issues. Features of the production systems for natural pork may vary among processors and groups of growers but, in general, sows and gilts are maintained and farrow on pasture or in pens with deep straw or other litter, and the pigs are raised and fed to processing weight in these or other less confined circumstances.

In direct response to these issues, both the National Pork Producers Council and individual operators worked together to establish and implement a set of environmentally positive guidelines for unit construction and operation.

It is not the purpose of this textbook to advance one form of ownership, one size range, or one production system over others. Rather, it is to describe major components and features of current pork production systems. The focus is on the broad pork production business of today and the future.

28.3   FARROW-TO-FINISH OPERATIONS

A majority of U.S. swine production enterprises are farrow-to-finish operations. These operations maintain a herd of sows and gilts, farrow the baby pigs, and feed these pigs to market weight of 230 to 280 pounds. There is considerable variation among farrow-to-finish producers in volume of pork produced, intensity of production (seasonal farrowing vs. farrowing a group of females every 6 weeks), and, of course, the labor required and the facility investment per pig marketed.

An increasing number of farrow-to-finish operations use physically separate facilities (Figure 28–4) for farrowing, a baby pig nursery, and finishing, but all three components of the operation can be done in a "production line" facility as shown in Figure 28–5 and Figure 28–6. Also, some producers farrow both spring and fall on pasture and finish the pigs in open pens or with little shelter. Table 28–2 illustrates a management schedule for such operations.

A two-litter-per-year pasture system operates on a 6-month cycle, with farrowing scheduled to utilize farm labor in other than planting and harvesting seasons. Pasture farrowing and other low-investment systems are usually secondary to crop enterprises and/or other animal enterprises on the farm. Buildings are simple in design and have few labor-saving devices.

Confinement farrow-to-finish systems may range from four farrowings per year, perhaps December and June for half the sows and February and August for the other half, to a group of sows farrowing every 6 weeks. The more intense the enterprise, the greater the investment. Custom-designed buildings may have slatted floors, allowing manure to drop to pits below, or a flush system to clean the floors. They likely have automatic, computer-controlled ventilation and temperature control systems and computer-controlled feeding systems.

Less land is required, of course, for the confinement buildings. However, considerable land is needed for distributing the manure. In fact, some states require that the manure management plan include a commitment of land to receive the manure, the acreage in accord with the year-long volume of nutrients in the manure. Because there may be legal liability, before approving a loan for facility construction lenders may require written commitments by nearby landowners to receive some of the manure.

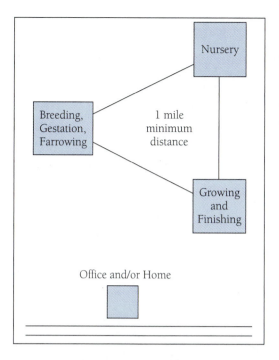

Figure 28–4
Photos are of high-quality gilts during gestations and a large litter of nursing pigs. Physical separation (shown in the drawing) of gestation/farrowing, nursery, and finishing units reduces the risk of any disease spreading throughout the entire operation.
(*Source:* Photos courtesy of Purdue University and Pig Improvement Company; drawing from *Pork Industry Handbook,* PIH-22.)

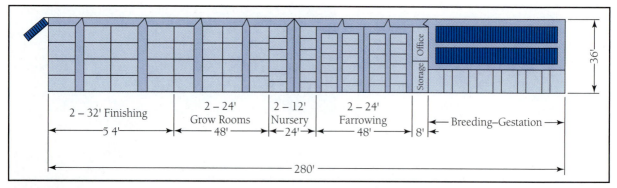

Figure 28–5

Floor plan for a production line system for a 100-sow herd. It has two 12-sow farrowing rooms, to accommodate farrowings every 6 weeks.
(*Source: Pork Industry Handbook*, PIH-22.)

Figure 28–6

A complete confinement swine production unit in Georgia. Note the lagoon for waste handling.
(Courtesy of Purdue University)

TABLE 28–2	Facilities Use Schedule for Nine Groups of 24 Sows, Farrow-to-Finish (Expressed as Day of Year)						
Sow Group	Breeding Begins	Move to Farrowing	Farrowing Room	Wean and Move to Nursery	Nursery Room	Enter Grow Finish	Sell by
1	0	110	A	141	A	172	296
2	16	126	B	157	B	188	312
3	32	142	A	173	A	204	328
4	48	158	B	189	B	220	344
5	64	174	A	205	A	236	360
6	80	190	B	221	B	252	376
7	96	206	A	237	A	268	392
8	112	222	B	253	B	284	408
9	128	238	A	269	A	300	424
1	144	254	B	285	B	316	440

Source: *Pork Industry Handbook*, PIH-75, 1996.

As in any business, a pork producer can benefit from establishing production and efficiency goals and comparing his or her operation "profile" with an industry standard or with the profiles of other operations. Table 28–3 provides some example performance profiles and some "attainable goals."

Some may assume that only intensive, high-investment operations can meet such goals. That is not the case. The labor requirement for a low-investment or seasonal furrowing operation may be higher, but with high-quality genetics, rations, and management such an operations can achieve some of those goals and be a profitable business.

Multiple-Site, Segregated Early Weaning, Split-Sex and Phase Feeding

An increasing number of producers have adopted the *multiple-site system* shown in Figure 28–4 and consisting of three locations: (1) a breeding-gestation farrowing site, (2) a nursery site; and (3) a growing-finishing site. Multiple-site operations generally result in healthier pigs. The all-in all-out concept is maintained by rooms, by building, and by site, resulting in fewer disease problems and improved feed efficiency. Pork producers also make most efficient use of time and facilities by establishing a "use schedule" similar to the example shown in Table 28–2.

Although many pigs are weaned at 3 to 5 weeks, **segregated early weaning** (SEW) at 10 to 14 days of age has become a popular practice among larger producers. It involves early group feeding of special diets and shielding the pigs from disease to help ensure continuous growth of healthy pigs. At 10 to 14 days of age, each litter of pigs is still receiving passive immunity to most diseases via the lactating sow. This immunity is enhanced by vaccinations given to the mother prior to or early in her gestation period.

> **Segregated early weaning (SEW)**
> A practice of weaning pigs at 10 to 14 days of age and isolating multiple litters from other age groups to reduce risk of contraction of disease.

TABLE 28–3	Examples of Production Profiles and Attainable Goals for Swine Operations

	Performance Profiles			
Performance Measure	Excellent	Average	Poor	Attainable Goal
Pigs weaned/litter (no.)				
Gilts	Over 9.5	7.6–9.5	Under 7.6	10.0
Sows	Over 10.5	8.5–10.5	Under 8.5	11.0
Avg. daily gain (lb)				
40 lb to market	Over 1.5	1.3–1.5	Under 1.3	1.7
Age at 230 lb (days)	Under 182	182–227	Over 227	165
Feed efficiency (lb feed/lb gain)				
40 lb to market	Under 3.0	3.0–3.5	Over 3.5	2.8
Loin muscle area at 230 lb (sq in.)	Over 5.5	4.7–5.4	Under 4.7	6.0
Fat depth, 10th rib, at 230 lb (in.)	Under 0.9	0.9–1.3	Over 1.3	0.8
Percent lean in carcass	Over 56	52–56	Under 52	57
Hogs discounted at market (percent)	Under 2	2–5	Over 5	1.5
Growing-finishing mortality (percent)	Under 2	2–4	Over 4	Under 2
Hogs sold/female/year (no.)	Over 19	15.5–19.0	Under 15.5	21.0

Source: *Adapted from the Pork Industry Handbook, PIH-100. Purdue University Cooperative Extension Service.*

SEW also involves the movement of pigs into a facility with other young pigs, all within an age range of 7 days. This facility is a highly insulated structure with temperature controlled initially at 85° to 86°F, then lowered 3 to 4 degrees each week for several weeks. Managed as an all-in all-out operation, the facility is carefully cleaned and sanitized between groups of pigs.

Much of the success of this early weaning system is specially formulated diets that enable the pigs to shift from their mother's milk to a solid diet and water. Highly skilled management also is required to ensure that pigs do not become contaminated from other pigs, workers, rodents, or equipment. The pigs remain in this facility for about 6 weeks and to a weight of about 50 pounds. Then they are moved as a group into separate growing-finishing facilities (see Color Plate D).

Two other practices employed in SEW, and in other operations where pigs will be grown together until market age, are *split-sex feeding* and *phase feeding*. Split-sex feeding refers to the placing of newly weaned barrows and gilts in separate groups until market age. The basis for this practice is that most gilts are leaner than barrows and their nutritional requirements will be different.

Phase feeding refers to adjusting or changing rations to best fit the nutrient requirement of pigs as they grow and develop. For example, a ration containing highly digestible protein and carbohydrate sources is provided for early-weaned pigs. Another ration is formulated for pigs after they have adapted to dry feed and are weighing at least 20 pounds, until they reach typical feeder pig weights of about 40 pounds. At that time, they then would go onto a sequence of two or more growing-finishing rations.

Pigs generally will remain in the growing-finishing facilities for about 120 to 127 days, when they should be ready for market at 240 pounds or more (see color insert).

28.4   FEEDER PIG PRODUCTION ENTERPRISES

Feeder pig production units require the maintenance of a breeding herd, usually crossbred gilts and sows that benefit from hybrid vigor (see Chapter 18) and boars. Because feeder pigs are usually sold at 30 to 60 pounds, feeder pig production requires relatively less feed in relation to labor or other costs. Emphasis is on making full use of buildings and skilled labor. Farrowings are scheduled as frequently as possible, within the limitations of disease control and proper breeding herd management.

The low-investment feeder pig unit is usually a relatively small enterprise (50 to 100 sows) designed to supplement rural family farm or nonfarm earnings by providing a way to gain cash income from the use of available labor and facilities. Low-investment feeder pig production often employs a central farrowing house (perhaps a remodeled building or a "pull-together" building on a concrete slab) and an open-sided sow-and-pig nursery. The breeding herd is usually maintained as groups in open-shed facilities (Figure 28–4) or on pasture. Pigs are sold at about 6 to 8 weeks of age.

The high-investment feeder pig unit usually maintains 200 or more sows and is operated by full-time swine staff. Sophisticated buildings and equipment reduce labor requirements and provide a controlled environment. Pigs usually move directly to feed-out or finishing operations, often by contract that includes a base-price arrangement. This avoids some marketing costs and the dangers of spreading disease among "pooled" pigs.

Some major pork-producing companies with feeding facilities or contract arrangements throughout the Corn Belt have chosen to locate their sow herd and feeder pig production in the southern part of the Corn Belt, perhaps central or southern Missouri. This latitude has the advantage of a more temperate climate for the sows herd and a lower energy cost for heating the facilities. Pigs then are generally moved north to finishing units.

28.5 PIG FINISHING OPERATIONS

Pig finishing operations do not require the maintenance of a breeding herd and avoid farrowing and nursery responsibilities. They depend, though, on a steady source of satisfactory feeder pigs at an equitable price.

To reduce health and management problems, when purchasing groups of feeder pigs, the buyer should try to purchase pigs that

- Are uniform in size and weight (e.g., 45 to 50 pounds each)
- Appear to be alert, vigorous, and free from injury or disease
- Are healed from castration
- Are free of parasites (internal and external)
- Have been tail docked
- Are produced and sold by reliable persons

Young pigs that are transported some distance and commingled with other groups of pigs are subjected to the stresses of fatigue, thirst, and hunger. Upon arrival, they also experience changes in facilities, pen mates, social order, and diet. To reduce these stresses, they should be kept as comfortable as possible in transit and upon arrival provided water and a specially formulated diet. Use of electrolytes in the water and an antimicrobial feed additive in the diet are helpful in reducing stress, diarrhea, and death losses.

With a feeder pig finishing operation there is more opportunity, compared to sow herd enterprises, to vary production volume in response to expected profit levels. The penalty for holding or reducing production is relatively modest; the operator can increase volume on short notice. Successful feeder pig finishers tend to be (1) short on labor and husbandry time or skills needed to manage a sow herd, but long on feed grain to grow and finish hogs to market weights; (2) skilled in buying and selling or in negotiating contracts; and (3) in need of steady income from periodic hog marketing or payment for contract feeding

The *low-investment feeder pig finishing system* uses simple, open-fronted buildings with concrete slabs or dirt lots. Building investments are low, but there are higher labor requirements and more problems with flies, odors, and waste disposal. Because facilities investment is low, the producer who buys pigs on the open market rather than by long-term contract can shut down more economically when there is little chance for profit. The *high-investment system* uses sophisticated equipment to reduce labor requirements and provide a controlled environment. However, management must utilize these facilities to capacity, because the high cost of owning them continues whether or not they are in use. Many using this system buy feeder pigs on a continuing basis directly from feeder pig producers, are shareholders in a feeder pig-producing cooperative or partnership, or feed hogs owned by another party, with payment on a per pig per day or per pound of gain for their labor, facilities, and utilities.

28.6   INVESTMENT

Starting in the hog business requires considerable capital regardless of type of unit or system chosen. The first investment, of course, is in the animals. Those considering the business usually visit hog production units, consult with extension specialists, develop a plan that suits their needs, then consult a bank or other credit source to seek financing. The credit source, as well as the operator, will need answers to the following questions: What proportion of the investment can the operator provide? Will the unit have competitive production costs? Will income meet debt (principal and interest) payments? Will there be adequate cash flow to meet daily costs of feed, veterinary expense, and other operating expenses in addition to debt payments? Swine buildings and swine equipment are both rather short-lived so must be amortized over a rather short period, perhaps 8 to 12 years.

Table 28–4 estimates the minimum competitive size and the practical limit in size for various production units. This table does not infer that smaller units fail to add to farm income or profit. Many smaller units exist to utilize available family labor, unused buildings or land, or some other resource with no alternative use. In such a situation, size of unit is set by the limiting resource. Even a very small unit may make a significant contribution to farm earnings.

A new production unit, however, must operate at sufficient volume to make efficient use of labor, buildings, equipment, and management skills. Table 28–4 suggests that a high-investment farrow-to-finish operation requires about 100 sows producing 1,500 pigs per year to pay for labor, management, environmentally controlled buildings, and supporting equipment for waste handling, feed processing, and emergency power generation. A feasible goal for the previously discussed 216 sow operation (nine groups of 24 sows), assuming a herd average of 2.1 litters per sow per year and nine pigs per litter weaned, is over 4,100 pigs per year. A typical investment for this type and size of operation is presented in Table 28–5.

TABLE 28–4	Minimum Competitive Size and Feasible Maximums for Various Pork Production Systems			

Production System	Minimum Competitive Size		Feasible Goal	
	Number of Sows	Pigs Produced per Year	Number of Sows	Pigs Produced per Year
Sow herd enterprises:				
Feeder pig production operations:				
Low-investment systems	36	540	150	2,250
High-investment systems	100	1,500	400	6,000
Farrow-to-finish operations:				
Two-litter pasture systems	25	375	150	2,250
Low-investment confinement systems	60	900	150	2,250
High-investment confinement systems	100	1,500	400	6,000
Feeder pig finishing enterprises:				
Low-investment systems	—	600	—	5,000
High-investment systems	—	1,500	—	15,000 +

Source: *Pork Industry Handbook, PIH-48.*

TABLE 28–5	Estimated Investment and Operating Costs for Nine Groups of 24 Sows, Farrow-to-Finish		

Facilities	Purchase Price	Life (years)	Annual Fixed Cost
Farrowing facilities:			
Equipment	$26,000	8	$4,108
Buildings	58,300	15	5,130
No. farrowing rooms	2		
No. crates per room	20		
Nursery facilities:			
Equipment	13,000	8	2,054
Buildings	29,200	15	2,570
Grower-finishing facilities:			
Equipment	36,000	8	5,688
Buildings	144,000	15	12,672
Gestation/breeding facilities:			
Equipment	16,200	8	2,560
Buildings	64,800	15	5,702
Subtotal:			
Equipment	$91,200		$14,410
Buildings	296,300		26,074
Support facilities:			
Lagoon/manure stg.	$11,000	20	$693
Manure disposal equipment	35,000	10	4,130
Pickup	18,000	5	4,824
Stock trailer	7,000	20	581
Feed mill and storage facilities	35,000	15	3,652
Subtotal	$106,000		$13,880
Total	$493,500		$54,364
Equipment	197,200		28,289
Buildings	296,300		26,074

Source: *Pork Industry Handbook, PIH-83, 1996.*

28.7 LABOR AND MANAGEMENT

Working successfully with animals requires a special kind of person, one who maintains a love for animals and life. In the case of a swine operation, it also requires considerable technical knowledge and skill. Where the total operation is handled by one or two people, perhaps within a general farm business, the operator needs to be proficient in breeding techniques, use of medications, equipment maintenance, feed handling, pest control, and manure handling, as well as record keeping and financial planning. Note in Table 28–6 the improvement in labor efficiency, labor costs, and production per person in larger, high investment systems.

For larger operations, perhaps involving a million dollars or more investment, several employees, and thousands of pigs produced annually, the manager must focus on the broad issues of personnel selection, training and supervision, measures of productivity and performance (of both the animals and the employees), genetic and feed

TABLE 28–6	Labor Requirements and Costs, and Per-Person Capacity for Various Pork Production Systems				
	Hours of Labor per Sow Unit of Production[a]		**Labor Cost**[b]	**Production per Person**	
Production System	*Direct*	*Total*		*Size of Sow Herd*	*Pigs per Year*
Sow herd enterprises:	*Per Sow Unit*[a]		*Per Pig*		
Feeder pig production operations:					
Low-investment systems	20	26	$11.62	150	2,250
High-investment systems	14	18	8.13	215	3,200
Farrow-to-finish operations:			*Per Cwt. Gain*		
Two-litter pasture systems	36	48	9.49	85	1,250
Low-investment confinement systems	34	45	8.97	90	1,300
High-investment confinement systems	22	28	7.61	135	2,050
Feeder pig finishing enterprises:	*Per 100 Pigs*[a]				
Low-investment systems	75	100	4.11	—	4,000
High-investment systems	60	80	3.31	—	5,000

[a]The sow is the unit for sow herd enterprise data; it denotes a mature female in production and includes a "supporting cast" of boars, replacement gilts, and progeny in various stages of growth with 14–16 market hogs sold yearly per sow unit. The unit for feeder pig finishing enterprises is 100 purchased pigs; it assumes that feeders are fed on a continuous basis, and for each 100 pigs fed out per year, only about one-third are on hand at any one time.

[b]Labor is charged at $9.00 per hour.

Source: *Adapted from* Pork Industry Handbook, *PIH-48.*

sources, and product quality, as well as on financial issues—input cost, contract terms, and payment for animals sold.

The crop producer who produces hogs as a secondary enterprise has a labor supply that varies from month to month so is likely to favor pasture or other production systems that permit flexibility in scheduling. However, where there is a constant labor supply, where hogs are the major enterprise, or where hogs supplement income from an outside, full-time job, the operator usually avoids the pasture system because labor demands are 50 to 100 percent greater per hog produced than with a highly automated, environmentally regulated system.

Producers often design their on-farm facilities to utilize a supply of available family labor (growing children, spouse, "off-season" labor not needed for crop production, etc.). Thus, there is a personal and family interest in enterprise success and profit.

28.8　FEED

Principles of nutrition, feed ingredients, nutrient requirements, ration formulations, feed processing, and least-cost rations are all discussed in Chapters 3 and 5. A producer generally will do a good job of feeding swine of any age or class by (1) buying feed from a reputable manufacturer and following the manufacturer's feeding and management suggestions, or (2) formulating the mixing rations using formulas and suggestions from the university and its extension service. However, the producer can make choices and decisions that influence feed cost. Because feed is the largest single production cost, economical feeding is necessary for a profitable enterprise.

<table>
<tr><td rowspan="2" style="background:#1a4c7c;color:white">TABLE
28–7</td><td colspan="3">Feed Requirements and Feed Conversion Rates
for Various Pork Production Systems</td></tr>
</table>

Production System	Feed per Unit of Production[a]		Feed Conversion (lb feed/ cwt. produced)
	Corn Equivalent (bu)	*Weight of Purchased Feed (lb)*	
Sow herd enterprises:	*Per Sow Unit[a]*		
Feeder pig production operations:			
Low-investment systems	60	1,130	474
High-investment systems	56	1,165	453
Farrow-to-finish operations:			
Two-litter pasture systems	202	2,350	400
Low-investment confinement systems	203	2,495	406
High-investment confinement systems	197	2,550	400
Feeder pig finishing enterprises:	*Per 100 Purchased Pigs[a]*		
Low-investment systems	960	10,650	394
High-investment systems	930	10,400	382

[a]*The sow is the unit for sow herd enterprise data; it denotes a mature female in production and includes a "supporting cast" of boars, replacement gilts, and progeny in various stages of growth with 14–16 market hogs sold yearly per sow unit. The unit for feeder pig finishing enterprises is 100 purchased pigs; it assumes that feeders are fed on a continuous basis, and for each 100 pigs fed out per year, only about one-third are on hand at any one time.*

Source: *Pork Industry Handbook, PIH-48.*

Feed represents about 50 percent of the production cost in a feeder pig production unit, about 60 percent in a farrow-to-finish unit, and about 70 percent in a feeder pig finishing unit. Table 28–7 shows feed requirements and efficiency data for the various basic production systems described in Section 28.3. Values given are averages; many producers are attaining better returns from their feed investment. Feed conversion is probably the best measure of animal performance and enterprise efficiency. It is calculated as total units (pounds or kilograms) of feed divided by total units of pork produced. In sow herd enterprises, weight gains in the breeding herd are taken into account because this weight eventually is sold for processing. In pig finishing enterprises, only feedlot gains are considered.

Data in Table 28–7 for sow herd enterprises include gestation and lactation feed, as well as an allowance for boars and for bringing replacement gilts to breeding age. All female breeding herd replacements are assumed to come from within the herd. Thus, feed conversion reflects such factors as disease problems, mortality, feed wastage, conception rates, litter size marketed, genetic ability to grow efficiently, and weight loss en route to market.

The feed requirement data in Table 28–7 are divided into two categories, feed grain (corn equivalent) and purchased feed (supplement and "creep" feed). If a feed grain other than corn is used, requirements in some cases would be adjusted. Grain sorghum is equivalent to corn on a weight basis; 1 pound of corn equals 1.14 pounds of oats, or 0.96 pound of wheat, or 0.94 pound of barley. These equivalencies may vary, depending on the proportion that grain is of the total ration. Figures for purchased feed are based on the use of a complete supplement containing 40 percent crude protein. Figures would not be affected significantly if supplemental protein, minerals, and vitamins were purchased separately.

The producer can choose from several alternatives in providing swine rations. One can grow or buy feed grains, purchase supplements, and mix the ration at the unit, have a feed supplier formulate and deliver complete rations to the bins or self-feeders, or handle rations in some intermediate way.

Data in Table 28–7 are based on the operator producing grain and buying a supplement containing all other ingredients needed, except for creep feed. Creep feed, for pigs prior to weaning, is usually a rather complex mix and is used in small amounts. Studies indicate that the operator using more than 100 tons of feed per year might profit from some type of "on-the-farm" feed preparation system. One hundred tons of feed will produce about 250 market pigs. The selection of a system for preparing rations is based on such things as (1) knowledge of nutrition; (2) time and/or labor available; (3) availability and cost of ingredients; (4) equipment available for measuring, grinding, and mixing; (5) storage facilities; (6) volume of feed used; and (7) personal preference.

28.9   GENETICS SOURCES AND BREEDING SYSTEMS

Mating systems and breeding programs were discussed in some detail in Chapter 20. The producer of purebred boars and gilts to be sold as replacements will select within the breed those animals excelling in traits of highest commercial value. Outstanding purebred animals are needed by other breeders and also for commercial crossbreeding programs.

Most commercial swine production enterprises, regardless of specific production system, follow a rotational crossbreeding program, as shown earlier in Figure 20–5 and 20–6. The crossbred sow usually will farrow more pigs per litter and produce more milk than the average of her parent breeds. The crossbred baby pig typically is more vigorous and healthier than purebred pigs in the same environment. Most production units retain replacement females from within the herd and maintain a reasonably high level of hybrid vigor in successive generations by introducing outstanding purebred boars of a different breed into the herd.

Boar herd production records, health history, and performance records are important in evaluating the potential boar. The same is true for gilts, where they are purchased as herd foundation animals or for a specific mating system in an established herd. In addition, behavioral traits, soundness of feet and legs, body conformation and type, pedigree records, and breed should be considered. Therefore, time is well spent in evaluation of seedstock suppliers and their genetic improvement program when selecting replacement boars.

As discussed in Chapter 2, with larger swine operations, more intense selection for economic traits, and recent advances in reproduction, several global companies have become prominent in providing breeding stock to the commercial swine industry. This follows the pattern of the poultry industry during the last half of the twentieth century.

28.10   MANURE AND ODOR MANAGEMENT

Recent federal legislation and regulations, and parallel action in many states, provide specific guidance to swine and other animal producers regarding the handling of manure.

All animal operations are specifically defined as animal feeding operations (AFOs), and those above certain sizes, according to numbers of animals or total animal weight, are considered a concentrated animal feeding operation (CAFO).

A swine operation with at least 2,500 swine weighing 55 pounds or more, or 10,000 swine weighing less than 55 pounds each, is a large CAFO.

An operation is a medium CAFO if it has 750 animals, each weighing 55 pounds, or 3,000 swine *and* a human-made ditch or pipe that carries manure or wastewater from the operation *or* the animals come into contact with surface water running through their confinement area.

Most states, through specified state agencies have authority to manage federal CAFO programs and to issue permits. A state also may have additional and more specific regulations and permitting requirements.

Minimum federal regulations for all CAFOs require the following as part of a producer's permit:

1. A written nutrient management plan.
2. Annual reports to the state permitting authority.
3. Maintenance of nutrient management records for at least 5 years.

Because so much phosphorus is carried in manure, its level in the manure, application rate on land, and its utilization by crops that land produces are key items in most manure management plans. Permits for large CAFOs require, for example, analyzing manure for nutrient content annually, analyzing the soil every 5 years where manure is applied, and avoiding application to any land within 100 feet of surface water. Requirements for medium CAFOs generally parallel those for large CAFOs. Nitrogen may be present in swine manure in larger proportion than phosphorus, but much of it may volatilize. Some states, however, require parallel data and procedures for the manure nitrogen. Additional requirements may be imposed by the state permitting authority and, if the state authority finds that *any* animal feeding operation is adding pollutants to surface water, it may require a CAFO permit.

Odor emissions—ammonia hydrogen sulfide, and other volatile compounds—from animal enterprises may be significant and sometimes offensive to workers and neighbors. However, as of this writing, odor is addressed in most states only indirectly by "separation distances," the minimum distance a swine facility must be from the nearest residence, park, or other public facility.

Odor intensity at any point is influenced not only by the unit emissions but also by wind direction and velocity. Little quantitative data exists from research on odor-contributing emissions, the distances they may travel under varied circumstances, or the physiological effect on people.

Even though considerable research has been done on potential methods of controlling odor, no method has proven foolproof. Observation and experience suggest odor and its negative effects, perceived or real, are minimized by one or more of the following:

1. Locate swine facilities downwind and a maximum distance from any residence or public facility, behind a hill, and out of general public sight.
2. Plant a shelterbelt of trees and shrubs to both detour the wind currents and reduce public visibility.
3. Use a closed manure tank or a manure pit under the swine building to ensure the liquid manure surface is not exposed to the atmosphere. Avoid use of an open lagoon; if a lagoon already exists, cover the manure surface with straw or other material that minimizes volatilization.

4. Notify neighbors before emptying the pit or tank and hauling the liquid manure. Keep equipment clean; avoid spills or drippings on the roadway.

5. Knife in the liquid manure, below the soil surface.

28.11   MARKETING

In the early twenty-first century, a producer's marketing plan for feeder pigs or finished hogs focuses on ensuring steady sales and on limiting price risk. The steady sales can be ensured by means of a long-term sales agreement (contract) and price risk can be limited either by use of the futures market (discussed earlier) or by terms of the agreement.

A decade earlier more than 80 percent of finished hogs were sold in what is called the cash market (or spot market), the price available or that could be negotiated with a buyer the day of sale. Should there be a large volume of market hogs, a large volume of fresh pork in storage, or processing workers on strike that day, the price would be low. There have been times of oversupply of finished hogs and no market; the processors couldn't use more hogs. The producer carried the price risk and if there were no market for finished hogs, they'd just need to be fed longer.

At present, surveys indicate less than 20 percent of finished hogs are sold on the cash market. More than 80 percent are marketed under a prearranged agreement with a processor or are owned by a processor or other party. Most long-term sale contracts don't *eliminate* price risk, they *limit* it. Contracts vary in their terms, and negotiation of those terms before contract signing is important. Delivery dates, weights and numbers, or ranges for each, and sometimes the time of day or night for delivery are specified. The price per pound of live weight or carcass weight is calculated according to some type of formula. The major input to the formula is some type of public market price, such as the "nearby live hog futures" price or a wholesale pork price, the day of delivery.

The formula for finished hogs usually includes a premium/discount provision for quality or leanness. It may also include an adjustment based on price of corn, a major production cost, or some other component.

Some agreements include a risk-sharing provision. For example, if the calculated price from the formula is between $50 and $60 per 100 pounds of live animal, the producer is paid that price. If the calculated price is above $60, the producer receives half the excess; if the calculated price is below $50, such as $46, the producer receives $48. In this case, the processor carries half the price risk below $50, but is compensated by retaining half the excess above $60. Pricing contracts for feeder pigs may, of course, be similarly structured.

A 2000 survey[1] indicated that many contracts with processors used a cash or spot price as basic input to the formula. With a smaller and smaller proportion of finished hogs actually being sold on the cash or spot market, the question arises, "How reliable is the spot or cash market in measuring the total pork supply and demand?" There is considerable evidence that the larger producers and the producers with, on the

---

[1] Lawrence, John D., and Glenn Grimes. *Production and Marketing Characteristics of U.S. Pork Producers.* Ames, Iowa: Iowa State University, 2000.

average, higher quality and more consistent quality hogs are either processor owned or priced on a formula. If that is the case, and most of the finished hogs being sold on the spot or cash market are of generally lower quality, the lower price established for such hogs unduly lowers a calculated formula price.

Other food-processing industries might be studied and considered by the pork industry, as the trend to contract formula pricing continues. In the vegetable industry virtually all production is contracted and there is no reliable cash or spot market for large volumes. Contract formulas may therefore be based, at least in part, on such items as labor, energy, or other production input costs, average retail prices for the finished product, or even the national Consumer Price Index (CPI).

28.12   PORK INDUSTRY ORGANIZATIONS

Three national organizations work on behalf of the pork industry. The National Pork Producers Council (NPPC), headquartered in Des Moines, Iowa, focuses on public policy issues such as federal legislation and regulations, working closely with and representing 44 state pork producer organizations. The NPPC is financed largely by membership fees and a voluntary check-off which, at this writing, is 10 cents per $100 animal sales.

The National Pork Promotion and Research Board (NPB), located in the NPPC national office, is financed largely by a federally legislated check-off of 40 cents per $100 sales. The NPB provides information to media and consumers, publishes fact sheets on production practices, organizes conferences, and sponsors research. The 2002 check-off funds financed $2.7 million in research on environmental issues, disease prevention, housing, pork quality, and other topics.

Although the two organizations are separate legal entities, they work closely together and, at this writing, members are considering a merger of the two. An annually scheduled national pork industry trade show, the National Pork Congress, attracts a large attendance of producers and other pork industry people, and features demonstrations, seminars, and displays of equipment, buildings, and management innovations.

The U.S. Meat Export Federation (USMEF), headquartered in Denver, Colorado, promotes export sales of beef and lamb, as well as pork. It is financed by the meat industry and, for certain programs, federal export promotion funds provided through the USDA's Foreign Agricultural Service. USMEF has offices in Singapore, Seoul, Tokyo, Osaka, Hong Kong, Shanghai, Taipei, Moscow, St. Petersburg, Mexico City, and London, plus representatives in several other regions.

QUESTIONS FOR STUDY AND DISCUSSION

1. In what states are most of the swine raised? Why?
2. Describe a total confinement swine enterprise, including types of buildings and how the animals are handled during the life cycle.
3. Define and explain the purposes of the following: multiple-site system, segregated early weaning, split-sex feeding, and phase feeding.

4. Some swine producers specialize in feeder pig production. Describe that enterprise, including the age or weight pigs are usually sold, typical litter size, and important financial and management considerations.

5. What are the major components of swine rations and about what percentage of the total cost of swine production is represented by feed?

6. Describe the breeding system that is most common in swine production, and tell why it is so popular.

7. How are most hogs that are ready for processing sold? How are most feeder pigs sold?

8. Describe how the work of the pork industry promotional organizations is financed.

The Cow Herd 29 and Ewe Flock

Intensive cattle and lamb feeding enterprises need a source of animals for feeding. Most intensive feeding operations are located where large volumes of grains—especially corn, sorghum, or barley—are produced. Animals being fed for slaughter gain more rapidly and produce higher quality carcasses when fed high-energy rations.

In contrast, cow herds and ewe flocks effectively utilize forages—grass, hay, or silage (see Color Plates B and C). Many such flocks and herds are maintained in the range and grassland areas such as the Sandhills of Nebraska, mountain meadows of the western states, Flint Hills of Kansas, "West River" country of South Dakota, rolling pasture land of Missouri, and flat grasslands of the Southeast where cotton formerly flourished. In some of these areas, unit acreages are relatively large, perhaps thousands of acres; in the semi-arid or arid regions 5 to 50 acres may be required per cow and 1 to 10 acres per ewe.

Some herds and flocks are maintained where there is also intensive cropping. Crop aftermath, such as cornstalks, and nonpermanent pasture that is part of a crop rotation system are the major feed sources.

Feeder cattle and feeder lamb production enterprises are businesses. The product is the calf or lamb, weaned and ready for purchase by one who will finish it for slaughter. Inputs to the business are capital for land, breeding stock, equipment, feed, and health expenses, as well as labor and management ability. Sheep numbers decreased during the last century whereas beef cow numbers increased. (See Figure 29–1.)

This chapter separately discusses feeder cattle and feeder lamb production, including geographic regions of production, types of enterprises, costs, nutrition, and management.

Upon completion of this chapter, the reader should be able to

1. Describe the feeder cattle and feeder lamb production enterprises.
2. Identify the areas of the United States where most such enterprises exist and explain why.
3. Differentiate among sizes of units in labor efficiency and compatibility with other farm or ranch enterprises.
4. Explain why feed is a major cost and differentiate among stages in the reproductive cycle in terms of feed required per animal.
5. Describe the breeding systems most commonly used in feeder production units.
6. List the steps to follow to reduce risk of illness at or following weaning.
7. Describe the marketing systems commonly used for feeders, as well as for cull cows or ewes.

8. Understand each of the following: body condition scoring, phases of the production and reproduction year of the cow and ewe, backgrounding, preconditioning (including vaccination programs), ewe and ram breeds, and accelerated lambing.

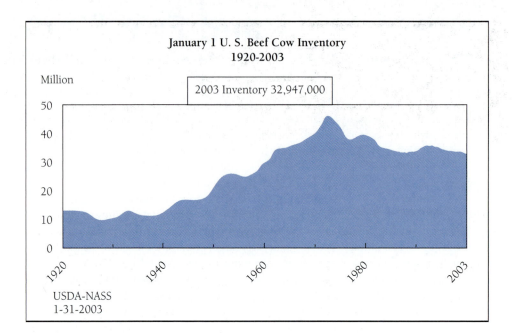

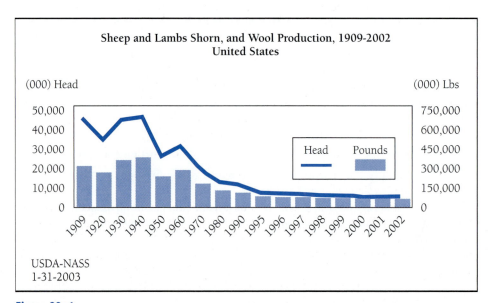

Figure 29–1

Trends in beef cow and sheep numbers during the last century show a sharp contrast. (*Source*: USDA, NASS.)

Figure 29–2
A cow herd and its calves on a ranch in the Pacific Northwest.
(Courtesy of Oregon State University)

29.1 FEEDER CALF PRODUCTION

The production of feeder calves is the goal of commercial (nonpurebred) cow-calf operations (see Color Plate C). It is a large and diverse business. Many people are involved in the business and many different types of feeder cattle, including straightbreds and crossbreds, are raised (Figure 29–2).

The investment required for land and cows is high relative to the value of calves at market time. Some operators in the western states lease federally owned land for grazing, thus reducing capital requirements. Because of the rugged and hardy nature of the beef cow and because cows spend a large portion of the year on range or pasture, building and equipment requirements are not large in relation to other agricultural enterprises. However, shrewd management is necessary for a cow-calf production enterprise to be profitable.

Feeder calf (and stocker) marketing is complex and market timing is critical. Marketing may be as important as production efficiency in achieving a profit. Most feeder calves are offered for sale in the spring and fall, having been born 6 to 8 months earlier. Good managers recognize this and try to avoid marketing their feeder cattle when the supply is large. Forward contracting, for delivery of feeders to feedlot operators at a later date, and the use of feeder futures contracts can be very helpful to "lock in" a sale price or level of profit early in a production cycle.

29.2 GEOGRAPHIC DISTRIBUTION OF COW HERDS

Beef cow herd (or cow-calf operations) in the United States can be characterized as large in number of operations but rather small in number of cows per operation. There were about 805,000 cow-calf operations in the United States in 2002 (Figure 29–3). About 78 percent of these were herds of 50 cows or less. However, the number of cows in these smaller herds represented only 29 percent of the total cow numbers.

On the other hand only 5.4 percent of all cow-calf operations were larger than 500 cows. These herds averaged about 820 cows per herd and accounted for 14.3 percent of all beef cows. Larger herds are primarily located in the Southwest, Great Plains, and western regions (Figure 29–4).

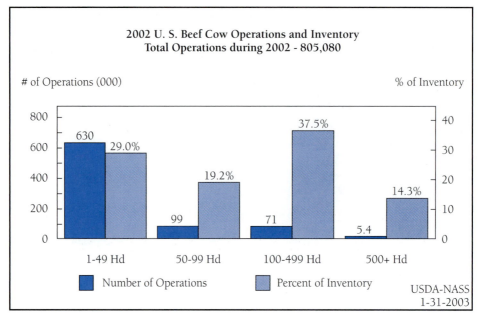

Figure 29–3

Beef cow operations and inventory for 2002. Note that the number of beef cow operations with less than 50 cows represents a very high proportion (about 71 percent) of the total number of operations. (*Source*: USDA, NASS.)

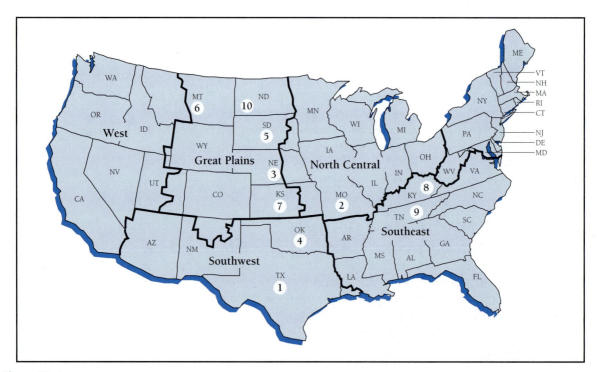

Figure 29–4

Beef cow numbers are highest in the Southwest and Great Plains regions. The 10 leading states and their rank in 1999 are shown. (*Source*: Compiled from *Cattle*, USDA, NASS, January 1999.)

There are five major regions of feeder calf production in the United States: the southeastern United States, the North Central region, the Southwest, the Great Plains, and the western states (Figure 29–4). There are differences among the areas in the general type of cattle produced and the system used to produce them. A characteristic common to all feeder-producing areas is an abundant and relatively inexpensive source of forage. The type of forage varies from area to area, but ruminants have the ability to utilize a wide variety.

In the southeastern states, cows and calves are often maintained on grass 10 months or more annually. Cool-season grasses are productive in the spring and fall, whereas Bermuda grass grows well during the summer heat. To provide grazing through the winter, winter rye is often interseeded into Bermuda grass pastures in the fall. Because the Southeast produces more feeder cattle than it uses in finishing operations, many are transported to the Corn Belt or plains states for finishing.

In the southern plains, there is considerable winter grazing of winter wheat, planted in the fall for grain harvest the following summer. The grazing actually benefits the grain crop, if rainfall is adequate. It forces the plant to stool or branch, increasing the number of wheat stalks and heads.

Most Corn Belt land is too valuable for grass production, so this area is a net importer of feeders. Because corn is the principal crop, there is a good feed supply, and the cattle feeder has a competitive advantage for finishing cattle. Cattle also may utilize corn stover, other crop residues, and silages (Figure 29–5). Some feeding operations in the Corn Belt are part of a diversified farming operation. These feeders prefer to purchase feeder calves in the fall, to use labor during an otherwise slack winter period, and to use crop residue. However, long and severe winters may offset part of the Corn Belt's competitive advantage.

High-quality feeder cattle are a tradition in the Corn Belt, and are often demanded whether produced in the Corn Belt or in other areas—the Plains, the West, or the Southeast.

The southern plains and western Corn Belt are considered the center of the commercial cattle-feeding industry. This region has characteristics that permit good feeding enterprises and good animal performance: ample supply of feed grains, abundant forages, and semiarid climate. With these characteristics and the advantages of large-volume feeding, cost of production can be lower than many smaller farmer feedlots. Even though the states of Nebraska, Kansas, Oklahoma, Texas, and South Dakota all rank among the top 10 states in beef cow numbers (Figure 29–4), these states still import feeder cattle from the Southeast and West.

Figure 29–5
Heifers and cows utilize crop residue in the fields after harvest. (*Source*: *Kansas Farmer*.)

Large numbers of feeder cattle are produced in the "West River" country of South Dakota, the Sandhills of western Nebraska, the Flint Hills of Kansas, the Osage area in eastern Oklahoma, and adjacent states like Missouri, where some land is not suited to intensive crop production and the native grasses are very productive under the climate conditions of the region.

The meat-processing industry has grown with the cattle-feeding industry in the Plains, and the two industries combined constitute a major portion of the area's economy.

Much of the western United States is arid or semiarid and the vegetation is sparse. It is not uncommon for ranchers to graze one cow unit (a cow and her calf, up to weaning age) on 50 or more acres of land, so cows in this area must be hardy and have the ability to travel long distances for grass and water. Even so, the region is well adapted to cattle production and contains some of the largest cow herds in the nation. Because it is a grain-deficient area, many of the feeder cattle produced there are shipped to the Plains and Corn Belt for finishing.

29.3   TYPES AND SIZES OF COW-CALF HERDS

In the production of feeder calves there is considerable variation in size of unit, in management systems, and in type of cattle involved. *Ranching* is considered a larger operation whereas *farming* comprises a smaller acreage and usually smaller numbers of cattle. Both may well be family-type operations. Ranches are more common in the West and on the Plains and farms in the Corn Belt and the Southeast. Most farmers are diversified and raise both crops and livestock. Some operators have a full-time nonfarm occupation and raise cattle for supplemental income. The number of cattle per farm in these cases usually is small. The production of feeder calves lends itself to many types of operations and from 5 or 10 cows to several thousand in one unit.

Cattle are used as scavengers in many areas and are a means of utilizing field crop residues and other low-quality roughages. As in other breeding herds, excellence in genetics, reproduction, nutrition, herd health, and close and timely management are required for a profitable cow-calf operation. Several performance measures are important to overall profitability (Table 29–1). Goals for profitability include

- High conception rates: 95 percent
- Percent of calves weaned: 95 percent
- Calf weaning weight: 550–600 pounds
- High production efficiency: low variable and fixed costs (see "Production efficiency" in Table 29–1)
- High-grade calves: top market prices

Profitability of beef operations is also related to herd size. For example, a summary of the Kansas Farm Management Association's 172 beef herds[1] indicated that the total cost per cow for 100 beef cows was 4 percent lower than for 50-cow herds. In addition, 200- and 300-cow herds had 4 and 6 percent lower costs per cow than the 100-cow herd, due largely to lower labor costs, depreciation, and interest on fixed assets.

---

[1] Muirhead, S. "Kansas farm management association summary." *Feedstuffs*. April 18, 1994.

TABLE 29-1	Example of Performance Measures for Determining the Management Profile of a Beef Cow-Calf Enterprise

| | Performance Level | | | |
Performance Measure	Excellent	Average	Poor	Your Herd
Conception rate (breeding season percent)	95	85	75	
Calves weaned:				
Percent of calves weaned	95	90	85	_____
Percent of calves weaned per cow exposed	90	77	64	_____
Calf weaning weight (lb):				
Avg. weaning weight	550	460	380	_____
Lb of calf weaned per cow exposed	496	352	242	_____
Production efficiency:				
Feed costs per cow:				
Hay cost[a]	$100.00	$100.00	$96.00	_____
Time fed (mo.)	4	5	6	_____
Tons required/cow	2.0	2.5	3.2	_____
Pasture cost[b]	$46.55	$54.00	$48.00	_____
Grazing time (mo.)	6	6	6	_____
Acres needed/cow	1.33	2.00	4.00	_____
Stalk and crop aftermath cost[c]	0	0	0	_____
Grazing time (mo.)	2	1	0	_____
Grain cost	$13.00	$10.00	0	_____
Protein supplement cost	$14.00	$5.60	0	_____
Salt and mineral cost	$12.60	$9.80	$4.00	_____
Total feed costs per cow	$186.15	$179.40	$148.00	_____
Nonfeed costs per cow[d]	$72.50	$48.00	$19.50	_____
Total variable costs per cow[e]	$258.65	$227.40	$167.50	_____
Feeder calf value per cow exposed (75¢ market price)	$338.25	$243.75	$167.25	_____
Returns above—				
Total variable costs per cow[e]	$79.60	$16.35	−$0.25	_____
Feed costs per cow	$152.10	$64.35	$19.25	_____
Breakeven price above—				
Total variable costs per pound[ef]	$0.58	$0.70	$0.76	_____
Feed costs per pound[f]	0.42	0.56	0.67	_____

[a]Hay requirements assume 5, 15, and 25% loss in harvesting, storage, and feeding for excellent, average, and poor producers respectively; hay value is based upon assumed quality of $50/ton, $40/ton, and $30/ton for excellent, average, and poor performance levels, respectively.

[b]Estimated pasture yields are 3 tons, 2 tons, and 1 ton/acre for excellent, average, and poor producers, respectively; costs are direct costs only plus $7.00/acre for fencing and water; costs based upon Purdue Ext. Publ. ID-68, Tables C-19 and C-20.

[c]No charge assessed for grazing of salvage cornstalks or crop aftermath.

[d]Nonfeed costs include: veterinary, breeding, insurance, bedding, hired labor, utilities, fuel, equipment repair, marketing, and transportation.

[e]Total costs, excluding investment overhead and family labor.

[f]Breakeven price assumes entire calf crop sold rather than sale of cull cows and fixed replacement rate of heifers.

Source: Adapted from Trouble-Shooting the Beef Cow-Calf Operation. ID-138, Purdue University.

TABLE 29–2	Estimated Annual Labor Requirement per Animal for Cow Herds or Grazing Operations				
	Beef Cow Herd				
Number	Calf Sold at Weaning (hr/cow)	Calf Sold as Yearling (hr/cow)	**Backgrounding of Steers (hr/steer)**	**Drylot Grazing Steers (hr/steer)**	**Wintering and Grazing Steers (hr/steer)**
Under 40	14.0	23.0	9.0	2.74	11.5
40–80	11.0	17.0	6.6	1.40	7.15
80–120	7.0	11.5	4.8	1.15	5.40
120–200	5.0	8.0	3.5	1.00	4.15
200–500			2.0	0.90	2.65
Over 500			1.6	0.85	2.20

Source: *Beef cow herd data from* Farm Management Guides, MF-266 (1981), MF-358 (1979), *and* MF-360 (1979). *Manhattan: Cooperative Extension Service, Department of Economics, Kansas State University.*

High-quality management and labor are needed during the breeding season, especially when artificial insemination is used, and during calving. During these periods the herd must be checked frequently. Time, labor, and supplemental feed are required during the winter months, but herd care during the winter normally is not a full-time job unless the operation is large. When cattle are on grass, little labor and expense are involved except for breeding and calving periods. Time is needed, of course, for hay harvest and maintenance jobs such as fence repair. The estimated labor requirement for selected cattle operations is presented in Table 29–2.

Some farmers and ranchers in the business of producing feeders do not maintain a cow herd but rather purchase weaned calves weighing 400 to 500 pounds and grow them on grass and cheap roughage until they weigh 700 to 800 pounds. These *stockers* then are sold as heavy feeders to be finished on a high-grain ration. This specialized system avoids the management of a cow herd and requires less labor, but carries higher financial risk. Dollars are invested a shorter time than with a cow herd, but possible fluctuations in cattle prices at the normal times of cattle purchase and sale cause the risk.

See Appendix B for the location of parts and terminology used in evaluation of cows and calves.

29.4   FEED AND OTHER COW-CALF HERD COSTS

During the grazing season in most areas, the only supplemental feed usually required is salt and minerals. Salt and mineral mixes generally are provided free choice in either block form or as loose salt in "salt boxes" placed throughout the grazing area, perhaps near water. Though salt and mineral blocks usually permit adequate consumption, many producers prefer a loose salt-mineral mixture (e.g., six parts trace mineral salt to four parts dicalcium phosphate) because cattle can more easily meet their requirements.

A beef cow requires about 5 tons of forage, on a dry-matter basis, per year. This includes feed for cows, bulls, replacement stock, and the calf to weaning. On the average, about 3 tons of "hay equivalent" per cow is consumed on pasture and about 2 tons

TABLE 29–3	Effect of Month of Birth on 190-Day Weight of Beef Calves		
Month of Birth	Number of Calves	Preweaning Rate of Gain (lb per day)	190-Day Weaning Weight (lb)
February	108	2.07	464
March	710	2.00	451
April	979	1.91	434
May	351	1.86	424
June	124	1.77	407

Source: *Minyard, J. A.,* Proceedings, The Range Beef Cow, *South Dakota State University, Agriculture Research and Extension Center, Rapid City, South Dakota, 1973.*

of forage per cow should be provided during the nongrazing season, as crop residue, hay, or silage. These will vary among areas, depending on climate.

During the winter, for animals scheduled to calve in the spring, it is best to divide the cow herd into three units: mature cows, heifers calving for the first time, and younger replacement heifers. If they are all fed together and if the requirements of the heifers are met, the cows will be overfed. If the ration is designed for the mature cows, the first-calf heifers and replacement heifers will be underfed. The result would likely be inadequate growth and there may be some delay in initiation of the estrus cycle in the case of replacement heifers. For this reason, a specialized business of developing heifers, from their own weaning through insemination through early pregnancy, has developed in the Corn Belt.

The nutritional goal for the cow herd is to maintain optimum productive performance. Cows need to be bred and all conceive within a 60-day breeding season so the calves will be born early (assuming early spring calving). The cow will more likely be on lush grass during lactation and produce heavier calves at weaning. The data in Table 29–3 illustrate the influence of birth date on 190-day weight of beef calves born in the spring under northern plains conditions.

Each region of the United States has its desirable or advantageous calving times. In areas of severe winter storms, a late spring calving date may mean fewer death losses from storms, even though the calves will be lighter at weaning. Fall calving has certain advantages—low labor and housing requirements at calving time and the calf is large enough by spring to utilize grass effectively. Also, fall-dropped calves are available to feedlots when there is a need, so there may be a sale price advantage.

Table 29–4 shows the nutritional regime needed in each of four periods in the beef cow year. In *period 1*, the first 82 days after calving, the cow is lactating heavily, undergoing uterine involution (when the uterus returns to its normal size and physiological state), and preparing for rebreeding. Thus, her nutrient needs are the highest of the year. Because the calf receives most of its nutrients from milk during this period, adequate energy and protein for the cow are essential to permit adequate milk production and calf growth. Lactating cows should receive sufficient protein from young forage, but additional energy may need to be provided from supplemental grain feeding.

TABLE 29–4	The Nutritional Requirements for a 1,100-Pound Cow in the Four Periods of a Cow's Year			

	Period			
	1 *Postcalving* *(Early Lactation)*	*2* *Early Gestation* *and Lactation*	*3* *Gestation*	*4* *Late Gestation* *(Precalving)*
Days in period	82	123	110	40
TDN (lb/day)	13–15[a]	11–12	9.5	11.2
Total NE$_M$ (Mcal/day)[b]	15.49	12.34	9.84	12.96
Total metabolizable protein (lb/day)	1.77	1.33	1.05	1.37
Calcium (g/day)	34	27	20	28
Phosphorus (g/day)	24	18	15	18
Vitamin A (IU/day)	38,000	38,000	25,000	26,000

[a]*Depends on milking ability, age, and condition.*

[b]*Includes 0.34 Mcal NE$_M$/lb of milk produced.*

Sources: *Adapted from* Nutritional Requirements of Beef Cattle, *7th ed., National Research Council, 1996; and* Cow Herd Health, *Kansas State University, 1977.*

During *period 2*, the cow is in the early stage of pregnancy and still lactating. Assuming she calved in the spring, she should be gaining weight to prepare for the coming winter. Over time, the calf will depend less on milk and consume more grass. Normally the cow is on grass and no supplemental feed is required unless a drought or unusual situation exists.

In *period 3*, the calf normally is weaned and the cow is in midgestation. She only needs to maintain herself and provide nutrients for the developing fetus. Nutritional needs are lowest, in terms of volume, during this period, and lower quality forages can be fed during this time.

Late gestation, or about 50 days prepartum, is considered the second most important period; the cow is preparing for lactation, and this is the time during which 70 to 80 percent of the total fetal growth occurs. As a guide, any underconditioned cows should receive sufficient nutrients to be gaining weight.

In addition to the factors mentioned, nutritional needs of the cow are influenced during these periods by cow size, level of milk production, condition of the cow, and environmental conditions.

Monitoring cows for body condition has become more widely used as a tool to reduce the incidence of extremely under- or overconditioned cows. (See also Section 15.2.) Cows are visually evaluated early in and through pregnancy, with emaciated cows receiving a score of 1 and obese cows a score of 9. An extensive study by Spitzer et al.[2] indicates that the well-fleshed but not overconditioned beef cow, with a score of 4, 5, or 6, gives birth to a heavier calf.

These researchers also found that cows with higher condition scores at calving and cows gaining weight after calving resulted in improved reproductive efficiency. Benefits were more cows in estrus and more cows pregnant in the 60-day breeding season. Also, calves nursing from cows that gain weight after calving were heavier at weaning time.

---

[2] Spitzer, J. C., et al. *J. Anim. Sci.* 73 (1995): 1251.

29.5 WEANING AND MANAGEMENT OF FEEDER CALVES FOR SALE

Calves are normally weaned 6 to 8 months after birth. In general, 205 days is a weaning age target used by many in the industry, and some producers like to compare their 205-day weights with those of other herds. Weaning is a stressful time for calves; they are separated from their dams for the first time. In addition, the heavier calves at weaning often are moved directly to a feedlot and that, too, is stressful to the calf.

Negative effects of these stresses can be minimized and calves will do better in the feedlot if they have been fully preconditioned 30 to 45 days before weaning. **Preconditioning** might include (1) castration and dehorning if needed; (2) worming; (3) treatment for grubs; (4) implanting with a growth stimulant; (5) vaccination for diseases such as infectious bovine rhinotracheitis (IBR), bovine virus diarrhea (BVD), bovine respiratory syncytial virus (BRSV), leptospirosis, pasturella infection, and blackleg; and (6) adapting to a feedlot finishing ration.

If these health and management procedures are accomplished 30 to 45 days before shipping, the calves will be protected from the stress of shipping and of the feedlot. Illness and loss can be significantly reduced. For more details on vaccination programs for preconditioned calves, see Box 29–1.

Fast-growing calves, which are relatively heavy at weaning, will likely go directly from the producer to the feedlot. They may weigh 600 pounds at weaning and have the genetic capacity to gain rapidly to a finished weight of 1,000 to 1,200 pounds.

Feeder cattle should be sorted by weight. When heavy, large-frame calves are mixed with more compact calves, the more compact calves will reach the Choice grade sooner and at lighter weight. By the time the large-frame animals reach the Choice grade, the compact animals may be overfinished (beyond Yield Grade 1 to 2).

Calves that are light at weaning—400 to 500 pounds—are more likely to be *backgrounded* before going to a feedlot. Backgrounding consists of feeding a growing ration to calves to weights of 650 to 700 pounds, a weight range better suited for the finishing feedlot. As discussed in Section 8.3, backgrounding provides the opportunity to feed homegrown feeds, take advantage of winter labor, and benefit from increased income through efficient weight gains made during this period (see Color Plate C). These calves should, of course, also be preconditioned.

> *Preconditioning*
> Preparing feeder calves for movement from their home ranch or farm to a feedlot by weaning, vaccinating, and other postweaning practices.

29.6 FEEDER CATTLE MARKETING

Where feeder cattle are sold and moved directly to a feedlot, there are relatively few marketing problems except for negotiating price and other sale terms and arranging transportation. The same generally is true when large lots of cattle move through special feeder auctions located in or adjacent to feeder producing areas. Demand by consumers and retailers for product traceback capability, in time will require animal identification and record of ownership from birth to product. This will reduce or eliminate a historic problem in movement of feeder cattle: the mixing of cattle from numerous sources and their movement through a series of auctions or dealers. This has caused even more stress on animals, and has led to a higher incidence of sickness and death in some feedlots.

Heifers not saved for breeding move with the steers to feedlots. The "inferior calf" that is culled usually will be sold through the normal channels, but will bring less money. Experience is necessary when purchasing feeders to detect inferior calves mixed with a larger group. They should be sorted out and rejected or purchased at a reduced price. (See Chapter 8.)

BOX 29–1

Alabama Value-Added Preconditioned Calves (Program A)

Preconditioned calves are calves that have been pre-pared for the stresses of weaning, shipment, and placement into commercial or custom feedlots. These calves are less likely to suffer from respiratory diseases and other ailments, and go onto feed more quickly, with cheaper, faster gains. Buyers are willing to pay more for preconditioned calves. The following vacci-nation program is one of three utilized in the Alabama Value-Added Calves program which has strengthened the link between buyers and feedlot operators.

Program A (Process before and at weaning, ship later)

● *1st vaccination 3 to 4 weeks before weaning*

  1. IBR-PI₃-BVD-BRSV
     Approved as safe for use in calves nursing cows.
  2. *Pasteurella haemolytica* leukotoxoid

  3. Brucellosis "calfhood vaccination" for heifers
     Needed to ship to some states
  4. 7-way clostridial (if needed)
  5. Leptospirosis (if needed)
● *2nd vaccination at weaning*
  1. IBR-PI₃-BVD-BRSV modified live
  2. *Pasteurella haemolytica* leukotoxoid
  3. 7-way clostridial (if needed)
  4. Leptospirosis (if needed)
● *Ship 4 to 6 weeks after weaning*

The figure below illustrates the diminishing immu-nity provided by the first vaccinations and the bene-fit of providing a second vaccination at weaning.

Source: Adapted from "Proper preparation of calves for shipment is essential." *Feedstuffs*, May 15, 1996 (authored by J. Floyd, extension veterinarian, Auburn University).

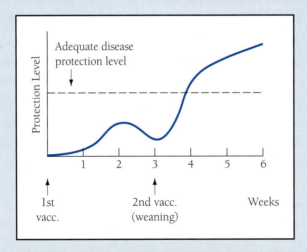

Cows culled from the herd because of age or other problems have generally gone directly to processing, but the increasing demand for hamburger and recent re-search on cull cows suggests a 3-month feeding period can increase their value suffi-ciently to make such feeding a profitable venture.

29.7   COW-CALF HERD MATING SYSTEMS

A review of mating systems, described in Chapter 20 (Sections 20.6 and 20.7) may be beneficial (though swine are used as the example) and 20.7 before proceeding with this section. Crossbreeding is very common in feeder production; heterosis can benefit both the feeder producer and the feedlot operator.

The two-breed cross has been popular in beef cow herds. A good example is mating Hereford bulls to Angus females. The calves produced are polled, have a black body and white face, and commonly are referred to as "black baldies." This system provides hybrid vigor only in the calf; replacement straightbred females generally are purchased from other breeders. This is a simple system, and the crossbred calf is in demand.

Two-breed and three-breed rotational crosses described in Sections 20.6 and 20.7 are increasingly popular because they provide heterosis in the replacement heifer that joins the cow herd—more rapid growth and milk production from that animal occurs.

In the two-breed rotational crossbreeding program, bulls of two breeds generally are maintained and each is mated to cows of the opposite breed. Offspring heifers (F_1 generation) saved for breeding are then mated to a bull of the breed opposite that of the heifer's sire (the same breed as their dam). For example, assume the two sires are Angus and Hereford. Heifers from an Angus sire and a Hereford dam are mated to a Hereford bull. Their offspring (F_2) are mated to an Angus bull. After the first generation, the replacement heifers are then mated to a sire of the breed genetically *least* related to the heifer, and some hybrid vigor results in each successive generation.

One could also describe this arrangement as maintaining two "herds," with herd 1 headed by an Angus bull and herd 2 headed by a Hereford bull. Heifers produced in herd 1 are moved to herd 2 for breeding, and vice versa. This system produces both a crossbred cow and a crossbred calf and achieves about 67 percent of maximum heterosis. A three-breed rotational cross will achieve about 86 percent of maximum heterosis.

Bull selection is very important, as the bull represents 50 percent of the gene source and 50 percent of the opportunity for genetic improvement in all offspring. Performance-tested bulls carry measures of their genetic value or potential. Major performance traits to consider are birth, weaning, and yearling weights, and growth rate and feed efficiency "on test," at the age their offspring would be in a feedlot. Bull feeding tests are normally about 140 days in duration.

A bull having a birth weight of 70 to 90 pounds, a weaning weight of 600 pounds or more at 205 days, a yearling weight of 1,100 or more, and gaining 3.5 pounds daily or more on test would be considered to have an excellent performance record. Such bulls usually sell for a premium and should sire vigorous and rapidly growing calves; typically they are used on mature and large-frame cows. With young heifers or small-frame cows, there is a risk that the calves would be large enough to cause some difficulty in parturition.

Performance-tested sires, used in a planned crossbreeding program, can provide both permanent genetic improvement and heterosis in the immediate next generation.

At the U.S. Meat Animal Research Center at Clay Center, Nebraska, research has been directed to classification of breed groups for crossbreeding (Table 29–5). Those crosses with more Xs had more rapid growth and were heavier at maturity, were more lean, reached puberty at a younger age, and produced more milk.

29.8   DISTRIBUTION AND TYPES OF EWE FLOCKS

Although the USDA has not collected annual data on numbers since 1999, it appears that sheep and lamb inventory has continued to decline, from about 13 million in 1991 to less than 8 million in 2002. Several factors have contributed to this decline: low per capita meat (lamb and mutton) demand, availability and price of artificial fibers to replace wool, problems with predators, labor availability and cost, and a decreasing processing and marketing infrastructure. However, in 2003 there remain an estimated 63,000 ewe flock or lamb feedlot producers.

TABLE 29–5	Breed Crosses Grouped by Biological Type on the Basis of Four Major Criteria			
Breed Group	**Growth Rate and Mature Size**	**Lean-to-Fat Ratio**	**Age at Puberty**	**Milk Production**
Jersey-X	X	X	X	XXXXX
Hereford-Angus-X	XX	XX	XXX	XX
Red Poll-X	XX	XX	XX	XXX
South Devon-X	XXX	XXX	XX	XXX
Tarentaise-X	XXX	XXX	XX	XXX
Pinzgauer-X	XXX	XXX	XX	XXX
Sahiwal-X	XX	XXX	XXXXX	XXX
Brahman-X	XXXX	XXX	XXXXX	XXX
Brown Swiss-X	XXXX	XXXX	XX	XXXX
Gelbvieh-X	XXXX	XXXX	XX	XXXX
Simmental-X	XXXXX	XXXX	XXX	XXXX
Maine-Anjou-X	XXXXX	XXXX	XXX	XXX
Limousin-X	XXX	XXXXX	XXXX	X
Charolais-X	XXXXX	XXXXX	XXXX	X
Chianina-X	XXXXX	XXXXX	XXXX	X

Source: *U.S. Meat Animal Research Center, Clay Center, NE.*

Almost 82.5 percent of the 5.31 million *breeding* sheep and replacement lambs in 1999 were located in the 17 westernmost states. The 10 leading states were ranked as follows:

1. *Texas*
2. *Wyoming*
3. *California*
4. *Utah*
5. *Montana*
6. *South Dakota*
7. *Idaho*
8. *New Mexico*
9. *Colorado*
10. *Iowa*

Figure 29–6 shows typical ewe flocks in the leading western states.

U.S. sheep production is classified into two management types: *range flocks* and *farm flocks*. Although exceptions exist, a north-south line through the plains states divides the United States into the eastern farm flock management area and the western range area (Figure 29–7).

Sheep are better suited than cattle for those range areas that have less grass and more brush and shrubs because sheep better utilize some of those latter plant species (see Color Plate B). Over two-thirds of the commercial range sheep producers also handle cattle, and some also graze goats. These animals differ in the vegetative species they use most efficiently, so a mix of animal species will use the range more effectively. The Edwards Plateau region of Texas is an example; the complementary effect on forage utilization from grazing cattle, sheep, and goats on the same land is significant.

Most range lambs are sold following weaning, for slaughter if summer range conditions have provided adequate nutrition, or for feeding. Because of lack of available

Figure 29–6
Most lambs are produced in flocks like these, in the mountain states and on the western plains.
(Courtesy of Oregon State University and *Western Livestock Journal*.)

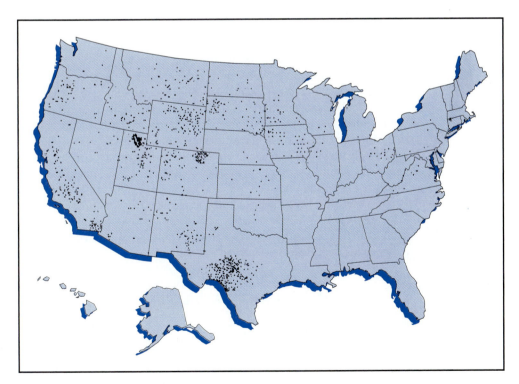

Figure 29–7
An approximation of the distribution of the 7.24 million total U.S. sheep population in 1999. One
dot equals 5000 head.
(*Source*: Adapted from USDA data.)

Table 29–6	Farm Flock Production Averages and Factors Affecting Profit

Seven-Year (1994–2001) Flock Averages

Conception rate	95.48%
Lambs born/ewe exposed (to the ram)	1.81
Lambs sold/ewe exposed	1.62
Pounds of lamb sold/ewe exposed	201.43
Weight/lamb sold, pounds	124.6
Price/cwt for lambs sold	$84.11
Feed cost/lamb sold	$24.17
Pounds of feed/pound of gain	3.01
Death loss, birth to market	10.71%
Pounds of wool/ewe sheared	9.1

Calculated Impact of Key Factors on Profit per Ewe

5% change in conception rate	$ 6.25
0.10 more lambs sold/ewe	$ 7.38
5 pounds heavier lamb at market	$ 4.56
$5 higher lamb price per cwt	$10.00
$0.25 higher price per pound of wool	$ 2.00
Increase ewe longevity from 5 to 6 years	$ 3.20

Source: *Adapted from flock data, Minnesota Community and Technical College, Pipestone.*

grains, range operators seldom finish their lambs in drylot. Almost 49 percent of all sheep marketed in 1999 were in predominantly feedlot states: California, Texas, and Colorado.

Corn Belt and eastern states generally are considered in the farm flock production area. Concentrate feeding is more prevalent, especially in finishing lambs but also in maintaining the ewe flock. Number of sheep per operation and acreage per operation are considerably lower in the farm flock area, most flocks consisting of less than 100 head. Because grains are in greater supply, most producers start lambs on a creep-feeding program at a young age and market a finished lamb at 3 to 5 months of age. Many will be marketed at 110 pounds at about 130 days of age. Few lambs from these flocks are sold as feeders. Ewe flocks are generally on pasture 4 to 8 months each year and during the remaining time utilize forage crop residue or are in drylot. Table 29–6 provides production data from a northern Corn Belt flock and the calculated impact of production improvements that might be achieved in flocks.

29.9 EWE FLOCK INVESTMENT AND LABOR

Capital investment needed for a ewe flock enterprise will vary. Where land, shelter, and facilities already exist and the sheep will utilize underemployed labor plus crop residue, the only major cost will be the ewes. In contrast, the purchase of land, construction of buildings, hiring of labor, and purchase or harvest of feed for the enterprise would result in significant cost. The sheep industry has long been characterized as one that can be entered with a lower initial investment, relative to value of product produced, than is the case with most other livestock industries. The main factors affecting profitability are *weaning percentage, death loss, age at culling,* and *production costs* (Table 29–6).

TABLE 29–7	Estimated Annual Labor Requirements per Ewe Under Farm Flock Conditions	
Number of Ewes	**Hours per Ewe**	
Under 35	7	
35–75	6	
75–100	5	
100–200	4	
200–300	3	
300–500	2	
Over 500	1.5	

Source: *Kansas State University Farm Management Guide MF-421, 1982.*

Production goals that ensure high returns per ewe[3] are

- High conception rates   95 + percent
- High lambing percentages   175 + percent
- Low lamb mortality   10 percent or less
- A strong marketing program
- High quality and yield of wool
- Longevity of breeding stock

Total labor needs, however, can be high in relation to value of product. Labor requirement varies with the stage in the ewes' reproductive cycle and also is influenced by the intensity of the sheep operation. Lambing is the time of highest labor need; regular observation is of utmost importance. An estimate of annual labor requirements under farm flock conditions is shown in Table 29–7.

Large sheep units in the western and some plains states involve vast acreages and large sheep numbers. Most of these units utilize government-owned grazing land, leased to the producer. Because land is not privately owned, no permanent facilities may exist, and because the area grazed may be vast, the band of sheep requires a mobile caretaker or herder. One herder usually takes care of a band of about 1000 ewes for the duration of the grazing season. Sheep herders have been scarce in recent years and in strong demand, especially since the United States placed a restriction on immigration of Spanish Basques as herders.

29.10   FEED AND OTHER EWE FLOCK COSTS

Feed costs can account for up to 60 percent of the total costs of a sheep operation. Body condition scoring and adjustments in rations should be made to ensure that most ewes stay within accepted weight ranges (Figure 29–8). The ewe can be expected to lose 5 to 7 percent of her body weight due to lactation demands and regain weight and body condition during late gestation.

---

[3] Adapted from North Central Regional Extension Publication 240 (revised 1993).

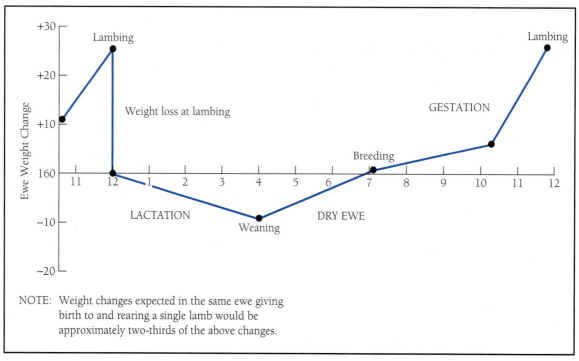

Figure 29–8
Typical weight changes for a 160-pound ewe giving birth to and rearing twins.
(*Source: Sheep Industry Handbook*, SID Nutrition.)

Guidelines for determining annual feed needs and feed costs for a 155-pound ewe in average condition at breeding time are as follows:[4]

- 7 months of pasture
- 700 pounds of high-quality legume or legume-grass hay
- 45 pounds of grain before lambing (flushing and late gestation)
- 90 to 120 pounds of grain and supplement for a 60-day lactation period (depending on nursing a single lamb or twins)

Feed costs for ewes managed under range conditions are lower. Lease grazing rates for federal ranges, which include land managed by the Bureau of Land Management (BLM) and National Forest Service (NFS), are a factor in this lower feed cost. Approximately 20 percent of the western range producers that handle 50 percent of the West's commercial ewes use some BLM or NFS land.

Because the majority of feeder lambs are produced on the open range, some distance from any headquarters facility, it is not customary to provide lambs a creep ration.

Other costs in addition to feed, including veterinary expense, vaccinations, utilities, marketing costs, ewe replacement costs, farm overhead costs, and interest, amount to about $37 per head, based upon mid-1990 costs of production.

---

[4] Adapted from North Central Regional Extension Publication 240.

29.11   SHEEP BREEDING SYSTEMS

Sheep breeds are often classified as ewe breeds or ram breeds. *Ewe breeds* are generally white-faced sheep that produce the finer wool. Examples are the Merino and Rambouillet (see Chapter 21). These breeds are adapted to range conditions and have high reproductive efficiency, wool production, milking ability, and longevity. Replacement ewes, both in range areas and for some farm flocks, generally are selected from these breeds or from crosses involving these breeds. *Ram breeds* tend to be of the meat type and are selected largely on growth rate, efficiency of gains, and carcass quality.

Traditionally, range sheep producers have perpetuated the ewe breeds for their flocks by using both ewes and rams of the ewe breeds, or at least ewe-breed rams on a sufficient portion of the flock to produce straightbred ewe lambs for flock replacements. Farm flock producers generally purchase replacement ewes of the ewe breeds from range producers and mate them to a sire of one of the black-faced ram breeds. These matings will tend to produce lambs with good growth rate and carcass quality, yet the ewes will produce relatively fine wool, will have clean, open faces, and will be adapted to grazing and open pasture conditions.

Sires from the Finn breed became popular during the 1970s to increase prolificacy in sheep flocks. Research at the USDA Meat Animal Research Center at Clay Center, Nebraska, indicated that Finn-cross ewes produced 134 lambs per 100 ewes bred, compared to 75 lambs per 100 ewes bred in Rambouillet-cross ewes.[5] More recently, producers have tended to reduce the percentage of Finn sheep by using Finn-cross rams, resulting in one-fourth Finn replacement ewes.

A breed known as the Dorper has been introduced into the United States via embryos brought from Africa and implanted in selected ewes. The Dorper breed is popular in Africa and known for its hardiness and adaptability to desert conditions, light fleece, and meat production. See Section 21.6 for other data on sheep breeds.

Some managers who have gained experience and achieved high production in the flock with once-a-year lambing have adopted the *accelerated lambing program*. This program involves ewes lambing more frequently than once each year, thereby providing a more uniform supply of lambs throughout the year. It allows the manager to take advantage of producing lambs during months when feed is more abundant and cheaper, and marketing when prices are higher. Accelerated lambing requires prolific ewes that can breed successfully at any season of the year. *Synchronization of estrus* also can be used to stimulate ewes to initiate the estrous cycle outside of the normal breeding season.

One system consists of *three lamb crops in two years*. All ewes would have the following 2-year schedule:

Mating	Lambing
May	October
January	June
September	February

You can see how accelerated lambing increases the intensity of production and the need for labor and management skills. A successful program results in more lambs and greater returns than once-a-year lambing.

---

[5] Laster, D. B., et al. *J. Anim. Sci.* 35 (1972): 79.

Another practice to increase the reproductive rate of the flock is to breed ewe lambs during their first year of age rather than wait an extra year. With excellent nutrition and management, most ewe lambs born early in the year develop sufficient size and will reach puberty at 9 to 10 months of age. Special nutrition and management must be provided to these younger ewes throughout pregnancy, lambing, lactation, and after lambing to ensure that they reach mature size.

Sheep breeders are well acquainted with the desired conformation. The parts of the sheep and their common terminology are presented in Appendix B.

29.12   SHEEP MANAGEMENT PROBLEMS

Predator control is a major problem in the sheep industry. The coyote is the most costly predator in the western and plains states, attacking and killing both ewes and lambs. In highly populated areas, stray dogs may inflict high death losses. Electric fences, night penning of sheep, trained dogs, and various predator-frightening devices have been used to reduce predation in farm flock areas, but range producers have limited means of preventing predation.

External parasites are a significant problem in some sheep units, especially farm flocks handled on limited acreage or in drylot (see Section 12.7).

The availability of experienced and trained personnel is a limiting factor in both range and large confinement enterprises.

Lamb death losses of approximately 15 to 20 percent are not uncommon in the industry. More intensified sheep enterprises can result in an increase in the occurrence of respiratory diseases, enterotoxemia, and also reproduction diseases that may affect the survivability of embryos or lambs at birth.

29.13   LAMB MARKETING

Most feeder lambs are sold at 60 to 75 pounds directly to feeders, or through dealers or auctions. Long distances between the range and many feedlots result in high transportation costs and considerable shrinkage, as well as some illness. Especially in farm flocks, lambs are given supplemental grain, and large-framed lambs are sold for slaughter at 110 to 130 pounds.

Some midwestern producers have marketed finished lambs through a marketing pool, to provide a larger group that can be transported more economically to a processor. In the eastern states some processors specialize in serving the ethnic market and provide an outlet for finished lambs from farm flocks in that region.

The low volume in recent decades has resulted in a very limited marketing and processing infrastructure for lambs. Processors have lacked a dependable Supply to service customers who demand lamb, and many long-time producers have become further away from any processor. The major problem for many midwestern producers has not been negotiating price; it has been to find a market.

No doubt as a consequence, a joint venture between the Mountain States Lamb Cooperative, a producer group with members in 10 western states, and B. Rosen & Sons, Inc., with plants in New York and Colorado, was formed in early 2003 to provide an integrated supply system from the ranch or farm to food service and retail. The joint venture, called MSR, indicated plans to offer a series of branded products, including a natural (no antibiotic and no hormone) line and an organic line, both source verified from farm to package.

The salvage value of cull ewes and rams is low in comparison to that of cows and sows. Therefore, it is financially advantageous for a sheep producer to maintain ownership of the ewe as long as she is regularly giving birth to and weaning healthy lambs.

QUESTIONS FOR STUDY AND DISCUSSION

1. Why is the production of feeder cattle for feedlots a specialized business? Describe the enterprise.

2. In recent decades the proportion of feeder cattle raised in the southeastern states has increased. Why has that occurred? List some of the unique features of feeder cattle production in the Southeast.

3. What are the major feeds used for a cow herd and in the production of feeder cattle? Estimate the weight of total feed required to produce a calf to weaning age.

4. What does preconditioning of calves for sale usually include?

5. Name and describe two mating systems that are used in calf production.

6. What percent of maximum heterosis can be achieved with the use of a three-breed rotational cross-mating system?

7. In what geographic region of the United States are most sheep raised? Why?

8. Describe a typical mating system used in range sheep production. What is meant by "ewe breeds" and "ram breeds"?

9. List three of the most important factors that affect the amount of profit earned in sheep production.

10. After lambing, how much weight loss would be expected in the ewe, and when should her body condition be restored?

11. Explain how an accelerated lambing program can be established. What are its advantages and requirements?

12. As a sheep breeder, would you try to breed ewe lambs in their first year of age? If so, why?

30

The Business of Dairying

Dairy production enterprises in the early twenty-first century range from a few milk cows that are part of a general farming operation to a herd of 3,000 or more cows where the sole business focus is milk production. The milk produced may be destined for packaging as fluid milk or for processing into ice cream, yogurt, cheese, or other products (see Color Plate A). Milk as a product is fully discussed in Chapter 31.

Dairy production is a long-term and a relatively high-investment business. A milk producer seeks a monetary return for labor, capital, feed, management ability, and the risk assumed.

This chapter discusses some of the factors that influence the profitability of dairying including location, size, investment needed, labor, feed, type of product desired, and other factors.

Upon completion of this chapter, the reader should be able to

1. List a series of characteristics or advantages of the dairy business.
2. Chart the recent trends in numbers of cows and average milk production per cow.
3. Explain why dairy cow herds are more common in certain regions of the United States.
4. List the major costs in a dairy production unit.
5. List major management practices necessary for a profitable dairy operation.
6. List the major breeds of dairy cows and dairy goats.
7. Chart a typical lactation curve for a dairy cow, indicating approximate pounds per day at several points on the curve.

30.1 CHARACTERISTICS AND TRENDS

To better understand modern dairying, it is helpful to list some of the characteristics of dairying. These include the following:

1. A continuous supply of personnel is necessary because the milking herd must be milked at least twice daily every day of the year, and the personnel must be dependable and high quality.
2. Dairying provides a steady source of income. Milk sold usually is paid for biweekly, which enhances cash flow.
3. Dairying is considered a long-time commitment because it requires a large investment per cow in building and equipment. Higher efficiency requires labor-saving equipment.

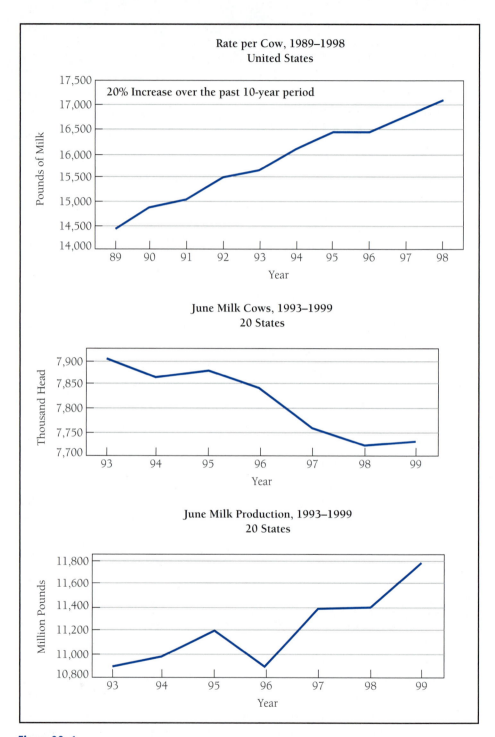

Figure 30–1

Milk production per cow continues to increase, the result of genetic selection for production and high-quality herd management, and the number of milking herds continues to decline. (*Source*: USA, NASS.)

4. Milk is a relatively perishable product and must be kept cold and rapidly transported to the processing plants. Great care is given to quality control in the production, handling, processing, and preservation of milk and dairy products.

5. Milk is a highly desired and respected food. It contains needed nutrients and is often described as "nature's most nearly perfect food." This lends stability to the demand for milk.

6. The dairy business is one of the most stable livestock enterprises, from an economic standpoint. Milk production does not fluctuate in long-term cycles as does beef, lamb, or pork production. Prices received for milk are more stable because of the more constant supply and demand, and also because of federal marketing orders that help to stabilize price.

7. The high-producing dairy cow converts energy and protein in grains and high-quality forages to human food quite efficiently, compared to other classes of domestic livestock.

The number of dairy cows on U.S. farms has steadily decreased from the peak of about 25 million in 1940 to less than 9 million in 2002. During this period, the number of farms with dairy cows as a source of income decreased from about 4.6 million to less than 100,000. For every 46 dairy farms in 1940, now there is only 1. However, the number of cows per farm and the productivity per cow have increased dramatically (Figure 30–1). Cows on DHI records, the national dairy record system, in 1940 averaged about 8,100 pounds of milk per lactation. Today DHI herds produce an average of over 22,700 pounds per cow (Table 30–1).

Fluid milk (whole milk) is now the product of commercial dairy farms, though a substantial amount of this is converted to other products after reaching the processing plant. Payments for both fluid and manufacturing milk are made according to a *component pricing system* (which includes pounds of protein, butterfat, and other solids).

Of interest in this discussion is the average production and composition of milk produced by cows of major dairy breeds and the relative number of cows of each breed. Table 30–1 provides breed summaries from all herds enrolled in the National Cooperative Dairy Herd Improvement Program in 2002.

Table 30–1	**Official Averages of Breeds Enrolled in the National Cooperative Dairy Herd Improvement Program, 2002**					
	Ayrshire	*Brown Swiss*	*Guernsey*	*Holstein*	*Jersey*	*Red & White*
Herds (no.)	144	292	189	18,835	1,127	33
Cows/herd	50	49	46	178	139	54
Total cows	7,226	14,192	8,616	3,356,044	156,638	1,797
Milk/cow (lb)	15,878	17,847	14,931	21,919	15,652	20,086
Protein (%)	3.16	3.36	3.37	3.05	3.59	3.02
Fat (%)	3.89	4.05	4.54	3.64	4.62	3.67

Source: *Adapted from USDA, ARS. Summary of DHI Herd Averages, 2002.*

Other characteristics of milk are important to the consumer. Flavor of milk can be altered by feed and management practices. (See Section 16.6.) Color of milk may vary among animals, and certainly does among breeds. Guernsey milk, for example, has a richer yellow color, preferred by some people and promoted by some who produce and sell Guernsey milk.

30.2   GEOGRAPHIC DISTRIBUTION, SIZE, AND TYPES OF UNITS

What is the best location for a dairy operation? This is influenced by markets, feed supply, labor, and perhaps climate.

Traditionally, closeness to a market—consumers or processors—has been a major consideration. This is because milk is relatively perishable compared to other foods and, because it is bulky (about 87.5 percent water), transportation costs are high. However, highly insulated milk-hauling tank trucks and the interstate highway system do allow moving milk hundreds of miles. (See Figure 30–2.)

Ample feed supply is essential for dairying. Dairy cows in milk have a high energy requirement and use large quantities of both grain and high-quality forage. Forages are usually grown fairly close to the dairy because they are bulky and expensive to transport.

Climate influences both feed supply and the productivity of dairy cows. All popular breeds of dairy cattle adapt well to cold weather. However, temperatures averaging over 85°F for several days, with high humidity, will decrease feed consumption and milk production. These effects can be minimized by good management, including shade and, in the case of confinement units, air movement and even sprinklers. These are discussed briefly in Chapters 6 and 11.

Dairying requires expensive equipment to meet sanitation requirements and minimize labor, and also requires considerable knowledge and skill. Therefore a dairy farm enterprise must be large enough to utilize the equipment and hold down per unit cost of production.

On many dairy farms the farm family supplies most of the labor, perhaps supplemented with hired labor who also help with other farming operations. However, large, specialized units of several thousand cows are dependent on hired labor. Such units are more prevalent in states such as California, Idaho, Arizona, Texas, and Florida, but exist in many states. Cost of labor varies considerably, according to wage rates and types of industries in the community.

Because labor is a significant portion of the costs in dairying, there has been much incentive to increase labor efficiency. Milking parlors, equipped with pipeline and automatic takeoff milking systems that route the milk directly to stainless steel bulk tanks, are an example (Figure 30–3).

The number of cows needed for an efficient operation depends, of course, on the volume of milk produced per cow, as well as on the degree of mechanization.

Table 30–2 shows the approximate number of dairy production units in each of several size groups in 2002. The data, plus the fact that there are significant economies of scale in dairy units, suggest that further concentration of milk production into fewer units will occur in future years.

For example, the 613 largest herds enrolled in DHI presented in Table 30–2 averaged 1,875 cows per herd and over 22,700 pounds of milk in 305 days. Their milk yield was 1,500 pounds more per cow than the 21,211 pound average of all DHI herds.

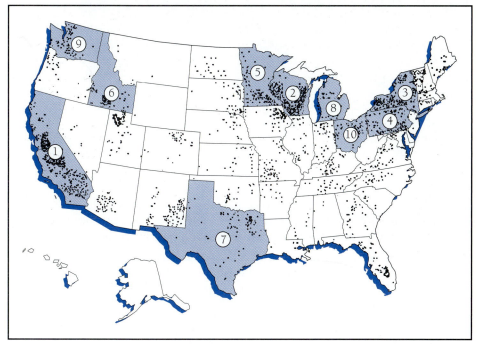

Figure 30–2

Although most dairy herds are fed stored feeds in confinement systems, some continue to utilize pastures and "intensive grazing systems." The lower map shows the 10 leading states in total milk production in 1999. Suburban population growth and expansion, and lower investment costs per cow has caused more and larger herds to relocate in northwestern states such as Idaho and Washington and in the midwestern states such as Indiana.

(Photo courtesy of Holstein Association and *Milk Production*; map source USDA, NASS, July 1999.)

Also, USDA data indicated that there were about 97,500 herds and less than 9 million dairy cows in 2001. Although the number of operations with herds above 500 cows milked 35 percent of all cows, they produced 39% of the milk.[1]

---

[1] *Source*: USDA, NASS.

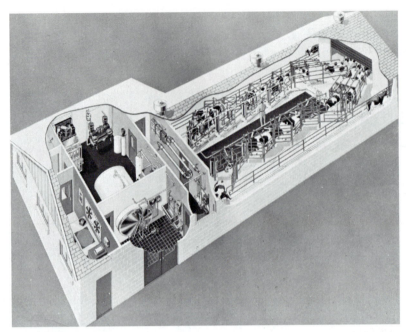

Figure 30–3
Milking parlors contribute to efficient use of labor in dairying.
(*Source*: WestfaliaSurge, Inc.)

Table 30–2	Sizes of Dairy Herds Enrolled in National Cooperative Dairy Herd Improvement Program, 2002			
	Number of Herds		**Percent of Total Cows**	
Herd Size	**1998**	**2002**	**1998**	**2002**
Less than 50	10,897	7,968	9.3	6.6
50–99	15,293	11,732	24.9	18.5
100–199	6,164	5,344	19.5	16.4
200–299	1,458	1,439	8.2	7.9
300–499	1,017	1,145	9.1	9.9
500–999	780	926	12.7	14.6
1,000 or more	429	613	16.3	26.0
All herds	36,038	29,167	100.0	100.0

Source: *Adapted from USDA, ARS. DHIA Summary of Herd Averages, August 1999 and 2002.*

Note in Table 30–2 the movement toward larger herds in the 3-year period between 1998 and 2002. Because the larger herds can make more efficient use of management talent, expensive equipment, and labor, they enjoy a competitive advantage over most smaller herds. An example is one 2,700-cow operation where cows are milked three times per day. The milk flows by pipe from the cow through a rapid chiller and directly to a parked tanker trailer. A truck tractor shuttles the trailers to a processing plant 70 miles away. One manager supervises about 25 employees (about 1 employee per 100 cows). All cows are milked in a "double 30" milking parlor handled by three workers each shift. The other employees handle feeding, insemination, calving, and moving the cows to and from the parlor.

As discussed earlier, because of expensive equipment and because a higher producing herd is not quickly or easily put together, dairying must be considered a long-term, high-investment business. The decision to milk cows must be accompanied by (1) the intention of dairying for a span of years, and (2) a decision to milk enough cows and produce enough milk to be efficient.

Cows are a major investment. Money is needed for buying or raising replacement heifers. Credit usually can be obtained for partial financing of a herd of cows, but bear in mind that a cow has a limited life, contrasted to real estate.

Most feed may be raised on the typical dairy farm but still represents considerable investment. Purchased supplements represent a small proportion of feed costs on most dairy farms, but the outlay of cash is significant for a large herd.

Generalizations on costs of buildings and equipment needed for profitable dairying certainly do not apply to every operation. However, facilities for a 200-cow unit with a double-10 milking parlor, refrigeration tanks, and accessory equipment approach $200,000 in the late 1990s. This does not include housing, feed-handling equipment, and other such necessary items. Economies of scale apply primarily to the milking facilities; investment in housing and feeding facilities are related almost directly to the number of cows.

Money for general operating costs—veterinary expenses, equipment repairs, power, and other such items—can be significant in a large operation. A business must provide a return on investment, as well as return for labor, management ability, and risk, in order to be fully satisfactory. The income from dairying is relatively unique in that it is continuous and provides good cash flow. Milk is picked up daily or on alternate days, and payment is usually made biweekly or monthly.

30.3   LABOR

Because labor represents a higher proportion of production costs in dairying than in most other livestock operations, it is important to manage the labor for top efficiency.

Labor on a dairy farm must be of high quality and well trained. A worker must be alert. Sanitation is important. Cows must be kept clean. Udder infections must be detected, if they exist, and care exercised to prevent spreading infections among cows. Equipment must be cleaned, teats cleaned, and many other precautions taken to ensure a high-quality product. Estrus detection, artificial insemination, and care and feeding of newborn calves require trained and attentive personnel.

It is implied in the above discussion that dairy production units are owner-operator units, with the operator owning the land, equipment, and cows, and carrying all risk. This is true in most cases, but there are other business and financial arrangements, such as corporations or partnerships.

30.4   FEED AND OTHER COSTS

Supply of feed was mentioned in Section 30.2 as a factor in dairy farm location. In most dairy operations, feed represents 45 to 60 percent of production costs.

Chapter 6 covered ruminant nutrition and presented most of the principles involved in feeding growing heifers and mature dairy cattle. Tables 6–1 and 6–2 gave examples of the nutrient requirements of dairy cattle according to stage of production. Sections 6.5 and 6.6 presented some considerations peculiar to feeding dairy cattle, and Table 6–7 gave an example of a complete ration for dairy cows. Because calves are functionally nonruminants until 4 to 6 weeks of age, they were discussed in Chapter 5.

Forage—pasture, hay, silage, and freshly chopped forage—is the foundation of most feeding systems for cows. Forage must be of high quality (relative feed value of 150 or higher) for maximum consumption and milk production. There is a limit to the capacity of a cow's digestive system. If forage is low in quality, a cow eats less, and the amount eaten is not efficiently digested.

Intensive grazing—where cows are restricted to a strip of pasture sufficient for one or a few days—makes excellent use of forage and can be practical with a smaller herd that is part of a general farming operation. Confining the cows to drylot and feeding the pasture forage freshly chopped also allows more efficient use of the forage, but cost of chopping and hauling, and weather problems, must be considered to determine if the practice is feasible.

Some dairy producers harvest the entire crop of forage as silage and/or hay, cutting at the best stage of growth, and then feed it daily to the cows in confinement. This takes advantage of the increased efficiency in forage production and utilization, compared to pasturing, while also eliminating the need for daily chopping of fresh forage.

Heifers should enter the milking herd between 22 and 24 months of age. A dairy producer may wish to save most heifers until they are old enough for some to be culled visually or on the basis of performance—essentially, all heifers born. This represents a large expense and considerable feed used for animals other than those in milk production (Figure 30–4).

Is it best to raise replacement heifers or buy them? As with other management problems, the answer depends on conditions. Those who raise their own indicate they do so because "home-grown heifers cost less money," or due to "more profitable use of land and labor," "less disease in my herd," and for other reasons. One

Figure 30–4
Cost of raising replacement heifers is significant because of both feed and labor. These calves on a dairy farm in Colorado should be about 2 years of age when they begin to produce milk. (Courtesy of Hoards Dairyman.)

important reason is that the producer who raises the heifers may do a better job of choosing the best heifers and also selecting the most appropriate sires to mate with the heifers.

Replacement heifers on a dairy farm represent a sizable investment. There are situations in which capital is limited and the capital available can be used more wisely for mature cows producing milk than for maintaining heifers. Feed, space, and labor might also be limited.

At the same time, there are some farmers who are in the business of raising replacement heifers. A certain farmer, for example, may not have milking facilities or an available market but have ample forage and other feed, as well as skill and experience in raising heifers. Such a farmer might purchase heifer calves, raise them to breeding age, artificially inseminate them, and sell them just before they calve. In some cases, these farmers raise replacement heifers for large dairies on a contract basis.

30.5   HANDLING MALE CALVES

Only a very small number of dairy bulls—those where production records of relatives indicate the animal has a very high breeding value, as discussed in a later section—are raised to maturity and used for artificial insemination or for natural service. Most bull calves are fed to produce veal or are further grown and fed for beef.

Most veal calves are raised on an all-milk-replacer diet under very controlled management conditions to about 350 to 420 pounds. They grow quite rapidly, reaching market weight within about 12 to 16 weeks of age. Most veal commands a premium price, well above beef prices, because of its tender meat.

Profitability of feeding calves for veal depends primarily on (1) purchase price or production cost at birth, (2) cost of milk replacer and other feed, plus feed conversion, (3) mortality rate and cost of medication, and (4) price received for veal. Other management factors to consider are days on feed, calves per cycle, and cycles per year. Because whole milk usually is too valuable to feed to calves, high-quality milk replacers, comprised largely of whey and other dairy byproducts and sold largely as dry powder, usually are used.

In some cases it is practical to raise bull calves for a longer feeding program, as bulls or as steers. Profitability varies, of course, depending on veal and beef prices and on feed supplies and costs.

As with feeding for veal, calves of the heavier breeds (Holstein, Brown Swiss) are more likely to be profitable when raised for prolonged feeding. These large-type cattle gain fast, utilize roughage effectively, and usually put on low-cost gains. The carcasses are more angular and carry less marbling than carcasses from beef animals, but they do yield a high proportion of lean meat.

30.6   SOUND DAIRY HERD MANAGEMENT PRACTICES

Dairying requires extremely good management to be efficient and profitable. Some important areas of herd management include

- Selecting for desired production and structural soundness, especially the udder and the feet and legs (see Figure 30–5)
- Maintaining excellence in reproductive and overall herd health to achieve a 12.5- to 13-month calving interval

Figure 30–5
In addition to selecting for high milk production, most dairy breeders place a high priority on strongly attached, capacious udders and straightness of feet and legs, illustrated by this group of Jersey cows.
(Courtesy of Agri-Graphics, Ltd.)

- Feeding according to milk production and body condition
- Using records for herd management and decision making
- Providing proper housing for comfort and cleanliness of the cow
- Using proper milking procedures that ensure high-quality milk and a healthy udder

Selection, calving interval, proper feeding, and use of records are discussed in the following paragraphs. Housing and proper milking procedures are discussed in Chapter 31. For more details on realistic dairy herd goals, see Box 30–1. Also, see Appendix B for the parts of the dairy cow and their terminology.

Dairy managers can be highly selective in their choice of sires available through artificial insemination companies. More than 75 percent of all dairy operations use AI and maintain a semen inventory. Through cooperative efforts of DHI organizations and AI companies, young sires from planned matings are sampled throughout the country. Daughter-herdmate comparisons are used to evaluate all young sires and to update evaluations on older sires.[2] Most dairy managers select about 80 percent of their semen needs from sires with progeny records and about 20 percent from young promising sires based upon their pedigree information.

Semen from only the top-ranking progeny-tested sires and young sires is made available through the AI companies. Sire summaries are made available twice yearly that rank bulls according to their *predicted transmitting ability (PTA)*. Some of the informa-

---

[2] Data provided by the National Association of Animal Breeders (NAAB) show that about 13 million units of semen are sold in the United States each year.

<table>
<tr><td colspan="2">BOX 30–1</td><td colspan="2">Realistic Goals for a Dairy Herd</td></tr>
</table>

The sale of milk represents about 90 percent of the income on most dairy operations. Increasing milk production per cow before expanding herd size is usually the best approach to increasing profit. Today, most dairy farmers also are controlling production costs more carefully and placing more emphasis on *net profit* per cow. The following are some management items with realistic goals for many Holstein herds.

Item	Herd Goals
Milk per cow (lb)	>21,000
Protein per cow (lb)	>660
Milk fat per cow (lb)	>740
Calving interval (days)	365–380
Days to first breeding	50–60
Days to conception	85–100
Services per conception, avg.	1.5
Cows pregnant on exam (%)	85–90
Average days nonlactating	45–60
Percent of all cows in milk	>90
Percent I.D. by sire	>90
Percent from proven sires	>80

Many herds today have attained or exceeded the above goals. Each herd manager should establish goals according to the present status of the herd and adjust the suggested goals accordingly. Establishing goals in other phases of the operation also can be helpful in planning the future direction of the overall dairy business.

Source: Compiled by the authors from Dairy Herd Improvement (DHI) record summaries.

tion provided includes: PTA for milk yield, milk fat and protein yield, milk fat and protein percentage, overall type, and udder composite scores. The average PTAs for active AI bulls are consistently higher than those for non-AI bulls (e.g., 750 to 1,200 pounds milk) so genetic progress is much greater using selected AI sires. Most dairy producers establish a minimum goal for PTA milk and protein yield, and PTA type. Also, because information is provided on the expected percentage of difficult first births, dairy managers can have greater confidence in breeding their heifers artificially by selecting sires with low difficult-birth scores.

A desired average calving interval for the dairy herd is 12.5 to 13 months. When cows conceive within 85 to 100 days after calving to meet this ideal calving interval, they have more days of productive life in the higher portion of the lactation curve (Figure 30–6). To achieve an acceptable overall calving interval for the herd, the incidence of uterine infections must be kept low. The herd should be on a sound reproductive health program with scheduled visits by a competent veterinarian.

Some larger dairy operations utilizing AI, such as the Arizona Jersey herd shown in Figure 30–7, use electronic aids to help detect cows as they approach the desired breeding date. Pressure sensors (attached to the rump) or pedometers (attached

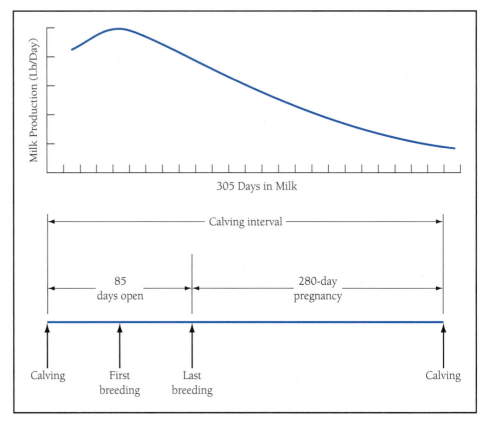

Figure 30–6

The cow's milk production and reproduction year. A cow with a 12-month calving interval would have a 305-day lactating period and a 60-day nonlactating period. A cow with a 12-month calving interval must be pregnant within 85 days.
(*Source*: Purdue University.)

Figure 30–7

A large Jersey herd utilizing open-elevated shades (in background) and fenceline feeding for automated total mixed rations. Photo at right shows pedometer that is attached to the ankle of all cows when they reach 40 days postpartum.
(Courtesy of Purdue University.)

to the ankle (see Figure 30–7), transmit increased estrus activity to the dairy office computer. By twice-daily monitoring, the herd manager determines which cows to observe more closely. Then, if primary signs of estrus are observed, these cows are bred.

Formulation of dairy rations was discussed briefly in Section 6.3, and as mentioned in this chapter, high-quality forage is very important for economical rations and high milk production. Today, most dairy herd rations are fed as **total mixed rations (TMR)**, where all components of the ration are carefully weighed and blended together. Also, most dairies will have cows grouped and fed according to their production level. For example, large herds may be fed and managed as

1. High-production group
2. Low-production group
3. First-lactation cows
4. Nonlactating cows and heifers approaching parturition

The amount to be fed to each group is calculated on the nutrients needed per head daily on a *dry-matter basis*. These calculations are converted to an *as-fed basis* when weighed, blended, and fed to the group.

To ensure adequate dry matter and energy intake, and sufficient physical fiber, most group rations also are calculated to have a certain forage-to-concentrate ratio. For example, the high-producing group may receive a 50:50 ratio to help ensure adequate energy intake. The low-producing group may receive a 60:40 ratio because less energy is needed and forages are more economical.

High-producing cows can be expected to consume 3.8 to 4.0 percent of their body weight in pounds of dry matter daily. For example, cows in the high-production group averaging 1,450 pounds of body weight could be expected to consume about 55 to 58 pounds of dry matter each day.

Another management tool used by many dairy farmers is *body condition scoring*. Fat covering is an excellent indicator of the amount of stored energy in dairy cows. Excessively fat cows are more apt to have calving difficulty and often have sluggish appetites after calving. Cows with insufficient body reserves produce less milk and are more prone to metabolic disorders and impaired reproductive performance. Therefore, body condition scoring is helpful in determining when to move cows into other production groups or to provide special attention to particular cows.

The most common system for dairy cows is a numerical system from 0 to 5 with half scores in between whole numbers (0 = extremely emaciated, 5 = very fat, based primarily on observance and feel of fat covering around the tailhead and rump area). To be effective, cows should be scored at calving time, peak production, midlactation, and when ready for the nonlactating (dry) period. Suggested time of scoring and desired scores are shown in Table 30–3.

Records are necessary in determining the present performance of the dairy herd, establishing attainable performance goals, and deciding how to reach these goals. Although private records are helpful in decision making, many dairy herds (about 36,000) are on the official **Dairy Herd Improvement (DHI)** record-keeping system. Dairy herds are enrolled in this program through state dairy associations and sponsorship by USDA and state extension services.

For most official DHI programs, a trained supervisor checks each farm monthly to obtain milk yields and samples from each cow (Figure 30–8). Samples are sent to a

Total mixed ration (TMR)
All-in-one ration. Ration components combined; theoretically, each bite of food would contain all nutrients in the correct proportions.

Dairy Herd Improvement (DHI)
USDA Extension Service record-keeping program for dairy cows and goats.

TABLE 30–3	Body Conditioning Scores for Dairy Cattle	
Time of Scoring	**Desired Score**	**Acceptable Range**
Cows:		
At calving	3.5	3.0–4.0
Peak lactation	2.0	1.5–2.0
Midlactation	2.5	2.0–2.5
Start of dry period	3.5	3.0–3.5
Heifers:		
6 months of age	2.5	2.0–3.0
Breeding (13–15 mo.)	2.5	2.0–3.0
At calving	3.5	3.0–4.0

Source: *Patton, R A., et.al.*, Body Condition Scoring—A Management Tool, *Michigan State University, 1988.*

Figure 30–8
The DHI supervisor checks each herd once a month to record milk weight and other data, and to collect a milk sample from each cow. (Courtesy of Kansas State University.)

central milk testing laboratory, generally within the state, where each sample is analyzed for protein, fat, and somatic cell content. These results, and other herd information, are compiled in a computer and usually are teleprocessed to one of the nine Dairy Records Processing Centers located in the United States. The newly compiled monthly reports are returned quickly to the DHI member.

Monthly DHI reports provide the herd manager with valuable management information such as (1) production, income, and feed cost summary; (2) current breeding and reproductive summaries; and (3) the rolling yearly herd average for milk, fat, and protein yield.

Two additional reports are provided to help in determining which animals to cull or keep. The Predicted Producing Ability (PPA) ranks each cow in the herd according to her expected deviation from herdmates; it is best used to determine the most productive cows in the next lactation. The Predicted Transmitting Ability (PTA) is the best indicator of a cow's transmitting ability, and both production and pedigree information are considered in its calculation.

Many on-farm computers have direct access to the Dairy Records Processing Center database. Direct access to records by telephone (DART) enables producers to update their records weekly or daily rather than monthly to help make timely management decisions. These include lists and dates such as cows to calve, cows to turn dry, cows to breed, and so on. Many managers have found that less time spent in doing chores and more time spent interpreting information from records has resulted in improved efficiency and profitability of the dairy operation.

30.7   MARKETING ANIMALS CULLED FROM THE HERD

Beef is a major byproduct of dairying. It is estimated that 15 to 20 percent of the beef consumed in the United States comes from animals that are primarily of dairy breeding. This figure includes the fed steers and bulls discussed in the previous section, but a major portion is contributed by cows culled from milking herds. Of the 35.7 million head of cattle slaughtered in 2001, 7.4 percent were cows culled from dairy herds.[3]

The marketing of cull cows for beef provides from 5 to 10 percent of the gross returns to the dairy producer. Refer to Section 22.3 for a discussion of supply and price trends for various classes and grades of slaughter cattle and Chapter 23 for discussions of appraisal for slaughter.

30.8   DAIRY GOAT MANAGEMENT

There has been a significant growth in the U.S. dairy goat industry in the last 30 years, reaching a peak about 1982, then sharply declining and leveling off in recent years to about 46,000 head. The five leading dairy goat states, based upon enrollment in DHI in 1998, were California, Wisconsin, Ohio, New York, and Oregon. These five states represent about 50 percent of all goats on official test.

Most goat milk produced under Grade A standards (Figure 30–9) is sold as fresh pasteurized milk . However, production of U.S. gourmet cheeses has become more popular in recent years, competing well with imported cheeses from Europe. Dairy goat breeds found in the United States, in order of numerical preference, are Nubian, Alpine, Saanen, LaMancha, Toggenburg, and Oberhasli. Top producers in all breeds have exceeded 4,000 pounds of milk per 305-day lactation, the record being 6,650 pounds produced by a Toggenburg. In 1998, the average of all breeds was 1,775 pounds of milk of 3.66 percent fat and 3.15 percent protein.

Does usually are bred at 10 to 12 months of age at 85 to 95 pounds. Estrous cycle length averages 21 days during the breeding season, usually from August through January, and is initiated by decreasing photoperiod. Birth of twins and triplets is most common. Most newborn kids weigh about 6 pounds.

Typically, does lactate for 8 to 10 months. Commercial dairies reduce seasonal milk production by delayed breeding of some does, artificial light control, and hormonal

> **Doe**
> A female goat.

---

[3] USDA, NASS.

Figure 30–9
Dairy goats being milked in a modern milking parlor.
(Courtesy of American Dairy Goat Association and The Coach Farm, Pine Plains, New York.)

treatment to trigger onset of the estrous cycle. Does produce the most milk during the third or fourth lactation, declining in later lactations. Studies have shown increased milk production related to incidence of triplets and quadruplets as compared to singles or twins. Production testing programs are available through DHI associations.

Facilities needed for dairy goats and dairy cows are similar. However, high, well-constructed fences are needed for both goat confinement and to keep out predators, dogs being the greatest problem. High-quality forage is very important because of nutrient content and the tendency of goats to refuse coarse stems. Loose housing systems and confinement in small areas are common. Housing that provides wet and/or drafty conditions is especially harmful to dairy goat health.

Kids can withstand cold temperatures if they are in a dry and draft-free area. Coccidiosis is a common problem in wet and unclean housing. Caprine arthritis and encephalitis (CAE) is the greatest cause of animal and milk loss. Many goat raisers pasteurize and administer the first colostrum, then feed pasteurized milk, to reduce the spread of CAE from dam to kid through nursing or feeding of raw milk. A rigid program for internal parasite control must be included in the raising and maintenance of a healthy herd.

QUESTIONS FOR STUDY AND DISCUSSION

1. Describe the need for quality personnel and labor on the dairy farm.
2. Why is dairying a relatively stable business? Compare cash flow of dairying to other enterprises.
3. Compare the quality and amount of forage needed for lactating dairy herds to that of beef cow herds.

4. Describe the trends in milk cow numbers, total milk production, and production per cow in the United States.

5. On what basis are most dairy producers being paid for their milk?

6. What features of a particular area of the United States would make dairying competitive with other enterprises?

7. Give examples to illustrate the specialization and large investment of modern dairying.

8. About what percentage of the total cost of producing milk does feed cost represent?

9. Define *TMR*. Explain "as-fed" versus "dry-matter" calculations.

10. What are advantages of growing herd replacements properly to calve at 22 to 24 months of age?

11. Describe fancy or formula-fed veal production. What are major factors that determine profitability of the business?

12. Name several types of information provided by DHI records.

13. Why and when should cows be "body condition" scored?

14. What is the typical amount of milk produced by dairy goats (a) per day and (b) per lactation?

Milk Handling 31 and Marketing

Because milk is a relatively perishable food, and because a high percentage is consumed in a relatively natural state, handling of milk to preserve its natural and desired characteristics is very important. Milk must be produced in a clean environment, using properly cleaned and sanitized equipment. It must be cooled rapidly, processed, and distributed to consumers in a rather short time. Where processing occurs, as in production of ice cream, cheese, or butter, more time elapses, but the processing must be initiated and completed before quality of the milk is impaired.

This chapter describes the handling and distribution of milk, from the time the cows are milked until the milk reaches the consumer. It includes discussions of sanitation ordinances, milking facilities, cooling and storage, equipment, marketing procedures and regulations, and consumption of dairy products.

Upon completion of this chapter, the reader should be able to

1. List six important characteristics of dairy facilities.
2. Contrast, with sketches or verbally, the different milking parlor designs.
3. Define *stanchions, tie-stalls*, and *free-stalls*.
4. Describe the functioning of a milking machine, including vacuum, pulsations, and flow of milk from the teat.
5. Describe how milk is handled, held, and transported after it leaves the cow.
6. Chart the normal 12-month pattern of U.S. milk production volume.
7. Differentiate between fluid and manufacturing milk and between Classes I, II, and III.
8. Define the term *federal market order* and describe the pricing system that functions within orders.
9. Define *pasteurization* and *homogenization*.
10. Contrast the per capita consumption trends for whole milk, skim or low-fat milk, and cheese over recent decades.

31.1 FACILITIES FOR THE DAIRY HERD

Housing arrangements for the dairy herd must mesh with other components of the total system. These areas include

Milking facilities	Herd replacement facilities
Feed storage and alleys	Waste management area
Free-stall or tie-stall housing	Exercise lots
Breeding and treatment area	Grassed observation pasture
Maternity area	Paved traffic alleys

Figure 31–1
Schematic of a modern parlor equipped with automatic milker takeoff units and clean-in-place equipment.
(Courtesy of Fair OAKS Dairy, IN.)

Properly designed and well-managed facilities are essential in the production of high-quality milk. The overall dairy system should fulfill the following criteria:

- Exceed all Grade A sanitation standards
- Be designed for efficient use of labor
- Be designed for flexibility, such as future expansion
- Provide a comfortable and easily cleaned environment for the cow
- Provide for safe and efficient movement of cattle, people, and machinery

> **Milking parlor**
> A special facility in which cows, goats, or sheep are milked.

The more common system for housing and milking of the dairy herd is the **milking parlor** and **free-stall housing** (Figures 31–1 and 31–2). Free-stalls allow the cow to enter freely, lie down, and leave, but not turn around. Cows step up from an alley behind the stalls, and, if properly designed, the cow's position ensures that urine and feces will drop into the alley. These and any wet bedding are removed at least daily with scraping vehicles, automatic scrapers, or by flushing with water.

> **Free-stall**
> Dairy barn and stall design that permits cows to choose a slightly elevated individual stall for resting.

Considerably less labor is required with free-stalls and a milking parlor than with *tie-stall* or *comfort-stall barns*. Feeding can be arranged to require less labor, with automated delivery systems (augers, belts) or by a drive-through system (Figure 31–2).

Udder, teat, and leg injuries occur less often in well-designed and properly managed free-stalls. Construction costs are less and the system is more flexible. For example, if herd size is increased, free-stall facilities often can be expanded and the additional cows milked in the existing parlor.

Figure 31–2
Free-stalls save bedding and labor. Cows stay cleaner and therefore less space per cow is required. Drive-through alley allows automation in feeding cows in production groups. (Courtesy of the USDA, South Dakota State University and Fair OAKS Dairy.)

Modern tie-stall barns generally house two rows of 30 to 45 cows. Most are arranged so cows face outward, with a central alley behind and a feed alley in front of the cows. Each cow is provided an individual stall and quickly learns to select her stall when entering the barn. Tie-stall barns also have other advantages. Each cow receives her individual feed allowance so feed intake can be better controlled, and cows can receive more individual attention. Some purebred breeders prefer tie-stalls because they can display and merchandise their animals more effectively. When cows are milked in-place in tie-stalls rather than in a milking parlor, the milking process is more strenuous; however, pipeline milking and automatic takeoff units have improved milking efficiency.

Sanitation is the primary consideration in milking facilities because selling milk depends on meeting sanitation requirements. Most states have adopted state laws that govern milking and milk-handling equipment and procedures; cities and counties also may have milk sanitation ordinances.

To ensure that the consumer receives a continuous supply of high-quality Grade A milk, every producer must pass initial inspection to receive a permit. Sanitation ordinances also require that all cows be free of diseases that are threats to human health.

Several on-site farm inspections are conducted annually to ensure that housing, milking facilities, and surroundings are properly maintained and clean. Cleanliness of the milk room and assurance of a safe water supply are carefully evaluated during each inspection.

Following each farm inspection, a *Dairy Farm Inspection Report* is posted in the milk room area. Failure to correct violations can result in suspension of the Grade A permit.

In addition to sanitary standards, as described above, milk samples are collected by the milk hauler and must meet certain chemical, bacteriological, and temperature standards.[1] Minimum standards and goals include the following:

Temperature	Cooled to 45°F or lower within 2 hours after milking.	*Goal:* 38° to 42°F
Bacterial	Not to exceed 100,000 per mL prior to commingling with other producers' milk.	*Goal:* 10,000 per mL or less
Somatic cells	Individual producer's milk not to exceed 750,000 per mL.	*Goal:* 200,000 per mL or less
Antibiotic	Zero tolerance	*Goal:* Negative

Most dairy producers sell milk of higher standards than required. For example, milk is quickly cooled to about 38°F and maintained at that temperature until picked up. Also, milk plants often impose higher requirements and provide quality incentives to help ensure milk of high quality.

31.2   MILKING MACHINES AND PROCEDURES

Essentially all dairy cows in the United States are milked by machines. Milking systems vary in design and cost, but all have four major components: milking unit, pulsation system, vacuum supply system, and milk flow system (Figure 31–3).

The milking unit consists of four teat cups connected to a "claw," suspended under the cow and sometimes supported or attached to an automatic detacher arm, and the connecting milk and air tubes.

All milk is drawn from the teat by partial vacuum. Vacuum also massages the teat and moves milk through the milk flow system. The *pulsation system* introduces vacuum and atmospheric pressure alternatively between the teat cup shell and liner, providing a *milking phase* followed by a *resting phase*. Pulsation rates range from 40 to 70 pulsations per minute, with the ratio of milking phase time to resting phase time ranging from 50:50 to 70:30.

In the milking phase, a vacuum of 12 to 15 inches of mercury is applied to the air space between the teat cup shell and liner (Figure 31–4). This results in a partial vacuum applied to the exterior of the teat, which opens the streak canal and, with proper milking procedures and milk let-down, removes the milk.

For the resting phase, the pulsator interrupts the vacuum and allows air to enter the space between the teat cup shell and liner; this allows closure of the liner around the teat. This action massages the teat and lets milk flow from the udder cistern and upper ducts into the teat cistern, so it can be withdrawn during the next milking phase. It also prevents congestion of blood in the lower portion of the wall of the teat.

The milk flow system, also under partial vacuum, consists of hoses and pipes that convey the milk from the cow to a "receiver jar." From this point, it is pumped to the bulk tank cooler in the milkroom. All pipelines are of sanitary design and constructed of proper size.

---

[1] The minimum standards are specified by *The Grade A Pasteurized Milk Ordinance (PMO)*. Additional standards may be imposed, and those shown are subject to change.

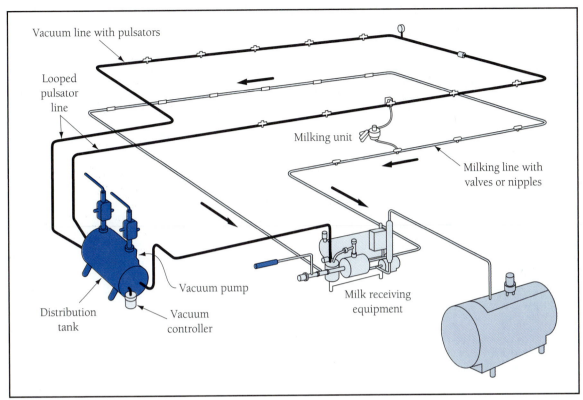

Figure 31–3

A machine milking pipeline system for a milking parlor. Note vacuum pump and line, plus the line that moves milk from the milk receiving equipment to the cooling and holding tank. (*Source*: WestfaliaSurge, Inc.)

To prepare the udder and ensure milk let-down, the milker first removes a few squirts of milk from each quarter and checks for abnormal milk. Also, this first milk usually is higher in bacteria count and is best discarded. Each teat is then cleaned with a warm sanitizing solution and dried with a single-service towel. Some producers dip the end of each teat in a germicide (approved by the FDA) to reduce risk of bacteria entry into the teat during milking. If predipped, each teat is dried after about 30 seconds with a single-service towel. Once the teats are thoroughly dry, and about 45 seconds after first stimulation, the teat cups are positioned on the teats, with care being taken to minimize air intake through the teat cups.

It is important that the cow be milked out quickly, usually in 3 to 6 minutes. In manual machine removal, the vacuum is shut off at the claw and the teat cups gently removed. Automatic detachers, which allow more cows to be handled by one person, sense when milk flow diminishes, automatically shut off the vacuum when flow ceases, and gently detach the teat cup cluster (Figure 31–5).

The final step in the milking procedure is dipping or spraying the lower one-third of each teat with germicide. This is especially important because the teat end is dilated and bacteria are more likely to enter the teat at the completion of milking.

Prior to use on the next cow, the teat cups should be rinsed and sanitized to reduce bacterial transfer from cow to cow. Many dairies have installed automatic

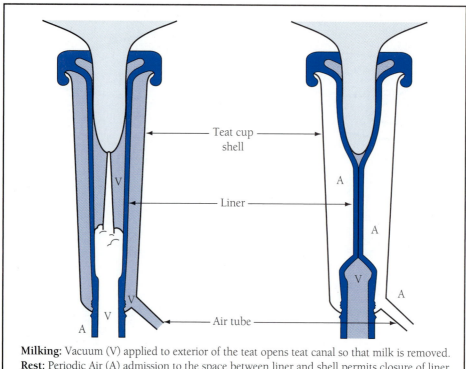

Milking: Vacuum (V) applied to exterior of the teat opens teat canal so that milk is removed. **Rest:** Periodic Air (A) admission to the space between liner and shell permits closure of liner for massage.

Figure 31–4
Status of teat cup during milking and resting phases.
(*Source*: WestfaliaSurge, Inc.)

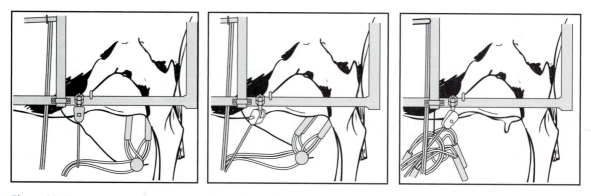

Figure 31–5
An automatic takeoff unit shuts off vacuum when milk flow ceases, then gently removes the milking cluster.
(*Source*: Westfalia Systemat, Elk Grove, IL.)

backflush systems that rinse the claw and teat cups with water, sanitize, rerinse, and remove excess water automatically.

As shown earlier in Figure 31–1, milking parlors are constructed so the milker has easy access to each cow's udder. Most provide for two groups of cows with several cows on each side of the milker, such as a double-8 or double-10. Today, most parlors are of the "herringbone" design.

In recent years, the "parallel" milking system has become more popular. In this system, groups of cows (usually 10 or more) enter each side and stand parallel to each other with their rear udders exposed to the milker. Each milking unit is attached from the rear of the cow rather than from the side. This system, as well as other newer systems, also allow each group of cows to "rapid-exit" when milking is completed.

The milking process becomes habitual. Both milkers and cows become accustomed to a standard routine. A good milker is observant, works quietly and swiftly, is gentle to the cows, and is dependable. Successful dairy managers try to attract and retain good milkers.

31.3   COOLING AND HOLDING MILK AT THE PRODUCTION UNIT

Milk should be cooled rapidly to about 38°F to prevent rapid multiplication of bacteria and an elevated lactic acid concentration. Cooling is done in refrigerated, stainless steel bulk tanks, equipped with an agitator to slowly mix the entering warm milk and speed the cooling process (Figure 31–6). Recording thermometers are common on many new milk cooler models. Heat recovery systems sometimes are used to capture the heat from milk cooling and use it to heat water for use in the parlor or milkroom. Also, some dairies use "plate coolers" that transfer heat from the warm milk to water. This procedure lowers milk temperature a few degrees before it enters the bulk tank in order to reduce the cost of cooling and also to provide warm water for utility use.

All milk cooling equipment is designed and installed with ease and effectiveness of cleaning in mind. Completely automatic **clean-in-place (CIP)** systems require less labor for cleaning. Rinse water and detergent solutions are circulated through the milking units, pipelines, and cooler tank without dismantling of the equipment. Proper cleaning procedures, whether performed manually or by CIP systems, include the following steps:

> **Clean-in-place (CIP)**
> Automated cleaning techniques used to clean milking, milk-handling, and milk storage equipment.

1. *Rinse* with warm water immediately after milking and before milk residues dry and adhere to surfaces. Hot water (above 140°F) should not be used because milk protein residues may become coagulated at such high temperatures.

2. *Clean* with a detergent and high-temperature water (about 120°F).

3. *Postrinse* with clean water to remove all traces of the detergent.

4. *Disinfect* prior to the next milking, to further ensure that all interior surfaces that are in contact with milk are free from living bacteria.

Collection of milk from the farm by bulk truck on alternate days is practiced most commonly. From some larger dairies, milk is moved daily. Milk haulers trained and licensed by the state board of health check the appearance and temperature of the milk,

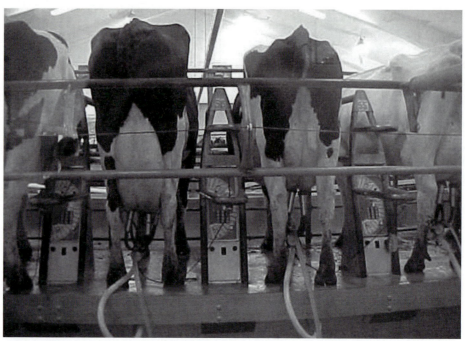

Figure 31–6
Most milk is handled in bulk. It moves from the cow by hose and stainless steel or glass
pipelines to bulk cooling tanks. From there it is pumped to a truck.
(Courtesy of WestfaliaSurge, Inc.: and Fair OAKS Dairy.)

measure the volume of milk, thoroughly mix the milk, and then collect a sample of milk
to be tested at the processing plant. The final step is to transfer the milk from the cooler
to the truck tank by hose.

It should be noted that in the entire process from milking to the delivery of milk
to the processing plant, milk always is contained in a *closed system*, is quickly cooled and
maintained at a low temperature, and is properly handled to help ensure a high-quality
product for the consumer.

31.4   MILK MARKETING

Most dairy farmers sell their milk through cooperative marketing associations or direct
to processor. Other methods of marketing, such as direct to consumer, involve only a
small proportion of milk production.

Cooperative milk marketing associations have existed for more than 65 years
in certain areas. They provide the producers more bargaining power because an association might control most of the milk produced in the area. The association can afford to
hire experienced people to deal with prospective purchasers and also to help producers
with production and sanitation problems.

Milk production is somewhat seasonal (Figure 31–7), whereas consumption is
less seasonal. During periods of surplus production, not all milk can be used for fluid
milk sales; some must be routed for manufactured products. Without cooperatives,
processors could cease to purchase from some producers, or buy milk from certain producers at a much lower price, for manufacturing. With cooperatives, a portion of the to-

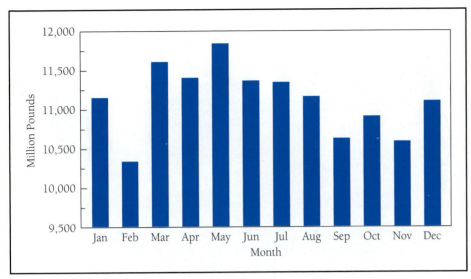

Figure 31–7
Seasonal milk production as reflected in the monthly milk production of all cows in the leading
20 states during 1997. U.S. milk production is usually lower during late winter (February) and
the fall months.
(*Source*: Adapted from USDA data.)

tal supply can be classed as surplus and used as manufacturing milk in peak seasons,
and each producer therefore sells some milk at a lower price. Most milk marketing associations have developed around large population centers.

Most milk is marketed and priced under "federal market orders." The purpose
of a federal order is to promote and maintain orderly milk marketing conditions for dairy
farmers and to ensure consumers an adequate supply of pure and wholesome milk
(Figure 31–8).

Each order may be different, but all contain a provision for establishing minimum prices that processors must pay producers for milk of each "use" classification. Second, all milk is pooled so that farmers share equitably in the higher price received for
milk that is used in fluid form as well as the lower prices received for manufacturing milk.

Processors or handlers of milk must pay the minimum prices, calculated from
provisions of the order, for each use classification, make accurate weights and tests, and
account for the way the milk is used. The order does not control how much the processor can buy, from whom, or at what price the processor may sell. Farmers may produce
and sell any amount of milk they wish under a federal order and may sell to anyone they
wish, independently or through a cooperative.

Milk for *fluid use* is placed in *Class I*, the highest price class. Class I use includes
whole, low-fat, skim, and flavored milk, milk drinks, and buttermilk. Grade A milk that
is used in manufactured products is placed in one or two lower price classes. In most
markets, *Class II* includes frozen desserts, cottage cheese, and cream. *Class III* includes
milk used for butter and cheese, in most instances. The quality of the milk is no different; the proportion placed in each class is simply dependent on the amount that can be
used for the Class I purposes, then for Class II purposes.

Because milk production is slightly higher in the spring, a larger percentage of
milk is diverted to manufacturing, and so is classed accordingly.

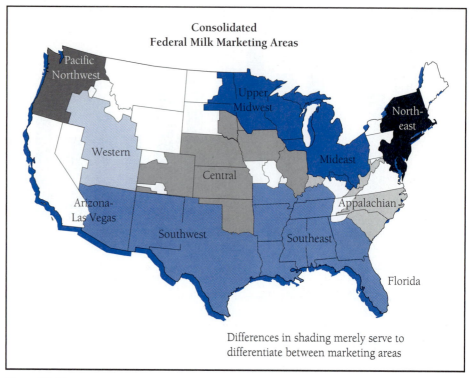

Figure 31–8
Eleven federal milk marketing order (FMMO) areas, October 1999. The previous 31 FMMOs were consolidated into 11 orders with similar milk class utilization.
(*Source*: USDA.)

Changes in the dairy processing industry have influenced the marketing of farm-produced milk. There are fewer processing plants, and those that exist handle more of the product. This centralization of processing means fluid milk or cream plants no longer exist in small towns.

31.5   MILK PRICING SYSTEMS

Milk prices paid to producers are usually quoted on a hundredweight (cwt) basis. The average price per cwt that a processor pays may be determined by (1) supply and de-mand—the competitive price required to attract milk to the plant, (2) a price set by bar-gaining between the processor and a producer association, or (3) a price determined by some previously agreed-upon formula, as in federal order markets.

Until 1996, the milk pricing system for most of the 33 federal milk marketing orders was based upon the Basic Formula Price (BFP). This system relied upon monthly surveys of the manufacturing-grade milk in the upper Midwest and used a price formula linked to market prices for butter, nonfat dry milk, and cheese for each price update.

The most recent change in marketing was passage of the 1996 Federal Agricul-tural Improvement & Reform (FAIR) Act. The purpose of this act was to (1) consolidate the 33 federal orders into 11 orders and (2) create a more equitable system for produc-ers and processors.

The average price received per 100 pounds of Grade A milk from 1992 to mid-1999 ranged from $12.00 in mid-1995 to a high of $18.00 in December 1998 (Figure 31–9).

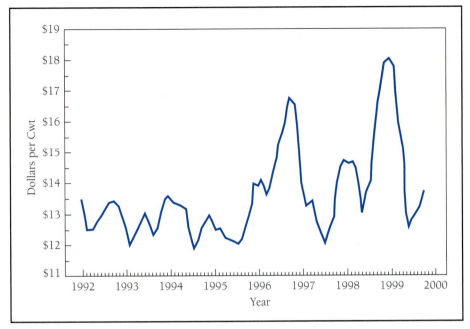

Figure 31–9
The average price per hundred pounds of Grade A milk since 1992. The variation in price is due primarily to seasonal effects upon the availability of milk and increased demand by the processors. (*Source*: USDA, NASS, July 1999.)

However, for the first six months of 1999, the average price had declined to $14.43 per hundred for Grade A milk. Some producers, with cows that produce milk with higher solids and who receive quality incentives for producing milk of the highest quality standards, receive an additional 50 to 75 cents more per hundred pounds of milk sold.

The newest system of milk pricing is referred to as *multiple component pricing*, or MCP. This system places a value on milk based on its most valuable components, protein, fat, and other solids, in accordance with their use in consumer-purchased products. In recent years, consumers continue to purchase more cheese and lowfat and skim milk. Under the MCP system, dairy farmers are provided a greater incentive to produce the components in milk of highest value, flavor, and nutrition.

Components and the typical values that determine a producer's milk price are illustrated by the following:

Components	Value per Component
Amount of protein	$1.50 to $1.90 per lb
Amount of milk fat	$0.60 to $0.95 per lb
Amount of other solids	$0.58 to $0.90 per lb
Price differential for Grade A[2]	$0.05 to $0.40 per lb
Somatic cell count—each 100,000 below 350,000 per mL	$0.038 to $0.075 per cwt

[2] A weighted price adjustment is made for the proportion of Class I milk actually used for fluid purposes and the remaining proportion used for Class II and Class III. A typical proportion of use might be 40, 15, and 45 percent for Classes I, II, and III, respectively.

With multiple component pricing, a greater value is placed on the pounds of milk solids sold, especially protein, than with previous milk pricing systems. Table 31–1 shows the differences between breeds in average protein and milk fat percentages. Under this system, the Holstein breed should continue to have an advantage in total milk yield. However, most cows in the other breeds should compensate for less milk yield by producing milk of higher value because of the higher solids content.

31.6   PROCESSING AND MANUFACTURING

About 38 to 40 percent of the fluid milk leaving production units is processed for consumption as whole milk, skim milk, or low-fat milk (1 to 2 percent fat). All milk is pasteurized, most by the high-temperature short-time (HTST) method, which involves heating to 161°F and holding at that temperature for 15 seconds in a continuous flow process, to destroy any disease-producing organism that may be in the milk.

Skim milk results from removing nearly all milk fat, usually by centrifugal force. In a centrifugal separator, the high-density water and nonfat solids move to the outside, the lighter cream or fat remains near the center, and the two leave the separator in separate streams. Most skim milk marketed contains added levels of nonfat milk solids (largely carbohydrate and protein) to provide both nutrients and consumer appeal.

The cream that is removed may vary in percentage of fat, depending on the intensity of the centrifugal force. Half-and-half contains about 12 percent, coffee cream about 20 percent, and whipping cream about 40 percent.

Milk intended for consumption as whole milk is usually homogenized. Milk as it comes from the cow, if left to stand, will separate gradually, with the cream rising to the top and the skim milk settling to the bottom. Homogenization is achieved by intense pressure and results in a rather permanent blending, breaking down the fat globules into minute particles dispersed throughout the milk. The number of fat globules in homogenized milk is about 10,000 times greater than in untreated milk.[3]

Each state prescribes the percentage of fat that should be in whole milk (usually 3.25 percent), and that level normally is achieved by blending milk from various sources that may vary in fat content. Blending of whole and skim milk is done to achieve the 1 or 2 percent product.

Ice cream or soft-serve mixes are also produced by blending to achieve the desired fat content, as well as by the addition of sugar and other ingredients.

Nearly all processes in modern dairy plants are continuous, in-line processes, except for the production of butter and most cheese. Butter is churned from cream (plus a small amount of salt) in a rotating drum that contains several paddles. The churning results in fat globules combining into successively larger units until a rather solid mass develops. The relatively nonfat liquid (buttermilk) is drained off, and the butter eventually is formed into bars or patties.

In cheese making, whole or skim milk, plus perhaps other ingredients, is placed in a tank with specific bacteria or enzymes. Rennet or some other product that will coagulate the protein is added. The mixture is stirred, the protein coagulates, and the remaining liquid, whey, is drained off. Fermentation occurs during the process described

---

3 *Dairy Handbook a Alfa-Laval.* Alfa-Laval AB, Dairy and Food Engineering Division. P.O. Box 1008, S-722 14 Vasteras, Sweden.

and, except for cottage cheese, continues during a specified storage time to achieve the distinct flavor and other characteristics desired in each type of cheese.

Many dairy byproducts such as dried skim milk, whey, and dried buttermilk are used as livestock feed ingredients. See Table 3–6 for nutrient composition of these feeds.

Dried skim milk, with only the water and fat removed, is very high in protein and highly digestible. The protein is also of high quality, containing significant levels of critical amino acids (see Section 3.4), so dried skim milk is valuable in feeds for non-ruminants—pigs, chickens, calves, and also pets. Dried buttermilk is similar in composition and value.

Whey, the byproduct of cheese manufacturing, contains a high proportion of digestible carbohydrate and is high in B vitamins, but is rather low in protein. The microorganisms that cause fermentation produce large quantities of B vitamins. Whey is also credited with having certain "unidentified growth factors" for livestock and poultry.

Other miscellaneous dairy byproducts, such as cheese rinds, outdated fluid milk, and liquid whey, also are used for livestock feeding. Dairy byproducts—especially dried whey and buttermilk—are used widely in human foods as well.

31.7   CONSUMPTION TRENDS

Trends in per capita consumption of dairy products are indicated in Table 31–1. Butter and fluid milk consumption have dropped; cheese consumption has increased.

Increased use of margarine as a spread and for cooking is no doubt the major reason for decreased butter consumption. More milk fat is used in nonbutter items than in butter. Such products include ice cream, nonbutter spreads, and other processed food items.

Other trends are evident in use of dairy products. Some people consume more skim milk and less whole milk. Emphasis on low-calorie diets to prevent obesity and heart disease, as well as out of a desire to keep a trim figure, have caused people to avoid some of the fat in the diet. Skim milk and low-fat milk supplies the protein, carbohydrates,

TABLE 31–1	Per Capita Consumption of Dairy Products (in pounds)			
	1978	**1988**	**1998**	**2003***
Fluid milk and cream	255.0	238.0	219.0	207.8
Butter	4.4	4.4	4.2	4.7
American cheese	9.6	11.6	12.0	—
Other hard cheese	7.3	12.0	16.5	—
Cottage cheese	4.7	3.8	2.7	—
Evaporated and condensed milk	7.5	7.9	2.2	—
Ice cream	17.7	17.8	19.5	—
Ice milk	7.7	7.8	7.9	—
Other frozen products	2.2	2.7	2.1	—
Dried milk products	6.0	7.2	7.3	—
Total human use (milk equivalent)	549	585	580	—

Sources: *Adapted from USDA, ERS Statistical Bulletin 965, July 1999; and* Agricultural Outlook, *July–August 1999.*
*FAPRI-UMC Report #03–03, April 2003.

TABLE 31–2	Nutritional Value of Fluid Milk Products (per 8-ounce serving)			
	Whole Milk	Skim Milk	Low-fat 2% Milk	Low-fat 1% Milk
Calories	150	85	121	102
Protein (g)	8	8	8	8
Fat (g)	8	<0.5	5	3
Percent of calories from fat	48	0	37	26
Calcium (mg)	291	302	297	300
Vitamin A (IU)	300	500	500	500

Source: *Adapted from American Dairy Association data.*

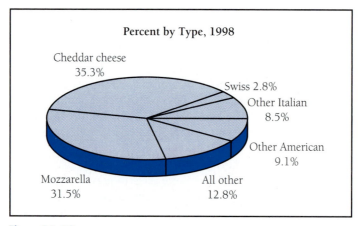

Figure 31–10
The most popular cheeses in the United States are the cheddar and Italian cheese/mozzarella varieties.
(*Source*: USDA, NASS, April 1999.)

minerals, and vitamins of milk with less fat (Table 31–2). Most dairies merchandise special low-fat milk, usually containing about 1 or 2 percent fat and fortified with vitamins. A "fat-free" dairy product advertised to possess the creamy texture and good taste of whole milk and containing less fat than skim milk (0.29 gram per 8-ounce serving) was introduced in the western states in 1996. Some consumers prefer this type of product to skim milk, perhaps due to the flavor contributed by the low amount of fat or to the low prestige rating of skim milk.

There has been a sharp increase in consumption of low-fat dairy desserts, including yogurt, promoted first by local and franchise drive-ins but sold also in restaurants and in stores for home use.

Technological developments and progressive merchandising have helped expand the use of cheese, especially in the cheddar and mozzarella types (Figure 31–10). A wide variety of cheeses are available in small packages—in self-service display counters, featured for Christmas gifts, and promoted for use in spreads, salads, and cooked foods. Cheeses are used more for variety, which American consumers now feel they can afford, than to supply needed protein as a meat substitute. Recognize, however, that cheese is a highly nutritious food, rich in high-quality protein.

Considerable dry nonfat or whole milk solids are used in manufactured foods such as pancake, muffin, and cake mixes, cereals, cookies, crackers, bread, cakes, and similar items. An increasing amount is used for cooking in the home. Whey, a byproduct of cheese manufacturing, is also used in certain manufactured foods.

New trends are on the horizon. More different consumer items utilizing dairy products will be developed and sold. The flavor characteristics of milk fat may be exploited. Milk fat may be heated to intensify the butter flavor, hydrolyzed to produce a cheese-type flavor, or heated with sugar to develop a butterscotch flavor. The protein and carbohydrate of milk also have distinctive characteristics that are being studied in the development of new products.

31.8   PROMOTION OF DAIRY PRODUCTS

Dairy products are promoted primarily by retailers, processors, the American Dairy Association (ADA), various state agencies, and breed associations.

There is some brand loyalty among consumers who purchase milk and other dairy products in stores. Because most fluid milk is sold through stores, advertising is aimed at promoting the brand name of the products, as well as increasing total purchase of dairy products. Wholesalers compete vigorously for the best location and the most space in self-service display coolers.

The National Dairy Promotion and Research Board, established by Congress in 1983 and comprised of 36 U.S. dairy leaders, carries out a coordinated program to maintain and expand markets and uses for fluid milk and dairy products. Its programs, financed by an assessment of up to 15 cents per hundredweight of milk sold, include advertising, nutrition research and education, product development, public relations, and program evaluation.

The American Dairy Association, a part of the United Dairy Industry Association, represents an industry effort in dairy product promotion. Financed by producers, it sponsors newspaper, magazine, radio, and television advertising. The association has cooperated with producers and processors of other foods for promotional purposes. Full-color ads featuring cottage cheese-and-peach salads, milk and cookies, or other combinations have been used widely and effectively.

In most states in which dairy production and processing are significant industries, state dairy associations promote dairy products in cooperation with the ADA and other groups. Funds may be obtained by donations, voluntary deductions from milk payment checks, or a tax allowed by the state legislature and voted by a majority of the dairy producers. A few states appropriate money for this purpose.

One significant effort of ADA staff has been the training of supermarket managers in managing their dairy departments to achieve maximum volume and profit.

Two other units of the United Dairy Industry Association, Dairy Research Inc. (DRINC) and the National Dairy Council, finance and conduct research in human nutrition, the nutritional properties of dairy products, utilization of dairy byproducts, the methods of processing, and other topics.

The purpose of this research is to eventually increase consumption and use of dairy products, as well as to provide more efficient processing of higher quality and more nutritious products. Most financing comes from dairy product processors and related industries.

QUESTIONS FOR STUDY AND DISCUSSION

1. How does free-stall housing differ from tie-stall housing?
2. Name the major components of the total dairy setup.
3. What are the major merits and limitations of tie-stall barns?
4. Name three quality standards required for production of milk sold for human consumption.
5. What is meant by pulsation rates and pulsation ratios of milking machines?
6. Explain two major differences between a "herringbone" and "parallel" milking parlor.
7. Describe proper preparation of the udder prior to attachment of the milking unit.
8. What are standards for cooling of milk, and how is it accomplished?
9. Provide four steps in proper cleaning of milking equipment from the end of one milking to the beginning of the next.
10. Why do many dairy owners belong to cooperatives?
11. What are the major factors that determine the price received per hundred pounds of raw milk?
12. What is accomplished by pasteurization and homogenization of milk in the processing plant?
13. Which dairy products are less popular and which are more popular today than 20 years ago?
14. What associations or groups within the United States have greatly helped in the promotion of dairy products?

32

Horses and Ponies

Horses and ponies are an increasingly important segment of the animal industry. Numbers have increased rapidly during recent decades, and with continued population growth and affluence in the United States, further increases are expected. That is, the horse and pony industry closely follows the economic trend.

This chapter will serve as an introduction to the industry, to breeds, and to horse and pony characteristics. Discussions of selection, management, breeding, training, and equitation are also included. Horse and pony nutrition and feeding is discussed in Chapter 7. The terms *horse* and *light horse* will be used often in the chapter to include both horses and ponies. Horses used in earlier years for farm tillage and field work and for hauling commodities and produce—large, muscular animals weighing 2,000 to 2,600 pounds—were called *draft horses*. The term *light horse* was used to describe smaller, less muscular, lightweight horses used for pleasure and for handling livestock (Figure 32–1).

Upon completion of this chapter, the reader should be able to

1. Describe the magnitude of the light horse industry, in terms of approximate numbers of animals and horse shows.
2. Distinguish between riding, race, and driving horses and list major breeds used for each.
3. List two or more breeds categorized as light breeds, draft breeds, and ponies.
4. List and describe the five major colors of horses.
5. Define conformation and list at least one desired trait related to the head, neck, shoulders, back, and feet and legs.
6. List and describe the three natural horse gaits.
7. Distinguish the differences in appearance of temporary and permanent teeth.
8. List at least two criteria for proper horse housing.
9. List the major types of rations based on the age and specific requirements of the horse.
10. Indicate the importance of forage in the horse's ration and criteria for selecting hay for horses.
11. Differentiate among pasture breeding, hand mating, and artificial insemination.
12. List several important vaccines typically included in a vaccination program.
13. Indicate when the mare becomes seasonally active sexually, and how the estrous cycle can be manipulated for earlier breeding.

Figure 32–1
Horses and ponies are valuable for both work and pleasure. The left photo shows a cutting horse in competition. The goal is to keep the calf isolated from the herd. At the right, a Hackney pony at a major show. (Courtesy of *Quarter Horse Journal* and Silver Summit Farm.)

14. List several factors that can cause lowered conception rates in mares.
15. List several sound management practices that help to ensure (a) successful foaling and (b) good foal health and survival.
16. List several basic guidelines for successful training of the horse.

32.1   SIZE AND SCOPE OF THE INDUSTRY

A high point in the U.S. horse population was reached in 1915 with an estimated 26.6 million head. Horse numbers steadily decreased to a low point of about 3 million in 1959. Since that time the population has increased to an estimated 6.9 million head and 7.1 million participants in 1998.[1] Major reasons for this increase include the popularity of horses for youth and adult leisure and for 4-H projects.

About 3.9 million families have light horses for riding and pleasure. The average number of people that ride horses annually is estimated at 30 million.[2] There are about 10,000 horse shows each year that contribute over $220 million annually to the economy.

An estimated half million horses are used as "cow ponies" in the United States. Most are where the cow herds are—on ranches of the Plains and Rocky Mountain area, in the southern Corn Belt, and in the Southeast. They are also used to "ride the pens" in commercial feedlots.

Horse racing involves slightly more than 1 percent of the horse population. Nearly 142 thousand races run annually in the United States. Of these, about 70,000 are Thoroughbreds that compete in about 56,000 races with the balance consisting of har-

---

[1] American Horse Council Foundation.
[2] United States Equine Marketing Association, P.O. Box 2529, Leesburg, VA 22075.

Figure 32–2
The left photo shows a girl and her horse in barrel race competition. At right, harness racers in tight formation.
(Courtesy of *Quarter Horse Journal* and Quarter Horse Association.)

ness and quarter horse races (Figure 32–2). Horse racing ranks as the second leading spectator sport, next to baseball, with annual attendance of about 70 million spectators.

Race attendance and the size of the purse largely depend on whether the state permits parimutuel wagering on the outcome of the race. Where it is permitted, dollars wagered are taxed, and the revenue to the taxing body—state or local governmental unit—is significant.

Although draft horse and mule numbers have diminished greatly, they are still used as a major source of power (Figure 32–3) on some small farms and are enjoyed by others as a hobby.

32.2   TYPES, COLORS, AND BREEDS

In most livestock and poultry breeds, all animals of one breed are of similar color, and color is nearly always a primary differentiating factor among breeds. This is not necessarily true in horses.

Horse owners often have strong preferences for coat color, and color is often a primary factor in describing or identifying individual horses. However, purchasing of horses based largely on color can restrict the number and quality of horses available. Also, a breeding program based primarily upon color can slow genetic progress in other traits. However, certain colors are discriminated against or not permitted in some breeds of horses.

The five basic horse colors of bay, black, brown, chestnut, and white are briefly described as follows:

- *Bay.* A mixture of red and yellow, much the color of a loaf of well-baked bread. A light bay shows more yellow, a dark bay more red. The darkest is the mahogany bay. Bays always have black points (mane, tail, and legs).
- *Black.* True black without light areas. A black horse almost invariably has black eyes, hoofs, and skin. The points are always black.

Figure 32–3
The matched team of Belgian horses at left is being sold at an auction in northern Indiana. On the right is pictured an exceptional Clydesdale horse. Clydesdales are known for their strength and stylish appearance as well as a draft animal for work, show, or simple pleasure.
(Courtesy of the Farmer's Exchange, New Paris, IN; Clydesdale Breeders of the United States; and Maureen Blaney.)

- *Brown.* Body brown, usually with light areas at muzzle, eyes, and flank. Many brown horses are mistakenly called black because they are so dark. The mane and tail are black.

- *Chestnut (sorrel).* The coat is basically red, varying from a bright yellowish red to rich mahogany red. The mane and tail are normally the same shade as the body. If the mane and tail are lighter, the horse is termed a flax or flaxen chestnut. The mane and tail are never black. Sorrel is differentiated by some breed associations as a light red color.

- *White.* The true white horse is born pure white and remains pure white throughout its lifetime. Very little, if any, seasonal change takes place in its coat color. Age does not affect it.

The major variations of color are as follows:

- *Dun and buckskin.* The dominant hair is some shade of yellow, which may vary from a pale yellow to a dirty canvas color, with mane, tail, skin, and hoofs grading from white to black. Duns often have a dorsal stripe. Buckskins usually have black points.

- *Gray.* Born blue or almost black, more and more white hairs come into the coat until the age of 8 or 10, when the horse will appear almost white. The dapple (small patches or spots of black) generally comes between the second and fifth year.

- *Palomino.* A golden color, varying from bright copper color to light yellow, with white mane and tail. True palominos have no black points. The breed description lists the ideal color as that of a "newly minted coin."

- *Pinto (calico or paint)*. A pinto is a spotted horse that has more than one color in or on its coat in large irregular patches or spots, usually on a white background.

- *Roan*. The coat carries white hairs intermingled with one or more basic colors. Most roans are combinations of bay, chestnut, or black, with white hairs intermingled. They are known as red, strawberry, or blue roan, respectively.

Haircoat patterns are also important in some breeds of horses.

- *Appaloosa*. Patterns may appear on any base color. Four distinguishing characteristics are (1) spotted coat patterns; two of the more distinctive are leopard (white with spots over entire body) and blanket (white blanket over his loin and hips, with or without spots); (2) mottled skin, especially around the muzzle and genitalia; (3) white sclera (outer ring) around the eye; and (4) vertically striped hooves.

- *Paint/pinto*. Pattern may appear on any base color and is a combination of white and colored markings. Terms for the two most common color patterns are *tobiano* and *overo*. (See Pinto, Table 32–1.)

TABLE 32–1	Origin and Color of Common Breeds of Horses and Ponies	

Types & Breeds	Origin	Color
Light Breeds		
American Albino	United States—White Horse Ranch, Naper, NE	Snow-white hair, pink skin, preferably dark eyes.
Quarter Horse	United States	Chestnut, sorrel, bay, and dun are most common, but may be palomino, black, brown roan, or copper-colored.
American Saddle Horse	United States—Fayette, KY	Bay, brown, chestnut, gray, black, or golden
Appaloosa	United States—Oregon, Washington, and Idaho	Variable, usually white over loin and hips, having dark round or egg-shaped spots thereon. Spots are less than four inches across.
Arabian	Saudi Arabia	Bay, gray, and chestnut; occasionally black; white marking on head and legs common. The skin is always dark.
Hackney	England—Norfolk and adjoining counties on eastern coast	Chestnut, bay, and brown; occasionally roan and black; white markings desirable.
Morgan	United States—New England states	Bay, brown, black, and chestnut; extensive white markings are uncommon.
Morocco Spotted	United States—from animals of Hackney and Saddle Horse breeding	Spotted. The secondary color must not be less than 10 percent.
Palomino	United States	Golden (color of newly minted gold coin, or very slightly lighter or darker). Mane and tail white, silver, or ivory, and having not more than 15 percent dark or chestnut hair in either.

(continued)

TABLE 32–1	Origin and Color of Common Breeds of Horses and Ponies—Continued	
Types & Breeds	**Origin**	**Color**
Pinto	United States—from horses brought in by Spanish	Preferably half color or colors and half white with many spots well placed. The two distinct pattern markings are *overo* (one white leg, head all or partially white) and *tobiano* (four white legs, head dark). Spots are usually more than four inches across.
Standardbred	United States	Bay, brown, chestnut, and black are most common, but gray, roan, and dun are found.
Tennessee Walking Horse	United States—middle basin of Tennessee	Sorrel, chestnut, black, roan, white, bay, brown, gray and golden; white markings on feet and legs are common.
Thoroughbred	England	Bay, brown, chestnut, and black, and less frequently, gray; white markings on the face and legs are common.
Draft Breeds		
Belgian	Belgium	Chestnut or roan with black points.
Clydesdale	Scotland	Bay, brown, chestnut, and roan, with splashes of white, especially on the face and lower legs.
Percheron	France	Black, gray, or dapple gray.
Shire	England	Bay, brown, black, and gray, with white markings.
Suffolk	England	Chestnut with white markings.
Ponies		
Pony of the Americas	United States—Mason City, IA	Similar to Appaloosa; white over the loin and hips, with dark round or egg-shaped spots.
Shetland Pony	Shetland Islands	All horse colors, solid or broken.
Welsh Pony	Wales	Any color except piebald (large amount of white on head) and skewbald (white markings on body).

Sources: *"Meet the Breeds,"* Pamphlet, Kentucky Horse Park, Lexington, KY, 1989, and numerous breed association publications.

Because the horse was domesticated some 5000 years ago, breeding and selection has resulted in a variety of types and breeds developed for different purposes. The types of horses are commonly categorized as *riding horses, race horses, driving horses,* and *draft horses.* The more typical breeds listed according to their primary use, approximate height, and weight are shown in Table 32–2.

Note the difference in conformation and size of the draft breeds that commonly stand 64 to 68 inches tall (16 to 17 hands) and may weigh a ton or more (Figure 32–3) to that of the pony. Most ponies, such as the Shetland shown in Figure 32–4, may be only 36 inches tall (9 hands) at the withers and weigh as little as 500 pounds. Also, the mule has become popular as an animal of pleasure and work. The mule foal, shown in Figure 32–4, is a hybrid resulting from the cross between two species of the Equidae family— the breeding of a male donkey (or jack) to a mare of the horse breeds.

Other differences in appearance and conformation of several breeds of horses and ponies are illustrated in Figure 32–4. Also, the origin and prevalent colors of most horse and pony breeds are presented in Table 32–1. The horse breeds categorized by type, principal use, and approximate height and weight are shown in Table 32–2.

TABLE
32–2

Types, Uses, Heights, and Weights of Horse Breeds

Type	Use	Breeds	Height (hands)[a]	Weight (lb)
Riding Horses	Three-gaited saddle horses	American Albino Horse American Saddle Horse Appaloosa Arabian Cleveland Bay Missouri Fox Trotting Horse Morgan Morocco Spotted Horse Palomino Pinto Thoroughbred	14.2–17	900–1,400
	Five-gaited saddle horses	American Saddle Horse	15.2–16.2	900–1,400
	Walking horses	Tennessee Walking Horse Grades, crossbreds, or following purebreds: Appaloossa Arabian Morgan	15–16	1,000–1,200
	Stock horses	Morocco Spotted Horse Palomino Pinto Quarter Horse Thoroughbred	14.1–16	1,000–1,300
	Polo mounts	Grades, crossbreds, and purebreds of all breeds, but predominantly of Thoroughbred breeding	14.2–15.2	1,000–1,250

(continued)

Table 32-2	Types, Uses, Heights, and Weights of Horse Breeds—Continued			
	Hunters and Jumpers	Grades, crossbreds, and purebreds of all breeds, but predominantly of Thoroughbred breeding	15.2–16.2	1,000–1,250
	Ponies for riding	Connemara Pony		
		Gotland Horse		
		Pony of the Americas	9–14.2	500–900
		Shetland		
		Welsh		
Race Horses	Running race horses	Thoroughbred	15.1–16.2	900–1,150
	Harness race horses (trotters and pacers)	Standardbred	14.2–15.2	900–1,200
	Quarter race horses	Quarter Horse	14.2–15.2	1,000–1,200
	Heavy harness horses	Hackney	14.2–16.1	900–1,300
	Fine harness horses	American Saddle Horse (predominantly, although other breeds are so used)	14.2–17	900–1,400
Driving Horses	Roadsters	Standardbred	14.2–15.2	900–1,200
	Ponies for driving:			
	Harness show ponies	Hackney		
	Heavy harness ponies	Shetland	9–14.2	500–900
		Welsh		
Draft Horses	Power, Work—especially in winters, parades (advertising)	Belgian	16–17	1,800–2,000
		Clydesdale	16–17	1,700–1,900
		Percheron	16–17	1,600–2,000
		Shire	16–18	2,000–2,200
		Suffolk	16	2,000–2,400

[a]Height denoted as hands and inches (14.2 hands = 14 hands and 2 inches, or 58 inches).

Source: Adapted from Horses! Horses! Horses! by Dr. M. E. Ensminger, 3699 E. Sierra Avenue, Clovis, CA 93612.

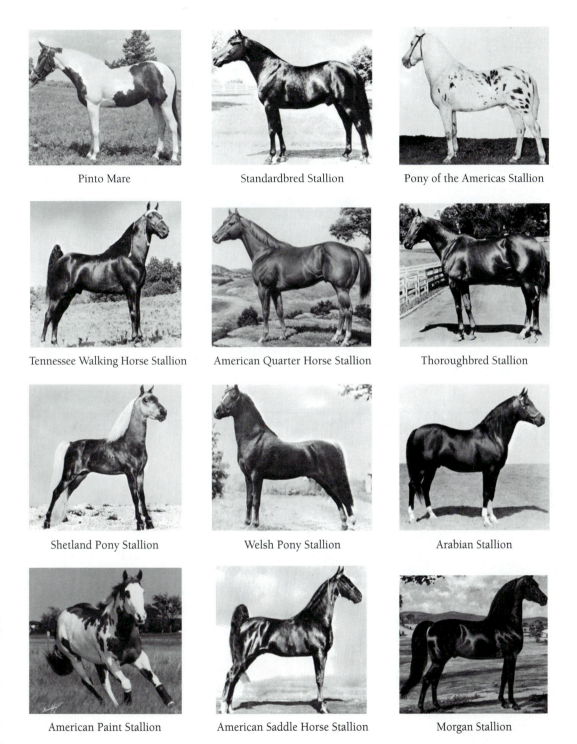

| Pinto Mare | Standardbred Stallion | Pony of the Americas Stallion |

| Tennessee Walking Horse Stallion | American Quarter Horse Stallion | Thoroughbred Stallion |

| Shetland Pony Stallion | Welsh Pony Stallion | Arabian Stallion |

| American Paint Stallion | American Saddle Horse Stallion | Morgan Stallion |

Figure 32–4
Common breeds of horses and ponies, and donkey and mule.
(Courtesy of USDA and American Donkey and Mule Society.)

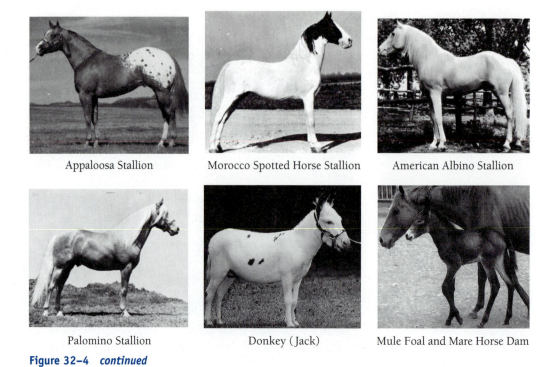

Appaloosa Stallion   Morocco Spotted Horse Stallion   American Albino Stallion

Palomino Stallion   Donkey (Jack)   Mule Foal and Mare Horse Dam

Figure 32–4   *continued*

Figure 32–5
The parts and desired conformation of the horse.
(*Source*: American Quarter Horse Association.)

32.3 CONFORMATION AND BODY PARTS

Desired performance of the horse, pony, or mule, whether it be for work (pulling, packing, roping), short- or long-distance racing, hunting and jumping, or for pleasure (trail rides, shows) is directly related to proper conformation.

Conformation is defined as an arrangement and blending of muscle, bone and other body tissue that results in a desired balance and pleasing appearance. Each breed has differences in conformation desired by breeders and owners. This desired or ideal conformation differs with the type of horse and its intended purpose. Breeders seek the desired conformation (form) to provide the desired performance (function).

For proper evaluation of conformation you must understand the desired function of the horse, all parts of the horse, and the function of the parts (Figure 32–5).

Some desired traits related to the horse's body structure and parts include

Head	ears squarely set and proportional to head
	eyes bright, large, properly located to the side to provide a rounded arch of vision
	nostrils—large
	jaw and muzzle—well defined
Neck	long and slender, proportional to length of head and should junction at the chest
Shoulder	proper angle (45–50°) in relation to the pasterns
Withers	sharp, prominent, slightly higher than hind quarters
Back	short top line
Barrel	ribbed high, adequate depth of heart, well proportioned
Hind quarters	square, full appearance when viewed from side, well proportioned under the body
Feet and legs	straight line from point of buttocks to hocks, parallel to cannon bone, and slightly behind the heel

Ideal leg structure is illustrated in Figure 32–6. Correct structure and position of feet and legs are especially important as they determine how smoothly the horse moves at each gait, and with speed, endurance, and safety in racing or work.

Also, an important structural factor in a smooth ride and in comfort and endurance for the horse is the slope and length of the shoulder and pastern. If they are too steep or short, there is little cushioning or spring each time a foot strikes the ground. If the slope is too flat or long, there is considerable strain on the pastern (Figure 32–7).

32.4 TEETH AND AGE DETERMINATION

The age of horses under 10 years of age can be determined by the teeth—the number, size, wear, and degree of forward angle (Figure 32–8). Temporary teeth are smaller and usually whiter than permanent teeth. A mature male usually has 40 teeth, a mature female 36, and a foal of either sex at 6 to 10 months has 24. Although not totally accurate, the appearance of the teeth and their degree of wear are commonly used to estimate age when records of a horse's birth date aren't available.

The number of teeth, size, amount of wear, and degree of forward angle are carefully inspected in estimating age. Factors causing differences in appearance and wear include the quality and form of feed, whether the horse has been largely stable or

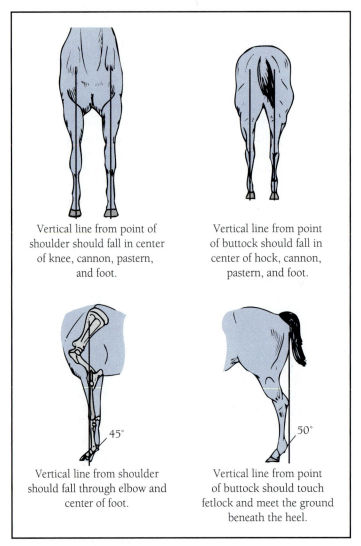

Vertical line from point of shoulder should fall in center of knee, cannon, pastern, and foot.

Vertical line from point of buttock should fall in center of hock, cannon, pastern, and foot.

45°

Vertical line from shoulder should fall through elbow and center of foot.

50°

Vertical line from point of buttock should touch fetlock and meet the ground beneath the heel.

Figure 32–6
Ideal leg structures.
(*Source*: Adapted from National 4-H Council.)

Figure 32–7
The ideal pastern slope is 45°.
(*Source*: National 4-H Council.)

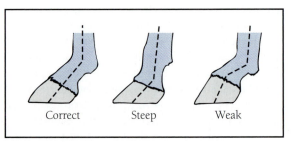

Correct   Steep   Weak

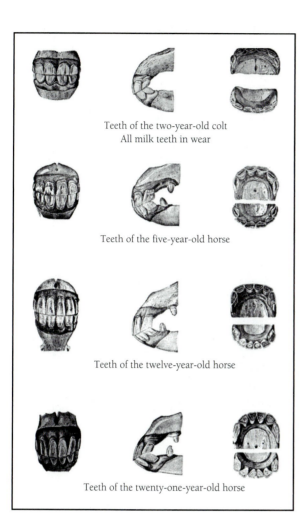

Figure 32–8
Front, side, and surface views of
incisors of a horse at various ages.
(*Source*: National 4-H Council.)

Teeth of the two-year-old colt
All milk teeth in wear

Teeth of the five-year-old horse

Teeth of the twelve-year-old horse

Teeth of the twenty-one-year-old horse

pasture fed, disease, degree of dental care, and heredity. For example, horses on sandy ranges or pastures usually will have teeth showing more wear than horses that are largely stable fed.

32.5   UNSOUNDNESS

Unsoundness refers to an injury or deviation in structure that limits the usefulness of the horse, from very slight to severe, and adversely affects its value. On the other hand, *blemishes* refer to less serious defects, such as a wire-cut injury (or subsequent scar), that do not affect soundness but may detract from a horse's appearance, thus reducing its value. Horses often have blemishes but are sound.

Because the purpose of a horse is movement, structural deformities, weaknesses, or injuries that impair or make walking or running painful are highly significant and economically very important. Space prevents listing and discussing all possible unsoundnesses and their causes, but common ones and their locations are shown in Figure 32–9.

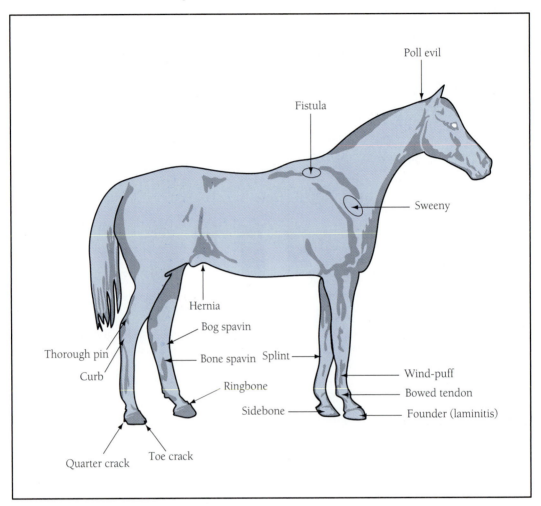

Figure 32–9
Terms and locations of some common unsoundness that may occur in the horse (or pony and mule).
(*Source*: Purdue University.)

Most unsoundnesses of the feet and legs are caused by injury or strain; a horse with a structural weakness is logically more susceptible to injury and strain. Unsoundnesses prevent a horse from performing at its maximum capability. Mares or stallions with unsoundnesses caused by structural weakness should not be considered for breeding purposes.

32.6 GAITS

A *gait* is a rhythmic movement of the feet and legs, natural or acquired by training. The three natural gaits are the walk, trot, and canter or gallop. The term *beat* is used to denote the number of movements in a gait. These are illustrated in Figure 32–10.

The *walk* is a natural, slow, flatfooted, four-beat gait—the latter meaning that each foot takes off from and strikes the ground at a separate interval. It should be springy, regular, and true.

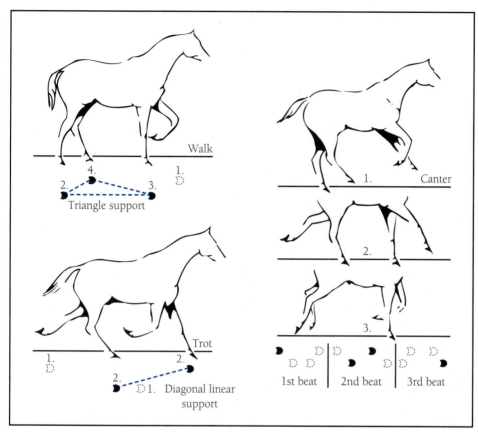

Figure 32–10

The three basic gaits of horses and ponies, showing foot and leg positions. Solid hoofprints are touching the ground.

(*Source: Horses! Horses! Horses!* by Dr. M. E. Ensminger, Clovis, CA.)

The *trot* is a natural, intermediate to rapid, two-beat, diagonal gait in which the front foot and the opposite hind foot leave and strike the ground simultaneously. For a brief moment all four feet are off the ground, and the horse seemingly floats through the air.

The *canter* is a slow- to medium-speed, restrained, three-beat gait. Two diagonal legs are paired, producing a single beat that falls between the successive beats of the other, unpaired legs.

The *gallop* is a fast, four-beat gait where the feet strike the ground separately—first one hind foot, then the other hind foot, followed very quickly by the forefoot on the same side as the first hind foot, then the other forefoot. The gallop is the natural fast gait of both the wild horse and the Thoroughbred race horse.

The *pace* is a fast, two-beat gait used in harness racing in which the front and hind feet on the same side start and stop simultaneously.

Other gaits are the *stepping pace,* which is a slower and modified pace, *foxtrot, running walk, rack* (formerly called *single-foot*), and *amble.* The rack and stepping pace are required of the five-gaited horse in addition to the three natural gaits.

In pleasure-horse shows, three-gaited and five-gaited horses compete on their ability to move smoothly and correctly at each of the required gaits. This, then, reflects the ability of the horse, the excellence of training, and the ability of the rider or driver.

Some conformation defects may affect a horse's feet or leg movements resulting in an altered gait, and in some cases where cases cause injury (such as cases where the hoof of one foot striking another).

32.7   HOUSING AND FENCING FOR THE HORSE

Many horses are kept in stalls or in small pens, much different from their natural environment where they roamed freely over the hills and plains. Many of the behavioral characteristics of horses are those needed for survival in a natural environment—long and strong legs, speed, grazing ability, and ability to fight with legs and teeth.

Considerable exercise is important for horses kept in stalls, for their health and vigor and to prevent development of bad habits or vices due to boredom. Open and luxuriant pasture can provide both exercise and a part of the nutritional needs.

Good management includes providing safe and functional facilities. Fences should be constructed of hardwood board, post and rail, metal pipe, cable, woven or smooth wire, or other smooth and strong material. Also, white vinyl rail fencing that is especially attractive and needs no painting is now commercially available. Barbed wire should not be used.

Stalls, runs, and pens should be free of sharp edges, protruding objects, or anything that might injure horses. Structures or circumstances that would permit a horse to get a foot caught, such as bars or open slats in stall fronts, should be avoided.

Adequate shelter from inclement weather or intense heat should be provided (Figure 32–11). For small family operations, simple economical housing is adequate, and with proper management can ensure comfort and healthy conditions for the horse. Two important housing factors are an adequate quality of well-distributed fresh air and control of humidity. Considerable heat and moisture produced by confined horses

Figure 32–11
An attractive horse facility that provides protection from inclement weather conditions and ease in housing management. (Courtesy of FBi Buildings.)

must be removed from the stable through proper ventilation design and stable management. Beginners sometimes tend to close up facilities in cold weather to best suit their comfort resulting in excessive humidity, insufficient fresh air and air movement, and strong odors.

As in all other livestock and poultry operations, appearance of the facilities provides a clue to the quality of management. Attractive and functional barns and fences and other facilities in good repair suggest to a visitor or client good management and care of the animals.

32.8   FEEDING GUIDELINES

Nutrition and feeding are discussed in depth in Chapter 7. Horses on pasture will graze intermittently—eat small amounts throughout the day. Feeding schedules in confinement should approximate the natural feeding habits. Horses may have access to hay on a continuing basis, and any grain fed should be in small amounts provided two or three times per day, not in one large portion once a day. Plenty of clean, fresh water should be available to the horses at all times.

Table 32–3 provides some typical guidelines that would approximate the needs of the horse. Note the changes in nutrient content of the rations as the young horse matures and has different requirements based upon the demands of its owner (maintenance, growth, gestation, lactation, or work).

TABLE 32–3	Horse Feeding Guidelines

Maintenance ration:
   9% crude protein in total ration
   2% of body weight per day as forage
   Usually do well on forage, mineralized salt block
      and water
Gestation ration:
   11% crude protein
   0.6% calcium and 0.5% phosphorus
   Increase concentrate to 1% of body weight/day
   Free access to water and good forage
Lactating ration:
   13–14% crude protein
   1 1/2% of body weight as concentrate
   Feed frequent small meals
   Calcium at 1% and phosphorus at 0.7%
Work ration:
   Depends on exercise level
   8–10% crude protein
   Energy is 25%, 50%, and 100% above maintenance
      for light, medium, and strenuous exercise
   Frequent meals
   1% of body weight is minimum/day for forage

Weanling ration:
   16% crude protein
   1% calcium and 0.8% phosphorus
   Limited feed concentrate 1 lb/day/month of age
   Access to free exercise, minerals, water, and good
      alfalfa
Yearling ration:
   14% crude protein
   0.6% calcium and 0.5% phosphorus
   1 1/2% of body weight/day as hay
   Plenty of exercise, water, and minerals

Source: 4–H Animal Science Meeting Guide Notebook, *Minnesota Extension Service.*

Also, note in Table 32–3 that forage constitutes 50 percent or more of the daily ration of the horse. As a rule, horses should consume no less than 1 percent of their body weight in hay (or hay equivalent from pasture) to ensure proper functioning of their digestive system. The wise horse manager is very selective to ensure that only high quality hay is purchased and fed. Green, leafy, early-cut forage, free from dust, mold and weeds, that has been properly harvested and stored should result in high-quality hay. When significant quantities are to be purchased, forage test results should be available to supplement visual appraisal.

32.9   HEALTH MANAGEMENT

As in other species, successful health management requires recognition of good signs of health—in order to act promptly when abnormal signs are observed. Emphasis also must be on a sound health program, or preventive disease control, rather than treatment of disease when they occur.

Horses are susceptible to numerous communicable diseases that should be controlled through a proper vaccination schedule, as agreed upon by the owner and veterinarian. Several of the more significant diseases and a vaccination schedule are shown in Table 32–4.

Most common horse diseases can be prevented by an initial vaccination followed by annual "booster shots" to maintain immunity. Such diseases include tetanus, strangles, influenza, eastern and western encephalomyelitis, and viral rhinopneumonitis (the latter especially affecting brood mares). Also, when horses are moved to another stable, the Coggin's test for equine infectious anemia should be performed.

To properly assess the health of the horse regular observation is essential—to become acquainted with the normal traits and behavior and to note deviations from normal. Important signs to check include changes in appetite, body condition and weight, fecal consistency and color, urine color and flow, excessive or lack of sweating, coughing, degree of alertness (or depression), or changes in hair coat.

Upon observation and suggestion of abnormal health, the horse's vital signs should be closely monitored; these include body temperature, heart rate, and respiration rate. Also, mucosal membrane color and skin pliability should be checked for abnormal changes.

Colic and *founder,* two major but noninfectious health problems in horses, are related to feeding and are discussed in detail in Chapter 7. Prevention of diseases and parasite infestations is necessary in close confinement situations. Horses need to be wormed regularly—about every 3 to 4 months—to prevent internal parasite infestation.

32.10   FEET AND DENTAL CARE

Foot care is an essential part of horse management; the foot is said to be the foundation of the horse. Foot neglect can reduce or prevent the horse's normal performance. Structural parts of the hoof are shown in Figure 32–12. The foot and hoof function together to support weight and absorb shock as the foot strikes the ground. Equally important is the movement of the *frog* portion of the hoof, which with each step creates force against the lateral cartilages. These, in turn, compress blood veins that force blood toward the heart.

TABLE 32–4		Vaccination Schedule for Horses				
Disease/ Vaccine	**Foals, Weanlings, Horses no vaccination history**	**Yearlings**	**Performance Horses**	**Pleasure Horses**	**Broodmares**	**Comments**
Tetanus Toxoid	First dose: 3–4 months Second dose: 4–5 months	Annual	Annual	Annual	Annual: 4–6 weeks prepartum	Booster at time of penetrating injury or surgery if last dose is not administered within 6 months; antitoxin administered if a cut or wound could harvest tetanus organism
Encephalomyelitis (EEE, WEE, VEE)	First dose: 3–4 months Second dose: 4–5 months	Annual: spring (2 doses, 3 weeks apart)	Annual: spring (2 doses, 3 weeks apart)	Annual: spring (2 doses, 3 weeks apart)	Annual: 4–6 weeks prepartum	Booster every 6 months in endemic areas; the antigen for VEE is available only as a combination vaccine with EEE and WEE, but is needed only when threat exists of a Venezuelan outbreak
Influenza	First dose: 6 months Second dose: 7 months Third dose: 8 months (repeat at 3 month intervals)	Every 3 months	Every 3 months	Annual: with added boosters prior to likely exposure (shows, etc.)	At least biannual with one booster 4–6 weeks prepartum.	A series of at least 3 doses is recommended for primary immunization of foals
Rhino-pneumonitis (EHV-1, EHV-4)	First dose: 5–6 months Second dose: 6–7 months Third dose: 7–8 months (repeat at 3 month intervals)	Every 3 months	Every 3 months	Optional: biannual if elected	5th, 7th, and 9th month of gestation (inactivated EHV-1 vaccine)	If primary series is started before 3 months of age, a 3-dose primary series is preferred. Vaccination of mares before breeding and 4–6 weeks prepartum is suggested. Some veterinarians suggest year-round protection
Rotavirus	Not approved for use in the foal				8th, 9th, and 10th month of gestation; booster 30–60 days of gestation; annual booster; consult with a veterinarian as to whether to repeat with a single booster or continue series	
Rabies	First dose: 3–4 months Second dose: 4–5 months	Annual	Annual	Annual	Annual: before breeding Rabies vaccination (recommended in endemic areas)	
Botulism	3-dose series at 30-day intervals; age at first injection dependent on local factors	Annual	Annual	Annual	Annual: 4 weeks prepartum	Only in endemic areas

Source: Adapted from the *American Association of Equine Practitioners*.

Figure 32–12

External structures of the hoof. The weight of the horse is not exerted on the sole but is borne by the hoof wall. Note the wedge-shaped frog that occupies the area between the bars, and expands the heels as the foot strikes the ground.
(*Source*: Purdue University.)

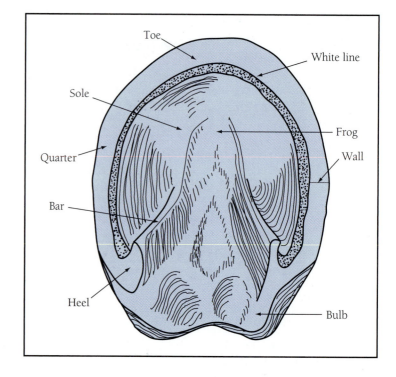

Proper regular trimming and shoeing of the hoof are essential for normal foot function. The feet of brood mares, foals, and riding horses should be trimmed every 4 to 6 weeks. Horses kept in stalls should have their feet checked and the bottoms of the hooves cleaned out daily, to prevent thrush.[3] Riding horses that are shod should have their shoes checked on a regular basis, and, depending on intensity of use, shoes may need to be replaced or tightened every 4 to 6 weeks. Only trained and experienced persons should be allowed to trim feet or to shoe, especially in the case of riding and performance horses.

Dental examinations and maintenance should be performed on a regular basis. *Floating,* the filing of sharp enamel points, should be a routine practice (on a 6-month or yearly basis) to facilitate chewing and reduce the chance of laceration of the tongue or cheeks. These sharp *points* develop from the grinding action of the jaws and appear on the outside of the upper and inside of the lower cheek teeth. As points develop, the chewing of food becomes uncomfortable, and, if severe enough, can result in lacerations of the tongue and cheek.

For prevention of dental problems, a biannual oral exam for older horses, and at least an annual oral exam for others, is recommended to detect and correct dental problems. Observation for abnormal signs should be a routine practice. These include holding the head awkwardly when eating, facial swelling, excessive dropping of feed when chewing, reduced feed intake, and foul breath.

---

[3] Thrush refers to a disease of the frog portion of the hoof that can be caused by many different organisms.

32.11 BREEDING MANAGEMENT OF THE MARE AND STALLION

Chapters 18 through 20 discuss, from a genetic standpoint, selection principles and breeding programs used for livestock and poultry. In addition to selection of conformation and personality traits, selection of both mares and stallions should be based upon performance records, and, if available, the performance records of their progeny. Conformation traits and performance ability are measurable with moderate to high heritability estimates. (See Table 18–2.)

In the case of horses, most selection for genetic improvement is *practiced on the sire side*. Owners of breeding stallions select males that best complement each mare in an effort to produce foals of excellent conformation. Similarly, mare owners select stallions of similar size that correct faults in the mare. In larger herds, especially in production of racing stock, conscious decisions are made as to which mares to breed. But in the case of pleasure horses and working stock, all mares are bred if pregnancy would not interfere unduly with their use.

Discussion of breeding methods here, then, is directed to the mechanics of mating, once the desired stallion has been identified. For review, refer to Chapter 14 where reproductive anatomy and physiology are discussed.

Breeding methods range from unattended pasture mating to artificial insemination of the mare. In pasture mating, a stallion is simply pastured with the mare; the mares usually are bred whenever they show signs of estrus. Usually only one stallion is kept with mares at a given time; two or more will tend to fight and can injure each other or the mares. Pasture mating is convenient and results in high conception rates, although accurate breeding records are more difficult to maintain.

A rather common practice in breeding horses is to maintain both the estrous (in heat) mare and the stallion under control of handlers. The mare is said to "stand for service," and the term used to describe the system is "hand mating." Artificial breeding is also used and has some distinct advantages, especially for large breeding studs; for example, it eliminates the need to move stallions or mares for breeding. The success of both hand breeding and artificial insemination depends largely on the ability of the manager to determine optimum time for breeding.

Where hand mating or artificial insemination are used, mares usually are teased with a stallion every other day during the breeding season to check for signs of estrus.

The mare is *seasonally polyestrous*. During the winter months the nonpregnant mare is in an inactive reproductive status referred to as *anestrus*, and will not respond to the stallion. However, the mare's estrous cycle becomes active in the spring and summer months.[4] Estrus is characterized by receptivity to the stallion. The entire estrous cycle of the mare is approximately 21 days, and length of heat ranges from 5 to 7 days. Ovulation usually occurs 24 to 36 hours before the end of estrus. Therefore, mares are more likely to conceive if bred near the end of the heat period. Some breeders hand mate on the third day of estrus and repeat every other day until the end of heat. Some can detect ovulation time by rectal palpation of the ovaries and use this as a guide to breeding time.

---

[4] Most mares also exhibit "foal heat" about 7 to 9 days after foaling. Some owners may desire to breed certain mares during foal heat, but not until a thorough examination by a veterinarian indicates she is free from lacerations, bruises, and infection.

Artificial insemination diminishes risk of diseases and infection transmission or injury to the mare or stallion. Semen can be evaluated prior to insemination, and more mares can be bred during a season as well as during the stallion's lifetime. Some breed associations do not permit registration of a foal conceived by artificial insemination, whereas others allow artificial insemination only when done on the premises by one natural service. Semen for insemination usually is collected by use of a collection dummy or mare. As the stallion mounts, the penis is guided into an artificial vagina where the ejaculated semen is collected. Insemination of the mare is by insertion of a catheter, pipette, or insemination syringe far into the cervix and deposition of the semen.

Some breeders choose to breed earlier in the year, but as mentioned earlier, the mare is seasonally anestrus during the winter months. Some extend the photoperiod starting in December by increasing artificial light to a 16-hour light day. This technique stimulates hormonal production bringing about estrous cycle activity and ovulation within 60 to 90 days.

Another technique that can be used to manipulate the estrous cycle is the use of hormone preparations such as prostaglandin F_2alpha. These mimic the effects of prostaglandins, causing regression of the corpus luteum, if present, and consequent reduced progesterone activity and shortened diestrus. Other hormones, such as human chorionic gonadotrophin, which possess a high LH activity can be administered IV to hasten ovulation. Progesterone and synthetic progestins also can be used to prevent mares from coming into estrus.

Low conception rates and most other breeding problems result from failure to synchronize breeding with ovulation time—due in part to poor observation of signs of estrus. Also, infections within the mare's reproductive tract contribute to lowered conception, therefore proper breeding hygiene methods are essential. These include proper washing of external genitalia and wrapping of the mare's tail to reduce introduction of infectious agents, and thorough washing and rinsing of the stallion's penis.

Nutrition prior to breeding and during pregnancy is also important (see Chapter 7). Body condition of both stallion and mare should be carefully monitored to ensure that neither obese nor thin conditions occur.[5]

32.12   BROOD MARE AND FOAL MANAGEMENT

Although the average gestation length is about 345 days, it will vary from 320 to 360 days, depending upon factors such as the breed, sex of the foal (colts are carried several days longer), and plane of nutrition.

The owner should anticipate the increased needs for energy, protein, minerals and vitamins during the last trimester. Nutritional requirements of the pregnant mare increase dramatically during the last third of gestation corresponding to the rapid increase in fetal development—about 50 percent of fetal growth occurs during the last trimester. Prior to that time, the mare's body condition generally can be maintained on good pasture, fresh water, and a supplemental trace mineral mixture. When pasture quality diminishes, mares may need additional grain and hay, and where forage is poor and scant also provide supplemental vitamins.

---

[5] Body condition scores have been established (Dr. D. R. Hennike, Texas A&M University) with a score of 1 representing a very poor condition, to a score of 9 for an extremely fat condition. Horses in good condition should score 5, a moderate condition.

Brood mares in late gestation (9th to 11th months) should be fed a balanced ration that maintains a body condition of 6.0 to 7.5. Texas A&M University research also suggests that feeding supplemental fat during this time and in early lactation results in improved reproductive efficiency after foaling and an enriched milk for the foal.

Pregnant mares should be on a vaccination schedule, such as provided in Table 32–4, which not only protects the mare but also provides for passive antibody immunity in the foal through the colostrum. Note that these vaccines are administered to the broodmare about 4 to 6 weeks prior to foaling.

Brood mares in late stages of pregnancy should not be subjected to strenuous work but enough exercise to maintain muscle tone. As foaling time approaches, the mare should be moved to a clean, disinfected stall, or, if weather permits, to a clean, grassy paddock. Maternity stalls should be well ventilated, about 150 to 170 square feet (approximately square), and the floor covered with a comfortable synthetic material or well bedded with clean straw. Sawdust, wood shavings, or other materials that tend to stick to the foal's nostrils should be avoided. Stalls also should be free from obstructions and partitioned to provide some isolation for the mare.

An initial sign of approaching parturition is a distended udder with a waxy discharge from the streak canals of the nipples during the last few days before foaling. Also, within about 7 to 10 days of foaling, the muscles and ligaments in the pelvic area begin to relax and result in a sunken appearance over the hips and around the buttock. Finally, the mare will show signs of nervousness, increase her movement in the stall, and urinate more frequently. She also may show signs of cramping and sweating. As these late signs appear, the mare should be prepared for foaling by wrapping her tail, and cleaning and disinfecting her udder and external genitalia.

The stages of parturition have been discussed in Chapter 14, and need not be further discussed in detail. However, certain sound management practices should be mentioned. These include being present in case of difficulties that could render harm to the mare or foal. Difficult foaling, or *dystocia,* requires veterinary or experienced attention. The caretaker also should note that the placenta has been expelled shortly after foaling. If not, it should be tied up or wrapped in plastic to reduce the risk of contamination until expelled or attended by a veterinarian.

Attention should be given to the newborn foal to ensure its mouth and nostrils are clear of mucous materials. The umbilical cord should be broken about 1 to 1 1/2 inches from the body and quickly dipped with a 7 to 10 percent iodine solution to prevent bacterial invasion into the body. If the mare didn't receive a tetanus booster about 30 days prior to foaling, the foal should be given a tetanus antitoxin injection. The foal should be strong enough and capable of nursing within the first 2 to 3 hours. Some foal may need to be assisted in nursing, and very weak foal may need to receive colostrum via a stomach tube.

In some cases, newborn foals have difficulty in eliminating the dark-colored fetal excrement, termed *meconium.* If not passed within about 12 hours, the foal is constipated and should receive an approved enema preparation.

Mares may also become constipated after foaling. Feeding a wheat bran mash-warm water mixture and high-quality alfalfa hay for 7 to 10 days can reduce the incidence or severity. The mare also should be regularly observed for signs of colic, which sometimes occurs due to continuing uterine contractions, or sometimes related to a retained placenta.

Lactating mares such as those shown in Figure 32–13 should receive additional nutrients from concentrate feeds to ensure adequate milk and prevention of excessive body weight loss.

Figure 32–13
Lactating mares with their foals on pasture. Supplemental concentrate feeds should be provided especially when pastures are not plentiful and high quality.
(Courtesy of Tennessee Walking Horse Breeders' and Exhibitors' Association.)

32.13   TRAINING

Training should begin early in the foal's life. Halter breaking, "gentling," picking up the feet, and brushing can and should be started while foals are nursing, and they should be halter broken by the time they are weaned. Usually little additional training is provided for the next 2 years as they grow and mature, but continued attention—brushing, handling, leading by halter—is worthwhile.

At about 2 years of age, further training usually is begun. Horses are sacked-out (taught to be content while a sack or blanket is put on their back, so they will accept a saddle later), bridled (so they become accustomed to a bit in their mouth and to reins), saddled, and taught to walk or stand saddled. They become accustomed to moving with weight on their backs. For training, a plain snaffle bit (simple bar in the mouth) or a rawhide or rope hackamore (loose loop around the nose) is used.

When the horse is familiar with and accustomed to the saddle and bridle, it can then be taught to "ground drive." A bitting rig (leather girth with loops for the reins) can be used or driving lines can be run through the stirrups of the saddle, with the stirrups tied to each other under the belly. The horse should be taught to turn, stop, and back up before riding is attempted. At the time of the first ride, it knows how to respond to nose or bit pressure.

A small round pen is best for the first few rides on the young horse. These rides should be used to let it learn how to move with the extra weight and to stand while being mounted and dismounted. The experience during the first ride often determines the horse's attitude toward future rides; the rider should attempt to make the experience as pleasant as possible.

After the horse is accustomed to the rider's weight, further training can begin. At this point, turning, stopping, and backing should be no problem, provided the rider's hands are kept low, approximating the direction of pressure the horse received during driving. The rider should first use direct reining (pulling the right or left rein in the desired direction). After several rides a combination of direct and neck reining can be used. In neck reining, the two reins are held in one hand. To turn the horse to the right, the right rein pulls on the right end of the bit and at the same time the left rein is pulled across the top

Figure 32–14
Patience, knowledge, and experience are essential characteristics of successful horse trainers. The goal of all horse training is to achieve a calm, well-mannered horse that can perform to the best of its ability.
(Courtesy of *Quarter Horse Journal*.)

of the left side of the neck, applying pressure at that point. The horse learns, with repeated experience, to move to the right when rein pressure is exerted to the upper left part of the neck. Eventually, only the weight of the rein at that point will cause the horse to turn. Neck reining usually is preferred because it is less work and there is less arm movement for the rider (Figure 32–14). It also causes less bit action in the horse's mouth. Leg pressure may also be included in the training process—the horse can be taught to "side pass" or move to the side, away from leg pressure, especially helpful when working livestock.

When the horse responds well to the basic rein and leg cues, more advanced maneuvers can be taught.[6] Most are combinations of simpler movements, and instruction utilizes the horse's previously acquired knowledge.

In early training, only close approximations of the desired response should be expected from the horse. As training progresses, more precision and responsiveness should be demanded. Horses vary in personality and speed of learning, and this variation is accommodated by successful trainers or by the attentive amateur. The trainer should use forms of reinforcement (reward) to enhance the horse's response to a certain stimulus.

Most horse owners can train their own horse successfully if training begins with the foal, by progressing gradually through the training steps outlined. Training of older horses is usually best left to experienced trainers.

32.14 EQUITATION

Equitation refers to proper body positions of the rider. Correct equitation enables the rider to cue the horse to certain maneuvers and at the same time not interfere with, and perhaps enhance, the horse's natural rhythm and balance. For example, the rider of a jumping horse must be forward in the saddle to provide balance while the horse goes

---

[6] A stimulus learned through repetitions or practice in training is referred to as a "conditioned" stimulus, or often referred to as a *cue*.

over a jump. The rider of a cutting horse, however, must put more weight to the rear because the horse's front end is usually low.

Equitation begins with mounting and dismounting. In western equitation, it is customary to mount from the left side. The reins should be held by the left hand at the top of the horse's neck, and held short enough to control the horse for mounting. The rider faces the rear or the side of the horse and places the left foot in the stirrup. The saddle horn is grasped with the right hand and, using spring or lift, the right leg is brought up and over the rear of the saddle. For dismounting, the process is reversed. The rider should be in complete control of the horse during mounting or dismounting.

In the saddle the rider should sit with head up, shoulders back, and back straight but not stiff (Figure 32–15). Contact with the horse should be through the rider's seat, inner thigh, and upper calf. Leg position can help maintain balance and upper body position. Stirrups can be adjusted to help maintain correct leg position. The rider's shoulders, hips, and heels should be in a straight line. Heels held low help the rider absorb the shock of each stride of the horse. Toes should be pointed forward or slightly turned out. The ball of the foot should be resting on the stirrup, with the boot heel nearly touching the stirrup.

Hand position and use are important parts of equitation. For western shows, the reins are held in one hand with only one finger or no fingers between the two reins. The reining hand is held in front of and above the saddle horn so the rider can control the horse with only small hand movements. English show riders hold reins with both hands. The reins and lower arm form a straight line from the horse's mouth to the rider's elbow.

Figure 32–15
Correct body position for a western rider. The rider should sit with shoulders back, knees bent, and heels down. The shoulders, hips, and back of the heels should be in a straight line. (*Source*: National 4-H Council.)

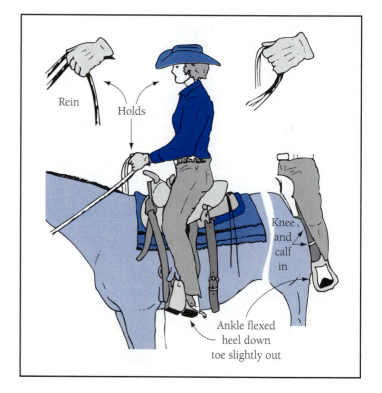

Rein

Holds

Knee and calf in

Ankle flexed heel down toe slightly out

In any kind of riding, the rider's hands should follow the movement of the horse's head, and this is achieved by flexing the elbow. Normally the upper arm will be down, in line with the upper body, but its position will change slightly as the rider "gives and takes" to follow the horse's head and neck movements. Riders should strive to develop "good hands," to cue the horse without interfering with performance.

QUESTIONS FOR STUDY AND DISCUSSION

1. About how many horses are there in the United States? Are most kept for pleasure or work? Give some examples.

2. There are five basic horse colors. List and describe them.

3. List and briefly describe five of the major breeds of horses in the United States. Categorize the breeds chosen above according to light, draft, or pony breeds.

4. Briefly discuss the importance of conformation and soundness to performance and value of the horse.

5. Name the three natural gaits and two acquired gaits in five-gaited horses.

6. What problems are likely to occur when horses are stabled in facilities? when closed too tightly—even in inclement cold weather?

7. How do protein and energy needs change with development and function of the horse?

8. What are characteristics of excellent quality hay?

9. Name two major disorders related to the feeding of horses.

10. What is the purpose of "floating" of teeth?

11. How often should horses (a) receive medication for parasites; (b) have vaccinations brought up-to-date; (c) have hooves trimmed; (d) have teeth floated?

12. Name several potential horse diseases that can be controlled through vaccines.

13. What signs exhibited by the mare enable the owner to know when she should be bred?

14. List the more important management practices that precede foaling.

15. List at least three newborn foal management practices that aid in its survival.

16. Define *equitation*. What does it include, and why is it important?

17. When should training usually begin, and what are the major steps and processes in training?

The Business 33 of Producing Eggs

An increase in U.S. per capita egg consumption from 232 eggs in 1990 to 253 eggs in 2002 suggests that egg production is a growing U.S. business with opportunity for more growth. As with other food businesses, consumers, through their purchases in retail stores and restaurants, control the future of the egg production business.

Consumers are increasingly sensitive to food safety and the influence of diet on their health, vitality, and longevity. Action by the American Heart Association in late 2000 indicating "an egg a day is OK" was a positive message to both consumers and the egg industry.

Seventy-three billion eggs were produced in the United States in 2002, and they had a value at the production unit of about $4.5 billion. After cartoning of shell eggs for retail (about 60 percent of total production) or processing into liquid form (at what are called breaker plants) for restaurant-ready or retail-ready product, or inclusion in packaged products, the total product value would be several multiples of that $4.5 billion.

In contrast to **broilers**, where about 15 percent of U.S. production is destined for export, less than 1 percent of eggs are exported. A few shell eggs are exported to Canada, Hong Kong, or other destinations, but processed egg products have more potential for export.

Egg production as a whole is highly concentrated, both geographically and among producers. The five top egg-producing states account for more than 50 percent of eggs produced, with the top 280 companies in early 2003 producing about 95 percent of the eggs. About 60 companies had more than 1 million layers; a dozen had more than 5 million each.

At the same time, there are a number of specialty egg operations that produce for specific markets; on the average, these are smaller operations. There is considerable demand for brown eggs (in certain regions), organic eggs, eggs produced by birds on "free range," or eggs with specific nutrient features brought about by special diets.

As a business, egg production focuses on converting grain and other feed ingredients, labor, and management skill into a relatively high-value product (see color Plate E). The business utilizes technology and products of the animal genetics, animal health, equipment, and construction businesses, as well as large volumes of electricity and fuel. It is a highly competitive business with, in many cases, large investment per unit. That fact requires precision management, careful personnel selection and training, and financial risk-limiting strategies.

These factors have resulted in steady improvement in the industry's productivity and efficiency. Between 1990 and 2002 the total U.S. layer flock productivity increased from 256.8 eggs per layer per year to 263.[1]

> **Broiler**
> A young chicken of either sex, being fed for meat production

---

[1] *Don Bell's Flock Projection and Economic Commentary,* November 5, 2002.

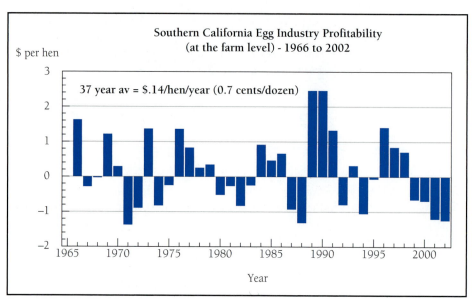

Figure 33–1

The calculated profitability of the southern California egg industry varies from year to year and quarter to quarter, and that is true for the rest of the country. Though a vast majority of the nation's eggs are now sold under some form of long-term agreement that may temper price fluctuations, this chart is based on spot or cash market prices and average production costs for California.

(*Source:* Don Bell, poultry specialist, University of California.)

As they leave the production unit, eggs can be described as a commodity, an item produced by many parties competing with the others to be the lowest cost producer and subject to rather large swings in total supply. Consumption, however, is relatively steady through the year and from year to year, so when supply goes up, price drops. Any industry that produces a commodity is subject to the risk of oversupply.

A majority of businesses in the egg industry are profitable most years, but profits fluctuate (Figure 33–1). There is probably more variation among producers in enterprise profitability than among years. As in other commodity businesses, when the industry is profitable, the more efficient producers will increase production more rapidly; the less efficient producers will lose market share.

Because of industry concentration, both geographically and in unit size, the industry is relatively vulnerable to the impact of any disease or food safety incident. An outbreak of avian influenza in several states and flocks early in this century caused a sharp drop in total production and a resultant increase in egg price.

Bird management focuses on both efficiency and bird welfare. Healthy birds that are comfortable produce more eggs. Housing and labor costs per bird or per dozen eggs are lower when birds are housed in cages. Bird cost (hatching, pullet growing, etc.) per dozen eggs is lower if flocks are molted (egg production interrupted for a rest) after a period of lay, then returned to production for a second and sometimes a third cycle. Most layer flocks are therefore housed in cages and are molted for a second cycle.

Both practices, induced molting and housing birds in cages, have been challenged by animal welfare interests in recent years (more intensely in Europe than in the

United States) and the U.S. egg industry has been especially progressive in developing and implementing science-based guidelines for each.

All of these issues will be discussed in this chapter. Upon completion of the chapter, the reader should be able to

1. Describe the recent trend in egg consumption and the proportion of eggs utilized as shell eggs versus processed product.
2. Describe the magnitude of the egg production industry and its geographic concentration.
3. Cite the reasons for the geographic shift in geographic density of egg production in recent decades.
4. List and differentiate among the four types of egg production operations.
5. Trace the typical life cycle of an egg-producing bird, from hatch through one or more production cycles.
6. Explain the influence of photoperiod on egg production.
7. Define *molting* and discuss economic and animal welfare issues related to it.
8. Discuss the issue of housing birds in cages versus floor or range handling of birds.
9. Describe the seasonal variation in egg demand and its general impact on price.
10. Describe how replacement pullets are raised.

33.1 GEOGRAPHIC DISTRIBUTION OF EGG PRODUCTION

Recent decades have seen a dramatic shift in where a high proportion of eggs are produced. Production by state in 2001 is shown in Figure 33–2 and the top 10 states for both 1981 and 2001, as well as hen numbers in 2003, are listed in Table 33–1. Note that Ohio, number 10 in 1981 with about 2.4 billion eggs produced, jumped to second in 2001, with 7.9 billion produced, a 230 percent increase. Iowa, producing more than 8 billion in 2001, wasn't in the top 10 list 20 years earlier. Its production had jumped 350 percent in 20 years. Alabama, Florida, and North Carolina had dropped off the top 10 list by 2001.

The 2003 hen numbers indicate the shift continues. By mid-2003 Nebraska had moved up to seventh in hen numbers.

Why the drastic changes? There appears to be several reasons: (1) The Southeast was no longer as competitive in labor cost at a time when many production facilities there were obsolete and ready for replacement, and the industry saw increased egg demand ahead. Lower labor costs several decades earlier had prompted the heavy production in the Southeast. (2) There was a lower cost corn, especially in the western Corn Belt. (3) More cropland received the manure from large production units. Corn Belt cropland also produced high yields, which utilized more phosphorus and nitrogen, the major nutrients in poultry manure. (4) Recruitment efforts increased by the Corn Belt states, to achieve more business activity and economic growth. States, cities, and counties provided tax and other incentives to relocate or establish new operations. (5) Both farmers and others in these states sought to increase local market demand for corn. Egg production, other animal operations, and ethanol plants were a part of the total effort.

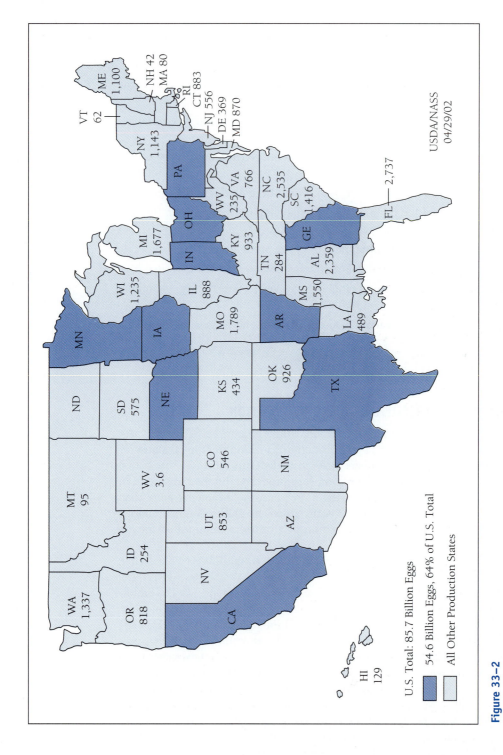

U.S. Total: 85.7 Billion Eggs

54.6 Billion Eggs, 64% of U.S. Total

All Other Production States

USDA/NASS
04/29/02

Figure 33–2

Egg production by states for 2001, with the top 10 states highlighted. State data don't necessarily describe area density. Texas is a very large state, and layer density is no doubt greater in northern Florida or in most North Carolina or Missouri counties than in Texas. (*Source:* USDA, NASS.)

Table 33–1	Egg Production by States, 1981 versus 2001, and Hen Numbers, 2003

| | Eggs Produced (millions) | | | | Hen Numbers (thousands) | |
| | 1981 | | 2001 | | 2003 | |
Rank	State	Number	State	Number	State	Number
1	CA	8,400	IA	8,676	IA	37,131
2	GA	5,578	OH	7,900	OH	28,478
3	PA	4,268	PA	6,662	PA	22,729
4	IN	4,093	IN	6,025	IN	21,970
5	AR	3,996	CA	5,996	CA	19,033
6	TX	3,224	GA	5,086	TX	14,017
7	AL	3,095	TX	4,734	NE	11,604
8	NC	3,078	AR	3,427	GA	10,996
9	FL	2,802	MN	3,112	MN	10,460
10	OH	2,431	NE	3,001	FL	10,205

Sources: *USDA, NASS and United Egg Producers. United Voices, July 20, 2003.*

California's drop in both state rank and total production can be traced to that state's rapid population growth, urban development, and higher land values. The population density makes it more difficult for an animal operation to avoid objections by neighbors to odor. In addition, the state and some counties had established increasingly stringent state environmental regulations that some operations could not meet.

Production in Pennsylvania and Indiana increased, but the increases were in approximate proportion to that of the United States and the states' rankings were unchanged. Their proximity to millions of consumers and to corn in the eastern Corn Belt, even in the face of their own population growth and environmental regulations, appeared to help stabilize these states' egg production industries.

The factors mentioned above underline the fact that total industries respond, in their locations and operations, to economic and other forces. Therefore, it should not be assumed the egg production industry 20 years from now will be unchanged.

33.2   TYPES AND SIZES OF PRODUCTION UNITS

Types of egg production units can be categorized as follows:

1. Commercial operations, producing for cartoning or processing. (See Figure 33–3.)

2. Breeder operations, producing eggs that will be hatched and the chicks become parents and grandparents of birds in the commercial operations.

3. Specialized units. Some may produce organic or "open range" eggs. Some may supply eggs to farmers' markets or deliver weekly to homes on a subscription arrangement. These operations, generally among the smaller, tend to serve a specific market sector.

4. Hobby operations. These may range from a few birds to supply eggs for home use to breeders who maintain rather pure lines of the historic breeds.

Figure 33–3
In most commercial units, feed distribution and egg gathering are computer controlled. Belts in front of each cage tier move the eggs to the ends of the row, where an up or down egg escalator puts the eggs on a wide gathering belt or conveyor. That gathering belt or conveyor extends through all houses of the complex and delivers eggs to a packing room.
(Courtesy of Purdue University.)

The vast majority of egg production is in *commercial operations,* generally comprised of several hundred thousand to several million white birds, housed in cages and producing white-shell eggs. The birds would be a developed strain of the Leghorn breed. A few operations of this size, especially in New England, produce some brown-shell eggs. Brown-shell eggs are produced by the Rhode Island Red, New Hampshire, and Plymouth Rock breeds.

A typical layer house may hold 125,000 to 175,000 birds, and has three to five tiers of cages on each side of three to five aisles. Upper tiers of cages are recessed and the backs of lower tiers are sheltered by curtains so that bird droppings from each cage go directly to a manure pit below all cages.

One production site may have 5 to 10 or more such houses, all connected to an egg-packing room and, in some cases, a cartoning or breaking operation. Flock schedules are staggered in order to maintain a steady egg volume and case weight for the total site or complex.

Both unit efficiency and individual bird performance are influenced by bird density, the amount of space provided each layer. In general, birds given more space produce more eggs, the eggs are larger, and bird mortality is lower. However, feed, building, and equipment costs per dozen eggs are higher and, in northern climates, heating costs may be higher because there is less bird-generated heat.

Table 33–2	Effects of Layers per Cage on Livability and Egg Production	
	Three Birds per Cage	*Four Birds per Cage*
Eggs/hen housed	246.4	233.9
Pounds feed/dozen	3.69	3.66
Egg weight/dozen (oz)	24.7	24.6
Egg mass/hen housed (lb)	32.3	30.4
Cracked eggs (%)	3.0	3.4
Mortality (%)	7.7	10.3

Source: *Bell, Donald D., and John Carey. Progress in Poultry, June 1988. University of California Cooperative Extension.*

Table 33–2 summarizes 5 years of research in North Carolina. It compares production results with three versus four birds in 216-square-inch cages (54 vs. 72 square inches per bird).

Cage sizes and numbers of birds per cage vary among the nation's layer operations, but for optimum bird welfare, a panel of scientists recommended in 2000 that layers be provided 67 to 86 square inches of usable space per bird. Other guidelines recommended included that cage bottom slope (necessary for eggs to roll out onto collection belts) not exceed 8°, that feeder space allow all birds to eat at the same time, and that ammonia concentration in the house atmosphere at the cages not exceed 25 ppm. Through the national United Egg Producers organization, a husbandry certification program is available for producers that meet or exceed the guidelines.

Nipple waterers are either suspended in each cage or in the wire partitions between cages, and a mechanical feed distribution system (a chain or flexible auger) and an egg-gathering belt run along the front of each tier of cages.

Continuous air movement is critical, to provide oxygen for the birds and to remove CO_2, ammonia, and other byproducts of bird metabolism and manure. For example, in one house for 125,000 birds, 50 exhaust fans, each 4 feet in diameter, provide a ventilating capability of 6 cubic feet of air exchange per bird per minute.

In a vast majority of commercial operations, the ventilation, egg gathering, and feed distribution systems are computer controlled. Parameters can be set or changed at the unit office and, perhaps, remotely by telephone. A backup generator is provided in case of interruption in electricity supply and a security system notifies personnel on site or elsewhere of any interruption or problem in the system.

Most of the larger egg-producing companies have layer complexes at several locations, often in several states. Many of these companies are privately owned, with financial data not publicly available. These private companies may range considerably in size, and may include general farm operations with large egg production enterprises.

A few of the major egg producers are public companies, or a subsidiary of a public company, with ownership stock available for public trading. Some are organized as cooperatives, usually with corn-producing farmers holding a majority of ownership.

Many of the larger egg production businesses also operate egg cartoning or breaking facilities, utilizing both the eggs their company produces and eggs produced by others. In most cases the latter eggs are covered by some form of long-term marketing

Table 33-3	The Five Largest U.S. Egg Production Companies, 2003	

Company	Million Layers	Headquarters Location
Cal-Maine Foods	20	Jackson, MS
Land O'Lakes	20	Minneapolis, MN
Rose Acre Farms	17.5	Seymour, IN
Michael Foods	14	Minneapolis, MN

Source: *United Egg Producers.*

Figure 33-4
Poultry breeder unit in Washington. The right photo shows electronic recording of individual bird egg production. (Courtesy of Pfizer H & N.)

agreement or contract. Feed for and financing of these contracted operations may be provided by the major company or, in some cases, the birds may be owned by the company. In 2003, the highest volume of eggs were controlled by four companies (Table 33–3).

Breeder operations are, of course, smaller in number and volume. Including the breeder flocks for broiler lines, they comprise about 15 percent of total egg production volume. (See Figure 33–4.)

Most breeder hen operations for layer lines are owned or franchised by three or four global layer genetics companies. Lines with high genetic value are developed through continued selection, usually in company-owned facilities. Most commercial pullets are "double crosses," the result of mating two "single crosses." After the testing of many different lines and line crosses, four grandparent lines are identified for the development and marketing of each commercial line.

The grandparent crosses usually are produced in genetic company facilities. Their offspring, selected roosters and pullets of the two single crosses and that will be parents of commercial pullets, typically are moved, after hatch, to the facilities of franchise operators. These franchises produce and hatch the fertile eggs, sort the newborn chicks by sex, vaccinate them, and place the day-old chicks in pullet-rearing units.

So that individual birds' egg numbers, egg weights, and quality traits can be recorded, individual layer cages may be used in these genetic companies' research and strain selection efforts. Once strains are selected, the flocks generally are housed in pens with wire or slat floors, and droppings go to a pit below. Nests are provided for nesting and egg laying, and the eggs gathered by hand.

At the franchise operation, where the fertile eggs are produced that will provide chicks for commercial production, considerable mechanization of the operation can be done.

The **specialty egg** production operations may range widely in size, numbers, or types of market to be served. A walk through a farmers' market or a Wild Oats retail food store will illustrate that variety. Specialty eggs make up less than 5% of the entire egg market.

> **Specialty egg**
> Eggs enriched or produced with possible benefits, i.e., high Vitamin K, Omega 3 fatty acids or cage free hens.

Birds may be housed in cages, on the floor in confinement (this represents 95 percent of layers), or provided open range. Eggs may be candled and cartoned by hand or mechanically. A price premium usually exists for these producers and, of course, is needed to offset the generally higher labor and other costs of production.

Hens that produce eggs identified as organic, free range, or cage free are generally housed on the floor, perhaps with access to an outside pen. Some type of bedding or litter, such as straw or ground corncobs, is provided to absorb the droppings and nests are provided for nesting and egg laying.

The *hobby* operations also may vary considerably. Some may be 4-H or FFA projects and some may be simply for home egg use. A number, however, are handled by what are termed breed fanciers—people who simply like to own and help maintain layer breeds that might otherwise disappear in this age of commercial production.

Merits of housing layers in cages versus the floor or bedding system may be debated. Healthy and comfortable birds generally are more productive and can have longer productive lives.

Commercial strains in use have been developed to be adapted to and highly productive in the cage system. Relative to layers of 30 or more years ago, they weigh less, produce more eggs per bird, and encounter fewer disease and health problems. The key to healthy, comfortable birds in either system is well-maintained equipment, attentive personnel, and top husbandry practices.

33.3   GROWING REPLACEMENT PULLETS

Replacement pullet operations are almost universally geared to the house size and the schedule of the layer operation to which the pullets are destined. The pullets may be owned and grown by the company that has the layer operation, by the franchise hatchery that mated the parents and hatched the chicks, or by a third party that specializes in this business. In the last two cases, there is usually a multiyear or multiflock agreement that specifies the genetic strain or line, the number of pullets to be raised, and the date, age, and weight range for delivery to the layer unit. The agreement will also specify an immunization schedule for the potential disease problems in the area.

If pullets are being raised for a 125,000-bird layer house, about 130,000 chicks will be placed in the growing house, taking into account some expected death loss and ensuring that there will be sufficient numbers of pullets delivered for the layer house to be fully occupied.

Each genetics company provides to the pullet grower a detailed management guide for its strain or strains of pullets. A guide for cage growing may include the following:

1. Keep house temperature at 85°F the first 2 days.
2. Keep relative humidity at 60 percent.
3. Put newspaper on the cage floor for the first week.
4. Place a handful of feed on the paper in each cage at chick arrival.
5. Provide 22 hours of light per 24 hours (1–2 foot-candle intensity) the first week.
6. Watch chicks carefully for water, feed, disease or other problems. Chirping may indicate bird discomfort.

Some replacement pullets are floor-reared and, in that case, the chicks are confined to a brooder (a hover that holds temperature at about 90°F), with about 500 chicks per brooder. Over time, the hover is gradually raised but, depending on season, supplemental heat is provided until birds are fully feathered. In other features, the guidelines are comparable to those for cage-reared pullets.

The chicks will have been sexed so that only pullet chicks (with a few exceptions) are delivered to the grower house. Any roosters will soon be evident, by the larger and brighter comb, and can be removed.

Specific ration guidance is given for each of two or more growth periods, including energy density, protein level, and levels of other nutrients. Feed consumption per 100 chicks and weight targets are so the operator can assess performance by weighing a sample of birds and reviewing feed consumption records each week. Where the pullets are being grown for another party, it is common for the layer unit manager to monitor the pullets' progress and perhaps visit the operation during the growing period.

The management guide for each strain outlines a detailed lighting program, in both hours and intensity according to age or weight of the strain. Initiation of egg laying is based upon the physiological effect of light stimulation. When light is transmitted to the eye, changes occur in the hypothalamus, increasing the hormonal action of the pituitary gland and, in turn, the follicles of the ovary.

As the pullets approach the time they will be moved to the layer unit (16 to 18 weeks of age), the lighting and feeding patterns in the pullet house can be adjusted to mesh with those of the destination layer house, so that the pullets can make an easier adjustment to their new environment after relocation. A basic rule is that the light should not be diminished after the birds are placed in the layer house.

The grower usually carries the responsibility to deliver the pullets and payment typically is based on the number of healthy pullets delivered and placed in the cages. At the grower unit, pullets are gathered and placed by hand in a crate, then crates are stacked on trucks and hauled to the layer unit. That distance could be a few feet or several hundred miles. A special crew is often contracted for the task of placing the birds in the layer cages. Depending on the season, weather conditions, distance, and number of trucks available, it may take 3 or 4 days to move a "house" of pullets. For a typical 125,000-bird house where birds move 300 miles, over a 3-day period the pullets are gathered and loaded in the evening, the trucks arrive at the layer house by dawn, and the birds are in their new cages by 10 A.M.

Table 33–4	Performance of Best and Poorest 25 Percent of U.S. Layer Flocks[a]		
Measurement	Best 25%	Poorest 25%	Difference
Eggs/hen housed	237.8	211.5	26.3
Eggs/100 hens/day	85.8	77.7	8.1
Mortality during lay period (%)	2.4	9.4	7.0
Daily feed/hen (lb)	0.195	0.227	0.032
Feed/dozen eggs (lb)	2.86	3.33	0.47
Average case weight (lb)	48.5	45.0	3.5

[a]Flocks hatched in 2002, data from 21 to 60 weeks of age.

Source: Bell, Don. The U.S. Egg Industry—Egg Production and Flock Productivity, 1990 to 2000. Davis: University of California–Davis, 2002.

33.4 MANAGING THE LAYER FLOCK

Because commercial layer operations are so similar—birds housed in cages with automated and computer-controlled equipment and all focused on a standard product, the egg—and because the genetics companies provide management guides designed for their individual strains, it is sometimes suggested that the operator needs to focus on only three things: feed, water, and air. That appears to make layer flock management rather simple and success automatic. Data in Table 33–4 suggest that is not the case. Note the tremendous range in eggs per hen housed, mortality, feed conversion, and other factors between the nation's top and bottom 25 percent of flocks hatched during 2002.

Layer performance certainly can be influenced by the level of management during pullet growing, but many factors in the laying house—ventilation, health maintenance, monitoring of feeders and waterers, removal of weak or dead birds, hours of light and light intensity, feeding frequency—can influence bird performance and unit profitability.

Air may be taken for granted in animals maintained on pasture or in open sheds. However, in environmentally controlled buildings with such dense populations as exist in layer houses, an entire flock can suffocate if exhaust fans fail to operate for several hours. That is why ventilation capacity, automatic fan operation, a backup generator, and a security system that notifies personnel of any malfunction are so critical.

At 70°F, 100 layers will consume more than 5 gallons of water per day; thus 125,000 birds in a house will need more than 6,000 gallons. A million layers will need 20,000 gallons, and more if the temperature is higher. This strongly suggests water storage capacity at the layer house or complex, to accommodate any interruption in water supply, and also a high-volume backup supply. Rural water districts can provide either the major or backup water supply in some cases.

Water needs to be high quality. For watering nipples in cages to operate consistently and without dripping, the water needs to be clean and free of dirt or other foreign matter. Either groundwater (from wells) or water from reservoirs can be used, but it should be tested for turbidity (lack of clarity), pH, nitrates, iron, fluoride, sulfate, phosphate, and, of course, bacteria. Acceptable levels of each usually are listed in management guides, and water filtering and treatment facilities are a normal part of many layer operations.

Rations are specifically formulated for each strain and for stage of lay, and genetics companies and feed or premix suppliers generally provide nutrition consultation and troubleshooting. Experienced personnel visit layer complexes on a regular basis and provide management advice and support.

Pullets entering the layer house no doubt will have been raised in cages with automatic feeding and water systems, but there is still an acclimation process for them to go through. In the layer house, they will likely be in cages of a different size, with different cage mates (so the "social order" will need to be newly established), and the water nipples may be located or function differently. The background noises—fans, feed chains, and egg-gathering belts—in the building will be different and the personnel will have different routines. However, the pullets will have been immunized according to an agreed-upon schedule, the duration of light and frequency of feeding, and even ration ingredients may have been coordinated with the pullet-rearing operation to minimize the stress of relocation.

For each commercial strain, a set of flock objectives, by week of age, is usually provided. To illustrate, excerpts from one such set are provided in Table 33–5. Note the eggs per hen housed objective increases from 0.63 at week 19, probably the first or second week in the layer house, to 6.39 at week 27, then tapers off to 4.56 at week 78. The standard 30-dozen case weight of 48 pounds is targeted at 36 weeks. The target cumulative eggs per hen housed at week 78 was 323.

A second set of objectives would exist for the second cycle, after molt. Unit personnel check egg and bird weights weekly by a sampling technique, other data are entered, and computer printouts chart flock performance in comparison with the performance targets provided by the strain source or by production company management. Two such charts, for flocks in their first or second lay cycles, are shown in Figure 33–5.

Pullet eggs are small; note that the case weight (30-dozen case) goes from 36.6 up to more than 52 pounds in Table 33–5. The standard case weight for egg marketing transactions is usually 48 pounds, with a modest premium for higher case weight and a significant deduct for a lighter case weight. The staggered scheduling of flocks in a complex allows blending of eggs from younger and older flocks by the computer-controlled egg-gathering system, so that eggs arriving at the packing machine will average 48-pound case weight.

Table 33–5	**Example Flock Management Objectives, Single Lay Cycle**				
Week	Eggs/Hen Housed	Cumulative Eggs/Hen Housed	Body Wt. (lb)	Case Wt. (lb)	Feed/day/ 100 Hens
19	0.63	0.63	3.11	36.6	16.5
23	5.99	17.02	3.45	41.8	22.0
27	6.39	42.36	3.62	44.8	22.9
36	6.11	98.54	3.75	48.0	22.9
50	5.64	180.53	3.75	51.0	23.1
64	5.11	255.61	3.75	52.1	23.8
78	4.56	323.01	3.75	52.4	23.8

Source: *DeKalb Poultry Research, Inc.*

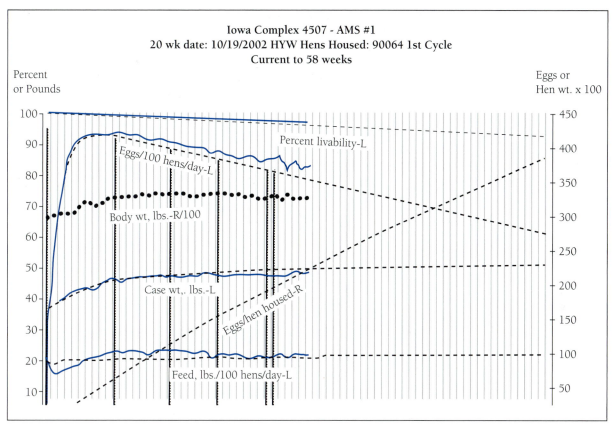

Figure 33–5
Performance charts for flocks in their first cycle (top) and second cycle, after molt (bottom), showing their performance from the 20th week of age through the week ending 6/21/03. The HYW at the top refers to the strain of bird. Labels have been inserted on the top chart to help interpretation. Note each refers to the left reference numbers (L) or the right reference numbers (R). Flock performance can be compared with the target lines established by company management.
Source: (AGRI-TECH, Sparboe Agricultural Corp.)

(continued)

33.5   INDUCED MOLTING

To the biologist or ornithologist, molt is the naturally occurring process of the replacement of feathers in birds. It tends to occur at a certain age and in a specific pattern for each species. Egg production usually slows or ceases during molt and, after molt, egg production, shell quality, and egg weight increase significantly from their levels just prior to molt. Molt can be induced at predetermined times, usually between 60 and 70 weeks of age, by reducing the photoperiod (changing the lighting schedule) and limiting feed intake for a period.

   Induced molt allows extending the life of the hen, keeping the layers for a second or second and third lay cycle, and reducing the number of 16- to 18-week-old replacement pullets that need to be purchased. Production essentially ceases during molt, of course, but most costs continue. This continued cost is generally more than

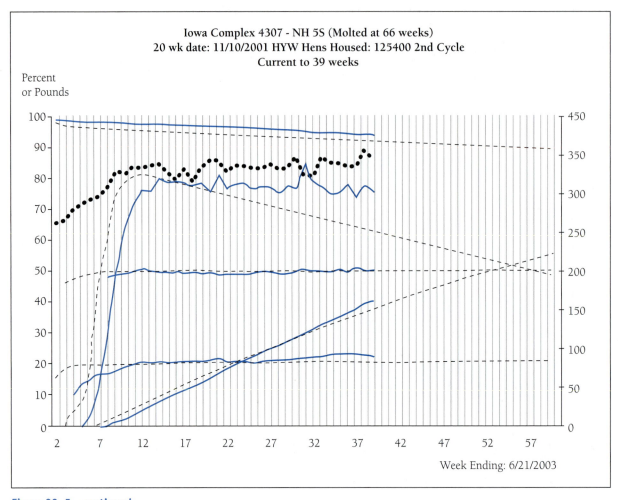

Figure 33–5—continued

Egg production company managers, as well as site managers, use charts such as these to monitor production volume and efficiency of each flock on a weekly basis. That helps them in refining flock management or in their choice of strains for later flocks.

offset by the savings in pullet costs and the better production following molt. (See Figure 33–6.) Molt is part of the management schedule in more than 85 percent of commercial layer flocks. Most flocks are molted once, a few twice (the second usually at 95 to 100 weeks of age). A planned molt schedule might be modified, of course, as a result of flock performance or health, or a shift in egg prices.

Although molt is a natural phenomenon, some animal welfare interests have challenged the practices associated with induced molt, especially the limiting of feed intake and the resultant weight loss in the birds. The 2000 report of the scientist panel on husbandry guidelines urged (1) that producers and researchers work together to develop alternatives to feed withdrawal for molting, (2) that producers ensure adequate and palatable feed, (3) that weight loss not compromise hen welfare, and (4) that mortality during molt not substantially exceed normal flock mortality. Research on alternatives for inducing molt is under way at several locations.

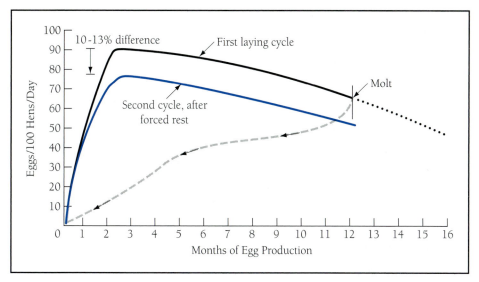

Figure 33–6
Typical effect of induced molt on egg production. If molt is not induced by about 12 months of lay, production declines to a nonprofitable level.
(*Source:* Ohio State University.)

33.6   HANDLING MANURE AND UNIT EMISSIONS

Poultry manure is an excellent nutrient source for crop production or for other fertilizing uses and its high level of calcium carbonate benefits soils that are low in pH. In some layer operations, the manure has become a significant income source. Considering the costs of nitrogen and other fertilizing elements in commercial fertilizer and of lime for correcting soil pH, manure from a layer operation applied to cropland early in the twenty-first century had a calculated value of approximately $30 per ton. The cost of loading, hauling, and applying to land within a 10-mile radius approximated $15 per ton.

The house ventilation system dries the manure to 30–35 percent moisture. This minimizes nitrogen loss to ammonia and also minimizes insect problems. Most manure pits below the cages are designed for a year or more of storage. Manure is usually removed annually and loaded on spreader trucks with a skid loader. In some units, a belt under each tier of cages moves the manure daily to spreader trucks or to an adjacent composting unit, for production and packaging of a higher value product for lawns or gardens. Layer operations with more than 82,000 hens on site for more than 45 days per year are subject to the CAFO permit requirements discussed in Chapter 28. For those relatively few operations that handle manure in liquid form, the threshold is 30,000 hens.

As with other animal operations, CAFO permits and manure management plans at this writing focus largely on avoiding discharge or leaching of excessive nitrogen or phosphorus to surface or groundwater. Phosphorus levels in the manure can be reduced by use of the enzyme phytase in rations. This enzyme increases bird utilization of dietary phosphorus, so less need be in the ration and less comes through in the manure.

It is anticipated that emissions of particulate matter as well as ammonia, hydrogen sulfide, and other contributors to odor will become a part of CAFO regulations and it certainly would be prudent for unit operators to monitor and hold down, to the extent feasible, problem emissions from their units. As in any animal operations, wise site selection, good facility design, and top management are critical to avoiding significant environmental problems and neighbors' concerns.

33.7   EGG MARKETING

Figure 33–7 shows monthly table egg production and midmonth prices for "Midwest Large" eggs. Most egg production is in commercial units that operate year-round and consumption varies little through the year. Price variations that occur tend to be the result of holiday demand, such as the Easter season, or temporary oversupply or shortage.

From the graphs in Figure 33–7 it is apparent that price varies more than supply. When the supply of a commodity increases, the percentage drop in price tends to be the square of the percentage increase in supply.

Because most eggs come from large, high-investment production units, most egg producers need to limit their price risk. Similarly, those who carton or break eggs for retail or further processing need to ensure a steady inflow of eggs to their operations. Therefore, well over 90 percent of the eggs from commercial production units are either owned by the processor or are "marketed" under some type of long-term marketing agreement.

Even though most agreements are confidential, an agreement between a producer who packs eggs on pallets and the cartoner or breaker that buys the eggs will likely include the following:

1. Time period covered—years or numbers of successive flocks.
2. Number of cases per year, or a range in numbers.
3. A formula for calculating price, and the source of inputs to that formula.
4. Case weight, usually 48 pounds, and premiums or discounts for higher or lower case weights.
5. How case weight is determined, by the pallet or the truckload.
6. Which party provides the packing materials—flats, pallets, and pallet wrapping.
7. Which party provides the transportation.
8. When title to the eggs is transferred, at the loading dock or at the receiving plant.
9. How soon after loading is payment made, and how—by check or by wire transfer.

The spot market, where price is negotiated at delivery, is used for a relatively small proportion of eggs; however, it plays a key role in most agreements. A private company, Urner Barry, has established itself as a major source of egg price information. Every trading day its personnel talk to buyers and sellers across the country asking, "What are you paying/receiving for large/medium eggs today?" Buyers and sellers subscribe to the Urner Barry daily reports and one or more (New York Large White, Midwest Breaker, or other) Urner Barry quotes may be stipulated inputs to the price formulas.

The marketing and pricing of organic or other eggs may, of course, be by long-term agreement or a price negotiated at the point of sale, and Urner Barry or other price reports may be used in these cases.

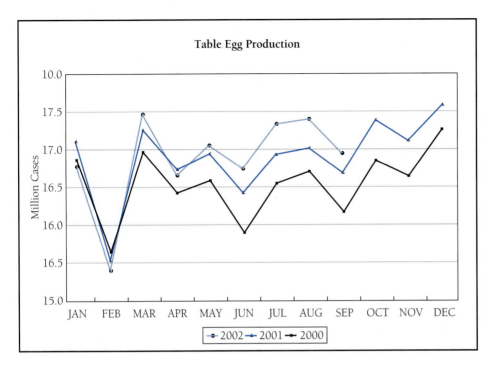

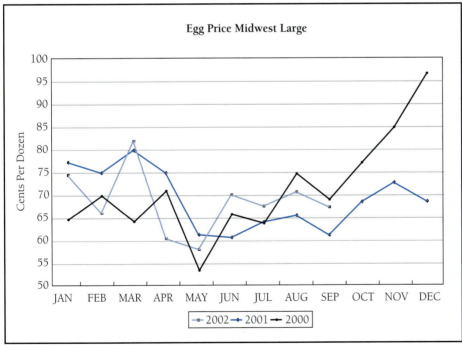

Figure 33–7

Table egg production by months in the top chart and the price of Midwest Large eggs by month in the bottom chart illustrate that percentage variation in price is greater than percentage variation in supply. Factors other than supply influence egg prices considerably. Easter and the Thanksgiving and Christmas seasons are times of higher egg demand.
(*Source:* United Egg Producers.)

QUESTIONS FOR STUDY AND DISCUSSION

1. List three of the top five states in egg production. Why did egg production move from the Southeast toward the Corn Belt in the 1990s?

2. What percent of laying hens in the United States are housed in cages? Why?

3. If you were manager of a unit where layers are housed in cages, how would you respond to those who raise concerns about this practice?

4. Of what breed are most of the commercial layer strains?

5. Explain the influence of photoperiod (daylight or artificial light) on egg production.

6. At what age are replacement pullets moved to a layer house?

7. Define *molting*. How is molting induced in a layer flock? Why is it used?

8. List five components that are a part of most long-term egg marketing agreements.

9. What does CAFO mean, and how is it related to egg production operations?

34

The Egg as a Product

This chapter focuses on the egg that is produced for food. The egg is a source of valuable nutrients; it can be a part of the main course, or used in salads, desserts, or snacks. The egg is subject to classification and grading by the USDA or private companies. It is an instrument of commerce, to be bought, sold, and promoted. It is a source of financial profit.

The egg has quality attributes—shell texture and shape, shell and yolk colors, and flavor—that are established by genetic and environmental factors. Some of these attributes are subject to change after the egg is laid and subjected to handling, storage, and processing. Sections that follow discuss the role each of these factors plays in egg quality.

Upon completion of this chapter, the reader should be able to

1. Sketch an egg and identify its major parts.
2. Contrast the physical and nutritional characteristics of the egg albumen and yolk.
3. List the USDA egg grades and the major factors that determine the grade of an egg.
4. List the major factors that influence egg quality.
5. Describe how eggs normally are handled after they leave the hen to ensure maintenance of quality.
6. List a series of products that contain processed eggs.
7. Describe the trend in U.S. per capita egg consumption.

34.1 CHARACTERISTICS AND ATTRIBUTES

Figure 34–1 shows the parts of an egg and also a cross section of the shell. The *yolk* develops, while attached to the hen's ovary, as a potential source of nutrients for a developing embryo, were the egg to become fertilized. It contains the germinal disc where fertilization may occur, although almost all table eggs are infertile.

The yellow color of the yolk is due to fat-soluble pigments, mostly xanthophylls transferred from the digestive tract via the blood and lymph supply and usually supplied by such feed as alfalfa meal and yellow corn. When compared to free-ranged hens, light-colored yolks are most common when birds in confinement are fed today's commercially prepared rations. Also, there are light-colored egg yolk markets (mayo and speciality breads), so some hens are fed diets low in xanthophylls, and the eggs become "chalky" in appearance.

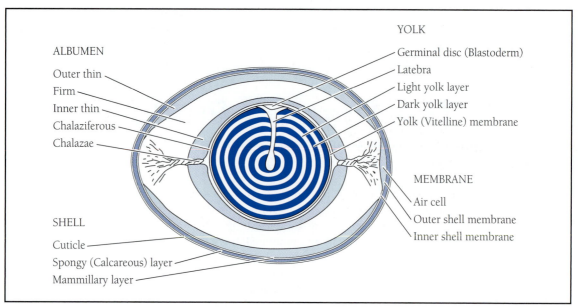

ALBUMEN
Outer thin
Firm
Inner thin
Chalaziferous
Chalazae

YOLK
Germinal disc (Blastoderm)
Latebra
Light yolk layer
Dark yolk layer
Yolk (Vitelline) membrane

MEMBRANE
Air cell
Outer shell membrane
Inner shell membrane

SHELL
Cuticle
Spongy (Calcareous) layer
Mammillary layer

Figure 34–1
The parts of an egg.
(*Source*: USDA.)

TABLE 34–1	**Average Composition of the Whole Egg**				

	Percent by Weight	**Percent**			
		Dry Matter	*Protein*	*Fat*	*Ash*
Whole egg	100	34.5	11.8	11.0	11.7
White	58	12	11.0	0.2	0.8
Yolk	31	52	17.5	32.5	2.0
		Calcium Carbonate	*Magnesium Carbonate*	*Calcium Phosphate*	*Organic Matter*
Shell	11	94.0	1.0	1.0	4.0

Source: *Data adapted from USDA, FSQS,* Agricultural Handbook No. 75, *1977 and No. 8, 1989.*

> **Albumen**
> A viscous protein that comprises most of the white of an egg.

The **albumen** (or egg white), thick and thin, includes all material between the vitelline yolk membrane and the inner shell membrane, and is secreted to surround the yolk in successive sections of the hen's oviduct. It is the thick (inner) albumen that stands higher and spreads less when the shell is broken. Composition of the yolk and white are shown in Table 34–1.

The two *shell membranes* are lacelike protein materials that surround the albumen and provide a protective barrier. The air cell forms between the two membranes at the large end of the egg.

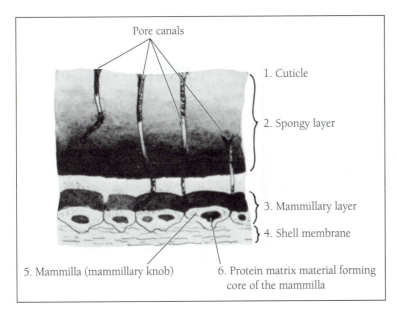

Figure 34–2
A magnified cross section of the eggshell. (*Source*: USDA.)

The inner layer of the *shell*, formed first, is apparently two or three layers of tiny round calcified pebbles (mammilla) deposited on the outer shell membrane. These are then covered and cemented together, except for certain *pores* (Figure 34–2), by a second layer, a calcareous spongy material of calcium carbonate. The third, outer, shell layer is a thin protein material, which tends to seal the shell. This outer layer is called the *cuticle*.

Shell shape and strength are important. An egg that is too long or has too large a circumference may be crushed, or squeezed and cracked, while being handled or moved in containers. Those that are especially short and round tend to roll faster, and can become cracked or can crack others on gathering belts or in graders. Uniformity of shape and texture helps appearance.

Shell texture and thickness are major quality attributes. The pores that admit oxygen for embryonic development in a fertilized egg are a detriment to a table egg; the simultaneous loss of carbon dioxide permits oxidation and quality deterioration.

Breeders of egg-type chicken strains measure shell strength by (1) micrometer measurement of shell cross section, (2) force required to break the shell, and (3) specific gravity. A weak or porous shell will admit more air, causing the egg to float in water, so the specific gravity is lower.

The proportion of thick albumen, and the degree to which the albumen remains thick until the egg is used, is a major quality trait. Consumers prefer thick albumen; when an egg is broken for frying or poaching, an egg with a thick albumen retains a small circumference and stands up high. This is measured by breaking out an egg onto a flat surface and measuring the height of the albumen. The height is expressed in relation to egg weight as Haugh units.

A thin albumen denotes lack of freshness, though albumen quality may also be influenced by other factors. Eggs laid late in the hen's production cycle usually have lower Haugh unit scores. Some pullets lay eggs with relatively thin albumen, especially when environmental temperature is high. (See Section 34.4.)

Consumers like a clear albumen, a high yolk, definite and uniform yolk color, and chalaza that are not prominent. *Chalaza* are spiral bands of thicker albumen, contained within the albumen, which help to maintain the yolk in its position so it doesn't stick to one side of the egg. Table eggs are scanned by strong lights to ensure freedom from "blood spots" (traces of blood resulting from capillary rupture at the time of ovulation) and "meat spots" (these are usually traces of tissue from the hen's ovary or oviduct) that infrequently occur as the egg is formed.

Eggs are especially nutritious. The albumen is very high in protein and the protein is of high quality because of the excellent balance of amino acids. The yolk contains fat and cholesterol. It also is high in protein and contains fat-soluble vitamins A, D, E, and K, and the elements iron, phosphorus, sulfur, copper, potassium, sodium, magnesium, calcium, chlorine, and manganese.

A large egg contains about 4.5 grams of fat (1.5 of which is saturated fat), and 213 milligrams of cholesterol, 6.25 grams of protein, and about 70 calories each. Many nutritionists believe that eggs can fit into a healthy, well-balanced diet when consumed in moderate amounts.

34.2   CLASSES AND GRADES

Shell eggs are classed according to size (ounces per dozen) and graded according to standards of quality by a plant grader and then certified by a USDA grader.[1] USDA specialists help institutional buyers (retailers, restaurants, schools, military, etc.) develop and prepare explicit shell egg specifications tailored to their requirements. USDA graders then provide certification, in the plant by sampling, that purchases comply with those specifications. An official grading certificate accompanies each shipment from the grading plant. Table 34–2 and Figure 34–3 summarize and illustrate weight classes.

The grade is an indicator of internal quality and the condition and appearance of the shell. Eggs are graded by being passed over a lighted background or by holding

TABLE 34–2	U.S. Weight Classes for Consumer Grades of Fresh Eggs		
Size or Weight Class	Minimum Net Weight per Dozen (oz)	Minimum Net Weight per 30 Dozen (lb)	Minimum Weight for Individual Eggs at Rate per Dozen (oz)
Jumbo	30	56	29
Extra large	27	50.5	26
Large	24	45	23
Medium	21	39.5	20
Small	18	34	17
Peewee	15	28	—

Source: *USDA, FSQS, Agricultural Handbook No. 75, 1972.*

---

[1] Material in this section describes grades and classes effective December 4, 1995. AMS/USDA.

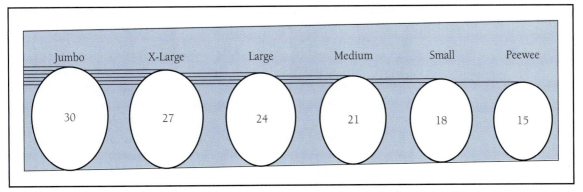

Figure 34–3
The size of shell eggs is based on minimum net weight expressed in ounces per dozen.
(*Source*: USDA.)

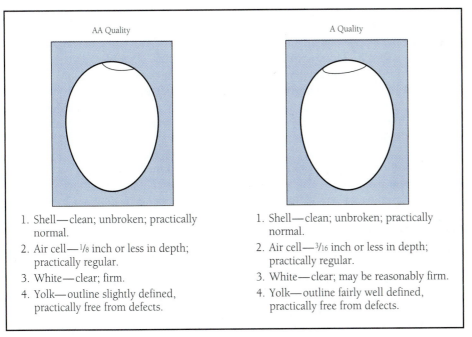

AA Quality	A Quality
1. Shell—clean; unbroken; practically normal.	1. Shell—clean; unbroken; practically normal.
2. Air cell—1/8 inch or less in depth; practically regular.	2. Air cell—3/16 inch or less in depth; practically regular.
3. White—clear; firm.	3. White—clear; may be reasonably firm.
4. Yolk—outline slightly defined, practically free from defects.	4. Yolk—outline fairly well defined, practically free from defects.

Figure 34–4
Illustration of differences in the air cell of a AA and A quality egg. The air cell of the fresh egg is minimal because very little moisture has been lost from the egg. Other standards are shown below the sketches.
(*Source*: USDA.)

up the large end to a box opening, behind which is a bright light. You can assess the size of air cell and the distinctness of yolk shadow outline and shape of yolk, and can detect yolk or albumen defects or germ development by rotating the egg before the candling light. (See Figure 34–4.)

The internal and external standards of individual eggs, by USDA grade, are listed below:

- *AA Quality.* The shell must be clean, unbroken, and practically normal. The air cell must not exceed 1/8 inch in depth, may show unlimited movement, and may be free or bubbly. The white must be clear and firm so that the yolk outline is only slightly defined when the egg is twirled before the candling light. The yolk must be practically free from apparent defects.

- *A Quality.* The shell must be clean, unbroken, and practically normal. The air cell must not exceed 3/16 inch in depth, may show unlimited movement, and may be free or bubbly. The white must be clear and at least reasonably firm so that the yolk outline is only fairly well defined when the egg is twirled before the candling light. The yolk must be practically free from apparent defects.

- *B Quality.* The shell must be unbroken, may be abnormal, and may have slightly stained areas. Moderately stained areas are permitted if they do not cover more than 1/32 of the shell surface if localized, or 1/16 of the shell surface if scattered. Eggs having shells with prominent stains or adhering dirt are not permitted. The air cell may be over 3/16 inch in depth, may show unlimited movement, and may be free or bubbly. The white may be weak and watery so that the yolk outline is plainly visible when the egg is twirled before the candling light. The yolk may appear dark, enlarged, and flattened, and may show clearly visible germ development but no blood due to such development. It may show other serious defects that do not render the egg inedible. Small blood spots or meat spots (aggregating not more than 1/8 inch in diameter) may be present.

Most eggs, in automated grading plants, are scanned electronically as the rows of rolling eggs pass over that lighted background. An egg that is dirty or unsound is not eligible for grading. In automated grading plants, such an egg is detected by the scanning equipment and an electronic signal causes the egg to be dropped off the packaging line.

Eggs that go to breakers, for further processing, are generally not graded. Rather, delivery contracts specify quality features and per dozen weights. Scanners similar to those used in cartoning lines detect and drop out any eggs with cracked shells or other defects.

When eggs are grouped, as in 1- or 3-dozen cartons, 30-dozen cases, or larger lots, reasonable tolerances for each grade are justified. A grade label and typical egg carton is shown in Figure 34–5. Figure 34–6 shows the "broken out" appearance of eggs of the three grades.

34.3   INHERITANCE AND EGG QUALITY

Heritabilities for major egg quality traits average about 0.60 for shape and weight and 0.35 for albumen quality and specific gravity. Presence or absence of yolk abnormalities also is inherited to a significant extent.

Because of the economic importance and relatively high heritabilities of these traits, breeders expend much effort in measuring these traits in experimental lines and varieties.

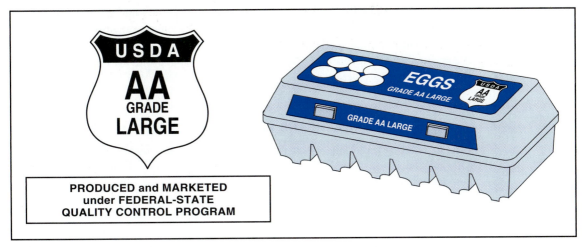

Figure 34–5
Official grade label (Grade AA Large) for egg cartons (left). Also note the USDA legend and labeling on a typical carton for Grade AA and Grade A eggs.
(*Source*: Adapted from USDA.)

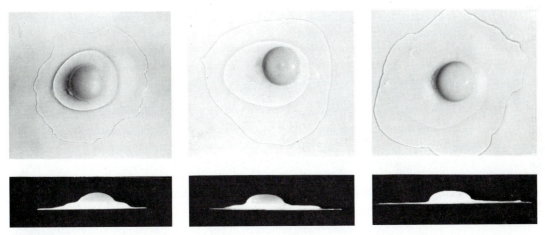

Figure 34–6
The pictures show the interior quality of eggs that meet the specification of individual shell eggs with respect to albumen and yolk quality for AA Quality, A Quality, and B Quality.
(Courtesy of USDA.)

34.4   EFFECT OF SEASON AND AGE OF HEN ON EGG QUALITY

Shell thickness and strength decline as the hen ages. This contributes to lower interior quality, especially after storage, and to increased percentages of cracked eggs. Because 25 to 40 percent of the calcium used in shell formation is drawn from skeletal stores rather than from the ration, the body may act to conserve skeletal calcium as the laying season progresses.

Cold temperatures and low relative humidity tend to cause thicker, denser shells; temperatures above 80°F and high humidity result in thinner shells, and egg size may be reduced slightly. Yolk mottling and blood spots tend to increase through the laying season and as warmer weather is reached, especially during the first year of lay.

With essentially all birds confined and receiving dry, mixed rations, there is little seasonal effect on yolk color. With birds on open range, yolks are much darker in the spring as birds eat green forages high in the xanthophylls.

34.5   EFFECT OF HEALTH AND NUTRITION OF THE HEN ON EGG QUALITY

All nutrients are important for egg production, but certain of them are more critical than others—that is, they are more likely to be deficient in natural rations and so are of greater concern in ration formulation. For example, a hen uses about 2.2 grams of calcium per egg and has additional needs for other body functions, and absorption from the digestive tract is less than 100 percent. A ration level of about 3.5 percent calcium normally is recommended, though higher levels sometimes are needed in hot weather when feed consumption is lower. When more than 6 percent calcium is used, production generally declines. Relationship to and amount of vitamin D_3 is also important.

Yolk pigmentation is subject to many nutritional influences. *Xanthophylls* are the main pigment, and the xanthophylls of yellow corn appear equally potent to the xanthophylls in alfalfa products in contributing to rich yolk color. *Chlorophyll* derivatives at relatively high concentrations have been known to cause greenish yolks.

Mottled yolks have resulted from feeding certain estrogens and from the use of cottonseed meal in layer rations. Some cottonseed meal contains *gossypol*, which causes an olive green or olive brown color in the yolk and, in some instances during egg storage, a mottling of the yolk color. Gossypol is toxic and can reduce egg production. For these reasons, only "degossypolized" cottonseed meals generally are used in poultry rations.

Albumen quality or chemical content appears to be influenced little by chemical composition of the ration. Much research has been directed to causing thicker albumen. Ammonium chloride in the ration does cause thicker albumen, but results in lower production and other undesirable side effects.

High concentrations of ammonia in the atmosphere of the laying hens or egg storage area have been shown to decrease the proportion of thick albumen and to increase yolk mottling.

Some diseases affect egg quality, as well as quantity, even after the birds appear to have recovered. Birds that have had bronchitis tend to produce thin-shelled and abnormally shaped eggs. This disease and certain others may also cause thin, watery whites. Certain drugs also will affect egg quality.

Birds maintained in a controlled environment and fed a standard, well-balanced ration generally will produce eggs with uniform and consistent quality characteristics, including yolk color, albumen thickness, and flavor.

34.6   SHELL EGG HANDLING AND PACKAGING

Quality requirements for Grade A or AA eggs dictate that cooling facilities exist at the production unit. The quicker egg temperature is lowered to 40 to 45°F or below, the less deterioration in internal quality results. Many states have laws requiring eggs to be stored at 45°F. The relative humidity of storage coolers is held at 70 to 85 percent, also to prevent deterioration and weight loss.

In many modern units, eggs are washed, scanned, sized, and packaged on line as they move by conveyor belt from the layer house and before being placed in a cooler.

All commercial eggs are washed in 90°F+ water with detergents and sanitizers, and the water is required to be 20°F warmer than the incoming eggs. If the washing solution is warmer than the egg, the solution is not absorbed into the egg. Such washing not only removes stains and dirt but also reduces bacterial population. However, eggs *must not* be allowed to soak in water. Soaking for as little as 1 to 2 minutes can facilitate microbial penetration into the egg's shell.

In generally smaller units, perhaps up to 1 million birds, 30-egg flats are stacked on pallets, the pallet shrink wrapped and moved to a cooler, then transported by refrigerated truck to grading or breaking plant.

After eggs are washed and properly rinsed in a sanitized solution, they are usually sprayed with an extremely light oil or oil emulsion; sometimes a special silicone-type product is used. This is to close pores in the shell and prevent further evaporation and deterioration; it also improves appearance.

The eggs are then sorted by size and cased or packaged in retail containers. Some suggest placing cases or cartons in the cooler for up to 24 hours prior to filling. Hot and dry cartons or cases can absorb up to a pound of water from 30 dozen eggs. Because of the weight loss, this could lower the size classification of a case or lot. A USDA grader on site samples the cartons or cases to ensure that the grade specifications in the buyer's purchase contract are met, and provides a certification with each batch to be shipped.

In handling eggs, the critical factors are rapid cooling, washing, rinsing, sanitizing, oiling, sizing, packaging, and rapid shipment to the retailer, with handling restricted to as few times as possible.

34.7   SHELL EGG DISTRIBUTION AND MERCHANDISING

Most eggs go from the grading plant to a retailer's or restaurant chain's distribution facility, or to a private distributor who serves such. With recent mergers of major retailers and resultant concentration, with some retailers handling more than 10 percent of U.S. retail food sales, greater efficiencies in egg movement can be achieved. With checkout counter sales data electronically aggregated, an egg supplier may receive daily orders not only on numbers of cartons of each weight and grade but also the sequence for loading the trucks. When these trucks arrive at the retailer's facility, those pallets destined to individual stores can be moved quickly to the retailer's distribution trucks.

Large egg producers in densely populated areas may sell directly to retail stores, delivering sized, packaged eggs to the stores 6 days a week. Price, volume, and other terms usually are established by contract; the contract may provide price fluctuations according to prices that exist at an agreed-upon market. Some eggs are sold at farmers' markets or are delivered to the consumer's door weekly on a subscription or continuing purchase basis.

34.8   EGG PROCESSING AND PRODUCTS

More than 30 percent of U.S. eggs go to breakers and that percentage is increasing. The term *breaker* is commonly used to describe the plant or the business that converts the shell egg to liquid egg—whites or yolks. The breaker plant may dry or freeze the product, or process it further into consumer-ready products. At this writing 71 U.S. egg-processing plants operate under the USDA/FSIS (Food Safety and Inspection Service) plant and process inspection program.

A majority of liquid egg products go to fast-food and other restaurants for use in scrambled eggs or other menu items (see Figure 34–7). Some is partially cooked in serving-size portions. Among other products made either in the breaker plants or in separate facilities that use the liquid or dried eggs are cake and pancake mixes, eggnog flavoring beads, egg rolls, chiffon pies, noodles, and mayonnaise.

Because the egg is such a concentrated source of available nutrients, it is also a good medium for growth of beneficial or harmful bacteria. These bacteria can flourish in the egg contents and products if they are not handled properly. Breaker plant equipment is virtually all stainless steel, and all facilities and equipment must be regularly steam cleaned and otherwise sanitized. The process must also be planned and fully monitored to avoid introduction of bacteria from any external source, and quality control laboratories regularly check samples from steps in the processing lines for pathogens, such as *Salmonella*, or other bacteria. The process must include pasteurization of finished product.

Many egg products are virtually identical to the fresh egg in nutritional value, flavor, or cooking properties. However, the industry produces many items that differ by increasing the proportion of whites or by blending or mixing with other ingredients, such as flavors, spices, and preservatives.

34.9 CONSUMPTION TRENDS

Per capita egg consumption in the United States increased from 234 in 1990 to 255 in 2002 and the trend continues upward. Two reasons seem apparent for the increase: (1) More processed, liquid eggs are used in fast-food outlets and as ingredients in packaged foods, and (2) consumers are more positive about the nutritional and health benefits of eggs.

Figure 34–7
Scrambled eggs with sausage and hash browns (left) and eggs in a breakfast sandwich (right) are both popular breakfast items at fast-food outlets today.
(Courtesy of Pfizer H & N.)

In recent years, fast-food outlets have featured quick breakfasts that are more nutritionally complete than a donut or sweet roll. An egg or egg and meat in a biscuit sandwich can be ordered at the drive-up window en route to work. A Mexican omelet and pancakes are common on fast-food menus as well as the menus of other restaurants. Eggs are a major ingredient in pancake mix. The proportion of eggs processed into liquid form, a high proportion for such eventual use, increased from 20 percent in 1990 to more than 31 percent in 2001.

Research in the 1960s and 1970s on the potential causes of human heart disease and blockage of the circulatory system indicated high levels cholesterol in the blood to be a likely cause. Because the egg yolk contains considerable cholesterol, most in the medical profession cautioned people about eggs, and consumption dropped sharply.

Continued research, however, has differentiated two general types of cholesterol—good (HDL) and bad (LDL)—and that a person's tendency to manufacture cholesterol (related to genetics and perhaps other factors) has more to do with blood cholesterol level than does cholesterol intake. Also, the egg industry has lowered the average cholesterol content in eggs from about 270 milligrams to about 220. Most of this reduction has been the result of genetic change in the nation's layer flock.

Whereas the American Heart Association had earlier advised that three eggs per week should be the limit, the Association changed its guideline in late 2000 to "an egg a day."

Egg consumption and consumer awareness of the egg and its positive features have been considerably enhanced by aggressive and effective industry organizations. For example, most consumers are well aware of the mind-catching phrase, "the incredible edible egg" used in advertisements, educational programs, and literature. The American Egg Board, funded by a national check-off on all egg production from companies with more than 75,000 layers, is the major egg promotion organization. The Board and the United Egg Producers, described in Chapter 33, jointly fund an Egg Nutrition Center, which focuses on gathering and providing scientifically accurate information on the relationship of eggs to the nation's health and nutrition. A panel of independent scientists guides its research and educational efforts.

In states with large egg industries, industry-financed state organizations promote egg consumption by such means as fair booths, serving omelets at public events, school demonstrations, and promotional literature.

New product development is also financed by some of these organizations, as well as by the food industry, the development work usually being done in university food science departments or industry laboratories and test kitchens. The American public seems always ready for a new and attractive food item.

QUESTIONS FOR STUDY AND DISCUSSION

1. Sketch a cross section of an egg and identify its major parts.
2. Compare the yolk and the white in appearance and nutritional characteristics.
3. Describe how eggs are handled and stored after they are laid, in order to maintain high quality.
4. Eggs for retail are graded for quality and classed for size. How is quality usually measured? What is the difference in weight per dozen between successive weight classes?

5. If you should break an old egg and a newly laid egg into a flat pan, how would they differ in appearance?

6. Describe the shell of an egg and its function. Of what is it composed? Why is the shell important?

7. What is the recommended temperature for storing eggs?

8. What percent of eggs are processed? What are some major uses of processed eggs?

9. Describe the trend in per capita consumption of eggs in the United States in recent decades.

The Business 35
of Commercial
Poultry Meat Production

Dramatic increases in production have occurred in the poultry industry, especially in the broiler business (see Color Plate F). These highly efficient birds specifically bred for meat production reach a market weight of about 5 pounds in 48 to 50 days, after having consumed less than 10 pounds of high-energy mixed ration. Note the more than five-fold increase in pounds of broilers marketed annually since 1961. (Figure 35–1 and Table 35–1).

The value of U.S. broilers produced in recent years has exceeded $16 billion annually and over 80 percent of the total poultry meat produced each year. The popularity of broilers in providing a high-quality economical source of meat for the American consumer has steadily increased to over 80 pounds per capita in 2002.

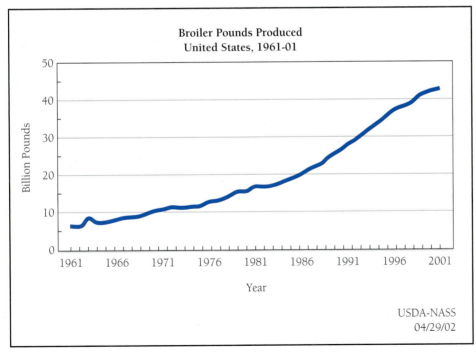

Broiler Pounds Produced
United States, 1961-01

USDA-NASS
04/29/02

Figure 35–1

The annual production of broilers (in pounds) since 1960 has increased from about 5 billion to more than 40 billion pounds.
(*Source:* USDA, NASS.)

Table 35–1	U.S. Broiler Industry Production and Consumption Trends				
	Broilers Produced		**Average Price Received by Producers (cents/lb)**	**Value of Production ($ millions)**	**Per Capita Consumption (lb)**
Year	*Million Birds*	*Million Pounds*			
1955	1,092	3,350	25.2	844	13.8
1965	2,334	8,111	15.0	1,218	29.6
1975	2,950	11,095	26.3	2,899	36.9
1985	4,470	18,810	30.1	5,668	56.1
1997	7,764	37,591	37.7	14,159	73.0
2000	8,390	39,884	40*	16,000*	80.5

Sources: *Brown, Robert H. Feedstuffs, October 1979, p. 94; Poultry and Egg Situation, May 1990; USDA, NASS. Poultry Production and Value, May 1996; USDA, ERS. Agricultural Outlook, January 1999;.*

* estimated

Table 35–2	U.S. Broiler Exports by Year	
Year	*1,000 Pounds*	*Percent of Production*
1970	93,707	1.2
1980	567,050	5.0
1990	1,165,180	6.1
2000	4,918,000	16.3
2001	5,555,000	18.0
2002	4,800,000	15.0

Source: *USDA, ERS.*

Rapid-gaining broiler strains, nutritional advances, labor-saving devices, assembly-line operations, vertical integration, production scheduling, and careful cost control allow mass production of a uniform product at relatively low cost. Price, relative to other meats, has been attractive. Increased broiler production is likely to continue for several years, and although the increase in per capita consumption in the United States may slow, industry growth in the immediate future will be influenced by export volume, now at about 15 percent of production. The export of broilers steadily increased during the 1990s and into 2001, especially to Russia, Hong Kong, Japan, Mexico, and China. There was a drop in 2002, however, resulting largely from a temporary ban in imports by Russia. (See Table 35–2.)

The production of turkeys rose steadily during the last half of the twentieth century and is now about 270 million birds annually (Figure 35–2) with a market value of about $3 billion per year. Consumption per person amounts to about 17.5 pounds per person each year. A small percentage of poultry meat consumed is derived from culled laying hens, a small number of roosters, and other fowl such as ducks, geese, pheasant, and quail.

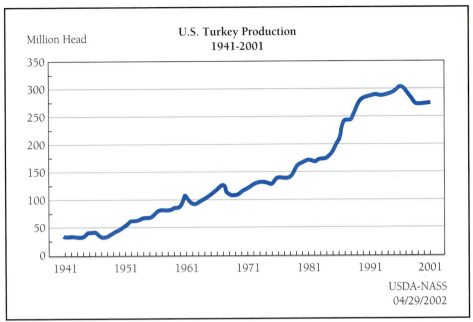

Figure 35–2
The annual production of turkeys in the United States has steadily increased since about 1950 and peaked at almost 303 million birds in 1996 and has been steady since.
(*Source:* USDA, NASS.)

Figure 35–3
Fried chicken being prepared in fast-food shop. Sale of broiler meat through fast-food shops has contributed to the increase in per capita poultry consumption.
(Courtesy of Pfizer H & N.)

The poultry industry is highly integrated, so that production is tied closely to retail sales—in supermarkets, restaurants, or ready-to-eat-chicken shops (Figure 35–3). Demand for poultry appears rather steady, day to day and week to week, so complete scheduling of all steps in the production and processing scheme can help to establish prices for all parties and reduce risk.

Increased per capita poultry consumption has been associated primarily with lower and lower product costs. Margins of profit in production and processing are small. Investments are high, because of the necessity of large volume. Disease outbreaks and their treatment can be very costly, so most producers have sophisticated disease prevention programs.

Broiler breeders are aggressively developing strains that may have genetic resistance to certain common diseases. They are also exerting effort toward increasing flavor of broiler meat by genetic selection. There is evidence this can be achieved.

Most of this chapter covers the broiler business. The development of this business is illustrated in Figure 35–1. Sections 35.6 and 35.7 discuss turkeys and other fowl.

Upon completion of this chapter, the reader should be able to

1. Appreciate the increased trends in poultry production and consumption brought about by efficiency throughout the industry.

2. Describe the broiler business in terms of geographic concentration and size of production units.

3. Define *integration* and describe how it functions in the broiler industry.

4. Explain how broilers normally are handled, from the time the chicks are delivered to the growing house until they depart for the processing unit.

5. Contrast turkey and broiler production, in terms of seasonality and bird size.

35.1 GEOGRAPHIC DISTRIBUTION AND SPECIALIZATION

Figure 35–4 shows where most broilers are produced. Production areas are concentrated for several reasons. The broiler business really began about 1923 in the Delmarva (Delaware–Maryland–Virginia) area. Other concentrated production areas developed early in Maine, Connecticut, and Rhode Island, near centers of population. During the 1950s, prompted by temperate climate, available labor, and increased grain production, broiler production expanded rapidly in the southern states. (See also Table 35–3.) Note that although most broilers are produced in the southern and eastern areas of the country, successful operations exist in other areas, especially areas with high grain production. Areas within states with lighter population densities and labor resources and fewer environmental pressures also are attractive for broiler production.

The turkey industry is less concentrated in specific regions of the United States (Figure 35–5) as indicated by the leading states of Minnesota, North Carolina, Arkansas, Virginia, Missouri, California, and Indiana.

35.2 BROILER UNITS: TYPES AND SIZES

Virtually all broilers are raised under a contract between a processor and/or feed supplier and the grower, or are owned by the processor. The contracting firm generally assumes most or all of the price risk and provides the day-old chicks and the feed. The grower furnishes the housing, equipment, fuel, and labor. The grower provides daily management, but generally is supervised by a firm representative who visits the premises on a regular basis. Litter (bedding), medicinals, and other inputs may be provided by either party, depending on the terms of the contract.

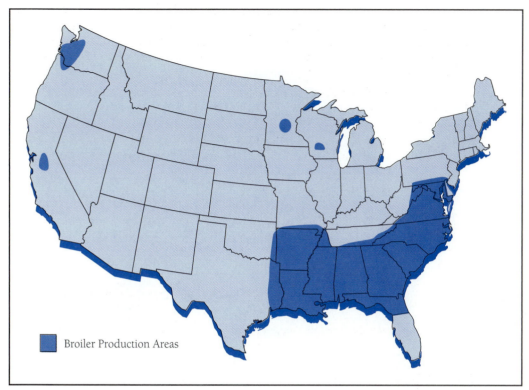

Figure 35–4
Broiler production is most heavily concentrated in the south and southeastern regions of the United States, and areas of the upper Midwest and the Pacific coastal states.
(*Sources:* Agri-Bio Corporation and USDA.)

Table 35–3	Top Eight States in Broiler Production, 2002

State	Billion Birds
Georgia	1.3
Arkansas	1.2
Alabama	1.1
Mississippi	770
North Carolina	735
Texas	588
Maryland	293
Delaware	251

Source: *USDA, ERS 2002.*

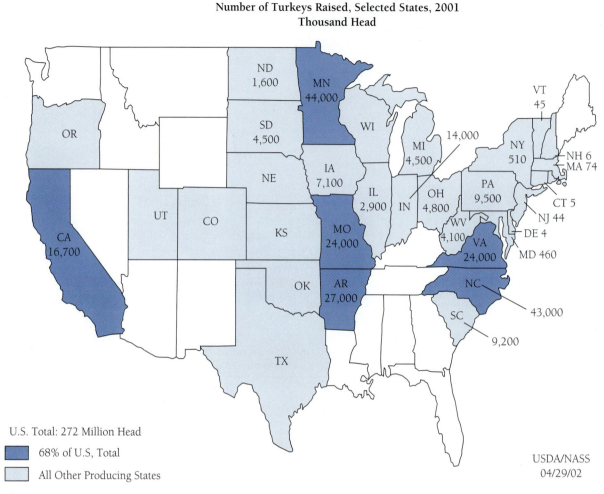

Number of Turkeys Raised, Selected States, 2001
Thousand Head

U.S. Total: 272 Million Head

68% of U.S, Total

All Other Producing States

USDA/NASS
04/29/02

Figure 35–5
States where most of the U.S. turkeys are raised. The darker shaded states produced 68% of the U.S. total in 2001.

In 1950, about 631 million broilers were marketed in the United States. Since that time broiler production has expanded over 13 times to more than 8.4 billion broilers marketed annually. Production units have steadily decreased in number but increased in capacity. These large, efficient operations and good management have resulted in an excellent cost per pound of broiler produced.

The top 10 broiler processing companies in 2002 ranked by average weekly ready-to-cook products in millions of pounds are listed in Table 35–4.

These 10 companies produced, slaughtered, and processed more than 70 percent of all broilers produced annually. The production and processing of a major proportion of the industry's output by a few companies is expected to continue in the future.

The number of broiler companies has decreased to 42 today—as the industry has continued to consolidate and further consolidation is expected.

The term *integrated* often is used to describe the poultry industry. The supplying of inputs (feed, birds, etc.), the growing, processing, and, in some cases, even the retailing are tied together by contractual agreements.

Table 35–4	Top Ten Broiler Companies, 2002, in Pounds of Ready-to-Cook Product		
Rank	Company	Weekly Average (million lb)	Processing Plants
1.	Tyson Foods, Inc.	148.84	106
2.	GoldKist, Inc.	61.53	24
3.	Pilgrims Pride Corp.	57.53	27
4.	Con Agra Poultry Co.	51.53	30
5.	Perdue Farms, Inc.	48.15	26
6.	Wayne Farms, LLC	29.15	26
7.	Sanderson Farms, Inc.	25.11	14
8.	Mountaire Farms	19.71	6
9.	Cagles, Inc.	16.41	8
10.	Foster Farms	15.54	16
	Top 10 total	473.27	283
	Industry total	663.13	343

Source: *WATT Poultry USA, January 2003.*

The major entrepreneur or contractor may be a processor who wants to ensure that a plant, with a processing capacity of perhaps several hundred thousand broilers a day, has a steady supply of birds for processing. It may be a feed company that contracts with both growers and a processor to ensure a good feed manufacturing volume. In a few instances the contractor is strictly an agent or entrepreneur, establishing contracts with all parties and seeking a return on contracting skill and on the risk assumed.

Among broiler firms in major broiler-producing areas, most birds are housed in what are termed "conventional houses." These generally are 400 to 500 or more feet long and 40 to 50 feet wide, with no internal posts, and with open sides, and curtains for use in cool or inclement weather (Figure 35–6). In most of these cases the ceilings are insulated. A few have fans to aid ventilation. Most have dirt floors; a few have concrete or asphalt. This type of house, of course, is most common in Georgia through Texas because it is lower in cost and appropriate for the climate.

Complete environmental control, with fully insulated, windowless houses, cement floors, and a fully automatic ventilation system, is more common in the northern states, including the Delmarva area. Brooder units provide supplemental heat the first days or weeks for each lot of broilers.

Litter used may be chopped straw, wood shavings or chips, sawdust, chopped bark, rice hulls, ground corncobs, or certain other inert material. The main factors in litter selection are availability, cost, and moisture-absorbing capacity. Most broiler producers believe fairly coarse litter is better, in that it provides a softer, cushion-type support for the birds, and air will move through it to keep it drier. Litter is placed in the building before a batch of chicks is delivered, usually at a depth of 3 to 5 inches. Sometimes the litter may be used two or three times before complete replacement.

Some broiler producers insist on fresh litter with each lot of chicks, primarily to avoid the spread of disease to the next batch. Others contend that old litter provides, in many cases, a low level of disease exposure so the birds build up a natural immunity early. Reuse of litter is obviously cheaper. Where it is reused, wet or decomposed areas, especially around waterers, are removed and the remainder fully stirred, by hand or

Figure 35–6
Brooding day-old broiler chicks, left. Right photo shows broilers nearing market weight in an open-style house with curtains.
(Courtesy of Pfizer H & N.)

Figure 35–7
Broiler chicks a few days old, utilizing a modern system of watering.
(Courtesy of Chore-Time, Milford, IN.)

mechanically. In some cases, it is shoved into a single pile to heat for several days. There usually is enough moisture to permit some fermentation and an increase in temperature to the point that some disease organisms are killed.

Brooders, feeders, and waterers often are suspended from the ceiling. At the end of the growing period, these can be lifted out of the way, birds penned up more easily for catching, and the building can be cleaned and disinfected mechanically and the litter removed or stirred.

A variety of mechanical feeders and waterers are used (Fig. 35–7). Storage for 3 to 7 days' feed supply usually is provided. Feed is augered to individual suspended feeders or is available to birds in open troughs equipped with slow-moving, endless link chains. Waterers may be open troughs with continuous water circulation or drinking units where water is released into a small cup by the birds' activating a small valve (Figure 35–7). Because of the large number of birds handled by one person, dependability of the equipment—freedom from breakdowns, water leakage, or other malfunction—is especially important.

Sufficient feed and water space for all birds is essential. With the rapid establishment of the social order, shy birds may otherwise be kept away and not eat or drink enough to gain efficiently.

35.3   BROILER MANAGEMENT

Day-old crossbred broiler chicks of both sexes are delivered to the unit, usually in a well-ventilated van or panel truck. The hatchery is usually a part of the integrated operation.

The birds are vaccinated before delivery. It is important that the house, litter, water, and heating units be ready for them. Feed is sometimes withheld 2 to 3 hours so that water is consumed first. The brooder area should be at about 90°F or slightly above.

The new chicks are confined to an interior portion of the house, usually by fine-mesh wire, hardware cloth, or cardboard "chickguard" 18 inches high, so they will be close to feed, water, and, if necessary, supplemental brooder heat, and so they will learn to eat and drink. Feed may be scattered first on newspaper or the chick box tops, or be available in small open feeders (Figure 35–6). Extremely small grit, granite, or other insoluble material may also be available. The grit is held in the gizzard to serve as a grinding agent in the digestive process.

After several days, the confinement area is enlarged, and eventually the confinement is removed and birds have access to the total building, including the mechanical feeders and waterers.

Density may range from 0.5 to 1 square foot per bird, and will be determined by expected relative profitability. Broiler prices, performance of different crosses at varied bird densities, cost of housing, and other factors will influence the density.

Research indicates that density could increase to as little as 0.4 square foot per chick without significant decrease in weight gain, and that successive increases in density will increase gain per unit of feed, likely due to reduced exercise.[1] Bird density in most commercial broiler units commonly ranges from 0.7 to 0.8 square foot per bird.

Most broiler management decisions have been made before broilers enter the house. Purchase and sales agreements, or a contract that covers the firm's and grower's responsibilities, have been established, equipment is functioning, litter is in place, and the ration has been formulated and delivered to the growing house.

Management during the growing period involves primarily (1) observing the birds to watch for disease symptoms or evidence of poor performance; (2) checking of feeding, watering, and ventilation equipment to ensure that it functions properly; and (3) maintaining a proper supply of feed and water.

Disease outbreaks generally are treated by vaccination or administration of drugs via feed or water, or by spray. Individual handling of birds generally would be far too costly.

Large broiler operation size and labor-saving equipment, combined with highly skilled management, have resulted in greatly reduced labor requirements. Georgia research[2] showed that operators of broiler flocks, averaging 122,000 birds with annual production of 732,000, were able to reduce the labor required per 1,000 birds to slightly under 3 hours for the entire growing period. Labor needs for operations equipped with mechanical starter feeders and nipple drinkers were about 60 percent as great as operations with conventional feeders and waterers.

---

[1] Adapted from *Scientific Aspects of the Welfare of Food Animals*. Council for Agr. Sci. and Tech. Report No. 91, November 1981.
[2] Lance, G. Chris. *Poultry Sci.* 74 (1995): 398.

35.4   CONTRACT GROWING

The principle of contract broiler growing is that the grower has land, labor, limited capital, and a limited desire or capacity to assume price risks. Yet if the grower's labor, land, and a certain amount of capital (buildings and equipment) are invested in the enterprise, there will be sufficient incentive to achieve good production, to manage the birds well.

Contracting firms provide chicks, feed, and experienced field personnel who visit each production unit on a regular schedule or who are on call to check for or to prescribe action to handle any problems that may develop.

Most contracts guarantee the grower a minimum income per 1,000 birds started. Contracts usually provide that the firm (processor, feed manufacturer, or other) and grower share equally the calculated profit above the minimum guarantee, or that higher payments per 1,000 broilers or per 100 pounds of broilers be paid the grower when prices of broilers or dressed carcasses exceed preestablished levels. The calculations, of course, may include charges or deductions for mortality, condemnations, hauling charges, medicines, and other expenses. In some contracts, the grower is paid a set rate per pound of broiler delivered to the plant, depending on the feed conversion rate.

To enable the grower to obtain financing or feel justified in investing in buildings and equipment, contracts usually cover several years. In some cases the firm advances credit to the grower for these investments.

Efficiency of integrated broiler production and processing systems varies, of course, with many factors, including the grower's management ability, quality of housing, capacity at each location, accessibility to good roads, and distance from feed mill, hatchery, and processor. Therefore, it is logical that contract terms take these factors into account.

In recent years, rapidly growing numbers of franchise drive-ins and restaurants that feature deep-fat-fried or barbecued chicken have become a part of the vertically integrated business. Certain of these retail groups have been organized by processors or growers.

35.5   FROM GROWING HOUSE TO PROCESSOR

Most broilers are processed at about 4.0–4.2 pounds and at about 40 to 42 days of age. Since the 1970s, the average weight of processed birds has steadily increased. Heavier birds generally are produced and processed more efficiently per pound of bird. There is less "down time" at the growing unit (usually 14 days between lots is scheduled for clean-out, disinfecting, repair of equipment, etc.), less labor is required to handle baby chicks and finished birds, and the chick cost is lower per pound of bird produced. In the plant, labor and processing costs are also lower per pound of bird or carcass.

The number of broilers processed fluctuates from month to month. Both producers and processors try to avoid heavy workloads during major holiday periods, and because of the consistent 7-week growing period this may cause the cyclic slaughter pattern illustrated in Figure 35–8. Price appears to be influenced more by other factors than by supply.

Feed and water usually are withdrawn about 12 hours before processing. This prevents undue wastage of feed that would be in the digestive tract at the time of processing and results in maximum carcass yield.

The processing date for each lot usually is predetermined, because broiler contracts schedule production to keep processing units busy and a steady supply of dressed birds on the market. Because the birds are quiet and more easily handled in the dark, they are usually loaded during the night or very early morning. Experienced crews work through the building in minimum light, pick up the birds by hand and place them in metal modules (crates). The modules are loaded by forklift onto a truck, usually 6,000 to 6,500 birds per truck.

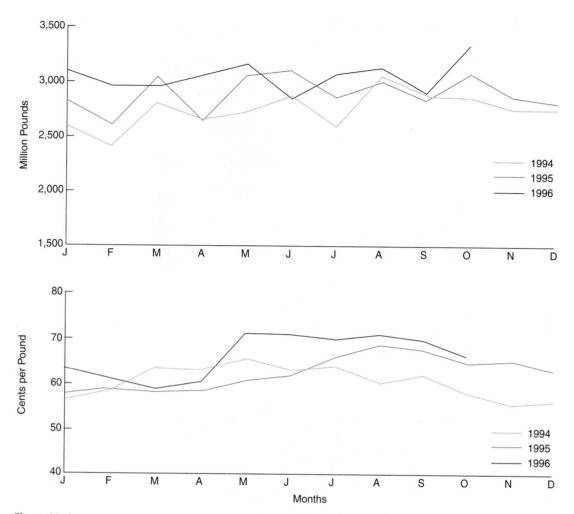

Figure 35–8
Changes in seasonal broiler slaughter volume (top) and prices (bottom) in recent years. (Adapted from USDA sources)

At the plant the trucks may be held in ventilated and heated or cooled sheds until the modules can be unloaded by forklift. The modules are emptied onto a wide belt where workers shackle the individual birds for the processing line. The workers are trained and supervised to minimize bruising and death loss. Poultry meat processing is discussed in Chapter 23.

35.6 TURKEY PRODUCTION

The turkey meat business parallels the broiler business, except that production is more seasonal, the growth period is much longer, hens and toms are usually grown separately, and a much higher proportion are grown on open range or in open shelters (see Color Plate F). The top 10 states in turkey production are shown in Table 35–5. Per capita consumption for 2002 was about 17.5 pounds in the United States (Table 35–6). Vertical integration from producer to processor is similar to the broiler industry, and the top 15 processors control more than 90 percent of all U.S. production. Several of these large processors are producer owned.

Table 35–5	Top Ten States in Turkey Production, 2001

State	Million Head
Minnesota	44,000
North Carolina	43,000
Arkansas	27,000
Virginia	24,000
Missouri	24,000
California	18,700
Indiana	14,000
Pennsylvania	9,500
South Carolina	9,200
Iowa	7,100

Source: *USDA, NASS.*

Table 35–6	Approximate U.S. Turkey Production and Consumption		
Year	Million Birds Raised	Million Pounds Raised	Pounds Consumed per Person
1990	282	6,043	17.5
1995	293	6,779	18.0
2000	270	6,934	18.0
2002	270	7,290	17.5

Source: *National Turkey Federation.*

Annual production and per capita consumption have been very consistent in recent years (Table 35–6). Though pounds consumed per person has changed little, less is handled as home cooked, whole birds and more is utilized as precooked turkey product, such as sliced breast meat used in sandwiches. Relatively more of the turkey hens, being lighter when finished—about 15.5 pounds at 14 weeks of age—are utilized as whole birds for roasting. Toms are marketed at a heavier weight—about 32 pounds at 18 weeks of age—and most are further processed into the precooked or "portion control" items such as sandwich meat, turkey franks, cutlets, or individual items such as turkey legs.

With more precooked product and steady demand through the year, production has become less seasonal. However, slightly higher placement of poults during the first 6 months of the year ensures more processed turkey for the holiday season, from Thanksgiving through New Year's Day.

The current turkey production industry in the United States and Canada traces its roots to an emigrant from England to Canada who, in 1927, brought with him some Broad-Breasted Bronze Turkeys. Their efficiency and meat quality, superior to native North American turkeys, prompted increased production and consumption of turkey meat. Over time, continued genetic selection and matings have resulted in more breast meat, meatier thighs, and white feathers. The latter means a uniformly white skin after the feathers are removed.

Figure 35–9
Tom turkeys approaching market weight.
(Courtesy of Nicholas Turkey Breeding
Farms.)

Feed efficiency for turkey poults ranges from slightly over 2 pounds of feed per pound of marketed bird for light broiler types to about 3.4 pounds of feed per pound of bird for toms of the heavier roaster types. There has been significant progress, as a result of selection and breeding systems within the major varieties, in developing efficient lines and strains of poults.

Delivered day-old poults usually are held in confinement with supplemental heat and facilities similar to those provided broilers. Birds may be kept in open pens or on open range year-round in the South and after warmer spring weather in the northern states until ready for market (Figure 35–9). In the north central states, however, a high proportion of poults are grown to market weight in complete confinement.

35.7   OTHER MEAT-PRODUCING FOWL

Chickens and turkeys from laying and breeding flocks provide about 5 percent of the total poultry meat supply. Being older, they provide less tender meat, which usually is used in soups, chicken pies, and other processed food items. Some hen chickens are sold at retail for stewing.

Capons are produced as specialty items. They are male chickens, surgically castrated when 10 days to 4 weeks of age. They are slaughtered at 5 to 6 months, weighing 10 to 14 pounds, and produce carcasses that are relatively tender, juicy, and flavorful.

About 20.7 million ducks were processed in federally inspected plants in 1996.[3] Most are produced in the northeastern states, the upper Midwest (especially Indiana), and California. Ducks are fed rations similar to those fed to broilers and turkey poults. They are processed at 6 to 8 weeks of age at a weight of 6 to 7 pounds.

Many geese are raised on farms in the north central states. Traditionally, South Dakota ranks first in goose production, at about 200,000 pounds yearly.[4] Most are raised by Hutterite colonies in relatively large flocks. Demand for geese parallels the seasonal

---

[3] USDA. *Agricultural Statistics*, 1998.
[4] *Poultry Digest* 38 (1979): 392.

demand for turkeys; they are often used at Thanksgiving and Christmas. The meat is usually juicier than turkey meat. Birds are normally 20 or more weeks of age and weigh 10 or more pounds when dressed.

Pheasant and quail have become rather popular in some regions as menu items in some of the finer restaurants and for home use. Birds are handled and processed in a manner similar to the way broilers are handled, but in smaller groups. Because less is known about nutrition and other aspects of management, production costs usually are higher.

The ratite industry has gained popularity in the United States in recent years. These include the ostrich from Africa, the emu from Australia, and the rhea from South America. Ratites are well known for their skins that can be processed into high-quality leather goods such as boots and purses. Breeders also promote ratite for their production of red meat. The U.S. ratite industry in the mid-1990s was estimated at 500,000 emus, 200,000 ostriches, and a much lower number of rhea, largely breeding flocks with only limited marketing of ratite meat.

The growth and development, behavior, and characteristics of ratites differ greatly from birds bred and raised for commercial poultry production. Most obvious is their size, as an adult ostrich may weigh 200 to 350 pounds and stand 7 to 10 feet tall. At 3 to 4 years of age ostriches reach sexual maturity and begin to breed. Typically, several hens mate with one male in the spring months. Incubation period is about 6 weeks. Chicks reach their full height at about 6 months of age. Ostriches are also known for their speed and are capable of running up to 40 miles per hour. Birds maintained for breeding may live as long as 50 years.

The meat, hides, and feathers of ostriches have commercial value when demand and markets are available. Ostrich meat has a texture and color similar to beef and is comparable in iron and protein content. Cooked lean cuts of ostrich meat contain considerably less fat than beef and chicken. Some restaurants have now added prepared ostrich meat dishes to their menus.

The emu also is a large bird—reaching full size within about 2 years at a height of 5 to 6 feet tall and weighing up to 150 pounds. The female breeds as early as 18 months of age and may continue to produce eggs for more than 15 years. The incubation period for eggs is about 50 days, the eggs being incubated by the male during this period.

Products derived from the emu include leather, meat, oil, and decorative egg shells. Emu meat is similar to that of the ostrich.

QUESTIONS FOR STUDY AND DISCUSSION

1. Most broilers are produced in contract operations. What does that mean? What are the common features of a typical broiler contract?
2. Why did contract broiler production become common in the southeastern states?
3. Describe how most broilers are produced, including type of housing, age when they enter and leave the rearing house, and types of rations.
4. Generally describe the mating system used in production of broiler stock. Are most of a single breed or crosses of several breeds? Is the breeding done by the broiler growers or by specialized breeders? What are some major goals of a breeding program?
5. At what age and weight are broilers usually processed? Briefly describe the process.
6. Name five species of poultry, other than chickens, that are raised for meat in the United States.

36
Aquaculture

Aquaculture refers to the growing and rearing of aquatic organisms including fish, mollusks (oysters, clams, mussels), crustaceans (crab, lobster, crayfish, shrimp), and aquatic plants. Aquaculture can be viewed as a form of animal agriculture, an enterprise where most of the organisms are grown for food and cultured in a water environment. In addition to providing food for human consumption, aquaculture also provides numerous species for sport, bait, and ornamental purposes.

Aquaculture has provided food for people throughout the world for thousands of years and has been traced back to fish farming by the Chinese over 3,000 years ago. Although some forms of U.S. aquaculture were practiced in the late 1800s, it was not a major industry until the 1950s when commercial catfish farming greatly expanded in the southern states. Through increased knowledge and advanced technology the aquaculture industry has steadily grown into an important segment of U.S. agriculture with commercial fish farming represented in every state.

A majority of the catfish and trout, about one-third of the salmon, and about half of the shrimp consumed in the United States is produced by aquaculturists. Aquaculture is a rapidly developing segment of American agriculture, and, for some farmers, provides a means of changing or diversifying their animal operation, as well as providing an investment opportunity for them or others.

When discussing modern animal agriculture, one almost always relates to the breeding and management of dairy, beef, swine, sheep, or horses, or more recently perhaps some alternative enterprises such as dairy goats, ratites, or even bison. Why do we include this chapter dealing with aquaculture? The answer is simple—aquaculture provides an important source of food in the diets of people throughout the world and undoubtedly will contribute more in the future with an expanding world population.

U.S. consumption of fish and seafood increased dramatically in the 1980s. Today about 14.5 pounds is consumed per person annually, on a boneless trimmed basis. Although less than beef, pork, or chicken, the consumption of fish and seafood is equal to or greater than turkey consumption, and far greater than veal or lamb and mutton. The increase in overall aquatic foods most likely is due to the health benefits they provide. More specifically, many aquatic organisms contain high levels of omega–3 or n3 fatty acids. These fatty acids have very specific bonds located at the third position of the fatty acid and, in turn, elicit unique properties. For instance, Eskimos eat diets very high in fat (e.g., whale blubber) and rarely develop cardiovascular disease (CVD). The same is true for many of the people who live and eat aquatic organisms along the Mediterranean Sea. Epidemiological studies have shown n3 fatty acids to be associated with a variety of health benefits. By eating more n3 fatty acids via aquatic organisms, there is a direct decrease in saturated fats and conversely an increase in polyunsaturated fatty acids (PUFA), especially n3, which in turn leads to improvements in blood lipid profiles, reduced risk of stroke, lowering of blood pressure, and a concomitant reduction in the risk of CVD.

Many of the characteristics of successful livestock and poultry production (Sections 1.8 and 1.9) and essentially all of the principles of animal science (e.g., genetics, nutrition, disease control) apply to successful aquaculture businesses. As in other agricultural products, the consistent production of a high-quality, marketable product is essential for a successful aquaculture business.

Today's more specialized aquaculturists must provide sufficient volume of a high-quality product to be sold through fish markets in the larger cities. The smaller producer often must develop a local market through promotion and consistent sale of a product desired by the consumer.

Most critical to successful aquaculture production is (1) the timely application of sound management practices concerning the species being grown, and (2) keeping the water environment of the species in excellent condition. Excellence in water quality requires control of water temperature within an acceptable range, and management practices to assure sufficient oxygen, proper level of nutrients, and control of pH, ammonia, and pollutants.

This chapter will discuss characteristics and establish principles of aquaculture, primarily as they apply to commercial fresh fish production.

Upon completion of the chapter, the reader should be able to

1. Identify the region of the world and countries that dominate the aquaculture industry.
2. Characterize the aquaculture industry in the United States.
3. Provide examples of mollusks and crustaceans cultured in the United States.
4. Categorize the more significant species of warm-water and cold-water fish produced in the United States.
5. Describe how water temperature changes affect the metabolic rate and activity of fish.
6. List several important factors that determine water quality.
7. List some visual signs of oxygen depletion in fish containments and factors that may cause it.
8. Briefly describe four methods of rearing fish.
9. Contrast differences in eating behavior of carnivorous and omnivorous fish species and how these differences affect the type of diet formulated.
10. Define a "complete diet" and the essential characteristics of a high-quality diet.
11. List several types of stress that may predispose fish to diseases and indicate the more important disease prevention practices.
12. List several factors to be considered before fish are harvested for market.
13. Provide the more significant points in development of a marketing plan.

36.1 GLOBAL AQUACULTURE

Global food fish production consists of both aquaculture and "captured" sources, those that grow wild in oceans and other bodies of water. Total production of captured finfish and shellfish amounted to about 92 metric tons in the mid 1990s; two-thirds were available as edible fish, and the balance processed into fishmeal and oil for industrial uses,

TABLE 36–1	Leading Aquaculture Countries and Projected Production Volumes in 2000		
Country	**Rank**	**Metric Tons (thousands)**	**Percentage of Global Share**
China	1	22,500	64.30
India	2	2,400	6.86
Japan	3	1,400	4.01
Korean Republic	4	1,220	3.50
Philippines	5	1,000	2.86
Indonesia	6	875	2.50
Thailand	7	680	1.94
United States	8	500	1.43
Bangladesh	9	475	1.36
Taiwan	10	300	0.86
Other countries		3,640	10.40
Total		34,990	

Source: *Adapted from FAO Fisheries Department* Review of World Aquaculture Outlook, *1999.*

and for animal feeds. Few if any captured fisheries are likely to increase production because of the limited supply of fish.

On the other hand, aquaculture contribution of global fish production has increased dramatically in the last few decades. For example, in the mid–1980s, aquaculture provided about one-eighth of the total global food fish. By 1995, its contribution had more than doubled to slightly more than one-fourth of the world's supply. With continued expansion and improved technology, aquaculture production should be able to provide most, and perhaps all, of the deficit between capture fisheries supplies, continued growth in global human population, and consumer demand.

Globally, China is by far the leading food fish producer providing over 64 percent of the total world production from aquaculture (Table 36–1). Chinese aquatic fish production ranks second only to pork production in providing food for human consumption. Carp is the most prevalent species of food fish in the world. Raised largely in China, carp totals about half of the world's fish production.

As shown in Table 36–1, most of the other leading countries in aquaculture production are Asian where fish make up a higher percentage of the human diet than other areas of the world. Although U.S. aquaculture is a relatively young industry and represents only about 2 percent of world production, it currently ranks eighth largest in the world. U.S. aquaculture production increased to about 500 metric tons in the year 2000 or almost 21 percent higher than in 1995. At that time annual U.S. production was estimated to be nearly over 413.4 metric tons and valued at over 729 million dollars (Table 36–2).

36.2 U.S. AQUACULTURE FOOD PRODUCTION

About 100 species of aquatic organisms are produced through cultured practices in the United States. The more important species, based upon quantity of production and dollar value, are listed in Table 36–2. Freshwater fish (primarily catfish, trout, tilapia, and

TABLE 36–2	Important Aquaculture Organisms in the United States, 1995			
	Quantity (Metric Tons)	Percent of Total Quantity	Value in Dollars (thousands)	Percent Total Value
Catfish	202,706	49.0	350,681	48.1
Oysters*	109,080	26.4	70,646	9.7
Crawfish	26,375	6.4	34,815	7.2
Trout	25,240	6.1	52,752	4.8
Salmon	14,106	3.4	75,504	10.4
Clams*	13,481	3.3	19,221	2.6
Baitfish	9,883	2.4	71,355	9.8
Tilapia	6,838	1.7	22,634	3.1
Hybrid St. Bass	3,772	0.9	21,161	2.9
Shrimp	1,000	0.2	8,820	1.2
Mussels	930	0.2	1,218	0.2
Sturgeon	20	0.0	290	0.0
Total	413,431		729,097	

*Shell weight included.

Source: Adapted from FAO 1997 data.

hybrid striped bass) represent about 60 percent of U.S. aquaculture production. The production of oysters, grown mainly along the coasts of the Northeast, Gulf of Mexico, and Pacific Northwest, and mussels and clams are second in U.S. production and amount to about one-third of total production.

U.S. aquaculture's production of human food can be categorized into the production of fish (primarily freshwater species), mollusks, and crustaceans. Fish production represents over 60 percent, mollusks about 33 percent, and crustaceans about 6 percent of all U.S. production.

Characteristics of most fish species grown for the U.S. food industry are presented in Table 36–3. Note the differences in growing and spawning temperatures among the species. For example, cold water temperatures are required for trout, walleye, and salmon, in contrast to the warmer water requirements of catfish, hybrid striped bass, and tilapia.

A fundamental knowledge of conformation and parts of the fish species better enables you to recognize differences in growth and appearance and in evaluating and describing their condition. The external parts of the fish and illustrations of some of the fish species commercially grown in the United States are shown in Figures 36–1 and 36–2.

Channel catfish are the most prevalent fish species and represent almost 50 percent of the total quantity, by weight, of U.S. aquaculture production. The meat of catfish is characterized as white, sweet, and mild. Catfish are warm water fish. Therefore they are largely produced in commercial ponds located in the Mississippi Delta region— Mississippi produces about 60 percent of all U.S. catfish, followed by Arkansas, Alabama, and Louisiana. About 177,000 acres of ponds are dedicated to the aquaculture of catfish in the United States. More details regarding catfish culture are presented in Section 36.4.

TABLE
36–3

Fish Species Commonly Grown for Food in the United States and Selected Requirements

Species	Production Potential	Ease of Culture	Marketing Potential	Temperature Requirements		Protein Feed Requirements	Most Common Production Systems
				Growout	Spawning		
Atlantic salmon *Salmo salar*	Moderate	Easy	High	50–62°F	42–50°F	45–50%	Raceways, net pens[a]
Channel catfish *Italurus punctatus*	Excellent	Easy	High	80–85°F	72–82°F	28–30%	Ponds, cages
Coho salmon *Oncorhynchus kisutch*	Moderate	Easy	High	48–58°F	45–55°F	45–50%	Raceways, net pens
Hybrid striped bass *Morone saxatilis x Morona chrysops*	Excellent	Moderate	Moderate	77–80°F	60–68°F	45–50%	Ponds, cages, Recirc. systems
Largemouth bass *Micropterus salmoides*	Excellent	Moderate	High	55–80°F	60–65°F	40%	Ponds
Rainbow trout *Oncorhynchus mykiss*	Good	Easy	High	50–60°F	50–55°F	45–50%	Raceways, net pens
Smallmouth bass *Micropterus dolomieui*	Moderate	Easy	Moderate	50–70°F	58–62°F	45%	Ponds
Tilapia *Tilapia, Sarotherodon, Oreochromis, spp.*	Good	Easy	Moderate	80–84°F	80–84°F	28–32%	Recirc. systems, cages
Walleye *Stizostedium vitreum*	Moderate	Difficult	High	72–77°F	48–55°F	40–50%	Ponds
Yellow perch *Perca llavescens*	Poor	Moderate	High	70–77°F	50–60°F	40–50%	Ponds, recirc. systems

[a]Net pens are similar to cages but are much larger in capacity.

Source: Indiana Aquaculture Plan, *Indiana Aquaculture Association and Purdue University, 1997.*

Figure 36–1
The external parts of the fish and correct terms. (*Source:* Purdue University.)

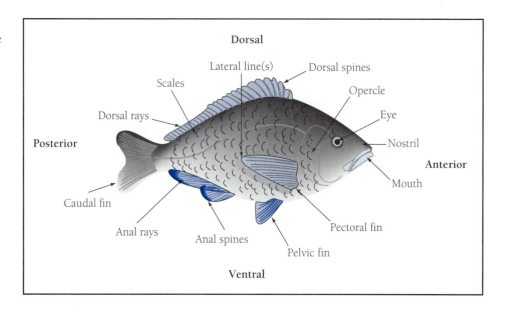

Hybrid striped bass

Channel catfish

Tilapia

Rainbow trout

Figure 36–2
Some food fish species commercially produced in the United States through aquaculture. (Courtesy of Illinois-Indiana Sea Grant Program.)

Tilapia, characterized as mild-tasting white-fleshed fish, is now being produced in most of the states. Tilapia is the generic name given to a variety of species in the *Cichlid* family. Two Tilapia species, the Nile and Blue, are the most commonly cultured species. A warm-water fish, Tilapia must be grown at temperatures above 80°F. Production of Tilapia has steadily escalated and expanded geographically because of its consumer popularity. Also, advanced technology in rearing systems that offer relatively inexpensive heat sources (e.g., solar greenhouses, improved insulation of buildings) has enabled Tilapia production to expand farther north into colder regions of the United States and closer to some larger city markets. U.S. Tilapia production amounts to about 15 million pounds annually. Consumer demand far exceeds present domestic Tilapia supply. Imports, primarily from Taiwan, provide an additional 50 million pounds each year to U.S. consumers.

The *Hybrid striped bass* is a cross between the "wild" striped bass family, a native of ocean salt waters along the east and west U.S. coasts, and the white bass, a fish that lives in fresh water. This hybrid was developed to replace the decline in numbers of striped bass harvested by commercial fisheries.

This new hybrid species is popular with American consumers because of its mild taste and firm texture. Its popularity also is rapidly increasing as a commercially grown food fish because of its faster growth rate, greater disease resistance, and higher survival rate than either parent species. Researchers have shown hybrid striped bass to be well suited to pond culture and are actively involved in research to improve other production techniques. Hybrid striped bass are marketed at 1.5 to 2.0 pounds at a higher market price than channel catfish.

Trout farming is most prevalent in Idaho where annual yield is valued at over $50 million annually. Idaho provides about 88 percent of all *Rainbow trout* produced in the United States. Its meat may be a white, pink, or orange color, and is mild with a nut-like flavor. Most trout are produced in raceways, a series of tanks through which water flows continuously, and further described in Section 36.4.

Atlantic salmon production amounted to only 3.4 percent of total U.S. aquaculture production, in volume, in 1995. However, because it commands an excellent retail price, salmon amounted to 10.4 percent of the total production in value. Traditionally, Atlantic salmon has been marketed through the "white tablecloth" restaurants because of its delicate flavor and mild taste. Today, producers also are promoting greater Atlantic salmon consumption through the targeting of midscale restaurants.

Oysters are grown primarily along the coasts of the Northeast, the Gulf of Mexico, and the Pacific Northwest. Oysters have been cultured in the United States for more than a century and are cultured by some because of their high popularity and value as seafood. Note that the quantity of oysters shown in Table 36–2 include shell weight that represents about 92 percent of the oyster. Actual edible oyster production approximated 8,773 metric tons in 1995.

Crayfish is the most prevalent crustacean produced in the United States. About 90 percent of all crayfish are produced in Louisiana, where they are typically grown as a single crop in earthen ponds, or as a double crop in a rice rotation system.

Another crustacean cultured in the United States is *freshwater shrimp*, sometimes referred to as *prawns*. About 70 percent of farmed shrimp are produced in Texas. They are grown in earthen ponds, similar in structure used in catfish farming, and also require the same high standards for water quality. More recently the trend in U.S. shrimp production has been in the testing and development of large-scale, water recirculating grow-out systems.

Today, only about 20 percent of the world's freshwater shrimp are grown in the Western Hemisphere. However, the United States leads in shrimp culture technology, especially in intensive hatchery production and management. Most seedstock shrimp are produced in hatcheries located in the southern coastal states of South Carolina, Florida, and Texas.

Today's more specialized aquaculturists must provide sufficient volume of a high-quality product to be sold through fish markets in the larger cities. The smaller producer often must develop a local market through promotion and consistent sale of a product desired by the consumer.

U.S. aquaculturists can be categorized into three grower groups.[1] Of these the most prominent are the well-established producers who specialize in the production of the more popular species such as trout, salmon, or catfish. The second group consists of those who are exploring different growing and culturing methods, and producing other species such as Tilapia, hybrid striped bass, or some form of shellfish such as shrimp or crayfish. The third group is those aquaculturists, in minimal production but experimenting in other species production, such as walleye, to determine their potential as a food fish in an aquaculture business.

36.3   WATER SOURCES AND FACTORS AFFECTING WATER QUALITY

No organism is more dependent upon their environment than those living in water (Figure 36–3). For example, as cold-blooded animals, the activity and growth of fish is largely dependent upon water temperature. They also are dependent upon a water supply that provides an adequate source of oxygen and nutrients, is maintained at an ac-

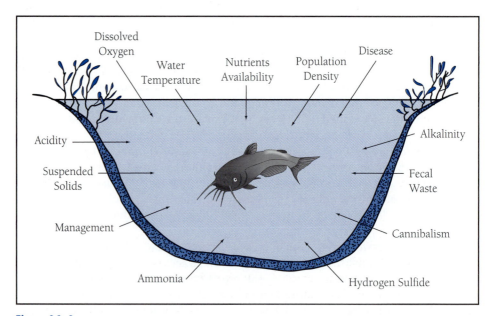

Figure 36–3
The aquatic environment of fish and many factors that may affect their health and performance. (*Source:* Illinois-Indiana Sea Grant Program.)

[1]Farmline. *Aquaculture Laying Groundwork for Future Growth,* May 1992.

ceptable pH, and accommodates the breakdown and elimination of the waste products of digestion. Several of the more important factors affecting the health and performance of fish are discussed in the following paragraphs.

Water Sources

Success in aquaculture requires an abundant source of clean fresh water. Depending upon the needs of fish species, sources can come from wells, springs, streams, lakes, storage ponds, or municipal water. Many aquaculture systems require very large quantities—such as trout farming in Idaho that depends upon an aquifer source of high-quality, cold spring water. Also, large quantities of fresh water are needed in shrimp farming that has a high stocking density and greater water requirement than most other forms of pond aquaculture.[2]

Water Temperature

The fish's metabolic rate increases as water temperatures rise, and decreases as water temperatures fall, and directly affects its activity, food consumption, and oxygen demand. For example, metabolic rate can be expected to double with every 18°F increase in water temperature.

As shown earlier in Table 36–3, each fish species requires rather specific and different water temperature ranges for spawning and growth. Temperatures above or below the desired range can result in reduced growth extending the time required to reach market size. Temperature extremes also result in increased stress and susceptibility to disease, and in extreme situations, can result in significant fish losses. Sudden water temperature changes, as sometimes occurs when fish are moved, can result in stressed or dead fish although these changes may not vary more than 10 to 15°F.

Another factor related to water temperatures and fish activity and performance is the *stratification* of water into temperature layers from top to bottom of the body of water as affected by environmental temperatures. These layers are referred to as the *epilimnion, thermocline,* and *hypolimnion,* respectively. The epilimnion, or upper layer, forms as days become warmer and is characterized as the warmer, lighter, less dense, oxygen-rich layer. The thermocline, or intermediate layer, possesses a notable and abrupt cooler temperature range. The bottom layer, or hypolimnion, is coolest and most dense, and therefore contains the least amount of oxygen.

Without mechanical aeration, stratification can occur because the warmer lighter upper layer may not mix with the cooler, heavier lower layers of water. This can result in extreme differences in dissolved oxygen availability to the fish. However, the alert manager is constantly aware of potential stratification conditions and will utilize mechanical aeration to prevent it.

Oxygen

High levels of dissolved oxygen are essential. In pond culture, *phytoplankton* (microscopic plant growth) are the principal source of dissolved oxygen through the process of photosynthesis, whereas dissolved oxygen losses are largely due to respiration by fish, *plankton* (small animals, bacteria, and plants), and diffusion of oxygen into the atmosphere. The amount of light that penetrates through the water is especially important in photosynthesis by the phytoplankton. Plant growth and rate of oxygen production decrease as discoloration and depth of water increase.

---

[2]Goldburg, R., and T. Triplett. *Murky Waters: Environmental Effects of Aquaculture in the United States.* The Environmental Defense Fund, 1997.

Excessive levels of organic matter entering the water also can upset oxygen balance through high oxygen usage in the decaying process. Sources include materials such as uneaten fish feed and nitrogenous waste products.

Phytoplankton and other aquatic plants, while producing oxygen during daylight hours are oxygen users during darkness, and when in excess can result in insufficient oxygen availability to the fish. Also, excessive plankton, usually associated with blue-green algae on the water surface, can die and the decaying process that follows can use large quantities of oxygen and cause lowered oxygen availability.

Fish must have an adequate supply of oxygen for survival and growth, and their needs depend somewhat upon their species, stage of development, and cultural conditions. About 4 ppm is considered essential for good health, growth, and reproduction. Commercial fish farmers monitor their ponds and other fish containments closely for oxygen depletion, visually and through use of oxygen meters (Figure 36–4). Dissolved oxygen is generally lowest at dawn due to the process of respiration and lack of photosynthesis during darkness. As a rule, supplemental aeration, shown in Figure 36–4, should be provided when dissolved oxygen decreases below 4 ppm. Visual signs of depletion include a darkening of water color, and swimming of fish near the surface during late night or early morning hours.

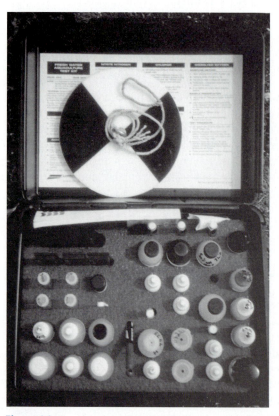

Figure 36–4
The desired water dissolved oxygen level of 4 to 5 ppm can be easily checked with an oxygen meter. If lower, water aeration should be provided by mechanical aerators.
(Courtesy of Illinois-Indiana Sea Grant Program.)

Carbon Dioxide

It is also important to understand the relationship between carbon dioxide (CO_2) content in the water and proper plant growth and fish health. In the daylight hours when photosynthesis is occurring, phytoplankton and aquatic plants use CO_2. Conversely, at night CO_2 is released during the process of respiration. Unusually high levels can occur with excessive plankton die-off. Levels greater than 20 ppm are considered excessive and adversely affect fish health. An ideal level of carbon dioxide is 5 ppm or less.

pH

Aquaculturists regularly monitor the **pH** (hydrogen ion concentration) of their fish water, especially when they suspect the carbon dioxide level to be higher than usual. Readings are best taken at daybreak when pH levels can be expected to be at a low point. A desirable pH range for best fish production is between 6.5 and 9.0. Typical pH levels will increase during the day and decrease during the night. High alkalinity waters with a pH of about 11, or acidic waters with a pH of 4, are considered to be lethal.

> **pH**
> An index of the acidity of a substance, the value of 7.0 being neutral, values above 7.0 being alkaline, and values below 7.0 being acid.

Nitrogenous Wastes

High levels of **ammonia**, should they occur in cultured fish water, are highly toxic to fish. Ammonia is a product of fish metabolism and bacterial decomposition of organic matter. It is excreted through the fish gills and feces into the water as un-ionized ammonia (NH_3) or ionized ammonia (NH_4). NH_3 is toxic at relatively low levels. Its increased concentration is related to the water pH and temperature, with highest concentrations after die-off of phytoplankton when carbon dioxide levels are higher than normal. When these conditions are likely, the water ammonia level should be monitored and can be easily accomplished using commercially available test kits.

> **Ammonia**
> NH_3, created by decomposition of organic matter in aquaculture water and in the process of nitrogenous breakdown by organisms such as those in the rumen.

Ammonia, through the nitrofication process by bacteria, is converted to *nitrite* (NO_2) which is detrimental to fish health. Nitrites enter the bloodstream across the gill membranes. When in excess, they then combine with the hemoglobin of the red blood cells to form *methemoglobin*, a compound that is unable to carry oxygen. Methemoglobin imparts a distinct brown color to the blood because of its low content of oxygen.

When ammonia levels are excessive, several management steps should be taken. These include correcting the pH levels by addition of agricultural lime, reduction or cessation of feeding, and flushing the system with new water if necessary.

Other Water Quality Factors

Water clarity also can affect fish health and growth performance. Problems can occur in water recirculation systems when suspended solids accumulate and the water becomes discolored. Suspended solids, in excess, can partially clog the gills of fish, reducing their ability to breathe and increasing their susceptibility to diseases. Suspended solids also can cause off-flavor problems in the meat of fish. Addition of flocculents and maintenance of a proper functioning filtration system are effective in reducing excessive levels of suspended solids.

Alkalinity and *water hardness* also can be related to decreased fish performance. Alkalinity (the buffering capacity of water) is expressed as ppm calcium carbonate. Hardness of water is largely affected by calcium and magnesium. Although they can differ greatly in some waters, alkalinity and hardness are similar in most waters. Some species such as crawfish and striped bass are adversely affected by *lack of water hardness*. As a rule, hardness should not be maintained above 50 ppm. Both alkalinity

> **Alkalinity**
> A measure of the pH buffering capacity.

and hardness can be adjusted upward by addition of some form of calcium and carbonates (e.g., lime, calcium chloride, or sodium bicarbonate) to the water.

Other water quality factors include *hydrogen sulfide* and *iron*. Hydrogen sulfide is most often related to accumulated food or fecal waste, or high levels of dissolved iron in the water source. Excessive iron forms red deposits that can settle on the gills of fish causing irritation.

36.4   METHODS OF FISH CULTURE

As the aquaculture industry has developed, four different production methods of rearing fish have evolved. These are (1) ponds, (2) raceways, (3) cages, and (4) water recirculation systems. Each will be briefly described in this section.

Levee-Type Ponds

> **Levee pond**
> A pond enclosed by an earthen dam.

The **levee pond** system is the most prevalent system used for fish production particularly in the southern portion of the United States where catfish farming is most concentrated (Figure 36–5). Within an operation, three types of ponds may exist to rear the fish through their stages of development—spawning, fingerling, and finishing.[3]

All commercial fishponds are constructed similarly but the size and number of each type will vary depending upon the overall size and objectives of each operation. In all cases, site location and topography are most important to ensure that (1) adequate water will be available, (2) chemical and physical properties of the soil are acceptable, (3) the site is not subject to flooding, and (4) the pond can be easily drained.

Pond shape also is an important factor to consider as it can greatly affect fish culture management and labor efficiency. Most levee-type fishponds are rectangular in design to provide ease in feeding and harvesting, and maintenance of high water quality. Irregular shaped ponds should not be constructed because of their difficulty in management.

Figure 36–5
Aerial and closeup views of catfish farm and levee pond system.
(Courtesy of Illinois-Indiana Sea Grant Program.)

---

[3]The stages of fish size and growth are: fry, between hatching size and about 1 inch in length; fingerlings, between 1 and 8 inches and ready for stocking; and finishing, those fish between fingerling and market size.

The layout and design of a typical levee-type pond complex is shown in Figure 36–6 and consists of four ponds served by a water well and roadways. Each pond is about 15 to 20 acres in size for economy in construction and best culture management. Ponds smaller than this acreage are more expensive per unit of size and larger units are more difficult to manage.

Regardless of their size, levee-type pond construction must consider (1) levee width, (2) amount of freeboard, and (3) depth and slope of levees (Figure 36–7). The main levee should be surfaced with gravel or crusted stone and be at least 20 feet wide. Side levees should be at least 16 feet wide. The wider main levee is necessary to accommodate the well that supplies water to the adjoining ponds and ease in traffic flow (feeding, harvesting, etc.).

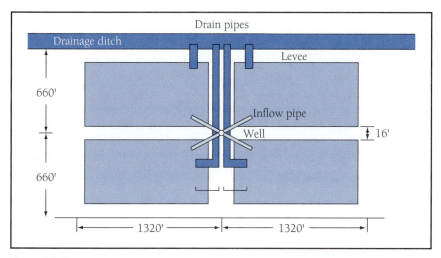

Figure 36–6
Sketch of a typical levee-type catfish pond.
(*Source:* Southern Regional Agricultural Center Publication #101.)

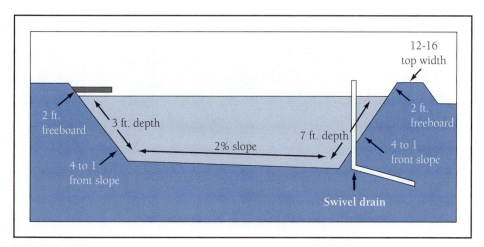

Figure 36–7
Cross section of a typical fish culture pond showing desired depth, slope of pond banks and floor, and width of levee.
(*Source:* Illinois-Indiana Sea Grant Program.)

> **Freeboard**
> Refers to the distance between the surface of the water in a pond and the top of the levee.

The term **freeboard** refers to the height of the levee above normal water level and should be between 1 and 2 feet. Excessive freeboard adds expense and hinders movement of equipment into and out of the water. Insufficient freeboard is more subject to erosion and overflow during heavy rains.

Levee slope refers to the ratio of horizontal distance in feet for each 1 foot in height. A satisfactory inside slope for most soil types is 3:1 when properly compacted.

The depth of commercial food fish ponds should be 2.5 to 3.5 feet in the shallow end and slope to a depth of not more than 6 feet in the deep end. A flat, sloping bottom is required for ease in harvesting and drainage.

Raceways

The raceway design consists of a series of long narrow channels through which there is a continuous flow of fresh oxygenated water. Trout and salmon are most commonly produced in raceway systems. A typical raceway will be about 90 feet long, 9 feet wide, and 3 feet deep (Figure 36–8). Larger raceways are usually constructed of concrete whereas smaller systems may be constructed of concrete, fiberglass, aluminum, soil, or other materials. A major advantage of raceways is that they enable close observation of the fish and ease in feeding and harvesting.

Natural water springs, such as the one shown in Figure 36–9, located above raceways provide an ideal source and high volume of cheap gravity-fed water. Energy costs increase substantially where the water source must be continuously pumped. Locations with an 8 to 10 percent slope enables water to flow through raceways by gravity. This water flow and a 1½- to 2-foot water drop between each unit provide sufficient water aeration.

Cages

Cage culture involves the suspension of cages in bodies of water that cannot be harvested by seining, or in some cases, they are used for convenience in management (Figure 36–10). This system also provides ease in fish observation, feeding, and harvesting.

Figure 36–8
Raceways are most commonly used in the culture of rainbow trout where a continuous supply of fresh cold water is required. (Courtesy of Illinois-Indiana Sea Grant Program.)

Figure 36–9
Natural springs provide a consistent source of water as needed in raceway systems. (Courtesy of Illinois-Indiana Sea Grant Program.)

Figure 36–10
Two types of fish cages suspended in water—both enable case in fish observation, feeding, and harvesting. (Courtesy of Illinois-Indiana Sea Grant Program.)

Farm ponds or barrow pits (as observed along some major highways) are excellent sites for cage fish production. However, if used for aquaculture, they must be free from contaminants such as crop and feedlot runoff, contain no wild fish, limited aquatic plants, and have an average depth of 4 to 5 feet. Cage production in these types of water containment also has the advantage of low initial investment when compared to start-up costs that include pond construction.

Figure 36–11
In recirculation systems, water is pumped through the tanks, then through filters (in background) that remove waste materials, and returned back through the tanks.
(Courtesy of Illinois-Indiana Sea Grant Program.)

Cage construction normally consists of a frame (e.g., PVC or plastic pipe) and plastic or fabric netting formed into a square, rectangular, or circular shape. An inside netting liner is suspended within each cage to contain the fish. Many systems also utilize a "fishing ring" that is attached to the upper inside portion of the cage and designed to keep feed from floating out of the cage. Each cage is equipped with flotation material to keep the upper portion above water level.

Water Recirculation Systems

In recent years, a fourth method of commercial fish production has been developed and is referred to as the water recirculation system. A biological filter, solids filter, pump, and culture tank are primary components of a recirculating system. This method uses tanks for growing fish (Figure 36–11) and recirculates about 90 percent of the water contained within a *closed loop system*. Fiberglass tanks consisting of several sizes and designs are commercially available and commonly used in recirculation systems.

Water recirculation systems require special equipment and skilled management to ensure that correct water temperature and quality are constantly maintained. Continuous mechanical aeration is essential for maintaining sufficient dissolved water oxygen. Also, special **biofilters** are essential for removal of metabolic waste and convert ammonia and nitrites (nitrogenous wastes) into relatively nontoxic nitrate—normally accomplished by natural biological processes in ponds and frequent water exchanges in raceways.

A major advantage of enclosed water recirculation systems is that they can be employed throughout the year in geographic areas where consistency in water temper-

> **Biofilter**
> A water filtration system that uses living organisms to alter or neutralize potentially harmful materials.

atures would be difficult to maintain. Recirculation systems provide for chilling or heating of water to best suit the species being cultured. A properly designed and managed water recirculation system also enables better disease and predator control than pond systems.

Major limitations of recirculation systems are the (1) high-capital requirements in equipment and structures, particularly in relation to volume of production, (2) costs involved in recirculating and maintaining water quality, and (3) high degree of management required of high-density fish production. Within the last 20 years several universities have conducted extensive research on water recirculation systems and research continues that focuses on the biological, engineering, and economic aspects of their use in commercial fish production.

In comparison to other systems, there are few commercial water recirculating systems in the United States but their use can be expected to increase as technology in their design and economic management improves.

36.5 FISH CULTURE MANAGEMENT PRACTICES

The important aspects of water temperature and water quality control have been discussed earlier in Section 36.3. Also to be successful, fish producers must properly manage the stocking, rearing, harvesting, and marketing of their fish.

Culture practices will vary depending upon the chosen fish species and method of production. This section deals with some of the more important management practices, with emphasis on those that have been successful on modern catfish farms.

Successful fish farming requires the stocking of growing facilities with healthy, high-quality fingerlings that are uniform in size. Each group of fingerlings to be stocked should be consistent in size but could average as small as 4 inches, or up to about 10 inches in length. These may be purchased from other commercial sources or produced at the farm location. If produced on the farm, the producer must have additional expertise and facilities to provide for spawning, hatching, and rearing fish to the fingerling stage. Almost 2 years are required to produce market size catfish from the spawning stage. Once fingerlings are stocked, about 200 days are needed to produce a marketable catfish.

A major factor that determines the length of the growing season is average water temperature. Catfish grow most rapidly at water temperatures of 75 to 85°F. Very little growth occurs at temperatures below 60°F.

In pond culture, the size of fingerlings when stocked, and the number (or pounds) stocked per surface acre of water also affect the length of growing season. As density per acre is increased daily growth performance usually decreases. Most catfish operations have stocking densities between 2,000 and 6,000 pounds per acre, although some have reported densities as high as 12,000 pounds per acre.

High-density stocking requires skilled management and regular monitoring of water quality, especially the dissolved oxygen concentration of the water. Managers must be able to act promptly and correctly when sudden changes occur in water quality or fish performance (i.e., decreased feeding activity).

Nutrition and Feeding

The feeding and nutrition of fish is affected by (1) species, (2) amount of natural and supplemental foodstuffs, (3) stocking density, and (4) water temperature and quality.

Carnivore
A flesh-eating animal.

Herbivore
Animal that normally eats vegetative materials.

Omnivorous
Living upon both plants and animals.

Fish species may be primarily **carnivorous**, **herbivorous**, or **omnivorous**. For example, trout are naturally carnivorous and prefer a diet of smaller fish or insects, and when cultured should receive animal protein in their diet. On the other hand, channel catfish are omnivorous and accept a wider variety of lower protein foodstuffs.

Research studies have determined that the recommended protein requirement of trout is 40 to 45 percent, whereas the optimal level for catfish is 30 to 32 percent. Most nutritional research has been conducted for these two species, although aquaculturists frequently use similar diet formulations for other species. When species data are limited, producers generally feed higher protein levels to the carnivorous species, and lower levels to omnivorous and herbivorous species.

Fish in a pond environment may have an opportunity to consume some aquatic vegetation or small animal life whereas fish reared in cages, raceways, or water recirculation systems are totally dependent upon the regular offering of a complete formulated ration. Although some producers offer a feed to supplement the natural habitat, most have operations that require a complete ration for optimal performance.

The growth rate of smaller fish may be adversely affected if producers maintain a high-stocking density. With crowding by larger fish at feeding time and limited feeding space, smaller fish have difficulty in obtaining an adequate amount of food.

When environmental temperatures or source of water result in low water temperature, the metabolic rate of fish also is low. As very little energy is expended in cold water, the appetite of fish also is low. Therefore, in the case of warm-water fish culture such as channel catfish, the producers monitor water temperature and reduce the amount of feed offered when water temperatures decrease to 40°F to 45°F.

As warm-water fish, channel catfish become more aggressive eaters when water temperature rises to about 60°F. Feed efficiency also improves as water temperature rises. Catfish producers can expect a feed conversion of about 4.0 pounds of feed per pound of gain at 60°F water temperature. Feed efficiency steadily improves to a ratio of about 1.5:1 at 80°F. Appetite and performance decrease as temperatures reach and exceed about 85°F. At water temperatures above 90°F, little if any feed should be made available.

Active aggressive-eating fish require high levels of dissolved oxygen. When oxygen levels drop too low, insufficient oxygen results in impaired respiration and metabolism. If these conditions develop, the producer must take steps to sufficiently aerate the water to improve performance, or in severe cases reduce fish losses.

Fish diet formulations should be based upon the most current National Research Council's (NRC) recommendations.[4] If so, they should contain the proper kinds and amounts of ingredients that result in a diet that is (1) adequate in energy, (2) high in digestibility, (3) adequate in protein amount and proper amino acid balance, and (4) palatable. Only high-quality protein feedstuffs should be used to assure the proper balance of amino acids. Rations also must be properly fortified with supplemental vitamins and minerals. Supplemental vitamin C should always be included, as fish are unable to synthesize sufficient amounts.

Fish have lower dietary energy requirements for protein synthesis than warm-blooded animals, in part because they do not have to maintain a constant body temperature. However, energy levels of diets is very important. Low levels can result in other nutrients not being used efficiently whereas extremely high energy levels limit the desired food intake for growth.

---

[4]National Research Council. *Nutrient Requirements of Fish*. Washington, DC: National Academy Press, 1993.

TABLE 36–4	Example of a Diet for Growing Catfish	

Ingredient	Percent of diet
Fish meal, menhaden	8.0
Soybean meal, 48% protein	48.3
Corn grain, ground	31.0
Rice bran (or wheat)	10.0
Dicalcium phosophate	1.0
Pellet Binder	1.5
Fat (sprayed)	0.05
Trace mineral mix*	0.05
Vitamin mix*	0.05
Ascorbic acid (Vitamin C)	0.03

*To provide recommended NRC levels.

Source: Swann, La Don, 1999. *Fish Nutrition Pamphlet.* Illinois-Indiana Sea Grant Program.

When formulating a "complete" diet for fish, feed mixtures must be formulated to meet the minimal daily requirement for each nutrient based upon the amount of feed to be consumed. Fish eat to satisfy their energy need. Therefore, the first step in formulation is to determine the energy level needed in each pound of the diet. Then the concentration of all other nutrients is determined and added at the desired percentage within each pound of feed. This procedure ensures that the fish will receive the proper level of all nutrients as they eat to fulfill their appetite.

Fish meal and soybean meal are excellent sources of protein in fish diet formulation because of their quantity of protein and amino acid balance. Other feeds such as cottonseed meal or blood meal are sometimes used. Corn and corn byproducts are common sources of energy because of their availability at economical cost and their high starch content. Starch also serves as a binding agent in pelleted and extruded feeds.

Fats and oils are added to some formulations as a concentrated source of energy and essential fatty acid content. In the processing of pellets they are not blended into the pellet as they prevent good binding, but are sprayed on the outside surface. This outer coating helps minimize the amount of fine particles, dustiness, and feed wastage. An example of a well-formulated catfish diet is presented in Table 36–4.

Two types of fish feeds are most commonly formulated and sold by feed manufacturers—floating and sinking pellets. Although floating pellets cost more, they are more water-stable than sinking pellets. Also, extruded pellets are less dense than compressed pellets and stay intact longer, a major benefit at times when fish are slow feeders. Floating pellets also enable the caretaker to better monitor fish feeding activity. For example, slow feeding activity can indicate poor water quality characteristics and is difficult to monitor when feed sinks quickly.

Some operations prefer to feed sinking pellets because of their low cost, or sometimes blend in enough floating pellets so that feeding activity can still be observed. Another factor to consider when purchasing fish feed is the size of pellet. Larger pellets of a size acceptable to the fish result in less expended energy in eating, and the pellet stays intact longer.

The feeding rate for fish also depends upon size and stage of growth. Most feed companies provide suggested feeding rates of their different formulations—most commonly provided on the feed tag attached to bagged feed. Automatic feeders are commonly used for fingerlings or smaller fish because of their need to feed more frequently. However, this method of feeding can result in overconsumption and reduced feed efficiency in larger fish. Therefore, producers prefer to feed larger fish once or twice daily. In the case of pond culture, feed is dispensed along all sides of the pond utilizing a vehicle-mounted mechanical feeder.

Research studies suggest that fish can experience "*compensatory growth,*" as is sometimes noted in cattle, sheep, or swine when they are deprived of sufficient nutrients for growth during an earlier period. In Arkansas studies, catfish not fed during the winter actually lost some weight, but then increased in size considerably more during the following summer months than catfish that received supplemental feed during the winter. Some researchers believe that winter-fed fish that gain some weight are more likely to withstand diseases or other stresses than if not fed during the winter months.

Feed is a major cost in rearing food fish and will commonly represent 58 to 60 percent of the variable costs of pond production. Most producers purchase commercial brands of fish diets, rather than formulating and mixing on the farm because of limited equipment or facilities.

To control feed costs, fish should be fed at the manufacturer's recommended rates, based on fish size and water temperature. Excessive feed is costly and increases the organic waste content of the water. This added waste contributes to increased oxygen demand necessary for bacterial breakdown and reduces the available oxygen for the fish. Because excessive feeding can occur at high water temperatures when fish eat less, it is recommended that fish not be fed when water temperatures exceed 90°F.

In contrast to catfish, trout are classified as a cold-water fish and their feeding management must be different. For example, their minimum temperature for growth is about 38°F. At this temperature or lower, trout are slow eaters and require only a maintenance diet. As water temperatures increase above 38°F, trout become active eaters. Their growth rate will increase until the water reaches about 65°F. Optimal water temperature for most efficient trout growth is in the range of 55 to 65°F.

Trout, other than fingerlings, are generally not hand-fed on commercial farms but receive feed upon demand from a mechanical feeder, or from feeders equipped with variable timers.

36.6   FISH HEALTH MANAGEMENT

As in other species of domesticated animals, the emphasis in fish culture must be upon disease prevention rather than disease treatment. Disease control requires that stresses be minimized throughout the life of the fish—from prefingerling to marketable size. This requires sound and timely management. Fish producers strive to keep every group of fish healthy so they reach market size on schedule, and are profitable. This requires that the fish remain free from extreme stresses and diseases throughout the growing period.

Fish afflicted by disease often have been predisposed to some form of stress for an extended period of time. Example of stresses include (1) poor water quality, (2) overstocking, (3) poor quality feeds, (4) environmental contamination, and (5) predation by birds or other animals.

As discussed in Chapter 12, diseases are classified as noninfectious (e.g., nutritional) and infectious diseases. Noninfectious diseases of fish include (1) metabolic problems related to nutrition, (2) chemical toxicity due to pesticides or other chemicals,

(3) environmental conditions related to water quality, and (4) physiological conditions (e.g., blood chemistry changes and organ dysfunction brought about by abrupt changes in/or poor management practices).

Infectious diseases in fish are caused by parasites, bacteria, fungi, and viruses.

Protozoa are the most common parasite affecting fish and multiply in large numbers on the skin or scale surfaces, or on the gills where they can create respiratory problems.

Most bacterial-related diseases are due to various stresses, as mentioned above. Of these, overcrowding and insufficient oxygen levels most commonly contribute to bacteria related diseases. There are large numbers and types of bacterial diseases that affect fish. Most are internal diseases although some locate on the gills and body surface.

Although fungal infections frequently occur in wildlife fish, they are not commonly encountered in fish culture. In this regard, there are only a few viral diseases in fish. But those that do occur can result in rapid and severe fish losses once they become established, especially in young fish.

Well-managed pond culture operations develop and implement a disease prevention plan that includes the following management practices:

- Regularly schedule draining of ponds to allow sufficient drying time to destroy parasites and parasite eggs.
- Carefully check and treat all new fish for any diagnosed diseases. A 2-week quarantine period should be provided before placing them with existing fish.
- Disinfect all equipment after each use.
- Maintain high water quality to include sufficient oxygen levels, correct pH, and proper temperature ranges.
- Avoid over- or underfeeding.
- Use only approved chemicals, as prescribed.

Early detection of potential disease problems requires regular observation of fish behavior, such as reduced eating or reduced vigor. Random fish samples should be checked for signs of parasites, such as sores or lesions. When fish show disease symptoms the assistance of a qualified fish pathologist or veterinarian should be sought to ensure correct diagnoses and treatment (Figure 36–12). Also, veterinary diagnostic laboratories are available in some states to assist in diagnosis and treatment of fish diseases.

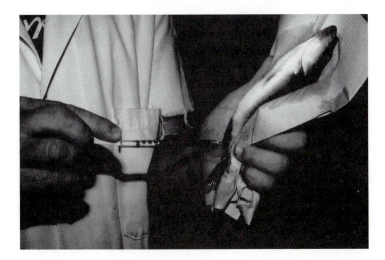

Figure 36–12
Fish that show disease symptoms should be sent to a diagnostic laboratory so proper disease identification can be determined and recommended treatment prescribed. (Courtesy of Illinois-Indiana Sea Grant Program.)

Figure 36–13
Seining of a small pond to harvest market-sized fish can be accomplished with hand labor. However, ponds consisting of several acres require the additional use of mechanized seining equipment.
(Courtesy of Purdue University.)

36.7 HARVESTING OF FOOD FISH

When fish reach market size they must be harvested and quickly processed, or moved as live fish quickly to the processor. Temporarily holding in water-filled storage tanks, or delays in being processed increase the risks of fish losses or lowered quality.

Preharvest checks should be made for good fish health, proper size, and possible presence of off-flavors that could result in rejection by the processor.[5] Harvesting should not start until all marketing details have been completed as harvesting, transporting, and processing must be performed in a timely manner.

Two types of harvesting are employed by producers, either partial or complete harvest. Partial harvest involves the use of larger mesh seines that enable grading by fish size. This method is commonly used by catfish producers and catches only those fish that have reached marketable size and weight (e.g., 1¼ pounds or larger) while allowing smaller fish to pass through the seine.

A complete harvest removes all fish from the containment. For example, the entire pond is seined, or with large bodies of water, drained down to concentrate the fish into a smaller area, then seined or netted. As the draining method requires considerable time and loss of water, it is usually restricted to smaller ponds, such as fingerling ponds, or for carnivorous species such as striped bass. However, it is a practical method of harvest in situations where larger ponds are difficult to seine because of their irregular shape, or structure, such as rocks or tree stumps. Newly constructed ponds should have smooth bottoms and be freed of obstructions to provide ease in harvesting.

Skilled farmers who have sufficient equipment and labor may choose to do their own harvesting (Figure 36–13). Others may employ a custom harvesting crew that provide all of the labor and equipment at a cost of a few cents per pound of harvested fish. On most commercial fish farms, power equipment is commonly used to pull and reel in the seine. Once concentrated, the fish are loaded using a truck-mounted powered boom

---

[5]In catfish pond culture, blue-green algae are responsible for many of the off-flavor problems, causing a musty flavor in the processed fish. Off-flavors are a costly problem because fish growers must continue to feed fish beyond the time when they are ready to be processed. The use of the chemical Diuron was approved in 1999 by the Environmental Protection Agency (EPA) for the control of blue-green algae in catfish ponds.

that lowers a basket into the water. The basket is filled, then lifted and emptied into an aerated water-filled tank truck, and quickly transported to the processor.

One very important management practice that should always be followed is not feeding fish 1 to 3 days prior to hauling to reduce the deposition of feed or fecal matter into the tank truck water.

36.8   MARKETING

Those fish producers who contract to produce fish on a time schedule are not concerned about locating an acceptable market or price. These growers already have a market when fish reach marketable size as established by earlier contractual agreement with the processor. However, those producers not under contract should determine the availability and preferred method of marketing before undertaking a fish farming business.

In the early stages of the business, every aquaculture producer should develop a marketing plan that

- Determines potential customers and competitors
- Projects estimated costs and returns to establish a breakeven price
- Anticipates future events such as consumer preferences
- Determines alternative courses of action
- Establishes production and marketing goals
- Periodically evaluates the performance of their aquaculture program

Fish can be sold wholesale through other businesses or be sold retail directly to the consumer. Large fish farmers usually have sufficient volume to market through direct wholesaling to processing plants, or through cooperatives, other wholesalers, or retailers. Marketing wholesale results in a lower price received by the farmer but marketing costs are lower than selling directly to the consumer.

Some market outlets prefer buying several species from a single source rather than one species from several sources, or purchasing only in large quantities. Small farmers may benefit through the pooling of their resources with other producers, enabling them to have greater access to these market outlets.

Selling direct to consumers is often a good marketing alternative for lower volume producers, or producers unable to provide a regular supply of market fish. The optimum marketable size (e.g., 1½ pound catfish) and product form (e.g., dressed with head intact, or fillets) must be determined in the establishment of a successful market. Both marketing costs and price received per pound are higher when selling fish direct.

Small producers also should analyze the opportunity to direct market to restaurants and other food stores. However, in addition to demanding a high quality product (Figure 36–14), most retail outlets insist upon consistency in supply and uniformity of product—requirements that are sometimes difficult to provide by the small producer.

36.9   PROCESSING INTO A CONSUMER-READY PRODUCT

As in the processing of other meat animals, fish-processing plants have become larger and more specialized, often purchasing and processing large quantities of a single species.

Figure 36–14
Fresh catfish fillets ready for consumer purchase are desired for their quality, flavor, and case in preparation.
(Courtesy of Purdue University.)

The processing of fish involves several steps before the product is ready for the consumer, as presented in the following sequence:

- Precheck for off-flavor
- Receiving and weighing of the live fish
- Initial steps of stunning and head removal
- Evisceration and skinning (or scaling)
- Washing, chilling, and inspection
- Fabrication (cutting into desired size, shape, and weight)
- Packaging and labeling
- Storing either refrigerated for rapid consumption, or frozen at correct temperatures
- Shipping of finished product to retailers

As in all other types of meat processing, fish-processing plants must implement proper sanitation and processing procedures that meet or exceed health regulations. (See Section 25.2.)

36.10 THE ATTRIBUTES OF FISH AS A FOOD

When properly prepared, fish provide the consumer a palatable high protein food. Also, fish is high in polyunsaturated fatty acids, B-vitamins, and trace minerals and is low in saturated fatty acids. Most fish, as purchased by the consumer, contain fewer calories per gram than typical cuts of red meat.

Of special interest to health-conscious people is that fish is relatively low in cholesterol and high in the *omega–3 fatty acids*. Omega–3's help lower blood cholesterol levels, particularly the low-density lipoprotein (LDL) cholesterols. They have been shown to reduce the risk of blocked arteries that lead to strokes or heart attacks, and reduce the risk of blood clotting.

As noted earlier in Section 36.2, the more popular farm-grown fish species possess a well-liked sweet mild flavor when properly grown and processed, and ultimately prepared by the consumer.

QUESTIONS FOR STUDY AND DISCUSSION

1. When did the aquaculture industry rapidly expand in the United States?

2. Name the leading country in aquaculture production and indicate the approximate percentage of the world's aquaculture production it provides.

3. Name two cold-water and two warm-water species of fish grown in the United States.

4. Which state leads in U.S. crawfish production?

5. How can stratification of water pose a problem in fish culture?

6. List at least four factors that affect water quality for fish.

7. What problems can organic matter runoff from feedlots, etc. cause in the production of fish?

8. Why would a high water pH reading in the early morning hours concern the fish producer?

9. How do high nitrate and nitrite levels affect fish health?

10. What is the desired levee-type pond shape and size for commercial catfish farming?

11. What species of fish are most commonly produced in raceways?

12. Briefly describe fish cages and list their advantages in fish management, as compared to pond rearing.

13. Compare the technology and management skills required of water recirculation systems to other systems of fish rearing.

14. As stated in this chapter, fish eat to satisfy their energy needs. Why should this be a primary consideration when formulating a complete fish diet?

15. What adjustment in feeding rate is needed when water temperatures are below or above the desired range of the fish species being cultured?

16. What is an excellent feed efficiency ratio for growing-finishing catfish?

17. Compare the merits of floating versus sinking type fish feeds.

18. Can fish experience compensatory growth? Explain.

19. List several important disease prevention practices in the culture of fish.

20. Describe the method of partial fish harvest.

21. Compare differences in marketing fish through contract farming, wholesale outlets, and direct retail sales.

22. What types of fatty acids are high in fish and seafood? Why are these important in human health?

37

Companion Animals

Upon completing this chapter, the reader should be able to

1. Describe the various types of animals that may be considered a pet.
2. Explain what makes certain animals more suitable for pets than others.
3. Understand the role of pets in human lives and what they provide.
4. Characterize the overall companion animal population.
5. Explain what contributes mostly to the overpopulation of animals.
6. Describe the role of a humane society.

37.1 WHAT IS A COMPANION ANIMAL?

This chapter provides an overview of companion animals. If asked when and where companion or pet animals originated, one theory is that the beginning occurred within the Western world, with farmers or village people in a quest to obtain food. This is evidenced by the discovery in northern Israel of a man and dog buried together dating back 12,000 years. The two were not simply tossed into the grave, but rather the man's hand was placed on the shoulder of the dog to perhaps symbolize a unique bond. As you will note in this chapter, the phrase *companion animal* can extend to a wide range of animals, but this chapter focuses primarily on cats, dogs, and show animals (Figures 37–1, 37–2, and 37–3).

Most companion animals may be best considered as those animals that have little or no economic function and thus are more commonly referred to as "pets." The range of animals that can be considered companion animals is extremely vast and applies to exotic as well as to domestic animals. Today, you might find an extremely wide range of exotic animals kept as "pets"; for example, former Iraqi dictator Saddam Hussein was known to keep tigers as pets. But these animals are considered far from "tame" or "domesticated." One of the larger issues in keeping exotic animals as pets is the lack of training on the owners' part to properly care for the animals. In general, exotic animals (e.g., skunks, lizards, deer, chimpanzees, poisonous snakes, raccoons, etc.) do not make the best pets. In fact, they can be dangerous and create increased liability for the owner in the event of a mishap. For several reasons, it is illegal to keep most of these animal as pets in the United States.

As with most animals, as exotic animals mature and their hormone balances change, they begin to display slightly different behaviors. Most mature exotic animals tend to become aggressive. Many owners find they can no longer keep this type of animal and typically encounter great difficulty in placing their animals in a new home. Furthermore, the demands for proper care and husbandry far exceed their expertise and consequently, many of these animals are released or donated to local animal care authorities.

Figure 37–1
A child and cat playing, a very common scene among pets and children.

Figure 37–2
Children playing with a goat.

Figure 37–3
A purebred boxer used for both companionship and show. (Purdue University)

Within the exotic pets category there is a group of animals that tend to be well maintained and thrive well—ornamental fish and birds. In general, the bright colors associated with these animals make them a favorite for professional offices and, to a lesser degree, homeowners. The benefits of owning exotic fish and birds include their visual appeal, their general ease of care, and the low commitment level required.[1] Exotic fish and birds basically require feeding and, if managed correctly, very little cleaning up and direct attention. On the other hand, cats and dogs greatly increase the level of commitment and require more care. Assuming that the owners are properly caring for their animals, both dogs and cats require routine visits to the veterinarian, ample exercise, proper nutrition, housing, and so on to maintain good health and well-being.

37.2 HUMANS AND COMPANION ANIMALS

As Americans, we love our pets and it shows in the numbers. In the United States there are approximately 61 million dogs and 74 million cats, and about 45 percent of Americans own one or the other. Most are treated extremely well. For instance, nearly 50 percent of the dogs and cats are given a Christmas gift, whereas 20–25 percent are given as an additional gift for a birthday. The dollars spent by Americans on their companion animals is sizable. Table 37–1 depicts the amount of dollars spent by Americans on their companion animals.

---

[1]In the context of this chapter, *commitment* refers to the amount of effort put forth by the business or homeowner toward the ornamental pet.

Table 37–1	American Spending on Pets	
Year	**Billions of Dollars**	
2003	$31.0 (estimate)	
2002	$29.5	
2001	$28.5	
1998	$23.0	
1996	$21.0	
1994	$17.0	

Since the early 1990s the pet industry has experienced substantial growth in pet supplies and services (grooming, animal care, etc.), which has in turn created opportunity for both existing and new companies. Much of the growth within this industry is a result of consumers having additional income to spend on their animals. Also, lifestyle trends and a shift in American desires toward pets have led to a significant increase in expenditures during the last decade. A growing number of pet-related companies have experienced consolidation among manufacturers, retailers, and service providers in order to remain competitive in this global economy. As stated, the economic strength of this industry is believed to be closely linked to the overall wealth of individuals. During more prosperous times, individuals are likely to *spend more for animal affection* and vice versa during the lean economic years. These actions clearly demonstrate the love shared between many owners and their pets.

Owners tend to select companion animals for (a) the enjoyment of playing with them, for animal affection, (b) companionship, (c) assistance in children learning, (d) companion communication, and (e) security. Young children typically are sheltered from animals because of the possibility of animals jumping on them, scaring them, or contaminating a food product. As a child matures, he or she often wants a pet. This role switches as children grow much older and leaves home. Then the parents often augment the companionship of their children with a pet.

In addition to these previously mentioned reasons, individuals may choose a dog or cat because of their ultimate need for that animal. For example, some animals may be acquired for work (police dogs, hunting dogs, and guide dogs), for show purposes, to make supplemental income (i.e., breeding dogs for resale), and for therapeutic work and research. The therapeutic work has not been "medically tested," but there are a large number of nursing facilities and homes where animals are being introduced to individuals to provide some comfort during their distress. Perhaps the treatment or satisfaction comes from the soothing effects of being less stressed through the bond relationship exhibited between an individual and a companion animal.

Animal research is not limited to farm animals or pets, but extends to specialized animals that mimic human characteristics such as monkeys and rodents. In general, research can be divided into two areas, medical research and nutrition. The human research side is driven primarily to find cures or solutions to human health diseases. In many cases, these trials require the sacrifice of animals to understand the mechanism of how specific genes affect diseases. Also, there are cases where investigators need to know the actions of specific drugs (i.e., those used in cancer patients). The overall mission is to utilize a biological system that might assist investigators in saving human lives.

In nutritional research, many animals are fed (a) diets that best estimate the nutritional requirement for that animal or (b) a product that may elicit increased growth without consuming additional feed. In both cases these factors are in constant change because of new genetic varieties of crops, discovery of new animals to be introduced into human populations, development of a variety of byproducts. For instance, byproduct from producing ethanol from grains is distiller grains. Their nutritional value is still being tested by researchers to discover how they may fit into society.

Determining nutritional needs many times involves making "educated guesses"; for instance, many researchers began feeding ostriches as they would feed domesticated turkeys. By challenging ostriches with this type of diet, researchers might recognize a problem associated with the development of a complete turkey diet. The turkey diet is used as a "benchmark," so the researchers have an initial starting point in their quest to design a suitable diet for an ostrich.

Regardless of research interest, many companies participate in a program to ensure humane treatment of laboratory animals. In fact, some review boards[2] are composed of faculty experts in the area, statisticians, veterinarians, and lay citizens. This type of review board brings a variety of people together for one common goal: humane treatment of laboratory animals.

In order to provide excellent facilities, many universities and private industries belong to the Association for Assessment and Accreditation of Laboratory Animal Care accreditation (AAALAC), which consists of a private nonprofit organization to promote humane treatment of animals in science through a voluntary accreditation program. At present, there are more than 650 companies, universities, hospitals, government agencies, and other research institutions that have earned and, in turn, demonstrated their commitment to responsible animal care and use. Participation is a volunteer effort, in addition to complying with the local, state, and federal laws that regulate animal research.

37.3   ANIMALS AND THE HUMANE SOCIETY

Despite the large number of conscientious cat and dog owners, many cats and dogs end up in shelters. Within the United States there are nearly 6,000 animal shelters and, according to the National Humane Society, nearly 6–8 million cats and dogs enter the shelter yearly (Figure 37–4), with nearly 50 percent being euthanized. In addition, U.S. shelters cost taxpayers approximately $2 billion annually. The high number of animals ending up in shelters primarily is attributed to the large number of feral (untamed wild) animals. Perhaps the biggest contributors to these feral animals are people who "dump" unwanted animals along roadsides or those who don't take responsibility for their animals' actions. Many of these abandoned animals will breed and multiply at a very fast pace. In addition, some owners allow the animals to run free. Many end up breeding with other strays or pets. This type of action greatly contributes to the already overpopulated number of unwanted animals. It is estimated that one cat and all her kittens over a lifetime are capable of producing nearly 420,000 offspring, whereas one female dog and subsequent litters could produce as many as 67,000 offspring. Hence, when you evaluate the entire population of abandoned cats and dogs across America this number

---

[2]Detailed protocol of the submitting investigator as well as his or her qualifications of the proposed work.

Figure 37–4
This shelter-found dog is a mixture of collie and Labrador retriever. As with many shelter animals, the dog was adopted as a pet by a responsible owner. This dog has turned out to be a wonderful companion. (Purdue University)

is staggering and contributes greatly to the 3–4 million currently being euthanized. This means that a dog or cat is euthanized approximately every 7.8 seconds. Experts believe the solution is widespread sterilization programs; only by spaying and neutering will the numbers of unwanted animals be reduced.

Where towns and cities have implemented a sterilization plan, a decline by 30 to 60 percent in potential companion animals has resulted. As with any program, such plans don't create themselves and require direction from committed individuals. More specifically, the successful programs require planning and coordination of many people to control pet population, from subsidized sterilization clinics to cooperative efforts involving local veterinarians to mass media educational campaigns. For example, North Central Indiana Spay and Neuter (NCISN) and rescue organization believes that euthanasia is unnecessary for loving companions. Moreover, the organization suggests that through proactive rescue, adoption efforts, and an affordable spay/neuter program, the need to eliminate animals via euthanasia could be avoided. Another major function of NCISN is to reduce the suffering, abandonment, and neglect of companion animals and ultimately reach a goal of no more homeless pets. With the assistance of compassionate people, NCISN has been responsible for the spay/neuter of over 5,000 animals. A major portion (2,100) have been relocated to new homes through efforts of dedicated volunteers.

In general, individuals do not want more legislation; however, it may be a necessary step to more directly impact the overpopulation of animals. That is, if legislation required that every pet adopted from a municipal or county shelter be sterilized within a certain period, then the number of animals released by individuals would not contribute to the rapid growth of unwanted animals. Also, laws that significantly increase the license fees for pets who have not been spayed or neutered should provide an incentive for new owners to sterilize their pets. As with any successful program, education is paramount and an essential part of solving this problem. More specifically, people need to realize the facts about pet overpopulation and sterilization. Along with an educational campaign, a subsidized spay or neuter fee would further encourage new owners to seek sterilization of their animal.

37.4   DOG AND CAT NUTRITION

The pet food industry is very large, totaling approximately $11.8 billion in sales for the United States and nearly $30 billion worldwide. This industry has realized positive growth at 4–6 percent annually and plays a key role in agriculture. More specifically, the pet food industry utilizes nearly 9 million tons of dry pet food, which accounts for many products from processing facilities (about 1.5 million pounds of poultry, swine, and beef). In addition, another 3.5 million tons of corn and 1 million tons of soybeans are utilized in diet formulations.

The first commercially prepared pet food was a dog biscuit introduced in England about 1860. Since that time, the pet food industry has rapidly expanded and continues to move at a record pace. Pet foods are available in canned, dry, and semimoist foods to meet a wide variety of nutritional needs and pet owners today have a tremendous choice of products. All reputable food companies have one thing in common: years of effort and research have gone into each and every commercially produced pet food product in the United States to ensure it provides the proper nutrition for the pet dog or cat. Since 1958, the Pet Food Institute (PFI) has been the primary voice of the U.S. pet food industry. This is a challenging task as the pet food industry is committed to providing clientele and their pets with products that meet the complex nutritional requirements of companion animals. With these complex mixtures and commitment to detail, there must be some governing "nonbias" group that oversees the accuracy of diets to ensure that consumers receive what they pay for and, more importantly, receive the nutritional profile expected for their animal. The Association of American Feed Control Officials (AAFCO) provides the mechanism for developing and implementing uniform and equitable laws, regulations, standards and enforcement policies for regulating the manufacture, distribution and sale of animal feeds; resulting in safe, effective, and useful feeds. AAFCO encourages intelligent solutions and innovative procedures and urges their adoption by member agencies, for uniformity.

As mentioned previously, companion animal owners have a wide range of pet foods to feed their animal. In daily management, animal owners need to consider the logistics of feeding their animal; for example, owners have the option of feeding them once daily, two to three times daily, or allow them free access to feed. Therefore, depending on the schedule, this may limit the major of feed type chosen (dry feed vs. canned) for that animal. Regardless of the option chosen, the owner needs to be sensitive to "overfeeding" or "underfeeding" the animal as well as keeping the proper ingredients in perspective for optimum health and performance. With proper health of your pet, you will be visiting a veterinarian on a routine basis, so during those visits ask him or her about your animal's weight. Managing an animal's weight should be extremely easy because the owner has full control over the diet and thus can easily regulate the intake. The amount and type of feed can be easily adjusted depending on the present condition (i.e., pregnancy, work, growth or maintenance). Tables 37–2 and 37–3 provide the recommended nutrients for dogs and cats, respectively.

Table
37–2

Nutrient Recommendations for Dogs

Nutrient	Units DM Basis	Growth, Reproduction Minimum	Adult Minimum	Maximum
Protein	%	22.00	18.00	
Arginine	%	0.62	0.51	
Histidine	%	0.22	0.18	
Isoleucine	%	0.45	0.37	
Leucine	%	0.72	0.59	
Lysine	%	0.77	0.63	
Methionine-cystine	%	0.53	0.43	
Phenylalanine-tyrosine	%	0.89	0.73	
Threonine	%	0.58	0.48	
Tryptophan	%	0.20	0.16	
Valine	%	0.48	0.39	
Fat[b]	%	8.0	5.0	
Linoleic Acid	%	1.0	1.0	
Minerals				
Calcium	%	1.0	0.6	2.5
Phosphorus	%	0.8	0.5	1.6
Ca:P ratio	%	1:1	1:1	2:1
Potassium	%	0.6	0.6	
Sodium	%	0.30	0.06	
Chloride	%	0.45	0.09	
Magnesium	%	0.04	0.04	0.3
Iron[c]	mg/kg	80	80	3,000
Copper[c]	mg/kg	7.3	7.3	250
Manganese	mg/kg	5.0	5.0	
Zinc	mg/kg	120	120	1,000
Iodine	mg/kg	1.5	1.5	50
Selenium	mg/kg	0.11	0.11	2
Vitamins				
Vitamin A	IU/kg	5,000	5,000	50,000
Vitamin B	IU/kg	500	500	5,000
Vitamin E	IU/kg	50	50	1,000
Thiamin[d]	mg/kg	1.0	1.0	
Riboflavin	mg/kg	2.2	2.2	
Pantothenic acid	mg/kg	10.0	10.0	
Niacin	mg/kg	11.4	11.4	
Pyridoxine	mg/kg	1.0	1.0	
Folic acid	mg/kg	0.18	0.18	
Vitamin B_{12}	mg/kg	0.022	0.022	
Choline	mg/kg	1,200	1,200	

[a]Presumes an energy density of 3.500 kcal ME/kg in accordance with regulation PF9. Rations greater than 4.000 kcal ME/kg should be corrected for energy density. Rations less than 3.500 kcal/kg should not be corrected for energy. Rations of low-energy density should not be adequate for growth or reproductive using data above.

[b]A true requirement for crude fat has not been established, the minimum level was based on recognition of crude fat as a source of essential fatty acids, as a carrier of fat soluble vitamins, to enhance palatability, and to supply an adequate caloric density.

[c]Because of poor bioavailability, iron from carbonate or iron oxide sources, and poor availability of copper from oxide, these sources should not be considered when determining minimum nutrient levels from these unsatisfactory iron and copper sources.

[d]Processing may destroy up to 90% of thiamin in the diet and should be considered in formulation to provide adequate amounts to meet nutrient levels after processing.

Source: Adapted from Feedstuffs, July 2001.

Table 37–3	**Nutrient Recommendations for Cats**

Nutrient	Units DM Basis	Growth, Reproduction Minimum	Adult Minimum	Maximum
Protein	%	30.00	26.00	
Arginine	%	1.25	1.04	
Histidine	%	0.52	0.52	
Isoleucine	%	1.25	1.25	
Leucine	%	1.20	0.83	
Lysine	%	1.10	1.10	
Methionine-cystine	%	0.62	0.62	
Phenylalanine-tyrosine	%	0.88	0.88	
Threonine	%	0.42	0.42	
Tryptophan	%	0.73	0.73	
Valine	%	0.62	0.62	
Crude fat[b]	%	9.0	9.0	
Linoleic acid	%	0.5	0.5	
Arachidonic acid	%	0.02	0.02	
Minerals				
Calcium	%	1.0	0.6	
Phosphorus	%	0.8	0.5	
Potassium	%	0.6	0.6	
Sodium	%	0.2	0.2	
Chloride	%	0.3	0.3	
Magnesium[c]	%	0.08	0.04	
Iron[d]	mg/kg	80	80	
Copper (extruded)[e]	mg/kg	15.0	5.0	
Copper (canned)[e]	mg/kg	5.0	5.0	
Manganese	mg/kg	7.5	7.5	
Zinc	mg/kg	75	75	2,000
Iodine	mg/kg	0.35	0.35	
Selenium	mg/kg	0.1	0.1	
Vitamins				
Vitamin A	IU/kg	9,000	5,000	750,000
Vitamin D	IU/kg	750	500	10,000
Vitamin E[f]	IU/kg	30	30	
Vitamin K[g]	—	—	—	—

[a]Presumes an energy density of 4,000 kcal ME/kg as determined in accordance with Regulation PF9 of AAFCO 2000. Rations greater than 4,500 kcal ME/kg should be corrected for energy density: rations less than 4,000 kcal ME/kg should not be corrected for energy. Rations of low-energy density should not be considered adequate for growth or reproductive needs based on comparison to the profiles alone.

[b]Although a true requirement for crude fat per se has not been established, the minimum level was based on recognition of crude fat as a source of essential fatty acids, as a carrier of fat-soluble vitamins, to enhance palatability, and to supply an adequate caloric density

[c]If the mean urine pH of cats fed ad libitum is not below 6.4, the risk of struvite urolithiasis increases as the magnesium content of the diet increases.

[d]Because of very poor bioavailability, iron from carbonate or oxide sources that are added to the diet should not be considered in determining the minimum nutrient level.

[e]Because of very poor bioavailability, copper from oxide sources that are added to the diet should not be considered in determining the minimum nutrient level.

[f]Add 10 IU Vitamin E above minimum level per gram of fish oil per kg of diet.

[g]Vitamin K does not need to be added unless diet contains greater than 25% fish on a dry matter basis.

(continued)

Table 37–3	Nutrient Recommendations for Cats—Continued			

Nutrient	Units DM Basis	Growth, Reproduction Minimum	Adult Minimum	Maximum
Thiamin[h]	mg/kg	5.0	5.0	
Riboflavin	mg/kg	4.0	4.0	
Pantothenic acid	mg/kg	5.0	5.0	
Niacin	mg/kg	60.0	60.0	
Pyridoxine	mg/kg	4.0	4.0	
Folic acid	mg/kg	0.80	0.80	
Biotin[i]	mg/kg	0.07	0.07	
Vitamin B_{12}	mg/kg	0.002	0.002	
Choline[j]	mg/kg	1,200	1,200	
Taurine (extruded)	g/kg	0.10	0.10	
Taurine (canned)	g/kg	0.20	0.20	

[h]Because processing may destroy up to 90% of the thiamin in the diet, allowances in formulation should be made to ensure the minimum nutrient level is met after processing.

[i]Biotin does not need to be added unless diet contains antimicrobial or antivitamin compounds.

[j]Methionine may be used to substitute for choline as a methyl donor at a rate of 3.75 parts for 1 part choline by weight when methionine exceeds 0.62%.

Official Publication AAFCO, 2000

Source: Adapted from Feedstuffs, July 2001.

37.5   BREEDS AND CHARACTERISTICS

The establishment of various breeds has been controlled by humans; however, some of the characteristics (behaviors) of dogs can be traced back to wild animals such as coyotes and other animals. Since 1884, the American Kennel Club (AKC), which is nonprofit (more information is available at **http://www.akc.org**), has established seven categories and a single miscellaneous group for dogs (Table 37–4). The categories are as follows: sporting, working, toy, hound, terrier, non-sporting, herding, and miscellaneous.

Sporting dogs are well liked and considered well-rounded companions and include pointers, retrievers, setters, and spaniels. These dogs have a remarkable instinct for water and woods, and many of these breeds actively continue to participate in hunting and other field activities. Owners for these dogs need to realize that most require regular and demanding exercise.

Hounds are well suited for hunting and continue to share that common ancestral trait. These dogs have an excellent ability to trail animals via smell, and strive to ultimately catch the animal or find what it is they are trailing. Within this category there is diversity which includes Pharaoh hounds, Norwegian elkhounds, Afghans, and beagles, among others. For example, a "Black and Tan" is well suited for running and baying coons, which means they will bark along the coon trail and then stand on the tree and continue to bark at the coon while hunters arrive. On the other hand, beagles do similar work but are well suited for chasing rabbits through tight brush and staying on the trail, and rarely bay a rabbit.

Table 37–4		Breeds of Dogs	

Sporting Dogs	Working Group	Toy Group
Brittany	Akita	Affenpinscher
Pointer	Alaskan Malamute	Brussels Griffon
German Shorthaired Pointer	Anatolian Shepherd	Cavalier King Charles Spaniel
German Wirehaired Pointer	Bernese Mountain Dog	Chihuahua
Chesapeake Bay Retriever	Boxer	Chinese Crested
Curly-Coated Retriever	Bullmastiff	English Toy Spaniel
Flat-Coated Retriever	Doberman Pinscher	Havanese
Golden Retriever	German Pinscher	Italian Greyhound
Labrador Retriever	Giant Schnauzer	Japanese Chin
English Setter	Great Dane	Maltese
Gordon Setter	Great Pyrenees	Manchester Terrier
Irish Setter	Greater Swiss Mountain Dog	Miniature Pinscher
American Water Spaniel	Komondor	Papillon
Clumber Spaniel	Kuvasz	Pekingese
Cocker Spaniel	Mastiff	Pomeranian
English Cocker Spaniel	Newfoundland	Poodle
English Springer Spaniel	Portuguese Water Dog	Pug
Field Spaniel	Rottweiler	Shih Tzu
Irish Water Spaniel	Saint Bernard	Silky Terrier
Nova Scotia Duck Tolling	Samoyed	Toy Fox Terrier
Retriever	Siberian Husky	Yorkshire Terrier
Spinone Italiano	Standard Schnauzer	
Sussex Spaniel		
Welsh Springer Spaniel		
Vizsla		
Weimaraner		
Wirehaired Pointing Griffon		

Hound Group	Terrier Group	Non-Sporting Group
Afghan Hound	Airedale Terrier	American Eskimo Dog
American Foxhound	American Staffordshire	Bichon Frise
Basenji	Terrier	Boston Terrier
Basset Hound	Australian Terrier	Bulldog
Beagle	Bedlington Terrier	Chinese Shar-pei
Black and Tan Coonhound	Border Terrier	Chow Chow
Bloodhound	Bull Terrier	Dalmatian
Borzoi	Cairn Terrier	Finnish Spitz
Dachshund	Dandie Dinmont Terrier	French Bulldog
English Foxhound	Irish Terrier	Keeshond
Greyhound	Kerry Blue Terrier	Lhasa Apso
Harrier	Lakeland Terrier	Löwchen
Ibizan Hound	Manchester Terrier	Poodle
Irish Wolfhound	(Standard)	Schipperke
Norwegian Elkhound	Miniature Bull Terrier	Shiba Inu
Otterhound	Miniature Schnauzer	Tibetan Spaniel
Petit Basset Griffon	Norfolk Terrier	Tibetan Terrier
Pharaoh Hound	Norwich Terrier	

(continued)

Table 37–4	Breeds of Dogs—Continued	

Hound Group	Terrier Group	Non-Sporting Group
Rhodesian Ridgeback	Parson Russell Terrier	
Saluki	Scottish Terrier	
Scottish Deerhound	Sealyham Terrier	
Whippet	Skye Terrier	
	Smooth Fox Terrier	
	Soft Coated Wheaten Terrier	
	Staffordshire Bull Terrier	
	Welsh Terrier	
	West Highland White Terrier	
	Wire Fox Terrier	

Herding Group	Miscellaneous	
Australian Cattle Dog	Beauceron	
Australian Shepherd	Black Russian Terrier	
Bearded Collie	Glen of Imaal Terrier	
Belgian Malinois	Neapolitan Mastiff	
Belgian Sheepdog	Plott Hound	
Belgian Tervuren	Redbone Coonhound	
Border Collie		
Bouvier des Flandres		
Briard		
Canaan Dog		
Cardigan Welsh Corgi		
Collie		
German Shepherd Dog		
Old English Sheepdog		
Pembroke Welsh Corgi		
Polish Lowland Sheepdog		
Puli		
Shetland Sheepdog		

Source: *Adapted from the American Kennel Club (AKC) Web site,* **http://www.akc.org**.

Working dogs were bred to perform specific tasks associated with work. For instance, the Siberian husky is used to pull sleds in order for people to travel across frozen and cold places; the Doberman pinscher may be used for protection in or outside a home. These dogs are considered quick learners and make very good companions. Because of their inherent size and strength, both huskies and Doberman pinschers are unsuitable as pets for most families, especially for those with small children. Moreover, the size and physical ability of these animals commands respect and by this virtue they must be properly trained for human safety.

Members of the terrier group have very distinctive personalities; that is, most would be considered energetic, feisty, and have little tolerance for other animals—including other dogs. In general, most terriers have wiry coats that require special attention to keep them groomed properly. In addition, dogs within this group span a considerable range from fairly small, as in the Norfolk, Cairn, or West Highland White Terrier, to the grand Airedale terrier. The ancestors of these modern terriers were bred to hunt and kill small animals. These dogs are not the most popular pets, and require owners with the determination to match their dogs' lively characters.

Toy dogs are small in stature, but are considered in many cases to be as tough as nails. Regardless of size, these little dogs will elicit a barking and feisty display when they run into something foreign to them. Toy breeds are very popular for apartment life as well as for individuals living in the city with small living spaces. In addition to being small, these dogs have relatively few inherent canine issues such as shedding, creating messes, and cost of care. Another positive aspect of toy breeds is training: it's easier to control a small 10-pound dog.

Dogs in the AKC's non-sporting category are not only diverse but also are considered sturdy with a wide range of personalities and appearances. For instance, the chow chow and bulldog are quite different from the dalmatian in temperament and appearance. This group has a varied collection in terms of size, coat, personality, and overall appearance.

In 1983, the herding group was created by the AKC. Members of this group were formerly in the working group and still share some of the physical abilities to control the movement of other animals. For example, the low-set corgi can move a herd of cows many times its size to pasture by leaping and nipping at their heels. Today, herding dogs are used mostly as house pets and rarely herd animals; however, through pure instinct, these dogs have been known to herd the children of the family. In general, herding dogs make excellent pets and respond very well to training.

When a particular breed is in judgment as to where it will exist, that breed is admitted to the Miscellaneous Class. Dogs in the Miscellaneous Class may compete and earn titles in AKC events; however, they are not eligible for championship points.

Currently, the Cat Fancier's Association, Inc. (CFA) recognizes 37 pedigreed breeds for showing in the Championship Class, and 4 breeds as Miscellaneous — American Bobtail, LaPerm, RagaMuffin, and Siberian. The objectives of CFA are (1) the promotion of the welfare of cats and the improvement of their breed; (2) the registration of pedigrees of cats and kittens; (3) the promulgation of rules for the management of cat shows; (4) the licensing of cat shows held under the rules of this organization; and (5) the promotion of interest of breeders and exhibitors of cats. In 1906, the CFA was formed and the first cat show licensed by CFA were held in Buffalo and another in Detroit. The first annual meeting for CFA was held in 1907 at Madison Square Garden, with dues totaling $155.25. Today they are well over $2 million. At present, the CFA conducts approximately 400 shows. Judges for these shows meet high qualification criteria and have completed a rigorous training program that well qualifies them to evaluate the show cat using CFA breed standards. To see numerous photographs and details about these breeds, visit **http://www.cfainc.org.**

QUESTIONS FOR STUDY AND DISCUSSION

1. What is a companion animal?
2. What are the leading companion animals in America?
3. Describe what owners want in companion animals.
4. Explain how some companion animals can be used for work-related activities.

5. Describe the overall population of wanted and unwanted companion animals.
6. How large is the pet food industry?
7. Who monitors or ensures quality for pet foods?
8. What are some general ingredients you might find in a typical dog and cat diet?
9. What makes a cat diet unique?
10. How many different breeds of dogs and cats are there?
11. Describe some of the general characteristics of the dog breeds.

 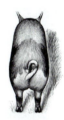

U.S. No. 1

U.S. No. 2

U.S. No. 3

U.S. No. 4

U.S. Utility

U.S. Slaughter Grades—Barrows and Gilts (USDA)

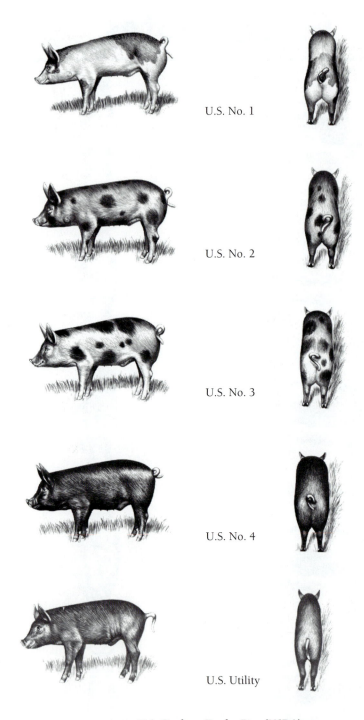

U.S. No. 1

U.S. No. 2

U.S. No. 3

U.S. No. 4

U.S. Utility

U.S. Grades—Feeder Pigs (USDA)

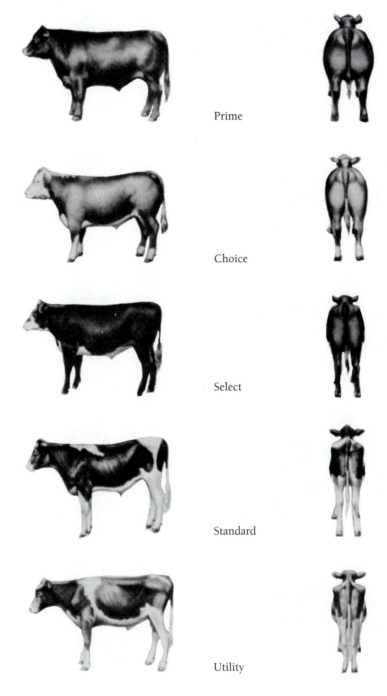

Prime

Choice

Select

Standard

Utility

U.S. Quality Grades—Slaughter Steers (USDA)

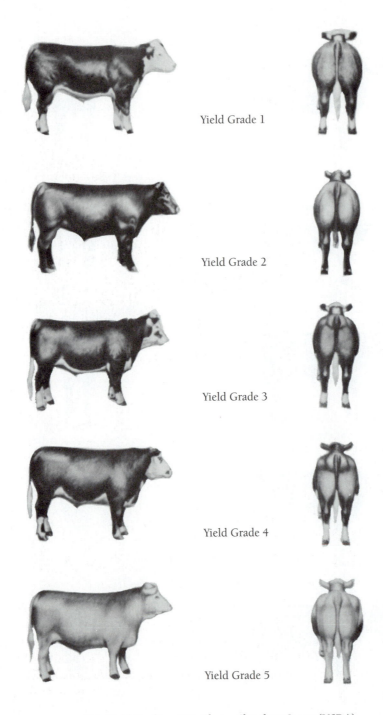

Yield Grade 1

Yield Grade 2

Yield Grade 3

Yield Grade 4

Yield Grade 5

U.S. Yield Grades—U.S. Choice Slaughter Steers (USDA)

Appendix
B
Illustrations of Animal Parts and Terms

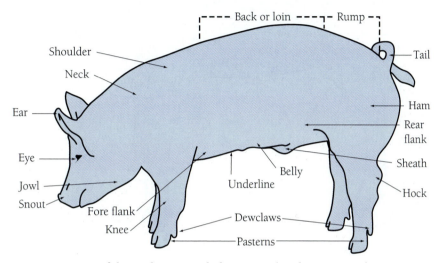

Parts of the Market Hog and Their Terms (Purdue University)

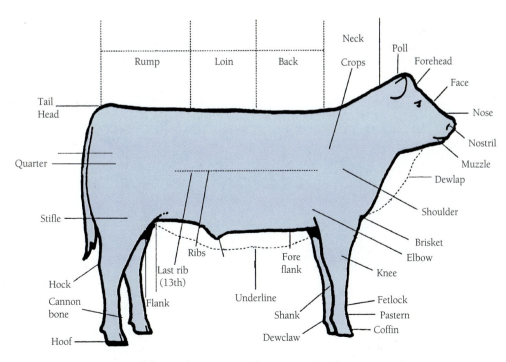

Parts of the Feeder Steer and Their Terms (Purdue University)

729

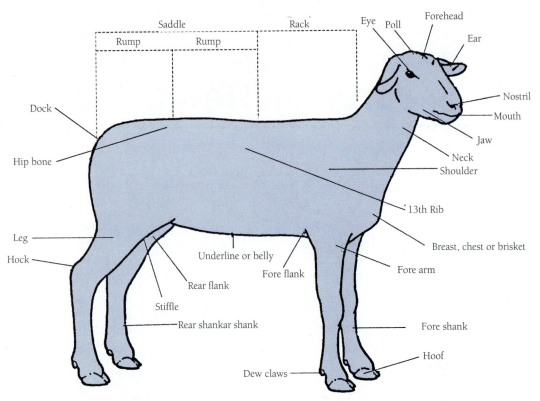

Parts of the Sheep and Their Terms (Purdue University)

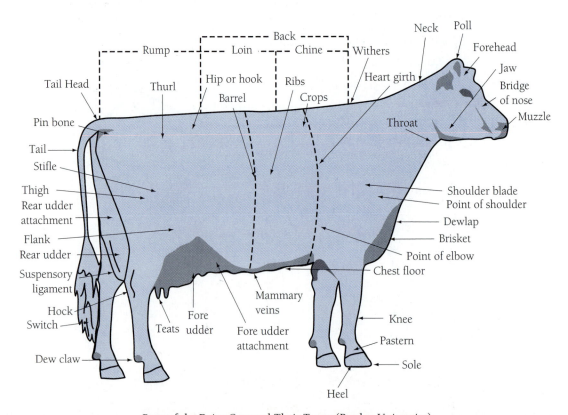

Parts of the Dairy Cow and Their Terms (Purdue University)

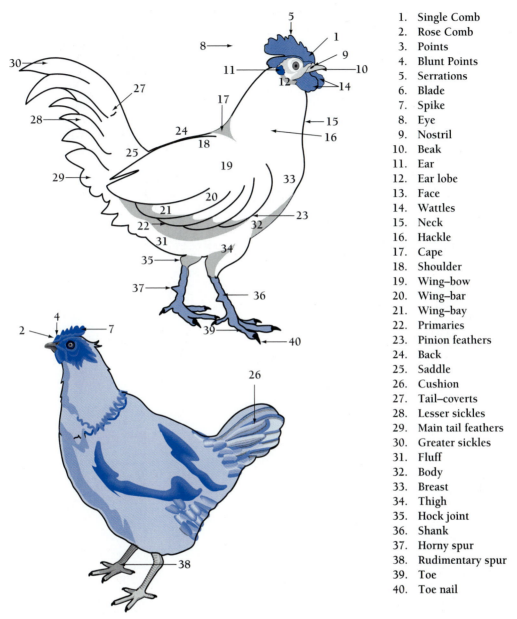

1. Single Comb
2. Rose Comb
3. Points
4. Blunt Points
5. Serrations
6. Blade
7. Spike
8. Eye
9. Nostril
10. Beak
11. Ear
12. Ear lobe
13. Face
14. Wattles
15. Neck
16. Hackle
17. Cape
18. Shoulder
19. Wing–bow
20. Wing–bar
21. Wing–bay
22. Primaries
23. Pinion feathers
24. Back
25. Saddle
26. Cushion
27. Tail–coverts
28. Lesser sickles
29. Main tail feathers
30. Greater sickles
31. Fluff
32. Body
33. Breast
34. Thigh
35. Hock joint
36. Shank
37. Horny spur
38. Rudimentary spur
39. Toe
40. Toe nail

Parts of the Chicken and Their Terms (Purdue University)

Glossary

This glossary is not meant to be a substitute for a dictionary nor for discussions presented in the body of the text. Rather, "working definitions" of some of the terms that may be new to some students are given. Definitions other than those given for each term might be equally accurate and complete. Refer to the Index for terms not given here.

Abomasum: The fourth compartment of the ruminant stomach where enzymatic digestion occurs. Often called the "true stomach."

Absorption: Movement of nutrients through the wall of the gastrointestinal tract and capillary walls into the circulatory system.

Acclimatization: Adaptation by an animal to several environmental factors over several days or weeks.

Acid detergent fiber (ADF): An indicator of relative digestibility of fibrous feeds.

Acidosis: A decrease of alkali in body fluids in proportion to the acid content; common in ruminants with sudden ration change and lowered rumen pH.

ADF: See *Acid detergent fiber*.

Adenosine triphosphate (ATP): A universal energy-transfer molecule. Upon oxidation of organic structures, the energy is captured and stored as ATP. A reverse reaction results in a release of energy for useful purposes.

ADG: See *Average daily gain*.

Adipose: Fat stored in loose connective tissue throughout the body.

Aeration: Addition of dissolved oxygen to water.

Agonistic: Aggressive animal behavior.

Albumen: A viscous protein that comprises most of the white of an egg.

Alkalinity: A measure of the pH buffering capacity.

Allantois: An intermediate fetal membrane that contributes to the formation of the umbilical cord and placenta.

Allele: Alternate gene(s) that occupies corresponding sites on homologous chromosomes.

Alveolus: A hollow follicle of cells. In the case of mammary tissue, a spherical structure with cells that manufacture milk and secrete it into the lumen of the alveolus. Pl., *alveoli*.

Amino acid: Nitrogen-containing component of a protein molecule.

Ammonia: NH_3 gas; created by decomposition of organic matter in aquaculture water and in the process of nitrogenous breakdown by organisms, such as those in the rumen.

Amnion: The innermost placental membrane surrounding the embryo and containing fluids that bathe the embryo.

Ampulla: An enlargement or dilation of a passageway, such as the innermost portion of the vas deferens in some male species, which is one of the accessory sex glands.

Anabolism: The building up of body tissues.

Androgen: A hormone associated with development of secondary sex characteristics in the male.

Anomaly: Deviation from normal structure, behavior, or appearance.

Anthelmintic: A substance that destroys or expels intestinal parasites.

Antibiotic: A compound that inhibits life. Each has some specificity, inhibiting only certain species or strains, such as certain kinds of bacteria.

Antibody: A chemical substance, in circulating fluids, colostrum, and milk, that contributes to immunity against a certain disease or infection.

Antigen: A substance that, when introduced into the body, brings about an immune response by a specific antibody.

Antioxidant: A compound that prevents oxidation. Used in mixed feeds to prevent rancidity or loss of vitamin potency.

Aquatic: Referring to animal or plant life in water.

Arthropod: A member of the phylum Arthropoda; primarily insects.

As-fed: An expression to denote nutrient content of a ration or feed when fed.

ATP: See *Adenosine triphosphate*.

Auction: A sale where successive bids are received and the animal is sold to the highest bidder.

Average daily gain (ADG): A measure of weight response of an animal or group of animals based on per-day performance.

Backfat: The fat over the back of a live hog or pork carcass.

Backgrounding: Growing of animals, typically young beef animals, for an extended period of time prior to placing them into a feedlot.

Bacteria: A class of single-celled organisms.

Bacteriostat: A compound that inhibits growth and reproduction of, or kills, certain bacteria.

Balanced: A term that applies to a ration or feed having all known required nutrients in the proper amount and proportion depending upon the requirements of the animal (e.g., maintenance, growth, reproduction).

Barrow: A male pig castrated before puberty.

Beget: To sire offspring.

Bile: Secretion of the liver stored in the gallbladder, if present, which emulsifies fats in the small intestine.

Biofilter: A water filtration system that uses living organisms to alter or neutralize potentially harmful materials.

Biological efficiency: Ratio of output to input of food materials (or nutrients).

Biosecurity: Refers to a planned system to prevent disease entry or cross-contamination within a herd or flock.

Blend price: Price paid a milk producer, based on the proportion of milk utilized in each price class, such as fluid and manufacturing.

Boar: Most commonly refers to a sexually mature, intact (uncastrated) male pig.

Bob veal: Slaughter calves under 3 weeks of age.

Bolus: A quantity of feed masticated and ready to be swallowed.

Boston shoulder: The top part of the shoulder of a pork carcass (sometimes called the Boston butt).

Bovine somatotrophin (BST): A peptide hormone produced through molecular biotechnology in recent years; when injected into lactating dairy cows, generally brings about a response in increased milk yield.

Boxed beef (pork): Precut portions of beef or pork prepared by the meat-processing plant for wholesale delivery.

Breed: A group of animals descended from common ancestry and possessing certain inherited characteristics that distinguish it from any other group.

Breed type: Distinctive features in which one breed differs from another.

Broiler: A young chicken, of either sex, being fed for meat production.

Broodfish: Mature fish kept for reproductive purposes.

Broodiness: The tendency of a hen to nest, to behave as if she were incubating eggs. Also used to describe behavior in other species.

BST: See *Bovine somatotrophin*.

Buck: Most commonly refers to a sexually mature, intact (uncastrated) male goat.

Bulk tank: Refrigerated tank that stores several hundred to thousands of gallons of milk on the farm until pickup.

Bull: Most commonly refers to a sexually mature, intact (uncastrated) male bovine.

Bullock: A term more currently used to indicate a young bull grown and fattened in a similar manner to beef steers and heifers.

Byproduct: A product of significantly less value than the major product. In beef cattle, the major product is meat; byproducts include the hide and other items.

Calf: Immature young of some large mammals; most commonly used to denote young cattle of either sex.

Calorie: The amount of heat required to raise 1 kg of water 1°C or 1 lb of water approximately 4°F.

Cannibalism: The tendency of animals to draw blood by pecking or biting other animals.

Capon: A male chicken grown for meat, castrated at about 6 weeks of age.

Carbohydrate: A class of nutrients, each composed of carbon, hydrogen, and oxygen, the latter elements present in a 2:1 ratio.

Carcass: The major portion of a meat animal remaining after slaughter. Varies among animals, but usually the head and internal organs have been removed. Skin and shanks are removed from cattle and sheep.

Carnivore: A flesh-eating animal.

Carotene: A precursor of vitamin A found largely in the yellow and green pigments of plants.

Caruncles: Disklike areas of the uterine endometrial lining which with the fetal cotyledons forms a strong connection known as the placentome.

Cashmere: Fine wool from the undercoat of certain breeds of goats.

Castration: Removal or permanent alteration of the testicles of a male animal.

Catabolism: The breaking down or degradation of body tissues.

Catalyst: A substance that speeds up a reaction without permanently entering into the reaction itself.

Cecum: A large pouch that is the forward part of the large intestine of a horse.

Cells of Leydig: Interstitial cells of the testicle that produce the male sex hormone or androgens.

Cellulose: A prevalent polysaccharide found in the fibrous portion of plants; digestible by ruminants because of cellulase produced by rumen microbes.

Centigrade: A scale of temperature with zero equal to the freezing temperature of water and 100 equal to the boiling temperature of water.

Cervix: The constricted necklike structure located between the vagina and uterine body.

Chalaza: A spiral band of thickened albumen that maintains the yolk in its position.

Chevon: Meat from the goat.

Cholesterol: A steroid needed in the structure of plasma membranes, synthesis of bile, and sex hormones and found in various parts of the body.

Chorion: The outermost placental membrane that makes contact with the maternal uterine endometrial lining.

Chromosomes: Body cell structures within the nucleus that contain hereditary materials (genes) and are microscopically visible only during cell division.

Chyme: A mixture of consumed feed, saliva, other enzyme-containing digestive juices, and/or microorganisms present in the digestive tract.

CIP: See *Clean-in-place*.

CL: See *Corpus luteum*.

Clean-in-place (CIP): Automated cleaning techniques employed in the cleaning of milking, milk-handling, and milk storage equipment.

Cloaca: The bulblike structure at the end of the digestive tract of birds; empties the digestive, urinary, and reproductive tracts.

Cloning: Making genetically identical copies of an individual.

Cock: Most commonly refers to a sexually mature, intact (uncastrated) male chicken.

Cockerel: A young male chicken.

Colic: Acute abdominal pain; pertaining to the colon.

Colon: The lower (posterior) portion of the large intestine.

Colostrum: First milk (lacteal secretion) secreted postpartum; contains antibodies for passive immunity of the newborn; also rich in nutrients.

Colt: A young horse or pony, generally under 1 year of age; sometimes refers to young males only.

Combing wool: Wool of staple length. Fibers are long enough that worsted fabric can be produced.

Comfort zone: The temperature range in which an animal will be comfortable and produce at a maximum level.

Compensatory gain: Gain at an above-normal rate following a period of little or no weight gain.

Complete feed: A nutritionally adequate feed for animals other than humans, compounded to be fed as the sole ration.

Concentrate: A feed high in digestible energy and low in fiber.

Concentrate feed: A feed used with another to improve the nutritive balance of the total ration.

Conception: Union of ovum and sperm; formation of the zygote.

Conduction: Transfer of heat through a medium; warm molecules impart kinetic energy upon contact with cooler molecules.

Conformation: The structure or shape of an animal or its carcass, or a cut from the carcass.

Congenital: Existing at birth; referring to hereditary defects or other defects that occurred during gestation.

Connective tissue: A type of supporting tissue that is made of protein and is especially strong. Some appear as long distinctive strands, others as a cementive mass between muscle cells.

Contagious: The characteristic of a disease that permits it to be readily transmitted from one animal to another.

Convection: Flow of heat through air or water.

Copulation: The act of sexual intercourse, where semen of the male is deposited in the reproductive tract of the female.

Corpus albicans: The term to denote tiny scars that remain after complete regression of earlier lutea.

Corpus luteum (CL): Active tissue that develops on the ovary at the site where an ovum has been shed. If conception does not occur, the tissue gradually regresses. If conception does occur, the tissue becomes functional, producing progesterone.

Correlation: The tendency for the rate of change of one variable to be associated with the rate of change of a second variable.

Cortex: The outer portion of an organ, such as ovary or kidney.

Cotyledon: Cup-shaped structure; areas of the fetal chorion that connect with the uterine caruncles of the maternal host.

Cow: A female bovine, usually after first pregnancy and parturition.

Cowper's glands: The paired accessory male sex glands posterior to the prostate and on each side of the urethra.

Creep feed: Feed provided young animals within an enclosure that excludes larger animals.

Cremaster muscle: The muscle that suspends the testicle and contracts or relaxes to enhance temperature regulation.

Crimp: Natural waves in wool fiber that provide characteristics of elasticity and resilience.

Critical temperature: The upper or lower temperature where an animal must increase or decrease the oxidation of energy sources to maintain body temperature.

Crossbred: An animal whose parents are of different breeds.

Crossbreeding: Mating animals of different breeds. A distinct type of outbreeding.

Crude protein: Nitrogen, present in feed, multiplied by 6.25. This factor is used because amino acids contain about 16 percent nitrogen. Crude protein includes non-protein nitrogen compounds.

Cryptorchid: A male animal in which one or both of the testicles remained in the body cavity and did not descend into the scrotum during fetal development.

Cull: To remove from a herd or flock, usually because of age, low performance, or an undesirable trait.

Cutability: The proportion of the weight of the carcass that is salable product after the original carcass has been trimmed or processed for sale.

Cwt: Abbreviation for hundredweight, 100 pounds.

Cytoplasm: The nonnuclear portion of a cell.

Dairy character: In a dairy cow, freedom from fleshiness or fatness, especially in the neck, withers, and thighs.

Dairy Herd Improvement (DHI): A USDA Extension Service record-keeping program for dairy cows and goats.

Dam: The mother of an animal.

Deoxyribonucleic acid (DNA): Double-stranded nucleic acid that is a component of chromosomes and that contains coded genetic information within its nucleotides.

Detergent: A compound that lowers surface tension. Used in mixed feeds because of apparent bacteriostatic effects. Also used for cleaning dairy equipment.

DHI: See *Dairy Herd Improvement.*

Digestion: The chemical and physical breakdown of nutrients in the gastrointestinal tract preparatory to absorption.

Digestion coefficient: The percentage of a nutrient that is absorbed from the digestive system.

Diploid: A cell having two full pairs of homologous chromosomes in its nucleus.

Direct marketing: Direct transaction between the livestock producer and the meat-processing establishment in the sale of animals.

Dissolved oxygen: The oxygen that is dissolved in water and required for animal life.

DNA: See *Deoxyribonucleic acid.*

DNA probe: A short section of single-stranded DNA (labeled with radioactivity) designed to be complementary to some characteristic part of the DNA to be tested.

Doe: A female goat.

Dominance (genic): The tendency of one gene to exert its influence over its partner after conception occurs and genes exist in pairs. There are varying degrees of dominance, from partial to complete to over-dominance.

Dominance (social): The tendency of one animal in a group to exert its social influence or presence over others in the group. See also *Social order* and *Pecking order.*

Dorsal: The upper or topside of the body of a fish.

Dress: The process of cleaning and preparing fish (or other meat animal products) for consumer use.

Dressing percent: Carcass weight divided by live weight and multiplied by 100. Usually the cold carcass weight is used.

Drove: A group of animals. Used more often in the case of hogs, sometimes with cattle.

E. coli: See *Escherichia coli.*

Effluent: Water discharged from an animal unit such as a livestock facility, or fishpond.

Elasticity of demand: The tendency of demand for a commodity to be influenced or changed by various factors. *Price elasticity* is the tendency of demand to change as price goes up or down. *Income elasticity* is the tendency of demand to change as the consumer's income goes up or down. Both are related to ability to buy.

Electrolyte: An ionizable substance in solution.

Emaciation: Abnormally thin body condition due to loss in body weight.

Embryo: The developing young, during pregnancy, from soon after zygote formation to placentation (or fetal stage).

Embryo transfer (ET): Five- to 6-day-old embryo from an elite female is transferred to one of the uterine horns of a recipient female—both horns if two embryos are transferred—most commonly done with cattle.

Emulsifier: A substance that improves the dispersion of tiny globules of one solution throughout the volume of another solution; often added to enhance mixing of fatty materials in liquid diets.

Endometrium: The inner layer of the uterine wall and body.

Endotoxin: A toxin produced within and by the animal.

Enterotoxemia: An acute disease within the intestine caused by toxins produced by the organism *Clostridium perfringens;* highly fatal and most common in sheep and cattle on concentrate rations.

Environment: All of the conditions to which an animal is exposed after conception.

Enzyme: An organic catalyst; it speeds or slows a chemical reaction without being used up in the reaction.

Epididymis: A tortuous tube leading from the testicle. A site of sperm storage and maturation.

Equitation: Body position and control actions of the rider of a horse.

Escherichia coli (E. coli): A bacterium that lives in the intestinal tract of most vertebrates. Much of the work using recombinant DNA techniques has been conducted with this organism because it has been genetically well characterized.

Estrogen: A female hormone that promotes development of the female reproductive tract and mammary tissue.

Estrous cycle: Physiological events occurring in the reproductive system of a female between one estrus period and another; about 17 days in sheep, 21 days in cattle, swine, and horses.

Estrus: The time when a female is "in heat" and will breed readily. Also called *estrus period.*

ET: See *Embryo transfer.*

Ether extract: Common expression for lipids (fats, oils) that are extractable with ether from feedstuffs.

Ethology: The scientific study of animal behavior in their natural or typical environments.

Evaporative cooling: A means of heat loss from the body (sweating, panting); warmer liquid molecules escape from the body via evaporation, resulting in a cooler body mass.

Evisceration: The process of internal organ removal from an animal, as in the preparation of meat for human consumption.

Ewe: A female sheep of any age.

Exotoxin: A toxin produced by an organism other than the host animal, such as organisms of the intestinal tract.

Eye muscle: The longissimus dorsi muscle of four-footed animals. The major muscle of a rib, loin steak, or chop.

F_1 generation: The first generation of offspring of a specific cross or mating.

Family: A group of animals within a breed that usually trace to some noted ancestor.

Farrow: The act of a sow or gilt giving birth to a litter of pigs.

Fat: A class of nutrients, each normally composed of glycerol and three fatty acids. The most potent energy source in rations. Contains the elements carbon, hydrogen, and oxygen; the hydrogen to carbon ratio is much higher than in carbohydrates.

FDA: See *Food and Drug Administration.*

Feed efficiency: Expression of feed (air-dry basis) required per unit of body weight gain (i.e., a feed efficiency of 2:1 indicates that 2 pounds of air-dry feed were required for 1 pound of weight gain).

Feeder: (a) A lamb, pig, or calf of a weight and size that needs to be grown and fattened before suitable for slaughter; (b) a person involved in the business of feeding livestock.

Fermentation: The conversion of one or more substances into more desirable substances through the actions of microorganisms (i.e., rumen fermentation).

Fertility: The capacity of a female or male to reproduce.

Fertilization: The union of the sperm with the egg.

Fetus: The prenatal animal after fetal membranes (placenta) become functional; beyond the embryo stage.

Fillet: A boneless piece of fish prepared for retail sales, or consumer use.

Filly: A young female horse or pony.

Fingerling: Fish ranging from 1 to 8 or 10 inches in length, beyond the "fry" stage, that are used in stocking grow-out facilities.

Flehmen: An action by a bull, boar, or ram associated with courtship and sexual activity. The lip curls upward and the animal inhales in the vicinity of urine or the female vulva.

Flock: A group of chickens, turkeys, sheep, or goats.

Flushing: The practice of feeding a higher amount of nutrients prior to and during the breeding season to improve ovulation rate; most commonly done in sheep management.

Foal: A newborn horse or pony.

Follicle (ovarian): The growth that appears on the surface of the ovary late in the estrous cycle and that contains the developing ovum.

Follicle-stimulating hormone (FSH): A hormone produced by the pituitary gland that promotes growth of ovarian follicles in the female and sperm in the male.

Food and Drug Administration (FDA): The federal agency in the Department of Health and Human Services that must approve all new drugs; its main purpose is to protect consumers from unsafe foods and fraudulent practices.

Forward contracting: Contracting at an early date to sell a product yet to be produced. A form of price protection.

Founder: Digestive malfunction in a horse, usually caused by excessive overeating. Symptoms include high temperature, foot deformity, and pain.

Four-way cross: Offspring from the mating of two single crosses.

Freeboard: Refers to the distance between the surface of the water in a pond and the top of the levee.

Free choice: Refers to individual feeds being available to animals so they can choose the proportion of each they prefer.

Free-stall: Dairy barn and stall design that permits cows to choose a slightly elevated individual stall for resting.

Fry: Minnows up to about 1 inch in length; the stage of fish growth between hatching and 1 inch of growth.

FSH: See *Follicle-stimulating hormone.*

Fungus: A group of plants without chlorophyll that reproduce by spores. Includes molds and can cause animal health problems.

Futures market: A market at which contracts for future delivery of a commodity are bought and sold.

Galactose: A hexose monosaccharide that, when linked with glucose, forms lactose in milk.

Gamma globulin: A protein contained in blood and in colostrum with general disease-fighting effects.

Gelding: A castrated male horse or pony.

Gender: The anatomical sex of an individual.

Gene: The simplest unit of inheritance. Physically, each gene is apparently a nucleic acid with a unique structure. It influences certain traits. Sometimes called a trait determiner.

Gene mapping: Determination of the relative locations of genes on a chromosome.

Genome: The complete genetic code of any individual, the genetic sum of its DNA.

Genotype: The genetic makeup of an animal. A listing of genes carried by the animal, for one or several traits.

Geothermal: A term indicating a system of building ventilation utilizing air drawn through ducts buried in the earth; provides a more uniform temperature throughout the year.

Gestation period: The duration of pregnancy.

Gilt: A female pig before farrowing. Some use the term until the second farrowing.

Glucocorticoid: Steroidlike substance(s) that influences metabolism, such as glycogen deposition, or provide's antiinflammatory effect; includes the stress hormones (cortisol and corticosterone, and epinephrine and norepinephrine).

Glycogen: A form of animal starch; polysaccharide formed in the liver that provides an immediate source of energy if needed.

GnRH: Gonadotropin-releasing hormone; controls the release of two hormones from the pituitary—follicle-stimulating hormone and luteinizing hormone.

Gonad: One of the two testes of a male or ovaries of a female; the organ that produces sex cells.

Gossypol: A yellow pigment in cottonseed or improperly processed cottonseed meal that is toxic to nonruminants, especially young chicks and pigs.

Grade: Any animal, not purebred, that possesses the major characteristics of a breed. Also, a category of animals or product sorted by quality.

Grade A milk: Milk produced by dairy farms under very rigid quality and sanitation standards; represents over 90 percent of all milk sold to milk plants.

Grease wool: Wool shorn from live sheep; unprocessed wool.

Greenchop: Forage harvested and fed in the chopped stage.

Gregarious: Tending to associate as a group such as a herd or flock.

Gross energy: The total energy content of a feed; the heat produced if it were burned.

HACCP: Hazard Analysis and Critical Control Point; a food safety program that requires at least one critical control point for each food safety hazard area where a risk is more likely to occur anywhere in the production and processing of foods.

Hand mating: A system where females (sows, ewes) are brought to the male of choice for the act of mating.

Haploid: Having only one complete set of chromosomes within the nucleus of a cell, such as the sex cells.

Heat increment: The energy used up in the consumption, digestion, and metabolism of a feed.

Hedge: A form of price protection, usually involving a commodities futures contract (e.g., selling beef carcass futures at the time feeder cattle enter a feedlot).

Heifer: A female bovine before calving. Some use the term until the second calving.

Heiferette: Used to describe a heifer that has calved once, perhaps prematurely, then was "dried up" and fed for slaughter.

Hen: A mature female chicken or turkey.

Herbivore: Animals that normally eat vegetative materials.

Herd: A group of animals, especially beef or dairy cattle or hogs.

Heredity: A study or description of genes passed from one generation to the next through sperm and ova. The heredity of an individual would be the genes received from the sire and dam via the sperm and ovum.

Heritability: The degree to which heredity influences a particular trait.

Heterosis: The tendency of offspring of a cross to perform better than the average of their parents.

Heterozygote: An animal whose genotype for a particular trait or pair of genes consists of unlike genes.

Home range: The territory an animal or group of animals normally occupy and in which they feel comfortable.

Homeostasis: Maintenance of physiological stability even though environmental conditions may change.

Homeotherm: An animal that utilizes or dissipates energy to maintain body temperature, usually in the range of 95° to 105°F. Typically called a warm-blooded animal.

Homologous: Pertaining to similar or corresponding structures within a biological system, such as homologous chromosomes (matching pairs, one of each parent).

Homozygote: An animal whose genotype for a particular trait (or pair of genes) consists of like genes.

Hybrid: The offspring of genetically dissimilar parents, whether cells, plants, or animals.

Hybrid vigor: The tendency of crossbred offspring to perform better in certain traits than the average of their parents; heterosis.

Hypoglycemia: Abnormally low levels of blood glucose.

Hypothalamus: The portion of the brain involved in release of certain hormones; also aids in regulation of appetite and body temperature.

ICSH: Interstitial cell stimulating hormone. Synonymous with the term LH, a hormone produced in the pituitary that stimulates testes production of testosterone.

Immunity: Resistance to an infectious agent or antigen.

Imprinting: The rapid and relatively irreversible learning that occurs within a few hours or days of birth.

Inbreeding: Mating animals that are related. Varies in degree, depending on degree of relationship.

Incisor: A front tooth, designed for cutting.

Index: A total merit score, usually calculated for breeding animals. The weight of each component of an index usually is determined by its heritability and economic importance.

Infection: Invasion of animal tissues or organs by a disease organism.

Infectious: The ability to be transmitted by disease; denoting a disease caused by a microorganism.

Infundibulum: A funnel-shaped passageway such as the innermost portion of the oviduct of the hen.

Ingesta: Contents of the digestive tract. Includes feed, digestive juices, bacteria, etc.

Inguinal: Relating to the groin region.

Insemination: The deposition of semen in the female reproductive tract.

Intact: A term sometimes used to describe a male animal that has not been castrated.

Integration: The joining together, by contracts or by ownership, of several components of an industry.

Invertebrate: An organism that does not possess a spinal cord but has a firm outer skeleton.

In vitro: In an artificial environment such as a laboratory rather than in a living body (in vivo).

Ionophore: A compound or substance sometimes used to alter rumen fermentation to improve efficiency of feed utilization.

Isometric: Refers to equal dimensions, as in growth of structures of the body.

Isthmus: A term denoting the relatively short section of the bird's reproductive tract where the shell membranes are produced, creating the shape of the egg.

Ketones: Potentially toxic substances created by only partial oxidation of fatty acids; can result in ketosis in cows, does, or ewes.

Kid: A young goat of either sex.

Kilogram: A metric system unit of weight, equal to 1,000 grams, or 2.205 pounds.

Kosher meat: Meat processed according to Hebrew law and considered fit for consumption by traditional Jews.

Lactation: The period of milk secretion. Usually begins at parturition and ends when offspring are weaned or, in the case of dairy cattle, when milking is stopped.

Lagoon: A human-made earthen storage pit designed to receive animal waste flushed from confinement facilities; second- or third-stage lagoon water is sometimes recycled back for later flushing.

Lamb: A sheep under 1 year of age.

Lateral: Refers to the side of a body, or object.

Lecithin: A feed additive classified as a stabilizer.

Levee pond: A pond enclosed by an earthen dam.

LH: See *Luteinizing hormone.*

Libido: Male sex drive.

Lignin: Indigestible material; not a carbohydrate, but in close association with cellulose as a component of the cell wall of plants and woody materials; steadily increases with plant maturity, thus reducing nutritive value.

Line: A group of animals descended from or related to a specific animal or source of genetic stock.

Linebreeding: A form of inbreeding to concentrate the genes of a particular individual into the pedigree.

Lipase: An enzyme present in gastric juice and pancreatic juice that acts upon fats to produce glycerol and fatty acids.

Lipids: A term that denotes fats, oils, and fatlike materials.

Lipoprotein: A compound of lipid and protein; lipids in blood are transported as lipoproteins.

Litter: (a) Bedding, often sawdust, chopped cobs, or shavings, used on the floor of some animal units; (b) group of pigs born of a female swine.

Locus: A specific site; commonly used to denote the area on a chromosome where a gene is located. Pl., *loci.*

Longevity: The tendency of a breeding animal to have a long productive life.

Luteinizing hormone (LH): A gonadotrophic hormone produced by the pituitary gland that causes rupture of ovarian follicles in the female (ovulation), and secretion of testosterone in the male.

Luteolysis: The degeneration or destruction of an ovarian corpus luteum (or lutea); Lutalyse™ is a commercially available prostaglandin used in estrus synchronization that causes luteolysis.

Lux: The international unit of illumination, equal to 0.0929 foot-candle or 1 lumen per square meter.

Lymph: Relatively clear fluid that is collected from tissues and enters lymphatic vessels; contains water, salts, proteins, and lymphocytes.

Magnum: A term that refers to the albumen-secreting portion of the oviduct of the egg-laying bird.

Malady: A disease or disorder of the body.

Mammary system: The udder, teats, and tissues associated with milk secretion, including the circulating fluids in the vicinity of the udder.

Marbling: The interspersion of fat particles in lean meat; intramuscular fat.

Mare: A female horse or pony approximately 2 years of age or older; designation from filly to mare status is somewhat dependent upon breed.

Margin: The difference between cost and sale price, usually per unit.

Market classes and grades: Standards established by the USDA that segregate animals, carcasses, and animal products into uniform groups, largely based on the best use of an animal and quality guidelines within that use.

Mastitis: An infection of the mammary gland, usually associated with abnormal milk.

Mature equivalent (ME): A term used to indicate an adjustment of performance records of an animal to the mature age of animals in its breed; provides a more accurate evaluation of records of all animals in test.

Mcal: Megacalorie. 1,000 kilocalories or 1,000,000 calories; common expression of energy in feeds.

ME: See *Mature equivalent.*

Meatiness: The degree of muscling or ratio of muscle to fat and bone.

Medial suspensory ligament: A primary support of the mammary system; heavy elastic tissue dividing the udder into halves; extends from the abdominal wall and to the base of the udder.

Medulla: The inner portion of a body organ.

Meiosis: Cell division early in the reproductive process, and in the formation of sperm and ova in the testicles and ovaries. Each pair of chromosomes in the cell being divided separates, and one member of each pair goes to each of the two newly formed cells.

Melengestrol acetate (MGA): A synthetic progesterone that inhibits estrus.

Metabolism: All physical and chemical processes occurring within a biological system.

Metabolites: Substances that are substrates for or are produced by metabolic processes or enzyme reactions.

MGA: See *Melengestrol acetate.*

Microingredients: Vitamins, minerals, antibiotics, drugs and other materials normally required in small amounts and measured in milligrams, micrograms, or parts per million (ppm).

Micronutrients: Nutrients added in extremely small quantities to rations.

Milking parlor: A special facility in which cows, goats, or sheep are milked.

Mitochondrion: A subcellular cytoplasmic structure that is the site of Kreb cycle activity, where most ATP synthesis occurs.

Mitosis: Cell division during normal growth of tissue. Each chromosome divides such that resulting new cells each have a full complement of chromosome pairs.

Mohair: Animal fiber of the Angora goat used for upholstery, rugs, and clothing; of greater length and less crimp than wool.

Mollusk: A member of the phylum Mollusca, consisting of all shellfish such as clams, snails, and oysters, but excluding crustaceans.

Molting: An interruption of egg laying, either natural or forced.

Monensin: An ionophore that increases the proportion of propionic acid in the rumen.

Monogastrics: The term denoting animals with a single stomach, in contrast to ruminants with four stomach compartments.

Monosaccharide: A single-sugar molecule; a carbohydrate.

Morphology: The form and structure of living things.

Motility: The ability to move about.

Mucus: A viscous liquid secreted by the mucous glands that contains a polysaccharide-protein material.

Mulefoot: Syndactylism. A hereditary defect where only one toe is present on the affected foot; may affect one or more feet; caused by a recessive gene.

Mutton: Meat from sheep older than 1 year of age.

Muzzle: The nose of cattle, horse, or sheep.

Mycoplasma: A genus of bacteria containing cells without a true cell wall; smaller than most bacteria but larger than viruses; many disease-causing species.

Myometrium: The muscular wall of the uterus.

National Research Council (NRC): Provides publications on nutrient requirements of different species.

NDF: See *Neutral detergent fiber.*

Near-infrared reflectance (NIR): A rapid method of obtaining quality estimates of forages; utilizes an NIR spectrometer.

Net energy: The energy of a feed that is available to the animal for growth, production, or work after digestion and metabolism.

Neutral detergent fiber (NDF): A measure of the cellulose, hemocellulose, lignin, and insoluble ash content of forages; an indicator of the intake potential within or among forages.

NFE: See *Nitrogen-free extract.*

NIR: See *Near-infrared reflectance.*

Nitrate: A compound that is converted from nitrite and is less toxic than ammonia and nitrite; formed through breakdown of nitrogenous materials.

Nitrite: A form of nitrogen converted by microbes from ammonia.

Nitrogen-free extract (NFE): A part of a feed consisting largely of sugars and starches.

Noninfectious disease: A nonpathogenic disease not transferred from one animal to another.

Nonprotein nitrogen (NPN): Refers to components in feed that contains nitrogen not incorporated into protein, such as urea.

Nonruminant: An animal without a functional rumen. Sometimes called a monogastric.

NPN: See *Nonprotein nitrogen.*

NRC: See *National Research Council.*

Nucleotide: A chemical component of the genetic material chromatin.

Nucleotide sequence: The sequence of the bases within the nucleic acid that determines what proteins will be made.

Nutrient: A chemical element or compound that is essential for normal body metabolism.

Obese: Excessively fat.

Off-flavor: An abnormal flavor in fish (or other food products), such as a rancid or muddy taste.

Offspring: Animals born to a parent; descendants, either the first or a later generation.

Olfactory: Pertaining to the sense of smell.

Omasum: The third compartment of the ruminant's stomach that receives material from the reticular-omasal orifice.

Omnivorous: Living upon both plants and animals.

Oogenesis: The process of formation and development of ova.

Outbreeding: Mating animals distinctly unrelated, usually with diverse type or production traits. Varies in degree, depending on the degree of divergence in type or production traits. See also *Crossbreeding* and *Linebreeding.*

Ovary: The female sex organ that produces ova after sexual maturity.

Oviduct: The tube leading from each horn of the uterus to the corresponding ovary.

Ovum: The female sex cell, produced on the ovary and carrying a sample half of the genes carried by the female in which it was produced. Pl., *Ova.*

Oxidative: Having the power or tendency to combine with oxygen.

Packer: Refers to the buyer of animals to be slaughtered and processed for meat.

Palatability: The property of being pleasing to the taste.

Palpation: A massaging or stroking of animal tissues.

Pampiniform plexus: An arrangement of arteries and veins above the testicle that aid in temperature regulation of the testicles.

Pancreas: A gland that produces and secretes pancreatic juice into the duodenum to enhance digestion.

Parasite: Animal life that lives on another organism.

Parturition: The act of giving birth—calving, lambing, farrowing, or foaling.

Pathogens: Disease organisms.

Pecking order: Social order; the tendency of animals to behave in an order of social dominance.

Pedigree: A record of ancestors.

Peptide: Two or more amino acids bonded together.

Per capita: Per person.

Performance testing: The use of testing and records in selecting breeding stock; sometimes animals from different producers are managed as a group at Central Testing Stations.

Peristalsis: A process by which muscular contractions propel food material through the digestive tract.

pH: An index of the acidity of a substance, the value of 7.0 being neutral, values above 7.0 being alkaline, and values below 7.0 being acid.

Pharynx: The upper portion of the throat; the junction of the respiratory and digestive tracts.

Phenotype: The characteristics that an animal shows or demonstrates. Includes both appearance and performance characteristics.

Photoperiod: The length of daylight or artificial light provided.

Photosynthesis: The process where plants convert carbon dioxide and other elements into simple carbohydrates with the production of oxygen, in the presence of light.

Phytoplankton or plankton: Microscopic plants or animals that live and grow in water.

Picnic: The lower part of the pork carcass shoulder.

Pituitary gland: A small endocrine gland located at the base of the brain that produces and secretes various hormones into the bloodstream. These hormones help regulate various body processes.

Placenta: A structure in mammals that provides the exchange of nutrients and waste materials between the maternal and fetal system.

Poikilotherms: Animals with little or no ability to maintain an even body temperature, so temperature is influenced largely by the environment. Referred to as cold-blooded animals.

Polyestrous: Refers to animals that have several estrous cycles per year or breeding season.

Pony: A small-type horse, under 14.2 hands (about 57 inches) in height at the withers.

Population: A defined segment or the totality of animals of a particular species.

Porcine somatotropin (PST): The type of somatotropin produced by swine.

Postpartum: After birth.

Postweaning: Time period from after offspring is weaned from dam until slaughter (or selection for animals entering the breeding herd).

Poult: A young turkey, male or female.

PPA: See *Predicted producing ability*.

Preconditioning: Preparing feeder calves for movement from their home ranch or farm to a feedlot by weaning, vaccinating, and other postweaning practices.

Predator: An animal that attacks or molests domestic animals, with the purpose of injury or death.

Predicted producing ability (PPA): An estimate of a dairy cow's ability to produce milk in future lactations as compared to her herdmates; useful in making culling decisions.

Predicted transmitting ability (PTA): A term that has replaced *predicted difference* and *cow index*; values predicting the genetic merit of dairy animals.

Pregnancy: The period from conception to birth.

Prenatal: Prior to birth.

Preweaning: Time period from birth to weaning of offspring.

Primal cuts: The most valuable cuts on a beef, pork, or lamb carcass.

Probe: A small metal ruler used to measure backfat; sometimes refers to an electronic sensor rather than a ruler.

Progeny: Offspring.

Progeny testing: Testing of offspring to determine which parents should be chosen for more extensive use in breeding programs.

Progesterone: A hormone produced by the corpus luteum on the ovary that aids in maintaining pregnancy.

Prolactin: An anterior pituitary hormone needed for synthesis of milk.

Prolific: Having a tendency to produce many offspring.

Prostaglandin: Certain organic acids of the body; involved in bringing about regression of the corpus luteum.

Prostate gland: A gland that surrounds the beginning of the urethra in the male.

Protein: A class of nutrients containing amino acids that may be present in feeds individually or combined. Contain the elements carbon, hydrogen, oxygen, and nitrogen, and usually sulfur.

Protozoa: Single-celled organisms that reproduce by fission. Many may cause animal diseases.

Proven sire (DHI): A bull with at least 10 daughters that have completed lactation records and that are out of dams with completed lactation records.

Proventriculus: An enlargement of the bird esophagus preceding the gizzard; the glandual or true stomach.

PST: See *Porcine somatotropin*.

PTA: See *Predicted transmitting ability*.

Puberty: The age when an animal is physiologically mature enough to be capable of sexual reproduction.

Pullet: A young female chicken, usually before she begins to lay.

Quality grades: Official standards established by the USDA for evaluating palatability; indicate characteristics of the carcass of market animals.

Raceway: A continuous flowing aquaculture water system, most often used in trout farming.

Radiation: The exchange of heat between two objects that are not touching; emission of heat rays from the warmer to the cooler object.

Ram: A sexually mature, intact (uncastrated) male sheep.

Random mating: Allowing selected animals to mate at random.

Ration: A combination of feeds, perhaps mixed together, that are fed to animals to meet nutrient requirements. May vary in quantity.

Ratite: An ostrich, emu, or rhea; birds known for their red meat and skins used for leather goods.

Reach: The difference between the average merit of a herd or flock, in one or several traits, and the average merit of those selected to be parents of the next generation; the selection differential.

Recessive: The tendency for a gene to be overshadowed by its partner in its influence on traits, after conception occurs and genes exist in pairs. See also *Dominance*.

Reciprocal cross: A mating where the breed of the sire and dam are reversed in relation to another mating. Sire A × Dam B is the reciprocal of Sire B × Dam A.

Registered: Recorded by a breed association.

Registration certificates: Papers showing that an animal has been recorded by the breed association as a purebred.

Repeatability: The tendency of animals to repeat themselves in certain performance traits in successive seasons, pregnancies, or lactations.

Reproduction efficiency: The tendency to produce large numbers of offspring in a given span of time.

Respiration: The process of plants and animals that converts chemical energy into other kinds of energy; oxygen is utilized and carbon dioxide is produced.

Reticulum: The second compartment of the ruminant stomach, where bacterial digestion continues. Has a honeycomb-textured lining, so is often called the "honeycomb," or "hardware" stomach.

Rickets: Abnormal bone development; bones easily bent or distorted.

Roughage: A feed low in digestible energy and high in fiber.

Rumen: The largest of the four stomach compartments in the adult ruminant. The site of active microbial digestion.

Ruminant: An animal with a functional rumen compartment in the stomach, plus three other compartments; a cud-chewing animal.

Runoff: Precipitation or excess liquid animal waste that does not infiltrate the soil and can potentially enter other water sources such as lakes, ponds, or streams.

Salinity: Refers to salt content in water.

Saturated fat: Fat with fatty acids having the maximum number of hydrogen atoms attached to them; no double bonds.

Scrotum: The saclike extension from the abdominal cavity of a male that contains the testicles.

Scrub: An animal having little or no improved breeding, or an animal of mixed, mongrel, or unknown breeding.

Secondary host: An animal that may host a disease organism, usually without developing symptoms of the disease itself.

Secretion: The release of material through the plasma membrane of a cell, such as milk secretion.

Sediment: Material such as uneaten food, soil, or fecal matter that settles to the bottom of water containments such as fish ponds or tanks.

Segregated early weaning (SEW): A practice of weaning pigs at 10 to 14 days of age and isolating multiple litters from other age groups to reduce risk of contraction of disease.

Seine: The netting material used to harvest fish.

Selection differential: The difference in a trait between the average merit of the population (usually a herd, flock, or segment thereof) and the average merit of those animals selected to be parents of the next generation.

Selection threshold: A trait line or standard established by a breeder, below which the breeder will reject potential breeding animals.

Semen: A mixture of sperm and accessory gland fluids produced by the testicle and accessory organs.

Seminal vesicles: Paired glands attached to the urethra, near the bladder, that produce fluids to carry and nourish the sperm.

Seminiferous tubules: Tubular structures within the testicles where spermatazoa are produced.

Service: The act of breeding or mating.

Settle: To conceive, after a female is bred.

SEW: See *Segregated early weaning*.

Shoat: A pig of either sex, usually between 60 and 160 pounds.

Shrinkage: Loss in weight. Usually used to describe weight loss when livestock is shipped. Weights usually are taken under comparable conditions before and after shipping.

Sib: Sibling; an animal with the same parents as another; a brother or sister.

Silage: Succulent feed produced from anaerobic storage and fermentation of forages.

Single cross: Offspring from the mating of two lines, often two inbred lines.

Sire: Male parent.

Social order: The tendency of animals to behave in an order of social dominance.

Solid mouth: A mouth with a full set of teeth that are in good condition.

Somatotropin: A protein hormone secreted by the pituitary gland in mammals. It directs lactation and growth. Because it is a protein, somatotropin is broken in the digestive tract if consumed in meat or milk.

Sonoscope: An electronic instrument that emits sound waves and measures the time required for the waves to bounce back from the junction between tissues; electronic probe.

Soundness: The health or physical integrity of an animal or some part of an animal.

Sow: A mature female pig, usually after one or two pregnancies.

Spay: To remove the ovaries or sever the oviducts leading to the uterus.

Spawning: The release of eggs from female fish.

Sperm: The male sex cell, produced by the testicle and carrying a sample half of the genes carried by the male in which it was produced.

Spermatic cord: The area between the testicles and the body that includes the vas deferens and the blood, lymphatic, and nerve supply to the testicles.

Spermatogenesis: Production of spermatozoa within the seminiferous tubules of the testicle.

Sphincter: An arrangement of circular muscular fibers, such as the sphincter surrounding the streak canal of the teat of the udder.

Squab: A young pigeon, sometimes rapidly grown for meat production.

Stag: A male pig, sheep, or bovine castrated after puberty and after secondary sex characteristics have developed.

Stallion: Most commonly refers to a sexually mature, intact (uncastrated) horse or pony.

Stanchion: A stall arrangement that constrains an animal, usually a milk cow, with parallel bars or pipes on each side of the neck.

Staple: Wool fibers sufficiently long to be used in worsted fabric.

Starch: An important nonfibrous polysaccharide; a major component of cereal grains and tubers that provides glucose for energy when digested.

Steer: A male bovine castrated before puberty.

Stocker: A term used to describe relatively light or thin feeder animals, or mature cows that are generally purchased to utilize low-quality roughage. In aquatic culture, applies to fish larger than fingerlings but smaller than food fish; usually greater than 8 to 10 inches.

Stover: Forage that remains after grain is removed.

Strain: A group of animals within a breed with characteristics that distinguish them from others in the breed.

Streak canal: The opening at the end of the teat held closed by circular sphincter muscles.

Stress: The sum of adverse physical, chemical, and emotional factors that results in physiological tensions within an animal. Conditions that cause animals to expend energy and experience physical strain or make them more susceptible to diseases.

Substrate: Material acted upon by an enzyme.

Superovulate: To cause the production and release of several ova at a single estrus, as performed in embryo transfer procedures.

Surfactant: See *Detergent*.

Synergistic: A condition where two or more structures or drugs work together in a coordinated effort.

TDN: See *Total digestible nutrient*.

TOBEC: A tradename for an electromagnetic scanning apparatus used to measure total body electrical conductivity as a means of measuring pork quality.

Tactile sense: Sense of touch.

Terminal market: A market where animals are gathered and sold by commission agents on behalf of the owner.

Territorialism: The tendency for animals to inhabit and protect a specific geographical region or domain.

Testicle: One of the paired male sex organs that produce sperm after sexual maturity. Pl., *testes* or *testicles*.

Therm: One thousand calories.

Thermogenesis: Heat generation within an animal, triggered by low environmental temperature.

Tie-stall: Refers to the type of barn or the method where animals are tied within an individual stall.

TMR: See *Total mixed ration*.

Tom: Most commonly refers to a sexually mature, intact (uncastrated) male turkey.

Total digestible nutrient (TDN): A measure of the digestible energy in a feed.

Total mixed ration (TMR): All-in-one ration. Ration components combined; theoretically, each bite of food would contain all nutrients in the correct proportions.

Toxemia: An illness or malady caused by a toxin.

Toxin: A harmful chemical of animal or bacterial origin.

Trace minerals: Mineral nutrients required by animals in micro amounts only (measured in milligrams per pound or smaller units).

Transgenic organism: An organism formed by the insertion of foreign genetic material into the reproductive cells.

Triglyceride: A compound composed of glycerol and three fatty acids.

Trimester: One-third of the length of a pregnancy.

Tunica dartos: The elastic tissue and muscle fibers within the scrotal wall that contract and relax to aid in temperature regulation of the testicles.

Turbidity: Muddy or cloudy water; caused by suspended soil particles or plankton.

Type: Referring to the structure or conformation of an animal, or the type of product it produces. Examples: Meat-type hog, wool-type sheep.

Unsaturated fatty acids: Fatty acids containing one or more double bonds; tend to be liquid and less stable than saturated acids.

Unsoundness: A defect in the structure or physical integrity of an animal or one of its parts.

Urethra: The portion of the urinary tract that carries urine from the bladder; in males, also provides passage for semen ejaculate.

Uterus: The female reproductive organ in which an embryo and fetus develop.

Variety: Poultry of specific genetic source or breeding.

Vas deferens: The passageway for sperm from the epididymis to the urethra.

Veal: Meat from a young milk- or milk-formula-fed calf; most are marketed at 350 to 425 pounds within 16 weeks of age, such as fancy veal.

Vector: The agent (e.g., plasmid, virus) used to carry new DNA into a cell; capable of transmitting disease.

Ventral: The underside of the body, such as the stomach.

Vertebrate: An organism with a segmented spinal column and inner skeleton.

VFAs: See *Volatile fatty acids*.

Vicuna: Animal fiber from a wild ruminant of the Andes; a fine, lustrous, wool-like undercoat.

Virus: A large and complex protein material that is capable of causing disease and which reproduces only inside a host cell.

Viscera: Usually refers to the organs of the abdominal cavity removed at slaughter, including stomach, intestines, liver, and other accessory organs. Heart and lungs may also be included. Part or all of the organs in the great cavity of the body. In fish, it includes the gills, heart, liver, spleen, stomach, and intestines.

Vitamin: One of a class of nutrients, each required in small quantities, that usually functions as a catalyst or catalyst component in body metabolism. Not needed as a source of energy or nitrogen.

Volatile fatty acids (VFAs): The group of volatile organic acids resulting from rumen fermentation; most prevalent are acetic, propionic, and butyric acids.

Wean: To cease to provide milk; to physically separate young from the nursing dam.

Wether: A young male sheep or goat castrated before puberty.

Wind chill index: A measure of the net chilling effect on animals of temperature and wind.

Withers: The area on an animal just behind the top of the shoulder.

Worsted fabric: High-quality woolen fabric comprised of long fibers that are small in diameter.

Yearling: Normally, an animal that is 1 year of age but not yet 2.

Yield: (a) Dressing percent of an animal slaughtered (see also *Dressing percent* and *Cutability*); (b) volume of milk produced per day or per lactation; (c) pounds of wool clipped per ewe; (d) percentage of clean wool after scouring.

Yield grade: A USDA grade that indicates the amount of trimmed, boneless major retail cuts that can be derived from a carcass; cutability.

Zoonoses: Diseases that can be transmitted between humans and animals.

Zooplankton: Microscopic animals that live and grow in water.

Zygote: The diploid cell resulting from the union of haploid female and male gametes.

Index